VIP, PACAP, AND RELATED PEPTIDES:
FROM GENE TO THERAPY

ANNALS OF THE NEW YORK ACADEMY OF SCIENCES
Volume 1070

VIP, PACAP, AND RELATED PEPTIDES: FROM GENE TO THERAPY

Edited by
Hubert Vaudry and Marc Laburthe

Published by Blackwell Publishing on behalf of the New York Academy of Sciences
Boston, Massachusetts
2006

Library of Congress Cataloging-in-Publication Data

VIP, PACAP, and related peptides : from gene to therapy / edited by
Hubert Vaudry and Marc Laburthe.
    p. ; cm. – (Annals of the New York Academy of Sciences,
ISSN 0077-8923 ; v. 1070)
    ISBN-13: 978-1-57331-550-0 (alk. paper)
    ISBN-10: 1-57331-550-8 (alk. paper)
    1. Neuropeptides. 2. Neuropeptides–Physiological effect. I. Vaudry,
Hubert. II. Laburthe, Marc. III. Series.
    [DNLM: 1. Neuropeptides. 2. Vasoactive Intestinal Peptide. 3.
Pituitary Adenylate Cyclase-Activating Polypeptide.
W1 AN626YL v.1070 2006 / WL 104 V811 2006]

    QP552.N39V57 2006
    612'.015756–dc22

                                                        2006016389

The *Annals of the New York Academy of Sciences* (ISSN: 0077-8923 [print]; ISSN: 1749-6632 [online]) is published 28 times a year on behalf of the New York Academy of Sciences by Blackwell Publishing, with offices located at 350 Main Street, Malden, Massachusetts 02148 USA, PO Box 1354, Garsington Road, Oxford OX4 2DQ UK, and PO Box 378 Carlton South, 3053 Victoria Australia.

Information for subscribers: Subscription prices for 2006 are: Premium Institutional: $3850.00 (US) and £2139.00 (Europe and Rest of World).
Customers in the UK should add VAT at 5%. Customers in the EU should also add VAT at 5% or provide a VAT registration number or evidence of entitlement to exemption. Customers in Canada should add 7% GST or provide evidence of entitlement to exemption. The Premium Institutional price also includes online access to full-text articles from 1997 to present, where available. For other pricing options or more information about online access to Blackwell Publishing journals, including access information and terms and conditions, please visit www.blackwellpublishing.com/nyas.

Membership information: Members may order copies of the *Annals* volumes directly from the Academy by visiting www.nyas.org/annals, emailing membership@nyas.org, faxing 212-888-2894, or calling 800-843-6927 (US only), or +1 212 838 0230, ext. 345 (International). For more information on becoming a member of the New York Academy of Sciences, please visit www.nyas.org/membership.

Journal Customer Services: For ordering information, claims, and any inquiry concerning your institutional subscription, please contact your nearest office:
UK: Email: customerservices@blackwellpublishing.com; Tel: +44 (0) 1865 778315; Fax +44 (0) 1865 471775
US: Email: customerservices@blackwellpublishing.com; Tel: +1 781 388 8599 or 1 800 835 6770 (Toll free in the USA); Fax: +1 781 388 8232
Asia: Email: customerservices@blackwellpublishing.com; Tel: +65 6511 8000; Fax: +61 3 8359 1120
Members: Claims and inquiries on member orders should be directed to the Academy at email: membership@nyas.org or Tel: +1 212 838 0230 (International) or 800-843-6927 (US only).

Printed in the USA.
Printed on acid-free paper.

Mailing: The *Annals of the New York Academy of Sciences* are mailed Standard Rate. **Postmaster:** Send all address changes to *Annals of the New York Academy of Sciences*, Blackwell Publishing, Inc., Journals Subscription Department, 350 Main Street, Malden, MA 01248-5020. Mailing to rest of world by DHL Smart and Global Mail.

Annals are available to subscribers online at the New York Academy of Sciences and also at Blackwell Synergy. Visit www.annalsnyas.org or www.blackwell-synergy.com to search the articles and register for table of contents e-mail alerts. Access to full text and PDF downloads of *Annals* articles are available to nonmembers and subscribers on a pay-per-view basis at www.annalsnyas.org.

The paper used in this publication meets the minimum requirements of the National Standard for Information Sciences Permanence of Paper for Printed Library Materials, ANSI Z39.48_1984.

ISSN: 0077-8923 (print); 1749-6632 (online)
ISBN-10: 1-57331-550-8 (paper); ISBN-13: 978-1-57331-550-0 (paper)

A catalogue record for this title is available from the British Library.

ANNALS OF THE NEW YORK ACADEMY OF SCIENCES
Volume 1070
July 2006

VIP, PACAP, AND RELATED PEPTIDES: FROM GENE TO THERAPY

Editors
HUBERT VAUDRY AND MARC LABURTHE

This volume is the result of the 7th **International Symposium on VIP, PACAP, and Related Peptides**, which was held on September 11–14, 2005, in Rouen, France.

CONTENTS

Financial assistance was received from:

Major Funders
- Conseil Régional de Haute-Normandie

Supporters
- Agglomération de Rouen
- Institut Fédératif de Recherches Multidisciplinaires sur les Peptides
- Institut National de la Santé et de la Recherche Médicale
- Municipalité de Rouen
- Science Action Haute-Normandie
- Technopole Chime-Biologie Santé
- Université Paris 7
- Université de Rouen

Contributors
- Affymetrix
- Applied Biosystems
- Bachem
- Beaufour Ipsen Pharma
- Ciphergen
- Debiopharm
- Dutcher
- Elsevier
- European Peptide Society
- Euroscreen
- Institut de Recherches Internationales Servier
- Johnson & Johnson
- Leica
- NeoMPS
- Promega
- Sigma
- VWR

Introduction

The 7th International Symposium on VIP, PACAP and Related Peptides was held in Rouen, France on September 11–14, 2005. It was followed by a satellite workshop (chairmen: Youssef Anouar, Lee E. Eiden, David Vaudry) entitled "Signaling Mechanisms of VIP, PACAP and Related Peptides: Contribution of Genomics, Proteomics and Bioinformatics" on September 15, 2005. Over 300 registered participants from all continents attended the symposium and/or the satellite workshop.

Vasoactive intestinal polypeptide (VIP) was discovered in 1971 by Sami Said, while he was working in the laboratory of the famous peptide chemist Victor Mutt, at the Karolinska Institute (*Science* 169:1217–1218) and pituitary adenylate cyclase-activating polypeptide (PACAP) was identified in 1989 by Akira Arimura and his co-workers at Tulane University, Belle-Chasse, LA (*Biochem. Biophys. Res. Commun.* 173:1271–1279). At the opening ceremony, the President of the University of Rouen awarded the Medal of Rouen University to Akira Arimura and Sami Said. During the symposium, Akira Arimura gave the opening lecture on the potential of PACAP for the treatment of renal failure associated with multiple myeloma while Sami Said, who is now working at the State University of New York, gave the closing lecture on the potential of VIP for the treatment of disorders of the lungs and pulmonary circulation. It was actually rewarding to see that the "fathers" of VIP and PACAP are still conducting outstanding research at the forefront of this rapidly expanding field. Gabriel Rosselin, a pioneer of gastrointestinal peptide research, who had organized the first symposium of this series in Strasbourg (Bischenberg), France, September 19–23, 1993, was the guest of honor of the 7th International Symposium on VIP, PACAP and Related Peptides.

VIP, PACAP, and related peptides (i.e., secretin, glucagon, glucagon-like peptides, growth hormone-releasing hormone, gastric inhibitory polypeptide, etc.) are undoubtedly among the most fascinating regulatory peptides. They belong to the largest family of bioactive peptides and thus provide a unique model for investigating the processes of molecular evolution that have led to the diversification of multigene families. VIP and PACAP are clearly implicated in a large array of physiological and pathophysiological processes related to development, growth, cancers, neuronal, endocrine, cardiovascular, respiratory, reproductive and digestive functions, immune responses, and circadian rhythms.

Five plenary lectures were given by prominent specialists, originating from all continents: Akemichi Baba, Osaka, Japan; Patricia Brubaker, Toronto, Canada; Bill K.C. Chow, Hong Kong, China; Rosa P. Gomariz, Madrid, Spain;

Ann. N.Y. Acad. Sci. 1070: xvii–xviii (2006). © 2006 New York Academy of Sciences.
doi: 10.1196/annals.1317.096

and Patrick Sexton, Parkville, Australia. Eleven state-of-the-art lectures, 26 oral communications and 96 poster presentations have addressed the role of VIP, PACAP, and related peptides in health and diseases. They were divided up by themes which included: cell signaling, neurobiology, functional contribution of gene knockout, immunity and inflammation, ligand receptor interactions, regulation of endocrine functions, neuroprotection, and clinical applications. The recent developments in immunology and potential value of VIP in treatment of inflammatory diseases were discussed in a Round Table on VIP/PACAP and Immunity. Among other recent breakthroughs in VIP/ PACAP research are the acknowledgment of VIP and PACAP or analogs as possible treatments for several diseases (inflammatory and neurodegenerative diseases) and the development of structural biology in the studies of receptors for VIP, PACAP, and related peptides that belong to the class II of G protein–coupled receptors.

The satellite workshop held after the symposium was dedicated to the signaling mechanisms of VIP, PACAP, and related peptides, with special emphasis on novel approaches to the study of signal transduction and transcriptional programs regulated by VIP, PACAP, and related peptides. The workshop included keynote presentations on transcriptional expression profiling and phosphoproteomic approaches applied to study signaling mechanisms, as well as several short communications and posters to illustrate the various molecular pathways involved in the pleiotropic effects of VIP, PACAP, and related peptides.

It is a pleasure to acknowledge the help of those individuals and organizations who made the symposium possible: the contributing scientists; the international committee; the scientific advisory board; the organizing committee and the staff of the conference; and the editorial department of the Academy, especially Ms. Linda H. Mehta. We would like to thank also all our colleagues and co-workers who have contributed to the organization of the symposium and satellite workshop, in particular Catherine Beau, Bruno Gonzalez, Jérôme Leprince, Margot Sauvadet, and Marie-Christine Tonon.

—HUBERT VAUDRY
Mont-Saint-Aignan, France

—MARC LABURTHE
Paris, France

Reminiscences of a Life in Science

I am grateful for the flattering picture that William Rostène made of me in his introduction. Even at my stage of life the ego benefits from recognition, but I am particularly grateful that his picture included not only me, but also the *group* of scientists that has collaborated with me. At the beginning of my career, I had no doubts my efforts would be successful. Pride, in part, led me to work long hours in the lab and to expect good results. But I was tempted to attribute exclusively to myself all the wonderful things I discovered.

Ego is a good catalyst for research, and ego makes it possible for most young scientists to succeed. Even later, ego is important for consolidating the vision you have of your work and for being efficient. Ego, however, entails some diplomatic work and sometimes open struggle, even with excellent friends. Now, with time, *j'ai mis de l'eau dans mon vin* as we say in French. That is to say, I have become wiser and more understanding. With the experience of serving on various committees, I find it more and more difficult to distinguish between the progresses of science that can be specifically ascribed to an individual and those which derive from his group and from the international scientific community in the background. I came to the conclusion that what makes good research possible is to set up a good team as soon as something new happens— and I was lucky to be part of good teams. At this stage, ego is not useful for progress.

What is the recipe for a good team? I do not have an answer. Affinity among the individuals making up the group is essential, of course, as well as the capacity of the entire group working together to be effective: the group should have sufficient coherence and shared motivation to take the research further. But there is something transcendent in the definition of a good team: it remains as difficult to find the right recipe as to discover how molecules create cyanobacteria—how cells join together to develop a coordinated function—or the evolutionary process results in structures that are more and more organized and coherent. Last but not least, the members of the team know when their work together is achieved, and they know how to split fruitfully when the time for splitting is reached. When you have had the chance to work in a good team, such an experience persists in your mind when the active pursuit of science becomes part of your past; this is what people mean when they talk about the "good moments of life"—as long as you agree that the term "good" does not necessarily mean "easy." I have to say to Dominique, Jean-Claude, Marc, William, Anne-Marie, Any, Claudine, and Denise, Shahin, and Christian, and more generally to people who have been working or are now working in the

Ann. N.Y. Acad. Sci. 1070: xix–xxi (2006). © 2006 New York Academy of Sciences.
doi: 10.1196/annals.1317.097

lab, that living and working with them during those years were part of these good moments in my life and I hope they feel the same. The time I spent with my other colleagues—in meetings like this one or visiting their labs or when they visited mine—has also been part of these good moments.

Of course, the opportunity to work in a lab involves dealing with some form of administration, and I would like to address that administrative interface, at least as it exists in Europe. Administrators will often say: "Look at what they are doing with 'our' money. These meetings are just a place where good friends meet. They just express their own self-satisfaction." They are inclined to conclude that developing huge programs like nuclear fusion or enormous telescopes or space stations is more worthwhile than funding such meetings and research. So we keep on struggling for biomedical research funding. Biomedicine, however, cannot evolve through applying the same methods as physics.

First, we must remember that it was easier to create a galaxy than to create life. It took six hundred million years to form the Earth as compared with nearly four *billion* years to create life, because it was achieved through endless trial and error: some developments succeed and some fail.

Second, to understand life's functions, we have to deal with millions of forms of RNA and proteins. Consequently, many groups of researchers are necessary, and methods from numerous disciplines must be applied to biomedical research.

To be competitive, in my view, we have to think in the manner of Theseus. This Greek hero had to find his way out of a labyrinth, and he was clever enough to find Ariadne and her clue. The Theseus method is an interesting approach, especially these days when we have to mount large research programs in order to compete with physics or spate research. We are now able to make maps of the labyrinth of life using nanotechnology, genomics, proteomics, bioinformatics, mass spectrometry synchrotrons, and so on. But suppose for a moment that Theseus had only *maps* of the labyrinth. If so, he probably would not have had enough oil in his lamp to check all the side routes on his way back and would have been killed by the Minotaur.

This is exactly why a meeting like this one is fruitful. It provides us with Ariadne's clue and helps those who draw maps of the labyrinth. You will see this process happening in the workshop that follows this meeting.

Now that computer rooms are available during congresses and that internet replaces the telephone, biomedicine also changes quickly. People live longer, at least in some places, and more and more diseases can be cured. The relationship of these changes with the progress in biologic research is not obvious to the layman, but this kind of progress is only possible because knowledge about life has moved forward. The cellular functions are now targeted to specific cellular networks. Everyone knows that behind a simple reaction there is a specific crossroad and many combinatorial stimulations that make it easy for the cells to adapt. It has become possible to understand how large functional groups react to the dynamics of the whole system and to acquire further insight into what makes cells live, evolve, or die.

An interesting side aspect of all this is that because of the coherence of life, general laws can be identified from a narrow field of research. This was illustrated in the past by my mentor Sol Berson. He was the pioneer of the nanotechnology techniques that our groups have applied for 40 years to assay active peptides in picoquantities and to provide evidence for many new receptors. Berson found these techniques through the observation of insulin binding to serum from insulin-treated diabetic patients.[1,2] My friend Martin Rodbell found G proteins while studying glucagons binding to liver plasma membranes.[5] So, good luck to the young scientists. Their work, which may appear to be narrowly focused, might end up having general consequences.

To conclude, I would like to express my high regard to Akira Arimura and offer him my sympathy for his suffering in the aftermath of the hurricane in New Orleans. I thank Sami Said and Dr. Arimura as well as the scientific committee for adapting the VIP, PACAP meetings; they have kept pace with the progress of biomedicine and remain a continuing part of its progress. I cannot conclude without mentioning my deeply cherished memories of my old friends Viktor Mutt and Christophe Yanaihara, and also of Daniélle Hui Bon Hoa who was a colleague in our lab and died so prematurely. I thank Hubert Vaudry and Marc Laburthe, who organized this meeting, for their kind invitation even though I am no longer active in science. Happily, I am still active in the pleasures of life. For that reason, I appreciated the choir and orchestra of the medical school of Rouen, directed by Pierre Déchelotte, the pixel display of Monet projected onto the cathedral, and the tour in Normandy, a region I love. Thanks and congratulations to all of you.

—GABRIEL ROSSELIN
Paris, France

Treatment of Renal Failure Associated with Multiple Myeloma and Other Diseases by PACAP-38

AKIRA ARIMURA, MIN LI, AND VECIHI BATUMAN

Department of Medicine, Tulane University School of Medicine,
New Orleans, Louisiana 70112, USA

ABSTRACT: Myeloma kidney injury is caused by the large amount of light chain (LC) of immunoglobulins produced by cancerous plasma cells through stimulation of proinflamatory cytokines like TNF-α and IL-6. PACAP-38 suppressed LC-stimulated cytokine production by tubular epithelial cells *in vitro* and *in vivo*, and prevented injury of these epithelial cells. The suppressive effect is comparable or greater than dexamethasone (dex). Although dex produces adverse side effects when it is given for a long time period, PACAP-38 is a natural and safe neuropeptide and no adverse effect has been reported when administered to produce significant biological effects. Furthermore, PACAP-38 suppressed growth of myeloma cells in culture and also suppressed production of their growth factor, IL-6, production from the bone marrow stromal cells that was stimulated by adhesion of myeloma cells. These findings render PACAP-38 worth evaluation as a safe and potent renoprotectant in myeloma kidney as well as a new antitumor agent for myeloma cells.

KEYWORDS: myeloma kidney; light chain; cytokine; PACAP-38; myeloma cells; PACAP receptors

INTRODUCTION

Renal failure is often associated with multiple myeloma (MM). It is characterized by cast formation in the tubules and tubulointerstitial injury, the entity called myeloma kidney. A large amount of light chain (LC) of immunoglobulin produced by myeloma cells is filtrated through the glomerulus and overflow the tubules. LC binds to Tamm Horsfall protein, forming casts that occlude tubules. LC undergoes endocytosis in the tubular epithelial cells, activating NFκB, stimulating production of proinflammatory cytokines that damage the

Address for correspondence: Akira Arimura, M.D., Ph.D., U.S.-Japan Biomedical Research Laboratories, Tulane University Hebert Research Center, Bldg. 30, 3705 Main St., Belle Chasse, LA 70037. Voice: 504-394-7199; fax: 504-394-7169.
    e-mail: arimura@tulane.edu

Ann. N.Y. Acad. Sci. 1070: 1–4 (2006). © 2006 New York Academy of Sciences.
doi: 10.1196/annals.1317.093

epithelial cells and tubulointerstitial tissue.[1,2] Although the pathophysiological mechanism has been elucidated, no effective treatment of myeloma kidney is available, apart from the limited use of steroids. Since pituitary adenylate cyclase-activating polypeptide (PACAP) is known to reduce cytokine production through suppression of NFκB in immune cells[3,4] and others, we evaluated PACAP-38 for possible beneficial effects for the treatment of myeloma kidney injury.

RESULTS AND DISCUSSION

First, immortalized human tubular cells were cultured to which human LC purified from the urine of a patient with MM was added. Addition of 50 μM LC stimulated production of tumor necrosis factor-α TNF-α and interleukin-6 (IL-6) time-dependently, reaching the plateau at 72 h.

PACAP-38 *dose* dependently suppressed LC-stimulated cytokine production as tested at 72 h. TNF-α production was suppressed by PACAP-38 at doses as low as 10^{-13} M. Dexamethasone (dex) also suppressed cytokine production.

LC-induced cytokine production is mediated by activation of mitogen-activated protein kinase (MAPKs).[2] PACAP-38 suppressed LC-stimulated p38 MAPK activation, but not extracellular signal regulated kinase (ERK) activation. PACAP-38 also suppressed LC-stimulated activation of p50 subunit of NFκB. P65 subunit of NFκB was activated in cultured tubular cells before addition of LC, but this level was also suppressed by PACAP-38. When LC was added to the tubular cell culture, the cells were detached, aggregated and underwent necrosis. However, these changes were prevented when PACAP-38 was added together with LC. Suppression of LC-stimulated TNF-α production by PACAP-38 was attenuated by either M65, a PAC1-specific inhibitor, or a VPAC1-specific inhibitor. Vasoactive intestinal peptide (VIP) also suppressed LC-induced TNF-α production, but to a lesser extent than PACAP-38, and VIP-induced suppression was greatly reduced by a VPAC1 inhibitor. These findings suggest that both PAC1- and VPAC1-receptors are involved in PACAP-induced suppression of cytokine production.

RT-PCR analysis of PACAP receptor mRNAs showed that tubular cells express both PAC1- and VPAC1-receptor mRNAs. Myeloma cells also express both PAC1- and VPAC1-receptor mRNAs. Human bone marrow stromal cells on which myeloma cells grow express both PAC1- and VPAC1-receptor mRNAs as well as VPAC2 receptor mRNA.

To examine the suppressive effect of PACAP *in vivo*, rats were administered intravenously LC alone or LC with PACAP for 3 days, and then TNF-α content in the kidney was determined. LC administration increased TNF-α content in the kidney several times, but simultaneous injection of PACAP-38 reduced the TNF-α level to nearly the control level. This suggests that administration of

PACAP-38 indeed reduces cytokine production stimulated by LC in the kidney.

It is known that renal failure is more closely related to tubulopathy than pure glomerulopathy.[5] In myeloma kidney, overflow of LC in the tubules stimulates cytokine production from the tubular epithelial cells, damaging the cells and interstitial tissues. If other proteins overflow the tubules, as seen in proteinuria, such as diabetic nephropathy, it could also stimulate cytokine production and damage the tubular epithelial cells and the interstitial tissue. If so, PACAP may also suppress such damage. This hypothesis was tested using diabetic rats induced by streptozotocin (SFZ). Some animals were subcutaneously implanted a miniosmotic pump to administer PACAP-38 i.v. continuously for 2 weeks. In STZ-treated rats, proteinuria occurred and the kidney was hypertrophied. Histological examination showed glomerular hypertrophy and vacuoles in the tubules. In the diabetic animals treated with PACAP-38, the glomerus looked normal and no vacuoles were seen in the tubules. TNF-α content in the kidney of diabetic rats was several times higher than that in control animals. But in diabetic rats treated with PACAP, the level was nearly the same as that in the control animals. These findings suggest that PACAP may also be used for preventing diabetic nephropathy.

On the other hand, PACAP is known to stimulate growth of some cancer cells.[6,7] If the treatment with PACAP stimulates growth of myeloma cells, its renoprotective effect would be greatly compromised. However, when human myeloma cells were cultured in the non-inactivating serum containing media, addition of PACAP-38 to the media dose dependently suppressed the growth of myeloma cells, as did dex. However, myeloma cells do not grow in peripheral blood, but grow in the bone marrow and adhere to the stromal cells. Adhesion of myeloma cells stimulates the release of growth factors such as IL-6 from stromal cells, enhancing growth of myeloma cells. PACAP-38 was found to suppress production of IL-6 stimulated by adhesion of myeloma cells. Therefore, PACAP suppresses growth of myeloma cells directly and also by affecting their internal milieu.

All of these findings render PACAP worth evaluation as a promising renoprotectant as well as a new antitumor agent in multiple myeloma.[8]

ACKNOWLEDGMENTS

This study was supported in part by Kaken American Foundation (AA) and by a Veterans Administration (VA) Merit Review Grant (VB).

REFERENCES

1. SENGUL, S., C. ZWIZINSKI, E.E. SIMON, *et al.* 2002. Endocytosis of light chains induces cytokines through activation of NF-κB in human proximal tubule cells. Kidney Int. **62:** 1977–1988.

2. SENGUL, S., C. ZWIZINSKI & V. BATUMAN. 2003. Role of MAPK pathways in light chain-induced cytokine production in human proximal tubule cells. Am J. Physiol. **284:** F1245–F1254.

3. DELGADO, M. & D. GANEA. 2001. Vasoactive intestinal peptide and pituitary adenylate cyclase-activating polypeptide inhibit nuclear factor-κ B-dependent gene activation at multiple levels in the human monocytic cell line THP-1. J. Biol. Chem. **276:** 369–380.

4. GANEA, D. & M. DELGADO. 2001. Neuropeptides as modulators of macrophage functions. Regulation of cytokine production and antigen presentation by VIP and PACAP. Arch. Immunol. Ther. Exp. (Warsz.) **49:** 101–110.

5. RISDON, R.A., J.C. SLOPER & H.E. DE WARDENER. 1968. Relationship between renal function and histological changes found in renal biopsy specimens from patients with persistent glomerular nephritis. Lancet **2:** 363–366.

6. BARRIE, A.P., A.M. CLOHESSY, C.S. BUENSUCESO, et al. 1997. Pituitary adenylyl cyclase-activating peptide stimulates extracellular signal-regulated kinase 1 or 2 (ERK1/2) activity in a Ras-independent, mitogen-activated protein Kinase/ERK kinase 1 or 2-dependent manner in PC12 cells. J. Biol. Chem. **272:** 19666–19671.

7. DOUZIECH, N., A. LAJAS, Z. COULOMBE, et al. 1998. Growth effects of regulatory peptides and intracellular signaling routes in human pancreatic cancer cell lines. Endocrine. **9:** 171–183.

8. ARIMURA, A., M. LI & V. BATUMAN. Potential protective action of pituitary adenylate cyclase activating polypeptide (PACAP-38) on in vitro and in vivo models of myeloma kidney injury. Prepublished on line as Blood First Edition Paper. October 4, 2005.

Clues to VIP Function from Knockout Mice

S.A. HAMIDI,[a,b] A.M. SZEMA,[a,b] S. LYUBSKY,[a,b] K.G. DICKMAN,[a,b]
A. DEGENE,[a,b] S.M. MATHEW,[a,b] J.A. WASCHEK,[c] AND S.I. SAID[a,b]

[a]SUNY, Stony Brook, New York 11794-8172, USA

[b]VA Medical Center, Northport, New York 11768, USA

[c]UCLA Medical Center, Los Angeles, California 90095, USA

ABSTRACT: We have taken advantage of the availability of vasoactive intestinal polypeptide (VIP) knockout (KO) mice to examine the possible influence of deletion of the VIP gene on: (a) airway reactivity and airway inflammation, as indicators of bronchial asthma; (b) mortality from endotoxemia, a model of septic shock; and (c) the pulmonary circulation. VIP KO mice showed: (a) airway hyperresponsiveness to the cholinergic agonist methacholine, as well as peribronchial and perivascular inflammation; (b) a greater susceptibility to death from endotoxemia; and (c) evidence suggestive of pulmonary hypertension.

KEYWORDS: VIP; lung; bronchial asthma; endotoxemia; septic shock; pulmonary circulation; knockout mice

INTRODUCTION

Over the years, a number of observations have suggested that vasoactive intestinal polypeptide (VIP) may play important roles in the regulation of lung and other vital organ function. Specifically, the observations relate to the role of the peptide in three clinical conditions or their experimental models: bronchial asthma, septic shock, and pulmonary hypertension. Such background information includes the following:

1. VIP is a potent relaxant of airway and pulmonary vascular smooth muscle, inhibitor of airway and pulmonary vascular smooth muscle proliferation, and a cotransmitter of the endogenous smooth muscle relaxant system.[1]
2. VIP has anti-inflammatory, immunomodulatory, and antiapoptotic properties,[2-4] and administration of either VIP or pituitary adenylate cyclase-activating polypeptide (PACAP), a closely related peptide that shares the

Address for correspondence: Sami I. Said, M.D., Pulmonary and Critical Care Medicine, SUNY Health Sciences Center, Stony Brook, NY 11794-8172. Voice: 631-444-1754; fax: 631-444-7502.
e-mail: sami.i.said@stonybrook.edu

Ann. N.Y. Acad. Sci. 1070: 5–9 (2006). © 2006 New York Academy of Sciences.
doi: 10.1196/annals.1317.035

same receptors with VIP, protects mice and rats against lung injury and endotoxin-induced lethality.[5,6]

3. Circulating levels of VIP are increased in clinical septic shock,[7] presumably as a defense mechanism.

The above findings suggested possible physiological roles for VIP as a modulator of bronchial asthma, septic shock and resultant multiorgan failure, and pulmonary hypertension.

We have now taken advantage of the availability of VIP knockout (KO) mice to examine the possible influence of deletion of the VIP gene on: (*a*) airway reactivity and airway inflammation, as indicators of bronchial asthma; (*b*) mortality from endotoxemia, a model of septic shock; and (*c*) the pulmonary circulation.

MATERIALS AND METHODS

Animals

VIP KO (VIP–/–) mice, prepared as described,[8] were backcrossed to the C57BL/6 strain. We bred the mice locally and confirmed the lack of the VIP gene by polymerase chain reaction (PCR) genotyping of tail-snips. Control, wild-type (WT) C57BL/6 mice were from Taconic Labs (Germantown, NY).

Airway Reactivity and Airway Inflammation

We examined bronchial reactivity in five male VIP KO and five male WT mice. The mice were anesthetized with pentobarbital, tracheostomized, and mechanically ventilated at a constant tidal volume. The cholinergic agonist methacholine was delivered as an aerosol, at four different concentrations, by an Aeroneb Nebulizer System (Buxco; Troy, NY). We recorded airway pressure continuously, and evaluated the degree of bronchoconstriction by increases in peak airway pressure that reflected increases in pulmonary resistance.

For evidence of airway inflammation, lungs were fixed, sectioned, and examined by a pathologist for the presence and intensity of inflammatory cell infiltrates.

Susceptibility to Endotoxemia

Five male and five female WT (15–17 weeks old) and five age-matched male and five female VIP KO mice were injected intraperitoneally with 1.0 mg lipopolysaccharide (LPS, *E. coli* serotype 0111:B4), and their survival was monitored over time.

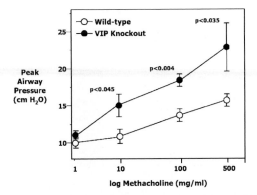

FIGURE 1. VIP KO mice were hyperresponsive to methacholine challenge. At three of the four concentrations tested, peak airway pressures reached significantly higher levels than in control mice.

Pulmonary Circulation

The thickness of pulmonary arteries was compared in lung sections from KO and WT mice.

Statistical Analysis

Results are expressed as means $\pm$ SEM. In each experimental model, the two groups of mice were compared by the unpaired, two-tailed Student's t-test, and P values were considered significant at or below 0.05.

RESULTS

Airway Reactivity and Airway Inflammation

Premethacholine levels of peak airway pressure were the same in both groups of mice, but reached significantly ($P < 0.05$) and considerably (32.4–46.2%) higher values in the VIP KO mice following three of the four doses (10, 10^2 and 5×10^2 mg/mL) of methacholine (FIG. 1).

Susceptibility to Endotoxemia

Male WT mice survived for 24.5 ± 0.7 h, while all male KO mice succumbed within 18 ± 1.1 h ($n = 5$, $P < 0.001$). The absence of the VIP gene, however, did not affect the lethality of LPS in female mice. Female mice in both groups died within 26.5 h after LPS treatment (FIGS. 2 and 3).

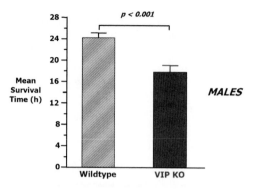

FIGURE 2. Age-matched male C57BL/6 WT and VIP KO mice were given 1 mg LPS (*E. coli* 0111:B4) by intraperitoneal injection, and survival was monitored over time. Values are means ± SEM, $n = 5$.

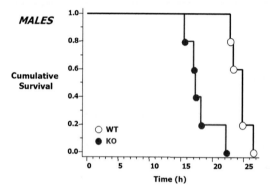

FIGURE 3. Kaplan–Meier plot of cumulative survival (fraction of mice alive versus time). Age-matched male C57BL/6 WT and VIP KO mice were given 1 mg LPS (*E. coli* 0111:B4) by intraperitoneal injection, and survival was monitored over time ($n = 5$ mice per group).

Pulmonary Circulation

In lung sections pulmonary vessels in KO mice appeared considerably thick-welled, compared to vessels in lungs from WT mice.

DISCUSSION

The rationale for examining VIP KO mice is that targeted deletion of the VIP gene, unless compensated for by other remaining genes, should result in impairment or loss of the functions attributed to VIP.

This study was designed to test our hypotheses that VIP is a major regulator of airway and pulmonary vascular smooth muscle function, a modulator of airway reactivity and inflammation, and a defender against acute lung injury and mortality from septic shock.

The results validate the original postulates on all three counts: Thus, VIP KO mice showed: (*a*) airway hyperresponsiveness to the cholinergic agonist methacholine, as well as peribronchial and perivascular inflammation; (*b*) a greater susceptibility to death from endotoxemia; and (*c*) evidence suggestive of pulmonary hypertension. It is notable that PACAP, which shares most or all of the actions of VIP, apparently did not make up for the absence of the VIP gene in these experimental models.

The findings therefore support the conclusion that in the absence of VIP, an important and essential modulator of major aspects of lung and other vital organ function, mice are likely to exhibit features of bronchial asthma, to die more readily of endotoxemia, and to develop pulmonary hypertension.

REFERENCES

1. SAID, S.I. 1991. Vasoactive intestinal polypeptide (VIP) in asthma. Ann. N. Y. Acad. Sci. **629:** 305–318.
2. VOICE, J.K. *et al.* 2002. Immunoeffector and immunoregulatory activities of vasoactive intestinal peptide. Regul. Pept. **109:** 199–208.
3. DELGADO, M. *et al.* 2004. The significance of vasoactive intestinal peptide in immunomodulation. Pharmacol. Rev. **56:** 249–290.
4. SAID, S.I. & K.G. DICKMAN. 2000. Pathways of inflammation and cell death in the lung: modulation by vasoactive intestinal peptide. Reg. Peptides **93:** 21–29.
5. DELGADO, M. *et al.* 1999. Vasoactive intestinal peptide (VIP) and pituitary adenylate cyclase-activation polypeptide (PACAP) protect mice from lethal endotoxemia through the inhibition of TNF-α and IL-6. J. Immunol. **162:** 1200–1205.
6. TANDON, R. *et al.* 1999. Protection by Vasoactive intestinal peptide (VIP) against endotoxin shock. Am. J. Respir. Crit. Care Med. **159:** A612.
7. BRANDTZAEG, P. *et al.* 1989. Elevated VIP and endotoxin plasma levels in human gram-negative septic shock. Reg. Peptides **24:** 37–44.
8. COLWELL, C.S. *et al.* 2003. Disrupted circadian rhythms in VIP- and PHI-deficient mice. Am. J. Physiol. **285:** R939–R949.

The Glucagon-Like Peptides

Pleiotropic Regulators of Nutrient Homeostasis

PATRICIA L. BRUBAKER

Departments of Physiology and Medicine, University of Toronto, Toronto on M5S 1A8, Canada

ABSTRACT: The glucagon-like peptides, GLP-1 and GLP-2, are cosecreted by intestinal L cells in response to nutrient ingestion. These peptides exert multiple effects on the gastrointestinal tract and pancreas to regulate the digestion, absorption, and assimilation of ingested nutrients, as well as providing feedback signals to the brain to modulate food intake. Tropic effects of GLP-1 and GLP-2 on their major peripheral target tissues, the beta cell and the intestinal epithelium, respectively, further enhance capacity for nutrient handling. When taken together, these findings demonstrate the diverse actions of the intestinal glucagon-like peptides to regulate nutrient homeostasis.

KEYWORDS: absorption; beta cell; digestion, glucagon; GLP-1; GLP-2; growth; insulin; intestine; motility; nutrient; satiety; secretion

INTRODUCTION

The proglucagon gene is expressed at high levels in pancreatic A and intestinal L cells, as well as in selected neurons of the hypothalamus and brain stem.[1] Tissue-specific posttranslational processing of proglucagon liberates a diversity of biologically active peptides, most notably glucagon in the A cells and the glucagon-like peptides (GLP-1 and GLP–2) in the L cells[2,3] (FIG. 1). The biological actions of glucagon to increase blood glucose levels through stimulation of hepatic glucose production have long been established.[4] However, it was not until 1987 and 1996, respectively, that biological activities were discovered for GLP-1 and GLP-2.[5,6] As discussed in this article, the GLPs are now known to exert pleiotropic effects on the gastroenteropancreatic axis, the brain, and peripheral tissues to modulate nutrient homeostasis at the levels of ingestion, digestion, absorption, and deposition. Through more recently described actions to enhance the growth and survival of both the intestinal mucos and the pancreatic islet, these peptides further promote positive energy balance.

Address for correspondence: Dr. P.L. Brubaker, Room 3366 Medical Sciences Building, University of Toronto, 1 King's College Circle, Toronto, ON M5S 1A8 Canada. Voice and fax: 1-416-978-2593. e-mail: p.brubaker@utoronto.ca

Ann. N.Y. Acad. Sci. 1070: 10–26 (2006). © 2006 New York Academy of Sciences.
doi: 10.1196/annals.1317.006

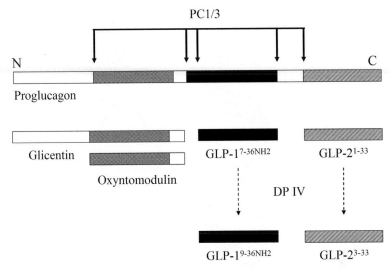

FIGURE 1. Schematic of GLP-1 and GLP-2 biosynthesis and degradation. GLP-1 and GLP-2 are synthesized as part of the larger proglucagon precursor molecule. Tissue-specific expression of prohormone convertase (PC) 1/3 in the intestinal L cells results in liberation of the biologically active GLPs, as well as of equimolar amounts of glicentin plus oxyntomodulin. The pancreatic hormone glucagon (shown in the checkerboard pattern) is not released from proglucagon in the L cell. After secretion of GLP-1 and GLP-2 from the L cell, both peptides are inactivated by dipeptidylpeptidase IV (DP IV)-mediated removal of the N-terminal two amino acids.[114,115]

REGULATION OF GLP-1 AND GLP-2 SECRETION BY NUTRIENTS

The major physiological stimulus for GLP-1 and GLP-2 secretion is ingestion of a meal (FIG. 2), consistent with important roles in the regulation of nutrient homeostasis (FIGS. 3 and 4). Despite the predominant localization of the L cell in the distal intestine,[7] intake of either glucose or fat increases release of GLP-1 and GLP-2 within 15–30 min, with a second peak of secretion occurring at 90–120 min after a meal.[8,9] We therefore postulated the existence of two pathways regulating GLP secretion, the first occurring immediately after nutrient ingestion, being mediated indirectly through activation of a neuro/endocrine mechanism, and the second occurring later, after transit of the nutrients through the intestinal lumen to interact directly with the L cell[10–12] (FIG. 2). A large number of studies in rodents have now demonstrated that placement of either glucose or fat directly into the upper gastrointestinal (GI) tract stimulates prompt rises in GLP release.[10,12–15] These effects are not on account of intestinal distension, and can be prevented by bilateral subdiaphragmatic vagotomy, as well as by administration of antagonists against both muscarinic (e.g., atropine and M1-specific) and gastrin-releasing

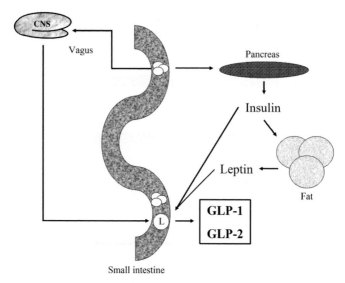

FIGURE 2. Schematic of indirect and direct regulation of GLP-1 and GLP-2 secretion by ingested nutrients. The indirect pathway is initiated by the presence of nutrients in the proximal small intestine, whereby a vagal pathway is activated leading to GLP secretion from the distal L cell. Absorption of ingested nutrients also results in increased levels of insulin and leptin, both of which contribute to enhanced GLP levels. The direct effects of nutrients on the L cell are temporally delayed as compared to the indirect effects, on account of transit of the ingested nutrients to the distal intestine. This pathway is mediated through direct actions of nutrients, and most notably long-chain mono- and polyunsaturated fatty acids, on the L cell.

peptide (GRP) receptors.[11,12,16,17] Similarly, infusion of atropine into humans prevents the early peak of GLP-1 secretion,[18] although the results of this study should be interpreted with caution, as atropine also delays gastric emptying, which may preclude activation of putative duodenal nutrient sensors. Nonetheless, consistent with the antagonist findings, GLP secretion is stimulated by bethanechol, M1 agonists, and GRP, both *in vivo* and *in vitro*, and is reduced in mice-lacking GRP receptors.[13,16,17,19] Hence, nutrient-dependent activation of the vagus stimulates the L cell via a cholinergic and "GRPergic"-dependent pathway. That the L cell being stimulated by this neural pathway actually resides in the distal intestine was demonstrated in studies showing complete abrogation of proximal nutrient-dependent GLP stimulation in rats following resection of the distal small and large intestine.[11] Importantly, although initial studies performed in rats demonstrated a role for the duodenal hormone glucose-dependent insulinotropic peptide (GIP) in activating the vagal pathway in response to proximal nutrients,[10,12] this has not been confirmed in humans.[20] Furthermore, although several other gut hormones and neurotransmitters have been shown to activate the L cell,[21,22] their role in regulating GLP

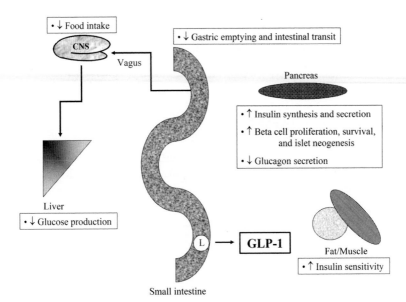

FIGURE 3. The biological actions of GLP-1 to modulate nutrient homeostasis include effects on the endocrine pancreas, GI tract, peripheral tissues, and the brain.

secretion in response to ingested nutrients currently remains elusive. Hence, many questions remain regarding the mechanism(s) underlying the early peak of GLP following a meal, particularly in humans.

Nutrient intake is also associated with release of two peripheral hormones with key roles in nutrient homeostasis, insulin, and leptin. Interestingly, recent studies have now demonstrated that both of these hormones stimulate GLP-1 release in association with activation of their respective receptors on the L cell[23,24] (FIG. 2). Consistent with these findings, leptin resistance has been demonstrated to decrease the effects of leptin on GLP-1 release,[23] and may therefore contribute to the reduced levels of GLP-1 that have been observed in obese individuals.[25] Similarly, GLP-1 levels have been reported to be low in patients with insulin resistance or type 2 diabetes, independent of both obesity[26,27] and clearance rates.[28] However, as found for leptin, a very recent study has now shown that insulin resistance can be induced in the L cell,[24] suggesting a possible explanation for the low levels of GLP-1 in such patients. Hence, the regulation of GLP-1 and GLP-2 secretion by other nutrient-dependent hormones may provide feedback to the L cell regarding nutrient homeostasis and energy stores at the level of the whole organism.

Finally, a number of studies have now clearly demonstrated that nutrients can exert direct, stimulatory effects on the L cell, suggesting a role in the late phase of GLP secretion following nutrient ingestion. Placement of glucose into the lumen of the ileum increases GLP release,[10,29] while recent

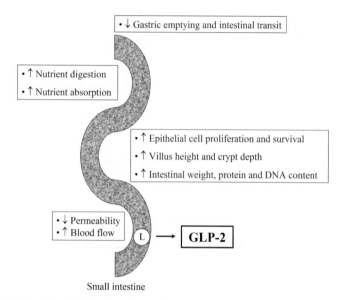

FIGURE 4. The biological actions of GLP-2 to modulate nutrient homeostasis are largely restricted to the effects on the GI tract.

studies by Gribble and Reimann have demonstrated that glucose stimulates GLP-1 secretion through both K_{ATP}- and sodium–glucose luminal transporter (SGLT)-dependent mechanisms.[30,31] However, the levels of glucose that actually reach the distal intestine are very low,[32] and glucose is therefore not likely to exert major effects on the L cell under physiological conditions. Similarly, although amino acids and peptones can also directly stimulate the L cell,[29,33] the majority of protein absorption by the GI tract occurs proximally.[34] In contrast, fatty acids reach the distal intestine at high levels,[35] and placement of fat into the ileum stimulates GLP release.[10] Furthermore, long-chain fatty acids, and particularly of mono- and polyunsaturated fatty acids, stimulate GLP-1 secretion when applied directly to the L cell, through mechanisms that appear to involve the newly described fatty acid receptor, GPR120[36] and/or protein kinase Cζ (PKCζ).[37,38] When taken together, these findings suggest an important physiological role for fatty acids in the direct regulation of the distal L cell by ingested nutrients.

BIOLOGICAL ACTIONS OF GLP-1 AND GLP-2 TO PROMOTE POSITIVE NUTRIENT BALANCE

Nutrient Digestion and Absorption

The major biological actions of GLP-2 occur at the level of the GI tract (FIG. 4). Administration of either GLP-2 or of a degradation-resistant analog (e.g., Gly^2-GLP-2; Fig. 1) to rodents for 10–14 days increases small and, to a lesser extent, large intestinal wet weight, in association with enhanced protein and DNA content.[6,39–43] The tropic effects of GLP-2 in the small intestine are largely restricted to the crypt-villus axis, concomitant with increased crypt cell proliferation and reduced villus apoptosis.[40,43–45] Although the importance of GLP-2 in intestinal development is not yet clear, on account of the current lack of a GLP-2 "null" mouse, a recent study has demonstrated that GLP-2 does play a role in the physiological regulation of intestinal tropic responses to fasting and refeeding in the adult mouse.[45] Furthermore, GLP-2 exerts mucosal protective effects in a variety of pathophysiological models, preventing and/or repairing the damage caused by injury or administration of noxious agents.[46–50] These effects are associated, in part, with a reduced inflammatory response of the bowel to injury,[48,50] possibly consequent to the ability of GLP-2 to reduce intestinal permeability.[51] Finally, GLP-2 has been demonstrated to enhance small bowel growth in rodent models of intestinal insufficiency, most notably following massive small bowel resection.[52–54]

GLP-2 has a number of additional actions on the GI tract to directly increase the digestion and absorption of nutrients (Fig. 4). Acute infusion (e.g., 1–4 h) of GLP-2, but not of GLP-1, into rats increases glucose uptake, through recruitment and/or activation of both SGLT-1 and glucose transporter-2.[55,56] As GLP-2 also acutely enhances blood flow in the intestine,[57] nutrient uptake may also be facilitated indirectly through this mechanism. Furthermore, chronic administration of GLP-2 (e.g., 10–14 days) increases expression of not only SGLT-1, but also several brush border digestive enzymes, including maltase and sucrase isomaltase.[39,52,54,58] As absorption of disaccharides and triglycerides, as well as of monosaccharides and amino acids is enhanced in rodents following chronic treatment with GLP-2,[39,52] these findings are consistent with effects of this hormone to increase both digestive enzyme activity and absorptive capacity in the small intestine.

When taken together, these studies have clearly demonstrated the unique actions of GLP-2 to enhance intestinal growth and prevent or repair damage, in association with an increased capacity of the gut to digest and absorb ingested nutrients. Consistent with these findings in rodents, patients with short bowel syndrome also demonstrate an enhanced ability for enteral nutrient ingestion after only 30-day treatment with GLP-2.[59] Clinical trials are now ongoing to determine the efficacy of a degradation-resistant analog of GLP-2 (Teduglutide) in the treatment of patients with short bowel syndrome, as well in individuals with inflammatory bowel (Crohn's) disease (http://www.npsp.com).

Interestingly, despite a number of studies on the 7-transmembrane, Gαs-coupled GLP-2 receptor in both heterologous and intestinal cell lines,[60–62] the exact mechanism of action of GLP-2 within the intestinal epithelium remains unclear. The GLP-2 receptor has been localized to a number of different cell types, including enteroendocrine cells, enteric neurons, and subepithelial myofibroblasts.[41,63,64] However, the one consistent pattern in all of these studies is the absence of receptor expression in the cells that respond to GLP-2, most notably the crypt and villus epithelial cells. Hence, it has been proposed that the tropic effects of GLP-2 are mediated indirectly, through one or more other growth factors.[63] One recent study has demonstrated a role for keratinocyte growth factor in mediating the tropic effects of GLP-2 on the colon[41]; however, the identity of the factor(s) responsible for GLP-2 action in the small intestine remains elusive. Insulin like growth factor-1 has very recently been identified as a mediator of the tropic effects of GLP-2 on the small and large intestine.[65] The existence of such indirect mediators of GLP-2 action is also consistent with a recent report that acute administration of GLP-2 is associated with activation of Akt in intestinal epithelial cells,[66] a signaling molecule that is associated with cell growth and survival, but that is not typically activated by $G\alpha s$-linked receptors.

Nutrient Deposition

In contrast to the relative restricted effects of GLP-2, the biological actions of GLP-1 to regulate nutrient homeostasis are extremely diverse (FIG. 3). The first, and arguably most important effect of GLP-1 to be discovered was its ability to stimulate insulin secretion in a glucose-dependent fashion.[5,67-70] Together with the duodenal hormone, GIP, GLP-1 functions as a physiological "incretin," or an intestinal hormone released in response to nutrients that enhances glucose-stimulated insulin secretion.[71] Indeed, in a large number of studies on mice lacking the GLP-1 receptor, the major phenotype is mild hyperglycemia consequent to impaired insulin release.[72-74] Importantly, and in contrast to the actions of many other insulin secretagogues, GLP-1 also enhances insulin biosynthesis,[75] thereby replenishing insulin stores, and suppresses glucagon release, possibly through a somatostatin-dependent mechanism.[67,76] Hence, the response of the endocrine pancreas to nutrient ingestion is integrated through the actions of GLP-1 to ensure both nutrient disposal and maintenance of normoglycemia.

The mechanism by which GLP-1 enhances glucose-dependent insulin secretion has been the subject of extensive investigation over the past decade. The beta cell expresses high levels of the 7-transmembrane, G protein-coupled GLP-1 receptor,[77] and activation of this receptor enhances the actions of glucose on the insulin secretory pathway, via direct effects on K_{ATP} channels, voltage-dependent calcium channels, and secretory granule exocytosis.[78] These findings are consistent with reports that the ligand binding to the GLP-1 receptor is coupled to activation of a cyclic adenosine $3',5'$-monophosphate (cAMP)/protein kinase A-dependent signaling.[77] However, the GLP-1 receptor can bind to multiple G proteins,[79] and consistent with this finding, GLP-1 has been found to stimulate proliferation and inhibit apoptosis in rodent beta cells[80-83] through an Akt- and PKCζ-dependent pathway.[84,85] Furthermore, recent studies have elucidated a novel role for GLP-1 as a beta cell tropic factor *in vivo*, enhancing beta cell growth and survival as well as stimulating islet neogenesis in a variety of different rodent models, including normal, prediabetic, and diabetic animals, as well as rodents with pancreatic insufficiency on account of streptozotocin treatment or partial pancreatectomy.[80–83,86–86] Hence, like GLP-2, GLP-1 exerts important tropic effects on its major target tissue, thereby further increasing its actions to promote a positive energy balance.

In addition to the endocrine pancreas, a number of peripheral tissues have also been reported to be targets of GLP-1 action. In particular, GLP-1 suppresses hepatic glucose production, as well as stimulates glucose uptake in muscle and fat,[89,90] thereby further promoting nutrient disposal and reducing glycemia. As the cloned GLP-1 receptor has not been identified in these tissues, it remains possible that an alternative GLP-1 receptor is responsible for these actions. However, quite interestingly, the actions of GLP-1 on the liver, albeit not on fat and muscle, have recently been reported to require the vagus nerve,[91,92] which is known to express the GLP-1 receptor.[93] Similarly, although the islet expresses high levels of the GLP-1 receptor, the biological actions of GLP-1 on insulin secretion have also been reported to require an intact sensory afferent nervous system.[91,94] Hence, as found for GLP-2, at least some of the physiological actions of GLP-1 appear be exerted indirectly, through the actions of other mediators.

Excitingly, the pleiotropic actions of GLP-1 on nutrient homeostasis, and on blood glucose levels in particular, have led to clinical trials for the use of this hormone in the treatment of patients with type 2 diabetes. The vast majority of studies to date has demonstrated that chronic administration of GLP-1, long-acting analogs, GLP-1 receptor agonists (e.g., exendin-4), or agents to reduce degradation of GLP-1 (and other incretins), leads to significant reductions in both glycemia and HbA1c levels, in rodent models of diabetes, as well as in patients with type 2 diabetes.[95–99] To date, there have been few reported side effects of such therapy, with transient nausea appearing to be most common.[100] The successful results of such trials have now led to the approval of exendin-4 (Exenatide; Byetta™ Amylin Pharmaceuticals, Inc., San Diego, CA) for use in patients with type 2 diabetes.[101]

Feedback Effects

Feedback loops within endocrine systems are a common mechanism by which a fine degree of control is exerted over circulating hormone levels. Consistent with this general model, both nutrient ingestion and the presence of nutrients in the intestinal lumen are subject to feedback regulation by GLP-1 and/or GLP-2.

Peripheral administration of GLP-1 or exendin-4 acutely suppresses food intake in both rodents and humans.[69,102–105] Recent studies have demonstrated that, although GLP-1 can cross the blood–brain barrier,[106] the effects of the circulating peptide on food intake are mediated through sensory afferent nerves, most likely the vagus.[102,104,105] Importantly, these effects occur at doses that do not induce behavioral effects or cause conditioned taste aversion.[104] Furthermore, other products of the intestinal L cell, including oxyntomodulin (FIG. 1) and peptide $YY^{3-36NH2}$ also suppress food intake, with peptide $YY^{3-36NH2}$ acting in a synergistic fashion with the GLP-1 receptor agonist, exendin-4.[104,107] Consistent with these acute effects to induce satiety, chronic administration of GLP-1 or exendin-4 also results in weight loss.[100,103,108] Of particular importance, patients with type 2 diabetes who are treated with exendin-4 exhibit this weight loss despite improved glycemic control.[100]

Finally, GLP-1 and, to a lesser extent, GLP-2, exert feedback effects on the GI tract to delay gastric emptying and reduce intestinal motility[15,109,110] (FIGS. 3 and 4). Oxyntomodulin and glicentin, which are cosecreted with GLP-1 and GLP-2 (FIG. 1), act in conjunction with these peptides to reduce both gastric acid secretion and GI motility.[111-113] Together, these effects contribute to the "ileal brake" effect, providing feedback to the proximal GI tract whereby the presence of nutrients in the lumen, particularly in the distal intestine, generates a hormonal signal leading to an increased time of exposure of those nutrients to the digestive and absorptive processes of the GI tract. However, of some relevance to this hypothesis, it has also been proposed that, through its effects to delay gastric emptying, GLP-1 also slows down the rate of entry of glucose into the circulation.[114] Hence, the ileal brake may actually serve multiple roles, whereby nutrient digestion and absorption are delayed in order to ultimately promote efficient nutrient uptake, with release of both GLP-1 and GLP-2 continuing until all of the luminal nutrients have been absorbed.

CONCLUSIONS

Through their effects to enhance nutrient digestion and absorption by the GI tract, to facilitate insulin secretion and, hence, nutrient disposal, and to promote the growth of both the intestinal epithelium and the beta cell, the glucagon-like peptides exert a wide array of biological actions resulting in positive energy balance. These peptides function within a classical endocrine feedback loop, being stimulated by nutrient ingestion and providing feedback signals to limit their secretion. Together, GLP-1 and GLP-2 provide an integrated mechanism by which nutrient homeostasis is regulated.

ACKNOWLEDGMENTS

This work was supported by grants from the Canadian Diabetes Association and the Canadian Institutes of Health Research. PLB is supported by the Canada Research Chairs Program.

REFERENCES

1. LEE, Y.C., P.L. BRUBAKER & D.J. DRUCKER. 1990. Developmental and tissue-specific regulation of proglucagon gene expression. Endocrinology **127:** 2217–2222.
2. DHANVANTARI, S., N.G. SEIDAH & P.L. BRUBAKER. 1996. Role of prohormone convertases in the tissue-specific processing of proglucagon. Mol. Endocrinol. **10:** 342–355.
3. ROUILLÉ, Y., G. WESTERMARK, S.K. MARTIN & D.F. STEINER. 1994. Proglucagon is processed to glucagon by prohormone convertase PC2 in αTC1-6 cells. Proc. Natl. Acad. Sci. USA **91:** 3242–3246.
4. LEFÈBVRE, P.J. 1995. Glucagon and its family revisited. Diabetes Care **18:** 715–730.
5. MOJSOV, S., G.C. WEIR & J.F. HABENER. 1987. Insulinotropin: glucagon-like peptide I(7-37) co-encoded in the glucagon gene is a potent stimulator of insulin release in the perfused rat pancreas. J. Clin. Invest. **79:** 616–619.
6. DRUCKER, D.J., P. EHRLICH, S.L. ASA & P.L. BRUBAKER. 1996. Induction of intestinal epithelial proliferation by glucagon-like peptide 2. Proc. Natl. Acad. Sci. USA **93:** 7911–7916.
7. EISSELE, R., R. GÖKE, S. WILLEMER, et al. 1992. Glucagon-like peptide-1 cells in the gastrointestinal tract and pancreas of rat, pig and man. Eur. J. Clin. Invest. **22:** 283–291.
8. D'ALESSIO, D., R. THIRLBY, E. LASCHANSKY, et al. 1993. Response of tGLP-1 to nutrients in humans. Digestion **54:** 377–379.
9. XIAO, Q., R. BOUSHEY, D.J. DRUCKER & P.L. BRUBAKER. 1999. Secretion of the intestinotropic hormone glucagon-like peptide-2 is differentially regulated by nutrients in humans. Gastroenterology **117:** 99–105.
10. ROBERGE, J.N. & P.L. BRUBAKER. 1993. Regulation of intestinal proglucagon-derived peptide secretion by glucose-dependent insulinotropic peptide in a novel enteroendocrine loop. Endocrinology **133:** 233–240.
11. ROBERGE, J.N., K.A. GRONAU & P.L. BRUBAKER. 1996. Gastrin-releasing peptide is a novel mediator of proximal nutrient-induced proglucagon-derived peptide secretion from the distal gut. Endocrinology **137:** 2383–2388.
12. ROCCA, A.S. & P.L. BRUBAKER. 1999. Role of the vagus nerve in mediating proximal nutrient-induced glucagon-like peptide-1 secretion. Endocrinology **140:** 1687–1694.
13. PERSSON, K., R.L. GINGERICH, S. NAYAK, et al. 2000. Reduced GLP-1 and insulin responses and glucose intolerance after gastric glucose in GRP receptor-deleted mice. Am. J. Physiol Endocrinol. Metab. **279:** E956–E962.
14. KNAPPER, J.M., A. HEATH, J.M. FLETCHER, et al. 1995. GIP and GLP-1(7-36)amide secretion in response to intraduodenal infusions of nutrients in pigs. Comp Biochem. Physiol C. Pharmacol. Toxicol. Endocrinol. **111:** 445–450.
15. SCHIRRA, J., M. KATSCHINSKI, C. WEIDMANN, et al. 1996. Gastric emptying and release of incretin hormones after glucose ingestion in humans. J. Clin. Invest **97:** 92–103.
16. ANINI, Y., T. HANSOTIA & P.L. BRUBAKER. 2002. Muscarinic receptors control postprandial release of glucagon-like peptide-1: in vivo and in vitro studies in rats. Endocrinology **143:** 2420–2426.

17. ANINI, Y. & P.L. BRUBAKER. 2003. Muscaranic receptors control glucagon-like peptide 1 secretion by human endocrine L cells. Endocrinology **144:** 3244–3250.

18. BALKS, H.J., J.J. HOLST, A. VON ZUR MÜHLEN & G. BRABANT. 1997. Rapid oscillations in plasma glucagon-like peptide-1 (GLP- 1) in humans: cholinergic control of GLP-1 secretion via muscarinic receptors. J. Clin. Endocrinol. Metab. **82:** 786–790.

19. ABELLO, J., F. YE, A. BOSSHARD, *et al.* 1994. Stimulation of glucagon-like peptide-1 secretion by muscarinic agonist in a murine intestinal endocrine cell line. Endocrinology **134:** 2011–2017.

20. NAUCK, M.A., M.M. HEIMESAAT, C. ORSKOV, *et al.* 1993. Preserved incretin activity of glucagon-like peptide 1 [7- 36 amide] but not of synthetic human gastric inhibitory polypeptide in patients with type-2 diabetes mellitus. J. Clin. Invest. **91:** 301–307.

21. DUMOULIN, V., T. DAKKA, P. PLAISANCIE, *et al.* 1995. Regulation of glucagon-like peptide-1-(7-36)amide, peptide YY, and neurotensin secretion by neurotransmitters and gut hormones in the isolated vascularly perfused rat ileum. Endocrinology **136:** 5182–5188.

22. PLAISANCIE, P., C. BERNARD, J.-A. CHAYVIALLE & J.-C. CUBER. 1994. Regulation of glucagon-like peptide-1-(7-36) amide secretion by intestinal neurotransmitters and hormones in the isolated vascularly perfused rat colon. Endocrinology **135:** 2398–2403.

23. ANINI, Y. & P.L. BRUBAKER. 2003. Role of leptin in the regulation of glucagon-like peptide-1 secretion. Diabetes **52:** 252–259.

24. LIM, G. & P.L. BRUBAKER. 2006. The effect of insulin resistance on glucagon-like peptide-1 (GLP-1) secretion in mouse enteroendocrine cells [abstract]. Diabetes **55:** 364.

25. RANGANATH, L.R., J.M. BEETY, L.M. MORGAN, *et al.*1996. Attenuated GLP-1 secretion in obesity: cause or consequence. Gut **38:** 916–919.

26. RASK, E., T. OLSSON, S. SODERBERG, *et al.* 2001. Impaired incretin response after a mixed meal is associated with insulin resistance in nondiabetic men. Diab. Care **24:** 1640–1645.

27. VILSBOLL, T., T. KRARUP, C.F. DEACON, *et al.* 2001. Reduced postprandial concentrations of intact biologically active glucagon-like peptide 1 in type 2 diabetic patients. Diabetes **50:** 609–613.

28. VILSBOLL, T., H. AGERSO, T. KRARUP & J.J. HOLST. 2003. Similar elimination rates of glucagon-like peptide-1 in obese type 2 diabetic patients and healthy subjects. J. Clin. Endocrinol. Metab **88:** 220–224.

29. DUMOULIN, V., F. MORO, A. BARCELO, *et al.* 1998. Peptide YY, glucagon-like peptide-1, and neurotensin responses to luminal factors in the isolated vascularly perfused rat ileum. Endocrinology **139:** 3780–3786.

30. REIMANN, F. & F.M. GRIBBLE. 2002. Glucose-sensing in glucagon-like peptide-1-secreting cells. Diabetes **51:** 2757–2763.

31. GRIBBLE, F.M., L. WILLIAMS, A.K. SIMPSON & F. REIMANN. 2003. A novel glucose-sensing mechanism contributing to glucagon-like peptide-1 secretion from the GLUTag cell line. Diabetes **52:** 1147–1154.

32. FERRARIS, R.P., S. YASHARPOUR, K.C.K. LLOYD, *et al.* 1990. Luminal glucose concentrations in the gut under normal conditions. Am. J. Physiol. Gastrointest. Liver Physiol. **259:** G822–G837.

33. REIMER, R.A., C. DARIMONT, S. GREMLICH, *et al.* 2001. A human cellular model for studying the regulation of glucagon-like peptide-1 secretion. Endocrinology **142:** 4522–4528.
34. MODIGLIANI, R., J.C. RAMBAUD & J.J. BERNIER. 1973. The method of intraluminal perfusion of the human small intestine. II. Absorption studies in health. Digestion **9:** 264–290.
35. LIN, H.C., X.T. ZHAO & L. WANG. 1996. Fat absorption is not complete by midgut but is dependent on load of fat. Am. J. Physiol **271:** G62–G67.
36. HIRASAWA, A., K. TSUMAYA, T. AWAJI, *et al.* 2005. Free fatty acids regulate gut incretin glucagon-like peptide-1 secretion through GPR120. Nat. Med. **11:** 90–94.
37. ROCCA, A.S. & P.L. BRUBAKER. 1995. Stereospecific effects of fatty acids on proglucagon-derived peptide secretion in fetal rat intestinal cultures. Endocrinology **136:** 5593–5599.
38. IAKOUBOV, R., C.I. WHITESIDE & P.L. BRUBAKER. 2004. Protein kinase C zeta activation and translocation are required for fatty acid-induced secretion of GLP-1 in intestinal endocrine L cells [abstract]. Eur. Assoc. Study Diab. 16A.
39. BRUBAKER, P.L., A. IZZO, M. HILL & D.J. DRUCKER. 1997. Intestinal function in mice with small bowel growth induced by glucagon-like peptide-2. Am. J. Physiol. Endocrinol. Metab. **272:** E1050–E1058.
40. BURRIN, D.G., B. STOLL, R. JIANG, *et al.* 2000. GLP-2 stimulates intestinal growth in premature TPN-fed pigs by suppressing proteolysis and apoptosis. Am. J. Physiol. Gastrointest. Liver Physiol. **279:** G1249–G1256.
41. ORSKOV, C., B. HARTMANN, S.S. POULSEN, *et al.* 2005. GLP-2 stimulates colonic growth via KGF, released by subepithelial myofibroblasts with GLP-2 receptors. Regul. Pept. **124:** 105–112.
42. GHATEI, M.A., R.A. GOODLAD, S. TAHERI, *et al.* 2001. Proglucagon-derived peptides in intestinal epithelial proliferation: glucagon-like peptide-2 is a major mediator of intestinal epithelial proliferation in rats. Dig. Dis. Sci. **46:** 1255–1263.
43. BURRIN, D.G., B. STOLL, X. GUAN, *et al.* 2005. Glucagon-like peptide 2 dose-dependently activates intestinal cell survival and proliferation in neonatal piglets. Endocrinology **146:** 22–32.
44. TSAI, C.H., M. HILL, S.L. ASA, *et al.* 1997. Intestinal growth-promoting properties of glucagon-like peptide-2 in mice. Am. J. Physiol. Endocrinol. Metab. **273:** E77–E84.
45. SHIN, E.D., J.L. ESTALL, A. IZZO, *et al.* 2005. Mucosal adaptation to enteral nutrients is dependent on the physiologic actions of glucagon-like peptide-2 in mice. Gastroenterology **128:** 1340–1353.
46. BOUSHEY, R.P., B. YUSTA & D.J. DRUCKER. 1999. Glucagon-like peptide 2 decreases mortality and reduces the severity of indomethacin-induced murine enteritis. Am. J. Physiol. Endocrinol. Metab. **277:** E937–E947.
47. BOUSHEY, R.P., B. YUSTA & D.J. DRUCKER. 2001. Glucagon-like peptide (GLP)-2 reduces chemotherapy-associated mortality and enhances cell survival in cells expressing a transfected GLP-2 receptor. Cancer Res. **61:** 687–693.
48. L'HEUREUX, M.-C. & P.L. BRUBAKER. 2003. Glucagon-like peptide-2 and common therapeutics in a murine model of ulcerative colitis. J. Pharm. Exp. Ther. **306:** 347–354.

49. PRASAD, R., K. ALAVI & M.Z. SCHWARTZ. 2000. Glucagonlike peptide-2 analogue enhances intestinal mucosal mass after ischemia and reperfusion. J. Ped. Surg. **35:** 357–359.

50. ALAVI, K., M.Z. SCHWARTZ, J.P. PALAZZO & R. PRASAD. 2000. Treatment of inflammatory bowel disease in a rodent model with the intestinal growth factor glucagon-like peptide-2. J. Pediatr. Surg. **35:** 847–851.

51. BENJAMIN, M.A., D.M. MCKAY, P.C. YANG, et al. 2000. Glucagon-like peptide-2 enhances intestinal epithelial barrier function of both transcellular and paracellular pathways in the mouse. Gut **47:** 112–119.

52. SCOTT, R.B., D. KIRK, W.K. MACNAUGHTON & J.B. MEDDINGS. 1998. GLP-2 augments the adaptive response to massive intestinal resection in rat. Am. J. Physiol. Gastrointest. Liver Physiol. **275:** G911–G921.

53. WASHIZAWA, N., L.H. GU, L. GU, et al. 2004. Comparative effects of glucagon-like peptide-2 (GLP-2), growth hormone (GH), and keratinocyte growth factor (KGF) on markers of gut adaptation after massive small bowel resection in rats. JPEN J. Parenter. Enteral Nutr. **28:** 399–409.

54. MARTIN, G.R., L.E. WALLACE & D.L. SIGALET. 2004. Glucagon-like peptide-2 induces intestinal adaptation in parenterally fed rats with short bowel syndrome. Am. J. Physiol Gastrointest. Liver Physiol **286:** G964–G972.

55. CHEESEMAN, C.I. 1997. Upregulation of SGLT-1 transport activity in rat jejunum induced by GLP-2 infusion in vivo. Am. J. Physiol. Regul. Integr. Comp. Physiol. **273:** R1965–R1971.

56. CHEESEMAN, C.I. & D. O'NEILL. 1998. Basolateral D-glucose transport activity along the crypt-villus axis in rat jejunum and upregulation induced by gastric inhibitory peptide and glucagon-like peptide-2. Exp. Physiol. **83:** 605–616.

57. GUAN, X., B. STOLL, X. LU, et al. 2003. GLP-2-mediated up-regulation of intestinal blood flow and glucose uptake is nitric oxide-dependent in TPN-fed piglets 1. Gastroenterology **125:** 136–147.

58. KITCHEN, P.A., A.J. FITZGERALD, R.A. GOODLAD, et al. 2000. Glucagon-like peptide-2 increases sucrase-isomaltase but not caudal-related homeobox protein-2 gene expression. Am. J. Physiol. Gastrointest. Liver Physiol. **278:** G425–G428.

59. JEPPESEN, P.B., B. HARTMANN, J. THULESEN, et al. 2001. Glucagon-like peptide 2 improves nutrient absorption and nutritional status in short-bowel patients with no colon. Gastroenterology **120:** 806–815.

60. YUSTA, B., R. SOMWAR, F. WANG, et al. 1999. Identification of glucagon-like peptide-2 (GLP-2)-activated signaling pathways in baby hamster kidney fibroblasts expressing the rat GLP-2 receptor. J. Biol. Chem. **274:** 30459–30467.

61. YUSTA, B., J. ESTALL & D.J. DRUCKER. 2002. Glucagon-like peptide-2 receptor activation engages bad and glucagon synthase kinase-3 in a protein kinase A-dependent manner and prevents apoptosis following inhibition of phosphatidylinositol 3-kinase. J. Biol. Chem. **277:** 24896–24906.

62. JASLEEN, J., N. SHIMODA, E.R. SHEN, et al. 2000. Signaling mechanisms of glucagon-like peptide 2-induced intestinal epithelial cell proliferation. J. Surg. Res. **90:** 13–18.

63. YUSTA, B., L. HUANG, D. MUNROE, et al. 2000. Enteroendocrine localization of GLP-2 receptor expression in humans and rodents. Gastroenterology **119:** 744–755.

64. BJERKNES, M. & H. CHENG. 2001. Modulation of specific intestinal epithelial progenitors by enteric neurons. Proc. Natl. Acad. Sci. USA **98:** 12497–12502.
65. DUBE, P.E., C.L. FORSE, J. BAHRAMI, *et al.* In Press. The essential role of insulin-like growth factor-1 in the intestinal tropic effects of glucagon-like peptide-2 in mice. Gastroenterology.
66. DUBE, P.E. & P.L. BRUBAKER. 2005. Glucagon-like peptide-2 acutely activates PKB/Akt and decreases apoptotic signaling in the mouse small intestine [abstract]. Gastroenterology **128:** 1295A.
67. KREYMANN, B., M.A. GHATEI, G. WILLIAMS & S.R. BLOOM. 1987. Glucagon-like peptide-1 7-36: a physiological incretin in man. Lancet **2:** 1300–1304.
68. NAUCK, M.A., A. SAUERWALD, R. RITZEL, *et al.* 1998. Influence of glucagon-like peptide 1 on fasting glycemia in type 2 diabetic patients treated with insulin after sulfonylurea secondary failure. Diabetes Care **21:** 1925–1931.
69. TOFT-NIELSEN, M.B., S. MADSBAD & J.J. HOLST. 1999. Continuous subcutaneous infusion of glucagon-like peptide 1 lowers plasma glucose and reduces appetite in type 2 diabetic patients. Diabetes Care **22:** 1137–1143.
70. GUTNIAK, M.K., B. LINDE, J.J. HOLST & S. EFENDIC. 1994. Subcutaneous injection of the incretin hormone glucagon-like peptide 1 abolishes postprandial glycemia in NIDDM. Diabetes Care **17:** 1039–1044.
71. CREUTZFELDT, W. 2005. The [pre-] history of the incretin concept. Regul. Pept. **128:** 87–91.
72. SCROCCHI, L.A., B.A. MARSHALL, S.M. COOK, *et al.* 1998. Identification of glucagon-like peptide 1 (GLP-1) actions essential for glucose homeostasis in mice with disruption of GLP- 1 receptor signaling. Diabetes **47:** 632–639.
73. FLAMEZ, D., P. GILON, K. MOENS, *et al.* 1999. Altered cAMP and Ca^{2+} signaling in mouse pancreatic islets with glucagon-like peptide-1 receptor null phenotype. Diabetes **48:** 1979–1986.
74. HANSOTIA, T., L.L. BAGGIO, D. DELMEIRE, *et al.* 2004. Double incretin receptor knockout (DIRKO) mice reveal an essential role for the enteroinsular axis in transducing the glucoregulatory actions of DPP-IV inhibitors. Diabetes **53:** 1326–1335.
75. DRUCKER, D.J., J. PHILIPPE, S. MOJSOV, *et al.* 1987. Glucagon-like peptide I stimulates insulin gene expression and increases cyclic AMP levels in a rat islet cell line. Proc. Natl. Acad. Sci. USA **84:** 3434–3438.
76. HELLER, R.S., T.J. KIEFFER & J.F. HABENER. 1997. Insulinotropic glucagon-like peptide I receptor expression in glucagon-producing α-cells of the rat endocrine pancreas. Diabetes **46:** 785–791.
77. THORENS, B. 1992. Expression cloning of the pancreatic β cell receptor for the gluco-incretin hormone glucagon-like peptide 1. Proc. Natl. Acad. Sci. USA **89:** 8641–8645.
78. GROMADA, J., K. BOKVIST, W.G. DING, *et al.* 1998. Glucagon-like peptide 1(7-36) amide stimulates exocytosis in human pancreatic β-cells by both proximal and distal regulatory steps in stimulus-secretion coupling. Diabetes **47:** 57–65.
79. MONTROSE-RAFIZADEH, C., P. AVDONIN, M.J. GARANT, *et al.* 1999. Pancreatic glucagon-like peptide-1 receptor couples to multiple G proteins and activates mitogen-activated protein kinase pathways in Chinese hamster ovary cells. Endocrinology **140:** 1132–1140.
80. XU, G., D.A. STOFFERS, J.F. HABENER & S. BONNER-WEIR. 1999. Exendin-4 stimulates both β-cell replication and neogenesis, resulting in increased β-cell mass and improved glucose tolerance in diabetic rats. Diabetes **48:** 2270–2276.

81. WANG, Q. & P.L. BRUBAKER. 2002. Glucagon-like peptide-1 treatment delays the onset of diabetes in 8 week-old db/db mice. Diabetologia **45:** 1263–1273.
82. LI, Y., T. HANSOTIA, B. YUSTA, *et al.* 2003. Glucagon-like peptide-1 receptor signaling modulates beta cell apoptosis. J. Biol. Chem. **278:** 471–478.
83. POSPISILIK, J.A., J. MARTIN, T. DOTY, *et al.* 2003. Dipeptidyl peptidase IV inhibitor treatment stimulates beta-cell survival and islet neogenesis in streptozotocin-induced diabetic rats. Diabetes **52:** 741–750.
84. WANG, Q., L. LI, E. XU, *et al.* 2004. Glucagon-like peptide-1 regulates proliferation and apoptosis via activation of PKB in pancreatic (INS-1) beta-cells. Diabetologia **47:** 478–487.
85. BUTEAU, J., S. FOISY, C.J. RHODES, *et al.* 2001. Protein kinase Czeta activation mediates glucagon-like peptide-1- induced pancreatic beta-cell proliferation. Diabetes **50:** 2237–2243.
86. DE LEON D.D., S. DENG, R. MADANI, *et al.* 2003. Role of endogenous glucagon-like peptide-1 in islet regeneration after partial pancreatectomy. Diabetes **52:** 365–371.
87. FARILLA, L., H. HUI, C. BERTOLOTTO, *et al.* 2002. Glucagon-like peptide-1 promotes islet cell growth and inhibits apoptosis in Zucker diabetic rats. Endocrinology **143:** 4397–4408.
88. TOURREL, C., D. BAILBE, M.J. MEILE & B. PORTHA. 2001. Glucagon-like peptide-1 and exendin-4 stimulate beta-cell neogenesis in streptozotocin-treated newborn rats resulting in persistently improved glucose homeostasis at adult age. Diabetes **50:** 1562–1570.
89. ALCÁNTARA, A.I., M. MORALES, E. DELGADO, *et al.* 1997. Exendin-4 agonist and exendin(9-39)amide antagonist of the GLP-1(7-36)amide effects in liver and muscle. Arch. Biochem. Biophys. **341:** 1–7.
90. MÁRQUEZ, L., M.A. TRAPOTE, M.A. LUQUE, *et al.* 1998. Inositolphosphoglycans possibly mediate the effects of glucagon- like peptide-1(7-36)amide on rat liver and adipose tissue. Cell Biochem. Funct. **16:** 51–56.
91. IONUT, V., K. HUCKING, I.F. LIBERTY & R.N. BERGMAN. 2005. Synergistic effect of portal glucose and glucagon-like peptide-1 to lower systemic glucose and stimulate counter-regulatory hormones. Diabetologia **48:** 967–975.
92. DARDEVET, D., M.C. MOORE, C.A. DICOSTANZO, *et al.* 2005. Insulin secretion-independent effects of glucagon-like peptide 1 (GLP-1) on canine liver glucose metabolism do not involve portal vein GLP-1 receptors. Am. J. Physiol Gastrointest. Liver Physiol. **289:** G 806–814.
93. NAKAGAWA, A., H. SATAKE, H. NAKABAYASHI, *et al.* 2004. Receptor gene expression of glucagon-like peptide-1, but not glucose-dependent insulinotropic polypeptide, in rat nodose ganglion cells. Auton. Neurosci. **110:** 36–43.
94. AHREN, B. 2004. Sensory nerves contribute to insulin secretion by glucagon-like peptide-1 in mice. Am. J. Physiol Regul. Integr. Comp Physiol **286:** R269–R272.
95. SZAYNA, M., M.E. DOYLE, J.A. BETKEY, *et al.* 2000. Exendin-4 decelerates food intake, weight gain, and fat deposition in Zucker rats. Endocrinology **141:** 1936–1941.
96. EGAN, J.M., G.S. MENEILLY & D. ELAHI. 2003. Effects of 1-mo bolus subcutaneous administration of exendin-4 in type 2 diabetes. Am. J. Physiol Endocrinol. Metab **284:** E1072–E1079.
97. HARDER, H., L. NIELSEN, D.T. TU & A. ASTRUP. 2004. The effect of liraglutide, a long-acting glucagon-like peptide 1 derivative, on glycemic control, body

composition, and 24-h energy expenditure in patients with type 2 diabetes. Diabetes Care **27:** 1915–1921.

98. ZANDER, M., S. MADSBAD, J.L. MADSEN & J.J. HOLST. 2002. Effect of 6-week course of glucagon-like peptide 1 on glycaemic control, insulin sensitivity, and beta-cell function in type 2 diabetes: a parallel-group study. Lancet **359:** 824–830.

99. AHREN, B., G. PACINI, J.E. FOLEY & A. SCHWEIZER. 2005. Improved meal-related {beta}-cell function and insulin sensitivity by the dipeptidyl peptidase-IV inhibitor vildagliptin in metformin-treated patients with type 2 diabetes over 1 year. Diabetes Care **28:** 1936–1940.

100. KENDALL, D.M., M.C. RIDDLE, J. ROSENSTOCK, et al. 2005. Effects of exenatide (exendin-4) on glycemic control over 30 weeks in patients with type 2 diabetes treated with metformin and a sulfonylurea. Diabetes Care **28:** 1083–1091.

101. 2005. Exenatide (Byetta) for type 2 diabetes. Med. Lett. Drugs Ther. **47:** 45–46.

102. NASLUND, E., B. BARKELING, N. KING, et al. 1999. Energy intake and appetite are suppressed by glucagon-like peptide-1 (GLP-1) in obese men. Int. J. Obes. Relat Metab Disord. **23:** 304–311.

103. RODRIQUEZ, D.F., M. NAVARRO, E. ALVAREZ, et al. 2000. Peripheral versus central effects of glucagon-like peptide-1 receptor agonists on satiety and body weight loss in Zucker obese rats. Metabolism **49:** 709–717.

104. TALSANIA, T., Y. ANINI, S. SIU, et al. 2005. Peripheral exendin-4 and peptide YY3-36 synergistically reduce food intake through different mechanisms in mice. Endocrinology **146:** 3748–3756.

105. BAGGIO, L.L., Q. HUANG, T.J. BROWN & D.J. DRUCKER. 2004. A recombinant human glucagon-like peptide (GLP)-1-albumin protein (albugon) mimics peptidergic activation of GLP-1 receptor-dependent pathways coupled with satiety, gastrointestinal motility, and glucose homeostasis. Diabetes **53:** 2492–2500.

106. KASTIN, A.J., V. AKERSTROM & W. PAN. 2002. Interactions of glucagon-like peptide-1 (GLP-1) with the blood-brain barrier. J. Mol. Neurosci. **18:** 7–14.

107. DAKIN, C.L., C.J. SMALL, R.L. BATTERHAM, et al. 2004. Peripheral oxyntomodulin reduces food intake and body weight gain in rats. Endocrinology **145:** 2687–2695.

108. NASLUND, E., N. KING, S. MANSTEN, et al. 2004. Prandial subcutaneous injections of glucagon-like peptide-1 cause weight loss in obese human subjects. Br. J. Nutr. **91:** 439–446.

109. BOZKURT, A., E. NASLÜND, J.J. HOLST & P.M. HELLSTRÖM. 2002. GLP-1 and GLP-2 act in concert to inhibit fasted, but not fed, small bowel motility in the rat. Reg. Pep. **107:** 129–135.

110. NAGELL, C.F., A. WETTERGREN, J.F. PEDERSEN, et al. 2004. Glucagon-like peptide-2 inhibits antral emptying in man, but is not as potent as glucagon-like peptide-1. Scand. J. Gastroenterol. **39:** 353–358.

111. KIRKEGAARD, P., A.J. MOODY, J.J. HOLST, et al. 1982. Glicentin inhibits gastric acid secretion in the rat. Nature **297:** 156–157.

112. JARROUSSE, C., C. CARLES-BONNET, H. NIEL & D. BATAILLE. 1993. Inhibition of gastric acid secretion by oxyntomodulin and its (19-37) fragment in the conscious? rat. Am. J. Physiol. Gastrointest. Liver Physiol. **264:** G816–G823.

113. PELLISSIER, S., K. SASAKI, D. LE NGUYEN, et al. 2004. Oxyntomodulin and glicentin are potent inhibitors of the fed motility pattern in small intestine. Neurogastroenterol. Motil. **16:** 455–463.

114. WILLMS, B., J. WERNER, J.J. HOLST, *et al.* 1996. Gastric emptying, glucose responses, and insulin secretion after a liquid test meal: effects of exogenous glucagon- like peptide-1 (GLP-1)-(7-36) amide in type 2 (noninsulin- dependent) diabetic patients. J. Clin. Endocrinol. Metab. **81:** 327–332.
115. PEDERSON, R.A., H.A. WHITE, D. SCHLENZIG, *et al.* 1998. Improved glucose tolerance in Zucker fatty rats by oral administration of the dipeptidyl peptidase IV inhibitor isoleucine thiazolidide. Diabetes **47:** 1253–1258.
116. TAVARES, W., D.J. DRUCKER & P.L. BRUBAKER. 2000. Enzymatic- and renal-dependent catabolism of the intestinotropic hormone glucagon-like peptide-2 in rats. Am. J. Physiol. Endocrinol. Metab. **278:** E134–E139.

Secretin: A Pleiotrophic Hormone

J.Y.S. CHU,[a] W.H. YUNG,[b] AND B.K.C. CHOW[a]

[a]Department of Zoology, Kadoorie Biological Science Building,
The University of Hong Kong, Pokfulam, Hong Kong

[b]Department of Physiology, Faculty of Medicine, The Chinese
University of Hong Kong, Shatin, Hong Kong

ABSTRACT: Secretin holds a unique place in the history of endocrinology and gastrointestinal physiology, as it is the first peptide designated as a hormone. During the last century since its first discovery, the hormonal effects of secretin in the gastrointestinal tract were extensively studied, and its principal role in the periphery was found to stimulate exocrine secretion from the pancreas. Recently, a functional role of secretin in the brain has also been substantiated, with evidence suggesting a possible role of secretin in embryonic brain development. Given that secretin and its receptors are widely expressed in multiple tissues, this peptide should therefore exhibit pleiotrophic functions throughout the body. The present article reviews the current knowledge on the central and peripheral effects of secretin as well as its therapeutic uses.

KEYWORDS: secretin; central and peripheral functions

INTRODUCTION

In the last century, there were several milestones in the research of secretin including identification, purification and structural determination, synthesis, receptor identification and characterization, and transcriptional regulation. Now it has been established that secretin acts as a gastrointestinal hormone regulating exocrine secretion from the pancreas, gall bladder, and stomach. More recently, it has also been demonstrated that secretin acts as a neuropeptide in the central nervous system (CNS), participating in neuronal signaling, behavioral modulation, as well as central–peripheral interactions. It is noteworthy, that more and more evidence is accumulated suggesting its physiological properties outside the brain–gut axis. Here, we intend to provide an updated review on the pleiotrophic functions of secretin in our body.

Address for correspondence: Dr. Billy K.C. Chow, Department of Zoology, The University of Hong Kong, Pokfulam Road, Hong Kong, SAR, PRC. Voice: 852- 22990850; fax: 852-25599114.
e-mail: bkcc@hkusua.hku.hk

Ann. N.Y. Acad. Sci. 1070: 27–50 (2006). © 2006 New York Academy of Sciences.
doi: 10.1196/annals.1317.013

In the Central and Peripheral Nervous Systems

The first report suggesting the neuroactive function of secretin in the nervous systems dates back to 1979, in which Propst *et al.* demonstrated the stimulatory effect of secretin on cyclic adenosine 3':5'-phosphate (cAMP) production in neuroblastoma–glioma hybrid cells.[1] Subsequent studies coincide with the finding and stated that secretin could also induce cAMP formation in cultured mouse brain cells,[2] rat brain slide preparations,[3] hypothalamic and hippocampal regions,[4] and in the rat superior cervical ganglion (SCG).[5] Since some of these stimulatory effects could be partially blocked by the use of a synthetic antagonist of secretin, secretin (5–27),[2] it therefore appears that secretin should have specific regulatory influence on the second messengers in specific neuronal cell bodies.

After these initial studies, as cAMP has been implicated in regulating tyrosine hydroxylase (TH), an enzyme catalyzing the rate-limiting step in the biosynthesis of catecholamine, secretin's effects on the activity of neuronal TH was proposed in the sympathetic nerve terminals. Results from Ip *et al.* suggested that in the rat SCG, secretin-evoked cAMP production in individual ganglia has a strong correlation with TH activity,[5] and subsequent studies by Schwarzschild *et al.* revealed that *in vitro* administration of secretin could stimulate TH activity in three autonomic end organs—the iris, submaxillary gland, and pineal gland—that are innervated by SCG neurons through sympathetic nerve endings.[6] Additionally, secretin was also found to increase TH activity in the hypothalamus,[7] sympathetic ganglia within the thoracic paravertebral ganglia,[8] and in the autonomic end organ innervated by the middle and inferior cervical ganglia, for example, the heart.[8] Together with the fact that secretin could (*a*) act synergistically with carbachol in regulating the synthesis of dihydroxyphenylalanine (DOPA),[9] (*b*) decrease dopamine turnover in the forebrain,[10] and (*c*) increase dopamine turnover in the median eminence,[10] it is generally believed that secretinergic pathways could be widely distributed among different types and regions of neurons and that secretin has regulatory effects on catecholamine neurotransmitter metabolism in the axon terminals of sympathetic neurons.

In addition, the direct roles of secretin as a neuromodulator, neurotransmitter, or neurotrophic factor have also been recognized. In the ventral spinal cord cultures, secretin at 1 μM concentration could promote neurite outgrowth, indicating its therapeutic potential in treating diseases involving degeneration and death of spinal motor neurons.[11] On the other hand, in both hypothalamic and cerebellar explants, we are able to demonstrate the K^+-evoked release of secretin, which is TTX and cadmium sensitive (unpublished).[10] The released secretin from the former region was thought to act in an autocrine manner to stimulate, through a cAMP-dependent pathway, *c-fos* and vasopressin (Vp) expression and somatodendritic release of Vp in magnocellular neurons. In the cerebellum, however, liberated secretin was shown to facilitate GABAergic

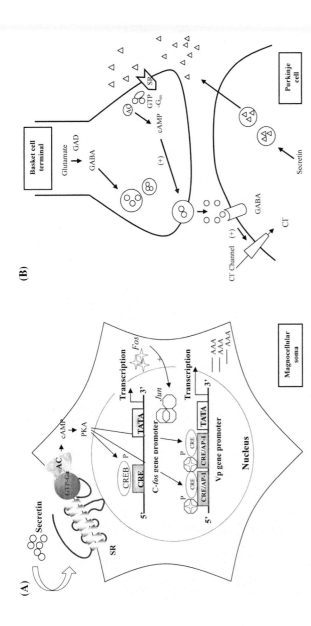

FIGURE 1. Diagrammatic illustration of the secretin functions in the hypothalamus and the cerebellum. (**A**) The mechanism of secretin-induced *c-fos* and Vp expression in the hypothalamic magnocellular neurons. Binding of secretin onto its receptor (SR) stimulates adenylyl cyclase (AC) and increases the intracellular cAMP level. Activated PKA then causes the phosphorylation of CREB in the nucleus as well as the transcription of *c-fos* immediate early gene. Subsequent dimerization of the product *Fos* with *Jun* protein allows its interaction with the CRE sequences in Vp promoter, hence upregulating the Vp gene expression. Through unknown pathway, cAMP could also enhance the stability of Vp mRNA in the cytoplasm by increasing the polyadenylate tail length. In addition, cAMP-induced PKA expression might also act directly to stimulate the Vp gene expression through CREB protein phosphorylation. (**B**) The function of secretin on basket cell-purkinje neuron axis. Secretin is released from the somatodendritic region of the Purkinje cell upon depolarization. The released secretin binds to its receptors on the presynaptic site, activating AC and increasing cAMP formation. The increased cAMP level facilitates GABA release from its vesicles in the presynaptic basket cell, which binds to the GABA$_A$ receptors on the Purkinje cell membrane and triggers the opening of C1 channels. This allows the entering of C1 into the Purkinje cells, leading to facilitation of IPSC. GTP-Gα_s = Stimulatory G protein; GAD = Glutamic acid decarbosylase.

inputs from basket cells to Purkinje neurons (FIG. 1).[12–14] In the nucleus of solitary tract (NTS), secretin excites NTS neurons through a voltage-independent membrane channel, nonselective cationic conductance (NSCC), which allows the passage of cations (Na^+, K^+, or Ca^{2+}) in various proportions.[15] These observations suggest a possible role of secretin in the central regulation of autonomic functions. Consistently, the central administration of secretin was found to induce the expression of *c-fos* in the area postrema, the dorsal motor nucleus, as well as the medial region of NTS and its relay station, including the parabrachial complex and medial and central amygdala. These findings thus provide an anatomical basis for the neuroregulatory role of secretin.[16] In addition, centrally administered secretin decreased blood pressure and increased bicarbonate output from the pancreas, indicating the central roles of secretin as a hypotension agent and in pancreatic secretion.[17,18] Finally, secretin has also been implicated in modulating the physiological and behavioral changes. It was shown to increase defecation, decrease novel object approaches, decrease open-field locomotor activity, and alter the respiration rate in rats.[19] In addition, it could also inhibit the pulsatile secretion of luteinizing hormone and prolactin in ovariectomized rats by activating the hypothalamic dopaminergic system.[7,20] These support a central role of the endogenously synthesized secretin.

Secretin has recently been demonstrated to enter the brain from the periphery through a nonsaturable transmembrane diffusion process in the vascular barrier and a saturable process in the choroids plexus.[21] Peripheral secretin, therefore, has the potential of passing through the blood–brain barrier to exert central effects. In agreement with this, peripheral-, intravenous- (i.v.), or intraperitoneal (i.p.)-administered secretin was able to (*a*) induce *c-fos* expression in discrete brain regions including amygdala, NTS, locus coeruleus, Barrington's nucleus, and arcuate nucleus[15,22,23]; (*b*) reestablish the communicative and affiliative interactions in autistic children[24]; and (*c*) modulate neural networks that are implicated in the acquisition and expression of emotional behaviors including fear and anxiety.[25] Therefore, it is plausible that secretin, either synthesized endogenously from the CNS or from the periphery, has diverse actions in the CNS.

In the Esophagus

Scanty data are available on the effect of secretin on the esophagus. Cohen *et al.* reported that secretin inhibits the rise in lower esophageal sphincter pressure (LESP) induced by gastrin during feeding while having little effect on resting sphincter pressure.[26] Miyata *et al.* reported that secretin has a long-acting effect on muscular relaxation of the lower esophageal sphincter in esophageal achalsia patients.[27] In addition, it has also been shown that HCl instillation into the duodenum produces an initial fall in the LESP with a corresponding rise in plasma secretin level.[28] In summary, secretin appears to act as a duodenal factor that regulates esophageal sphincter competence.

In the Cardiovascular System

In both cats and dogs, secretin was shown to raise the systemic arterial pressure and cardiac output,[29] with preferential direction of the blood flow to stomach, small intestine, pancreas, and kidney, but not to the heart and lung.[30] Nevertheless, in dogs having acute left ventricular failure, secretin injection was found to cause a general distribution of the increased cardiac output to the renal, carotid, femoral, and superior mesenteric arteries, leading to a restoration of blood flow to normal levels.[31] It could also enhance the left ventricular performance,[32] and consistently, clinical injection of pharmacological doses of secretin in patients with depressed cardiac function augmented left ventricular performance by means of arteriolar dilation.[33] Hence, it is possible that secretin could regulate the distribution of blood flow and have merit in treating acute left ventricular failure.

Aside from regulating cardiac output and peripheral blood flow, secretin also exerts various physiological actions in the cardiac system. Studies with secretin administered via the cannulated sinus node artery suggested that the peptide has direct action on atrial rate and contractility, as it could produce dose-related positive inotropic and biphasic chronotropic effects after injection.[34] Furthermore, it could activate adenylate cyclase in the heart in the presence of guanosine 5'-triphosphate (GTP),[35] even though its efficacy varies from species to species; it is active only in the rat but not in the guinea pig, rabbit, dog, and monkey.[36] Finally, secretin was also demonstrated to (*a*) increase stroke volume while decreasing the total systemic resistance in humans,[37] (*b*) stimulate the contractility rate of spontaneously beating right atrium in rats,[38] (*c*) potentiate the Ba^{2+} currents in isolated adult rat ventricular myocytes,[39] and (*d*) modify the effect of L-NAME on coronary permeability in both intact and diabetic rats.[40]

In the Lung

Human secretin receptor was originally isolated from lung cells.[41] It was found to be expressed predominantly on the basolateral membrane of the epithelial cells within the tertiary bronchus.[42] Physiologically, secretin could stimulate adenylate cyclase in surgically removed human lung tissue as well as within the lung membrane isolated from the rat, mouse, and guinea pig.[43,44] In addition, it could also stimulate the NPPB-sensitive channel-mediated Cl^- efflux from bronchial epithelial cells and cause a concentration-dependent relaxation of carbachol-induced precontracted bronchial smooth muscle,[42] indicating its role in bronchorelaxation via the regulation of epithelial ionic composition. Finally, secretin was also shown to produce a seven-fold increase in the release of bombesin/gastrin releasing peptide from small cell carcinoma of the lung (SCLC).[45] Since these peptides have been implicated in stimulating colony formation of SCLC, secretin might also be capable of altering the growth of SCLC cells.

In the Stomach

Secretin has long been implicated in regulating diverse gastric functions in several species, including the rat,[46,47] cat,[48] dog,[48–51] and humans.[49,52] In a physiological dose, it could act as an enterogastrone to stimulate gastric pepsin secretion,[48] inhibit gastric motility,[53] delay gastric emptying by decreasing the intragastric pressure as well as the stomach–duodenum pressure gradient,[54] and inhibit both meal- and pentagastrin-stimulated gastric acid secretion from parietal cells.[50,55] The action of secretin on gastric acid and motility inhibition is likely mediated through a neurohormonal mechanism. The neural pathways that mediate the relaxation and inhibitory action of secretin on forestomach muscle strips and gastric acid secretion, respectively, were demonstrated by Kwon et al.,[56] Li et al.[57] and Li et al.[58] Experimental evidence from these groups indicated that subdiaphragmatic perivagal capsaicin treatment in conscious rats could prevent inhibitory actions elicited by a physiological dose of secretin while vagotomy could abolish the inhibitory effect of secretin on gastric motility. These studies, therefore, suggest the action of secretin through vagal afferent pathways originating from the gastroduodenal mucosa.[57] Besides, they also demonstrated that vagal input could modulate the density of the secretin receptor-binding sites in the stomach, which might additionally play a permissive role in the stimulatory effect of secretin on the local release of SS, a paracrine mediator in inhibiting gastric acid secretion.[56] Finally, Li et al. demonstrated that intra-arterial injection of secretin at physiological doses could elicit responses in the nodose ganglia neurons that terminate in the mucosa,[58] and Yang et al. showed secretin receptor gene expression in nodose ganglia.[23] These findings provide the functional and anatomical basis to support systemic secretin in inhibiting gastric motility via the vagal afferent pathways.

Secretin also acts via hormonal mechanisms to regulate gastric motility and acid secretion. Intravenous infusion of the peptide inhibits pentagastrin release at the G cell level in the antrum.[46] Besides, it could also stimulate the release of SS and prostaglandins E2, which could inhibit the pentagastrin-induced gastric acid secretion and histamine-sensitive enzymes, respectively, at the parietal cell level.[59–61] Finally, there is also a report suggesting the involvement of secretin in reducing gastric histamine release in modulate acid secretion.[62] In summary, the enterogastrone effects of secretin are dually mediated by both neuronal and hormonal mechanisms.

On top of the enterogastrone's effect, secretin also stimulates adenylate cyclase activity in the mucosal membrane isolated from the corpus of rat stomach.[63] Besides, it could also inhibit the mucosal blood flow,[64] increase the activity of chief cells in human gastric mucosa,[65] enhance the activity of gastric lipase via the influence on gastric acid secretion,[66] stimulate the rapid postnatal growth and enzyme composition of developing stomach from rats,[67] and decrease the resting tension of longitudinal and circular muscles of the fundus.[68] Finally, secretin also (*a*) increases the paracellular resistance in gastric mucosa through an Src-mediated pathway; (*b*) potentiates the effect of epidermal growth factor on transepithelial resistance[69]; and (*c*) stimulates mucus and bicarbonate secretion to protect the gastric mucosa.[70]

In the Pancreas

Among all the described functions of secretin, the best established are its effects on the pancreas, in which secretin augments both exocrine and endocrine secretion, while its principal effect is to stimulate bicarbonate, water, and electrolytes from pancreatic ductal epithelial cells.[71] The secretin-stimulated secretion of pancreatic bicarbonate-rich fluid is in response to gastric acid and fat entering the duodenum. This is important in neutralizing the acid chyme in order to provide an optimal intraluminal pH for the normal functioning of pancreatic and intestinal enzymes involved in protein, lipid, and carbohydrate digestion. After the release of secretin from intestinal S cells, it acts directly or indirectly on five subcellular components to regulate bicarbonate secretion (FIG. 2).

1. The secretin receptor: it is localized on the basolateral membrane of ductal and centroacinar cells. Upon activation, it causes an increase in cytosolic concentration of cAMP and PKA activity that activates the Cl^-/HCO_3^- antiport situated on the apical membrane.
2. The Cl^-/HCO_3^- antiport: activation of this exchanger results in the secretion of HCO_3^- into ductal lumen while importing an equimolar of chloride ions.[72]
3. The H^+-ATPases: upon secretin stimulation, H^+-ATPases-containing cytoplasmic tubulovesicles are stimulated to incorporate into the basolateral membrane. This allows active electrogenic transportation of accumulated H^+ into the interstitial fluid.[73,74]
4. The cystic fibrosis transmembrane conductance regulator (CFTR): it is essential for the recirculation of Cl^- back to the glandular lumen, and hence, the maintenance of ionic gradient for bicarbonate ion release.[72,75]
5. The enzyme carbonic anhydrase: it is responsible for the bidirectional hydration of CO_2, which is locally produced as a metabolic product or diffused from the extracellular fluid into the alkalinized ductal cells.

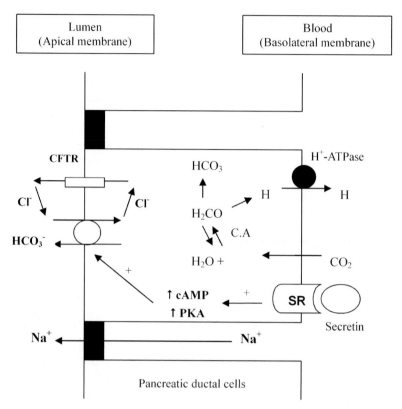

FIGURE 2. A current model of secretin-stimulated pancreatic bicarbonate secretion. Carbon dioxide diffuses into the ductal cell and is hydrated by carbonic anhydrase (C.A.) to carbonic acid, which then dissociates to form protons and bicarbonate ions. Accumulated bicarbonate is secreted into the ductal lumen through Cl^-/HCO_3^- antiport at the apical membrane. The cycling rate of this exchanger depends on the availability of the luminal Cl^-, which is in turn controlled by the CFTR. C.A. = Carbonic Anhydrase.

Secretin-evoked bicarbonate secretion into the duct lumen is accompanied with transepithelial movement of Na^+ and K^+, to a lesser extent, through the paracellular pathway to maintain electrical neutrality. The accumulation of electrolytes is followed by passive water movement into the duct lumen along the osmotic gradient and is therefore regulated by plasma osmolality.[76] In agreement with this, changes in plasma osmolality exhibited a negative correlation with changes in secretin-induced bicarbonate secretion, with Vp participating in fine tuning this process.[77] For instance, Vp released in elevated plasma osmolality could act on the duct cells, via intracellular Ca^{2+}, to reduce the secretin-stimulated pancreatic secretion.[77,78]

Aside from plasma osmolality and Vp, secretin-induced bicarbonate secretion is also regulated by other mechanisms: (*a*) arterial pH and carbon dioxide determine the incorporation of cytoplasmic tubulovesicles into the basolateral membrane of duct cells[79]; (*b*) insulin inhibits the bicarbonate output via cholinergic mechanisms[80]; (*c*) CCK and neurotensin interact with secretin to synergistically stimulate pancreatic bicarbonate secretion[81]; (*d*) substance P inhibits the induced bicarbonate output through inactivation of the Cl^-/HCO_3^- antiport[82]; (*e*) SS inhibits secretin-promoted exocrine secretion[83]; and (*f*) GR-Pergic neurons enhance the evoked exocrine secretion upon excitation.[84]

Secretin also performs various functions in the pancreas aside from stimulating pancreatic bicarbonate output. In the acinar cells, secretin could provoke two completely independent signaling pathways. At low concentration ($\sim 10^{-10}$ M), it stimulates adenylyl cyclase (AC) activity, whereas at high concentration ($> 10^{-8}$ M), it elevates the intracellular Mg^{2+} and activates the hydrolysis of phosphatidyl-inositol biphosphate (PIP_2), which are linked to intracellular Ca^{2+} and PKC isoforms.[85–87] Stimulation of the former pathway leads to a modest increase in amylase secretion, whereas activation of the later one stimulates the electrogenic Cl^- efflux in acinar cells through the slowly activating voltage-dependent K^+ current.[88] This allows the sensitization, and hence, potentiation of acinar cells to CCK- and caerulein-stimulated amylase release.[89–92]

Aside from regulating amylase secretion, secretin also increases tight junction permeability in CAPAN-1 pancreatic ductal carcinoma cells via a cAMP-dependent pathway.[93] In addition, it enhances polyamine synthesis and ornithine decarboxylase activity and potentiates the pancreatotrophic effects of CCK.[94] Consistent with this idea, secretin has been suggested to play a role in the rapid postnatal growth of the pancreas.[67,95] In addition, secretin was found to suppress the endocrine release of glucagon from pancreatic A cells[96] while inducing a dose-dependent increase in the basal insulin secretion[97] as well as the glucose-induced insulin release from the pancreatic B cells.[98,99] Since it could also improve glucose tolerance in a way that is not related to its mild insulinotropic effects in human adults,[100] insulin response to a secretin load was once suggested to be useful in providing a pertinent clinical guide for the selection of an appropriate therapy for patients with diabetes mellitus.[101] Finally, secretin was also found to potentiate the caerulein-induced cell injury/death as well as the redistribution of f-actin in pancreatic acinar cells.[92]

In the Intestine

There is a considerable body of evidence suggesting the involvement of secretin in intestinal physiology. Using semiquantitative RT-PCR, Wells *et al.* have recently demonstrated a role of secretin in upregulating the pepsinogen C gene in the rat intestinal cell, IEC-6.[102] In addition, the peptide has also been shown to inhibit gastrin-induced activation of ornithine decarboxylase in the same cell and stimulate a benzamil-sensitive and

5-nitro-2-(3-phenylpropylamino) benzoic acid-sensitive ion transport across the apical membrane of human intestinal epithelial Caco-2 cells.[103,104] Moreover, secretin could trigger, via its receptor on the smooth musculature, the relaxation of rat ileal and colonic longitudinal muscle,[105] and dog duodenum after vagotomy.[106] Finally, secretin has also been shown to (a) stimulate the secretion of bicarbonate, protein, mucus, and epidermal growth factor from the Brunner's glands[107–109]; (b) affect the maturation of gastrointestinal functions in suckling rat[110]; (c) inhibit the absorption of water, sodium, and glucose from the dog jejunum and the rat ileum[111,112]; and (d) increase the weight, DNA, and protein content of the rat small intestine.[113]

In the Liver

Functions of secretin within the liver are associated with bile production and release. Bile is produced by hepatocytes and secreted into the bile ducts where its volume and composition are modified by the transport of water and solutes from the ductal epithelial cells, cholangiocytes. In the large intraheptic bile duct units (diameter larger than 15 μm), secretin interacts with its receptor on the epithelial cells to stimulate the cAMP/PKA pathway, leading to the opening of CFTR channels by phosphorylation.[114] As a result, bicarbonate-rich bile is produced by the extrusion of Cl^- ions and subsequent activation of the Cl^-/HCO_3^- exchanger.[115,116] Nevertheless, the response of bile secretion upon secretin treatment differs markedly among mammalian species: a marked rise in flow is observed in humans, cat, dog, pig, and sheep, while there is no response observed in rat.[117]

Secretin-induced bile secretion is accompanied by various events in the cholangiocytes, including transepithelial movement of water into the ductal bile, which is also tightly regulated by secretin. In isolated rat cholangiocytes, exposure of secretin leads to a dose-dependent increase in membrane permeability to water by evoking the vesicular translocation of aquaporin-1 (AQP-1) from the intracellular compartment to the apical membrane.[118] This process was found to be microtubule-dependent as pretreatment of cholangiocytes with colchicine could completely abolish vesicular fusion.[119] Recent studies in cultured polarized mouse cholangiocytes demonstrated the coexpression of CFTR and Cl^-/HCO_3^- exchanger in AQP-1-containing vesicles and that these channel proteins are co-redistributed with AQP-1 to the apical membrane upon secretin treatment.[120] In addition, secretin also stimulates the translocation of the Na^+-dependent bile acid transporter to the plasma membrane of cholangiocytes, resulting in cholehepatic shunting of conjugated bile acids.[121] Thus, secretin-induced trafficking of vesicles in the cholangiocytes is the key mechanism by which secretin exerts its effects to increase the secretion of water, bicarbonate, and electrolytes as well as the absorption of biliary bile acids. Aside from these, secretin may also enhance excretion of cytotoxic bile acids in animals having cholestasis,[122] and dose-dependently inhibits motility of the proximal and distal segments of the bile duct component of the sphincter of Oddi in order to regulate bile flow into the duodenum.[123]

In the Adrenal Gland

The PC12 cell is a rat catecholamine-secreting pheochromocytoma cell line derived from the adrenal medulla, and secretin was found to stimulate the catecholamine biosynthetic pathway by increasing the TH activity via a PKA-dependent pathway.[124] Consistent with this, Mahapatra *et al.* demonstrated an induction of up to 26% of the total catecholamine release from PC12 cells upon secretin administration.[125] Additionally, secretin could also activate the expression of chromogranin A, a major soluble component of the secretory vesicles in catecholaminergic cells. In addition, secretin has also been shown to enhance the production of a mineralocorticoid, aldosterone, in dispersed rat zona glomerulosa cells, and inhibit ACTH-enhanced release of a glucocorticoid, corticosterone, from dispersed zona fasciculata–reticularis cells.[126]

In the Kidney

The renal function in early studies[127] with impure preparations of secretin suggested a diuretic action of secretin opposing the antidiuretic effects of Vp and SS.[128] Subsequent studies by Baron *et al.*,[129] Barbezat *et al.*,[130] Viteri *et al.*,[131] and Waldum *et al.*[132] consistently indicated the diuretic activity of secretin, while studies by Charlton *et al.*[133] showed an antidiuretic function of this peptide. The findings of the former groups suggested that secretin could increase the renal excretion of water, bicarbonate, sodium, potassium, or calcium in normal human subjects and dogs, with a significant rise in the urinary pH. The latter group, however, showed that i.v. injection of secretin into rats could decrease the urine output through activating the adenylate cyclase in the outer medulla of kidney, thus mimicking the effect of Vp in this region. In addition, they also showed that the antidiuretic effect of secretin is as potent as Vp in homozygous Vp-deficient Brattleboro rats. In agreement with these findings, our laboratory has recently found that rat secretin receptor is co-localized with the aquaporin-2 water channel in the cuboidal epithelial cells in the collecting duct, and secretin receptor knockout mice exhibited a phenotype with distended urinary bladder (unpublished data). Today, although the effect of secretin in urinary output remains controversial, the peptide is likely involved in regulating water homeostasis through its receptor in the kidney. In addition, secretin could increase renal blood flow,[134,135] peritubular capillary and interstitial hydrostatic pressures,[136] single-nephron glomerular filtration rate, and glomerular plasma flow.[137] It is therefore possible that secretin may participate in various renal functions.

In the Male Reproductive System

Only a few reports have contributed toward explaining the role of secretin in the male reproductive system. In the Leydig cells, secretin enhances testosterone and progesterone release,[138] whereas in the epididymis, secretin regulates the transport of anion and fluid by stimulating the short-circuit current through concurrent electrogenic Cl^- and HCO_3^- secretion.[139] The latter effect is reminiscent of the actions of secretin in the pancreatic and bile ducts, where proximally released secretin acts in an autocrine manner on its receptor at the apical membrane.

In the Female Reproductive Cycle

The physiological significance of secretin in the female reproductive system has not been examined in great depth. Nevertheless, there are data suggesting a role of secretin in the hormonal changes during a spontaneous menstrual cycle. In ovariectomized estrogen-primed rats, secretin injection into the preoptic nucleus could strikingly increase the circadian rise of luteinizing hormone release.[140] This is coherent with the putative positive correlations between plasma secretin and serum estradiol that primes the preovulatory LH surge,[141] and the inhibitory role of secretin on prolactin release during follicular and luteal phases of the menstrual cycle.[142] Taken together, secretin may play a stimulatory role in ovulation.

Plasma secretin levels fluctuate during pregnancy. It elevates in late pregnancy, with maximum levels at 36 weeks, and progressively falls to nonpregnant levels 5 days after delivery.[143] As there is a decrease in plasma secretin during oxytocin infusion in postterm pregnancies,[144] secretin may play a role in preterm pregnancy to prevent early expulsion of the fetus by counteracting the steady increase of oxytocin during pregnancy.

In Fasting

Secretin may also be involved in other physiological actions, such as the body metabolism during starvation. In humans and dog, several reports have revealed, for example, a dramatical eight fold increase in plasma secretin concentration from day 1 to 3 of fasting.[145–148] This increment in secretin level is several times greater than that seen after the intraduodenal load of acid and could be suppressed by an i.v. infusion of glucose or refeeding,[148] suggesting that the trigger is unrelated to the normal release mechanism of the peptide and that secretin could be pertinent to lipid metabolism. Consistent with this notion, synthetic porcine secretin was found to act as a potent lipolytic agent on rat adipose tissues of and on mouse fat cells.[149] In addition, it was also found to be more potent than glucagon in stimulating AC and lipase activity, resulting in lipolysis in adipose tissue and isolated fat cells.[150] Finally, secretin could also suppress basal and insulin-stimulated lipogenesis in fat cells.[151] These data, therefore, established a metabolic role of secretin particularly during fasting in mammals.

Others

Secretin could act in a concerted manner with insulin in stimulating glycolysis and net calcium efflux in the muscle compartment of the human forearm.[152] It could also increase the flow rate of thoracic duct lymph in dogs[153] and inhibit the Na^+/K^+ cotransport system in the red blood cells of geese.[154] Finally, intracerebroventricular (i.c.v.) injection of secretin could produce hypothermia with a decrease in the interscapular brown adipose tissue thermogenesis,[155] suggesting a role of secretin in thermoregulation.

THERAPEUTICAL APPLICATIONS

Autism

Horvath *et al.*[24] first described the beneficial effects of secretin on the behavioral deficits that characterize autism. They found that secretin treatment could lead to a marked improvement in speech, eye contact, and attention in three autistic children, thus initiating the popularity of using secretin as a treatment for this disorder. Although the efficacy of secretin in treating this neurological disorder remains a controversial issue, Kuntz *et al.*[156] and Toda *et al.*[157] provided experimental evidence elucidating how secretin could potentially alleviate autistic symptoms. In their studies, secretin administered peripherally increased extracellular glutamate and GABA levels in rat hippocampus, which could counteract the autism-related hypoglutamate conditions. Moreover, i.v. secretin could promote metabolism of serotonin and dopamine in the CNS in autistic subjects, and this could potentially offset the dysfunctioned serotonin system implicated in autism-related aggression. Secretin is transiently expressed in serotoninergic-mesencephalic neurons during neonatal brain development[158] and it can alter the activity of neurons involved in behavioral conditioning in stress adaptation and visceral reflex reactions.[16] Together with the observation that autistic subjects usually display a striking loss of Purkinje cells in the cerebellum,[159] this further supports the view that there is a possible relation between secretin and the etiology of a subgroup of autistic patients. This idea, however, demands further investigations.

Schizophrenia

Secretin usage for the treatment of schizophrenia was first reported by Sheitman *et al.*,[160] who demonstrated clinically relevant, but transient, improvements in symptoms after secretin infusion. In addition, patients receiving a single dose of secretin were consistently rated as more improved with the clinical global impression scale than those treated with placebo. Since i.p. injection of secretin has recently been shown to dose-dependently reverse the phencyclidine-induced impairments in prepulse inhibition,[25] the potential use of secretin in the treatment of schizophrenia merits further investigations.

Duodenal Ulcer

In 1966, Grossman[161] speculated that secretin would be effective in treating duodenal ulcer by stimulating the release of alkaline pancreatic juice, while suppressing the gastric secretion of acid. Since then, some studies have shown an impairment in secretin release in patients with duodenal ulcer,[162,163] linking the defect in endogenous secretion of secretin with the onset of duodenal ulcer. Secretin administration could inhibit the bombesin-evoked gastrin release and could reduce the epigastric pain in patients with duodenal ulcer.[164,165] Complete healing of duodenal ulcer was observed in the respective patients with a continuous treatment of secretin for 2 to 3 weeks[165]; hence, secretin is a therapeutically promising agent for treating duodenal ulcers.

EPILOGUE

The discovery of a blood-borne peptide, secretin, in 1902 had provided the impetus to establish endocrinology as a specialist field. Since then, the biology of this peptide has been studied most extensively in the gastrointestinal tract, providing mechanistic explanations to its transcriptional regulation, release, and bioactivities. Recently, other aspects of secretin including its structure, distribution, and extragastrointestinal functions have also been investigated. However, little is known regarding the physiology of secretin in nonvertebrate species. Characterization of the structure and function of secretin and its receptor in amphibian, avian, and fish species should be instrumental in a thorough understanding of the physiological evolution of this peptide. To fully appreciate the pleiotropic actions of secretin in our body, further research should therefore be devoted to elucidating the physiology of this ligand–receptor pair in a comparative and evolutionary perspective, which may provide clues to answering questions regarding the biological, biochemical, physiological, and pharmacological characteristics of this unique hormone.

REFERENCES

1. PROPST, F., L. MORODER, E. WUNSCH, et al. 1979. The influence of secretin, glucagon and other peptides, of amino acids, prostaglandin endoperoxide

analogues and diazepam on the level of adenosine 3',5'-cyclic monophosphate in neuroblastoma x glioma hybrid cells. J. Neurochem. **32:** 1495–1500.

2. VAN CALKER, D., M. MULLER & B. HAMPRECHT. 1980. Regulation by secretin, vasoactive intestinal peptide, and somatostatin of cyclic AMP accumulation in cultured brain cells. Proc. Natl. Acad. Sci. USA **77:** 6907–6911.

3. FREMEAU, R.T., JR., R.T. JENSEN, C.G. CHARLTON, *et al.* 1983. Secretin: specific binding to rat brain membranes. J. Neurosci. **3:** 1620–1625.

4. KARELSON, E., J. LAASIK & R. SILLARD. 1995. Regulation of adenylate cyclase by galanin, neuropeptide Y, secretin and vasoactive intestinal polypeptide in rat frontal cortex, hippocampus and hypothalamus. Neuropeptides **28:** 21–28.

5. IP, N.Y., C. BALDWIN & R.E. ZIGMOND. 1985. Regulation of the concentration of adenosine 3',5'-cyclic monophosphate and the activity of tyrosine hydroxylase in the rat superior cervical ganglion by three neuropeptides of the secretin family. J. Neurosci. **5:** 1947–1954.

6. SCHWARZSCHILD, M.A. & R.E. ZIGMOND. 1989. Secretin and vasoactive intestinal peptide activate tyrosine hydroxylase in sympathetic nerve endings. J. Neurosci. **9:** 160–166.

7. BABU, G.N. & E. VIJAYAN. 1983. Plasma gonadotropin, prolactin levels and hypothalamic tyrosine hydroxylase activity following intraventricular bombesin and secretin in ovariectomized conscious rats. Brain Res. Bull. **11:** 25–29.

8. SCHWARZSCHILD, M.A. & R.E. ZIGMOND. 1991. Effects of peptides of the secretin-glucagon family and cyclic nucleotides on tyrosine hydroxylase activity in sympathetic nerve endings. J. Neurochem. **56:** 400–406.

9. IP, N.Y. & R.E. ZIGMOND. 2000. Synergistic effects of muscarinic agonists and secretin or vasoactive intestinal peptide on the regulation of tyrosine hydroxylase activity in sympathetic neurons. J. Neurobiol. **42:** 14–21.

10. FUXE, K., K. ANDERSSON, T. HOKFELT, *et al.* 1979. Localization and possible function of peptidergic neurons and their interactions with central catecholamine neurons, and the central actions of gut hormones. Fed. Proc. **38:** 2333–2340.

11. IWASAKI, Y., K. IKEDA, Y. ICHIKAWA, *et al.* 2001. Vasoactive intestinal peptide influences neurite outgrowth in cultured rat spinal cord neurons. Neurol. Res. **23:** 851–854.

12. LEE, S.M., L. CHEN, B.K. CHOW, *et al.* 2005. Endogenous release and multiple actions of secretin in the rat cerebellum. Neuroscience **134:** 377–386.

13. NG, S.S., W.H. YUNG & B.K. CHOW. 2002. Secretin as a neuropeptide. Mol. Neurobiol. **26:** 97–107.

14. YUNG, W.H., P.S. LEUNG, S.S. NG, *et al.* 2001. Secretin facilitates GABA transmission in the cerebellum. J. Neurosci. **21:** 7063–7068.

15. YANG, B., M. GOULET, R. BOISMENU, *et al.* 2004. Secretin depolarizes nucleus tractus solitarius neurons through activation of a nonselective cationic conductance. Am. J. Physiol Regul. Integr. Comp. Physiol **286:** R927–R934.

16. WELCH, M.G., J.D. KEUNE, T.B. WELCH-HORAN, *et al.* 2003. Secretin activates visceral brain regions in the rat including areas abnormal in autism. Cell Mol. Neurobiol. **23:** 817–837.

17. HASHIMOTO, R. & F. KIMURA. 1987. The depressor effect of secretin administered either centrally or peripherally in conscious rats. Regul. Pept. **19:** 233–242.

18. CONTER, R.L., M.T. HUGHES & G.L. KAUFFMAN, JR. 1996. Intracerebroventricular secretin enhances pancreatic volume and bicarbonate response in rats. Surgery **119:** 208–213.

19. CHARLTON, C.G., R.L. MILLER, J.N. CRAWLEY, et al. 1983. Secretin modulation of behavioral and physiological functions in the rat. Peptides **4:** 739–742.

20. WEICK, R.F., K.M. STOBIE & K.A. NOH. 1992. Effect of [4Cl-D-Phe6,Leu17]VIP on the inhibition of pulsatile LH release by VIP and related peptides in the ovariectomized rat. Neuroendocrinology **56:** 646–652.

21. BANKS, W.A., D. UCHIDA, A. ARIMURA, et al. 1996. Transport of pituitary adenylate cyclase-activating polypeptide across the blood-brain barrier and the prevention of ischemia-induced death of hippocampal neurons. Ann. N. Y. Acad. Sci. **805:** 270–277.

22. GOULET, M., P.J. SHIROMANI, C.M. WARE, et al. 2003. A secretin i.v. infusion activates gene expression in the central amygdala of rats. Neuroscience **118:** 881–888.

23. YANG, H., L. WANG, S.V. WU, et al. 2004. Peripheral secretin-induced Fos expression in the rat brain is largely vagal dependent. Neuroscience **128:** 131–141.

24. HORVATH, K., G. STEFANATOS, K.N. SOKOLSKI, et al. 1998. Improved social and language skills after secretin administration in patients with autistic spectrum disorders. J. Assoc. Acad. Minor. Phys. **9:** 9–15.

25. MYERS, K., M. GOULET, J. RUSCHE, et al. 2004. Inhibition of fear potentiated startle in rats following peripheral administration of secretin. Psychopharmacology (Berl.) **172:** 94–99.

26. COHEN, S. & W. LIPSHUTZ. 1971. Hormonal regulation of human lower esophageal sphincter competence: interaction of gastrin and secretin. J. Clin. Invest. **50:** 449–454.

27. MIYATA, M., T. SAKAMOTO, T. HASHIMOTO, et al. 1991. Effect of secretin on lower esophageal sphincter pressure in patients with esophageal achalasia. Gastroenterol. Jpn. **26:** 712–715.

28. HONGO, M., A. ISHIMORI, A. NAGASAKI, et al. 1980. Effect of duodenal acidification on the lower esophageal sphincter pressure in the dog with special reference to related gastrointestinal hormones. Tohoku J. Exp. Med. **131:** 215–219.

29. ROSS, G. 1970. Cardiovascular effects of secretin. Am. J. Physiol **218:** 1166–1170.

30. KITANI, K., Y. SUZUKI & R. MIURA. 1978. Differences in the effects of secretin and glucagon on the blood circulation of unanesthetized rats. Acta Hepatogastroenterol. (Stuttg.) **25:** 470–473.

31. GUNNES, P. & O. REIKERAS. 1988. Peripheral distribution of the increased cardiac output by secretin during acute ischemic left ventricular failure. J. Pharmacol. Exp. Ther. **244:** 1057–1061.

32. GUNNES, P., O. REIKERAS & I. LYGREN. 1986. Secretin infusion in acute ischemic left ventricular failure: effects on myocardial performance and metabolism in a closed-chest dog model. J. Pharmacol. Exp. Ther. **239:** 915–918.

33. GUNNES, P. & K. RASMUSSEN. 1986. Haemodynamic effects of pharmacological doses of secretin in patients with impaired left ventricular function. Eur. Heart J. **7:** 146–149.

34. CHIBA, S. 1976. Effect of secretin on pacemaker activity and contractility in the isolated blood-perfused atrium of the dog. Clin. Exp. Pharmacol. Physiol. **3:** 167–172.

35. CHATELAIN, P., P. ROBBERECHT, P. DE NEEF, et al. 1980. Secretin and VIP-stimulated adenylate cyclase from rat heart. I. General properties and structural requirements for enzyme activation. Pflugers Arch. **389:** 21–27.

36. CHATELAIN, P., P. ROBBERECHT, M. WAELBROECK, *et al.* 1983. Topographical distribution of the secretin- and VIP-stimulated adenylate cyclase system in the heart of five animal species. Pflugers Arch. **397:** 100–105.
37. GUNNES, P., H.L. WALDUM, K. RASMUSSEN, *et al.* 1983. Cardiovascular effects of secretin infusion in man. Scand. J. Clin. Lab. Invest. **43:** 637–642.
38. ROBBERECHT, P., P. DE NEEF, M. WAELBROECK, *et al.* 1984. Secretin-induced changes in rate, contractility and adenylate cyclase activity in rat heart atria. Pflugers Arch. **401:** 1–5.
39. TIAHO, F. & J.M. NERBONNE. 1996. VIP and secretin augment cardiac L-type calcium channel currents in isolated adult rat ventricular myocytes. Pflugers Arch. **432:** 821–830.
40. SITNIEWSKA, E.M. & R.J. WISNIEWSKA. 2001. Influence of secretin and L-NAME on vascular permeability in the coronary circulation of intact and diabetic rats. Regul. Pept. **96:** 105–111.
41. PATEL, D.R., Y. KONG & S.P. SREEDHARAN. 1995. Molecular cloning and expression of a human secretin receptor. Mol. Pharmacol. **47:** 467–473.
42. DAVIS, R.J., K.J. PAGE, G.J. DOS SANTOS CRUZ, *et al.* 2004. Expression and functions of the duodenal peptide secretin and its receptor in human lung. Am. J. Respir. Cell Mol. Biol. **31:** 302–308.
43. ROBBERECHT, P., K. TATEMOTO, P. CHATELAIN, *et al.* 1982. Effects of PHI on vasoactive intestinal peptide receptors and adenylate cyclase activity in lung membranes. A comparison in man, rat, mouse and guinea pig. Regul. Pept. **4:** 241–250.
44. TATON, G., M. DELHAYE, J.C. CAMUS, *et al.* 1981. Characterization of the VIP- and secretin-stimulated adenylate cyclase system from human lung. Pflugers Arch. **391:** 178–182.
45. MOODY, T.W. & L.Y. KORMAN. 1988. The release of bombesin-like peptides from small cell lung cancer cells. Ann. N. Y. Acad. Sci. **547:** 351–359.
46. KOWALEWSKI, K. 1975. The effect of secretin on pentagastrin-stimulated secretion of gastric pepsin and acid in rats. Arch. Int. Physiol. Biochim. **83:** 255–259.
47. STOLLMAIER, W. & P.O. SCHWILLE. 1992. Endogenous secretin in the rat—evidence for a role as an enterogastrone but failure to influence serum calcium homeostasis. Exp. Clin. Endocrinol. **99:** 169–174.
48. STENING, G.F., L.R. JOHNSON & M.I. GROSSMAN. 1969. Effect of secretin on acid and pepsin secretion in cat and dog. Gastroenterology **56:** 468–475.
49. ITOH, Z., S. TAKEUCHI, I. AIZAWA, *et al.* 1975. The negative feedback mechanism of gastric acid secretion: significance of acid in the gastric juice in man and dog. Surgery **77:** 648–660.
50. CHEY, W.Y., M.S. KIM, K.Y. LEE, *et al.* 1981. Secretin is an enterogastrone in the dog. Am. J. Physiol. **240:** G239–G244.
51. LAFONTAINE, M., G.B. CADIERE, M.C. WOUSSEN-COLLE, *et al.* 1983. The role of secretin in the control of gastric secretion and emptying in the dog. Arch. Int. Physiol. Biochim. **91:** 459–463.
52. YOU, C.H. & W.Y. CHEY. 1987. Secretin is an enterogastrone in humans. Dig. Dis. Sci. **32:** 466–471.
53. SUGAWARA, K., J. ISAZA, J. CURT, *et al.* 1969. Effect of secretin and cholecystokinin on gastric motility. Am. J. Physiol. **217:** 1633–1638.
54. VALENZUELA, J.E. 1976. Effect of intestinal hormones and peptides on intragastric pressure in dogs. Gastroenterology **71:** 766–769.

55. HALTER, F., L. WITZEL, B. KOHLER, et al. 1975. Secretin snuff: evaluation of the action on pentagastrin-stimulated gastric acid secretion and on pancreatic bicarbonate production. Scand. J. Gastroenterol. **10:** 81–85.
56. KWON, H.Y., T.M. CHANG, K.Y. LEE, et al. 1999. Vagus nerve modulates secretin binding sites in the rat forestomach. Am. J. Physiol. **276:** G1052–G1058.
57. LI, P., T.M. CHANG & W.Y. CHEY. 1998. Secretin inhibits gastric acid secretion via a vagal afferent pathway in rats. Am. J. Physiol. **275:** G22–G28.
58. LI, Y., X. WU, H. YAO, et al. 2005. Secretin activates vagal primary afferent neurons in the rat: evidence from electrophysiological and immunocytochemical studies. Am. J. Physiol Gastrointest. Liver Physiol. **289:** 6745–6752.
59. SOLOMON, T.E., G. VARGA, N. ZENG, et al. 2001. Different actions of secretin and Gly-extended secretin predict secretin receptor subtypes. Am. J. Physiol. Gastrointest. Liver Physiol. **280:** G88–G94.
60. CHUNG, I., P. LI, K. LEE, et al. 1994. Dual inhibitory mechanism of secretin action on acid secretion in totally isolated, vascularly perfused rat stomach. Gastroenterology **107:** 1751–1758.
61. SHIRATORI, K., S. WATANABE & T. TAKEUCHI. 1993. Role of endogenous prostaglandins in secretin- and plaunotol-induced inhibition of gastric acid secretion in the rat. Am. J. Gastroenterol. **88:** 84–89.
62. GERBER, J.G. & N.A. PAYNE. 1996. Secretin inhibits canine gastric acid secretion in response to pentagastrin by modulating gastric histamine release. J. Pharmacol. Exp. Ther. **279:** 718–723.
63. THOMPSON, W.J., L.K. CHANG & E.D. JACOBSON. 1977. Rat gastric mucosal adenylyl cyclase. Gastroenterology **72:** 244–250.
64. FLETCHER, D.R., A. SHULKES & K.J. HARDY. 1985. The effect of neurotensin and secretin on gastric acid secretion and mucosal blood flow in man. Regul. Pept. **11:** 217–226.
65. STACHURA, J., K.J. IVEY, A. TARNAWSKI, et al. 1981. Fine-morphology of chief cells in human gastric mucosa after secretin. Scand. J. Gastroenterol. **16:** 713–720.
66. OLSEN, O., M. WOJDEMANN, B. BERNER, et al. 1998. Secretin and gastric lipase secretion. Digestion **59:** 655–659.
67. POLLACK, P.F., J.G. WOOD & T. SOLOMON. 1990. Effect of secretin on growth of stomach, small intestine, and pancreas of developing rats. Dig. Dis. Sci. **35:** 749–758.
68. LI, W., T.Z. ZHENG & S.Y. QU. 2000. Effect of cholecystokinin and secretin on contractile activity of isolated gastric muscle strips in guinea pigs. World J. Gastroenterol. **6:** 93–95.
69. CHEN, M.C., T.E. SOLOMON, S.E. PEREZ, et al. 2002. Secretin regulates paracellular permeability in canine gastric monolayers by a Src kinase-dependent pathway. Am. J. Physiol. Gastrointest. Liver Physiol. **283:** G893–G899.
70. SCHUSDZIARRA, V. & N. WEIGERT. 1987. Somatostatin and opioids, secretin and VIP–protectors of the mucosa?. Z. Gastroenterol. **25**(Suppl 3): 56–63.
71. GREGORY, R.A. 1974. The Bayliss-Starling lecture 1973. The gastrointestinal hormones: a review of recent advances. J. Physiol. **241:** 1–32.
72. GRAY, M.A., J.R. GREENWELL & B.E. ARGENT. 1988. Secretin-regulated chloride channel on the apical plasma membrane of pancreatic duct cells. J. Membr. Biol. **105:** 131–142.

73. GROTMOL, T., T. BUANES, T. VEEL, *et al.* 1990. Secretin-dependent HCO3- secretion from pancreas and liver. J. Intern. Med. Suppl. **732:** 47–51.

74. VILLANGER, O., T. VEEL & M.G. RAEDER. 1995. Secretin causes H+/HCO3- secretion from pig pancreatic ductules by vacuolar-type H(+)-adenosine triphosphatase. Gastroenterology **108:** 850–859.

75. RAEDER, M.G. 1992. The origin of and subcellular mechanisms causing pancreatic bicarbonate secretion. Gastroenterology **103:** 1674–1684.

76. KONTUREK, S.J., R. ZABIELSKI, J.W. KONTUREK, *et al.* 2003. Neuroendocrinology of the pancreas; role of brain-gut axis in pancreatic secretion. Eur. J. Pharmacol. **481:** 1–14.

77. KITAGAWA, M., T. HAYAKAWA, T. KONDO, *et al.* 1990. Plasma osmolality and exocrine pancreatic secretion. Int. J. Pancreatol. **6:** 25–32.

78. KO, S.B., S. NARUSE, M. KITAGAWA, *et al.* 1999. Arginine vasopressin inhibits fluid secretion in guinea pig pancreatic duct cells. Am. J. Physiol. **277:** G48–G54.

79. BUANES, T., T. GROTMOL, T. LANDSVERK, *et al.* 1988. Effects of arterial pH and carbon dioxide on pancreatic exocrine H+/HCO3- secretion and secretin-dependent translocation of cytoplasmic vesicles in pancreatic duct cells. Acta Physiol. Scand. **133:** 1–9.

80. HOWARD-MCNATT, M., T. SIMON, Y. WANG, *et al.* 2002. Insulin inhibits secretin-induced pancreatic bicarbonate output via cholinergic mechanisms. Pancreas **24:** 380–385.

81. CHEY, W.Y. & T. CHANG. 2001. Neural hormonal regulation of exocrine pancreatic secretion. Pancreatology **1:** 320–335.

82. HEGYI, P., M.A. GRAY & B.E. ARGENT. 2003. Substance P inhibits bicarbonate secretion from guinea pig pancreatic ducts by modulating an anion exchanger. Am. J. Physiol. Cell Physiol. **285:** C268–C276.

83. LEE, E., U. GERLACH, D.Y. UHM, *et al.* 2002. Inhibitory effect of somatostatin on secretin-induced augmentation of the slowly activating K+ current (IKs) in the rat pancreatic acinar cell. Pflugers Arch. **443:** 405–410.

84. PARK, H.S., H.Y. KWON, Y.L. LEE, *et al.* 2000. Role of GRPergic neurons in secretin-evoked exocrine secretion in isolated rat pancreas. Am. J. Physiol. Gastrointest. Liver Physiol. **278:** G557–G562.

85. TRIMBLE, E.R., R. BRUZZONE, T.J. BIDEN, *et al.* 1986. Secretin induces rapid increases in inositol trisphosphate, cytosolic Ca2+ and diacylglycerol as well as cyclic AMP in rat pancreatic acini. Biochem. J. **239:** 257–261.

86. TRIMBLE, E.R., R. BRUZZONE, T.J. BIDEN, *et al.* 1987. Secretin stimulates cyclic AMP and inositol trisphosphate production in rat pancreatic acinar tissue by two fully independent mechanisms. Proc. Natl. Acad. Sci. USA **84:** 3146–3150.

87. WISDOM, D.M., P.J. CAMELLO, G.M. SALIDO, *et al.* 1994. Interaction between secretin and nerve-mediated amylase secretion in the isolated exocrine rat pancreas. Exp. Physiol. **79:** 851–863.

88. KIM, S.J., J.K. KIM, H. PAVENSTADT, *et al.* 2001. Regulation of slowly activating potassium current (I(Ks)) by secretin in rat pancreatic acinar cells. J. Physiol. **535:** 349–358.

89. GARDNER, J.D. & M.J. JACKSON. 1977. Regulation of amylase release from dispersed pancreatic acinar cells. J. Physiol. **270:** 439–454.

90. SINGH, J., R. LENNARD, G.M. SALIDO, *et al.* 1992. Interaction between secretin and cholecystokinin-octapeptide in the exocrine rat pancreas in vivo and in vitro. Exp. Physiol. **77:** 191–204.

91. BOLD, R.J., J. ISHIZUKA, C.M. TOWNSEND, JR, *et al.* 1995. Secretin potentiates cholecystokinin-stimulated amylase release by AR4–2J cells via a stimulation of phospholipase C. J. Cell Physiol. **165:** 172–176.

92. PERIDES, G., A. SHARMA, A. GOPAL, *et al.* 2005. Secretin differentially sensitizes rat pancreatic acini to the effects of supramaximal stimulation with caerulein. Am. J. Physiol Gastrointest. Liver Physiol. **289:** 6713–6721.

93. ROTOLI, B.M., O. BUSSOLATI, V. DALL'ASTA, *et al.* 2000. Secretin increases the paracellular permeability of CAPAN-1 pancreatic duct cells. Cell Physiol. Biochem. **10:** 13–25.

94. HAARSTAD, H. & H. PETERSEN. 1989. Short- and long-term effects of secretin and a cholecystokinin-like peptide on pancreatic growth and synthesis of RNA and polyamines. Scand. J. Gastroenterol. **24:** 721–732.

95. JOEKEL, C.S., M.K. HERRINGTON, J.A. VANDERHOOF, *et al.* 1993. Postnatal development of circulating cholecystokinin and secretin, pancreatic growth, and exocrine function in guinea pigs. Int. J. Pancreatol. **13:** 1–13.

96. HISATOMI, A. & R.H. UNGER. 1983. Secretin inhibits glucagon in the isolated perfused dog pancreas. Diabetes **32:** 970–973.

97. AHREN, B. & I. LUNDQUIST. 1981. Effects of vasoactive intestinal polypeptide (VIP), secretin and gastrin on insulin secretion in the mouse. Diabetologia **20:** 54–59.

98. KOFOD, H., B. HANSEN, A. LERNMARK, *et al.* 1986. Secretin and its C-terminal hexapeptide potentiates insulin release in mouse islets. Am. J. Physiol. **250:** E107–E113.

99. KOFOD, H. 1986. Secretin N-terminal hexapeptide potentiates insulin release in mouse islets. Regul. Pept. **15:** 229–237.

100. SHERWOOD, N.M., S.L. KRUECKL & J.E. MCRORY. 2000. The origin and function of the pituitary adenylate cyclase-activating polypeptide (PACAP)/glucagon superfamily. Endocr. Rev. **21:** 619–670.

101. FUNAKOSHI, A., K. MIYAZAKI, H. SHINOZAKI, *et al.* 1985. Changes in insulin secretion after secretin administration and the implications in diabetes mellitus. Endocrinol. Jpn. **32:** 473–479.

102. WELLS, M., B. BROWN & J. HALL. 2003. Pepsinogen C expression in intestinal IEC-6 cells. Cell Physiol. Biochem. **13:** 301–308.

103. WANG, J.Y., S.A. MCCORMACK, M.J. VIAR, *et al.* 1994. Secretin inhibits induction of ornithine decarboxylase activity by gastrin in duodenal mucosa and IEC-6 cells. Am. J. Physiol. **267:** G276–G284.

104. FUKUDA, M., A. OHARA, T. BAMBA, *et al.* 2000. Activation of transepithelial ion transport by secretin in human intestinal Caco-2 cells. Jpn. J. Physiol. **50:** 215–225.

105. ANDERSSON, A., F. SUNDLER & E. EKBLAD. 2000. Expression and motor effects of secretin in small and large intestine of the rat. Peptides **21:** 1687–1694.

106. PETZOLD, A., A. OTTINGER, J. HAMMER, *et al.* 1991. Effect of gastrointestinal hormones on duodenal motility in acute experiments. Gastroenterol. J. **51:** 81–92.

107. KIRKEGAARD, P., O.P. SKOV, P.S. SEIER, *et al.* 1984. Effect of secretin and glucagon on Brunner's gland secretion in the rat. Gut **25:** 264–268.

108. MOORE, B.A., G.P. MORRIS & S. VANNER. 2000. A novel in vitro model of Brunner's gland secretion in the guinea pig duodenum. Am. J. Physiol. Gastrointest. Liver Physiol. **278:** G477–G485.

109. OLSEN, P.S., P. KIRKEGAARD, S.S. POULSEN, *et al.* 1994. Effect of secretin and somatostatin on secretion of epidermal growth factor from Brunner's glands in the rat. Dig. Dis. Sci. **39:** 2186–2190.

110. HARADA, E. & B. SYUTO. 1993. Secretin induces precocious cessation of intestinal macromolecular transmission and maltase development in the suckling rat. Biol. Neonate **63:** 52–60.

111. HIROSE, S., K. SHIMAZAKI & N. HATTORI. 1986. Effect of secretin and caerulein on the absorption of water, electrolytes and glucose from the jejunum of dogs. Digestion **35:** 205–210.

112. PANSU, D., A. BOSSHARD, M.A. DECHELETTE, *et al.* 1980. Effect of pentagastrin, secretin and cholecystokinin on intestinal water and sodium absorption in the rat. Digestion **20:** 201–206.

113. HOANG, H.D., J.G. WOOD, L.J. BUSSJAEGER, *et al.* 1988. Interaction of neurotensin with caerulein or secretin on digestive tract growth in rats. Regul. Pept. **22:** 275–284.

114. FITZ, J.G. 2002. Regulation of cholangiocyte secretion. Semin. Liver Dis. **22:** 241–249.

115. ALPINI, G., C.D. ULRICH, J.O. PHILLIPS, *et al.* 1994. Upregulation of secretin receptor gene expression in rat cholangiocytes after bile duct ligation. Am. J. Physiol. **266:** G922–G928.

116. ALPINI, G., S. GLASER, W. ROBERTSON, *et al.* 1997. Large but not small intrahepatic bile ducts are involved in secretin-regulated ductal bile secretion. Am. J. Physiol. **272:** G1064–G1074.

117. HUBEL, K.A. 1972. Secretin: a long progress note. Gastroenterology **62:** 318–341.

118. MARINELLI, R.A., L. PHAM, P. AGRE, *et al.* 1997. Secretin promotes osmotic water transport in rat cholangiocytes by increasing aquaporin-1 water channels in plasma membrane. Evidence for a secretin-induced vesicular translocation of aquaporin-1. J. Biol. Chem. **272:** 12984–12988.

119. MARINELLI, R.A., P.S. TIETZ, L.D. PHAM, *et al.* 1999. Secretin induces the apical insertion of aquaporin-1 water channels in rat cholangiocytes. Am. J. Physiol. **276:** G280–G286.

120. TIETZ, P.S., R.A. MARINELLI, X.M. CHEN, *et al.* 2003. Agonist-induced coordinated trafficking of functionally related transport proteins for water and ions in cholangiocytes. J. Biol. Chem. **278:** 20413–20419.

121. ALPINI, G., S. GLASER, L. BAIOCCHI, *et al.* 2005. Secretin activation of the apical Na+-dependent bile acid transporter is associated with cholehepatic shunting in rats. Hepatology **41:** 1037–1045.

122. FUKUMOTO, Y., F. MURAKAMI, A. TATEISHI, *et al.* 2002. Effects of secretin on TCDCA- or TDCA-induced cholestatic liver injury in the rat. Hepatol. Res. **22:** 214–222.

123. AL JIFFRY, B.O., J.M. JOBLING, A.C. SCHLOITHE, *et al.* 2001. Secretin induces variable inhibition of motility in different parts of the Australian possum sphincter of Oddi. Neurogastroenterol. Motil. **13:** 449–455.

124. WESSELS-REIKER, M., R. BASIBOINA, A.C. HOWLETT, *et al.* 1993. Vasoactive intestinal polypeptide-related peptides modulate tyrosine hydroxylase gene expression in PC12 cells through multiple adenylate cyclase-coupled receptors. J. Neurochem. **60:** 1018–1029.

125. MAHAPATRA, N.R., M. MAHATA, D.T. O'CONNOR, *et al.* 2003. Secretin activation of chromogranin A gene transcription. Identification of the signaling pathways in cis and in trans. J. Biol. Chem. **278:** 19986–19994.

126. NUSSDORFER, G.G., M. BAHCELIOGLU, G. NERI, et al. 2000. Secretin, glucagon, gastric inhibitory polypeptide, parathyroid hormone, and related peptides in the regulation of the hypothalamuspituitary-adrenal axis. Peptides **21:** 309–324.
127. DRAGSTEDT, C.A. & S.E. OWEN. 1931. The diuretic action of secretin preparations. Am. J. Physiol.— Legacy **97:** 276–281.
128. LONDONG, W., V. LONDONG, R. MUHLBAUER, et al. 1987. Pharmacological effects of secretin and somatostatin on gastric and renal function in man. Scand. J Gastroenterol. Suppl. **139:** 25–31.
129. BARON, D.N., F. NEWMAN & A. WARRICK. 1958. The effects of secretin on urinary volume and electrolytes in normal subjects and patients with chronic pancreatic disease. Experientia **14:** 30–32. Ref type: generic.
130. BARBEZAT, G.O., J.I. ISENBERG & M.I. GROSSMAN. 1972. Diuretic action of secretin in dog. Proc. Soc. Exp. Biol. Med. **139:** 211–215.
131. VITERI, A.L., J.W. POPPELL, J.M. LASATER, et al. 1975. Renal response to secretin. J. Appl. Physiol. **38:** 661–664.
132. WALDUM, H.L., J.A. SUNDSFJORD, U. AANSTAD, et al. 1980. The effect of secretin on renal haemodynamics in man. Scand . J. Clin. Lab. Invest. **40:** 475–478.
133. CHARLTON, C.G., R. QUIRION, G.E. HANDELMANN, et al. 1986. Secretin receptors in the rat kidney: adenylate cyclase activation and renal effects. Peptides **7:** 865–871.
134. LAMEIRE, N., R. VANHOLDER, S. RINGOIR, et al. 1980. Role of medullary hemodynamics in the natriuresis of drug-induced renal vasodilation in the rat. Circ. Res. **47:** 839–844.
135. FADEM, S.Z., G. HERNANDEZ-LLAMAS, R.V. PATAK, et al. 1982. Studies on the mechanism of sodium excretion during drug-induced vasodilatation in the dog. J Clin. Invest. **69:** 604–610.
136. MERTZ, J.I., J.A. HAAS, T.J. BERNDT, et al. 1983. Effects of secretin on peritubular capillary physical factors and proximal fluid reabsorption in the rat. J. Clin. Invest. **72:** 622–625.
137. MARCHAND, G.R. 1986. Effect of secretin on glomerular dynamics in dogs. Am. J. Physiol. **250:** F256–F260.
138. KASSON, B.G., P. LIM & A.J. HSUEH. 1986. Vasoactive intestinal peptide stimulates androgen biosynthesis by cultured neonatal testicular cells. Mol. Cell Endocrinol. **48:** 21–29.
139. CHOW, B.K., K.H. CHEUNG, E.M. TSANG, et al. 2004. Secretin controls anion secretion in the rat epididymis in an autocrine/paracrine fashion. Biol. Reprod. **70:** 1594–1599.
140. KIMURA, F., N. MITSUGI, J. ARITA, et al. 1987. Effects of preoptic injections of gastrin, cholecystokinin, secretin, vasoactive intestinal peptide and PHI on the secretion of luteinizing hormone and prolactin in ovariectomized estrogen-primed rats. Brain Res. **410:** 315–322.
141. HOLST, N., T.G. JENSSEN, P.G. BURHOL, et al. 1989. Plasma gastrointestinal hormones during spontaneous and induced menstrual cycles. J. Clin. Endocrinol. Metab. **68:** 1160–1166.
142. HOLST, N., T.G. JENSSEN, P.G. BURHOL, et al. 1991. Prolactin response to secretin during the spontaneous menstrual cycle in women. Gynecol. Obstet. Invest. **31:** 37–41.
143. HOLST, N., T.G. JENSSEN, P.G. BURHOL, et al. 1989. Plasma secretin concentrations during normal human pregnancy, delivery, and postpartum. Br. J Obstet. Gynaecol. **96:** 424–427.

144. HOLST, N., T.G. JENSSEN, P.G. BURHOL, *et al.* 1988. Gastrointestinal regulatory peptides during oxytocin infusion in post-term pregnancies. Acta Physiol. Scand. **132:** 23–27.
145. MASON, J.C., R.F. MURPHY, R.W. HENRY, *et al.* 1979. Starvation-induced changes in secretin-like immunoreactivity of human plasma. Biochim. Biophys. Acta **582:** 322–331.
146. BELL, P.M., R.W. HENRY, K.D. BUCHANAN, *et al.* 1985. Cimetidine fails to suppress the rise in plasma secretin during fasting. Regul. Pept. **10:** 127–131.
147. MANABE, T., Y. TANAKA, K. YAMAKI, *et al.* 1987. The role of plasma secretin during starvation in dogs. Gastroenterol. Jpn. **22:** 756–758.
148. THUESEN, B., O.B. SCHAFFALITZKY DE MUCKADELL, J.J. HOLST, *et al.* 1987. The relationship of secretin and somatostatin levels in plasma to glucose administration and acid secretion during fasting. Am. J. Gastroenterol. **82:** 723–726.
149. RUDMAN, D. & A.E. DEL RIO. 1969. Lipolytic activity of synthetic porcine secretin. Endocrinology **85:** 214–217.
150. RODBELL, M., L. BIRNBAUMER & S.L. POHL. 1970. Adenyl cyclase in fat cells. 3. Stimulation by secretin and the effects of trypsin on the receptors for lipolytic hormones. J. Biol. Chem. **245:** 718–722.
151. NG, T.B. 1990. Studies on hormonal regulation of lipolysis and lipogenesis in fat cells of various mammalian species. Comp. Biochem. Physiol. B. **97:** 441–446.
152. CHISHOLM, D.J., G.A. KLASSEN, J. DUPRE, *et al.* 1975. Interaction of secretin and insulin on human forearm metabolism. Eur. J. Clin. Invest. **5:** 487–494.
153. RAZIN, E., M.G. FELDMAN & D.A. DREILING. 1962. The hormonal regulation of thoracic ductal lymph flow. The effect of secretin and related hormones on the thoracic ductal flow and composition in dogs. J. Surg. Res. **2:** 320–331.
154. LYGREN, I., H.L. WALDUM, P.G. BURHOL, *et al.* 1982. Secretin exerts inhibition of the 8BrcAMP-stimulated 86Rb influx into avian red blood cells. Regul. Pept. **4:** 221–225.
155. SHIDO, O., Y. YONEDA & T. NAGASAKA. 1989. Changes in brown adipose tissue metabolism following intraventricular vasoactive intestinal peptide and other gastrointestinal peptides in rats. Jpn. J. Physiol. **39:** 359–369.
156. KUNTZ, A., H.W. CLEMENT, W. LEHNERT, *et al.* 2004. Effects of secretin on extracellular amino acid concentrations in rat hippocampus. J. Neural. Transm. **111:** 931–939.
157. TODA, Y., K. MORI, T. HASHIMOTO, *et al.* 2004. Efficacy of secretin for the treatment of autism. No To Hattatsu **36:** 289–295.
158. LOSSI, L., L. BOTTARELLI, M.E. CANDUSSO, *et al.* 2004. Transient expression of secretin in serotoninergic neurons of mouse brain during development. Eur. J. Neurosci. **20:** 3259–3269.
159. KOVES, K., M. KAUSZ, D. RESER, *et al.* 2002. What may be the anatomical basis that secretin can improve the mental functions in autism? Regul. Pept. **109:** 167–172.
160. SHEITMAN, B.B., M.B. KNABLE, L.F. JARSKOG, *et al.* 2004. Secretin for refractory schizophrenia. Schizophr. Res. **66:** 177–181.
161. GROSSMAN, M.I. 1966. Treatment of duodenal ulcer with secretin: a speculative proposal. Gastroenterology **50:** 912–913.
162. BLOOM, S.R. & A.S. WARD. 1975. Failure of secretin release in patients with duodenal ulcer. Br. Med. J **1:** 126–127.

163. OHARA, H., S. INABE, S. YABU-UCHI, *et al.* 1978. Secretin secretion in patients with duodenal ulcer, chronic pancreatitis and diabetes mellitus. Gastroenterol. Jpn. **13:** 21–27.
164. KISFALVI, I. 1979. Inhibition of bombesin-stimulated gastric acid secretion by secretin, glucagon and caerulein in patients with duodenal ulcer. Digestion **19:** 315–321.
165. DEMLING, L., W. DOMSCHKE, S. DOMSCHKE, *et al.* 1975. Treatment of duodenal ulcer with a long-acting synthetic secretin: a pilot trial. Acta Hepatogastroenterol. (Stuttg.) **22:** 310–313.

VIP–PACAP System in Immunity

New Insights for Multitarget Therapy

R.P. GOMARIZ,[a] Y. JUARRANZ,[a] C. ABAD,[a] A. ARRANZ,[a] J. LECETA,[a] AND C. MARTINEZ[b]

[a]Department of Cell Biology, Faculty of Biology, Complutense University, 28040 Madrid, Spain

[b]Department of Cell Biology, Faculty of Medicine, Complutense University, 28040 Madrid, Spain

ABSTRACT: Our research about VIP/PACAP and the immune system goes back to 1990 when our group described the expression of VIP on lymphocytes for the first time. Since this year, using three models of disease, septic shock, rheumathoid arthritis, and Crohn's disease, we are trying to contribute with new pieces to the puzzle of immunity to approach the use of VIP/PACAP system as a therapeutic agent. In 1999 we established that the first step in the beneficial effect of the VIP/PACAP system exerts consists in its potent anti-inflammatory action. Thus, VIP and PACAP inhibit the expression and release of proinflammatory cytokines and chemokines, and enhance the production of the anti-inflammatory factors. These effects were reported both *in vitro* and *in vivo*, are mediated by the presence of PAC1, VPAC1, and VPAC2 receptors, in the three models of diseases used. The next step was that the system favors Th2 responses versus Th1 contributing to the remission of illness as rheumatoid arthritis or Crohn's disease by blocking the autoimmune component of these diseases. Because it appears that inflammatory processes requires more than blockade of a single mediator, new therapies blocking several components of both the infection- and the autoimmunity-induced inflammation cascades should be an interesting focus of attention. In this sense, at present we are trying to dissect new aspects of the potential therapeutic of the VIP/PACAP system in the control of CC and CXC chemokine and their receptors, coagulation factors, adhesion molecules, acute phase proteins, and osteoclastogenesis mediators as well as in the modulation of the expression of Toll-like receptors. Our more recent data open a hopeful door for the therapeutic use of VIP/PACAP in humans.

KEYWORDS: neuroimmunomodulation; inflammation; VIP; PACAP; immune system; toll-like receptors; chemokines

Address for correspondence: R.P. Gomariz, Department of Cell Biology, Faculty of Biology, Complutense University, 28040 Madrid, Spain. Voice: 34-913944971; fax: 34-913944981.
e-mail: gomariz@bio.ucm.es

Ann. N.Y. Acad. Sci. 1070: 51–74 (2006). © 2006 New York Academy of Sciences.
doi: 10.1196/annals.1317.031

INTRODUCTION

Together with genetic and environment factors, the maintenance of health depends on the regulatory interactions between the three essential systems involved in homeostasis: nervous, endocrine, and immune systems. In 1987, N.H. Spector[1] named the science that explores the interactions between these three systems as neuroimmunomodulation. This fascinating discipline represents the new sign of an old science described by Jankovic.[2] He defined it as a modern reflection in neurosciences and immunosciences of the ideas and experiences of philosophers and ingenious observers of ancient Egypt, Greece, China, India, and other civilizations, who thought that the mind is involved in the defense against disease. A significant landmark for the behavior of this framework is the sharing of common ligand-receptor-effector molecular systems that were soon demonstrated and accepted for the endocrine and nervous systems. The immune system began to be considered as a member of this homeostatic framework when Blalock and Smith demonstrated in 1980 that immune cells produce adrenocorticotropic hormone (ACTH) and endorphins.[3] Herein, we review the current knowledge of the physiological role of vasoactive intestinal peptide (VIP)–pituitary adenylate cyclase-activating polypeptide (PACAP) in immunity, to further speculate, that advances in the understanding of this role may allow us to approach their use in human therapy.

VIP–PACAP FRAMEWORK IN IMMUNITY

To date, the VIP–PACAP system consists of two peptides and three receptors of the family[2] of G protein–coupled receptors (GPCRs). In relation to peptides, Said and Mutt in 1970 reported, for the first time, to the scientific community, the presence of VIP in lung, that was later isolated from small intestine,[4] in an article entitled with the premonitory title of "Polypeptide with broad biological activity: isolation from small intestine." Twenty years later, in 1989 PACAP was isolated from ovine hypothalamic extracts and it has been described as two amidated forms: PACAP-27 and PACAP-38.[5] VIP and PACAP were found widely distributed in the organism as neuroendocrine modulators.[6–10] Both peptides were also, more recently, assigned to the immune system, exerting very novel and interesting functions on innate and acquired immunity. In this system, VIP and PACAP storage and gene expression appear exactly in the same lymphoid subpopulations, double- and single-positive thymocytes and T and B lymphocytes from peripheral lymphoid organs.[11–14] Moreover, VIP is secreted to the lymphoid microenvironment after stimulation with agents that induce antigenic stimulation, inflammation, and apoptosis.[15] Thereafter, it has been described that VIP is produced by type 2 T cells following antigenic stimulation.[16] Thus, today we know that VIP and PACAP are peptides with a broad distribution in nervous, endocrine, and immune systems of complex living organisms, which show a highly conserved sequence, a similar gene structural organization, and a high degree of homology between the peptides and their precursors. These facts support the idea that VIP and PACAP genes have originated from a common ancestral sequence after gene duplication.[17] With regard to receptors, VIP and PACAP regulate numerous biological activities through the binding to three plasma membrane receptors belonging to the subfamily II within the superfamily of GPCR: VIP receptors type 1 and 2 (VPAC1 and VPAC2, respectively) that bind VIP and PACAP with equal affinity, and PACAP receptor (PAC1) that is PACAP selective.[10,18–20] To date, eight variants produced by alternative splicing of the transcript have been described for PAC1, seven of them coupled to the activation of adenylate cyclase and inositol/phosphate/phospholipase C systems,[21,22] and the eighth variant that stimulates an L-type calcium channel.[23]

In the immune system, Guerrero *et al.*,[24] using binding techniques, described for the first time in 1981 the VPAC1 receptor in human peripheral blood lymphocytes. Later, it was reported in human monocytes,[25] murine lymphocytes,[26] and rat alveolar macrophages.[27] Moreover, gene expression of VPAC1 receptor has been demonstrated in T and B murine lymphocytes subpopulations from peripheral lymphoid organs, in single- and double-positive CD4$^+$CD8$^+$ murine thymocyte subsets and in peritoneal macrophages.[12,28–30] With regard to VPAC2, its expression has been reported in lymphocytes and macrophages, being inducible and detected in lymphocytes only after stimulation through the T cell receptor (TCR)-associated CD3 molecule, and in macrophages after lipopolysaccharide (LPS) stimulation.[28–30] VPAC2 receptor is detected in mononuclear cells by immunohistochemical techniques 2 days after the detection of VPAC1 receptor, at sites of inflammation and antigen recognition.[31] However, the constitutive expression of VPAC2 receptor has been reported in human lymphoid cell lines. The PAC1 receptor is a PACAP selective receptor; in immune cells, this receptor is expressed only in macrophages,[32] binding VIP and PACAP with the same affinity and activating the IP/PLC pathway. Lymphocytes lack PAC1 receptor expression. Although it is still unknown which splice variant of PAC1 is present in macrophages, it seems clear that this receptor plays a crucial role in VIP and PACAP regulation of several agents involved in inflammation, such as interleukin-6 (IL-6).[33] In all, in the immune system, lymphocytes express VPAC1 and VPAC2 and macrophages express the PAC1 receptor and the two VPAC receptors. So, redundancy between VIP and PACAP in the immune system, in terms of both share their cellular expression pattern of ligand-receptors, together with the fact, as we discuss below, that this system performs similar functions, would suggest that the gene duplication could also have affected the control regions of the common ancestral gene, and could explain the pivotal role of the VIP–PACAP network in the control of immune homeostasis and disease.

NEW IMMUNE FUNCTIONS FOR OLD PEPTIDES

With this molecular substratum formed by two peptides and three receptors, our group and others are trying to dissect since 1990 their physiological meaning in the maintenance of the healthy status of the organism under the constant pressure of pathogen insult. We have tested the potential therapeutic effect of VIP–PACAP family in several models of both infection-induced inflammation and autoimmune-induced inflammation. The infection-induced inflammation has been studied in the LPS murine model of endotoxic shock and the autoimmune-induced inflammation in the murine model of Crohn's disease (CD), the trinitrobenzene sulfonic acid (TNBS)-induced colitis as well as in rheumatoid arthritis (RA) both, in human and in its murine model, the collagen-induced arthritis (CIA).

Endotoxic Shock

Endotoxin, which is found in the outer membrane of gram-negative bacteria, has been implicated in a variety of pathologies ranging from relatively mild (fever) to lethal (septic shock, organ failure, and death) as the etiological agent. Infections caused by gram-negative bacteria constitute one of the major causes of sepsis or septic shock that results from the inability of the immune system to limit bacterial spread during the ongoing infection. While normally helping to eradicate pathogens from a local infection of peripheral tissues, inflammation in sepsis develops into a systemic syndrome with multiple manifestations, such as hypotension, tissue injury, increased vascular permeability, disseminated intravascular coagulation, and ultimately, multiorgan failure and shock.[34]

Rheumatoid Arthritis

RA is an autoimmune disease of unknown etiology, characterized by the presence of an inflammatory synovitis accompanied by the destruction of joint cartilage and bone. Generally, it affects 1% of population and about twice as many women as men. Although RA can develop in childhood, in most cases it develops between the ages of 25 and 50 years. The drugs and agents currently used to treat RA have multiple effects, some of which are undesirable, and in long term these treatments do not prevent joint damage.[35]

CIA is a murine experimental disease model induced by immunization with type II collagen (CII). Because it shares a number of clinical, histological, and immunologic features with RA,[36] we have used the CIA model to study the potential effect of VIP on the pathogenesis of arthritis. It has been described that RA have two deleterious components: an inflammatory and a T helper 1

(Th1) component. Besides CIA model, we have used synovial tissue from human patients with RA to study VIP effect on humans.

Inflammatory Bowel Disease

CD and ulcerative colitis (UC) are examples of inflammatory bowel disease (IBD), and are multifaceted chronic autoimmune disorders with unknown etiology; to date, there is no known cure. IBD is thought to occur as a result of an inappropriate immune response to environmental factors in a genetically predisposed host, and it has become increasingly clear that cytokines play an important role in this process.

IBD affects 121 out of every 100,000 people in Europe.[37] The establishment of CD involves two kinds of overlapping immunological responses on the basis of the production of proinflammatory factors by both resident intestinal and recruited immune cells, and the activation of CD4 T cells with a Th1 profile.[38]

A frequently used murine model is based on the intrarectal administration of TNBS, which randomly haptenates the proteins in the colon mucosa and triggers an inflammatory and Th1 response similar to that in CD.[39] Current therapy of CD is not sufficient and is about two strategies, the long-term treatment with nonspecific anti-inflammatory drugs, and the more specific treatment with either recombinant anti-inflammatory cytokines (interferon-γ [IFN-γ], IL-10, IL-11), or specific monoclonal antibodies against proinflammatory cytokines (tumor necrosis factor-α [TNF-α], IL-12) and their receptors (TNFR, IL-6R).

Both strategies often trigger undesirable and potentially serious side effects. In summary, using these models, we have been trying to dissect since 1999 the potential therapeutic effect of VIP–PACAP network, describing the following regulatory immune functions.

Balance of the Generation of Inflammatory Factors

Although the VIP–PACAP system has been extensively reported to regulate a wide variety of functions in different systems,[6–8] the first function described in the immune system was its role in inflammation.[40] Different *in vitro* and later *in vivo* studies demonstrated its potent anti-inflammatory action.[28–30,41]

Inflammation is a self-defensive reaction aimed to eliminate or neutralize injurious stimuli, and restoring tissue integrity. The first studies demonstrated the anti-inflammatory properties of the VIP–PACAP system based on their ability to inhibit several macrophage functions, including phagocytosis, respiratory burst, and chemotaxis,[42] and also by inhibiting T cell proliferation and decreasing lymphocyte migration.[43] In this sense, VIP has been reported to inhibit the production of proinflammatory cytokines, such as TNF-α, IFN-γ, IL-6, and IL-12, to reduce the activity of inducible nitric oxide synthase (iNOS), and to enhance the production of the anti-inflammatory cytokines IL-10 and IL-1Ra.[29,41] TABLE 1 summarizes the effect of VIP–PACAP system on the balance of inflammatory mediators in different models of inflammatory/autoimmune disorders,[29,41,44–50] at mRNA and/or protein levels.

Besides, we have recently studied, for the first time, the inhibitory effect on mRNA production in cytokine receptors in a murine model of CD, including IL-1R2, IL-2Rγ, IL-6Ra, IL-12Rb, IL-15Rα, and IL-7R.

These results demonstrate that this system is able to reduce the synthesis of not only cytokines, but also their receptors.[44] In this study, we also pointed out the role of VIP–PACAP network in the production of chemokines and their receptors (TABLE 1). Chemokines are multifunctional mediators that promote immune responses, stem cell survival, development, and homeostasis, as well as triggering chemotaxis and angiogenesis.[51] These molecules constitute a family of structurally related low-molecular-weight (8–10 kDa) proteins, defined on the basis of their amino acid composition, and specifically on the presence of a conserved tetracysteine motif within the N terminus,[52–54] in four subfamilies named CXC, CC, C, and CX3C. To date,[43] chemokines have been described as those, which mediate their biological effects by binding to specific cell-surface protein-coupled receptors that at present consist of 19 receptors.

Chemokines and their receptors have been implicated in the physiopathology of a number of inflammatory diseases, playing a crucial role directing the migration of leukocytes toward sites of inflammation. Thus, they are important potential therapeutic targets because of their central role in the cell recruitment, because inappropriate cell recruitment is a hallmark of all autoimmune, allergic, and inflammatory diseases. Data about the effect of VIP–PACAP system in chemokine production have been reported on different animal models (TABLE 1).

As the table shows, VIP–PACAP regulate different chemokines involved in both monocyte and neutrophil chemotaxis, such as CCL2, CCL5, CCL9, CXCL1, CXCL2, CXCL3, CXCL8, and CX3CL1. Moreover, our group has recently demonstrated, for the first time, the involvement of this system in the downregulation of two chemokine receptors as CCR1 and CXCR2 in a murine model of CD[44] amplifying its enormous anti-inflammatory potential. CCR1 and CXCR2 are the main and the most ubiquitous receptors of CC and CXC families involved in the recruitment of monocytes and neutrophils, respectively. CCR1 has 10 ligands of the CC family that include CCL3, CCL5-7, CCL9, CCL13-16, and CCL23. Besides, CXCR2 has at least seven ligands of the CXC chemokine family, among them are CXCL1-3 and CXCL5-8.

As inflammation causes a vast amount of human morbidity and mortality, the function of the VIP–PACAP system in the prevention of inflammation by interfering with cellular recruitment through the balance of cytokines and chemokines and their receptors may represent an important tool to be applied to therapy.

TABLE 1. VIP–PACAP network regulation of inflammatory chemokines, cytokines, and their receptors in disease

		Septic shock	RA	CD	Delayed-type hypersensitivity	Neurodegenerative disorders
Chemokines	CCL2 (MCP-1)	↓	↓	↓	?	↓
	CCL5 (RANTES)	↓	↓	-	↓	↓
	CCL9 (MIP-1γ)	?	?	↓	?	↓
	CXCL1 (MIP-2α)	↓	↓	↓	?	↓
	CXCL2 (MIP-2β)	↓	↓	-	?	↓
	CXCL3 (GRO-γ)	↓	?	-	?	↓
	CXCL8 (IL-8)	↓	↓	-	?	?
	CX3CL1 (Fractalkine)	?	?	↓	?	?
Chemokine receptors	CCR1	?	?	↓	?	?
	CXCR2	?	?	↓	?	?
Cytokines	IL-1b	↓	↓	↓	↓	↓
	IL-1Rα	↑	?	?	?	?
	IL-2	?	?	↓	↓	?
	IL-6	↓	↓	↓	?	↓
	IL-10	↑	↑	↑	↑	?
	IL-11	?	↓	-	?	?
	IL-15	?	?	↓	?	?
	IL-17	?	↓	↓	?	?
	IL-21	?	?	↓	?	?
	IL-25	?	?	↓	?	?
	TGF-β	?	?	↓	?	?
	TNF-α	↓	↓	↓	?	↓
Cytokine receptors	IL-1R2	?	?	↓	?	?
	IL-2Rγ	?	?	↓	?	?
	IL-6Rα	?	?	↓	?	?
	IL-12Rβ	?	?	↓	?	?
	IL-15Rα	?	?	↓	?	?
	IL-17R	?	?	↓	?	?

Inhibition of the Adhesion Molecules Production

The early phase of endotoxic shock is characterized by inflammatory cytokine upregulation that leads to activation of both immune effectors: cells and vascular endothelium. Neutrophils are among the early immune cells to arrive at sites of infection, where they initiate antimicrobial and proinflammatory functions, which serve to contain infection. Activated neutrophils bind to vascular endothelial cells expressing adhesion molecules, migrate outside the vessel, and infiltrate tissue. The adherent infiltrating neutrophils are activated to release proteases, such as elastase and free radicals from granulocytes leading to organ damage. Thus, upon endotoxin stimulation, neutrophils, and adhesion molecules are strongly upregulated, being involved in the evolution of this infection-induced inflammation. The reduction of the inflammatory markers, such as infiltrating neutrophils, intercellular adhesion molecule-1 (ICAM-1), and vascular cell adhesion molecule-1 (VCAM-1) represents another important check-point to control the disease.[55] Using mice deficient in PAC1 and a PAC1 antagonist, we have demonstrated that VIP and PACAP as well as the PAC1 receptor are involved in the reduction of neutrophil recruitment in different target organs, such as intestine and liver, as well as in the downregulation of sICAM-1 serum levels and in intestine ICAM-1 and VCAM-1 mRNA expression.[56] Besides, in previous studies using PAC1−/− mice we have demonstrated a key anti-inflammatory function for this receptor in the protection of lethal endotoxemia through the inhibition of IL-6 production.[33] IL-6 cytokine is an important molecular marker of the development of endotoxic shock acting at different stages of the inflammatory cascade and elevated IL-6 serum concentrations have been described in this disorder. Studies using IL-6 deficient mice have indicated that sIL-6R/IL-6 complex amplifies leukocyte recruitment at sites of inflammation, by augmenting the local production of chemokines and modulating the expression of endothelial ICAM-1 and VCAM-1 to bind neutrophils.[57] Moreover, a region mediating IL-6 and IFN-γ responsiveness in the ICAM-1 promotor has been described.[58] As it has not been detected the presence and expression of PAC1 in endothelial cells so far, effects of VIP–PACAP framework on the reduction of neutrophil infiltration and ICAM-1 and VCAM-1 expression could be at least partly explained by the VIP-mediated inhibition of IL-6.

Downregulation of the Production of Coagulation Factors

Fibrinogen is a 340-kDa complex dimeric protein with each subunit composed of three no identical polypeptide chains (α, β, and γ) that is secreted into the blood after its synthesis in the liver. It is one of the factors implicated in disseminated intravascular coagulation, a disorder associated with septic shock that affects the function of the clotting system, which leads to multiple organ dysfunction syndrome. Induced expression of fibrinogen in inflammation is considered as a major cardiovascular risk factor since it is involved in blood clot formation.[59] Using mice deficient in PAC1 receptor, we have demonstrated that VIP and PACAP as well as the PAC1 receptor are involved in the reduction of the mRNA levels of β-fibrinogen.[56] Thus, after endotoxemia induction, increased levels of β-fibrinogen in both wild-type and PAC1−/− mice are produced, and VIP–PACAP treatment resulted in a significant downregulation of β-fibrinogen expression in wild-type mice. In contrast, both peptides failed to reduce β-fibrinogen mRNA expression in PAC1–/– mice indicating that the inhibition of VIP–PACAP is mainly mediated through PAC1. Plasma levels of three fibrinogen polypeptides rise coordinately in response to many inflammatory conditions. In cultured liver cells the addition of IL-6 can mimic this response. Furthermore, an IL-6 response element of the gamma chain has been identified in the promotor of human fibrinogen. In our model, the inhibitory effect of VIP and PACAP on fibrinogen expression was correlated with the reduction of IL-6 production,[59] suggesting that PAC1 deficiency leads to suppression of the negative control mechanism allowing the modulation of fibrinogen induction in endotoxemia, so that it protects mice from multiple organ failure, at least partially via the inhibited production of IL-6. On the other hand, although VPAC1 is the major receptor expressed in liver, PAC1 receptor has also been described in liver membranes, suggesting that both peptides may also reduce β-fibrinogen directly through its interaction with hepatocytes.[60,61]

Reduction of the Production of Acute-Phase Proteins

The acute-phase (AP) response is a nonspecific inflammatory reaction of the host that occurs shortly after any tissue injury. The response includes changes in the concentration of plasma proteins called acute-phase proteins (APPs), some of which increase its concentration. Thus, serum concentration of APP provide valuable diagnostic information for the detection, prognosis, or monitoring of septic shock. The acute-phase serum amyloid A (SAA) proteins are multifunctional apolipoproteins, which are involved in cholesterol transport and metabolism, and in modulating numerous immunological responses in inflammation and the AP response to infection. During the first stages of inflammation, these apolipoproteins have a crucial role in this process, however, its maintained overproduction may have negative clinical consequences,[62] and therefore effective temporal control is desirable. We have demonstrated that treatment with VIP drastically reduced serum SAA levels in the murine models of CD[63] and RA (manuscript in preparation) mainly through VPAC1. According to these results, in the murine model of endotoxic shock using PAC1-deficient mice, we have shown that SAA levels were significantly reduced in the serum of both wild-type and PAC1–/– mice after treatment with VIP and PACAP. SAA can be induced by either IL-1β or TNF-α whereas IL-6 has little effect by itself although it can synergize with them.[64] Furthermore, it looks as if there were a significant correlation between TNF-α levels and SAA concentrations in patients with sepsis.[65] In this sense, we have previously demonstrated that both peptides inhibit *in vivo* TNF-α production through VPAC1 receptor without involvement of PAC1. In all, the VIP–PACAP system is able to reduce the production of SAA through VPAC1 receptor.

Favor Th2 Versus Th1 Pathways

T cells, in particular CD4$^+$ T cells, have been implicated in mediating many aspects of autoimmune inflammation. These cells, upon antigenic stimulation, differentiate into Th1 and Th2 effectors cells, characterized by specific cytokine profiles and functions. Factors that influence their differentiation include antigen-presenting cell (APC) type, the expression of the costimulatory molecules expressed by them, and the cytokine microenvironment.[66] In recent years, VIP and PACAP have been shown to induce a Th2 response by stimulating the release of Th2 cytokines/chemokines and inhibiting the production of Th1 cytokines/chemokines in different models of inflammatory/autoimmune diseases (reviewed in Ref. 41) (TABLE 2). Moreover, VIP and PACAP preferentially induce the expression of costimulatory molecules related to Th2 differentiation (i.e., B7.2) in resting and Ag-stimulated macrophages as well as in dendritic cells *in vitro* and *in vivo*.[50,67] An additional mechanism could be the preferential VIP–PACAP-induced prevention of clonal deletion of Th2 against Th1 cells after antigenic stimulation, resulting in the generation of Th2 effectors and memory cells.[68] Several clinical observations suggest a pathogenic Th1 drive in RA leading to the perpetuation of chronic inflammation and recent data support that CD reflects an excessive Th1 response. Therefore, a shift in the balance of Th1/Th2 effectors cells toward anti-inflammatory Th2 cells would be expected to be clinically beneficial. It is interesting to note that recent studies have demonstrated that VIP–PACAP seem to be able to modulate the Th1/Th2 balance in the murine models of CIA and TNBS-induced colitis by inhibiting the production of IFN-γ (Th1 cytokine) and increasing IL-4 levels (Th2 cytokine).[45,63,69] In addition, another marker of CIA model is the Ig isotype switching that is directed by Th1 or Th2 cytokines in a different way (IFN-γ and IL-4 induce IgG2a and IgG1 synthesis, respectively). VIP and PACAP treatment produces a reduction in IgG2a and an increase in IgG1 confirming the induction of a Th2 response.[45,69] Finally, the last advance in our understanding of VIP effect on Th1/Th2 balance has been revealed by a microarray analysis in the TNBS-colitis model[44] studying for the first time the involvement of chemokine and cytokine receptors. TNBS-treated mice show an increase of Th1-related molecules, such as the chemokines/ receptors CCL3/CCR1, CCL4/CCR5, CX3CL1, CCR9, and CXCR3, implicated in Th1 cell recruitment as well as the Th1 inducing IL-12 system and IL-18 and its receptor (IL1-Rrp), which are related to the pathogenesis of CD.

VIP induces Th2 response versus Th1 with the inhibition of IFN-γ production and the related chemokines and cytokines, but the upregulation of IL-4 and IL-10 (TABLES 1 and 2).[29,41,44-50] In conclusion, VIP and PACAP interference with the activation and generation of Th1 cells and their secreted cytokines represents another important step in their beneficial effect for the treatment of inflammatory/autoimmune diseases.

TABLE 2. VIP–PACAP regulation of Th1/Th2 chemokines, cytokines, and their receptors in disease

		Septic shock	RA	CD	Delayed-type hypersensitivity	Neurodegenerative disorders
Chemokines	CCL3 (MIP-1α)	↓	↓	↓	↓	↓
	CCL4 (MIP-1β)	↓	?	↓	↓	↓
	CCL22 (MDC)	?	?	-	?	↓
	CXCL10 (IP-10)	?	?	-	?	↓
Chemokine receptors	CCR1	?	?	↓	?	?
	CCR4	?	?	↓	?	?
	CCR5	?	?	↓	↓	?
	CCR9	?	?	↓	?	?
	CXCR3	?	?	↓	?	?
Cytokines	IL-4	↑	↑	↑	↑	↑
	IL-12	↓	↓	↓	?	↓
	IL-18	?	↓	↓	?	?
	IFN-γ	↓	↓	↓	?	↓
Cytokine receptors	IL-1Rrp	?	?	↓	?	?

Protection from Bone Erosion

RA is characterized by progressive joint destruction. Osteoclasts involved in bone resorption are under control of both systemic factors and an array of local factors, such as cytokines, inflammatory mediators, and growth factors. The receptor activator of nuclear factor-κB ligand (RANKL) and its receptor activator of nuclear factor-κB (RANK) are essential for the development and activation of osteoclasts, appearing to be the pathogenic principle that causes bone and cartilage destruction in arthritis. Osteoprotegerin (OPG), a member of the TNF-receptor family expressed by osteoblasts, has documented effects on the regulation of bone metabolism (FIG. 1). OPG inhibits bone resorption and binds with strong affinity to its ligand RANKL, thereby preventing RANKL from binding to its receptor RANK. Several cytokines, such as IL-1β, TNF-α, IL-6, IL-11, or IL-17 produced by synovial tissue affect the presence of RANK, RANKL, and OPG in RA. IL-17 is a cytokine produced by a subset of activated memory Th1/Th0 cells[70] that has been shown to be an important osteoclast differentiation factor that induces RANKL expression leading to bone erosion in arthritis.[71] IL-11 also supports osteoclast formation by increasing RANKL expression in a STAT activation- dependent mechanism.[72] VIP has been shown to regulate several bone cell functions, in a way that affects bone resorbing activity of isolated osteoclasts and osteoclast formation[73] as well as osteoblast anabolic processes.[74] These effects are mediated by the presence of different VIP receptors in both types of bone cells since VPAC1 has been detected in osteoclasts[75] whereas VPAC2 is expressed in osteoblasts and VPAC1 is induced in advanced cultures of this cell type.[76] *In vitro* studies have demonstrated contradictory results: while VIP has been shown to promote the formation of mineralized nodules in cultures of osteoblasts,[74] it induces a transient inhibition and a delayed stimulation of osteoclast activity.[77] Using the CIA model of arthritis we have studied the expression of different mediators implicated in bone homeostasis, such as iNOS, cyclooxygenase-2 (COX-2), RANK, RANKL, OPG, IL-11, and IL-17.[78]

CIA mice treated with VIP present a decrease in mRNA expression of IL-17 and IL-11 in the joints. The ratio of RANKL/OPG decreased drastically after VIP treatment in the joint, which was correlated with an increase of OPG circulating levels of CIA mice treated with VIP. In addition, VIP treatment decreased the presence of mRNA expression of RANK, iNOS, and COX-2.

Most of the osteoclastogenic factors present in RA joints are thought to act indirectly enhancing RANKL expression by altering the RANK/RANKL/OPG system, which is the final regulator of bone resorption.[70] RANK is expressed on the surface of hematopoietic osteoclasts progenitors that belong to the monocyte/macrophage lineage, and also on mature osteoclasts. In arthritis, osteoclast precursors that express RANK recognize RANKL produced by osteoblasts/stromal cells or Th1 lymphocytes or dentritic cells, and differentiate into osteoclasts.[79] In this sense, we have reported a high level of RANK expression in the joints of arthritic mice, probably induced by the recruitment of osteoclast precursors caused by the local production of chemotactic chemokines by monocytes.[80] VIP decreased the expression of RANK in the joints of CIA mice to the levels detected in control nonarthritic mice. This effect may be due to the inhibition of RANK synthesis or alternatively to the inhibition of monocyte recruitment, since we have previously reported that VIP inhibits the local expression of the monocyte chemoattractant chemokines CCL3 (MIP-1) and CCL2 (MCP-1). VIP has been reported to inhibit the expression of RANKL and RANK induced by vitamin D in mouse bone marrow cultures.[73] Our results showed that VIP reduces the expression of RANK and RANKL in the joints of arthritic mice, and may account for the boneprotective properties of VIP in RA. On the other hand, its effects on the expression of OPG further support the postulated bone-protective property of VIP. In this way the balance RANKL-RANK/OPG that determines the erosive nature of RA is greatly reduced by VIP, accounting for the bone protection achieved by the treatment. The molecular mechanisms associated with these functions involved a reduction in the activity of the transcription factor NF-κB and a change in the activity of AP-1.[78] In all, the protective VIP effect on bone destruction in CIA could be due to different mechanisms that are not mutually exclusive (FIG. 1).[78] The first one could be the reduction of proinflammatory cytokines and other mediators involved in differentiation and activation of osteoclast precursor cells, together with an increase of anti-inflammatory cytokines. The second way could be the

(A) Bone erosion during RA

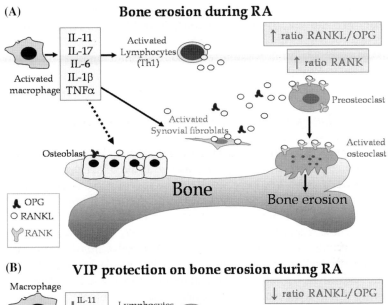

(B) VIP protection on bone erosion during RA

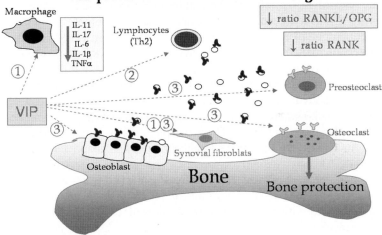

FIGURE 1. (A) Possible mechanisms involved in bone erosion in RA. Activated macrophages produces proinflammatory cytokines that induce lymphocytes Th1 and synovial fibroblasts to produce RANKL, and reduce OPG production, by osteoblasts which also express RANKL, and finally increase RANK expression in the surface of osteoclast precursors. All of them drive to an increase in ratio of RANKL/OPG and RANK in the joint that transforms preosteoclasts to activated osteoclasts, involved in bone erosion in RA. **(B)** After VIP treatment in the mouse model of RA, there is a decrease in the ratio RANKL/OPG in the joint, an increase in OPG serum levels, and a decrease in RANK expression in the joint. This protective effect could be due to different mechanisms that are not mutually exclusive: (*a*) VIP reduction of proinflammatory cytokines involved in differentiation and activation of osteoclasts precursor cells; (*b*) VIP induction of Th2 response; and (*c*) a direct effect on OPG, RANK, or RANKL expression on skeletal tissue, fibroblast, or immune cells present in the inflamed joint.

modification of cell type present in the joint induced by VIP. Finally, the third mechanism could be a VIP direct effect on OPG, RANK, or RANKL expression on skeletal tissue, fibroblast, or immune cells present in the inflamed joint. Our results support again the possibility of the therapeutic application of VIP in the treatment of human RA.

Balance of the Expression of Toll-Like Receptors

Two exciting discoveries guided the field of immunology to its present status of accelerated advancement, contributing to the understanding of innate and adaptive immune response. In 1996 the toll protein, at first described as a protein involved in the establishment of dorsoventral polarity in the developing Drosophila embryo,[81] was probed to be required to develop an effective immune response against fungus.[82] In 1998, TLR4 was identified as the LPS receptor required for mice to develop effective immune responses to gram-negative bacteria.[83] Today, it is known that toll-like receptors (TLR) are plasma–membrane- or intracellular compartment-expressed receptors, acting as sensors for invading pathogens, which initiated host defense responses in all multicellular organisms studied to date.[84–86] They are conserved proteins with an extracellular leucine-rich domain and an intracellular toll/IL-1 receptor-like (TIR) domain. Up to now, 11 TLRs have been identified in humans and 13 in mouse and TLRs 1–9 have conserved sequences between both species.[87]

Lipids, carbohydrates, nucleic acids, and various proteins collectively referred to as pathogen-associated molecular patterns (PAMPs) are among the structures recognized by TLRs. These receptors are essential to initiate an innate immune response and for the establishment of an adaptive response.[87–89] TLR2 and TLR4 are cell surface receptors involved in infection-induced and autoimmune-induced inflammatory diseases, such as sepsis, RA, and IBD.[84] TLR4 is essential for LPS signaling but requires an additional molecule, MD-2, which associates with the extracellular portion of TLR4. In addition, TLR4 has been involved in the recognition of endogenous ligands, such as hsp60, fibronectin, and multiple host proteins.[86] TLR2 recognizes a vast array of microbial components including lipoproteins from pathogens, peptidoglycan anchors from malaria-causing parasites, zymosan from fungi, and forms of LPS structurally different from those recognized by TLR4.[90] TLR2 and TLR4 can signal through the adaptor protein MyD88 as well as through MyD88 independent pathways, resulting in the activation of various intracellular signaling cascades, leading to the activation of NF-κB and IRF transcription factors' family members that finally induce the production of inflammatory mediators.[87] Since the stimulation of TLR2 and TLR4 mediates the expression in the cascade of inflammation-related genes, as well as the induction of an adaptive immune response of the Th1 type,[87] our hypothesis was that TLRs could be involved in the pathogenesis of inflammatory/autoimmune diseases, such as IBD and RA, and therefore VIP could downregulate, at least in part, the inflammatory response by modulating TLR expression.

TLR Expression

TLR and VIP in Gut Inflammation

The gut contains a diverse population of nonpathogenic, commensal bacteria whose contribution to gastrointestinal health and disease is now recognized. This microflora plays an important role in the development and expansion of lymphoid tissues and in the maintenance and regulation of gut immunity. A critical feature of the mucosal immune system is the ability to discriminate between harmful pathogens and the harmless members of the commensal flora, which is achieved in part, by TLRs.[91] The appropriate activation of TLRs has been demonstrated as an essential component of host immunity against pathogens, and it has been recently demonstrated that it is also vital for immune homeostasis.[92] TLR2 and TLR4 represent the main gut sensors to recognize molecular products from gram-positive and gram-negative bacteria, respectively. Several authors have described the constitutive expression of TLR2 and TLR4 in human and murine intestinal epithelial cells, as well as their modulation after gut inflammation.[93–95]

There is also evidence of the upregulation of TLR2 and TLR4 expression in macrophages from human-inflamed mucosa as well as of the association of a TLR4 polymorphism with both CD and UC in patients. In mice, it has been recently reported, that TLR2 and TLR4 were constitutively expressed in intestinal mucosa, being TLR4 upregulated in dextran sodium sulphate (DSS)-induced colitis, an animal model of human UC,[96] and in TNBS-induced colitis a mice model of CD.[95] Recently, our group has evaluated for the first time, the expression of TLR2 and TLR4 through the development of TNBS-induced colitis, showing the highest levels of messenger and protein expression on day 5, at the peak of the inflammatory parameter, whereas a decrease was seen on day 7, at the beginning of the chronic phase. VIP treatment exerted an inhibition of TLR2 and TLR4 expressions at both mRNA and protein levels from days 1 to 7. This effect had a similar time course triggering a general downregulation at colon local level with the involvement of both epithelial and mononuclear cells. Furthermore, VIP acted at systemic level in lymph nodes regulating the cellular traffic, reducing significantly the TNBS-increased proportion of dendritic cell (DC) and macrophage (MØ) and restoring the number of cells to the level of control animals. In addition, VIP exerted its effect by modulating the number of TLR2- and TLR4-positive DC, MØ, and lymphocytes. Our results described a new role of VIP in the gut microenvironment, as an effective modulator of the initial steps of acute inflammation, acting both at local and systemic levels, and leading to the restoration of the homeostasis lost after an established inflammatory disease.

TLR and VIP in RA

Although the ligands involved in TLR activation in RA are unclear, TLR signaling seems to be involved in its pathogenesis as shown by genetically deficient mice models.[97–99] The expression and function of TLR in RA tissues or cells have been studied by different groups. Higher expression of TLR2 and TLR4 in synovial tissues has been found in RA patients compared to osteoarthritis (OA) or healthy individuals.[100] However, previous studies have failed to demonstrate differences between RA and OA at the mRNA level of TLR2, TLR4, and TLR9 in synovial fibroblasts derived from joints whereas in synovial tissues they observed a weak expression of TLR2 mRNA in tissue sections of patients affected by OA compared to patients with RA.[101] Besides, it has been found that cultured synovial fibroblasts from patients with RA or OA express low levels of TLR2 and TLR9 mRNA and no differences between RA and OA were found in this study.[102] In cultured RA and OA fibroblast-like synoviocytes (FLS), bacterial peptidoglycans (PGs) and bacterial lipopeptide (BLP) but not CpG oligodeoxynucleotides, can activate, partially through TLR2, the expression of integrins, metalloproteases, and proinflammatory cytokines and chemokines.[102] Moreover, the expression of TLR2 was upregulated by PGs but the expression of TLR9 was not upregulated by CpG oligodeoxynucleotides in RA and OA FLS.[99] With regard to TLR4 data are scarce. It has been described that TLR4 is involved in the production of IFN-γ and CXC chemokine family in synovial cells.[103] We have recently studied in cultured FLS from patients with RA and OA the effect of VIP on basal, TNF-α, or LPS-induced expression of TLR2, TLR4, and MyD88, as well as its effects on TLR4-mediated production of CCL2 and CXCL8 chemokines. TLR2, TLR4, and MyD88 mRNA expression was increased in RA- compared to OA-FLS. The largest increase was observed for TLR4 and it was also upregulated at protein level by RA-FLS. TLR4 and MyD88 mRNA and proteins were induced by LPS and TNF-α in RA-FLS. VIP downregulated the induced, but not the constitutive expression, of TLR4 and MyD88 in RA-FLS. Furthermore, VIP treatment decreases CCL2 and CXCL8 chemokines production in response to TLR4 activation with LPS in RA-FLS.

In conclusion, our results about the modulation of TLR2 and TLR4 in IBD and RA by VIP could be explained by two possible mechanisms. The first one would be the secondary reduction of TLR2 and TLR4 caused by the VIP-mediated decrease of inflammatory mediators, such as IL-1β and IFN-γ, which have been described that synergize with bacterial products contributing to the amplification of TLR.

The other possible mechanism would involve the well-known negative effects of VIP signaling on NF-κB activation. In this sense, we have described in CIA model that VIP treatment *in vivo* prevented NF-κB nuclear translocation through the inhibition of IkBa phosphorylation/degradation.[78] Our results in

RA support this second mechanism because FLS are unable to produce TNF-α and other proinflammatory cytokines, such as IFN-γ and IL-1α.

The promising conclusion of these results on IBD and RA is that VIP causes a downregulation of TLR expression approaching the constitutive levels, suggesting a new therapeutic potential for the VIP–PACAP system, which would control the first stages of infection-induced and autoimmune-induced inflammatory diseases.

CONCLUDING REMARKS

The repertoire of immune functions of the VIP–PACAP system has expanded since its discovery as a component of the neuroimmuno network. This multifunctional system modulates key actions in the development of both infection- and autoimmune inflammation with an emerging role in the control of CC and CXC ligand-receptor chemokine binome as well as in the modulation of the expression of TLR opening a hopeful door for their therapeutic use in humans (FIG. 2).

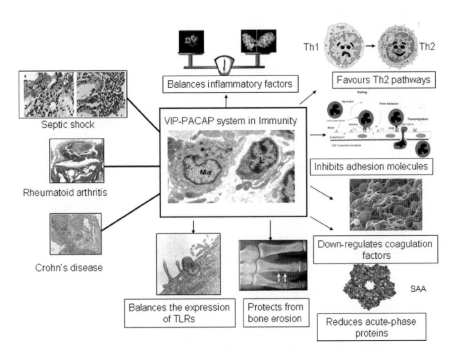

FIGURE 2. VIP–PACAP system in immunity, regulatory immune functions in septic shock, RA, and CD.

ACKNOWLEDGMENTS

This work was supported by grants BFI 2002-03489 from Ministerio de Ciencia y Tecnología (Spain), G03/152 from Fondo de Investigación Sanitaria (Spain), and a predoctoral fellowship from Ministerio de Ciencia y Tecnología (to A.A).

REFERENCES

1. SPECTOR, N.H. 1987. Epilogue. *In* Neuroimmune interactions: proceedings of the second international workshop on neuroimmunomodulation. B.D. Jancovic, B.M. Markovic & N.H. Spector, Eds.: Ann. N.Y. Acad. Sci. **496:** 750–751.
2. JANKOVIC, B.D., B.M. MARKOVIC & N.H. SPECTOR, EDS.: 1987. Neuroimmune interactions: proceedings of the second international workshop on neuroimmunomodulation. Ann. N.Y. Acad. Sci. **496:** 750–751.
3. BLALOCK, J.E. & E.M. SMITH. 1980. Human leukocyte interferon: structural and biological relatedness to adrenocorticotropic hormone and endorphins. Proc. Natl. Acad. Sci. USA **77:** 5972–5974.
4. SAID, S.I. & V. MUTT. 1970. Long acting vasodilator peptide from lung. Science **169:** 1217–1218.
5. MIYATA, A., A. ARIMURA, R.R. DAHL, *et al.* 1989. Isolation of a novel 38 residue-hypothalamic polypeptide which stimulates adenylate cyclase in pituitary cells. Biochem. Biophys. Res. Commun. **164:** 567–574.
6. HENNING, R.J. & D.R. SAWMILLER. 2001. Vasoactive intestinal peptide: cardiovascular effects. Cardiovasc. Res. **49:** 27–37.
7. SHERWOOD, N.M., S.L. KRUECKL & J.E. McRORY. 2000. The origin and function of the pituitary adenylate cyclase-activating polypeptide (PACAP)/glucagon superfamily. Endocr. Rev. **21:** 619–670.
8. DOGRUKUL, A.K.D., F. TORE & N. TUNCEL. 2004. Passage of VIP/PACAP/secretin family across the blood-brain barrier: therapeutic effects. Curr. Pharm. Design. **10:** 325–340.
9. GONZALEZ, B.J., M. BASILLE, D. VAUDRY, *et al.* 1998. Pituitary adenylate cyclase-activating polypeptide. Ann. Endocrinol. **59:** 364–405.
10. VAUDRY, D., B.J. GONZALEZ, M. BASILLE, *et al.* 2000. Pituitary adenylate cyclase-activating polypeptide and its receptors: from structure to functions. Pharmacol. Rev. **52:** 269–324.
11. GOMARIZ, R.P., M.J. LORENZO, L. CACICEDO, *et al.* 1990. Demonstration of immunoreactive vasoactive intestinal peptide (IR-VIP) and somatostatin (IR-SOM) in rat thymus. Brain Behav. Immun. **4:** 151–161.
12. GOMARIZ, R.P., M. DELGADO, J.R. NARANJO, *et al.* 1993. VIP gene expression in rat thymus and spleen. Brain Behav. Immun. **7:** 271–278.
13. LECETA, J., C. MARTINEZ, M. DELGADO, *et al.* 1994. Lymphoid cell subpopulations containing vasoactive intestinal peptide in the rat. Peptides **15:** 791–799.
14. ABAD, C., C. MARTINEZ, J. LECETA, *et al.* 2002. Pituitary adenylate cyclase-activating polypeptide expression in the immune system. Neuroimmunomodulation **10:** 177–186.
15. MARTINEZ, C., M. DELGADO, C. ABAD, *et al.* 1999. Regulation of VIP production and secretion by murine lymphocytes. J. Neuroimmunol. **93:** 126–138.

16. DELGADO, M. & D. GANEA. 2001. Is vasoactive intestinal peptide a type 2 cytokine? J. Immunol. **166:** 2907–2912.
17. OKHUBO, S., C. KIMURA, K. OGI, *et al.* 1992. Primary structure and characterization of the precursor to human pituitary adenylate cyclase-activating polypeptide. DNA Cell. Biol. **11:** 21–30.
18. HARMAR, A., A. ARIMURA, I. GOZES, *et al.* 1998. International Union of Pharmacology: XVIII. Nomenclature of receptors for vasoactive intestinal peptide and pituitary adenylate cyclase-activating polypeptide. Pharmacol. Rev. **50:** 265–270.
19. LABURTHE, M. & A. COUVINEAU. 2002. Molecular pharmacology and structure of VPAC receptors for VIP and PACAP. Regul. Pept. **108:** 165–173.
20. COUVINEAU, A., C. ROUYER-FESSARD, D. DARMOUL, *et al.* 1994. Human intestinal VIP receptor: cloning and functional expression of two cDNA encoding proteins with different N-terminal domains. Biochem. Biophys. Res. Commun. **200:** 769–776.
21. SPENGLER, D., C. WAEBER, C. PANTALONI, *et al.* 1993. Differential signal transduction by five splice variants of the PACAP receptor gene. Nature **365:** 170–175.
22. PANTALONI, C., P. BRABET, B. BILANGES, *et al.* 1996. Alternative splicing in the N-terminal extracellular domain of the pituitary adenylate cyclase-activating polypeptide (PACAP) receptor modulates receptor selectivity and relative potencies of PACAP-27 and PACAP-38 in phospholipase C activation. J. Biol. Chem. **271:** 146–2215.
23. CHATTERJEE, T.K., R.V. SHARMA & R.A. FISHER. 1996. Molecular cloning of a novel variant of the pituitary adenylate cyclase-activating polypeptide (PACAP) receptor that stimulates calcium influx by activation of the L-type calcium channels. J. Biol. Chem. **271:** 33226–33232.
24. GUERRERO, J.M., J.C. PRIETO, F.L. ELORZA, *et al.* 1981. Interaction of vasoactive intestinal peptide with human blood mononuclear cells. Mol. Cell. Endocrinol. **21:** 151–160.
25. WIIK, P., P.K. OPSTAD & A. BOYUM. 1985. Binding of vasoactive intestinal polypeptide (VIP) by human blood monocytes: demonstration of specific binding sites. Regul. Pept. **12:** 145–163.
26. OTTAWAY, C.A. & G. GREENBERG. 1984. Interaction of vasoactive intestinal peptide with mouse lymphocytes: specific binding and the modulation of mitogen responses. J. Immunol. **132:** 417–423.
27. SAKAKIBARA, H., K. SHIMA & S.I. SAID. 1994. Characterization of vasoactive intestinal peptide receptors on rat alveolar macrophages. Am. J. Physiol. **267:** L256–L262.
28. GOMARIZ, R.P., C. ABAD, C. MARTINEZ, *et al.* 2001. Vasoactive intestinal peptide, pituitary adenylate cyclase-activating polypeptide and immune system: from basic research to potential clinical application. Biomedical Rev. **12:** 1–9.
29. GOMARIZ, R.P., C. MARTINEZ, C. ABAD, *et al.* 2001. Immunology of VIP: a review and therapeutical perspectives. Curr. Pharm. Design. **7:** 89–111.
30. DELGADO, M., C. ABAD, C. MARTINEZ, *et al.* 2002. Vasoactive intestinal peptide in the immune system: potential therapeutic role in inflammatory and autoimmune diseases. J. Mol. Med. **80:** 16–24.
31. KALTREIDER, H.B., S. ICHIKAWA, P.K. BYRD, *et al.* 1997. Upregulation of neuropeptides and neuropeptide receptors in a murine model of immune inflammation in lung parenchyma. Am. J. Respir. Cell. Mol. Biol. **16:** 133–144.

32. Pozo, D., M. DELGADO, C. MARTINEZ, *et al.* 1997. Functional characterization and mRNA expression of pituitary adenylate cyclase activating polypeptide (PACAP) type I receptors in rat peritoneal macrophages. Biochim. Biophys. Acta **1359:** 250–262.

33. MARTINEZ, C., C. ABAD, M. DELGADO, *et al.* 2002. Anti-inflammatory role in septic shock of pituitary adenylate cyclase-activating polypeptide receptor. Proc. Natl. Acad. Sci. USA. **99:** 1053–1058.

34. RIEDEMANN, N.C., R.F. GUO & P.A. WARD. 2003. The enigma of sepsis. J. Clin. Invest. **112:** 460–467.

35. TAYLOR, P.C., R.O. WILLIANS & M. FELDMANN. 2004. Tumour necrosis factor alpha as a therapeutic target for immune-mediated inflammatory diseases. Curr. Opin. Biotechnol. **6:** 557–563.

36. MYERS, L.K., E.F. ROSLONIEC, M.A. CREMER, *et al.* 1997. Collagen-induced arthritis, an animal model of autoimmunity. Life Sci. **61:** 1861–1872.

37. LOFTS, E.V. 2004. Clinical epidemiology of inflammatory bowel disease: incidence, prevalence and environmental influences. Gastroenterology **126:** 1504–1517.

38. BOUMA, G. & W. STROBER. 2003. The immunological and genetic basis of inflammatory bowel disease. Nat. Rev. Immunol. **3:** 521–533.

39. NEURATH, M.F., I. FUSS, B.L. KELSALL, *et al.* 1995. Antibodies to interleukin 12 abrogate established experimental colitis in mice. J. Exp. Med. **1:** 1281–1290.

40. DELGADO, M., C. MARTINEZ, D. POZO, *et al.* 1999. Vasoactive intestinal peptide (VIP) and pituitary adenylate cyclase-activation polypeptide (PACAP) protect mice from lethal endotoxemia through the inhibition of TNF- alpha and IL-6. J. Immunol. **162:** 1200–1205.

41. DELGADO, M., D. POZO & D. GANEA. 2004. The significance of vasoactive intestinal peptide in immunomodulation. Pharmacol. Rev. **56:** 249–290.

42. DE LA FUENTE, M., M. DELGADO & R.P. GOMARIZ. 1996. VIP modulation of immune cell function. Adv. Neuroimmunol. **6:** 75–91.

43. DELGADO, M., M. DE LA FUENTE, C. MARTINEZ, *et al.* 1995. Pituitary adenylate cyclase-activating polypeptides (PACAP27and PACAP38) inhibit the mobility of murine thymocytes and splenic lymphocytes: comparison with VIP and implication of cAMP. J. Neuroimmunol. **62:** 137–146.

44. ABAD, C., Y. JUARRANZ, C. MARTINEZ, *et al.* 2005. cDNA array cytokines, chemokines and receptors involved in TNBS-induced colitis: homeostatic role of VIP. Inflamm. Bowel Dis. **7:** 674–684.

45. DELGADO, M., C. ABAD, C. MARTINEZ, *et al.* 2001. Vasoactive intestinal peptide prevents experimental arthritis by downregulating both autoimmune and inflammatory components of the disease. Nat. Med. **7:** 563–568.

46. GRIMM, M.C., R. NEWMAN, Z. HASSIM, *et al.* 2003. Cutting edge: vasoactive intestinal peptide acts as a potent suppressor of inflammation in vivo by trans-deactivating chemokine receptors. J. Immunol. **171:** 4990–4994.

47. KODALI, S., W. DING, J. HUANG, *et al.* 2004. Vasoactive intestinal peptide modulates Langerhans cell immune function. J. Immunol. **173:** 6082–6088.

48. VOICE, J.K., C. GRINNINGER, Y. KONG, *et al.* 2003. Roles of vasoactive intestinal peptide (VIP) in the expression of different immune phenotypes by wild-type mice and T cell-targeted type II VIP receptor transgenic mice. J. Immunol. **170:** 308–314.

49. GOETZL, E.J., J.K. VOICE, S. SHEN, *et al.* 2001. Enhanced delayed-type hypersensitivity and diminished immediate-type hypersensitivity in mice lacking the inducible VPAC2 receptor for vasoactive intestinal peptide. Proc. Natl. Acad. Sci. USA **98:** 13854–13859.

50. DELGADO, M., E. GONZALEZ-REY & D. GANEA. 2004. VIP/PACAP preferentially attract Th2 effectors through differential regulation of chemokine production by dendritic cells. FASEB J. **18:** 1453–1455.

51. ROMAGNANI, P., L. LASAGNI, F. ANNUNZIATO, *et al.* 2004. CXC chemokines: the regulatory link between inflammation and angiogenesis. Trends Immunol. **25:** 201–209.

52. BAGGLIONI, M., B. DEWALD & B. MOSER. 1997. Human chemokines: an update. Ann. Rev. Immunol. **15:** 675–705.

53. ROLLINS, B.J. 1997. Chemokines. Blood **90:** 909–928.

54. ZLOTNIK, A. & O. YOSHIE. 2000. Chemokines: a new classification system and their role in immunity. Immunity **12:** 121–127.

55. HENNINGER, D.D., J. PANES, M. EPPIHIMER, *et al.* 1997. Cytokine-induced VCAM-1 and ICAM-1 expression in different organs of the mouse. J. Immunol. **158:** 1825–1834.

56. MARTINEZ, C., Y. JUARRANZ, C. ABAD, *et al.* 2005. Analysis of the role of PAC1 receptor in neutrophil recruitment, acute-phase response and nitric oxide production in septic shock. J. Leukoc. Biol. **77:** 729–738.

57. ROMANO, M., M. SIRONI, C. TONIATTI, *et al.* 1997. Role of IL-6 and its soluble receptor in induction of chemokines and leukocyte recruitment. Immunity **6:** 315–325.

58. CALDENHOVEN, E., P. COFFER, J. YUAN, *et al.* 1994. Stimulation of the human intercellular adhesion molecule-1 promoter by interleukin-6 and interferon-gamma involves binding of distinct factors to a palindromic response element. J. Biol. Chem. **269:** 21146–21154.

59. HEINRICH, P.C., J.V. CASTELL & T. ANDUS. 1990. Interleukine-6 and acute phase response. Biochem. J. **265:** 621–636.

60. GUIJARRO, L.G., N. RODRIGUEZ-HENCHE, E. GARCIA-LOPEZ, *et al.* 1995. Receptors for pituitary adenylate cyclase-activating peptide in human liver. J. Clin. Endocrinol. Metab. **80:** 2451–2457.

61. ROBBERECHT, P., P. GOURLET, A. CAUVIN, *et al.* 1991. PACAP and VIP receptors in rat liver membranes. Am. J. Physiol. **260:** G97–G102.

62. UHLAR, C.M. & A.S. WHITEHEAD. 1999. Serum amyloid A, the major vertebrate acute-phase reactant. Eur. J. Biochem. **265:** 501–523.

63. ABAD, C., C. MARTINEZ, M.G. JUARRANZ, *et al.* 2003. Therapeutic effects of vasoactive intestinal peptide in the trinitrobenzene sulfonic acid mice model of Crohn's disease. Gastroenterology **124:** 961–971.

64. JENSEN, L.E. & A.S. WHITEHEAD. 1998. Regulation of serum amyloid A protein expression during the acute-phase response. Biochem. J. **334:** 489–503.

65. FRATELLI, M., M. ZINETTI, G. FANTUZZI, *et al.* 1997. Time course of circulating acute phase proteins and cytokines in septic patients. Amyloid: Int. J. Exp. Clin. Invest. **4:** 33–97.

66. LIEUW, F.Y. 2002. Th1 and Th2 cells: a historical perspective. Nat. Rev. Immunol. **2:** 55–60.

67. DELGADO, M., J. LECETA, R.P. GOMARIZ, *et al.* 1999. VIP and PACAP stimulate the induction of Th2 responses by upregulating B7.2 expression. J. Immunol. **163:** 3629–3635.

68. DELGADO, M., J. LECETA & D. GANEA. 2002. Vasoactive intestinal peptide and pituitary adenylate cyclase-activating polypeptide promote in vivo generation of memory Th2 cells. FASEB J. **16:** 1844–1846.
69. ABAD, C., C. MARTINEZ, J. LECETA, *et al.* 2001. Pituitary adenylate cyclase-activating polypeptide inhibits collagen-induced arthritis: an experimental immunomodulatory therapy. J. Immunol. **167:** 3182–3189.
70. ROMAS, E., M.T. GILLESPIE & T.J. MARTIN. 2002. Involvement of receptor activator of NFkB ligand and tumor necrosis factor-a in bone destruction in rheumatoid arthritis. Bone **30:** 340–346.
71. LUBBERTS, E., L. VAN DEN BERSSELAAR, B. OPPERS-WALGREEN, *et al.* 2003. IL-17 promotes bone erosion in murine collagen-induced arthritis through loss of the receptor activator of NF-kB ligand/osteoprotegerin balance. J. Immunol. **170:** 2655–2662.
72. WALTON, J.K., J.M. DUNCAN, P. DESCHAMPS, *et al.* 2002. Heparin acts synergistically with interleukin-11 to induce STAT3 activation and in vitro osteoclast formation. Blood **100:** 2530–2536.
73. MUKOHYAMA, H., M. RANSJÖ, H. TANIGUCHI, *et al.* 2000. The inhibitory effects of vasoactive intestinal peptide and pituitary adenylate cyclase-activating polypeptide on osteoclast formation are associated with upregulation of osteoprotegerin and downregulation of RANKL and RANK. Biochem. Biophys. Res. Commun. **271:** 158–163.
74. LUNDBERG, P., I. BOSTRÖM, H. MUKOHYAMA, *et al.* 1999. Neuro-hormonal control of bone metabolism: vasoactive intestinal peptide stimulates alkaline phosphatase activity and mRNA expression in mouse calvarial osteoblasts as well as calcium accumulation mineralized bone nodules. Regul. Pept. **85:** 47–58.
75. RANSJÖ, M., A. LIE, H. MUKOHYAMA, *et al.* 2000. Microisolated mouse osteoclasts express VIP-1 and PACAP receptors. Biochem. Biophys. Res. Commun. **274:** 400–404.
76. LUNDBERG, P., I. LUNDGREN, H. MUKOHYAMA, *et al.* 2001. Vasoactive intestinal peptide (VIP)/pituitary adenylate cyclase-activating peptide receptor subtypes in mouse calvarian osteoblasts: presence of VIP-2 receptors and differentiation-induced expression of VIP-1 receptors. Endocrinology **142:** 339–347.
77. LUNDBERG, P., A. LIE, A. BJURHOLM, *et al.* 2000. Vasoactive intestinal peptide regulates osteoclast activity via specific binding sites on both osteoclasts and osteoblasts. Bone **27:** 803–810.
78. JUARRANZ, Y., C. ABAD, C. MARTINEZ, *et al.* 2005. Protective effect of vasoactive intestinal peptide on bone destruction in the collagen-induced arthritis model of rheumatoid arthritis. Arthritis Res. Ther. **7:** R1034–R1045.
79. JONES, D.H., Y.Y. KONG & J.M. PENNINGER. 2002. Role of RANKL and RANK in bone loss and arthritis. Ann. Rheum. Dis. **61:** 32–39.
80. SZEKANECZ, Z., J. KIM & A.E. KOCH. 2003. Chemokines and chemokine receptors in rheumatoid arthritis. Semin. Immunol. **15:** 15–21.
81. ST JOHNSTON, D. & A.R.D. NUSSLEIN-VOLH. 1992. The origin of pattern and polarity in the Drosophila embryo. Cell **68:** 201–219.
82. LEMAITRE, B., E. NICOLAS, L. MICHAUT, *et al.* 1996. The dorsoventral regulatory gene cassette spatzle/Toll/cactus controls the potent antifungal response in Drosophila adults. Cell **86:** 973–983.
83. POLTORAK, A., X. HE, I. SMIRNOVA, *et al.* 1998. Defective LPS signalling in C3H/HeJ and C57BL/10ScCr mice: mutations in TLR4 gene. Science **282:** 2085–2088.

84. O'NEILL, L.A.J. 2004. TLRs: professor Mechnikov, sit on your hat. Trends Immunol. **25:** 687–693.
85. BEUTLER, B.. 2004. Interferences, questions and possibilities in toll-like receptor signalling. Nature **430:** 257–263.
86. TAKEDA, K., T. KAISHOT & S. AKIRA. 2003. Toll-like receptor. Ann. Rev. Immunol. **21:** 335–376.
87. DUNNE, A. & L.A.J. O'NEILL. 2005. Adaptor usage and toll-like receptor signalling specificity. FEBS Lett. **579:** 3330–3335.
88. PASARE, C. & R. MEDZHITOV. 2004. Toll-like receptors and acquired immunity. Semin. Immunol. **16:** 23–26.
89. UNDERHILL, D.M. & A. OZINSKY. 2002. Toll-like receptors: key mediators of microbe detection. Curr. Opin. Immunol. **14:** 103–110.
90. RIFKIN, I.R., E.A. LEADBETTER, L. BUSCONI, *et al.* 2005. Viglianti G, Marshak-Rothstein A. Toll-like receptors, endogenous ligands and systemic autoimmune disease. Immunol. Rev. **204:** 27–42.
91. CARIO, E. 2005. Bacterial interactions with cells of the intestinal mucosa: toll-like receptors and NOD2. Gut **54:** 1182–1193.
92. RAKOFF-NAHOUM, S., J. PAGLINO, F. ESLAMI-VARZANEH, *et al.* 2004. Recognition of commensal microflora by toll-like receptors is required for intestinal homeostasis. Cell **18:** 229–241.
93. CARIO, E., I.M. ROSENBERG, S.L. BRANDWEIN, *et al.* 2000. Lipopolysaccharide activates distinct pathways in intestinal epithelial cell lines expressing toll-like receptors. J. Immunol. **164:** 966–972.
94. SUZUKI, M., T. HISAMATSU & D.K. PODOLSKY. 2003. Gamma interferon augments the intracellular pathway for lipolysaccharide (LPS) recognition in human intestinal epithelial cells through coordinated up-regulation of LPS uptake and expression of the intracellular toll-like receptor 4-MD-2 complex. Infect. Immun. **71:** 3503–3511.
95. GOMARIZ, R.P., A. ARRANZ, C. ABAD, *et al.* 2005. Time-course expression of toll-like receptors 2 and 4 in inflammatory bowel disease and homeostatic effect of VIP. J. Leuk. Biol. **78:** 491–502.
96. ORTEGA-CAVA, C.F., S. ISHIHARA, M.A. RUMI, *et al.* 2003. Strategic compartmentalization of toll-like receptor 4 in the mouse gut. J. Immunol. **170:** 3977–3985.
97. CHOE, J.Y., B. CRAIN, S.R. WU, *et al.* 2003. Interleukin 1 receptor dependence of serum transferred arthritis can be circumvented by toll-like receptor 4 signalling. J. Exp. Med. **197:** 537–542.
98. ZHAI, Y., X.D. SHEN, R. O'CONNELL, *et al.* 2004. Cutting edge: TLR4 activation mediates liver ischemia/reperfusion inflammatory response via IFN regulatory factor 3-dependent MyD88-independent pathway. J. Immunol. **173:** 7115–7119.
99. BOULE, M.W., C. BROUGHTON, F. MACKAY, *et al.* 2004. Toll-like receptor 9-dependent and -independent dendritic cell activation by chromatin-immunoglobulin G complexes. J. Exp. Med. **199:** 1631–1640.
100. RADSTAKE, T.R., M.F. ROELOFS, Y.M. JENNISKENS, *et al.* Expression of toll-like receptors 2 and 4 in rheumatoid synovial tissue and regulation by proinflammatory cytokines interleukin-12 and interleukin-18 via interferon-gamma. Arthritis Rheum. **50:** 3856–3865.
101. SEIBL, R., T. BIRCHLER, S. LOELIGER, *et al.* 2003. Expression and regulation of toll-like receptor 2 in rheumatoid arthritis synovium. Am. J. Pathol. **162:** 1221–1227.

102. KYBURZ, D., J. RETHAGE, R. SEIBL, *et al.* 2003. Bacterial peptidoglycans but not CpG oligodeoxynucleotides activate fibroblasts by toll-like receptor signaling. Arthritis Rheum. **48:** 642–650.
103. PROOST, P., S. VERPOEST, K. VAN DE BORNE, *et al.* 2004. Synergistic induction of CXCL9 and CXCL11 by toll-like receptor ligands and interferon-gamma in fibroblasts correlates with elevated levels of CXCR3 in septic arthritis synovial fluids. J. Leuk. Biol. **75:** 777–784.

New Insights into the Central PACAPergic System from the Phenotypes in PACAP- and PACAP Receptor-Knockout Mice

HITOSHI HASHIMOTO, NORIHITO SHINTANI, AND AKEMICHI BABA

Laboratory of Molecular Neuropharmacology, Graduate School of Pharmaceutical Sciences, Osaka University, Suita, Osaka 565-0871, Japan

ABSTRACT: Pituitary adenylate cyclase-activating polypeptide (PACAP) is a structurally highly conserved neuropeptide and displays pleiotropic activity, including functioning as a neurotransmitter, neuromodulator, and neurotrophic factor. A series of recent experiments, including genetic manipulation of PACAP and its receptors, has led to better understanding of both normal and pathological processes in which PACAP has been proposed to play a role, and sheds light on previously uncharacterized functions of endogenous PACAP. The aim of this article is to briefly review the recent advances in understanding the role of PACAP in the central nervous system from PACAP- and PACAP receptor-deficient mice, particularly with respect to behavioral and neurological features, including psychomotor behavior, feeding, stress responses, memory performance, ethanol sensitivity, chronic pain, and circadian rhythms. This article also discusses their potential involvement in human diseases.

KEYWORDS: Adcyapl; animal model; knockout mouse; neuropeptide; PACAP

INTRODUCTION

Pituitary adenylate cyclase-activating polypeptide (PACAP) was originally isolated as a novel hypothalamic neuropeptide by Arimura's group in 1989, based on its ability to stimulate adenylate cyclase in rat anterior pituitary cell cultures.[1] PACAP exists in two biologically active forms, PACAP-38 and the C-terminally truncated PACAP-27. PACAP-27 has an amino acid sequence identity of 68% with vasoactive intestinal polypeptide (VIP) and of 37% with secretin, indicating that PACAP is a member of the VIP/glucagon/growth

Address for correspondence: Akemichi Baba, Laboratory of Molecular Neuropharmacology, Graduate School of Pharmaceutical Sciences, Osaka University, 1-6 Yamadaoka, Suita, Osaka 565-0871, Japan. Voice: 81-6-6879-8180; fax: 81-6-6879-8184.
e-mail: baba@phs.osaka-u.ac.jp

Ann. N.Y. Acad. Sci. 1070: 75–89 (2006). © 2006 New York Academy of Sciences.
doi: 10.1196/annals.1317.038

hormone-releasing hormone (GHRH)/secretin superfamily. PACAP is present not only in various areas of the central nervous system, including the hypothalamus and other brain regions, but also in peripheral tissues, such as testicular germ cells, pituitary gland lobes, and the adrenal medulla. PACAP is implicated in neurobiological functions, such as neurotransmission, neural plasticity, and neurotrophy (for reviews, see Refs. 2 and 3).

Molecular cloning studies have shown that these diverse activities of PACAP are mediated by three subtypes of class B (secretin-like) G protein–coupled receptors, one PACAP-specific (PAC1) receptor and two receptors that are shared with VIP (VPAC1 and VPAC2), whose signaling mechanisms mainly involve activation of adenylate cyclase, as well as depending on receptor subtypes and splice variants, and phospholipase-C cascades.[2-4]

Because appropriately selective low-molecular-weight PACAP receptor ligands are not available, studies on the physiological role of PACAP and each receptor subtype in function of the central nervous system have been hampered. Recently, several groups, including our own, have independently produced mice with targeted mutations of PACAP,[5-8] VIP,[9] and their receptors, PAC1 receptor[10-12] and VPAC2 receptor.[13,14] These mutant mice have not only led to a better understanding of the physiological roles of endogenous PACAP, but have also revealed some unexpected roles for PACAP.

The aim of this article is to review the recent advances in understanding the roles of PACAP from PACAP- and PACAP receptor-deficient mice, particularly with respect to behavioral and neurological functions. In addition, we aim to discuss their potential involvement in human diseases. Although this article cannot be encyclopedic, it attempts to discuss illustrative aspects of the central PACAPergic system. For more detailed information, a number of excellent reviews on PACAP are available.[2-4,15-27]

INCREASED POSTNATAL MORTALITY RATE IN PACAP-DEFICIENT MICE

PACAP-deficient mice from a mixed C57Bl/6J and 129/Ola genetic background generated by our own group are born in the expected Mendelian ratios, however, more than a half of PACAP-deficient mice die before weaning at around 3 weeks of age.[6] This agrees with the observation by Gray et al. who found that >90% of their knockout mice died a slow, wasting death or died suddenly in the second postnatal week, associated with dysfunction of lipid and carbohydrate metabolism.[5] The discrepancy in the survival rates between the mutants from the two laboratories may reflect their genetic backgrounds.[28] Indeed, our mutant mice that had been repeatedly backcrossed to the C57Bl/6J mice almost die before weaning, while mutants backcrossed to the ICR mice mostly survive to adulthood,[28] suggesting a considerable difference in the contribution of background genes to the mortality of PACAP-deficient mice.

In addition, environmental factors contribute to the higher mortality. Gray et al. reported that the survival rate of PACAP-deficient mice increased to 76%

when the environmental temperature was raised to 24°C instead of 21°C.[29] PACAP is expressed in several respiratory-related regions of the nervous system.[30] Cummings *et al.* hypothesized that the higher neonatal mortality in PACAP-deficient mice may be due to defective respiratory control and demonstrated a critical role for PACAP in respiratory control during periods of neonatal stress.[31] Collectively, these findings indicate that interaction of environmental stress with PACAP deficiency results in increased susceptibility to sudden death.

To ascertain whether this high mortality is reversed by overexpression of PACAP in peripheral tissue (such as pancreas), a genetic cross between PACAP-deficient mice[6] and transgenic mice overexpressing PACAP in pancreatic β-cells[32] has been performed, and the survival rate of their progeny examined.[33] PACAP overexpression, however, did not affect the survival rate of PACAP-deficient mice. Neuronal expression of PACAP may be a prerequisite for reversing the high mortality in PACAP-deficient mice.

PAC1 receptor-deficient mice also show a high postnatal mortality.[11,12,34] PAC1 receptor-deficient mice from a C57BL/6 background nearly all die during the second postnatal week from rapidly developing heart failure (right ventricle dilation) as a consequence of increased pulmonary arterial pressure. These findings demonstrate a crucial role of PAC1 receptor-mediated signaling for the maintenance of normal pulmonary vascular tone during early postnatal life.[34]

REDUCED FERTILITY IN PACAP-DEFICIENT MICE

Surviving PACAP-deficient female mice exhibit reduced fertility.[35] Although the mutant females exhibit no obvious defects in the length of their estrous cycles, their mating frequency is significantly reduced. In addition, maternal crouching behavior of the mutant females tends to decrease compared to wild-type females.

PACAP has been shown to be involved in the circadian and episodic release of luteinizing hormone (LH) from the pituitary *in vivo*.[36] Furthermore, endogenous PACAP has been shown to act as a pivotal modulator for synchronization of behavior with hypothalamic control of ovulation.[37] It is, therefore, plausible that the PACAPergic system is crucially implicated in female reproductive functions.

PAC1 receptor-deficient mice display decreased fertility[38] and delayed affiliative behavior.[39] PACAP and PAC1 receptor signaling have been suggested to affect different components of sexual behavior and social affiliation via regulation of pheromone processing, sensory circuitry, and endocrine maturation.[39] Further, it has been reported that VPAC2 receptor-deficient mice at the older age of 31 weeks exhibit diffuse seminiferous tubular degeneration with hypospermia and reduced fertility.[13]

PSYCHOMOTOR ABNORMALITIES IN PACAP- AND PAC1 RECEPTOR-DEFICIENT MICE

PACAP-deficient mice display remarkable behavioral abnormalities providing evidence that PACAP plays a previously uncharacterized role in the regulation of psychomotor behavior.[6] When placed into a novel environment, such as an open field, the mutants display significantly increased locomotor activity with minimal habituation to the environment, and less time engaged in licking and grooming behavior. The mutants also show explosive jumping behavior in the open field and increased exploratory behavior. These behavioral abnormalities may be due to perturbation of monoamine neurotransmission because serotonin metabolite 5-hydroxyindoleacetic acid is slightly decreased in the cerebral cortex and striatum of PACAP-deficient mice, and hyperactive behavior is ameliorated by the antipsychotic drug haloperidol.[6] In addition, the jumping behavior is suppressed by drugs that elevate extracellular serotonin, such as the selective serotonin reuptake inhibitors (see Shintani *et al.*, this volume).[40] However, the mechanistic basis and pathophysiological significance remain unclear.

Recently, Ogawa *et al.* showed that PACAP depletion does not affect the monoaminergic nervous system during early development at embryonic day (E)10.5 and E12.5.[41] Therefore, it is conceivable that defects in late development or an acute role for PACAP in psychological function may contribute to behavioral consequences of PACAP deficiency.

PAC1 receptor-deficient mice have been demonstrated to exhibit elevated locomotor activity and reduced anxiety-like behavior.[42] Furthermore, these mutants display markedly abnormal social behavior, indicating that PAC1 signaling is an important factor in the development and/or functioning of neural pathways associated with pheromone processing and regulation of social interaction.[39]

IMPAIRED FOOD INTAKE IN PACAP-DEFICIENT MICE

It has been shown that PACAP-deficient mice display a reduced intake of normal chow, but not a high-fat diet, and that PACAP directly activates orexigenic neuropeptide Y neurons.[43] A previous study shows that intracerebroventricular injection of PACAP increases neuropeptide Y mRNA in the arcuate nucleus, a feeding center, and decreases mRNA of corticotropin-releasing hormone, an anorectic peptide, in the paraventricular nucleus in rats.[44] These are the likely reasons for the reduced food intake in PACAP-deficient mice. Indeed, neuropeptide Y mRNA levels are significantly decreased in PACAP-deficient arcuate nuclei.[43]

Consistently, VPAC2 receptor-deficient mice display decreased food intake.[13] In these mutants, however, body weight is also reduced, and when adjusted for differences in body weight, the difference in food intake between mutant and wild-type mice is not statistically significant.

ROLES OF PACAP IN STRESS RESPONSES

Accumulating evidence suggests that PACAP is protective against various stressors, such as hypoglycemia and cold exposure. Hamelink *et al.* reported that PACAP is needed to couple epinephrine biosynthesis to secretion during insulin-induced hypoglycemia, and postulate that PACAP functions as an "emergency response" co-transmitter in the sympathoadrenal axis.[7] In addition, Gray *et al.* have shown that adaptive thermogenesis is impaired by insufficient norepinephrine stimulation of brown adipose tissue in cold-stressed, PACAP-deficient mice.[29]

Several lines of evidence have further implicated PACAP in the regulation of the hypothalamo-pituitary-adrenal (HPA) axis.[15] In PACAP-deficient mice, it has been shown that trimethyltin (TMT)-induced elevation of plasma corticosterone levels is absent, while basal corticosterone levels are not significantly different from those in wild-type mice (see Morita *et al.*, this volume).[45] These results demonstrate that endogenous PACAP is crucial for plasma corticosterone release, at least under conditions of TMT-mediated stress. Gray *et al.* have reported that some PACAP-deficient mice pups have normal corticosterone levels, whereas others have extremely high levels.[5] Hamelink *et al.* have shown that the diurnal rhythm of plasma corticosterone at rest is equivalent in their PACAP-deficient and wild-type mice, and that the acute corticosterone responsiveness to insulin administration is unimpaired in the mutant mice.[7] Collectively, these findings indicate that PACAP is involved in a variety of stress responses through modulation of the sympathoadrenal and/or HPA axes.

MEMORY PERFORMANCE IN THE PACAP MUTANTS

Memory in both vertebrates and invertebrates involves alteration in the efficiency of synaptic transmission, otherwise known as long-term potentiation (LTP) and long-term depression. In *Drosophila* memory mutant *amnesiac* with impaired memory retention, loss-of-function mutations have been identified in the *amnesiac* gene that codes for predicted mature peptides sharing limited homology with vertebrate PACAP, PACAP-related peptide, and GHRH.[46] In submammalian species investigated so far, PACAP and a GHRH-like peptide are located on the same precursor.[17,18]

Motivated by this finding, Otto *et al.* analyzed possible alterations in learning and memory in their PAC1 receptor-deficient mice, and showed a deficit in contextual fear conditioning, a hippocampus-dependent, associative learning paradigm.[12] They also showed that LTP is impaired in the mossy fiber-CA3 synapses. Extracellular recording in hippocampal slices has demonstrated that 0.05-nM PACAP-38 induces long-lasting facilitation of the basal transmission of CA1 synapses, while a high dose (1 μM) induces a long-lasting depression therein, showing that PACAP-38 modulates CA1 synaptic transmission in a dose-dependent manner.[47] *In vivo* LTP studies in the dentate gyrus of PACAP-heterozygous mutant mice and PAC1 receptor-mutant mice using two states of tetanus ("suprathreshold" and "at threshold") have shown that LTP depends on applied tetanus as well as the PACAP gene dose.[48] It is suggested that PACAP plays an important modulatory role in learning and memory, similar to that seen in *amnesiac* flies.

REDUCED RESPONSES TO ETHANOL IN PACAP-DEFICIENT MICE

Drosophila learning mutants that have mutations in components of the cAMP cascade—*rutabaga* (adenylyl cyclase) and *DCO* (protein kinase A catalytic subunit)—display increased sensitivity to ethanol. This is also the case for *amnesiac*, since Moore *et al.* have identified *cheapdate* as a mutant with enhanced sensitivity to ethanol, and revealed that *cheapdate* is allelic to *amnesiac*.[21,49,50] Therefore, the question of whether PACAP deficiency may be associated with altered sensitivity to ethanol has been addressed, and it has been shown that ethanol-induced hypothermic and hypnotic effects are significantly reduced in PACAP-deficient mice.[51] Because PAC1 receptor-deficient mice show normal sensitivity to ethanol,[42] lack of signal transmission of PACAP through the VPAC receptors may explain the reduced sensitivity to ethanol in PACAP-deficient mice.

In contrast to that in the *Drosophila* mutant *amnesiac*, PACAP deficiency in mammals causes reduced sensitivity to ethanol. However, these data suggest a connection between ethanol sensitivity and PACAP or *amnesiac* products. Functional comparison between PACAP and *amnesiac* may provide insights into the possible mechanisms of action and evolution of this peptidergic family.

LOSS OF INFLAMMATORY AND NEUROPATHIC PAIN IN PACAP-DEFICIENT MICE

PACAP-immunoreactive neural elements have been detected in the spinal dorsal horn and dorsal root ganglia,[52,53] suggesting a role for PACAP in the modulation of pain transmission. Several animal models have been studied to evaluate the role of PACAP in pain, but whether PACAP is nociceptive or anti-nociceptive is still controversial.[16] Recently, it has been demonstrated that inflammatory pain and neuropathic pain disappear in PACAP-deficient mice, and that upregulation of PACAP in the superficial layer of the spinal cord enhances the functional coupling of neuronal nitric oxide synthase (nNOS) and *N*-methyl-D-aspartate (NMDA) receptors and subsequent NO production, which is essential for chronic pain to occur.[54] PAC1 receptor-deficient mice also have a substantial, 75% decrease in chronic nociceptive response.[55] These results demonstrate that the PACAPergic system is a key element in pain hypersensitivity in the spinal cord with potential therapeutic relevance for relief of chronic neuropathic pain. The functional coupling of NMDA receptors and nNOS by upregulation of PACAP is the first *in vivo* example of late-onset, transcription, and activity-dependent, central sensitization involved in the maintenance of chronic pain.[54]

CIRCADIAN RHYTHMICITY IN PACAP AND ITS RECEPTOR MUTANTS

PACAP is co-expressed with glutamate in a subset of retinal ganglion cells included in the retinohypothalamic tract,[56] and the PACAP-containing retinal ganglion cells are identical to the subset of cells expressing melanopsin, a circadian photopigment.[57] PACAP exerts different effects on the photoentrainment of circadian rhythms depending on the applied dose.[20] PAC1 receptor-deficient mice show abnormalities in light-induced phase shift. Light stimulation at early nighttime results in larger phase delays, while that at late nighttime leads to phase delays instead of phase advances. These processes are dissociated from light-induced changes in the clock gene *per* and *c-fos* expression in the suprachiasmatic nucleus.[58]

In PACAP-deficient mice, the light-induced phase advance, but not phase delay, is significantly attenuated, whereas *c-fos* expression is attenuated in the opposite way, again suggesting that the induction of *c-fos* might not be essential for the phase shift.[59] However, Colwell *et al.* have shown, in their PACAP-deficient mice, that both light-induced phase delays and advances of the circadian system are attenuated.[8] The phenotypic difference between the two mutant strains may be due to differences in light intensity; the former study employing dimmer light for photic stimulation than the latter.

VPAC2 receptor-deficient mice have been shown to lose behavioral circadian rhythm under constant dark conditions and circadian expression of the clock genes, indicating that the VPAC2 receptor has crucial roles in circadian function.[14] Furthermore, studies in VIP- and VPAC2 receptor-deficient mice reveal that VIPergic signaling coordinates daily rhythms in the suprachiasmatic nucleus and behavior by synchronizing a small population of pacemaking neurons and maintaining rhythmicity in a subset of nonpacemaking neurons.[60]

NEURONAL DEVELOPMENT

Evidence suggests that PACAP plays diverse roles in mammalian neurogenesis,[2,3,23,61] therefore, it is conceivable that developmental defects may contribute some of the plethora of knockout phenotypes for PACAP and its receptors. PACAP and PAC1 receptor mRNA is expressed in the mouse nervous system from E9.5.[62] PAC1 receptor mRNA is expressed at very high levels in ventricular zones throughout the neuraxis, while PACAP mRNA is primarily expressed in postmitotic parenchymal tissue.[63]

Cultures of E13.5 rat cortical precursors produce PACAP as an autocrine signal to elicit cell cycle withdrawal, inducing the transition from proliferation to neuronal differentiation.[64] Intracerebroventricular injection of PACAP into E15.5 rats inhibits precursor mitosis and neurogenesis.[61] Furthermore, it has been shown that PACAP antagonists increase cell cycle exit, then increase apoptosis, of neuroblast cultures from E3.5 chick brain, indicating that endogenous PACAP is required to inhibit apoptosis and maintain full proliferative activity during early brain development.[65] In rat and mouse cerebellar granule neuron precursors, PACAP inhibits Sonic Hedgehog (Shh)-induced proliferation, while PACAP stimulates mitosis in the absence of Shh, suggesting that PAC1 receptors serve as sensors of environmental cues and coordinates brain neurogenesis.[66]

An *in vitro* neuronal culture model of embryonic stem (ES) cell differentiation has shown that PAC1 and VPAC receptors are functionally expressed in mouse ES cells and embryoid body-derived cells and that PACAP and VIP induce differentiation of ES cells into a neuronal phenotype.[67] In addition, it has been demonstrated that PACAP, VIP, and their receptor mRNAs are differentially expressed in ES cells, ES cell-derived, neural stem cell-enriched cultures, and differentiated neurons,[68] and that PACAP inhibits self-renewal and induces differentiation in neural precursor cells *in vitro* (see Hirose *et al.*, this volume).[69] These findings raise the possibility that PACAP is involved in embryonic development even at a very early stage, at least in part, via autocrine or paracrine mechanisms of action.

POTENTIAL INVOLVEMENT IN HUMAN DISEASES AND FUTURE OUTLOOK

This article reviews recent advances in understanding the role of PACAP by the study of behavioral and neurological consequences of PACAP- or PACAP receptor deficiency, and discusses aspects of the central PACAPergic system. Studies of PACAP- or PACAP receptor-deficient mice have led to a better understanding of the physiological roles of endogenous PACAP and have revealed some new roles for PACAP, which have not been adequately addressed. These studies have also provided new insights into the mechanisms of action involved in a number of physiological processes. Although it is difficult to readily extrapolate these insights to human disorders, the findings may be relevant for predicting, at least in part, putative developmental and/or neuroplastic abnormalities by assessing common genetic and/or signaling pathways.

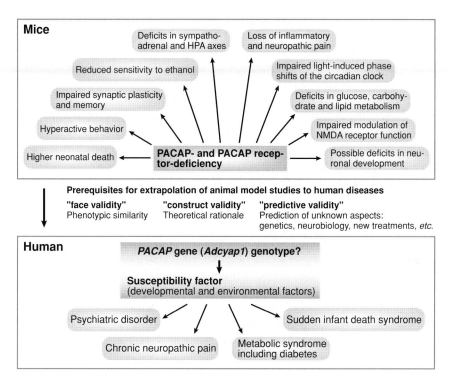

FIGURE 1. Conceptual framework for extrapolating the findings of animal model studies to human diseases.

FɪɢᴜʀE 1 shows a conceptual framework for extrapolating the findings from animal model studies to human diseases.

A possible linkage of certain psychiatric diseases with the PACAP gene *Adcyap1* has been suggested. Fine-scale mapping of a locus for severe bipolar mood disorder on chromosome 18p11.3 suggests that *Adcyap1* residing at 18p11.32[70,71] is located close to a bipolar disorder risk locus.[72] In addition, inositol depletion by lithium, a front-line treatment for bipolar disorder, has been demonstrated to be associated with coordinated upregulation of PACAP and the enzyme that processes the PACAP precursor, raising the possibility that PACAP may be a therapeutic effector of lithium in bipolar disorder.[73]

Studies in two related patients with a partial trisomy 18p revealed three copies of the PACAP gene and elevated plasma PACAP levels. These patients suffered from severe mental retardation and hematological abnormalities, although whether the former defect was a consequence of PACAP overexpression remains to be determined.[71]

The finding of respiratory control defects in PACAP-deficient neonates raises the possibility of a link between mutations in genes encoding components of the PACAP signaling pathways and sudden infant death syndrome in humans.[31,74]

Although a detailed review of peripheral roles of PACAP are beyond the scope of this article, PACAP has potent insulinotropic action as a pancreatic islet neuropeptide and insulin-sensitizing properties,[75,76] as well as playing a significant role in carbohydrate and lipid metabolism.[5] As discussed above, PACAP deficiency is associated with higher neonatal death.[5,6,11,12] The wasted appearance of PACAP-deficient neonates resembles some debilitating illnesses.[5] Studies in transgenic mice overexpressing PACAP in pancreatic β-cells suggest that PACAP has a significant role in streptozotocin-induced diabetes (a model of type 1 diabetes)[32] and lethal yellow agouti (KK*Ay*) mice (a genetic model for type 2 diabetes).[77] ZAC/PLAGL1, whose product regulates both apoptosis and cell cycle arrest, as well as the PAC1 receptor transcription,[78] is a strong candidate gene for transient neonatal diabetes.[79] Furthermore, an association between type 2 diabetes and two novel single nucleotide polymorphisms in *Adcyap1* has been found, although they may not be a major influence on susceptibility to type 2 diabetes.[80]

Many diseases are multifactorial, reflecting complex interactions between environmental and genetic factors, however, such pathogenesis remains poorly understood. Animal models, therefore, provide useful tools for investigating the mechanisms underlying human diseases. PACAP- as well as PACAP receptor-deficient mice will provide such models for human diseases relevant to their respective phenotypes, such as psychiatric diseases, chronic neuropathic pain, metabolic syndrome, and sudden infant death syndrome, with a phenotypic similarity (face validity) and hopefully theoretical rationale (construct validity) (FIG. 1).

Identification of genes influencing disease susceptibility will contribute to better understanding of the molecular mechanisms underlying pathogenesis, a necessary prerequisite for the rational design of new treatments. As discussed above, several linkage studies, as well as the multiple deficient phenotypes *per se*, imply that the *Adcyap1* genotype might be a susceptibility factor for some human diseases. Mutant mice may also provide useful models for predicting unknown aspects of human diseases, such as their genetics, neurobiology, or new treatments (predictive validity).

ACKNOWLEDGMENTS

Our research has been supported, in part, by: Grant-in-Aid for Exploratory Research from the Ministry of Education, Culture, Sports, Science and Technology of Japan; Grants-in-Aid for Scientific Research from Japan Society for the Promotion of Science; and by grants from Sankyo Foundation of Life Science and Taisho Pharmaceutical Co. Ltd.

REFERENCES

1. MIYATA, A., A. ARIMURA, R.R. DAHL, *et al.* 1989. Isolation of a novel 38 residue-hypothalamic polypeptide which stimulates adenylate cyclase in pituitary cells. Biochem. Biophys. Res. Commun. **164:** 567–574.
2. ARIMURA, A. 1998. Perspectives on pituitary adenylate cyclase activating polypeptide (PACAP) in the neuroendocrine, endocrine, and nervous systems. Jpn. J. Physiol. **48:** 301–331.
3. VAUDRY, D., B.J. GONZALEZ, M. BASILLE, *et al.* 2000. Pituitary adenylate cyclase-activating polypeptide and its receptors: from structure to functions. Pharmacol. Rev. **52:** 269–324.
4. HARMAR, A.J., A. ARIMURA, I. GOZES, *et al.* 1998. International Union of Pharmacology. XVIII.Nomenclature of receptors for vasoactive intestinal peptide and pituitary adenylate cyclase-activating polypeptide. Pharmacol. Rev. **50:** 265–270.
5. GRAY, S.L., K.J. CUMMINGS, F.R. JIRIK & N.M. SHERWOOD. 2001. Targeted disruption of the pituitary adenylate cyclase-activating polypeptide gene results in early postnatal death associated with dysfunction of lipid and carbohydrate metabolism. Mol. Endocrinol. **15:** 1739–1747.
6. HASHIMOTO, H., N. SHINTANI, K. TANAKA, *et al.* 2001. Altered psychomotor behaviors in mice lacking pituitary adenylate cyclase-activating polypeptide (PACAP). Proc. Natl. Acad. Sci. USA **98:** 13355–13360.
7. HAMELINK, C., O. TJURMINA, R. DAMADZIC, *et al.* 2002. Pituitary adenylate cyclase-activating polypeptide is a sympathoadrenal neurotransmitter involved in catecholamine regulation and glucohomeostasis. Proc. Natl. Acad. Sci. USA **99:** 461–466.
8. COLWELL, C.S., S. MICHEL, J. ITRI, *et al.* 2004. Selective deficits in the circadian light response in mice lacking PACAP. Am. J. Physiol. Regul. Integr. Comp. Physiol. **287:** R1194–R1201.
9. COLWELL, C.S., S. MICHEL, J. ITRI, *et al.* 2003. Disrupted circadian rhythms in VIP- and PHI-deficient mice. Am. J. Physiol. Regul. Integr. Comp. Physiol. **285:** R939–R949.
10. HASHIMOTO, H., N. SHINTANI, A. NISHINO, *et al.* 2000. Mice with markedly reduced PACAP (PAC(1)) receptor expression by targeted deletion of the signal peptide. J. Neurochem. **75:** 1810–1817.
11. JAMEN, F., K. PERSSON, G. BERTRAND, *et al.* 2000. PAC1 receptor-deficient mice display impaired insulinotropic response to glucose and reduced glucose tolerance. J. Clin. Invest. **105:** 1307–1315.
12. OTTO, C., Y. KOVALCHUK, D.P. WOLFER, *et al.* 2001. Impairment of mossy fiber long-term potentiation and associative learning in pituitary adenylate cyclase activating polypeptide type I receptor-deficient mice. J. Neurosci. **21:** 5520–5527.
13. ASNICAR, M.A., A. KOSTER, M.L. HEIMAN, *et al.* 2002. Vasoactive intestinal polypeptide/pituitary adenylate cyclase-activating peptide receptor 2 deficiency in mice results in growth retardation and increased basal metabolic rate. Endocrinology **143:** 3994–4006.
14. HARMAR, A.J., H.M. MARSTON, S. SHEN, *et al.* 2002. The VPAC(2) receptor is essential for circadian function in the mouse suprachiasmatic nuclei. Cell **109:** 497–508.

15. NUSSDORFER, G.G. & L.K. MALENDOWICZ. 1998. Role of VIP, PACAP, and related peptides in the regulation of the hypothalamo-pituitary-adrenal axis. Peptides **19:** 1443–1467.
16. DICKINSON, T. & S.M. FLEETWOOD-WALKER. 1999. VIP and PACAP: very important in pain? Trends Pharmacol. Sci. **20:** 324–329.
17. MONTERO, M., L. YON, S. KIKUYAMA, *et al*. 2000. Molecular evolution of the growth hormone-releasing hormone/pituitary adenylate cyclase-activating polypeptide gene family. Functional implication in the regulation of growth hormone secretion. J. Mol. Endocrinol. **25:** 157–168.
18. SHERWOOD, N.M., S.L. KRUECKL & J.E. MCRORY. 2000. The origin and function of the pituitary adenylate cyclase-activating polypeptide (PACAP)/glucagon superfamily. Endocr. Rev. **21:** 619–670.
19. FILIPSSON, K., M. KVIST-REIMER & B. AHREN. 2001. The neuropeptide pituitary adenylate cyclase-activating polypeptide and islet function. Diabetes **50:** 1959–1969.
20. HANNIBAL, J. 2002. Neurotransmitters of the retino-hypothalamic tract. Cell Tissue Res. **309:** 73–88.
21. HASHIMOTO, H., N. SHINTANI & A. BABA. 2002. Higher brain functions of PACAP and a homologous Drosophila memory gene amnesiac: insights from knockouts and mutants. Biochem. Biophys. Res. Commun. **297:** 427–431.
22. MORETTI, C., C. MENCACCI, G.V. FRAJESE, *et al*. 2002. Growth hormone-releasing hormone and pituitary adenylate cyclase-activating polypeptide in the reproductive system. Trends Endocrinol. Metab. **13:** 428–435.
23. WASCHEK, J.A. 2002. Multiple actions of pituitary adenylyl cyclase activating peptide in nervous system development and regeneration. Dev. Neurosci. **24:** 14–23.
24. GANEA, D., R. RODRIGUEZ & M. DELGADO. 2003. Vasoactive intestinal peptide and pituitary adenylate cyclase-activating polypeptide: players in innate and adaptive immunity. Cell Mol. Biol. **49:** 127–142.
25. HARMAR, A.J. 2003. An essential role for peptidergic signalling in the control of circadian rhythms in the suprachiasmatic nuclei. J. Neuroendocrinol. **15:** 335–338.
26. LI, M. & A. ARIMURA. 2003. Neuropeptides of the pituitary adenylate cyclase-activating polypeptide/vasoactive intestinal polypeptide/growth hormone-releasing hormone/secretin family in testis. Endocrine **20:** 201–214.
27. SOMOGYVARI-VIGH, A. & D. REGLODI. 2004. Pituitary adenylate cyclase activating polypeptide: a potential neuroprotective peptide. Curr. Pharm. Des. **10:** 2861–2889.
28. SHINTANI, N., S. TOMIMOTO, H. HASHIMOTO, *et al*. 2003. Functional roles of the neuropeptide PACAP in brain and pancreas. Life Sci. **74:** 337–343.
29. GRAY, S.L., N. YAMAGUCHI, P. VENCOVA & N.M. SHERWOOD. 2002. Temperature-sensitive phenotype in mice lacking pituitary adenylate cyclase-activating polypeptide. Endocrinology **143:** 3946–3954.
30. HANNIBAL, J. 2002. Pituitary adenylate cyclase-activating peptide in the rat central nervous system: an immunohistochemical and in situ hybridization study. J. Comp. Neurol. **453:** 389–417.
31. CUMMINGS, K.J., J.D. PENDLEBURY, N.M. SHERWOOD & R.J. WILSON. 2004. Sudden neonatal death in PACAP-deficient mice is associated with reduced respiratory chemoresponse and susceptibility to apnoea. J. Physiol. **555:** 15–26.

32. Yᴀᴍᴀᴍᴏᴛᴏ, K., H. Hᴀsʜɪᴍᴏᴛᴏ, S. Tᴏᴍɪᴍᴏᴛᴏ, *et al.* 2003. Overexpression of PACAP in transgenic mouse pancreatic β-cells enhances insulin secretion and ameliorates streptozotocin-induced diabetes. Diabetes **52:** 1155–1162.
33. Sʜɪɴᴛᴀɴɪ, N., H. Hᴀsʜɪᴍᴏᴛᴏ, K. Tᴀɴᴀᴋᴀ, *et al.* 2004. Overexpression of PACAP in the pancreas failed to rescue early postnatal mortality in PACAP-null mice. Regul. Pept. **123:** 155–159.
34. Oᴛᴛᴏ, C., L. Hᴇɪɴ, M. Bʀᴇᴅᴇ, *et al.* 2004. Pulmonary hypertension and right heart failure in pituitary adenylate cyclase-activating polypeptide type I receptor-deficient mice. Circulation **110:** 3245–3251.
35. Sʜɪɴᴛᴀɴɪ, N., W. Mᴏʀɪ, H. Hᴀsʜɪᴍᴏᴛᴏ, *et al.* 2002. Defects in reproductive functions in PACAP-deficient female mice. Regul. Pept. **109:** 45–48.
36. Sᴢᴀʙᴏ, E., A. Nᴇᴍᴇsᴋᴇʀɪ, A. Aʀɪᴍᴜʀᴀ & K. Kᴏᴠᴇs. 2004. Effect of PACAP on LH release studied by cell immunoblot assay depends on the gender, on the time of day and in female rats on the day of the estrous cycle. Regul. Pept. **123:** 139–145.
37. Aᴘᴏsᴛᴏʟᴀᴋɪs, E.M., R. Lᴀɴᴢ & B.W. O'Mᴀʟʟᴇʏ. 2004. Pituitary adenylate cyclase-activating peptide: a pivotal modulator of steroid-induced reproductive behavior in female rodents. Mol. Endocrinol. **18:** 173–183.
38. Jᴀᴍᴇɴ, F., N. Rᴏᴅʀɪɢᴜᴇᴢ-Hᴇɴᴄʜᴇ, F. Pʀᴀʟᴏɴɢ, *et al.* 2000. PAC1 null females display decreased fertility. Ann. N.Y. Acad. Sci. **921:** 400–404.
39. Nɪᴄᴏᴛ, A., T. Oᴛᴛᴏ, P. Bʀᴀʙᴇᴛ & E.M. Dɪᴄɪᴄᴄᴏ-Bʟᴏᴏᴍ. 2004. Altered social behavior in pituitary adenylate cyclase-activating polypeptide type I receptor-deficient mice. J. Neurosci. **24:** 8786–8795.
40. Sʜɪɴᴛᴀɴɪ, N., H. Hᴀsʜɪᴍᴏᴛᴏ, K. Tᴀɴᴀᴋᴀ, *et al.* 2006. Serotonergic inhibition of intense jumping behavior in mice lacking PACAP (Adcyap1-/-). Ann. N. Y. Acad. Sci. This volume.
41. Oɢᴀᴡᴀ, T., T. Nᴀᴋᴀᴍᴀᴄʜɪ, H. Oʜᴛᴀᴋɪ, *et al.* 2005. Monoaminergic neuronal development is not affected in PACAP-gene-deficient mice. Regul. Pept. **126:** 103–108.
42. Oᴛᴛᴏ, C., M. Mᴀʀᴛɪɴ, D.P. Wᴏʟғᴇʀ, *et al.* 2001. Altered emotional behavior in PACAP-type-I-receptor-deficient mice. Brain Res. Mol. Brain Res. **92:** 78–84.
43. Nᴀᴋᴀᴛᴀ, M., D. Kᴏʜɴᴏ, N. Sʜɪɴᴛᴀɴɪ, *et al.* 2004. PACAP deficient mice display reduced carbohydrate intake and PACAP activates NPY-containing neurons in the rat hypothalamic arcuate nucleus. Neurosci. Lett. **370:** 252–256.
44. Mɪᴢᴜɴᴏ, Y., K. Kᴏɴᴅᴏ, Y. Tᴇʀᴀsʜɪᴍᴀ, *et al.* 1998. Anorectic effect of pituitary adenylate cyclase activating polypeptide (PACAP) in rats: lack of evidence for involvement of hypothalamic neuropeptide gene expression. J. Neuroendocrinol. **10:** 611–616.
45. Mᴏʀɪᴛᴀ, Y., D. Yᴀɴᴀɢɪᴅᴀ, N. Sʜɪɴᴛᴀɴɪ, *et al.* 2006. Lack of trimethyltin (TMT)-induced elevation of plasma corticosterone in PACAP-deficient mice. Ann. N.Y. Acad. Sci. This volume.
46. Fᴇᴀɴʏ, M.B. & W.G. Qᴜɪɴɴ. 1995. A neuropeptide gene defined by the Drosophila memory mutant amnesiac. Science **268:** 869–873.
47. Rᴏʙᴇʀᴛᴏ, M., R. Sᴄᴜʀɪ & M. Bʀᴜɴᴇʟʟɪ. 2001. Differential effects of PACAP-38 on synaptic responses in rat hippocampal CA1 region. Learn. Mem. **8:** 265–271.
48. Mᴀᴛsᴜʏᴀᴍᴀ, S., A. Mᴀᴛsᴜᴍᴏᴛᴏ, H. Hᴀsʜɪᴍᴏᴛᴏ, *et al.* 2003. Impaired long-term potentiation in vivo in the dentate gyrus of pituitary adenylate cyclase-activating polypeptide (PACAP) or PACAP type 1 receptor-mutant mice. Neuroreport **14:** 2095–2098.
49. Bᴇʟʟᴇɴ, H.J. 1998. The fruit fly: a model organism to study the genetics of alcohol abuse and addiction? Cell **93:** 909–912.

50. MOORE, M.S., J. DEZAZZO, A.Y. LUK, et al. 1998. Ethanol intoxication in Drosophila: genetic and pharmacological evidence for regulation by the cAMP signaling pathway. Cell **93:** 997–1007.

51. TANAKA, K., H. HASHIMOTO, N. SHINTANI, et al. 2004. Reduced hypothermic and hypnotic responses to ethanol in PACAP-deficient mice. Regul. Pept. **123:** 95–98.

52. MOLLER, K., Y.Z. ZHANG, R. HAKANSON, et al. 1993. Pituitary adenylate cyclase activating peptide is a sensory neuropeptide: immunocytochemical and immuno-chemical evidence. Neuroscience **57:** 725–732.

53. ZHANG, Q., T.J. SHI, R.R. JI, et al. 1995. Expression of pituitary adenylate cyclase-activating polypeptide in dorsal root ganglia following axotomy: time course and coexistence. Brain Res. **705:** 149–158.

54. MABUCHI, T., N. SHINTANI, S. MATSUMURA, et al. 2004. Pituitary adenylate cyclase-activating polypeptide is required for the development of spinal sensitization and induction of neuropathic pain. J. Neurosci. **24:** 7283–7291.

55. JONGSMA, H., L.M. PETTERSSON, Y. ZHANG, et al. 2001. Markedly reduced chronic nociceptive response in mice lacking the PAC1 receptor. Neuroreport **12:** 2215–2219.

56. HANNIBAL, J., M. MOLLER, O.P. OTTERSEN & J. FAHRENKRUG. 2000. PACAP and glutamate are co-stored in the retinohypothalamic tract. J. Comp. Neurol. **418:** 147–155.

57. HANNIBAL, J., P. HINDERSSON, S.M. KNUDSEN, et al. 2002. The photopigment melanopsin is exclusively present in pituitary adenylate cyclase-activating polypeptide-containing retinal ganglion cells of the retinohypothalamic tract. J. Neurosci. **22:** RC191 (1–7).

58. HANNIBAL, J., F. JAMEN, H.S. NIELSEN, et al. 2001. Dissociation between light-induced phase shift of the circadian rhythm and clock gene expression in mice lacking the pituitary adenylate cyclase activating polypeptide type 1 receptor. J. Neurosci. **21:** 4883–4890.

59. KAWAGUCHI, C., K. TANAKA, Y. ISOJIMA, et al. 2003. Changes in light-induced phase shift of circadian rhythm in mice lacking PACAP. Biochem. Biophys. Res. Commun. **310:** 169–175.

60. ATON, S.J., C.S. COLWELL, A.J. HARMAR, et al. 2005. Vasoactive intestinal polypep-tide mediates circadian rhythmicity and synchrony in mammalian clock neurons. Nat. Neurosci. **8:** 476–483.

61. SUH, J., N. LU, A. NICOT, et al. 2001. PACAP is an anti-mitogenic signal in devel-oping cerebral cortex. Nat. Neurosci. **4:** 123–124.

62. SHEWARD, W.J., E.M. LUTZ, A.J. COPP & A.J. HARMAR. 1998. Expression of PACAP, and PACAP type 1 (PAC1) receptor mRNA during development of the mouse embryo. Brain Res. Dev. Brain Res. **109:** 245–253.

63. JAWORSKI, D.M. & M.D. PROCTOR. 2000. Developmental regulation of pituitary adenylate cyclase-activating polypeptide and PAC(1) receptor mRNA expression in the rat central nervous system. Brain Res. Dev. Brain Res. **120:** 27–39.

64. LU, N. & E. DICICCO-BLOOM. 1997. Pituitary adenylate cyclase-activating polypep-tide is an autocrine inhibitor of mitosis in cultured cortical precursor cells. Proc. Natl. Acad. Sci. USA **94:** 3357–3362.

65. ERHARDT, N.M. & N.M. SHERWOOD. 2004. PACAP maintains cell cycling and inhibits apoptosis in chick neuroblasts. Mol. Cell Endocrinol. **221:** 121–134.

66. NICOT, A., V. LELIEVRE, J. TAM, et al. 2002. Pituitary adenylate cyclase-activating polypeptide and sonic hedgehog interact to control cerebellar granule precursor cell proliferation. J. Neurosci. **22:** 9244–9254.

67. CAZILLIS, M., B.J. GONZALEZ, C. BILLARDON, *et al.* 2004. VIP and PACAP induce selective neuronal differentiation of mouse embryonic stem cells. Eur. J. Neurosci. **19:** 798–808.
68. HIROSE, M., H. HASHIMOTO, N. SHINTANI, *et al.* 2005. Differential expression of mRNAs for PACAP and its receptors during neural differentiation of embryonic stem cells. Regul. Pept. **126:** 109–113.
69. HIROSE, M., H. HASHIMOTO, J. IGA, *et al.* 2006. Inhibition of self-renewal and induction of neural differentiation by PACAP in neural progenitor cells. Ann. N.Y. Acad. Sci. This volume.
70. HOSOYA, M., C. KIMURA, K. OGI, *et al.* 1992. Structure of the human pituitary adenylate cyclase activating polypeptide (PACAP) gene. Biochim. Biophys. Acta. **6:** 199–206.
71. FRESON, K., H. HASHIMOTO, C. THYS, *et al.* 2004. The pituitary adenylate cyclase-activating polypeptide is a physiological inhibitor of platelet activation. J. Clin. Invest. **113:** 905–912.
72. MCINNES, L.A., S.K. SERVICE, V.I. REUS, *et al.* 2001. Fine-scale mapping of a locus for severe bipolar mood disorder on chromosome 18p11.3 in the Costa Rican population. Proc. Natl. Acad. Sci. USA **98:** 11485–11490.
73. BRANDISH, P.E., M. SU, D.J. HOLDER, *et al.* 2005. Regulation of gene expression by lithium and depletion of inositol in slices of adult rat cortex. Neuron **45:** 861–872.
74. STAINES, D.R. 2004. Is sudden infant death syndrome (SIDS) an autoimmune disorder of endogenous vasoactive neuropeptides? Med. Hypotheses **62:** 653–657.
75. YADA, T., M. SAKURADA, K. IHIDA, *et al.* 1994. Pituitary adenylate cyclase activating polypeptide is an extraordinarily potent intra-pancreatic regulator of insulin secretion from islet beta-cells. J. Biol. Chem. **269:** 1290–1293.
76. NAKATA, M., S. SHIODA, Y. OKA, *et al.* 1999. Insulinotropin PACAP potentiates insulin-stimulated glucose uptake in 3T3 L1 cells. Peptides **20:** 943–948.
77. TOMIMOTO, S., H. HASHIMOTO, N. SHINTANI, *et al.* 2004. Overexpression of pituitary adenylate cyclase-activating polypeptide in islets inhibits hyperinsulinemia and islet hyperplasia in agouti yellow mice. J. Pharmacol. Exp. Ther. **309:** 796–803.
78. CIANI, E., A. HOFFMANN, P. SCHMIDT, *et al.* 1999. Induction of the PAC1-R (PACAP-type I receptor) gene by p53 and Zac. Brain Res. Mol. Brain Res. **69:** 290–294.
79. KAMIYA, M., H. JUDSON, Y. OKAZAKI, *et al.* 2000. The cell cycle control gene ZAC/PLAGL1 is imprinted—a strong candidate gene for transient neonatal diabetes. Hum. Mol. Genet. **9:** 453–460.
80. GU, H.F. 2002. Genetic variation screening and association studies of the adenylate cyclase activating polypeptide 1 (ADCYAP1) gene in patients with type 2 diabetes. Hum. Mutat. **19:** 572–573.

Complexing Receptor Pharmacology

Modulation of Family B G Protein–Coupled Receptor Function by RAMPs

PATRICK M. SEXTON,[a,b] MARIA MORFIS,[a,b] NANDA TILAKARATNE,[a] DEBBIE L. HAY,[c] MADHARA UDAWELA,[a,b] GEORGE CHRISTOPOULOS,[a] AND ARTHUR CHRISTOPOULOS[a,b]

[a]Howard Florey Institute, [b]Department of Pharmacology, The University of Melbourne, Victoria 3010, Victoria, Australia

[c]School of Biological Sciences, University of Auckland, Symonds Street, Auckland, New Zealand

ABSTRACT: The most well-characterized subgroup of family B G protein–coupledreceptors (GPCRs) comprises receptors for peptide hormones, such as secretin, calcitonin (CT), glucagon, and vasoactive intestinal peptide (VIP). Recent data suggest that many of these receptors can interact with a novel family of GPCR accessory proteins termed *receptor activity modifying proteins (RAMPs)*. RAMP interaction with receptors can lead to a variety of actions that include chaperoning of the receptor protein to the cell surface as is the case for the calcitonin receptor-like receptor (CLR) and the generation of novel receptor phenotypes. RAMP heterodimerization with the CLR and related CT receptor is required for the formation of specific CT gene-related peptide, adrenomedullin (AM) or amylin receptors. More recent work has revealed that the specific RAMP present in a heterodimer may modulate other functions such as receptor internalization and recycling and also the strength of activation of downstream signaling pathways. In this article we review our current state of knowledge of the consequence of RAMP interaction with family B GPCRs.

KEYWORDS: receptor activity modifying protein; G protein–coupled receptor; calcitonin; CGRP; adrenomedullin; vasoactive intestinal polypeptide

INTRODUCTION

The superfamily of seven-transmembrane domain G protein–coupled receptors (GPCRs) is the largest group of receptor proteins; estimated to comprise

Address for correspondence: Patrick M. Sexton Ph.D., Howard Florey Institute, Level 2, Alan Gilbert Building, The University of Melbourne, 161 Barry Street, Carlton South, 3053, Victoria, Australia. Voice: 61-3-8344-1954; fax: 61-3-9347-0446.

e-mail: p.sexton@hfi.unimelb.edu.au

Ann. N.Y. Acad. Sci. 1070: 90–104 (2006). © 2006 New York Academy of Sciences.
doi: 10.1196/annals.1317.076

between 1% and 5% of the entire repertoire of human genes. These proteins serve as receptors for a diverse range of ligands ranging from photons through amino acids, amines, cations, and small peptides to cytokines, glycoproteins, and proteases. The majority of these receptors are related on a sequence similarity and evolutionary basis to the rhodopsin family of receptors, termed *family A* or *class I*. Two additional major GPCR families exist; termed *family B (class II)* and *family C (class III)*.[1] Family B receptors can be subclassified into three groups according to sequence similarity; those with GPCR proteolytic sites, those with cysteine-rich domains, such as frizzled or smoothened, and those that lack these domains.[2] This latter group includes the receptors for many peptide hormones and it is the interaction of this receptor subfamily with receptor activity modifying proteins (RAMPs) that is the primary focus of this article. These receptors have ~30–50% homology with each other and bind peptide hormones whose activity resides within sequences that range in length from 27 to ~50 amino acids. Peptide ligands for family B receptors include secretin, calcitonin (CT), amylin, CT gene-related peptide (CGRP), vasoactive intestinal peptide (VIP), pituitary adenylate cyclase-activating peptide (PACAP), glucagon, glucagon-like peptide-1 (GLP-1), parathyroid hormone (PTH) and its related protein (PTHrP), gastric inhibitory peptide (GIP), growth hormone releasing hormone (GHRH), and corticotropin-releasing factor (CRF).[3]

RAMPs

Although believed to be GPCRs, the molecular identity of receptors for CGRP, amylin, and adrenomedullin (AM) remained obscure for many years, despite numerous attempts to clone these receptors. The identity of the CGRP receptor was eventually resolved in 1998 with the discovery of a family of three novel accessory proteins (RAMPs).[4] Attempts to expression clone the CGRP receptor revealed a cDNA, that when expressed in oocytes, engendered a functional response to CGRP. This DNA encoded a single transmembrane spanning protein of 148 amino acids, later termed *RAMP1*, and not the expected seven-transmembrane domain structure of a GPCR. However, expression experiments in mammalian cells demonstrated that, by itself, RAMP1 could not reconstitute a CGRP receptor and that cotransfection of RAMP1 with the calcitonin receptor-like receptor (CLR) was required for the formation of a CGRP receptor.[4]

RAMP Structure

Depending on the species, there is a predicted N-terminal signal peptide of between 22 and 35 amino acids (RAMPs 1 and 3) or 24-60 amino acids

(RAMP2). This is linked in tandem to an extracellular amino terminus of about 90 residues in RAMPs 1 and 3, while RAMP2 sequences are generally longer, normally containing an additional 13 amino acids. There then follows a transmembrane domain spanning $\sim$22 amino acids and finally a C-terminal domain, usually consisting of 9 residues (FIG. 1).[5] All three RAMPs have four highly conserved cysteine residues that presumably form disulphide bonds, while the peptide sequences of RAMPs 1 and 3 contain an extra pair of cysteines. The known mammalian RAMP1s appear not to be glycosylated, and they lack consensus sites for N-glycosylation. RAMP2 and RAMP3 have multiple glycosylation sites. The majority of the known RAMP3s have four potential glycosylation sites (N29, N58, N71, and N104 in man) although only the fourth site (N104) is absolutely conserved. The equivalent of the fourth glycosylation site is also found in all known RAMP2s; in some mammalian species there is also an equivalent of the second site (i.e., N58 of RAMP3).[5]

The CLR

The CLR has $\sim$55% sequence identity with the calcitonin receptor (CTR) and had been earlier touted as a CGRP receptor,[6] but in many cell lines expression of CLR did not produce a CGRP responsive receptor.[4,7,8] As mentioned above, CLR indeed, has a role in the generation of CGRP receptors, with co-expression of CLR and RAMP1 forming a classical CGRP$_1$ receptor phenotype that was indistinguishable from that of the endogenous CGRP receptor expressed in SK-N-MC cells.[4] The two additional RAMPs were subsequently identified through database searching and named RAMP2 and RAMP3, respectively.[4] The three RAMPs share $\sim$30% sequence identity ($\sim$50% homology) (FIG. 1). Remarkably, cotransfection of RAMP2 or RAMP3 with CLR induced receptors with distinct pharmacology—that of AM receptors, which displayed equivalent pharmacology to endogenously expressed AM receptors in NG-108-15 cells.[4,9] It was this ability of a classical seven-transmembrane domain GPCR, to completely switch receptor specificity upon association with a group of novel proteins that provided a paradigm shift in our understanding of the molecular basis of GPCR phenotype, highlighting a novel mechanism that engenders much greater diversity in receptor repertoire than previously imagined.

OTHER FAMILY B RECEPTORS

The CT Receptor

It is now evident that RAMPs can interact with receptors other than CLR. CLR has the greatest sequence homology with the CTR and the potential

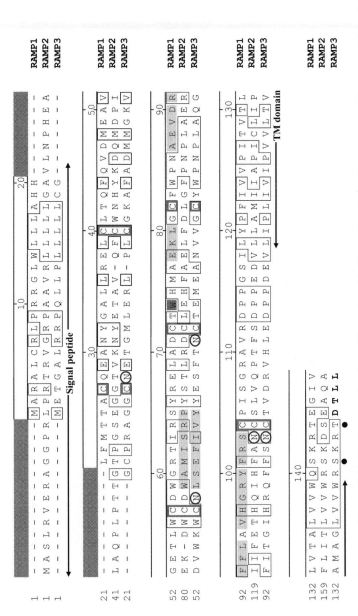

FIGURE 1. Alignment of human RAMPs. Conserved amino acids are boxed in black. Cysteine residues conserved between all three RAMPs, along with those conserved exclusively between RAMP1 and RAMP3, are boxed in dark grey. Tryptophan 74 of RAMP1, found to be important for BIBN4096BS affinity[33] is shaded in dark grey. Consensus phosphorylation sites are indicated with black filled circles. Consensus sites for N-linked glycosylation are circled in black. The consensus PDZ-binding domain in RAMP3 is depicted in bold type. Regions of RAMP1, RAMP2, or RAMP3 implicated in folding, dimerization, or receptor function are highlighted with grey shading.[27,38] (Reproduced with permission from Udawela *et al.*[24])

for interaction of this receptor with RAMPs was investigated soon after their discovery. Expression of amylin receptor phenotypes requires the co-expression of RAMPs with the CTR gene product.[10,11] However, as seen for CLR, the phenotypes induced by individual RAMPs are distinct. In COS-7 (African Green Monkey Kidney) or rabbit aortic endothelial cells, RAMP1 and RAMP3 induced amylin receptors that differed in their affinity for CGRP, while RAMP2 was relatively ineffective in inducing amylin receptor pheno-type. More recent work revealed that RAMP2 can also induce an amylin recep-tor phenotype, which is distinct from either the RAMP1 or RAMP3 induced receptors,[12,13] however, the efficacy of RAMP2 in creating this phenotype was highly dependent upon the cellular backgound and the isoform of CTR used in the study.

In humans, the major CTR variants differ by the absence (CTRa) or presence (CTRb) of a 16 amino acid insert in the first intracellular domain, with the CTRa being the more commonly expressed form and the variant used for initial studies with RAMPs.[10,11,14,15] Unlike CTRa, the CTRb variant is capable of strong induction of amylin receptor phenotype when co-expressed with any of the three RAMPs in COS-7.[12] These two human CTR isoforms differ in their ability to activate signaling pathways (presumably due to an effect on G protein coupling) and to internalize in response to agonist treatment,[16] which may suggest a role for G proteins in the ability of RAMPs to alter receptor phe-notype. Consistent with this hypothesis, RAMP2, which only weakly produced an amylin receptor from the CTRa receptor in COS-7 cells, potently gener-ated an amylin receptor from both CTR isoforms following cotransfection into CHO-P (Chinese Hawster Ovary-P) cells.[12]

Beyond CLR and CTR

Both the CLR and CTR share many conserved structural features with other family B GPCRs. It is thus possible that additional family B GPCRs are partners for RAMPs. Indeed, a recent screen for the capacity of receptors to translocate RAMPs to the cell surface, as a marker of RAMP-receptor interaction, has identified that 6 out of the 10 family B GPCRs tested, exhibit a demonstrable interaction with at least one RAMP. In addition to the CTR and CLR,[4,10,12] the VIP/pituitary adenylate cyclase activating peptide (VPAC) 1 receptor inter-acted strongly with all three RAMPs, while the PTH1, and glucagon receptors (GRs) translocated only RAMP2 and the PTH2 receptor translocated RAMP3 but not RAMPs 1 and 2.[17] The VPAC2, GHRH, GLP-1, and GLP-2 receptors did not significantly modify the distribution of any of the RAMPs (FIG. 2).[17] In addition to providing substantial evidence for a broader role for RAMPs, this work also revealed significant specificity for formation of RAMP-receptor pairs.

(A)

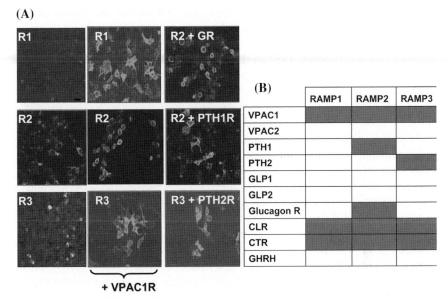

(B)

	RAMP1	RAMP2	RAMP3
VPAC1	▪	▪	▪
VPAC2			
PTH1		▪	
PTH2			▪
GLP1			
GLP2			
Glucagon R		▪	
CLR	▪	▪	▪
CTR	▪		▪
GHRH			

+ VPAC1R

FIGURE 2. Translocation of RAMPs by family B GPCRs. **(A)** Confocal images of cell-surface-expressed RAMPs. Wild-type RAMPs are relatively poorly expressed at the cell surface in the absence of receptor co-expression (left hand panels). Cotransfection of the VPAC1 receptor enables efficient translocation of all 3 RAMPs (middle panels), while the GR translocates RAMP2 (upper right panel), the PTH1 receptor RAMP2 (middle right panel), and the PTH2 receptor, RAMP3 (lower right panel). **(B)** Summary table of receptor interaction with RAMPs. Shaded boxes represent receptors that increase cell-surface expression of individual RAMPs. Blank boxes indicates no change in the level of cell-surface-expressed RAMP. (Adapted from Christopoulos *et al.*[17])

MOLECULAR MECHANISMS OF RAMP ACTION

Ramps as Receptor Chaperones

For CLR, heterodimerization with a RAMP is required for it to be efficiently transported to the cell surface of mammalian cells. In this sense, RAMPs have a chaperone role.[4] Epitope tagging of CLR has enabled information on its transport to the cell surface to be elucidated and shows that each RAMP can elicit translocation of CLR from intracellular compartments to the cell surface. Fluorescence-associated cell-sorting data and images from confocal microscopy illustrate that there is much more CLR at the cell surface when RAMPs are present.[4,7,18] RAMP interaction with the CLR also leads to modification of the terminal glycosylation of the receptor.[4,19] Although this modulation of CLR glycosylation was originally thought to be specific to RAMP1, it also occurs with RAMP2, and it is only the modified CLR that forms the functional AM_1 receptor and expresses at the cell surface.[19] It is likely that this modulation of glycosylation contributes to the cell-surface expression of the heterodimeric complex, however, the change in glycosylation of the receptor is not likely to play a role in generation of receptor phenotype; fully functional receptors are seen when RAMP and receptor are expressed in insect cells where no change in glycosylation status occurs.[20] It is currently unclear why the CLR does not express at the cell surface in the absence of RAMPs, however there is evidence that certain receptors may require association with other proteins for them to pass "quality-control checkpoints" during biosynthesis (reviewed by Bulenger *et al.*[21]). The other family B GPCRs that interact with RAMPs do not require heterodimerization to be transported to the cell surface and thus the chaperone role of RAMPs appears to be limited within this receptor family. Likewise, the glycosylation pattern of the CTR is not routinely altered when the receptor is coexpressed with RAMPs.[10,22]

Very recent work has demonstrated that RAMPs are required for the cell-surface expression and terminal glycosylation of the calcium-sensing receptor, a family C GPCR. In COS-7 cells, in the absence of RAMPs, the receptor remains in the endoplasmic reticulum (ER) as a 150 KDa, partially glycosylated receptor. RAMPs 1 and 3 (but not RAMP2) allow translocation to the Golgi and then to the cell surface, where the receptor then exists as a 175 KDa protein.[23] Furthermore, in HEK-293 (Human Embryonic Kidney) cells, where transfected receptor is expressed at the cell surface, siRNA knockdown of endogenous RAMP1 prevents this expression.[23] This work has significant parallels with the effects on CLR transport and glycosylation[4] and implies that RAMPs may have a chaperone role for other, as yet unidentified, receptors.

While knowledge of the role of RAMPs as receptor chaperones is evolving, a more consistent story has emerged for the cellular transport of RAMPs themselves, at least for RAMP1. In the absence of a receptor partner, RAMP1 exists as a homodimer and is retained in the ER. Heterodimerization with a receptor partner enables RAMP1 to transit through the Golgi and express at the cell surface, with a parallel decrease in the homodimeric form of the RAMP.[24] C-terminal deletion studies have identified a short sequence, located between Ser[141] and Thr,[144] critical for the retention of RAMP1 in the ER in the absence of receptor co-expression.[25] In contrast, studies with N-terminally epitope tagged RAMP2 or RAMP3 reveal significant cell-surface expression of these proteins even in the absence of receptor coexpression, although the extent to which this occurs appears to be influenced by the tag that is introduced and/or whether an artificial signal sequence is encoded.[17,26,27] The capacity of RAMPs 2 and 3 to be independently expressed at the cell surface appears to be dependent on their glycosylation status. As discussed above, both RAMP2 and RAMP3 are glycosylated, and mutation of these sites to prevent glycosylation abolishes receptor-independent (but not receptor-dependent) cell-surface expression.[26] Conversely, introduction of glycosylation sites into RAMP1 can allow receptor-independent cell-surface localization of this protein.

Ramps as Determinants of Ligand Binding Specificity

The classic function attributed to RAMPs, is their ability to switch the pharmacology of CLR, thus providing a novel mechanism for modulating receptor specificity; the CLR/RAMP1 complex is a high-affinity CGRP receptor but in the presence of RAMP2, CLR specificity is radically altered, the related peptide AM being recognized with the highest affinity while the affinity for CGRP is reduced ~100-fold. CLR/RAMP3 receptors are intriguing in that while AM is the highest affinity peptide, CGRP is recognized with moderate, rather than low affinity. Indeed, depending on the species and the form of CGRP (β versus α) the separation between the two peptides can be as little as 10-fold.[28]

The pharmacology of the CLR/RAMP1 complex and its sensitivity to the antagonist $CGRP_{8-37}$ closely matches an extensive literature relating to a "$CGRP_1$" receptor (as reviewed in Hay *et al.*[29]) and this is supported by data with the small molecule antagonist BIBN4096BS.[30–32] BIBN4096BS is a high-affinity antagonist of the $CGRP_1$ receptor that acts at the interface between CLR and RAMP1, with Trp^{74} of RAMP1 playing a key role its high-affinity binding.[33] Indeed, International Union of Pharmacology (IUPHAR) nomenclature has now ratified CLR/RAMP1 as the $CGRP_1$ receptor and CLR/RAMP2 and CLR/RAMP3 as AM_1 and AM_2 receptors, respectively.[31]

Similar to its functional interaction with the CLR, RAMP interaction with CTRs engenders novel receptor specificity with induction of multiple forms of amylin receptor. The pharmacological nature of the amylin receptor that is formed by RAMP and CTR co-expression appears to be quite variable and dependent upon host-cell environment, perhaps as a consequence of G protein and endogenous CLR/RAMP complement.[10,12] However, the current IUPHAR consensus is that there are three amylin receptors (AMY_{1-3}) generated by the interaction of the CTR with RAMPs 1–3, respectively.[31] Several of these subtypes have now been relatively well characterized. In terms of amylin affinity, the results are reasonably consistent across studies showing that amylin only binds with high affinity to RAMP-complexed CTRs. Similarly, human CT only appears to bind with high affinity to "free" CTRs.[34]

Some CTRs, when co-expressed with RAMPs (e.g., $AMY_{1(a)}$) have relatively high affinity for CGRP and may contribute to some aspects of reported non-$CGRP_1$ receptor pharmacology.[31,34,35] Indeed, there is evidence supporting the hypothesis that the $AMY_{1(a)}$ receptor underlies many reports of $CGRP_2$-like pharmacology.[34,35] $CGRP_2$ receptors are characterized by weak antagonism by the CGRP antagonist fragment, $CGRP_{8-37}$ and the $AMY_{1(a)}$ exhibits this behavior.[34] Furthermore, as significant amylin binding occurs in the vas deferens, a prototypical $CGRP_2$-receptor-expressing tissue, there may be *in situ* support for this hypothesis.[36] Recent pharmacological studies may now provide a basis for the separation of $CGRP_1$ and $AMY_{1(a)}$-based CGRP receptors with careful use of a spectrum of agonists and antagonists. However, in order to make robust correlations with possible molecular entities, the key receptor components need to be identified in tissues.

The full extent to which RAMPs act to modify receptor-binding specificity remains to be explored, although investigations to date with other family B receptors do not appear to support modulation of the binding site as a ubiquitous consequence of RAMP interaction. The most work to date in this regard has been performed on the VPAC1 receptor where co-expression of any of the three RAMPs failed to modify the binding affinity of a range of both high- and low-affinity agonists.[17]

Initial work using chimeras of RAMP1 and RAMP2 suggested that the RAMP N terminus was the principal domain involved in induction of CGRP and AM receptor phenotypes from the CLR.[8] Recent work suggests that, in addition to modulating phenotype, the RAMP N-terminal domain is sufficient to maintain a functional interaction between these two proteins. Expression of the RAMP1 extracellular domain (ECD) as a chimera with the transmembrane domain of the platelet-derived growth factor receptor enabled the generation of fully functional CGRP receptors that were only ~10-fold less potent than wild-type receptors.[37] Expression of the ECD alone also allowed formation of receptors that were weakly responsive to CGRP. Moreover, as seen in earlier studies,[25] the ECD alone was sufficient to promote terminal glycosylation of the CLR, and cotransfection of the CLR with the RAMP1 ECD caused secretion of the latter protein into the media.[37] Nonetheless, these data were not entirely consistent with earlier work where a marked loss of CGRP potency was found with only partial deletion of the transmembrane domain of RAMP1 (even though it was well expressed at the cell surface), which was paralleled by a significant reduction in the stability of the RAMP1-CLR heterodimer.[25]

A role for the transmembrane domain in stabilizing the interaction between RAMP and its interacting receptors is supported by studies of the interaction of RAMPs with CTRs, where differences in strength between RAMP1 and two interactions are evident. Analysis of N-terminal domain-swap chimeras between RAMP1 and RAMP2 provided supportive evidence for the predominant role of the RAMP ECD in defining receptor phenotype. In contrast, the level of amylin binding induced with wild-type RAMPs was paralleled by the chimeras according to the transmembrane domain/C terminus present.[13] Recent mutagenesis and deletion studies on the N terminus of RAMP1 have identified a stretch of aromatic residues within RAMP1 that are likely to be important for interaction between RAMP1 and the CLR, independent of the induction of CGRP phenotype. Mutation of Phe[93], Tyr[100] or Phe[101] and to a lesser extent, His[97] to alanine led to parallel losses in cell-surface expression of RAMP1 and CGRP binding, but did not alter the EC_{50} value for CGRP-dependent cyclic adenosine 3′,5′-phosphate (cAMP) response.[38] In contrast, mutation of Leu[94] to alanine led to increased cell-surface expression and CGRP binding, suggesting that leucine in this position provides steric hindrance to the interaction of RAMP1 and CLR. These data are supportive of components of the RAMP1 ECD being important for stable interaction between RAMP1 and CLR.

For RAMP2 interaction with the CLR to form the AM_1 receptor, at least one important region has been identified, comprising residues 77-101; amino acids 86-92 within this region play a particularly significant role as defined by deletion experiments.[27] Deletion of the corresponding region in RAMP3 (residues 59-65) had a similar effect on generation of a functional phenotype[27] and, as there is no sequence identity in this region of RAMPs, the results suggest either an important structural role in RAMPs or an allosteric effect of this region on CLR conformation. For RAMP1, deletion of the amino acids 101-103 abolishes induction of $CGRP_1$ receptor phenotype from the CLR, but substitution of individual amino acids in this segment with alanine do not alter the potency of CGRP,[38] again implying possible disruption of a required conformation of the protein.

RAMPs in Receptor Regulation

In an initial study it was observed that the type of RAMP did not affect CLR internalization or its targeting to a degradative pathway,[18] although more recent work indicates that this is not always the case (see below). CLR and RAMPs were co-localized and were internalized together following agonist stimulation and it is likely that they remain complexed throughout the life cycle of the proteins. Subsequent investigation of CL/RAMP1 complexes yielded similar results and further showed that the internalization was probably β-arrestin and dynamin dependent.[39]

Recent investigations of the trafficking properties of RAMP2 and RAMP3 have highlighted the significance of sequence divergence in the C-terminal tail of the proteins. Unlike RAMP1 or RAMP2, the C-terminal four amino acids of RAMP3 comprise a PDZ (Post synaptic density, Disc large protein and ZO = 1 proteins)-binding domain.[4] It appears that this domain may be responsible for the unique trafficking properties that have recently been reported for AM_2 receptors, which are able to recycle as opposed to AM_1 receptors that do not possess a PDZ domain and are targeted to a degradative pathway.[40] In HEK-293 cells all CLR/RAMP complexes are subject to targeting for degradation following internalization.[18] However, overexpression of N-ethylmalemide Sensitive Factor (NSF), selectively switched CLR/RAMP3 toward recycling.[40] A similar function of RAMP3 in AM receptor recycling was also shown in rat mesangial cells and Rat-2 fibroblasts.[40] Furthermore, only CLR/RAMP3 internalization may be regulated by NHERF (Na^+/H^+ Exchange Regulatory Factor)-1 binding to the RAMP3 PDZ domain.[41] These differential properties of RAMP2 and RAMP3 may underlie, at least in part, the evolutionary need for two AM receptors, which otherwise appear to have many similarities. Furthermore, such differences could contribute to the reported inconsistency in the literature of whether or not AM receptors are subject to agonist-stimulated desensitization (reviewed by Hay et al.[42]).

RAMPS as Modifiers of Receptor Signaling

The VPAC1 receptor strongly translocates all three RAMPs to the cell surface but its pharmacology, in terms of agonist binding, does not appear to be modified by their presence.[17] In contrast, RAMP2 overexpression enabled augmentation of VPAC1 receptor-mediated phosphatidylinositol (PI) hydrolysis relative to cAMP production, the latter being unaltered. The potency of the response (EC_{50} of VIP) was unchanged but the maximal PI hydrolysis response was elevated in the presence of RAMP2 (FIG. 3). It has been speculated that this may reflect a change in compartmentalization of the receptor signaling complex.[17] Such augmentation was not evident for the interaction of the VPAC1 receptor with RAMP1 or RAMP3; in these cases the outcome of heterodimerization may be more subtle or involve the modification of different receptor parameters such as trafficking.

There has been relatively little work examining the signaling consequences of CTR-RAMP interaction. However, no changes in maximal capacity for PI hydrolysis or cAMP production were observed with different CTR-RAMP

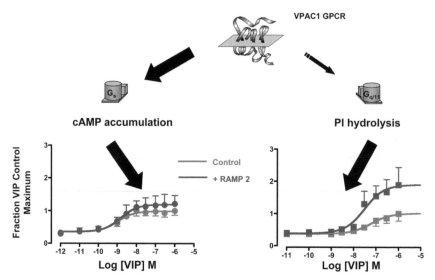

FIGURE 3. Modulation of VPAC1 receptor signaling by RAMP2. The VPAC1 receptor couples to multiple signaling pathways. VIP causes a potent stimulation of cAMP accumulation (left hand panel) and a weaker activation of PI hydrolysis (right hand panel). Co-expression of the receptor with RAMP2 leads to a specific enhancement of agonist-mediated PI hydrolysis without modulating cAMP production. (Adapted from Christopoulos et al.[17])

combinations in comparison to the CTR alone.[17] More recently, the relative coupling of AMY receptors to cAMP generation, intracellular Ca^{2+} mobilization and extracellular signal regulated kinase (ERK) 1/2 activation in transfected COS-7 cells has been evaluated. In contrast to the strong (>20-fold) increase in potency of AMY to generate cAMP, there was only a weak increase in AMY potency for Ca^{2+} and ERK1/2 signaling (less than fivefold),[43] suggesting that the CTR/RAMP-based AMY receptors are relatively less-well coupled to G_q than the isolated CTR. As discussed above, it is also evident that the generation of high-affinity AMY binding for CTR-RAMP2 complexes is particularly sensitive to experimental conditions. This relates to the splice variant of the CTR used and cellular background[12] and most likely reflects differences in coupling to signal transduction apparatus as overexpression of $G\alpha$ protein subunits can modulate induction of the AMY phenotype.[44]

For CLR-based receptors, efficient signalling (at least for Gs-mediated cAMP accumulation) also requires expression of an additional protein termed *receptor component protein (RCP)*.[45] Antisense knockdown of RCP expression strongly attenuates ligand-induced cAMP production of both CGRP and AM receptors.[46] RCP interacts with the second intracellular loop of CLR and may influence the stability of CL-RAMP complexes,[47] although it is unclear if there is a direct association of RCP with RAMPs or whether RCP influences coupling of receptors to alternate effector pathways.

CONCLUSION

The discovery of RAMPs has led to evolution of our understanding of how receptor diversity is implemented. Recent data now supports a broader role for RAMPs through interaction with additional GPCRs to CLR and CTR. Furthermore the spectrum of receptor function modified by RAMP interaction has been extended to include modulation of signalling efficacy and receptor regulation in addition to their more recognized action as chaperones and phenotypic modifiers. In common with ever-increasing examples of GPCR homo- and heterodimerization, it is no longer possible to consider GPCRs as isolated units. The challenge is now to understand the properties of each of these types of multimeric proteins, particularly their relevance outside of cultured cell systems and to explore these for potential discovery of selective pharmaceutical intervention that these complexes are likely to allow.

ACKNOWLEDGMENTS

Patrick M. Sexton and Arthur Christopoulos are Senior Research Fellows of the National Health and Medical Research Council of Australia (NHMRC). Maria Morfis is a NHMRC Dora Lush Biomedical Research Scholar. This work is funded in part by NHMRC project grant 299810 and the Ian Potter Neuropeptide Laboratory.

REFERENCES

1. FOORD, S.M. 2002. Receptor classification: post genome. Curr. Opin. Pharmacol. **2:** 561–566.
2. FOORD, S.M., S. JUPE & J. HOLBROOK. 2002. Bioinformatics and type II G-protein-coupled receptors. Biochem. Soc. Trans. **30:** 473–479.
3. SEXTON, P.M. Ed. 2002. Receptors and Channels, Special Issue on Class II G Protein Coupled Receptors. **8:** 135–255.
4. MCLATCHIE, L.M., N.J. FRASER, M.J. MAIN, et al. 1998. RAMPs regulate the transport and ligand specificity of the calcitonin-receptor-like receptor. Nature **393:** 333–339.
5. HAY, D.L., D.R. POYNER & P.M. SEXTON. 2005. GPCR modulation by RAMPs. Pharmacol. Ther., **109:** 173–197.
6. ELSHOURBAGY, N.A., J.E. ADAMOU, A.M. SWIFT, et al. 1998. Molecular cloning and characterization of the porcine calcitonin gene-related peptide receptor. Endocrinology **139:** 1678–1683.
7. HALL, J.M. & D.M. SMITH. 1998. Calcitonin gene-related peptide—a new concept in receptor-ligand specificity. Trends Pharmacol. Sci. **19:** 303–305.
8. FOORD, S.M. & F.H. MARSHALL. 1999. RAMPs: accessory proteins for seven trans-membrane domain receptors. Trends Pharmacol. Sci. **20:** 184–187.
9. FRASER, N.J., A. WISE, J. BROWN, et al. 1999. The amino terminus of receptor activity modifying proteins is a critical determinant of glycosylation state and ligand binding of calcitonin receptor-like receptor. Mol. Pharmacol. **55:** 1054–1059.
10. CHRISTOPOULOS, G., K.J. PERRY, M. MORFIS, et al. 1999. Multiple amylin receptors arise from receptor activity-modifying protein interaction with the calcitonin receptor gene product. Mol. Pharmacol. **56:** 235–242.
11. MUFF, R., N. BUHLMANN, J.A. FISCHER, et al. 1999. An amylin receptor is revealed following co-transfection of a calcitonin receptor with receptor activity modifying proteins-1 or -3. Endocrinology **140:** 2924–2927.
12. TILAKARATNE, N., G. CHRISTOPOULOS, E.T. ZUMPE, et al. 2000. Amylin receptor phenotypes derived from human calcitonin receptor/RAMP coexpression exhibit pharmacological differences dependent on receptor isoform and host cell environment. J. Pharmacol. Exp. Ther. **294:** 61–72.
13. ZUMPE, E.T., N. TILAKARATNE, N.J. FRASER, et al. 2000. Multiple ramp domains are required for generation of amylin receptor phenotype from the calcitonin receptor gene product. Biochem. Biophys. Res. Commun. **267:** 368–372.
14. KUESTNER, R.E., R. ELROD, F.J. GRANT, et al. 1994. Cloning and characterization of an abundant subtype of the human calcitonin receptor. Mol. Pharmacol. **46:** 246–255.
15. GORN, A.H., H.Y. LIN, M. YAMIN, et al. 1992. Cloning, characterization, and expression of a human calcitonin receptor from an ovarian carcinoma cell line. J. Clin. Invest. **90:** 1726–1735.
16. MOORE, E.E., R.E. KUESTNER, S.D. STROOP, et al. 1995. Functionally different isoforms of the human calcitonin receptor result from alternate splicing of the gene transcript. Mol. Endocrinol. **9:** 959–968.

17. CHRISTOPOULOS, A., G. CHRISTOPOULOS, M. MORFIS, *et al.* 2003. Novel receptor partners and function of receptor activity-modifying proteins. J. Biol. Chem. **278:** 3293–3297.
18. KUWASAKO, K., Y. SHIMEKAKE, M. MASUDA, *et al.* 2000. Visualization of the calcitonin receptor-like receptor and its receptor activity-modifying proteins during internalization and recycling. J Biol Chem **275:** 29602–29609.
19. HILAIRET, S., S.M. FOORD, F.H. MARSHALL, *et al.* 2001. Protein-protein interaction and not glycosylation determines the binding selectivity of heterodimers between the calcitonin receptor- like receptor and the receptor activity-modifying proteins. J. Biol. Chem. **276:** 29575–29581.
20. ALDECOA, A., R. GUJER, J.A. FISCHER, *et al.* 2000. Mammalian calcitonin receptor-like receptor/receptor activity modifying protein complexes define calcitonin gene-related peptide and adrenomedullin receptors in Drosophila Schneider 2 cells. FEBS Lett. **471:** 156–160.
21. BULENGER, S., S. MARULLO & M. BOUVIER. 2005. Emerging role of homo- and heterodimerization in G-protein-coupled receptor biosynthesis and maturation. Trends Pharmacol. Sci. **26:** 131–137.
22. LEUTHAUSER, K., R. GUJER, A. ALDECOA, *et al.* 2000. Receptor-activity-modifying protein 1 forms heterodimers with two G- protein-coupled receptors to define ligand recognition. Biochem. J. **351:** 347–351.
23. BOUSCHET, T., S. MARTIN & J.M. HENLEY. 2005. Receptor activity modifying proteins are required for forward trafficking of the calcium sensing receptor to the plasma membrane. J. Cell Sci. **148:** 4709–4720.
24. UDAWELA, M., D.L. HAY & P.M. SEXTON. 2004. The receptor activity modifying protein family of G protein coupled receptor accessory proteins. Semin. Cell Dev. Biol. **15:** 299–308.
25. STEINER, S., R. MUFF, R. GUJER, *et al.* 2002. The transmembrane domain of receptor-activity-modifying protein 1 is essential for the functional expression of a calcitonin gene-related peptide receptor. Biochemistry **41:** 11398–11404.
26. FLAHAUT, M., B.C. ROSSIER & D. FIRSOV. 2002. Respective roles of calcitonin receptor-like receptor (CRLR) and receptor activity-modifying proteins (RAMP) in cell surface expression of CRLR/RAMP heterodimeric receptors. J. Biol. Chem. **277:** 14731–14737.
27. KUWASAKO, K., K. KITAMURA, K. ITO, *et al.* 2001. The seven amino acids of human RAMP2 (86) and RAMP3 (59) are critical for agonist binding to human AM receptors. J. Biol. Chem. **276:** 49459–49465.
28. HAY, D.L., S.G. HOWITT, A.C. CONNER, *et al.* 2003. CL/RAMP2 and CL/RAMP3 produce pharmacologically distinct AM receptors: a comparison of effects of AM22-52, CGRP8-37 and BIBN4096BS. Br. J. Pharmacol. **140:** 477–486.
29. HAY, D.L., A.C. CONNER, S.G. HOWITT, *et al.* 2004. The pharmacology of CGRP responsive receptors in cultured and transfected cells. Peptides **25:** 2019–2026.
30. DOODS, H., G. HALLERMAYER, D. WU, *et al.* 2000. Pharmacological profile of BIBN4096BS, the first selective small molecule CGRP antagonist. Br. J. Pharmacol. **129:** 420–423.
31. POYNER, D.R., P.M. SEXTON, I. MARSHALL, *et al.* 2002. International Union of Pharmacology. XXXII. The Mammalian Calcitonin Gene-Related Peptides, AM, AMY, and Calcitonin Receptors. Pharmacol. Rev. **54:** 233–246.
32. MORFIS, M., A. CHRISTOPOULOS & P.M. SEXTON. 2003. RAMPs: 5 years on, where to now? Trends Pharmacol. Sci. **24:** 596–601.

33. MALLEE, J.J., C.A. SALVATORE, B. LEBOURDELLES, *et al.* 2002. Receptor activity-modifying protein 1 determines the species selectivity of non-peptide CGRP receptor antagonists. J. Biol. Chem. **277:** 14294–14298.
34. HAY, D.L., G. CHRISTOPOULOS, A. CHRISTOPOULOS, *et al.* 2005. Pharmacological discrimination of calcitonin receptor - receptor activity modifying protein complexes. Mol. Pharmacol. **67:** 1655–1665.
35. KUWASAKO, K., Y.-N. CAO, Y. NAGOSHI, *et al.* 2004. Characterization of the human calcitonin gene-related peptide receptor subtypes associated with receptor activity-modifying proteins. Mol. Pharmacol. **65:** 207–213.
36. POYNER, D.R., G.M. TAYLOR, A.E. TOMLINSON, *et al.* 1999. Characterization of receptors for calcitonin gene-related peptide and adrenomedullin on the guinea-pig vas deferens. Br. J. Pharmacol. **126:** 1276–1282.
37. FITZSIMMONS, T.J., X. ZHAO & S.A. WANK. 2003. The extracellular domain of receptor activity-modifying protein 1 is sufficient for calcitonin receptor-like receptor function. J. Biol. Chem. **278:** 14313–14320.
38. KUWASAKO, K., K. KITAMURA, Y. NAGOSHI, *et al.* 2003. Identification of the human receptor activity-modifying protein 1 domains responsible for agonist binding specificity. J. Biol.Chem. **278:** 22623–22630.
39. HILAIRET, S., C. BELANGER, J. BERTRAND, *et al.* 2001. Agonist-promoted internalization of a ternary complex between calcitonin receptor-like receptor, receptor activity-modifying protein 1 (RAMP1), and beta-arrestin. J. Biol. Chem. **276:** 42182–42190.
40. BOMBERGER, J.M., N. PARAMESWARAN, C.S. HALL, *et al.* 2005. Novel function for receptor activity modifying proteins (RAMPs) in post-endocytic receptor trafficking. J. Biol. Chem. **280:** 9297–92307.
41. BOMBERGER, J.M., W.S. SPIELMAN, C.S. HALL, *et al.* 2005. RAMP isoform-specific regulation of AM receptor trafficking by NHERF-1. J. Biol. Chem. **280:** 23926–23935.
42. HAY, D.L., D.R. POYNER & D.M. SMITH. 2003. Desensitisation of AM and CGRP receptors. Reg. Peptides **112:** 139–145.
43. MORFIS, M., T.D. WERRY, A. CHRISTOPOULOS, *et al.* 2005. Calcitonin and amylin signalling via calcitonin receptor and calcitonin receptor-RAMP complexes. Presented at the Gordon Research Conference on Molecular Pharmacology, Il Ciocco, Italy, May 10.
44. SMYTH, K., N. TILAKARATNE, M. MORFIS, *et al.* Influence of Gα protein subtype and expression level on receptor phenotypes generated from calcitonin receptor and RAMP interaction.[Abstract P1–26]. Proc. Aust. Soc. Clin. Exp. Pharmacol. Toxicol. **9:** 75.
45. LUEBKE, A.E., G.P. DAHL, B.A. ROOS, *et al.* 1996. Identification of a protein that confers calcitonin gene-related peptide responsiveness to oocytes by using a cystic fibrosis transmembrane conductance regulator assay. Proc. Natl. Acad. Sci. USA **93:** 3455–3460.
46. EVANS, B.N., M.I. ROSENBLATT, L.O. MNAYER, *et al.* 2000. CGRP-RCP, a novel protein required for signal transduction at calcitonin gene-related peptide and AM receptors. J. Biol. Chem. **275:** 31438–31443.
47. LOISEAU, S.C. & I.M. DICKERSON. 2004. CGRP receptor component protein (RCP); a multi-protein complex required for G protein-coupled signal transduction. Neuropeptides **38:** 115.

The Three-Dimensional Structure of the N-Terminal Domain of Corticotropin-Releasing Factor Receptors

Sushi Domains and the B1 Family of G Protein–Coupled Receptors

MARILYN H. PERRIN,[a] CHRISTY R. R. GRACE,[b] ROLAND RIEK,[b] AND WYLIE W. VALE[a]

[a] Clayton Foundation Laboratories for Peptide Biology, The Salk Institute, La Jolla, California 92037, USA

[b] Structural Biology Laboratory, The Salk Institute, La Jolla, California 92037, USA

ABSTRACT: The corticotropin-releasing factor (CRF) receptors, CRF-R1 and CRF-R2, belong to the B1 subfamily of G protein–coupled Receptors (GPCRs), including receptors for secretin, growth hormone-releasing hormone (GHRH), vasoactive intestinal peptide (VIP), pituitary adenylate cyclase-activating polypeptide (PACAP), calcitonin, parathyroid hormone (PTH), glucagon, and glucagon-like peptide-1 (GLP-1). The peptide ligand family comprises CRF, Ucn 1, 2, and 3. CRF plays the major role in integrating the response to stress. Additionally, the ligands exhibit many effects on muscle, pancreas, heart, and the GI, reproductive, and immune systems. CRF-R1 has higher affinity for CRF than does CRF-R2 while both receptors bind Ucn 1 equally. CRF-R2 shows specificity for Ucns 2 and 3. A major binding domain of the CRFRs is the N terminus/first extracellular domain (ECD1). Soluble proteins corresponding to the ECD1s of each receptor bind CRF ligands with nanomolar affinities. Our three-dimentional (3D) nuclear magnetic resonance (NMR) structure of a soluble protein corresponding to the ECD1 of CRF-R2β (1) identified its structural fold as a Sushi domain/short consensus repeat (SCR), stabilized by three disulfide bridges, two tryptophan residues, and an internal salt bridge (Asp65–Arg101). Disruption of the bridge by D65A mutation abrogates ligand recognition and results in loss of the well-defined disulfide pattern

Address for correspondence: Marilyn H. Perrin, The Clayton Foundation Laboratories for Peptide Biology, The Salk Institute, 10010 N. Torrey Pines Road, La Jolla, CA 92037. Voice: 858-453-4100, ext. 1497; fax: 858-552-1546.
e-mail: perrin@salk.edu

Ann. N.Y. Acad. Sci. 1070: 105–119 (2006). © 2006 New York Academy of Sciences.
doi: 10.1196/annals.1317.065

and Sushi domain structure. NMR analysis of the ECD1 in complex with astressin identified key amino acids involved in ligand recognition. Mutation of some of these residues in the full-length receptor reduces its affinity for CRF ligands. A structure-based sequence comparison shows conservation of key amino acids in all the B1 subfamily receptors, suggesting a corresponding conservation of a Sushi domain structural fold of their ECD1s.

KEYWORDS: CRF receptor; structure; ECD1; Sushi domains; B1 GPCR

INTRODUCTION

The isolation from the hypothalamus of corticotropin-releasing factor (CRF), was based on its function as an ACTH secretagogue.[1] CRF is the primary activator of the hypothalamic-pituitary-adrenal axis (HPA) and serves to integrate not only the endocrine, but also the autonomic and behavioral responses to stress. Subsequently, the CRF ligand family has been increased by the cloning of the three related peptides urocortins (UCNs) 1, 2, and 3, the latter two also known as stresscopin-related peptide and stresscopin, respectively.[2–5] The roles of the CRF ligand family continue to expand and currently include actions on the GI tract, pancreas, and muscle as well as on the reproductive, cardiovascular, and immune systems.

The actions of the ligands are initiated by binding and activation of CRF receptors. In rodents there is a single type I receptor, CRF-R1,[6–8] and two forms of the type II receptor, CRF-R2α and CRF-R2β, arising from alternative splicing.[9–12] In humans, two splice variants for CRF-R1 have been cloned[6] and transcripts for other splice variants have been reported[13]; three splice variants for CRF-R2 have been cloned.[14,15] Receptors orthologous to those in mammals have been identified in amphibia, fish, and birds.

The ligands and receptors display unique mutual specificities: CRF binds with higher affinity to CRF-R1 than to CRF-R2; Ucn 1 binds with equally high affinity to both receptor types; Ucn 2 and Ucn 3 are highly specific for CRF-R2.

The CRF system also includes two soluble binding proteins, CRF-BP[16] and sCRF-R2α,[17] the first being encoded by a distinct gene, while the second results from alternative splicing of the CRF-R2α gene. The CRF-BP binds CRF and Ucn 1 with high affinity. Interestingly, the ligand specificity of sCRF-R1α is more like that of CRF-R1 than like that of CRF-R2 in that CRF and Ucn 1 are bound with higher affinity than is Ucn 2 and Ucn 3 is bound with very low affinity.

The expression of CRF-R1 is widespread in the central nervous system as well as many peripheral tissues such as skin, gonads, GI tract, adrenal, and immune system.[18] The type I receptor is the major anterior pituitary receptor mediating the CRF-stimulated release of ACTH and thus the activation of the

HPA axis. Transgenic mice in which CRF-R1 is disabled exhibit a blunted stress response and have been termed "mellow mice."[19]

In the rodent, the expression of CRF-R2α is largely confined to the central nervous system; CRF-R2β is expressed in peripheral tissues such as epididymis, GI tract, heart, vasculature, and skeletal muscle.[18] The type II receptor expressed in the heart mediates the effects of the urocortins on cardioprotection[20] and a putative type II receptor in the pancreas is presumed to mediate the effects of urocortin on pancreatic hormone release.[21]

The CRF receptors belong to the B1 subfamily of G protein–coupled receptors (GPCRs) known as the secretin family which includes receptors for parathyroid hormone (PTH), calcitonin, calcitonin gene-related peptide, adrenomedullin, glucagon, glucagon-like peptide-1 (GLP-1), growth hormone-releasing hormone (GHRH), pituitary adenylate cyclase-activating polypeptide (PACAP), glucose-dependent insulinotropic peptide, and vasoactive intestinal peptide (VIP). The CRF receptors couple to G_s with subsequent activation of adenylate cyclase producing an increase in intracellular of cyclic adenosine 3',5'-phosphate (cAMP) and protein kinase A activity. The receptors also couple to G_q, and activation of protein kinase C, resulting in phosphatidyl inositol hydrolysis and an increase in intracellular calcium and diacyl glycerol. In some tissues there is also ligand-stimulated activation of mitogen-activated protein (MAP) kinase.

Distinct Regions of Receptors and Ligands Govern Binding and Activation

CRF Ligands

The CRF peptide ligands consist of 38–40 amino acids. The observation that deletion of the first 12 amino acids converts an agonist into an antagonist led to the development of high-affinity CRF antagonists such as astressin[22] and antisauvagine-30.[23] The development of the antagonists suggested that the C-terminal region of the ligand is not involved in signal transduction but rather, binds to the receptor and blocks binding of an agonist and subsequent signal transduction. Recent data on even more extensively N-truncated astressin-like antagonists[24] leads to the conclusion that the C-terminal 12 residues are sufficient for high-affinity binding.[25]

In order to investigate the region of the peptide involved in receptor signaling, a tethered-peptide receptor was created in which the N-terminal or first extracellular domain (ECD1) of CRF-R1 was replaced by the first 16 amino acids of CRF.[26] When expressed in mammalian cells, this chimeric receptor generated a continuous signal as measured by the accumulation of intracellular cAMP. The sempiternal signaling of this tethered-ligand receptor suggests that the N-terminal region of the ligand comprising, for example, the first 16 amino acids of CRF, is sufficient for activating the adenylate cyclase signaling pathway. Interestingly, a similar tethered-peptide PTH receptor is also constitutively active.[27,28]

CRF Receptors

The two types of CRF receptors are highly homologous and display ~70% sequence conservation in many regions, for example, the third intracellular loop. The majority of sequence differences occur in the extracellular domains, notably in the ECD1. Many studies using chimeric or mutant receptors have shown that the extracellular domains contribute to the binding and ligand selectivity[29–31] and further, that the ECD1 constitutes a major peptide-binding domain.[30,32–36]

The juxtamembrane receptor domain, that is, the receptor excluding the ECD1, also binds peptide ligands, albeit with significantly lower affinity.[35] The data from the constitutively active tethered-ligand receptor show that the juxtamembrane domain is involved in transducing the signaling response.

Soluble Proteins Corresponding to ECD1 of CRF Receptors

Further support for the key role of the ECD1 as a major binding domain derives from data showing that soluble proteins corresponding to the ECD1s of either CRF-R1 or CRF-R2 bind CRF ligands with nanomolar affinities.[37–39] Transfection of COS cells with cDNA encoding amino acids 1-119 of CRF-R1 resulted in secretion of the corresponding protein into the medium. Following enrichment of the protein by immunoaffinity chromatography, N-terminal sequencing revealed that the first amino acid of the secreted protein is Ser24. This result confirmed the proposal that the first 23 amino acids serve as a signal peptide. The soluble protein corresponding to the ECD1 of CRF-R1 binds astressin and Ucn 1 with nanomolar affinity ($K_d \sim 10$ nM).[37]

In order to obtain milligram quantities of the soluble protein, a bacterial expression system was chosen. The cDNA encoding amino acids 24-119 of hCRF-R1, ECD1-CRF-R1, (expressed as a thioredoxin fusion protein, containing an S-tag sequence for purification purposes and a thrombin cleavage site) was used to transform the Origami strain of *E. coli*. Following thrombin cleavage, the expressed protein was enriched by affinity chromatography and purified by high-performance liquid chromatography (HPLC). Biochemical characterization, using tryptic digestion and mass spectrometry, showed that the six cysteines form disulfide bonds with the following pattern: 1–3, 2–5, and 4–6.[37] This same disulfide pattern is observed in the soluble proteins corresponding to the ECD1s of the receptors for PTH[40] and GLP-1,[41] both of which are B1 receptor family members.

In similar fashion, a soluble protein, ECD1-CRF-R2β, comprising amino acids 27-133 of mCRF-R2β was expressed and purified from *E. coli*. Biochemical characterization showed that the disulfide pattern is the same as that determined for ECD1-CRF-R1.[39] The ECD1-CRF-R2β binds astressin, Ucn 1, and Ucn 2 with nanomolar affinities ($K_d \sim 10$ nM),[39] CRF with lower affinity (Kd ~ 100 nM), and Ucn 3 with even lower affinity (Kd > 200 nM). Because preliminary experiments showed that the yield of ECD1-CRF-R2β was greater than that of ECD1-CRF-R1, the former protein was chosen for structural studies.

Based on the activity of the tethered-ligand receptor chimera and taking into account data on the ECD1 as a major binding domain, a model was proposed for CRF receptor activation.[26] This model, which has now been extended to apply to the activation of all B1 receptors,[42] envisions receptor activation to involve two steps: First, the ECD1 of the receptor captures the ligand through high-affinity binding of its C-terminal region. Second, the bound ligand, now positioned in an appropriate proximity, presents its N-terminal region to the juxtamembrane domain of the receptor, initiating signal transduction.

Three-Dimensional Structure of the ECD1 of CRF-R2β

The importance of the ECD1 for CRF receptor-ligand recognition serves as the impetus for obtaining the three-dimensional (3D) structure of that domain. In order to achieve this, nuclear magnetic resonance (NMR) spectroscopy was chosen because of the frequent difficulties in obtaining crystals of large proteins. For NMR spectroscopy, milligram amounts of protein, isotopically labeled with ^{15}N and ^{13}C, are required. For the structural studies, a protein corresponding to amino acids 39-133 of mCRF-R2β was expressed in, and purified from *E. coli*, as described above, using minimal media supplemented with $(^{15}NH_4)_2SO_4$ and ^{13}C-D-glucose.[43]

Structural Fold of ECD1-CRF-R2β

Backbone assignments were obtained from triple resonance experiments while nuclear Overhauser enhancement spectroscopy (NOESY) experiments yielded the distance restraints.[43] The structure determination revealed the presence of two antiparallel pairs of β-sheets involving residues 61–66, 69–72, 79–84, and 98–102. The polypeptide fold is stabilized by three disulfide bonds between cysteine residues 45–70, 60–103, and 84–118 and by a core composed of an internal salt bridge between Asp65 and Arg101 sandwiched between two typtophan residues (71 and 109). The N and C terminii flanking the core are disordered, whereas the central region is highly ordered in all the conformers. These structural properties are shown in FIGURE 1A, in which the lowest energy conformer is presented as a ribbon diagram. The representation in FIGURE 1B is a superposition the 20 conformers that are consistent with the NMR data. The structure is a globular one with well-defined secondary structure and a

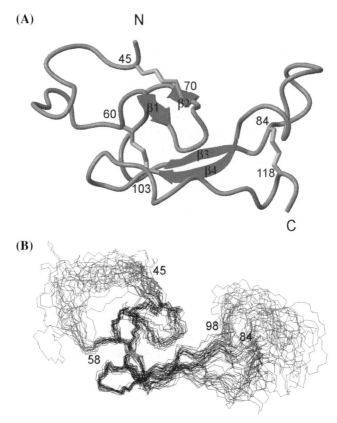

FIGURE 1. 3D structure of the ECD1-CRF-R2β. (**A**) Ribbon diagram of the lowest energy conformer. (**B**) Superposition of 20 conformers. Only amino acids 44-119 are shown.

large compact hydrophobic core. The disulfide pattern is the same as that determined biochemically for the soluble proteins corresponding to ECD1s of both CRF-R1 and CRF-R2β.

The core is surrounded by a layer of residues, Thr69, Val80, and Arg82, that show strict conservation and by other residues, Thr63, Ser74, and Ile67, which show conservative substitutions among the CRF receptors (FIG. 2 A). The other conserved residues are Pro72 and Pro83, which probably serve to end the β-strands and Gly77, Asn106, and Gly107, which are in the hinge regions and may be important for their relative orientation. Another set of conserved residues, Gly90, Tyr93, Asn94, and Thr96, are in the disordered loop between the β3 and β4 strands. The disordered loop comprising residues 39-58 is highly variable in sequence (FIG. 2 A) among the CRF receptor family members.

Analysis of the NMR data by the Dali program identified the structural fold of the highly ordered region as a short consensus repeat (SCR), or Sushi

(A)

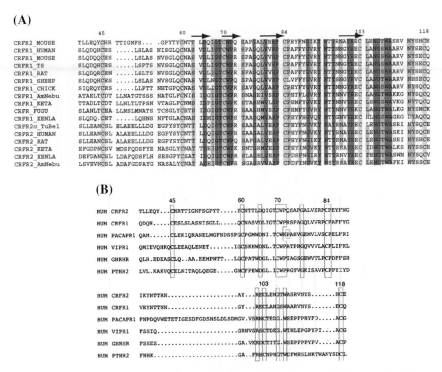

FIGURE 2. Sequence comparisons. **(A)** Sequence alignment of the ECD1s of CRF receptors. **(B)** Sequence alignment of the ECD1s of related B1 receptors.

domain, sometimes also called a Complement Control Protein, CCP, module.[44] The β-sheets and the four cysteines are characteristic of a Sushi domain, a well-known structural motif found more than 140 times in over 20 proteins, including at least 12 in the complement system, such as CD55. The 3D structure closest to that of ECD1-CRF-R2β is the first Sushi domain of β2-glycoprotein.[45] The Sushi domains are widely recognized for their roles in protein-protein interactions.

Peptide Hormone Binding Surface on the ECD1

In order to obtain structural information on the binding interface, NMR chemical-shift perturbation experiments were used to study the interaction of astressin with the ECD1-CRF-R2β. There was a change in the NMR spectrum of the ECD1-CRF-R2β bound to astressin compared to the spectrum of the free ECD1-CRF-R2β (Fig. 3 A). The largest chemical-shift changes were seen in the segments that include residues 67–69, 90–93, 102–103, and 112–116. These residues are situated between the edge of the tip of the first β-sheet and

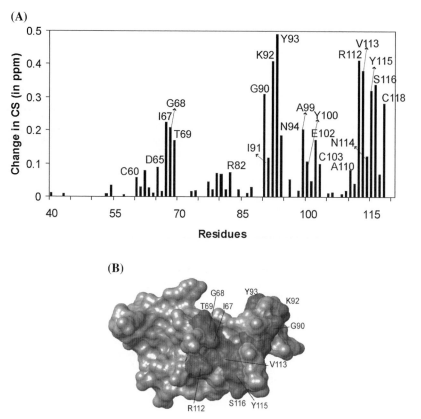

FIGURE 3. (A) The changes in the normalized chemical shifts following binding of astressin to the ECD1-CRF-R2β. (B) A surface representation of the ECD1-CRF-R2β showing the proposed residues involved in binding astressin.

the edge of the palm of the second β-sheet as shown in a surface representation in FIGURE 3 B. This surface of residues is proposed to constitute the peptide-binding interface. The chemical shift changes in the disordered C-terminal loop region, including residues 85–98 represent binding-induced folding.

Other CRF Receptors

Sequence comparison of the ECD1s of CRF-R1 and CRF-R2β from different species (FIG. 2 A) shows the conservation of many residues that play key roles in the structure. Many of those residues are the same ones that are proposed to interact in the binding of astressin (FIG. 3 B), so that the surface defined by those residues overlaps to a large extent with the binding surface. The coincidence of these two surfaces suggests that the ligand-binding surface is conserved across all the CRF receptors.

TABLE 1. Effect of mutation of residues in the ECD1 in the full-length receptor on the binding of astressin

Receptor	K_d(Astressin) nM	$\Delta\delta$ (ppm)
CRF-R2β	1.1 (0.7–2.0)	—
[K92Q]CRF-R2β	2.6 (1.9–3.6)	0.4
[R112E]CRF-R2β	8.0 (4.3–17.4)	0.4
[I67E]CRF-R2β	130 (85–191)	0.2
[Y115R]CRF-R2β	>200	0.3
[D65A]CRF-R2β	>200	0.1

K_d = dissociation constant; $\Delta\delta$ = change in NMR chemical shift of the indicated residue subsequent to binding of astressin to the ECD1-CRF-R2β.

Full-Length Receptor

The residues I67, K92, R112, Y115 show the largest NMR chemical-shift changes following binding of astressin (FIG. 2 A). These residues are proposed to lie on the surface of the ECD1. What relevance is this binding surface for ligand recognition by the full-length receptor? In order to address this question, those residues were mutated in the full-length CRF-R2β and the effect of the mutations on the affinity for astressin was determined. For example, the mutation R112E in the full-length receptor produced in a sevenfold decrease in the affinity for astressin, while the mutations I67E and Y115R result in >100-fold decrease in affinity for astressin. These data are summarized in TABLE 1. The decrease in affinities, as a result of the mutations, suggests that the binding surface in the full-length receptor is similar, if not identical to that determined for the isolated ECD1.

Role of Salt Bridge

Interestingly, even though Asp65 is not a surface residue in the ECD1 structure, the mutation D65A has a large effect on the binding of astressin to the full-length receptor (TABLE 1). In order to understand the reason for the importance of this salt bridge, NMR and biochemical methods were used to study the structure and ligand recognition of ECD1-CRF-R2β in which the salt bridge was disrupted. The protein, [D65A]ECD1-CRF-R2β (^{15}N-labeled), in which the mutation is presumed to disrupt the salt bridge, was expressed and purified from *E. coli*. The NMR spectrum of the mutant protein showed a collapse of the wild-type spectrum into the random coil chemical-shift region indicating a large conformational change from the well-defined Sushi domain structure toward that of a random coil state (Perrin *et al.*, to be published). One possible explanation is that the structure of the mutant is destabilized into a metastable state comprising many different conformations that exchange with one another over the time scale of NMR measurements. Biochemical characterization of the disulfide arrangement of [D65A]ECD1-CRF-R2β using mass spectrometric methods, showed no unique disulfide arrangement. Finally, [D65A]ECD1-CRF-R2β displayed no high-affinity binding to astressin. Thus, the internal salt-bridge in the ECD1 of the CRF receptor appears to play mainly a structural role and its presence is necessary for the high-affinity ligand recognition of both the isolated ECD1 and of the full-length receptor.

Model for CRF Receptor Activation

The 3D structure provides insight into the proposed mechanism of receptor activation. It is suggested from the structures of the ECD1 and of astressin 2B (Riek *et al.*, to be published) that the peptide interacts with the binding surface as shown in FIGURE 4 A. The structure of the ECD1 shows a cluster of positive charges (Arg47, Arg82, and Arg97) on the "back" of the protein's surface. Examination of the sequences of the extracellular domains 2–4 shows that they contain a cluster of negative charges. These considerations provide a rationale for the arrangement, shown in FIGURE 4 B, in which the positively charged back surface of the ECD1 is oriented towards the negatively charged extracellular domains 2–4. A consideration of the relative orientation of the ECD1 and the juxtamembrane domains of the receptor together with the 3D structures of the ECD1 and of astressin provides a molecular basis for the proposed two-step model of the receptor activation proposed. As indicated in FIGURE 4 B, the exposed binding surface of the ECD1 of the CRF receptor binds the C-terminal region of the peptide. The peptide, thus captured, is in position in the correct proximity to present its N terminus to the remainder of the receptor, in order to initiate signaling.

The B1 Receptor Subfamily

The sequences of the ECD1s of some B1 receptors are shown in FIGURE 2 B, in which the conserved residues are highlighted. These conserved residues are presumed to be responsible for maintaining the structural integrity of the ECD1. For example, the two tryptophan and six cysteine groups are strictly conserved as are the two prolines proposed to be important for ending the β-sheets and two glycines. In addition, there are many other residues that show only conservative changes. All the B1 subfamily receptors have either aspartic/glutamic acid and arginine/lysine residues in analogous positions for creating the structurally important internal salt bridge. Disruption of the salt bridge by mutation should produce deleterious effects on binding and signaling. Indeed, such effects have been reported for some of the B1 receptors.

The GHRH Receptor

It has been known for more than 20 years that a dwarf mouse, known as the *little* mouse, displays a phenotype characterized by a hypoplastic pituitary

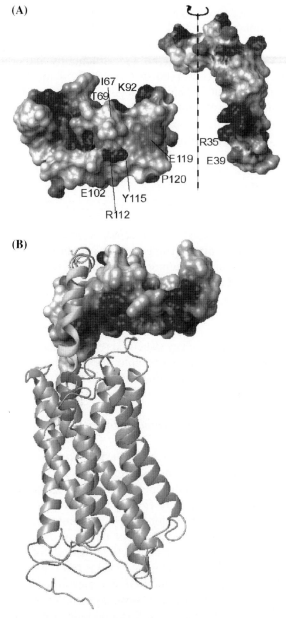

FIGURE 4. Proposed model of CRF receptor activation. (**A**) Surface representation of the ECD1-CRF-R2β and astressin (unpublished) showing the relative orientation of the peptide and major binding surface. (**B**) Schematic of the ECD1 oriented in the full-length receptor using rhodopsin as a model for the transmembrane domains. The model for the peptide includes an extension to include the N-terminal segment important for receptor activation.

and defective response to GHRH.[46] The cloning of its pituitary receptor revealed a single mutation, [D60G], in its ECD1.[46] This mutant GHRH receptor displays reduced ligand-induced signal transduction. The sequence comparison (FIG. 2B) shows that all the structurally important residues of the CRF receptor are conserved in the ECD1 of the GHRH receptor. Thus, the mutation [D60G], being analogous to the mutation [D65A] in CRF-R2β, would disrupt the internal salt bridge and result in a loss of the structural integrity of the ECD1 and of the GHRH binding surface. The consequences would then be compromised ligand recognition and impaired signal transduction.

Other B1 Receptors

Mutating D69 in the calcitonin receptor (corresponding to D65 in CRF-R2β) disrupted not only signaling but also the interaction of the receptor with RAMP1.[47] In the secretin receptor, the mutations [D49R] or [R83D] (corresponding to D65 or R101 in CRF-R2β) impaired binding and signaling.[48] In the VIP receptor, D68 (corresponding to D65 in CRF-R2β) is essential for binding VIP, and also for maintaining the constitutive activity of the [H178R] mutation.[49]

Thus, the conservation of the key structural residues suggests that the structural fold of the ECD1 of all the B1 receptors will be the same as that of the CRF receptor, namely a SCR/Sushi domain.

Extracellular Domain Interactions

Structures of extracellular domains of cell surface proteins are modular in nature and serve to facilitate protein-protein interactions. Two common structural modules are the SCR/Sushi and the epidermal growth factor (EGF) domains. Both motifs are found in many proteins, notably those in the complement system such as CD55 and its receptor CD97,[50] the latter being a member of the EGF-7TM GPCR family. The ECD1 of CD97 has multiple EGF modules.[51,52] The X ray structure of the extracellular domain of CD55 identified four SCR/Sushi domains.[53] The binding of CD55 to CD97 results directly from an interaction between their Sushi and EGF domains, respectively.[51]

Subsequent to our identification of a Sushi domain as the structural feature of the ECD1 of the CRF receptor, two Sushi domains were identified in the ECD1 of the GABA_B 1a receptor, another GPCR.[54] Interestingly, it was found that a soluble protein corresponding to the ECD1 of the GABA_B 1a receptor binds to the extracellular matrix protein fibulin-2,[54] which belongs to a family of proteins that are characterized by tandem arrays of EGF domains.[55]

Thus, the identification of a Sushi domain in the structure of the ECD1 of the CRF receptor raises the possibility of interactions of CRF receptors with other receptors and proteins that contain EGF structural modules. More generally, similar interactions of EGF module-containing proteins with the other receptors in the B1 subfamily may also exist.

CONCLUSIONS

The determination of the 3D NMR solution structure of the ECD1 of the CRF receptor has identified the structural fold adopted by this major binding domain as a SCR/Sushi domain. The structural fold is one that is found in many proteins and is responsible for their interactions with other proteins. The 3D structure identifies the peptide-binding surface on the ECD1 that appears to be relevant to ligand recognition of the full-length receptor. The structure also provides a molecular rationale for the conservation of key residues in the ECD1s of all the CRF receptors, as well as of all related B1 subfamily receptors.

The role of the internal salt bridge formed by a pair of anionic and cationic residues in ECD1 of the CRF receptor appears to be predominantly a structural one. The conservation of the corresponding residues in the ECD1s of the related receptors suggests that a corresponding salt bridge is a crucial structural component in their ECD1s also. Indeed, the compromised pituitary response to GHRH in the "*little*" mouse may now possibly be understood on the basis of the loss of the Sushi domain structure as a result of the disruption of the crucial salt bridge by mutation [D60A] in the ECD1 of its GHRH receptor.

Importantly, the 3D structure has provided important insight into the molecular model for the two-step process of peptide activation of the CRF receptor and its related B1 family members. Finally, the structure determination has opened up the possibility of novel interactions of CRF receptors, as well as of the other related receptors in the B1 subfamily, with proteins that contain extracellular EGF structural modules. The future promises to hold many surprises regarding the roles and interactions of CRF receptors and their B1 family relatives.

ACKNOWLEDGMENTS

We thank Dr. J. Rivier and Dr. J. Gulyas for peptide synthesis, and M. DiGruccio, R. Kaiser, and J. Vaughan for technical assistance. This work was supported by The Robert J. and Helen C. Kleberg Foundation, the Foundation for Research and NIDDK P01 DK26741. Roland Riek is a Pew Scholar. Wylie W. Vale is a FFR Senior Investigator and a consultant to, and has an equity interest in, Neurocrine Biosciences Inc.

REFERENCES

1. VALE, W., J. SPIESS, C. RIVIER & J. RIVIER. 1981. Science **213:** 1394–1397.
2. VAUGHAN, J., C. DONALDSON, J. BITTENCOURT, *et al.* 1995. Nature **378:** 287–292.
3. REYES, T.M., K. LEWIS, M.H. PERRIN, *et al.* 2001. Proc. Natl. Acad. Sci. USA **98:** 2843–2848.
4. LEWIS, K., C. LI, M.H. PERRIN, *et al.* 2001. Proc. Natl. Acad. Sci. USA **98:** 7570–7575.
5. HSU, S.Y. & A.J. HSUEH. 2001. Nat. Med. **7:** 605–611.
6. CHEN, R., K.A. LEWIS, M.H. PERRIN & W.W. VALE. 1993. Proc. Natl. Acad. Sci. USA **90:** 8967–8971.
7. VITA, N., P. LAURENT, S. LEFORT, *et al.* 1993. FEBS Lett. **335:** 1–5.
8. CHANG, C.-P., I.R.V. PEARSE, S. O'CONNELL & M.G. ROSENFELD. 1993. Neuron **11:** 1187–1195.
9. LOVENBERG, T.W., C.W. LIAW, D.E. GRIGORIADIS, *et al.* 1995. Proc. Natl. Acad. Sci. USA **92:** 836–840.
10. PERRIN, M., C. DONALDSON, R. CHEN, *et al.* 1995. Proc. Natl. Acad. Sci. USA **92:** 2969–2973.
11. KISHIMOTO, T., R.V. PEARSEII, C.R. LIN & M.G. ROSENFELD. 1995. Proc. Natl. Acad. Sci. USA **92:** 1108–1112.
12. STENZEL, P., R. KESTERSON, W. YEUNG, *et al.* 1995. Mol. Endocrinol. **9:** 637–645.
13. PISARCHIK, A. & A. SLOMINSKI. 2004. Eur. J. Biochem. **271:** 2821–2830.
14. ARDATI, A., V. GOETSCHY, J. GOTTOWICK, *et al.* 1999. Neuropharmacology **38:** 441–448.
15. KOSTICH, W.A., A. CHEN, K. SPERLE & B.L. LARGENT. 1998. Mol. Endocrinol. **12:** 1077–1085.
16. POTTER, E., D. BEHAN, W. FISCHER, *et al.* 1991. Nature **349:** 423–426.
17. CHEN, A.M., M.H. PERRIN, M.R. DIGRUCCIO, *et al.* 2005. Proc. Natl. Acad. Sci. USA **102:** 2620–2625.
18. HAUGER, R.L., D.E. GRIGORIADIS, M.F. DALLMAN, *et al.* 2003. Pharmacol. Rev. **55:** 21–26.
19. BALE, T.L. & W.W. VALE. 2004. Annu. Rev. Pharmacol. Toxicol. **44:** 525–557.
20. BRAR, B.K., A.K. JONASSEN, A. STEPHANOU, *et al.* 2000. J. Biol. Chem. **275:** 8508–8514.
21. LI, C., P. CHEN, J. VAUGHAN, *et al.* 2003. Endocrinology **144:** 3216–3224.
22. GULYAS, J., C. RIVIER, M. PERRIN, *et al.* 1995. Proc. Natl. Acad. Sci. USA **92:** 10575–10579.
23. HIGELIN, J., G. PY-LANG, C. PATERNOSTER, *et al.* 2001. Neuropharmacology **40:** 114–122.
24. RIJKERS, D.T., J.A. KRUIJTZER, M. VAN OOSTENBRUGGE, *et al.* 2004. Chembiochem **5:** 340–348.
25. YAMADA, Y., K. MIZUTANI, Y. MIZUSAWA, *et al.* 2004. J. Med. Chem. **47:** 1075–1078.
26. NIELSEN, S.M., L.Z. NIELSEN, S.A. HJORTH, *et al.* 2000. Proc. Natl. Acad. Sci. USA **97:** 10277–10281.
27. SHIMIZU, M., P.H. CARTER & T.J. GARDELLA. 2000. J. Biol. Chem. **275:** 19456–19460.
28. MONTICELLI, L., S. MAMMI & D.F. MIERKE. 2002. Biophys. Chem. **95:** 165–172.

29. LIAW, C.W., D.E. GRIGORIADIS, M.T. LORANG, *et al.* 1997. Mol. Endocrinol. **11:** 2048–2053.
30. DAUTZENBERG, F.M., G.J. KILPATRICK, S. WILLE & R.L. HAUGER. 1999. J. Neurochem. **73:** 821–829.
31. SYDOW, S., A. FLACCUS, A. FISCHER & J. SPIESS. 1999. Eur. J. Biochem. **259:** 55–62.
32. DAUTZENBERG, F.M., J. HIGELIN, O. BRAUNS, *et al.* 2002. Mol. Pharmacol. **61:** 1132–1139.
33. DAUTZENBERG, F.M., J. HIGELIN & U. TEICHERT. 2000. Eur. J. Pharmacol. **390:** 51–59.
34. PERRIN, M.H., S. SUTTON, D.B. BAIN, *et al.* 1998. Endocrinology **139:** 566–570.
35. HOARE, S.R., S.K. SULLIVAN, D.A. SCHWARZ, *et al.* 2004. Biochemistry **43:** 3996–4011.
36. WILLE, S., S. SYDOW, M.R. PALCHAUDHURI, *et al.* 1999. J. Neurochem. **72:** 388–395.
37. PERRIN, M.H., W.H. FISCHER, K.S. KUNITAKE, *et al.* 2001. J. Biol. Chem. **276:** 31528–31534.
38. HOFMANN, B.A., S. SYDOW, O. JAHN, *et al.* 2001. Protein Sci. **10:** 2050–2062.
39. PERRIN, M.H., M.R. DIGRUCCIO, S.C. KOERBER, *et al.* 2003. J. Biol. Chem. **278:** 15595–15600.
40. GRAUSCHOPF, U., H. LILIE, K. HONOLD, *et al.* 2000. Biochemistry **39:** 8878–8887.
41. WILMEN, A., B. GOKE & R. GOKE. 1996. FEBS Lett. **398:** 43–47.
42. HOARE, S.R. 2005. Drug Discov. Today **10:** 417–427.
43. GRACE, C.R., M.H. PERRIN, M.R. DIGRUCCIO, *et al.* 2004. Proc. Natl. Acad. Sci. USA **101:** 12836–12841.
44. BORK, P., A.K. DOWNING, B. KIEFFER & I.D. CAMPBELL. 1996. Q. Rev. Biophys. **29:** 119–167.
45. SCHWARZENBACHER, R., K. ZETH, K. DIEDERICHS, *et al.* 1999. EMBO J. **18:** 6228–6239.
46. LIN, S.-C., C.R. LIN, I. GUKOVSKY, *et al.* 1993. Nature **364:** 208–213.
47. ITTNER, L.M., F. LUESSI, D. KOLLER, *et al.* 2004. Biochem. Biophys. Res. Commun. **319:** 1203–1209.
48. DI PAOLO E., J.P. VILARDAGA, H. PETRY, *et al.* 1999. Peptides **20:** 1187–1193.
49. GAUDIN, P., J.J. MAORET, A. COUVINEAU, *et al.* 1998. J. Biol. Chem. **273:** 4990–4996.
50. HAMANN, J., B. VOGEL, G.M. VAN SCHIJNDEL & R.A. VAN LIER. 1996. J. Exp. Med. **184:** 1185–1189.
51. LIN, H.H., M. STACEY, C. SAXBY, *et al.* 2001. J. Biol. Chem. **276:** 24160–24169.
52. HAMANN, J., C. VAN ZEVENTER, A. BIJL, *et al.* 2000. Int. Immunol. **12:** 439–448.
53. LUKACIK, P., P. ROVERSI, J. WHITE, *et al.* 2004. Proc. Natl. Acad. Sci. USA **101:** 1279–1284.
54. BLEIN, S., R. GINHAM, D. UHRIN, *et al.* 2004. J. Biol. Chem. **279:** 48292–48306.
55. TIMPL, R., T. SASAKI, G. KOSTKA & M.L. CHU. 2003. Nat. Rev. Mol. Cell. Biol. **4:** 479–489.

Hedgehog Signaling

New Targets for GPCRs Coupled to cAMP and Protein Kinase A

JAMES A. WASCHEK,[a] EMANUEL DICICCO-BLOOM,[b]
ARNAUD NICOT,[b,c] AND VINCENT LELIEVRE[a,d]

[a]Mental Retardation Research Center, Semel Institute for Neuroscience, Jonsson Cancer Center, David Geffen School of Medicine, University of California at Los Angeles, California 90095, USA

[b]Department of Neuroscience and Cell Biology, University of Medicine and Dentistry of New Jersey/Robert Wood Johnson Medical School, Piscataway, New Jersey 08854, USA

[c]INSERM U732, Hopital St Antoine, 75012 Paris, France

[d]INSERM U676, Hôpital Robert Debré, 75019 Paris, France

ABSTRACT: Hedgehog (HH) is a secreted protein named for the bristle phenotype observed in *Drosophila* embryos that lack the corresponding gene. Three homologs have been characterized in vertebrates, all which have critical roles in the development of multiple organ systems. Moreover, these proteins regulate stem cell production and activation during tissue repair after injury, and appear to drive proliferation in a variety of type of tumors, including those arising in the brain, foregut, lung, breast, pancreas, stomach, and prostate. Early evidence from *Drosophila*, and later work in vertebrates established the cAMP/protein kinase A (PKA) pathway as a major pathway which opposes HH signaling, doing so by phosphorylating intracellular signaling mediators and targeting them for degradation. Thus, it seems possible that ligands which activate G protein–coupled receptors (GPCR) may act in some cases to oppose or enhance HH signaling. We studied a possible interaction of pituitary adenylyl cyclase-activating peptide (PACAP) with sonic hedgehog (SHH) in the developing cerebellum, where both PACAP and SHH are know to act. PACAP and the PAC1-specific agonist, maxadilan, were found to completely block the proliferative action of SHH on developing cerebellar granule neurons. It remains to be determined if HH/GPCR antagonistic interactions play additional important roles in development, plasticity, tissue repair, cancer, and other processes.

KEYWORDS: hedgehog; cancer; seven transmembrane; vasoactive intestinal peptide (VIP)

Address for correspondence: James A. Waschek, NRB 345, UCLA, 635 Charles E. Young Dr. South, Los Angeles, CA 90095-7332. Voice: 310-825-0179; fax: 310-206-5061.
e-mail: jwaschek@mednet.ucla.edu

Ann. N.Y. Acad. Sci. 1070: 120–128 (2006). © 2006 New York Academy of Sciences.
doi: 10.1196/annals.1317.089

INTRODUCTION

More than 7 years ago, we and others showed that mRNAs for pituitary adenylate cyclase-activating peptide (PACAP) and one of its receptors, PAC1, exhibited widespread expression in the embryonic mouse and rat neural tubes at the onset of neurogenesis.[1–5] Interestingly, PAC1 receptor gene transcripts were localized specifically to the "ventricular zone," the part of the neural tube that contains the proliferating progenitor cells that give rise to nearly all embryonic and most postnatal neurons, astrocytes, and oligodendrocytes.[3] When put into tissue culture, cells isolated from the neural tube at this stage responded to PACAP with a robust stimulation of cAMP production, indicating that these cells indeed expressed functional PACAP receptors.[3] Because no other functional cAMP-inducing ligand/receptor signaling pairs had been described in the early embryonic neural tube, this placed PACAP in a potentially unique position among signaling molecules. However, what could be the function of a cAMP signaling system in the neural tube at such an early stage of development? A potential explanation came from an unexpected field of study: *Drosophila* genetics. Here, in a screen for mutants that affected segment polarity, a gene regulating this process was identified called hedgehog (HH), which encoded a secreted factor. Three mammalian homologs were then discovered. One of these, named sonic hedgehog (SHH), was found to be an important regulator of neural patterning and neural cell proliferation. Referring again to *Drosophila*, genetic studies revealed that cAMP-dependent protein kinase A (PKA) was strongly antagonistic to HH-regulated processes. Some of the first evidence for this was that loss of PKA in flies that lacked HH resulted in a paradoxical induction of both HH target genes and HH-dependent patterning events (i.e., loss of PKA mimicked HH action).[6] Conversely, overexpression of PKA counteracted the effects of HH pathway activation. The interaction of these two pathways has since been a subject of active investigation in both *Drosophila* and vertebrates. For example, overexpression of dominant negative PKA in the dorsal neural tube of zebrafish embryos gave the same phenotype as SHH overexpression (neural tube ventralization).[7] Conversely, the phenotypic change induced by HH overexpression in zebrafish was suppressed by injection of mRNA encoding a constitutively active catalytic subunit of PKA.[8] This latter point strongly implied that one role of a cAMP/PKA-inducing system in the developing nervous system could be to counteract or delimit the actions of SHH. A model for PKA antagonism of the HH signaling in vertebrates is presented in FIGURE 1.

GPCR/SHH INTERACTION IN THE MOUSE NEURAL TUBE AND CEREBELLUM

The above model and the aforementioned studies in zebrafish suggested that G protein–coupled receptors (GPCR) that are positive coupled to a cAMP and

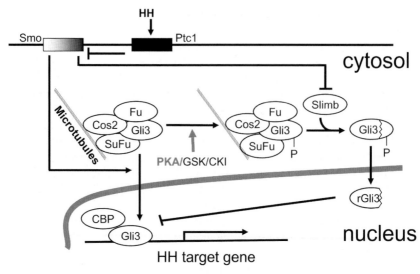

FIGURE 1. A model of HH signaling and PKA antagonism in vertebrates (modified from Johnson and Scott, Ref. 9). In the absence of HH, its cell membrane receptor Patched1 (Ptc1) inhibits constitutive smoothened (Smo) activity to keep the pathway quiescent. In the cytosol, a signaling protein complex containing Gli2 or Gli3 (homologs of *Drosophila* cubitus interruptus (Ci)) is held in check by way of its physical interaction with microtubules. Fused (Fu, a serine/threonine kinase) and Cos2 (a kinesin-like molecule) are components of this complex. In the presence of HH, Ptc1 activity is inhibited, allowing Smo to act. Smo is thought to induce hyperphosphorylation and dissociation of the signaling complex from microtubules. This allows Gli2 or Gli3 to translocate to the nucleus. There it associates with the cAMP-responsive element binding protein coactivator protein (CBP), binds to consensus sequences on target genes, and activates their transcription. In *Drosophila*, PKA appears to phosphorylate Ci, priming it for subsequent phosphorylation by glycogen synthase kinase (GSK) and a form of casein kinase I (CKI). Phosphorylation is thought to target Ci for proteolysis to transcription repressor form via interaction with Slimb, a factor with homology to vertebrate proteins that target proteosomes. Similar PKA-induced phosphorylation of Gli2 and Gli3 and cleavage seems to occur in vertebrates. Gli1 also has homology to CI. Gli1 gene expression is activated in many cell types by HH proteins via binding of Gli2 and/or Gli3 and accessory proteins to its promotor. The gene encoding Gli1 is therefore a target gene for HH proteins.

PKA induction could be in a position to counteract HH signaling. To begin to test the hypothesis that PACAP might do this in the neural tube, we took advantage of the fact that gli1 had been identified as a target gene commonly activated by SHH. We isolated cells from the mouse embryonic neural tube and found that PACAP downregulated gli1 gene expression,[3] providing the first evidence of an interaction between GPCR and HH signaling. Another location where SHH is known to act is in the external granular layer (EGL)

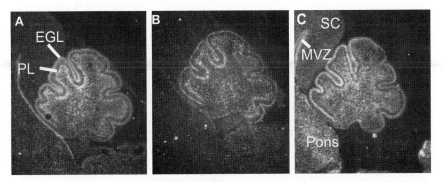

FIGURE 2. *In situ* hybridization showing PAC1 (**A**), Ptc (**B**), and gli1 (**C**) gene expression in postnatal day 1 rat brain. Dark field views of photomicrographs of sagittal sections of hindbrain and cerebellar (CB) are shown. Hybridization signals appear as white silver grains, which brightly label the EGL and Purkinje layers (PL). SC = spinal cord; MVZ = medulla ventricular zone. (Data are extracted from Reference 10 with permission from the *Journal of Neuroscience*.)

of the developing cerebellum. This is the proliferative zone that gives rise to the granule neurons of the cerebellum. Unlike most other germinal centers in the brain, this proliferative zone is on the outer surface of the structure, and is primarily active in the postnatal period. SHH is known to be a strong mitogen for cerebellar granule neural precursors (CGNPs) in the EGL. To determine if PACAP might be positioned to antagonize the mitogenic action of SHH in the EGL, we first determined if PAC1 receptor gene expression was co-localized with mRNAs for Ptc1 (the SHH receptor) and the SHH target gene gli1 in the cerebellar EGL.[10] Gene transcripts for PAC1, Ptc1, and gli1 were found to be very prominent in the EGL by postnatal day 1 in rats (FIG. 2). Transcripts for SHH were found to be present in the Purkinje cell layer, as previously reported by other groups,[11–14] but not in other areas of the developing cerebellum. PACAP gene expression was detected diffusely in the Purkinje layer and more strongly in the deep cerebellum nuclei.

The above *in situ* hybridization data show that mRNAs for receptors for PACAP and SHH are localized in the EGL, where CGNPs are actively proliferating. We established cultures of postnatal day 6.5 mouse CGNPs to determine the effects of PACAP and SHH on the proliferation of these cells.[10] PACAP, 10 nM, blocked the mitogenic effect of SHH in these cells by approximately 85% (FIG. 3). These antiproliferative effects were mimicked by the PAC1-specific agonist, maxadilan, and by the cAMP-inducing drug, forskolin.[10] Taken with the fact that SHH and PACAP receptors are co-expressed on the cells of the EGL of the cerebellum *in vivo*, the data suggest that a biologically relevant action of PACAP on these cells is to inhibit their proliferation, possibly by antagonizing SHH signaling.

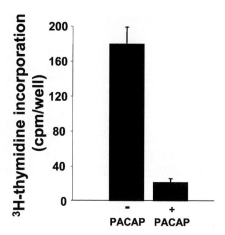

FIGURE 3. SHH-dependent proliferation of mouse cerebellar granule precursors is blocked by PACAP. Cerebellar granule precursors were isolated from mouse brain and incubated for 24 h in a medium containing SHH (3 μg/mL) as the sole mitogen. Inclusion of 10 nM PACAP resulted in an 85% reduction DNA synthesis. Under these conditions, cultures did not exhibit appreciable DNA synthesis in the absence of SHH or in the presence of PACAP alone. (Data are extracted from Reference 10.)

Other GPCRs have been proposed to regulate SHH activity in the cerebellar EGL. The chemokine receptor CXCR4 is expressed in the EGL, whereas one of its ligands, SDF-1α, is expressed in the pia matter. SDF-1 was found to synergize with SHH to induce the proliferation of cultures CGNPs, apparently via $G\alpha_i$.[15] It also appears that the metabotropic glutamate receptor type 4 (mGlu4), which mainly acts via G_i or G_o subtypes, has the capacity to interact with HH signaling to regulator CGNP proliferation and/or differentiation.[16]

HH SIGNALING IN CANCER

The first evidence indicating that the HH pathway could be important in the pathogenesis of cancer was that Gorlin's syndrome (also called basal cell nevus syndrome) was found to result from a mutation in a single copy of the Ptc1 gene. Gorlin's syndrome is a hereditary disease characterized by large size, and an increased incidence of tumors, especially basal cell carcinoma, and rhabdomyosarcoma, and medulloblastoma. Interestingly, the latter tumor arises from the cerebellar EGL, the germinal center in which we proposed a PACAP/SHH interaction (Ref. 10 and see above). Gorlin's syndrome has been accurately modeled in mice. Like in humans, mice with a mutation in a single copy of Ptc1 exhibit larger size and an increased incidence of medulloblastoma and rhabdomyosarcoma.[17] Interestingly, mutation of both Ptc1 alleles is lethal, with embryos dying in midgestation die due to massive overgrowth of the neural tube and other defects.

HH PATHWAY OVERACTIVITY IN DIVERSE TYPES OF TUMORS

Given the important actions of the HH pathway in the growth and development of many organs, it is not surprising that this pathway has recently been shown to play a role in the growth of a variety of human tumors. There are two main types of information which can indicate that the HH pathway might be overactive in tumors. The first is that HH target genes are expressed in tumors at higher levels than in nontumor tissue. The most reliable of these may be gli1, which is expressed in only a small minority of cell types in mature organisms, and seems to be a HH target during development in most if not all cell types that express the signaling components. Although gli1 could be a target of other mitogenic signaling pathways, none have as yet been identified. An independent type of evidence which can implicate HH-driven growth is that the rate of proliferation of derived cell lines is sensitive to the cyclopamine, an agent which acts to block HH signaling at the level of the co-receptor smoothened (Smo). Disadvantages of this pharmacological approach are that the specificity of this agent is not completely certain, and that cyclopamine may fail to block the pathway if activating mutations are present on signaling molecules downstream of Smo, for example SuFu or one of the gli genes. Nonetheless, using the dual criteria of high HH target gene expression and responsiveness of derived cell lines to cyclopamine, it appears that overactive HH signaling contributes to the growth rates of numerous cancer types (TABLE 1). However, unlike tumors associated with Gorlin's syndrome, overactive HH signaling in many of these tumors appears to be due to the autocrine production of one or

TABLE 1. Cancer types that may be driven by overactive HH signaling

Cancer type	Number of tumor samples positive/total analyzed	Reference
Esophageal	14/22, 4/6	18,19
Gastric	63/99, 6/6	19,20
Prostate (high grade)	42/59, 6/6	21,22
Breast	52/52	23
Glioma	12/12	24
Pancreas	6/6, 31*	19,25
Small cell lung cancer	5/10	26

NOTE: Criteria for inclusion in this table are both overexpression of HH target genes and cyclopamine responsiveness of derived cell lines. *Mean values of Indian HH mRNA in 31 tumor samples were 35-fold than in normal pancreatic tissue.

more HH proteins. This listing of tumors types in TABLE 1 is likely to expand because this is still a very new area of investigation.

POTENTIAL INTERACTION OF GPCRs AND HH SIGNALING IN CANCER

The fact that HH signaling is commonly antagonized by PKA suggests that Gs-coupled GPCRs could represent potential targets for the treatment of HH-overactive tumors. In this respect, it has been known for many years that VIP and PACAP receptors are expressed on many different types of tumors, including colon, breast, pancreas, glioma, prostate, lung (both large and small cell), pituitary, and lymphoma/leukemia (reviewed in Ref. 27). This set of tumor types clearly overlaps with those exhibiting overactive HH signaling. Thus, it would be of interest to determine the frequency of this overlap in individual tumor samples, and to study the interaction of these pathways in cell lines derived from the different kinds of tumors.

HH/GPCR PATHWAY INTERACTION IN STEM CELLS BIOLOGY, PLASTICITY, AND TISSUE REPAIR

SHH and PACAP and/or VIP have recently been shown to potentially be involved in several other growth-related processes, although their interactions in these processes have yet to be investigated. For example, SHH and PACAP have independently been shown to regulate adult neural stem cell behavior,[28,29] and axonal pathfinding.[30,31] In other work, SHH, VIP, and PACAP mRNAs have all been shown to be strongly induced in brain stem facial motor neurons after peripheral facial nerve axotomy,[32-34] suggesting that the interaction might be important in some aspect of neuron survival or regeneration.

REFERENCES

1. SHEWARD, W.J., E.M. LUTZ & A.J. HARMAR. 1996. Expression of pituitary adenylate cyclase activating polypeptide receptors in the early mouse embryo as assessed by reverse transcription polymerase chain reaction and in situ hybridisation. Neurosci. Lett. **216:** 45–48.
2. SHUTO, Y., D. UCHIDA, H. ONDA & A. ARIMURA. 1996. Ontogeny of pituitary adenylate cyclase activating polypeptide and its receptor mRNA in the mouse brain. Regul. Pept. **67:** 79–83.
3. WASCHEK, J.A., R.A. CASILLAS, T.B. NGUYEN, et al. 1998. Neural tube expression of pituitary adenylate cyclase-activating peptide (PACAP) and receptor: potential role in patterning and neurogenesis. Proc. Natl. Acad. Sci. USA. **95:** 9602–9607.

4. Sheward, W.J., E.M. Lutz, A.J. Copp & A.J. Harmar. 1998. Expression of PACAP, and PACAP type 1 (PAC1) receptor mRNA during development of the mouse embryo. Dev. Brain Res. **109:** 245–253.

5. Zhou, C.J., S. Shioda, M. Shibanuma, *et al.* 1999. Pituitary adenylate cyclase-activating polypeptide receptors during development: expression in the rat embryo at primitive streak stage. Neuroscience **93:** 375–391.

6. Li, W., J.T. Ohlmeyer, M.E. Lane & D. Kalderon. 1995. Function of protein kinase A in HH signal transduction and Drosophila imaginal disc development. Cell **80:** 553–562.

7. Epstein, D.J., E. Marti, M.P. Scott & A.P. McMahon. 1996. Antagonizing cAMP dependent protein kinase A in the dorsal CNS activates a conserved Sonic HH signaling pathway. Development **122:** 2885–2894.

8. Hammerschmidt, M., M.J. Bitgood & A. McMahon. 1996. Protein kinase A is a common negative regulator of HH signaling in the vertebrate embryo. Genes Dev. **10:** 647–658.

9. Johnson, R.L. & M.P. Scott. 1998. New players and puzzles in the HH signaling pathway. Curr. Opin. Genet. Dev. **8:** 450–456.

10. Nicot, A., V. Lelievre, J. Tam, *et al.* 2002. Pituitary adenylate cyclase-activating polypeptide and sonic HH interact to control cerebellar granule precursor cell proliferation. J. Neurosci. **22:** 9244–9254.

11. Wechsler-Reya, R.J. & M.P. Scott. 1999. Control of neuronal precursor proliferation in the cerebellum by sonic HH. Neuron **22:** 107–114.

12. Dahmane, N. & A. Ruiz i Altaba. 1999. Sonic HH regulates the growth and patterning of the cerebellum. Development **126:** 3089–3100.

13. Traiffort, E., D. Charytoniuk, L. Watroba, *et al.* 1999. Discrete localizations of HH signaling components in the developing and adult rat nervous system. Eur. J. Neurosci. **11:** 3199–3214.

14. Wallace, V.A. 1999. Purkinje-cell-derived sonic HH regulates granule neuron precursor cell proliferation in the developing mouse cerebellum. Curr. Biol. **9:** 445–448.

15. Klein, R.S., J.B. Rubin, H.D. Gibson, *et al.* 2001. SDF-1 alpha induces chemotaxis and enhances Sonic hedgehog-induced proliferation of cerebellar granule cells. Development **128:** 1971–1981.

16. Canudas, A.M., V. Di Giorgi-Gerevini, L. Iacovelli, *et al.* 2004. PHCCC, a specific enhancer of type 4 metabotropic glutamate receptors, reduces proliferation and promotes differentiation of cerebellar granule cell neuroprecursors. J. Neurosci. **24:** 10343–10352.

17. Goodrich, L.V., L. Milenkovic, K.M. Higgins & M.P. Scott. 1997. Altered neural cell fates and medulloblastoma in mouse patched mutants. Science **277:** 1109–1113.

18. Ma, X., T. Sheng, Y. Zhang, *et al.* 2006. Hedgehog signaling is activated in subsets of esophageal cancers. Int. J. Cancer **118:** 139–148.

19. Berman, D.M., S.S. Karhadkar, A. Maitra, *et al.* 2003. Widespread requirement for Hedgehog ligand stimulation in growth of digestive tract tumours. Nature **425:** 846–851.

20. Ma, X., K. Chen, S. Huang, *et al.* 2005. Frequent activation of the hedgehog pathway in advanced gastric adenocarcinomas. Carcinogenesis **26:** 1698–1705.

21. Sheng, T., C. Li, X. Zhang, *et al.* 2004. Activation of the hedgehog pathway in advanced prostate cancer. Mol. Cancer **3:** 29.

22. SANCHEZ, P., A.M. HERNANDEZ, B. STECCA, *et al.* 2004. Inhibition of prostate cancer proliferation by interference with SONIC HEDGEHOG-GLI1 signaling. Proc. Natl. Acad. Sci. USA **101:** 12561–12566.
23. KUBO, M., M. NAKAMURA, A. TASAKI, *et al.* 2004. Hedgehog signaling pathway is a new therapeutic target for patients with breast cancer. Cancer Res. **64:** 6071–6074.
24. DAHMANE, N., P. SANCHEZ, Y. GITTON, *et al.* 2001. The Sonic Hedgehog-Gli pathway regulates dorsal brain growth and tumorigenesis. Development **28:** 5201–5212.
25. KAYED, H., J. KLEEFF, S. KELEG, *et al.* 2004. Indian hedgehog signaling pathway: expression and regulation in pancreatic cancer. Int. J. Cancer **110:** 668–676.
26. WATKINS, D.N., D.M. BERMAN, S.G. BURKHOLDER, *et al.* 2003. Hedgehog signalling within airway epithelial progenitors and in small-cell lung cancer. Nature **422:** 313–317.
27. LELIÈVRE, V., N. PINEAU & J. WASCHEK. The biological significance of PACAP and PACAP receptors in human tumors. *In* Pituitary Adenylate Cyclase-Activating Polypeptide. H. Vaudry, Ed.: 361–399 H. Kluwer Academic Publishers. Norwell, MA.
28. LAI, K., B.K. KASPAR, F.H. GAGE & D.V. SCHAFFER. 2003. Sonic hedgehog regulates adult neural progenitor proliferation in vitro and in vivo. Nat. Neurosci. **6:** 21–27.
29. MERCER, A., H. RONNHOLM, J. HOLMBERG, *et al.* 2004. PACAP promotes neural stem cell proliferation in adult mouse brain. J. Neurosci. Res. **76:** 205–215.
30. GUIRLAND, C., K.B. BUCK, J.A. GIBNEY, *et al.* 2003. Direct cAMP signaling through G-protein–coupled receptors mediates growth cone attraction induced by pituitary adenylate cyclase-activating polypeptide. J. Neurosci. **23:** 2274–2283.
31. CHARRON, F., E. STEIN, J. JEONG, *et al.* 2003. The morphogen sonic hedgehog is an axonal chemoattractant that collaborates with netrin-1 in midline axon guidance. Cell **113:** 11–23.
32. ZHOU, X., W.I. RODRIGUEZ, R.A. CASILLAS, *et al.* 1999. Axotomy-induced changes in pituitary adenylate cyclase activating polypeptide (PACAP) and PACAP receptor gene expression in the adult rat facial motor nucleus. J. Neurosci. Res. **57:** 953–961.
33. ARMSTRONG, B.D., Z. HU, C. ABAD, *et al.* 2003. Lymphocyte regulation of neuropeptide gene expression after neuronal injury. J. Neurosci. Res. **74:** 240–247.
34. AKAZAWA, C., H. TSUZUKI, Y. NAKAMURA, *et al.* 2004. The upregulated expression of sonic hedgehog in motor neurons after rat facial nerve axotomy. J. Neurosci. **24:** 7923–7930.

Effect of VIP on TLR2 and TLR4 Expression in Lymph Node Immune Cells During TNBS-Induced Colitis

ALICIA ARRANZ,[a] CATALINA ABAD,[a] YASMINA JUARRANZ,[a]
MARTA TORROBA,[a] FLORENCIA ROSIGNOLI,[a] JAVIER LECETA,[a]
ROSA PÉREZ GOMARIZ,[a] AND CARMEN MARTÍNEZ[b]

*Department of Cell Biology, Faculty of Biology,[a] Faculty of Medicine[b]
Complutense University, 28040 Madrid, Spain*

ABSTRACT: Toll-like receptors (TLRs) are a family of pattern recognition receptors (PRRs), which recognize numerous molecules collectively named pathogen-associated molecular patterns, with an essential role in inflammatory conditions and connecting innate and acquired immune responses. Moreover, a new function of TLRs in the intestinal mucosa has been described. Under homeostatic conditions, TLRs act to protect the intestinal epithelium; but when homeostasis is disrupted, TLRs appear deregulated. Disruption of intestinal homeostasis occurs in disorders, such as Crohn's disease (CD). Trinitrobenzene sulfonic acid (TNBS)-induced colitis is a murine model of human CD and vasoactive intestinal polypeptide (VIP) exerts a beneficial effect, by decreasing both inflammatory and autoimmune components of the disease. Recently, we have demonstrated the constitutive expression of TLR2 and TLR4 at mRNA and protein levels in colon extracts and their upregulation in TNBS-treated mice as well as the effect of VIP treatment, approaching control levels. However, the systemic effect is little known. The present results demonstrate a beneficial role of VIP, restoring homeostatic conditions through the regulation of both lymphoid cell traffic and TLR2/4 expression on macrophages (MØ), dendritic cells (DCs), and CD4 and CD8 T lymphocytes.

KEYWORDS: TLR; lymph node; inflammation; homeostasis; VIP

INTRODUCTION

Toll-like receptors (TLRs) represent a primary line of defense against invading pathogens being ascribed to innate immune system and connecting innate and acquired immunity. Eleven TLRs have been identified in humans and

Address for correspondence: Carmen Martínez, Ph.D., Department of Cell Biology, Faculty of Medicine, Complutense University, 28040 Madrid, Spain. Voice: 34-91-3944971; fax: 34-91-3944981.
e-mail: cmmora@bio.ucm.es

Ann. N.Y. Acad. Sci. 1070: 129–134 (2006). © 2006 New York Academy of Sciences.
doi: 10.1196/annals.1317.001

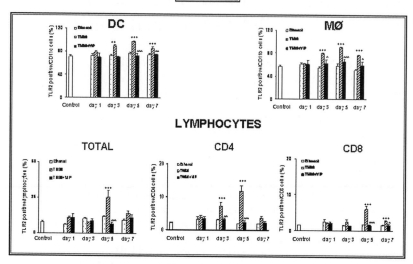

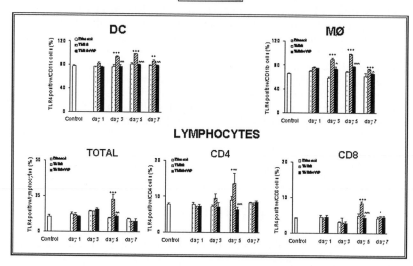

FIGURE 1. Flow cytometry analysis of TLR2 and TLR4 expression in MLN during TNBS-induced colitis, and the effect of VIP treatment. Colitis was induced by rectal instillation of TNBS in 50% ethanol and 1 nmol of VIP was injected intraperitoneally (i.p.) on alternate days. Cell suspensions from MLN extracted from control, TNBS, and VIP-treated animals were collected at different time points and processed for immunofluorescence staining by using a PE-conjugated TLR2 or TLR4 monoclonal antibody. Double labeling was

13 in mice, being TLRs1–9 conserved between species. All share a common cytosolic domain (toll/interleukin-(IL)-1 receptor) and an extracellular leucine rich repeat region that mediates recognition of exogenous and endogenous ligands. TLR4 and TLR2 belong to the group of TLR membrane surface receptors, TLR4 is essential for lipopolysaccharide (LPS) signaling and TLR2 recognizes a vast array of microbial components being both involved in inflammatory bowel diseases (IBD).[1,2] It has been described in different models of IBD both the constitutive and the inflammation-induced upregulation of TLR2 and TLR4 in the intestinal mucosa[3–6] resulting in the expression of a cascade of inflammation-related genes. In a recent report, we have shown the beneficial effect of vasoactive intestinal peptide (VIP) regulating the expression of these receptors in TNBS-induced colitis,[7] a murine model of human Crohn's disease (CD). We also demonstrated the VIP-mediated reduction of two crucial cytokines in the development of the disease, namely IL-1β and interferon-γ (IFN-γ). In the present article, we examined the immunological effects of VIP treatment on trinitrobenzene sulfonic acid (TNBS) induced upregulation of TLR2 and TLR4 in several types of immune cells from mesenteric lymph nodes (MLNs) that are essential for their traffic from/to sites of colitis.

RESULTS AND DISCUSSION

MLNs are the draining nodes of the intestinal tract wherein the presence of dendritic cells (DCs), as well as activated macrophages (MØ), leads to a secondary cellular immune response favoring T cells traffic into inflamed tissue. Thus, we argued about which particular populations were selectively regulated by TNBS-induced colitis and VIP treatment. To this end, we used two rat anti-mouse CD11c and CD11b monoclonal antibodies to label mainly DCs and MØ, respectively, whereas T lymphocyte subsets were labeled with monoclonal antibodies against CD4 and CD8 in the electronically gated lymphocyte population. Thus, we evaluated the proportion of cells bearing TLR2 and TLR4 by flow cytometry as an appropriate methodology for their quantization. As FIGURE 1 shows, around 70% of DCs and 60% of MØ constitutively express TLR2 and TLR4, and TNBS resulted in an increase of the proportion

---

also performed using these antibodies with either fluorescein isothiocynate (FITC)-conjugated rat anti-mouse CD11b or FITC-conjugated rat anti-mouse CD11c to detect mainly MØ and DCs, respectively. Lymphocyte subsets were labeled with monoclonal antibodies against CD4 and CD8 in the electronically gated lymphocyte population distinguished by their different forward scatter (FSC) versus side scatter (SSC) profiles. Immunofluorescence staining was analyzed by FACS calibur. Each result is the mean ± SEM of three separated experiments (three mice/group/experiment). $*P < 0.05$, $**P < 0.01$, $***P < 0.001$ analysis of variance (ANOVA) with respect to TNBS-treated mice. $^{++}P < 0.01$, $^{+++}P < 0.001$ (ANOVA) with respect to control animals.

of double-positive cells between days 3 and 7 with maximal percentages on day 5. The time course profiles for both TLRs were similar. Despite abundant data about TLRs expression on DCs and MØ, studies in immune cells during inflammatory process and in lymphocytes subpopulations are scarce. In recent years, it has been described both TLR2 and TLR4 expression on mouse T cells[8] and T and B cells from human PBMC,[9,10] as well as T subsets and B cells from feline lymphoid tissues.[11] In the present study, lymphocytes also expressed both TLR2 and TLR4 but had much lower percentage of positive cells compared to DCs and MØ, being significantly increased in TNBS mice on day 5. Our results about the expression of TLRs on lymphocytes suggest a role of these receptors in acquired immunity. Regarding T subsets, the number of CD4 cells expressing TLR2 and TLR4 was almost twofold higher than CD8 cells. The percentage of CD4 cells bearing TLR2 was upregulated in TNBS mice on days 3 and 5 ($\approx$ three- to five-fold, respectively) whereas, regarding CD8 cells, it increased between days 5 and 7 with maximal percentages on day 5 (threefold over control values). TNBS administration increased the proportion of CD4 and CD8 cells expressing TLR4 only on day 5. Although the functional relevance of TLR2 and TLR4 differential changes in TNBS mice in the subsets evaluated is not yet defined, upregulation of number of TLR2- and TLR4-positive cells suggests that inflammation initiated by TNBS causes a deregulation of TLRs that could contribute to the disruption of intestinal homeostasis observed in TNBS-induced colitis. Importantly, VIP treatment reduced significantly the proportion of TLR2- and TLR4-positive cells in all populations checked, restoring constitutive levels. Since TLR2 and TLR4 mediate a signaling cascade leading to the production of inflammatory factors, VIP could be involved in the blockade of the first stages of inflammation.

On the other hand, it is well established that the arrival of MØ and DC from inflamed gut within regional lymph nodes, which results in the migration of lymphocytes, contributes to intestinal inflammation. In this sense, the regulatory effect of VIP could also be on account of a modulation of the number of the different subpopulations recruited in MLNs . We investigated in TNBS-induced colitis, the effect of VIP on cellular traffic of DCs, MØ, and lymphocytes subsets in MLN. VIP treatment reduced significantly the TNBS-increased proportion of DCs and MØ on days 5 and 7, restoring the cellular number to values of control animals. However, TNBS treatment reduced significantly the percentage of T subsets at these time points whereas VIP counterbalanced this effect. FIGURE 2 summarizes the cellular composition of MLN in TNBS-induced colitis and VIP effect on day 5. Transit through lymph nodes (LN) is considered a crucial event in the immune response, since it is in these specialized sites where lymphocytes encounter antigens in association with antigen-presenting cells (APC), as DCs and MØ. As a consequence of these interactions, lymphocytes become functionally competent and migrate to sites of inflammation. TNBS-induced inflammation is characterized in the early stages of the disease, by a massive infiltration of neutrophils and

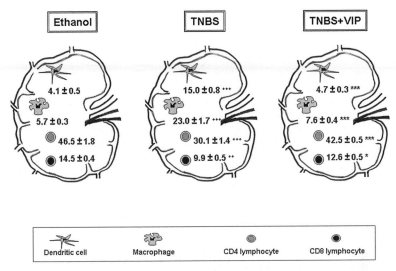

FIGURE 2. Summary of VIP effect on cellular traffic through MLNs of TNBS-treated animals on day 5. Colitis was induced as described in the legend of FIGURE 1 and mice were treated i.p. with 1 nmol of VIP on alternate days. Cell suspensions were prepared from four MLNs, selected for CD11c, CD11b, CD4, or CD8 phenotype and then subjected to flow cytometry. Numbers indicate the percentage of selected subpopulation into the cell suspension. Each result is the mean ± SEM of $n = 9$ from three separated experiments (three mice/group/experiment). $*P < 0.05$, $***P < 0.001$ (ANOVA) with respect to TNBS-treated mice. $^{++}P < 0.01$, $^{+++}P < 0.001$ (ANOVA) with respect to control animals treated with ethanol.

MØ producing high levels of proinflammatory cytokines followed by a T cell infiltration.[12] Our results reveal that VIP could selectively modulate cell recruitment into lymph nodes, which could prevent the pathological effects of TNBS.

In conclusion, our study supports and expands previous results obtained in the colon microenvironment, demonstrating the beneficial role of VIP in TNBS-induced colitis at systemic level. The present results show that VIP reestablishes the homeostatic condition through two complementary mechanisms. On the one hand, VIP modulates the immune cell traffic restoring the cellular composition in the MLN. On the other hand, VIP approaches the expression of TLR2 and TLR4 on DCs, MØ, and CD4$^+$ and CD8$^+$ lymphocytes to constitutive levels.

ACKNOWLEDGMENTS

This work was supported by grants BFI 2002–03489 from Ministerio de Ciencia y Tecnología (Spain) and a predoctoral fellowship from Ministerio de Ciencia y Tecnología (to A.A.).

REFERENCES

1. DUNNE, A. & L.A.J. O'NEILL. 2005. Adaptor usage and toll-like receptor signalling specificity. FEBS Lett. **579:** 3330–3335.
2. PASARE, C. & R. MEDZHITOV. 2004. Toll-like receptors and acquired immunity. Semin. Immunol. **16:** 23–26.
3. STROBER, W., I. FUSS & R.S. BLUMBERG. 2002. Immunology of mucosal models of inflammation. Annu. Rev. Immunol. **122:** 495–549.
4. ORTEGA-CAVA, C.F., S. ISHIHARA, M.A. RUMI, et al. 2003. Strategic compartmentalization of toll-like receptor 4 in the mouse gut. J. Immunol. **170:** 3977–3985.
5. FRANCHIMONT, D., S. VERMEIRE, H. EL HOUISNI, et al. 2004. Deficient host-bacteria interactions in inflammatory bowel disease? The toll-like receptor (TLR)-4 ASP299gly polymorphism is associated with Crohn's disease and ulcerative colitis. Gut **53:** 987–992.
6. ABREU, M.T., M. FUKATA & M. ARDITI. 2005. TLR signalling in the gut in health and disease. J. Immunol. **174:** 4453–4460.
7. GOMARIZ, R.P., A. ARRANZ, C. ABAD, et al. 2005. Time-course expression of toll-like receptors 2 and 4 in inflammatory bowel disease and homeostatic effect of VIP. J. Leukoc. Biol. **78:** 491–502.
8. MATSUGUCHI, T., K. TAKAGI, T. MUSIKACHAROEN, et al. 2000. Gene expressions of lipopolysaccharide receptors, toll-like receptors 2 and 4, are differently regulated in mouse T lymphocytes. Blood **95:** 1378–1385.
9. HORNUNG, V., S. ROTHENFUSSER, S. BRITSCH, et al. 2002. Quantitative expression of toll-like receptor 1–10 mRNA in cellular subsets of human peripheral blood mononuclear cells and sensitivity to CpG oligodioxynucleotides. J. Immunol. **18:** 4531–4537.
10. ZAREMBER, K.A. & P.J. GODOWSKI. 2002. Tissue expression of human toll-like receptors and differential regulation of toll-like receptor mRNAs in leukocytes in response to microbes, their products, and cytokines. J. Immunol. **168:** 554–561.
11. GLICERIO, I., S. NORDONE, K.E. HOWARD, et al. 2005. Toll-like receptor expression in feline lymphoid tissues. Vet. Immunol. Immunopathol. **106:** 229–237.
12. ABAD, C., C. MARTINEZ, M.G. JUARRANZ, et al. 2003. Therapeutic effects of vasoactive intestinal peptide in the trinitrobenzene sulfonic acid mice model of Crohn's disease. Gastroenterology **124:** 961–971.

Immunocytochemical Distribution of VIP and PACAP in the Rat Brain Stem

Implications for REM Sleep Physiology

ABDEL AHNAOU,[a,b] LAURENT YON,[c] MICHEL ARLUISON,[d] HUBERT VAUDRY,[c] JENS HANNIBAL,[e] MICHEL HAMON,[a] JOELLE ADRIEN,[a] AND PATRICE BOURGIN[a,f]

[a] UMR 677 INSERM/UPMC, GHU Pitié-Salpêtrière, 75013 Paris, France

[b] Johnson & Johnson Pharmaceutical Research and Development, N. V. Turnhoutseweg 30, B-2340 Beerse, Belgium

[c] INSERM U413, Institut Fédératif de Recherche Multidisciplinaire sur les Peptides (IFRMP23), Université de Rouen, 76821 Mont-Saint-Aignan France

[d] Neurobiologie des Signaux intercellulaires CNRS UMR 7101/UPMC, 7 quai Saint Bernard 75252 Paris, CEDEX 05 France, Institut des Neurosciences, CNRS UMR 7102/UPMC, 75005 Paris, France

[e] University Department of Clinical Biochemistry, Bisbebjerg Hospital, Bisbebjerg Bakke 23, DK-2400 Copenhagen NV, Denmark

[f] Stanford University, Department of Psychiatry and Department of Biological Sciences, Stanford, California 94305, USA

ABSTRACT: Recent evidence indicates that pituitary adenylate cyclase-activating polypeptide (PACAP) and vasoactive intestinal polypeptide (VIP) might play an important role in rapid eye movement sleep (REMS) generation at the pontine level in rats. We have thus examined the immunohistochemical distribution of VIP and PACAP in the pontine and mesencephalic areas known to be involved in REMS control in rats. A dense network of VIP-immunoreactive cell bodies and fibers was found in the dorsal raphe nucleus. A large number of PACAP-positive perikarya and nerve fibers was observed in the area known as the REMS induction zone within the pontine reticular formation (PRF). The present results provide an anatomical basis to our previous functional data, and suggest that PACAPergic mechanisms within the PRF play a critical role in long-term regulation of REMS.

Address for correspondence: Patrice Bourgin, M.D., Ph.D., Stanford University, Department of Biological Sciences, 371 Serra Mall, Stanford, CA 94305-5020. Voice: 650-723-5882; fax: 650-725-5356.

e-mail: bourgin@stanford.edu

Ann. N.Y. Acad. Sci. 1070: 135–142 (2006). © 2006 New York Academy of Sciences.

doi: 10.1196/annals.1317.095

KEYWORDS: pontine reticular formation; immunohistochemistry; pituitary adenylate cyclase-activating polypeptide; vasoactive intestinal polypeptide; rapid eye movement sleep; dorsal raphe nucleus

INTRODUCTION

A role for Vasoactive intestinal polypeptide (VIP) and pituitary adenylate cyclase-activating polypeptide (PACAP) in rapid eye movement sleep (REMS) regulation has been early suggested by studies reporting REMS enhancement following intracerebroventricular (i.c.v.) injection of VIP or PACAP.[1-3] Only a few studies, however, have examined the effects of direct intracerebral administration of VIP and PACAP on REMS.[4-7] These latter reports allowed the identification of brain areas involved in this sleep-promoting effect, i.e., the dorsal raphe nucleus[4] (DR) and the area within the pontine reticular formation (PRF), known as the "REMS induction zone"[5,6] in rat, and the amygdala, a structure strongly connected to the pontine generators of REMS, in cat.[7]

REMS regulation is known to be under the control of a reciprocal interaction between "REM off" and "REM on" structures located in the brain stem.[8] The "REM off" structures, i.e., the DR, that contains serotoninergic neurons, and the locus caeruleus (LC), that contains noradrenergic neurons, and the "REM on" structures, i.e., the pontine tegmentum and the PRF, are silent and active, respectively, during REMS (FIG. 1). The pontine tegmentum comprises the laterodorsal and pedunculopontine nuclei (LDT and PPT), two areas that provide a dense cholinergic innervation to the PRF[9] (FIG. 1). There is now ample evidence that cholinergic mechanisms at the pontine level are essential to the generation of REMS. In particular, direct local microinjection of carbachol (cholinergic agonist) allowed the identification of an "REMS induction zone" defined as the caudal part of the oral pontine reticular nucleus (cPnO) and the adjacent dorsal subcaeruleus nucleus (subCD) in rats.[10-12]

Interestingly, we found that a single microinjection of a low dose of VIP or PACAP into the REMS induction zone promotes sustained REMS for up to 12 days.[5,6] This long-term enhancement of REMS, which depends on VIP-PACAPergic/muscarinic receptor interactions,[13,14] is well suited for the study of both functional aspects and molecular mechanisms of this state of vigilance.

Both VIP and PACAP increase the intracellular content of cyclic AMP and act through activation of specific common receptors.[15] In order to determine the respective roles of each peptide as REMS promotor, and to establish the physiological relevance of their shared pharmacological effect, it is necessary to investigate whether endogenous VIP and/or PACAP actually exist in the areas targeted by the microinjections.

Here, we conducted an immunohistochemical study of VIP and PACAP in the rat brain stem in order to determine their respective distributions in the REMS induction zone and in "REM on" and "REM off" structures.

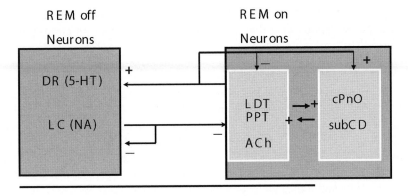

	VIP		PACAP	
	cell bodies	fibers	cell bodies	fibers
cPnO	-	±	++	++
subCD	-	±	++	++
LDT	±	±	-	+
PPT	-	+	-	+
DR	++++	+++	-	-
LC	-	+	+++*	++

FIGURE 1. The reciprocal model of "REM off"–"REM on" interaction[8] with the distribution of VIP- and PACAP-immunoreactive perikarya and fibers as determined in the components of the model. *Observed in colchicine-treated animals only. cPnO = caudal part of the oral pontine reticular nucleus; subCD = dorsal subcaeruleus nucleus; LDT = laterodorsal tegmentum; PPT = pedunculopontine tegmentum; DR = dorsal raphe nucleus; LC = locus caeruleus.

MATERIALS AND METHODS

Adult male Sprague Dawley rats, weighing 250–300 g, were used in the present study. The animals were handled in agreement with the ethical rules for experimentation on laboratory animals (DHEW Publication; 80–23; Office of Science and Health Reports, DRR/NIH, Bethesda, MD, USA; 1980). One day before sacrifice, half of the animals received an i.c.v. injection of colchicine. Briefly, 80 μg colchicine in 4 μL of phosphate buffered saline (PBS, pH 7.4) was injected over 10 min into the lateral ventricle under pentobarbital anesthesia (stereotaxic coordinates: P: 0.9 mm, L: 1.5 mm, V: −3.5 mm under the brain surface). Twenty-four hours later, the animals were deeply anesthetized (50 mg/kg of sodium pentobarbital) and perfused via the left ventricle with 100 mL of PBS followed by one of the fixative solutions described below. Brains were then removed and processed for immunostaining.

Three different antibodies were used with specific fixative conditions and dilutions as determined in previous experiments. *(1)* Anti-VIP antibodies (INC-STAR Corp., Stillwater, MN); dilution, 1:8000; fixation, 4% paraformaldehyde + 0.2% picric acid; 50-μm-thick free-floating coronal sections (vibratome). *(2)* Anti-PACAP27 antibodies (Peninsula, San Carlos, CA); dilution, 1:10,000; fixation, 3% paraformaldehyde + 1% glutaraldehyde; 15 μm-thick coronal sections on slides after cryoprotection (cryostat). *(3)* Monoclonal antibody against PACAP27 and PACAP38[16]; dilution, 1:5; fixation, 4% paraformaldehyde, 15-μm-thick coronal sections on slides after cryoprotection (cryostat). The latter two antibodies yielded comparable results except for the staining of nerve fibers that was greater using the monoclonal anti-PACAP antibody.

Tissue sections were pretreated with 0.1% H_2O_2 in PBS for 30 min to inhibit endogenous peroxidase activities. The sections were subsequently incubated overnight at 4°C with the primary antiserum supplemented with 1% normal goat serum and 0.2% Triton X-100. Then, the sections were incubated with a biotinylated secondary antibody (Vector, dilution, 1:250) for 2 h at room temperature. Thereafter, immunohistochemical staining was performed using the avidin-biotin-peroxidase method and H_2O_2-diamino-benzidine for revelation.

To control the specificity of the immunostaining, the primary antibodies were preabsorbed with an excess of respective peptide, VIP or PACAP. Immunostaining was not modified when the primary antibody directed against VIP or PACAP was preabsorbed with the heterologous peptide.

RESULTS

The distribution of PACAP- and VIP-immunoreactive elements appeared to be heterogeneous within brain stem structures known to participate in REMS regulation (FIG. 1). No VIP-positive cell bodies were visualized in the PRF, subCD, LDT, PPT, or LC nuclei. In contrast, numerous VIP cell bodies were observed in the DR nucleus, mainly in its caudal part. These neurons were fusiform, mainly uni- or bipolar, with short dendritic profiles, and were oriented in the dorsoventral direction within the DR (FIG. 2 B). Scarce VIP immunostained nerve fibers were observed in the PRF, subCD, LDT, PPT, and LC nuclei. They appeared as long varicose fibers and sometimes occurred individually (FIG. 2 D). A much greater number of short thin nerve fibers immunostained for VIP were identified in the DR where they exhibited a dorsoventral orientation (FIG. 2 B).

In sharp contrast with that found with anti-VIP antibodies, anti-PACAP antibodies labeled neuronal perikarya in the PRF, notably in the cPnO and adjacent subCD corresponding to the "REMS induction zone" (FIG. 1). These PACAP-immunoreactive neurons, of various sizes, were mostly multipolar, with very long dendrites and many of them corresponded to gigantocellular

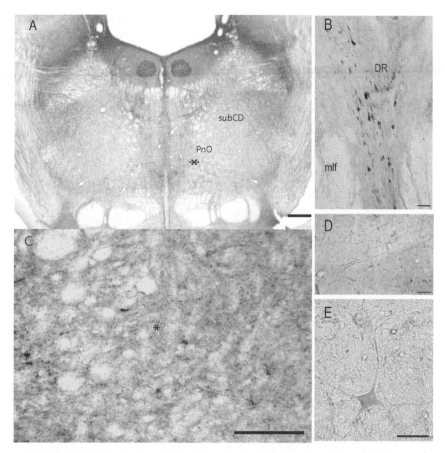

FIGURE 2. Photomicrographs of coronal sections showing VIP and PACAP immunoreactivity in REMS control areas. **(A)** PACAP immunoreactivity at level 8.8 mm posterior to bregma; **(B)** VIP immunoreactivity in the DR; **(C)** PACAP immunoreactivity in the cPnO; **(D)** scarce VIP positive nerve fibers in the cPnO; **(E)** PACAP-positive perikaryon with long dendrites in the cPnO. Scale bars: 500 μm in A, 25 μm in B, 100 μm in C, 50 μm in D, 50 μm in E. Abbreviations as in FIGURE 1.

nerve cell bodies (FIG. 2 E). A high density of PACAP-immunoreactive nerve fibers of various sizes without any particular orientation was also observed in the same area (FIG. 2 C). In rats treated with colchicine, a large number of PACAP-positive neurons were also visualized in the LC (FIG. 1).

DISCUSSION

Here, we report, for the first time, the presence of PACAP, but not VIP, in the REMS induction zone, an area defined as the caudal part of the PnO and the adjacent subCD nucleus. This PACAPergic labeling corresponds to both *(a)* fibers, suggesting the existence of a naturally occurring release of PACAP in this area, and *(b)* cell bodies with long dendritic ramifications, raising the question of the functional role of these cells, in particular whether they belong to the "REM on" cell component. Thus, further experiments are needed in order to measure the local release of PACAP and the firing rate of PACAP-positive cells *in vivo* across the sleep wakefulness cycle. The presence of PACAP but not VIP is also functionally coherent with the existence of specific binding sites with higher affinity for PACAP than VIP in the PRF.[6] Retrograde labeling studies have to be performed in order to determine the origin of the PACAP-positive fibers present in the REMS induction zone. Anterograde labeling studies from the LC would also be needed in order to determine the projection sites of the PACAP-positive cells located in this structure. Indeed, previous reports[17,18] and our observation of PACAP-positive cells in the LC of colchicine-treated animals suggest that PACAP might modulate wakefulness, in addition to its role in REMS regulation.

A large number of VIP-positive cell bodies as well as a dense VIPergic innervation were observed in the DR as previously reported.[19] In rats, microinjection of VIP into the DR, where serotoninergic neurons that project to the forebrain are located, has been shown to produce an increase in slow wave sleep (SWS) at low doses, and to promote REMS at larger doses.[4] VIP injected i.c.v. during the light period enhances REMS whereas it increases both SWS and REMS when injected during the dark period.[1,2] Altogether, these data suggest that VIPergic mechanisms within the DR may play a critical role in sleep wakefulness regulation, possibly through an inhibition of the serotoninergic system.

Our data do not support the hypothesis of a direct modulation of the "REM on" component by the VIPergic system. At variance with a previous report,[20] we did not detect any VIP-immunoreactive neuron in the pontine tegmentum. Only a very low density of VIP- and PACAP-immunoreactive fibers could be observed in this latter structure.

In conclusion, our immunocytochemical data showing that PACAP-positive cell bodies and fibers do exist in the REM sleep induction zone provide further support to the idea that PACAPergic mechanisms within the PRF play a critical role in long-term regulation of REMS.[3,6,14]

ACKNOWLEDGMENTS

Abdel Ahnaou was supported by a fellowship from the Fondation pour la Recherche Medicale. This work was granted by the Institut National de la Santé et de la Recherche Médicale and the Synthelabo European Sleep Research Program.

REFERENCES

1. Riou, F., R. Cespuglio & M. Jouvet. 1981. Hypnogenic properties of the vasoactive intestinal polypeptide in rats. C. R. Seances Acad. Sci. III. **293:** 679–682.
2. Obal, F., Jr., G. Sary, P. Alfoldi, *et al.* 1986. Vasoactive intestinal polypeptide promotes sleep without effects on brain temperature in rats at night. Neurosci. Lett. **64:** 236–240.
3. Fang, J., L. Payne & J.M. Krueger. 1995. Pituitary adenylate cyclase activating polypeptide enhances rapid eye movement sleep in rats. Brain Res. **686:** 23–28.
4. El Kafi, B., R. Cespuglio, L. Leger, *et al.* 1994. Is the nucleus raphe dorsalis a target for the peptides possessing hypnogenic properties? Brain Res. **637:** 211–221.
5. Bourgin, P., C. Lebrand, P. Escourrou, *et al.* 1997. Vasoactive intestinal polypeptide microinjections into the oral pontine tegmentum enhance rapid eye movement sleep in the rat. Neuroscience **77:** 351–360.
6. Ahnaou, A., M. Basille, B. Gonzalez, *et al.* 1999. Long-term enhancement of REM sleep by the pituitary adenylyl cyclase-activating polypeptide (PACAP) in the pontine reticular formation of the rat. Eur. J. Neurosci. **11:** 4051–4058.
7. Simon-Arceo, K., I. Ramirez-Salado & J.M. Calvo. 2003. Long-lasting enhancement of rapid eye movement sleep and pontogeniculooccipital waves by vasoactive intestinal peptide microinjection into the amygdala temporal lobe. Sleep. **26:** 259–264.
8. Pace-Schott, E.F. & J.A. Hobson. 2002. The neurobiology of sleep: genetics, cellular physiology and subcortical networks. Nat. Rev. Neurosci. **3:** 591–605.
9. Jones, B.E. 1990. Immunohistochemical study of choline acetyltransferase-immunoreactive processes and cells innervating the pontomedullary reticular formation in the rat. J. Comp. Neurol. **295:** 485–514.
10. Bourgin, P., P. Escourrou, C. Gaultier & J. Adrien. 1995. Induction of rapid eye movement sleep by carbachol infusion into the pontine reticular formation in the rat. Neuroreport **6:** 532–536.
11. Marks, G.A. & C.G. Birabil. 1998. Enhancement of rapid eye movement sleep in the rat by cholinergic and adenosinergic agonists infused into the pontine reticular formation. Neuroscience **86:** 29–37.
12. Fenik, V., V. Marchenko, P. Janssen, *et al.* 2002. A5 cells are silenced when REM sleep-like signs are elicited by pontine carbachol. J. Appl. Physiol. **93:** 1448–1456.
13. Bourgin, P., A. Ahnaou, A.M. Laporte, *et al.* 1999. Rapid eye movement sleep induction by vasoactive intestinal peptide infused into the oral pontine tegmentum of the rat may involve muscarinic receptors. Neuroscience **89:** 291–302.
14. Ahnaou, A., A.M. Laporte, S. Ballet, *et al.* 2000. Muscarinic and PACAP receptor interactions at pontine level in the rat: significance for REM sleep regulation. Eur. J. Neurosci. **12:** 4496–4504.
15. Vaudry, D., B.J. Gonzalez, M. Basille, *et al.* 2000. Pituitary adenylate cyclase-activating polypeptide and its receptors: from structure to function. Pharmacol. Rev. **52:** 269–324.
16. Hannibal, J., J.D. Mikkelsen, H. Clausen, *et al.* 1995. Gene expression of pituitary adenylate cyclase activating polypeptide (PACAP) in the rat hypothalamus. Regul. Pept. **55:** 133–148.
17. Tajti, J., R. Uddman & L. Edvinsson. 2001. Neuropeptide localization in the "migraine generator" region of the human brainstem. Cephalalgia **21:** 96–101.

18. HANNIBAL, J. 2002. Pituitary adenylate cyclase-activating peptide in the rat central nervous system: an immunohistochemical and in situ hybridization study. J. Comp. Neurol. **453:** 389–417.
19. PETIT, J.M., P.H. LUPPI, C. PEYRON, *et al.* 1995. VIP-like immunoreactive projections from the dorsal raphe and caudal linear raphe nuclei to the bed nucleus of the stria terminalis demonstrated by a double immunohistochemical method in the rat. Neurosci. Lett. **193:** 77–80.
20. SUTIN, E.L. & D.M. JACOBOWITZ. 1988. Immunocytochemical localization of peptides and other neurochemicals in the rat laterodorsal tegmental nucleus and adjacent area. J. Comp. Neurol. **270:** 243–270.

Microiontophoretically Applied PACAP Blocks Excitatory Effects of Kainic Acid *in Vivo*

TAMAS ATLASZ,[a] ZSOMBOR KŐSZEGI,[a] NORBERT BABAI,[a] ANDREA TAMÁS,[b] DORA REGLŐDI,[b,c] PETER KOVÁCS,[a] ISTVAN HERNÁDI,[a] AND ROBERT GÁBRIEL[a,d]

[a]*Department of Experimental Zoology and Neurobiology, University of Pécs, 7624 Pécs, Hungary*

[b]*Department of Anatomy, University of Pécs, 7624 Pécs, Hungary*

[c]*Neurohumoral Regulations Research Group of the Hungarian Academy of Sciences, University of Pécs, 7624 Pécs, Hungary*

[d]*Adaptational Biology Research Group of the Hungarian Academy of Sciences, University of Pécs, 7624 Pécs, Hungary*

ABSTRACT: Pituitary adenylate cyclase-activating polypeptide (PACAP) has been shown to be neuroprotective in animal models of different brain pathologies, including focal and global cerebral ischemia. The application of glutaminergic excitotoxin kainic acid (KA), similar to ischemic events, may lead to neurodegeneration. In the present article, we investigated the effects of microiontophoretic application of PACAP on the excitatory effects of KA. During recording-maintained spontaneous activity of single neurons, we microiontophoretized KA, which was followed by the application of PACAP-38. We found that PACAP could block the excitatory effects of KA in several brain areas (cortex: 89%, hippocampus: 36%, and thalamus: 50%). Moreover, we detected a lower level excitatory effect of PACAP alone (41%). The present results may explain the neuroprotective effects of PACAP observed in experimental models of glutamate (GLU)–receptor-mediated degenerative processes.

KEYWORDS: electrophysiology; extracellular recording; cortex; hippocampus; thalamus; neuroprotection

INTRODUCTION

Pathological increase of extracellular glutamate (GLU) concentration plays a key role in many neurological diseases and different GLU analogs have been

Address for correspondence: Dora Reglődi, Department of Anatomy, University of Pécs, 7624 Pécs, Szigeti u 12, Hungary. Voice: +36-72-536001; ext.: 5398; fax: +36-72-536393.
e-mail: dora.reglodi@aok.pte.hu

Ann. N.Y. Acad. Sci. 1070: 143–148 (2006). © 2006 New York Academy of Sciences.
doi: 10.1196/annals.1317.002

employed to study the processes leading to neurodegeneration and its prevention by neuroprotection.[1,2] Pituitary adenylate cyclase-activating polypeptide (PACAP) has been proposed as neuroprotective agent in several models of such conditions both *in vivo*[3,4] and *in vitro*.[5–8] At the molecular level, PACAP elevates cyclic adenosine $3',5'$-monophosphate (cAMP) levels and initiates protein kinase A-related mechanisms through a specific PAC1 receptor.[9] PACAP has been shown to be protective in few *in vivo* models of excitotoxic injury induced by GLU agonists: in ibotenate-induced cortical lesion in newborn hamsters and in monosodium GLU-induced retinal degeneration.[10,11] However, little is known about the *in vivo* effects of PACAP-exerted excitability on neurons of the central nervous system.[12] Therefore, the aims of the present study were (*a*) to examine the effect of microiontophoretically applied PACAP on maintained, baseline firing rate of neurons in different brain areas and (*b*) to investigate whether PACAP reduces neuronal excitation elicited by glutaminergic agonist kainic acid (KA) in anesthetized rats, *in vivo*.

MATERIALS AND METHODS

Adult male Long Evans rats (Charles River Laboratories, Budapest, Hungary) were studied during the experiments ($n = 11$). The animals were anesthetized by an intraperitoneal (i.p.) injection of ketamine (100 mg/kg) and were fixed in a stereotactic apparatus. The coordinates for target brain areas were AP: -3.6 mm, ML: 2.6 mm from bregma, and V: 0.5–7 mm from dura. At this position, cortical, hippocampal, and thalamic neurons could be recorded in a single vertical track.

Four-barreled electrodes (Carbostar-4, Kation-Scientific, Minneapolis, MN, USA) were used for recording and microiontophoresis. The impedance (measured at 50 Hz) was 0.2–0.6 MΩ for the central recording channel and, 5–150 MΩ for the drug channels, respectively. One of the drug channels (filled with 0.5M NaCl) was used for the application of continuous balancing current, while the other two capillaries were filled with one of the following substances (all dissolved in distilled water): KA (Sigma, 50 mM, ejection current: -5 to -30 nA), GLU (Fluka, 50 mM, -5 to -30 nA), and PACAP-38 (Tocris, 100 μM, -50 to $+50$ nA). Single-unit activity was recorded and passed to an IBM-compatible microcomputer through an analogue digital converter interface (Power 1401, CED, Cambridge, UK). Spike sorting and data analysis were performed by Spike2 software (CED). Frequency histograms were built and neuronal activity was displayed in cycles per seconds (cps, Hz). A neuron was considered responsive to treatment when its firing rate changed by $\pm20\%$ respective to its baseline level. Statistical analysis between pre- and post–treatment firing activity was performed using Student's paired and unpaired *t*-tests. The threshold for significance was set to $P < 0.01$. Changes of neural activity were also expressed as "normalized effects" (FIG. 2), indicating

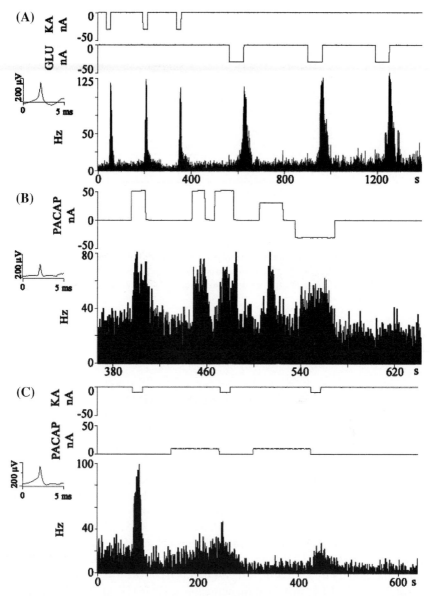

FIGURE 1. Effects of microiontophoretically applied KA, GLU, and PACAP on spontaneous neuronal activity in cortical, hippocampal, and thalamic neurons. (**A**) Control situation, excitatory effects on thalamic neuron induced by KA (50 mM) and GLU (50 mM). (**B**) PACAP (100 μM), ejected with either positive or negative current, increases firing rate of a thalamic neuron. (**C**) Microiontophoresis of PACAP (100 μM) on a thalamic neuron blocks KA-induced (50 mM) excitation. Hz, hertz; nA, nanoampere; s, second; ms, millisecond; μV, microvolt. Initial (barrel) concentrations of neuroactive compounds are indicated in brackets.

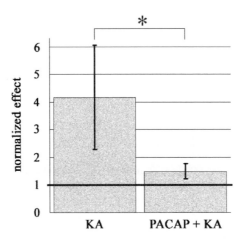

FIGURE 2. Summary diagram showing the normalized effects of microiontophoretically applied KA (50 mM) alone, or KA (50 mM) and PACAP (100 μM) on neuronal
activity. Initial (barrel) concentrations of neuroactive compounds are indicated in brackets.
The horizontal bold line indicates the spontaneous neuronal activity. KA: $n = 69$; PACAP
$+ KA: n = 40$. *$P < 0.01$.

the changes of the neurons' baseline firing rates (mean $\pm$ SEM) induced by
KA and PACAP treatments.

RESULTS AND DISCUSSION

In control situations, individually applied KA and GLU immediately excited
neurons studied in all brain regions (69 neurons of 11 animals). KA-evoked
responses were faster than those of GLU (Fig. 1 A). In line with earlier investigations,[12] we showed that PACAP alone exerted excitatory effect on hippocampal neurons and also stimulated cortical and thalamic neurons. Ejection
of PACAP-38 alone increased the firing rate of cortical ($n = 10/18$, 56%),
hippocampal ($n = 7/14$, 50%), and thalamic ($n = 12/38$, 32%) neurons. After
terminating the PACAP application, the neuronal activity returned to control
levels (Fig. 1 B). PACAP pretreatment was able to block the effects of KA
(Figs. 1C and 2) in 58% of recorded neurons (cortex: $n = 16/18$ 89%; hippocampus: $n = 5/14$, 36%; and thalamus: $n = 19/38$, 50%). PACAP-reduced
KA evoked excitation of neuronal firing and this effect was longlasting. The
above described action of PACAP may provide a useful neurophysiological
background for new potential treatment schedules in animal models of excitatory neurodegeneration and, possibly, in the treatment of similar symptoms
in humans. In addition, it is in line with the present observations, that a 1-h
exposure of cerebellar granule cells to PACAP induced a maximum survival of
cultured neurons and that longer incubations did not further increase survival.[13]

Similarly, in a rat model of focal cerebral ischemia, a 7-day pretreatment with PACAP resulted in no further reduction of the infarct size compared to a single preischemic bolus injection.[14] Acute treatment with PACAP seems to induce an immediate activation of a specific biochemical pathway, which promotes longlasting survival.[13]

In summary, the present study provides evidence that microiontophoretically applied PACAP on single neurons can reduce neuronal excitation induced by a potent GLU receptor agonist, such as KA.

ACKNOWLEDGMENTS

The authors thank Dóra Molnár for technical assistance.

This work was supported by OTKA T046589/61766 and the Hungarian Academy of Sciences.

REFERENCES

1. PEDERSEN, V. & W.J. SCHMIDT. 2000. The neuroprotective properties of glutamate antagonists and antiglutamatergic drugs. Neurotox. Res. **2:** 179–204.
2. DANYSZ, W. & C.G. PARSONS. 2002. Neuroprotective potential of ionotropic glutamate receptor antagonists. Neurotox. Res. **4:** 119–126.
3. SOMOGYVARI-VIGH, A. & D. REGLODI. 2004. Pituitary adenylate cyclase activating polypeptide: a potential neuroprotective peptide. Curr. Pharm. Des. **10:** 2861–2889.
4. TAMÁS, A. *et al.* 2004. Effects of pituitary adenylate cyclase activating polypeptide in retinal degeneration induced by monosodium- glutamate. Neurosci. Lett. **372:** 110–113.
5. SHOGE, K. *et al.* 1999. Attenuation by PACAP of glutamate-induced neurotoxicity in cultured retinal neurons. Brain Res. **839:** 66–73.
6. MORIO, H. *et al.* 1996. Pituitary adenylate cyclase activating polypeptide (PACAP) is a neurotrophic factor for cultured rat cortical neurons. Ann. N.Y. Acad. Sci. **805:** 476–481.
7. SAID, S.I. *et al.* 1998. Glutamate toxicity in the lung and neuronal cells: prevention or attenuation by VIP and PACAP. Ann. N.Y. Acad. Sci. **865:** 226–23.
8. RABL, K. *et al.* 2002. PACAP inhibits anoxia-induced changes in physiological responses in horizontal cells in the turtle retina. Regul. Pept. **109:** 71–74.
9. VAUDRY, D. *et al.* 2000. Pituitary adenylate cyclase activating polypeptide and its receptors: from structure to functions. Pharmacol. Rev. **52:** 269–324.
10. GRESSENS, P. *et al.* 2000. VIP and PACAP 38 modulate ibotenate-induced neuronal heterotopias in the newborn hamster neocortex. J. Neuropathol. Exp. Neurol. **59:** 1051–1062.
11. BABAI, N. *et al.* 2005. Degree of damage compensation by various PACAP treatments in monosodium glutamate-induced retina degeneration. Neurotox. Res. **8:** 227–233.
12. DI MAURO, M., S. CAVALLARO & L. CIRANNA 2003. Pituitary adenylate cyclase-activating polypeptide modifies the electrical activity of CA1 hippocampal neurons in the rat. Neurosci. Lett. **337:** 97–100.

13. VAUDRY, D. *et al*. 1998. Pituitary adenylate cyclase activating polypeptide stimulates both c-fos gene expression and cell survival in rat cerebellar granule neurons through activation of the protein kinase A pathway. Neuroscience **84:** 801–812.
14. REGLODI, D. *et al*. 2002. Effects of pretreatment with PACAP on the infarct size and functional outcome in rat permanent focal cerebral ischemia. Peptides **23:** 2227–2234.

Search for the Optimal Monosodium Glutamate Treatment Schedule to Study the Neuroprotective Effects of PACAP in the Retina

NORBERT BABAI,[a] TAMAS ATLASZ,[a] ANDREA TAMÁS,[b]
DORA REGLŐDI,[b,c] GABOR TÓTH,[d] PETER KISS,[b]
AND ROBERT GÁBRIEL[a,e]

[a]Departments of Experimental Zoology and Neurobiology, University of Pécs, 7624 Pécs, Hungary

[b]Department of Anatomy, University of Pécs, 7624 Pécs, Hungary

[c]Neurohumoral Regulations Research Group of the Hungarian Academy of Sciences, University of Pécs, 7624 Pécs, Hungary

[d]Department of Medical Chemistry, University of Szeged, 6701 Szeged, Hungary

[e]Adaptational Biology Research Group of the Hungarian Academy of Sciences, University of Pécs, 7624 Pécs, Hungary

ABSTRACT: We have previously shown the protective effects of pituitary adenylate cyclaseactivating polypeptide (PACAP) in monosodium glutamate (MSG)-induced retinal degeneration. In the present article, we have investigated the optimal model for examining this neuroprotective effect. One MSG treatment on postnatal (P) days 1 or 5, in spite of leading to same ultrastructural changes, does not cause enough damage to study neuroprotection. When retinas were treated three times with MSG, the entire inner retina degenerated. Neuroprotection with PACAP was achieved with at least two treatments. Evidence suggests that PACAP provides protection against excitotoxicity, therefore, it may be a useful agent in reducing excitotoxic damage in the retina.

KEYWORDS: retina; MSG; PACAP; plexiform layers; ribbon synapses; degeneration

INTRODUCTION

Pathological increase of glutamate levels plays a key role in neuronal damage in many diseases.[1] In the eye, several pathological conditions can be

Address for correspondence: Dora Reglődi, M.D., Ph.D., Department of Anatomy, University of Pécs, 7624 Pécs, Szigeti u 12. Hungary. Voice: +36-72-536001; ext.: 5398; fax: +36-72-536393.
e-mail: dora.reglodi@aok.pte.hu

Ann. N.Y. Acad. Sci. 1070: 149–155 (2006). © 2006 New York Academy of Sciences.
doi: 10.1196/annals.1317.003

mimicked by experimentally elevating extracellular glutamate concentrations.[2] The damaging effects of monosodium glutamate (MSG) on the retina have long been known. In accordance with others, we have also demonstrated that systemic treatment of neonatal rats with MSG leads to the destruction of the entire inner retina.[3] Pituitary adenylate cyclase-activating polypeptide (PACAP) and its receptors are present in the retina,[4,5] and the protective effects of both PACAP and vasoactive intestinal peptide (VIP) have been demonstrated in some retinal pathological conditions.[6–8] PACAP also attenuates glutamate toxicity in the retina *in vitro*,[9] in addition to similar effects observed in neuronal cultures.[10,11] Recently, we have shown that intravitreous injection of PACAP ameliorates MSG-induced retinal degeneration in neonatal rats.[3] The aims of the present article were (*a*) to examine the ultrastructural changes caused by MSG treatment and in correlation with these changes (*b*) to find the optimal model for investigating the neuroprotective effect of PACAP in MSG-induced retinal degeneration, that is, when MSG already exerts its damaging effects but those can still be readily reversed by PACAP treatment.

MATERIALS AND METHODS

Newborn Wistar rats were injected subcutaneously (s.c.) with 2 mg/g bodyweight MSG.[3] The following treatments were performed: one time injection on postnatal day (P) 1; one time injection at P5; and three times injection at P1, P5, and P9. In PACAP-treated pups, 100 pmol PACAP[3] in 5 μL saline was injected into the vitreous of one eye with a Hamilton syringe at the time of (*a*) the first, (*b*) the first two, or (*c*) all three MSG injections.

At P21, retinas were processed for histological analysis as previously described.[3] Briefly, immediately after removal, eyes were dissected in ice-cold phosphate buffer (PB) and fixed in 4% paraformaldehyde dissolved in 0.1 M PB. Sections of 2 μm were stained with toluidine blue. Samples for measurements derived from at least six tissue blocks prepared from at least three animals ($n = 2$–5 measurements from one tissue block). The following parameters were measured on digital photographs: cross-section of the retina from the outer limiting membrane to the inner limiting membrane; the width of the inner plexiform layer (IPL) from the bottom of perikarya of the last cell row in the inner nuclear layer (INL) to the top of perikarya in the ganglionic layer (GCL); and the number of cells/100 μm section length in the GCL. Statistical comparisons were made using the analysis of variance (ANOVA) test followed by Neuman Keuls posthoc analysis. Electron microscopy was performed on tissues fixed with 4% paraformaldehyde supplemented with 1% glutaraldehyde dissolved in 0.1 M PB. After washing in PB, tissue samples were treated with 1% OsO_4 in PB and embedded for routine electron microscopic examination.

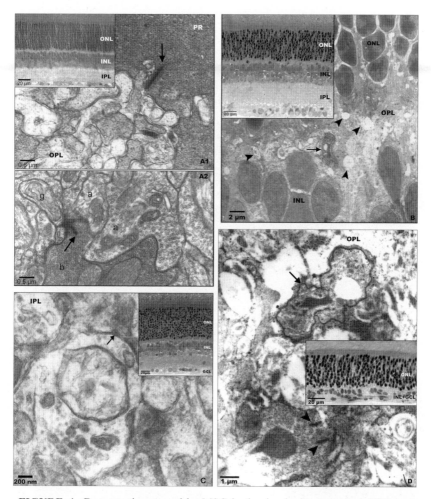

FIGURE 1. Degeneration caused by MSG in the developing rat retina. ONL: outer nuclear layer, OPL: outer plexiform layer, INL: inner nuclear layer, IPL: inner plexiform layer. (**A**) Control retina, black *arrow* shows intact ribbon synapses in the OPL (**A 1**; PR: photoreceptor cell) and in the IPL (**A 2**; b: bipolar cell terminal, a: amacrine cell dendrite, g: ganglion cell dendrite). (**B**) One time MSG treatment at P1; *arrowheads*: degenerative figures in the OPL; *arrow*: degenerating cell debris. (**C**) One time MSG treatment at P5; *arrow*: synapse in the IPL. (**D**) Three times MSG treatment; *arrows*: degenerating cell; *arrowheads*: ribbons in the photoreceptor cell. Inserts: light microscopic appearance of the retinas.

RESULTS AND DISCUSSION

Compared to normal retinas (FIG. 1 A), one time MSG treatment either at P1 or P5 did not cause discernible alteration at the light microscopic level, but

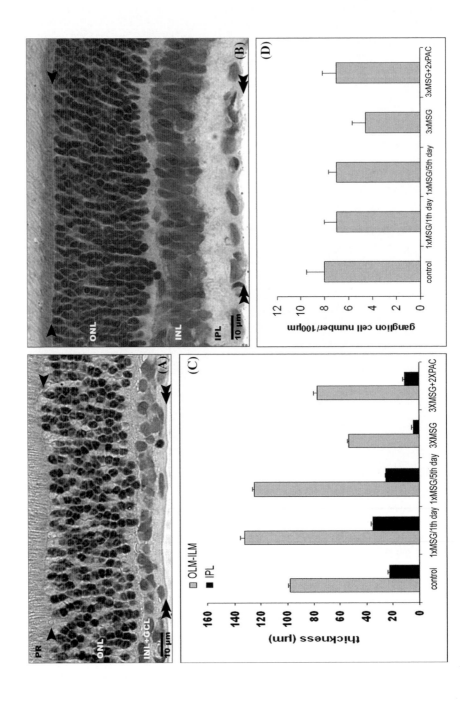

some degenerative processes were apparent with electron microscope (FIG. 1 B, C). The photoreceptor cells as well as the other cellular layers were normal. Though the plexiform layers seemed to be undisturbed in the light microscope, both the outer plexiform layer (OPL) and the IPL showed signs of damage at the ultrastructural level (FIG. 1 B, C). In the OPL, several holes could be found in the tissue around the photoreceptor cell terminals and scattered in the OPL. The ribbon synapses seemed unaffected, the cytoplasm of the cells (that of both the photoreceptors and the interneurons) was normal (FIG. 1 B). Degenerating cell debris was found in the OPL, but signs of increased microglial activity were not seen (FIG. 1 B). These findings correlate well with the facts that (*a*) mostly the inner retinal neurons contain ionotropic glutamate receptors,[12] therefore photoreceptors should remain unaffected and (*b*) in our earlier study, we have shown that only the inner retina degenerates after repeated MSG treatment.[3] However, it is surprising that even after one MSG treatment the OPL showed several signs of initial degenerative processes, while the IPL was almost unaffected. All types of synapses described in earlier studies were present.[13] Degenerative signs were only occasionally seen. There was little difference between the animals treated with MSG at P1 (when new cells are still being generated) or P5 (when most intensive synaptogenesis occurs).

The entire IPL disappeared in retinas with three times MSG treatment and the INL and GCL seemed fused (FIG 1 D). At the ultrastructural level, no remnants of the IPL (e.g., ribbon synapses of bipolar cells and serial amacrine cell synapses) were found. Only a structurally disturbed OPL could be studied, where the photoreceptor synapses were still present (FIG. 1 D), along with numerous degenerative structures (holes, cell debris, extracellularly located filamentous material, etc.). Such an extensive damage was not expected in the OPL, since only little alteration could be seen at the light microscopic level. Interestingly, in short-term excitotoxic insults the OPL and IPL seemed to be equally prone to damage.[14] The explanation for this may be that in our case the animals were killed long after the actual insult, therefore, there might have been enough time for structural rearrangement.[15]

FIGURE 2. Neuroprotecive effect of PACAP in MSG-induced degeneration; demonstration in light microscopic photographs and diagrams. **(A)** The INL and GCL are fused in the three times MSG-treated retina. **(B)** Three times MSG + two times PACAP-treated retina. The layers, including the IPL, are well visible. The ONL seems undisturbed, while the other layers of the retina are reduced. Both the width of the IPL and the distance between the outer (*arrowhead*) and inner (*double arrowhead*) limiting membranes (OLM-ILM) is the smallest in the three times MSG-treated preparations. In the PACAP-treated retina, these parameters are close to the control results. **(C)** Measurements of the IPL width and the OLM–ILM distance. **(D)** Number of cells in the ganglion cells layer/100 μm section length. In the three times MSG-treated retina, the cell number is decreased compared to the control and the other treatment schedules. Two times PACAP treatment counteracts the damaging effect of three times MSG treatment.

In the above model, neuroprotection with PACAP could be achieved with at least two treatments at the time of the first and second MSG application (FIG. 2 A, B). One PACAP application was not enough to exert protection; three applications did not provide significantly better protection than the two times treatment (not illustrated). This qualitative picture is well reflected in the measured morphometric parameters. (FIG. 2 C, D).

In summary, one time MSG treatment does not cause enough damage to study neuroprotection in our model, in spite of leading to some ultrastructural changes. At least two times MSG treatment is necessary to exert large-scale degeneration in the retina. Evidence also suggests that PACAP provides protection against excitotoxicity in the retina. This effect is likely to be mediated by PAC1 receptors coupled to high activity of cAMP production.[16] Topic PACAP application therefore may be a possible treatment for conditions caused by excitotoxic damage in the retina.

ACKNOWLEDGMENTS

This work was supported by OTKA T046589/61766, T048848, and the Hungarian Academy of Sciences.

REFERENCES

1. SUCHER, N.J., S.A. LIPTON & E.B. DREYER. 1997. Molecular basis of glutamate toxicity in retinal ganglion cells. Vision Res. **37:** 3483–3493.
2. VIDAL-SANZ, M. *et al.* 2000. Death and neuroprotection of retinal ganglion cells after different types of injury. Neurotox. Res. **2:** 215–227.
3. TAMÁS, A. *et al.* 2004. Effects of pituitary adenylate cyclase activating polypeptide in retinal degeneration induced by monosodiumglutamate. Neurosci. Lett. **372:** 110–113.
4. SEKI, T. *et al.* 2000. Gene expression for PACAP receptor mRNA in the rat retina by in situ hybridization and in situ RT-PCR. Ann. N. Y. Acad. Sci. **921:** 366–369.
5. IZUMI, S. *et al.* 2000. Ultrastructural localization of PACAP immunoreactivity in the rat retina. Ann. N. Y. Acad. Sci. **921:** 317–320.
6. SILVEIRA, M.S. *et al.* 2002. Pituitary adenylate cyclase activating polypeptide prevents induced cell death in retinal tissue through activation of cyclic AMP-dependent protein kinase. J. Biol. Chem. **277:** 16075–16080.
7. TUNCEL, N. *et al.* 1996. Protection of rat retina from ischemia-reperfusion injury by vasoactive intestinal peptide (VIP): the effect of VIP on lipid peroxidation and antioxidant enzyme activity of retina and choroid. Ann. N. Y. Acad. Sci. **805:** 489–498.
8. SEKI, T. *et al.* 2003. Pituitary adenylate cyclase activating polypeptide (PACAP) protects against ganglion cell death after cutting of optic nerve in the rat retina [abstract]. Regul. Pept. **115:** 55.
9. SHOGE, K. *et al.* 1999. Attenuation by PACAP of glutamate-induced neurotoxicity in cultured retinal neurons. Brain Res. **839:** 66–73.

10. SAID, S.I. *et al*. 1998. Glutamate toxicity in the lung and neuronal cells: prevention or attenuation by VIP and PACAP. Ann. N. Y. Acad. Sci. **865:** 226–237.
11. SHINTANI, N. *et al*. 2005. Neuroprotective action of endogenous PACAP in cultured rat cortical neurons. Regul. Pept. **126:** 123–128.
12. THORESON, W.B. & P. WITKOVSKY. 1999. Glutamate receptors and circuits in the vertebrate retina. Prog. Retin. Eye Res. **18:** 765–810.
13. DOWLING, J.E. 1987. The Retina. An Approachable Part of the Brain. Belknap Press, Cambridge, MA.
14. KLEINSCHMIDT, J., C.L. ZUCKER & S. YAZULLA. 1986. Neurotoxic action of kainic acid in the isolated toad and goldfish retina: II. Mechanism of action. J. Comp. Neurol. **254:** 196–208.
15. ROMANO, C., M.T. PRICE & J.W. OLNEY. 1995. Delayed excitotoxic neurodegeneration induced by excitatory amino acid agonists in isolated retina. J. Neurochem. **65:** 59–67.
16. NILSSON, S.F. *et al*. 1994. Characterization of ocular receptors for pituitary adenylate cyclase activating polypeptide (PACAP) and their coupling to adenylate cyclase. Exp. Eye Res. **58:** 459–467.

Can PACAP-38 Modulate Immune and Endocrine Responses During Lipopolysaccharide (LPS)-Induced Acute Inflammation?

AGNIESZKA BARANOWSKA-BIK, WOJCIECH BIK,
EWA WOLINSKA-WITORT, MAGDALENA CHMIELOWSKA,
LIDIA MARTYNSKA, AND BOGUSLAWA BARANOWSKA

*Neuroendocrinology Department, Medical Centre of Postgraduate Education,
01 – 813 Warsaw, Poland*

ABSTRACT: Pituitary adenylate cyclase-activating polypeptide (PACAP) shows a potential anti-inflammatory activity and interacts with the endocrine system. The aim of the present article was to evaluate the effects of PACAP38 on the endocrine and immune systems during acute inflammation. Rats used in the experiments, divided into four groups, were given intraperitoneal injection of, respectively 0.9% NaCl, LPS, PACAP38, and LPS+PACAP38. Hormone (pituitary, adrenal, and thyroid) and cytokine (TNF-α, IL-6, IL10) concentrations were measured 2 and 4 h after the injection. Treatment with LPS + PACAP, as compared to LPS, caused TNF-α and corticosterone to decrease and T4 to increase after 2 h. These data suggest that PACAP modulates both the endocrine and immune responses in this model of septic shock.

KEYWORDS: PACAP-38; immune system; endocrine system; acute inflammation

INTRODUCTION

Pituitary adenylate cyclase-activating polypeptide 38 (PACAP38), a 38-amino acid peptide, belongs to the glucagon secretin superfamily.[1] PACAP has been implicated as a regulator of the immune reaction during acute and chronic inflammation (e.g., rheumatoid arthritis).[2–4] Direct inhibition of proinflammatory cytokine production in response to PACAP was noted in macrophages. PACAP-stimulated upregulation of anti-inflammatory secretion during lipopolysaccharide (LPS)-induced acute inflammation (a model of

Address for correspondence: Agnieszka Baranowska-Bik, M.D., Neuroendocrinology Department, Medical Centre of Postgraduate Education, Marymoncka 99, 01 – 813 Warsaw, Poland. Voice: +48 22 56 93 850; fax: +48 22 56 93 859.
e-mail: zncmkp@op.pl

Ann. N.Y. Acad. Sci. 1070: 156–160 (2006). © 2006 New York Academy of Sciences.
doi: 10.1196/annals.1317.004

septic shock) has also been observed. In addition, PACAP protected mice from LPS-induced lethal endotoxemia.[3,4]

The endocrine system has been reported to participate in the processes accompanying acute inflammation. The action of the endocrine systems includes the mobilization of the hypothalamo–pituitary–adrenal axis and the attenuation of the hypothalamo–pituitary–thyroid axis.[5]

The aim of the present article was to evaluate the effects of PACAP38 administration on the endocrine and immune systems during LPS-induced acute inflammation. Serum concentrations of proinflammatory (tumor necrosis factor-α [TNF-α] and interleukin-6 [IL-6]) cytokines, anti-inflammatory (IL10) cytokines, and pituitary, adrenal and thyroid hormones were estimated so as to determine whether PACAP could change the hormonal activity and, simultaneously, act as an anti-inflammatory agent in acute inflammation.

MATERIALS AND METHODS

Male Wistar-Kyoto rats (250–300 g) used in the experiment were divided into four groups: group 1 (20 control animals), group 2 (19 animals), group 3 (14 animals), and group 4 (14 animals) received intraperitoneally 150 μL 0.9% NaCl, 600 μg LPS, 30 nmol PACAP38 and 600 μg LPS + 30 nmol PACAP38, respectively. Two to four hours after injection, the animals were decapitated. The trunk blood was collected and the serum was separated and stored at –20°C. All experimental procedures and protocols were approved by the First Warsaw Ethic Committee for Experiments on Animals. The serum concentrations of IL-6 and TNF-α were measured after 2 h, and IL10 after 4 h, using ELISA kits. The serum T3, T4, thyroid-stimulating hormone (TSH), and corticosterone concentrations were estimated after 2 and 4 h by radioimmunoassay (RIA) methods using commercial kits. The results were analyzed by the Mann–Whitney test to determine significant differences between the groups. The results were expressed as the mean ± SEM, and the statistical significance was accepted at $P < 0.05$. The correlation coefficient between cytokines and hormones was calculated according to the Spearman test.

RESULTS

Treatment with LPS, as compared to 0.9% NaCl (control group), significantly increased TNF-α and IL-6 serum concentrations after 2 h and IL10 concentrations after 4 h ($P < 0.001$, $P < 0.001$, $P < 0.001$, respectively; FIG. 1). Furthermore, LPS administration decreased serum concentrations of TSH after 2 h (2.63 versus 1.17 ng/mL; $P < 0.01$) and after 4 h (1.62 versus 1.06; $P < 0.05$) without changing T3 and T4 values. These results suggest that the activity of the hypothalamo–pituitary–thyroid axis decreases during acute

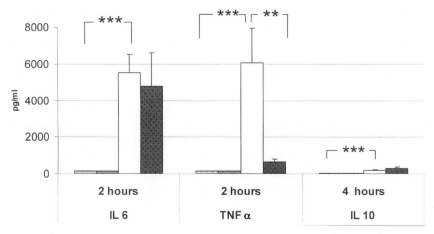

FIGURE 1. Serum cytokine concentration after 2 and 4 h. *** $P < 0.001$; ** $P < 0.01$.

inflammation as in the euthyroid sick syndrome where TSH values are below the normal range and T3, T4 concentrations remain unchanged at the onset of the severe illness.

As expected, LPS administration produced an increase in the corticosterone concentration after 2 (1831 versus 552 nmol/L; $P < 0.001$) and 4 h (1275 versus 638 nmol/L; $P < 0.05$). Administration of PACAP-38, as compared to 0.9% NaCl (control group), increased corticosterone concentration (1040 versus 638 nmol/L; $P < 0.05$) after 4 h.

The results obtained in the LPS + PACAP38 group showed an increase in TNF-α secretion after 2 h (compared to the control group). However, this effect was smaller than that seen in response to LPS ($P < 0.01$; FIG. 1). Moreover, LPS + PACAP38 treatment caused a reduction in corticosterone secretion induced by LPS after 2 h (LPS + PACAP versus LPS; $P < 0.05$; FIG. 2). Higher concentrations of T4 after 2 h were found in the group treated with LPS + PACAP38, as compared to the LPS-treated group (30.86 versus 22.22 nmol/L; $P < 0.05$). To assess a possible correlation, statistical analyses were done in all groups. A positive correlation between TNF-α and corticosterone ($R = 0.75$; $P < 0.001$), and that between IL-6 and corticosterone ($R = 0.69$; $P < 0.001$) were found. The correlations between T3, T4, TSH, and cytokines were not significant.

DISCUSSION

The effects of PACAP38 on macrophage and lymphocyte secretion during inflammation have been described as follows. PACAP38 attenuates production of proinflammatory agents (IL-6, IL12, TNF-α, TNFγ) and chemokines (MIP, IL8, RANTES); moreover, it could stimulate anti-inflammatory

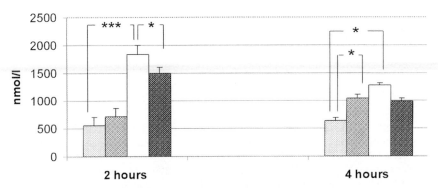

FIGURE 2. Serum corticosterone concentrations after 2 and 4 h. *** $P < 0.001$, * $P < 0.05$ NaCl PACAP38 LPS LPS + PACAP38.

cytokines (IL10).[6,7] In addition, the inhibition of iNOS synthesis in response to PACAP treatment has also been observed.[6,8] The present study has shown that PACAP38, administered in an average dose (30 nmol per animal), can modulate TNF-α production. An increase in TNF-α after 2 h was seen in the case of LPS + PACAP38. However, this effect was found to be much less significant than that caused by LPS. PACAP38 did not affect the IL-6 and IL10 concentrations since the levels of these cytokines were comparable both in LPS and LPS + PACAP38 groups. This suggests that there is a probable dose-dependency in anti-inflammatory effects of PACAP.

Previous works have demonstrated the changes in the endocrine system that result from acute inflammation reactions.[5] An increase in serum proinflammatory cytokines stimulates the synthesis of glucocorticoids through a direct action on the adrenal cortex. Additionally, indirect action on steroidogenesis via stimulation of corticotropin releasing hormone (CRH) and adrenocorticotrophic hormone (ACTH) secretion has also been reported.[5]

In our study, PACAP treatment, in combination with LPS, resulted in a lower, compared to that induced by LPS, concentrations of corticosterone after 2 h. Together with findings of a strong positive correlation between proinflammatory cytokines and corticosterone, the above results lead us to the conclusion that modulation of corticosterone secretion is mostly related to PACAP38-dependent suppression of proinflammatory cytokine production.

A significant increase in T4 concentration after 2 h, but not after 4 h, was observed in the PACAP38 + LPS-treated group as compared to the LPS-treated group. However, PACAP failed to changed TSH secretion since no differences in TSH concentration were found. These results suggest that PACAP modulates T4 secretion via other mechanisms than the hypothalamo–pituitary–thyroid axis.

We conclude that PACAP38 is a regulator of immune and endocrine responses during acute inflammation, but its effect seems to be short-lived.

ACKNOWLEDGMENT

This study was supported by scientific program no. 501-2-2-25-28/00.

REFERENCES

1. VAUDRY, D., B.J. GONZALEZ, M. BASILLE, *et al.* 2000. Pituitary adenylate cyclase-activating polypeptide and its receptors: from structure to function. Pharmacol. Rev. **52:** 269–324.
2. SHERWOOD, N.M., S.L. KRUECKL & J.E. MCROY. 2000. The origin and function of PACAP /glucagon superfamily. Endocr. Rev. **21:** 619–670.
3. DELGADO, M., C. MARTINEZ, D. POZO, *et al.* 1999. VIP and PACAP protect mice from lethal endotoxemia through the inhibition of TNF-α and IL-6. J. Immunol. **162:** 1200–1205.
4. DELGADO, M., C. ABAD, C. MARTINEZ, *et al.* 2003. PACAP in immunity and inflammation. Ann. N. Y. Acad. Sci. **992:** 141–157.
5. BASEDOVSKY, H.O. & A. DEL REY. 1996. Immune–Neuro–Endocrine interactions: facts and hypothesis. Endocrine Rev. **17:** 64–102.
6. GANEA, D. & M. DELGADO. 2001. Neuropeptides as modulators of macrophage functions. Regulation of cytokine production and antigen presentation by VIP and PACAP. Archiv. Immunol. et Therap. Exp. **49:** 101–110.
7. DELGADO, M., E.J. MUNOZ-ELIAZ, R.P. GOMARIZ & D. GANEA. 1999. VIP and PACAP enhance IL10 production by murine macrophages: in vitro and in vivo studies. J. Immunol. **162:** 1707–1716.
8. DELGADO, M., E.J. MUNOZ-ELIAS, R.P. GOMARIZ & D. GANEA. 1999. VIP and PACAP prevent inducible nitric oxide synthase transcription in macrophages by inhibiting NF-κB and IFN regulatory factor activation. J. Immunol. **162:** 4685–4696.

The Glucagon–Miniglucagon Interplay

A New Level in the Metabolic Regulation

DOMINIQUE BATAILLE, GHISLAINE FONTÉS, SAFIA COSTES, CHRISTINE LONGUET, AND STÉPHANE DALLE

INSERM U 376, CHU Arnaud-de-Villeneuve, 34295 Montpellier Cedex 05, France

ABSTRACT: Miniglucagon (glucagon 19–29) is the ultimate processing product of proglucagon, present in the glucagon-secreting granules of the α cells, at a close vicinity of the insulin-secreting β cells. Co-released with glucagon and thanks to its original mode of action and its huge potency, it suppresses, inside the islet of Langerhans, the detrimental effect of glucagon on insulin secretion, while it leaves untouched the beneficial effect of glucagon on glucose competence of the β cell. At the periphery, miniglucagon is processed at the surface of glucagon- and insulin-sensitive cells from circulating glucagon. At that level, it acts *via* a cellular pathway which uses initial molecular steps distinct from that of insulin which, when impaired, are involved in insulin resistence. This bypass allows miniglucagon to act as an insulin-like component, a characteristic which makes this peptide of particular interest from a pathophysiological and pharmacological point of views in understanding and treating metabolic diseases, such as the type 2 diabetes.

KEYWORDS: Miniglucagon; glucagon; proglucagon processing; insulin; type 2 diabetes

INTRODUCTION

The control of the glucose metabolism relies on both regulatory hormones (mainly insulin) and counterregulatory hormones (such as glucagon). From glucagon, is produced another peptide, miniglucagon or glucagon (19–29), which acts against its mother-hormone, and adds to the metabolic regulation novel steps that are to be taken into account in understanding (and possibly treating) metabolic diseases, such as type 2 diabetes.

Address for correspondence: Dominique Bataille, INSERM U 376, CHU Arnaud-de-Villeneuve, 34295 Montpellier Cedex 05, France. Voice: +33 4 67 41 52 30; fax: +33 4 67 41 52 22.
e-mail: bataille@montp.inserm.fr

Ann. N.Y. Acad. Sci. 1070: 161–166 (2006). © 2006 New York Academy of Sciences.
doi: 10.1196/annals.1317.005

PROGLUCAGON

Proglucagon[1] is the archetype of a multifunctions precursor,[2] which by tissue-specific posttranslational processing, furnishes the organism with signal molecules implicated in various regulatory mechanisms. In contrast to intestinal L cells that produce oxyntomodulin, glicentin, GLP-1, and GLP-2, the type of processing present in the endocrine pancreas, which appeared probably late during evolution, isolates the glucagon sequence from the octapeptide common to oxyntomodulin and glicentin, completely changing its biological specificity. However, glucagon is not the single peptide produced from proglucagon in α cells, since a further step in processing liberates a peptide displaying original and unexpected, if not surprising, properties.

MINIGLUCAGON

This peptide corresponds to the 11-amino acid C-terminal fragment of glucagon (glucagon 19–29). Its existence was uncovered[3] as being the peptide responsible for the observed glucagon effect on the hepatocyte plasma membrane calcium pump.[4] Since that time, we have observed that it is produced from circulating glucagon (few percent is transformed during a pass through an organ) by an enzymatic system at the level of the Arg_{17}-Arg_{18} basic doublet at the surface of glucagon-sensitive target cells.[5] Immediately after its local production and action, it is very quickly cleared from circulation. Accordingly, this peptide, carried to its target tissues by its mother hormone glucagon, is not a hormone itself and may act only locally, by modulating the actions of the mother hormone.

In those (glucagon- and miniglucagon-sensitive) tissues, it displays biological effects at extremely low doses (picomolar or lower), at least three order of magnitude below the active doses of glucagon.[6] The amount of miniglucagon produced (3–4% of the glucagon concentrations in α cells, a similar proportion in peripheral target tissues) is largely compensated by the huge difference between the active doses of the respective peptides. Miniglucagon is cosecreted with glucagon under hypoglycemic physiological conditions.[7]

Miniglucagon blocks the effect of all insulin secretagogues that use calcium entry into β cells, such as glucose, glucagon, GLP-1, or sulfonylurea.[6,7] It acts through a repolarization of the β cell plasma membrane, closing the voltage-sensitive calcium channels whose opening is necessary for secretion, while leaving untouched the cyclic adenosine $3',5'$-monophosphate (cAMP) levels.[6]

It is produced both in the glucagon secretory granules and at the surface of target tissues by the miniglucagon-generating endopeptidase (MGE), recently identified[8] as a metalloendoprotease (NRDc) that cleaves at the N-terminus

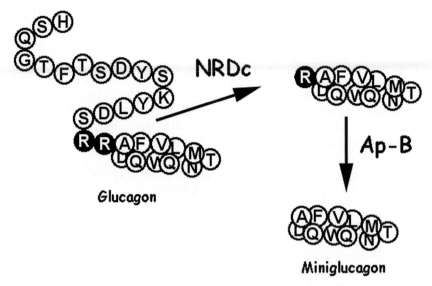

FIGURE 1. Processing of glucagon into miniglucagon by N-arginine dibasic convertase (NRDc) and aminopeptidase-B (Ap-B).

of arginine residues followed by suppression of the remaining basic residue(s) by a specific aminopeptidase (FIG. 1).

GLUCAGON–MINIGLUCAGON RELATIONSHIPS INSIDE THE ISLET

It is noteworthy that a *bona fide* partnership exists between α and β cells thanks to the fact that 35–50% of the latter have DIRECT contacts with α cells.[9] During the interprandial state when the hyperglycemic action of glucagon is necessary, the cosecreted miniglucagon "turns the calcium tap off" in the β cell, rendering impossible any nonphysiological effect of glucagon. This explains that, while exogenous glucagon stimulates insulin release,[10] endogenous glucagon, fortunately does not.[11] It remains to explain why glucagon has receptors at the surface of the β cell.[10] The glucagon receptors are coupled to adenylate cyclase, the cAMP produced stimulates protein kinase A (PKA) that activates molecules implicated in the trafficking of secretory granules, as well as nuclear targets which control gene expression necessary for regulating the secretory machinery. Glucose competence, that is, the ability to respond properly to glucose, may be maintained, or even recovered when partially lost, by peptides, such as glucagon,[12] that increase the cyclic AMP–PKA pathway in the β cell. Recent data[13,14] strongly suggest that the p44/42 MAP kinases (ERK1/2) are implicated in those mechanisms, since they activate nuclear transcription factors which regulate genes necessary for the β cell survival and functions, such as the antiapoptotic factor Bcl-2 or proinsulin. Furthermore,[14] ERK 1/2 phosphorylate proteins present at the surface of secretory granules, favoring the insulinosecretory effect of the awaited postprandial glucose wave (glucose competence). Since glucagon is able to stimulate ERK 1/2 in the β cell (13) in a pure cAMP/PKA manner, thus independently of calcium entry, it may exert its beneficial actions on the neighboring β cells (maintenance of the β cell mass and glucose competence) during the interprandial states, actions untouched by miniglucagon.

GLUCAGON–MINIGLUCAGON RELATIONSHIPS AT THE PERIPHERY

Secreted glucagon is transported to its target tissues *via* circulation. In contrast, its very short half-life precludes a hormonal status for miniglucagon. Thus, any observed effect of miniglucagon on a tissue remote from the islets is on account of a local production from circulating glucagon through MGE present at the cell surface. The possible importance of miniglucagon at the periphery was recently supported by the observation that a perfusion of the peptide together with a glucose load in vigil rats leads to a smaller insulin response, without any change in glycemia.[15] These observations led us to hypothesize the existence of an effect of the peptide on peripheral glucose-utilizing tissues, favoring or mimicking insulin action. We observed[15] that miniglucagon, as insulin does, induces translocation of the Glut-4 glucose transporter to the

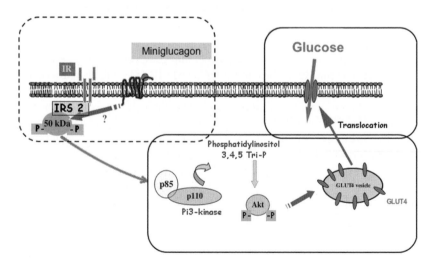

FIGURE 2. Mode of action of miniglucagon on glucose transport in 3T3-L1 adipocytes. The plain frames highlight the pathways common with insulin, and the dotted frame the original pathway.

plasma membrane of 3T3 adipocytes by use of the distal steps used by insulin (PI_3-kinase and Akt/PKB). On the other hand, the early steps differ: instead of using phosphorylation of the insulin receptor (IR) and of the insulin receptor substrate-1 (IRS-1), the miniglucagon action relies on phosphorylation of a 50-kDa protein that forms with the nonphosphorylated IR/IRS-2 (and not IRS-1) complex a signaling platform that passes the miniglucagon message to the PI_3-kinase/Akt/Glut-4 pathway (FIG. 2). Since a similar mechanism exists in the muscle (unpublished observations), miniglucagon appears as an insulin partner, acting in a direction opposite to that of its mother hormone, glucagon. Accordingly, any change in the MGE activity at the surface of target cells will set differently the glucagon–miniglucagon balance and, as a consequence, the glucagon versus insulin relationship, of major importance in regulating the catabolic–anabolic balance. The secondary processing of glucagon into miniglucagon appears thus as a new level in the metabolic regulation, which should be taken into account in the understanding of metabolic diseases, such as type 2 diabetes.

REFERENCES

1. BATAILLE, D. 1996. Preproglucagon and its processing. *In* Handbook of Experimental Pharmacology. P.J. Lefèbvre, Glucagon III, Eds.: 31–51. Springer, Heidelberg.
2. CHRÉTIEN, M. & N.G. SEIDAH. 1984. Precursor polyproteins in endocrine and neuroendocrine systems. Int. J. Pept. Protein Res. **23:** 335–341.
3. MALLAT, A., C. PAVOINE, M. DUFOUR, *et al.* 1987. A glucagon fragment is responsible for the inhibition of the liver Ca^{2+} pump by glucagon. Nature **325:** 620–622.
4. LOTERSZTAJN, S., R.M. EPAND, A. MALLAT & F. PECKER. 1984. Inhibition by glucagon of the calcium pump in liver plasma membranes. J. Biol. Chem. **259:** 8195–8201.
5. BLACHE, P., A. KERVRAN, M. DUFOUR, *et al.* 1990. Glucagon (19–29), a Ca^{2+} pump inhibitory peptide, is processed from glucagon in the rat liver plasma membrane by a thiol endopeptidase. J. Biol. Chem. **265:** 21514–21519.
6. DALLE, S., P. SMITH, P. BLACHE, *et al.* 1999. Miniglucagon (glucagon 19–29): a potent and efficient inhibitor of secretagogue-induced insulin release through a Ca^{2+} pathway. J. Biol. Chem. **274:** 10869–10876.
7. DALLE, S., G. FONTÉS, A.-D. LAJOIX, *et al.* 2002. Miniglucagon (glucagon 19–29): a novel regulator of the pancreatic islet physiology. Diabetes **51:** 406–412.
8. FONTÉS, G., A.-D. LAJOIX, F. BERGERON, *et al.* 2005. Miniglucagon-generating endopeptidase, which processes glucagon into miniglucagon, is composed of NRD-convertase and aminopeptidase-B. Endocrinology **146:** 702–712.
9. ORCI, L., F. MALAISSE-LAGAE, M. RAVAZOLLA, *et al.* 1975. A morphological basis for intercellular communication between alpha- and beta-cells in the endocrine pancreas. J. Clin. Invest. **56:** 1066–1070.
10. KAWAI, K., C. YOKATO, S. OHASHI, *et al.* 1995. Evidence that glucagon stimulates insulin secretion through its own receptor in rats. Diabetologia **38:** 274–276.

11. MOENS K., V. BERGER, J.M. AHN, *et al.* 2002. Assessment of the role of interstitial glucagon in the acute glucose secretory responsiveness of in situ pancreatic beta-cells. Diabetes **51:** 669–675.
12. HUYPENS, P., Z. LING, D. PIPELEERS & F. SCHUIT. 2000. Glucagon receptors on human islet cells contribute to glucose competence of insulin release. Diabetologia **43:** 1012–1019.
13. DALLE, S., C. LONGUET, S. COSTES, *et al.* 2004. Glucagon promotes CREB phosphorylation via activation of ERK1/2 (extracellular signal-related kinase 1/2) in MIN6 cell line and isolated islets of Langerhans. J. Biol. Chem. **279:** 20345–20355.
14. LONGUET, C., C. BROCA, S. COSTES, *et al.* 2005. ERK1/2 (p44/42 MAP kinases) phosphorylate synapsin I and regulate insulin secretion in the beta MIN6 cell line and islets of Langerhans. Endocrinology **146:** 643–645.
15. FONTÉS, G., T. IMAMURA, C. ILIC, *et al.* 2004. Miniglucagon: an unexpected insulin's partner in the peripheral effects of the hypoglycemic hormone [abstract]. Diabetologia **47(S1):** A9.

Effects of VIP and VIP–DAP on Proliferation and Lipid Peroxidation Metabolism in Human KB Cells

MICHELE CARAGLIA,[a] ALESSANDRA DICITORE,[b] GAIA GIUBERTI,[b]
DIANA CASSESE,[b] MARILENA LEPRETTI,[b] MARIA CARTENÌ,[c]
ALBERTO ABBRUZZESE,[b] AND PAOLA STIUSO[b]

[a]Experimental Pharmacology Unit Experimental Oncology Department
National Cancer Institute Fondazione G. Pascale, 801331 Naples, Italy

[b]Department of Biochimica and Biophysic, Second University of Naples,
80138 Naples, Italy

[c]Department of Sperimental Medicine, Second University of Naples,
80138 Naples, Italy

ABSTRACT: In the present study, we have utilized the transglutaminase
(TGase) enzyme to modify the primary structure of VIP with diamminopropane (DAP) at the level of the Gln16. We have investigated the conformational stability of VIP and VIP–DAP in solution by limited proteolysis
experiments. The VIP–DAP appears to be more resistant to the proteolytic attack of trypsin, thus indicating that the derivatization in position
16 is able to stabilize the structure of the peptide. However, we have studied their role in cell cycle modulation and antioxidant activity in the
oropharyngeal epidermoid carcinoma KB cells.

KEYWORDS: limited proteolysis; vasoactive intestinal peptide; vasoactive
intestinal peptide analogs; conformational analysis; intracellular lipid
peroxidation; cell cycle modulation

INTRODUCTION

Vasoactive intestinal peptide (VIP) is a 28-amino acid long peptide and is
a member of the secretin–glucagon family of regulatory peptides involved in
the modulation of numerous biological functions.[1] VIP is a potent relaxant
of human airway smooth muscle and of pulmonary blood vessels,[2–4] inhibits
the release of macromolecules from secreting glands,[5,6] functions as neurotransmitter of the nonadrenergic–noncholinergic inhibitory nervous system,[7–9]

Address for correspondence: Paola Stiuso, Dipartimento di Biochimica e Biofisica, Seconda Uni-
versità di Napoli, Via Costantinopoli 16, 80138 Napoli, Italy. Voice: +39-081-5667577; fax: +39-081-
5665863.
e-mail: paola.stiuso@unina2.it

Ann. N.Y. Acad. Sci. 1070: 167–172 (2006). © 2006 New York Academy of Sciences.
doi: 10.1196/annals.1317.007

produces anti-inflammatory effects,[10,11] modulates T cell and B cell prolifer-ation,[12] and inhibits human natural killer cell activity. At least two different G protein–coupled receptors, named VPAC1 and VPAC2, mediate these ac-tions. Studies on VIP are often aimed at creating synthetic analogs able to bind to its receptors and to act with a selective physiological effect. Robberecht *et al.* demonstrated the unexpected importance of Gln16 in the central region of the secretin family peptides for its interaction with the receptor N-terminal domain. We have used the transglutaminase (TGase) enzyme to modify the primary structure of VIP with diamminopropane (DAP) at the level of the Gln16.[13–17] VIP–DAP acts as structural VIP agonist, both *in vitro* and *in vivo*. In particular, the IC50 and EC50 values from binding experiments with hu-man and rat VPAC1 and VPAC2 receptors indicate that VIP–DAP has higher affinity and potency than VIP.[13] In the present study, we have investigated the conformational stability of VIP and VIP–DAP and their role in cell cycle modulation and antioxidant activity.

MATERIALS AND METHODS

Cell Culture and Cell Cycle Assay

The oropharyngeal epidermoid carcinoma KB cells, obtained from the American Type Tissue Culture Collection (Rockville, MD, USA) were grown in DMEM supplemented with heat inactivated 10% fetal bovine serum (FBS), 20 mM HEPES, 100 U/mL penicillin, 100 μg/mL streptomycin, 1% L-glutamine, and 1% sodium pyruvate. The cells were grown in a humidified atmosphere of 95% air/5% CO_2 at 37°C. Cells were centrifuged and directly stained in a propidium iodide (PI) solution (50 μg PI in 0.1% sodium citrate, 0.1% NP40, pH 7.4) for 30 min at 4°C in the dark. Flow cytometric analysis was performed using a FACScan flow cytometer (Becton Dickinson, San Jose, CA) interfaced with a Hewlett Packard computer (mod.310) for data analysis. To evaluate cell cycle, PI fluorescence was collected as FL2 (Log scale) by the CellFIT software (Becton Dickinson). The data were acquired after analysis of at least 20,000 events.

Intracellular Lipid Peroxidation (Malondialdehyde Assay)

The binding of thiobarbituric acid to malondialdehyde-bis-(dimethyl-acetal)1,1,3,3-tetramethoxypropan (MDA) formed during lipid peroxidation results in a chromogenic complex. Briefly, the homogenate was obtained by lysing 1–5 10^6 cells with 10% ice-cold TCA. The lysis mixture was centrifuged at 800 g for 10 min. Aliquots (1 mL) of supernatant were added to 1 mL of 0.6% 2-thiobarbituric acid (TBA) and heated in a boiling water bath for 10 min. The

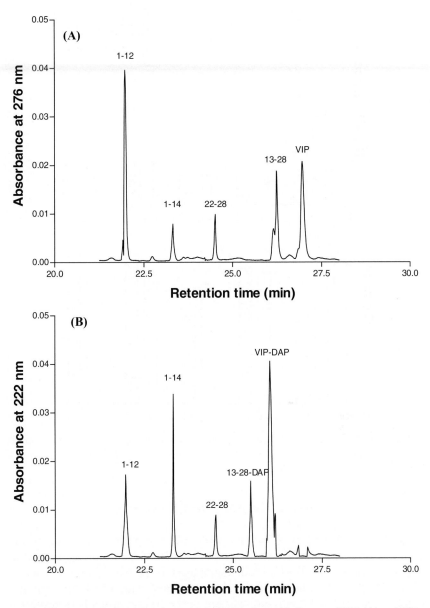

FIGURE 1. RP–HPLC elution pattern by C18 column of aliquots of VIP (**A**) and VIP–DAP (**B**) reacted with trypsin after 5 min of incubation.

samples were cooled and the color read at 535 nm. The concentration of MDA was determined by comparison with a standard reference curve. The TBA and MDA were purchased from Sigma AG (Sigma-Aldrich srl, Milano, Italy).

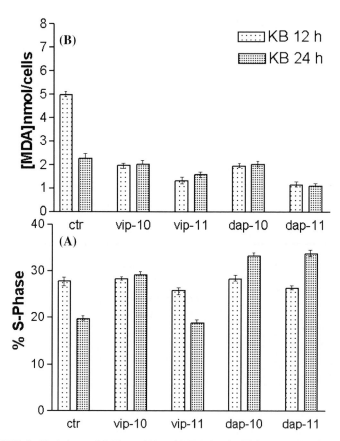

FIGURE 2. Variations of S-Phase (**A**) and MDA levels (**B**) in KB cells after 12 and 24 h of incubations with VIP and VIP–DAP at 10^{-10} and 10^{-11} M. Results are expressed as MDA nmol/ total number cells.

RESULTS AND DISCUSSIONS

The structural stability of VIP and VIP–DAP was performed by proteolysis with trypsin at an enzyme:substrate ratio of 1:1000. The proteolytic mixture was analyzed by RP–HPLC chromatography (FIG. 1). After 5 min of reaction, 32% of VIP–DAP is not hydrolyzed as compared with 17.8% for VIP. It is evident that VIP–DAP is more resistant to the tryptic proteolysis respect to VIP and this result can be explained either by steric hindrance that may prevent the optimal binding of the proteolytic enzyme or by a structural change on the peptide conferring it more rigidity. The structural analysis on VIP–DAP by mass spectrometry (data not shown) has shown that the derivatization of Gln^{16} protects Met^{17} from oxidation. We have analyzed the VIP–DAP role on the cell cycle modulation and antioxidant activity. This peptide at 10^{-10}–10^{-11}M concentration induces cell proliferation and modulates cell cycle of KB cells (causing an increase of the cells in S phase) at 12 and 24 h of incubation (see FIG. 2 A). Moreover, we have evaluated the effect of VIP–DAP on the lipid peroxidation by cytosolic MDA assay. In KB cells VIP–DAP induces a more pronounced decrease of MDA production at 10^{-11}M if compared with equimolar concentration of VIP after 24 h of incubation. Our data suggest that the introduction of a positive charge (DAP group) at the level of Gln[16] of VIP by TGase reaction increases both its biological functions (cell cycle modulation and antioxidant activity) and its stability to proteolysis.

REFERENCES

1. SHERWOOD, N.M., S.L. KRUECKL & J.E. MCRORY. 2000. The origin and function of the pituitary adenylate cyclase-activating polypeptide (PACAP)/glucagon superfamily. Endocr. Rev. **21:** 619–670.
2. MATSUZAKI, Y., Y. HAMASAKI & S.I. SAID. 1980. Vasoactive intestinal peptide: a possible transmitter of nonadrenergic relaxation of guinea pig airways. Science **210:** 1252–1253.
3. SAGA, T. & S.I. SAID. 1984. Vasoactive intestinal peptide relaxes isolated strips of human bronchus, pulmonary artery, and lung parenchyma. Trans. Assoc. Am. Physicians **97:** 304–310.
4. GRONEBERG, D.A., J. SPRINGER & A. FISCHER. 2001. Vasoactive intestinal polypeptide as mediator of asthma. Pulm. Pharmacol. Ther. **14:** 391–401.
5. COLES, S.J., S.I. SAID & L.M. REID. 1981. Inhibition by vasoactive intestinal peptide of glycoconjugate and lysozyme secretion by human airways in vitro. Am. Rev. Respir. Dis. **124:** 531–536.
6. MARTIN, S.C. & T.J. SHUTTLEWORTH. 1996. The control of fluid-secreting epithelia by VIP. Ann. N. Y. Acad. Sci. **805:** 133–147.
7. CAMERON, A.R., C.F. JOHNSTON, C.T. KIRKPATRICK & M.C.Q. KIRKPATRICK. 1983. The quest for the inhibitory neurotransmitter in bovine tracheal smooth muscle. J. Exp. Physiol. **68:** 413–426.
8. COOKE, H.J. 2000. Neurotransmitters in neuronal reflexes regulating intestinal secretion. Ann. N. Y. Acad. Sci. **915:** 77–80.
9. NUSSDORFER, G.G., M. BAHCELIOGLU, G. NERI & L.K. MALENDOWICZ. 2000. Distribution, functional role, and signaling mechanism of adrenomedullin receptors in the rat adrenal gland. Peptides **21:** 309–324.
10. SAID, S.I. 1991. VIP as a modulator of lung inflammation and airway constriction Am. Rev. Respir. Dis. **143:** S22–S24.
11. DELGADO, M., C. ABAD, C. MARTINEZ, *et al.* 2002. Vasoactive intestinal peptide in the immune system: potential therapeutic role in inflammatory and autoimmune diseases. J. Mol. Med. **80:** 16–24.
12. ISHIOKA, C., A. YOSHIDA, H. KIMATA & H. MIKAWA, 1992. Vasoactive intestinal peptide stimulates immunoglobulin production and growth of human B cells. Clin. Exp. Immunol. **87:** 504–508.
13. DE MARIA, S., S. METAFORA, V. METAFORA, *et al.* 2002. Transglutaminase-mediated polyamination of vasoactive intestinal peptide (VIP) Gln16 residue modulates VIP/PACAP receptor activity. Eur. J. Biochem. **269:** 3211–3219.

14. HOLTMANN, M.H., E.M. HADAC & L. MILLER. 1995. Critical contributions of amino-terminal extracellular domains in agonist binding and activation of secretin and vasoactive intestinal polypeptide receptors. Studies of chimeric receptors. J. Biol. Chem. **270:** 14394–14398.

15. GOURLET, P., J. P. VILARDAGA, P. DE NEEF, *et al.* 1996. The C-terminus ends of secretin and VIP interact with the N-terminal domains of their receptors. Peptides **17:** 825–829.

16. NICOLE, P., L. LINS, C. ROUYER-FESSARD, *et al.* 2000. Identification of key residues for interaction of vasoactive intestinal peptide with human VPAC1 and VPAC2 receptors and development of a highly selective VPAC1 receptor agonist. Alanine scanning and molecular modeling of the peptide. J. Biol. Chem. **275:** 24003–24012.

17. ONOUE, S., A. MATSUMOTO, Y. NAGANO, *et al.* 2004. Alpha-helical structure in the C-terminus of vasoactive intestinal peptide: functional and structural consequences. Eur. J. Pharmacol **485:** 307–316.

The Delayed Rectifier Channel Current I_K Plays a Key Role in the Control of Programmed Cell Death by PACAP and Ethanol in Cerebellar Granule Neurons

HÉLÈNE CASTEL,[a] DAVID VAUDRY,[a] YAN-AI MEI,[b]
THOMAS LEFEBVRE,[a] MAGALI BASILLE,[a] LAURENCE DESRUES,[a]
ALAIN FOURNIER,[c] HUBERT VAUDRY,[a] MARIE-CHRISTINE TONON,[a]
AND BRUNO J. GONZALEZ[a]

[a] *INSERM U413, Laboratory of Cellular and Molecular Neuroendocrinology, European Institute for Peptide Research (IFRMP 23), University of Rouen, 76821 Mont-Saint-Aignan, France*

[b] *Department of Physiology, School of Life Science, Fudan University, Shanghai 200433, China*

[c] *INRS-Institut Armand Frappier, Université du Québec, H9R 1G6 Pointe-Claire, Canada*

ABSTRACT: Alcohol exposure during development causes severe brain malformations, and thus, identification of molecules that can counteract the neurotoxicity of ethanol deserves high priority. Since activation of potassium (K^+) currents has been shown to play a critical role in the control of programmed cell death, we have investigated the effects of ethanol and PACAP on K^+ currents in cultured cerebellar granule cells using the patch-clamp technique in the whole cell configuration. In the presence of the fast-inactivating I_A current blocker 4-AP, a focal application of ethanol (200 mM) in the vicinity of granule cells provoked a robust hyperpolarization and a marked increase of the delayed rectifier I_K current. Addition of PACAP (0.1 μM) in the bath solution prevented ethanol-induced membrane hyperpolarization and suppressed the stimulatory effect of ethanol on I_K current. These data suggest that ethanol alters neuronal survival, at least in part, through activation of I_K, and that PACAP abolishes ethanol-induced cerebellar granule cell death via inhibition of I_K.

KEYWORDS: cerebellar granule cell; potassium current; PACAP; ethanol; membrane potential

Address for correspondence: Hubert Vaudry, INSERM U413, Laboratory of Cellular and Molecular Neuroendocrinology, European Institute for Peptide Research (IFRMP 23), University of Rouen, 76821 Mont-Saint-Aignan, France. Voice: 33-235-14-6624; fax: 33-235-14-6946.
e-mail: hubert.vaudry@univ-rouen.fr

Ann. N.Y. Acad. Sci. 1070: 173–179 (2006). © 2006 New York Academy of Sciences.
doi: 10.1196/annals.1317.008

INTRODUCTION

Prenatal exposure to ethanol causes impairment of brain development associated with behavioral and cognitive deficits known as fetal alcohol syndrome (FAS).[1,2] The cerebellum is one of the brain areas that is most sensitive to ethanol during development, with loss of both Purkinje neurons[3] and granule cells.[4] Pituitary adenylate cyclase-activating polypeptide (PACAP) and its receptors are actively expressed in the rat cerebellar cortex during postnatal development.[5-7] In particular, high levels of the PACAP-specific receptor PAC1-R are found in the external granule cell layer, a germinative matrix that gives rise to cerebellar interneurons.[8] *In vitro* studies have shown that incubation of cultured cerebellar granule cells with PACAP promotes cell survival and stimulates neurite outgrowth.[9,10] Previous reports have revealed that PACAP prevents cerebellar granule neurons from ethanol-induced apoptotic cell death.[11,12] Since there is strong evidence that K^+ channel-mediated signals play an important role in the control of programmed cell death,[13] we have investigated the possible action of ethanol and PACAP on resting membrane potential and K^+ current in cultured cerebellar granule neurons.

EFFECTS OF ETHANOL AND PACAP ON GRANULE CELL MEMBRANE POTENTIAL AND I_K CURRENT

The effects of ethanol and PACAP on K^+ currents have been studied on granule cells cultured for 5–10 days in a 25 mM $[K^+]$-containing medium in the presence of serum by using the patch-clamp technique in the whole cell configuration. In the presence of the Na^+ channel blocker tetrodotoxin and the fast-inactivating I_A current blocker 4-aminopyridine (4-AP), cerebellar granule cells exhibited a mean resting membrane potential of -58 ± 3.2 mV ($n = 16$) and a delayed rectifier K^+ (I_K) current amplitude of 234 ± 18 pA ($n = 58$). As shown in FIGURE 1, a 5-s ejection of ethanol (200 mM) elicited a marked hyperpolarization (FIG. 1 A). Exposure of granule cells to PACAP (0.1 μM) strongly attenuated the ethanol-induced hyperpolarization (FIG. 1 A). To investigate the possibility of a direct modulation of delayed rectifier outward K^+ channels, I_K currents were evoked by 500-msec depolarizing pulses from –50 mV to +50 mV at 15-s intervals. Ejection of ethanol (200 mM) significantly ($P < 0.001$) enhanced I_K by $116 \pm 2.5\%$ ($n = 7$). This activation process was immediate and control level was recovered within only 15s of washout (FIG. 1 B). Addition of PACAP (0.1 μM) in the bath solution abolished the stimulatory effect of ethanol on I_K currents (FIG. 1 C).

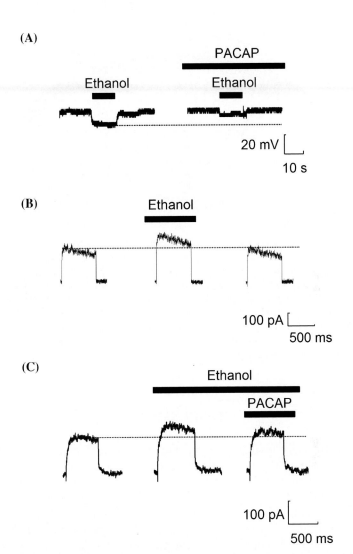

FIGURE 1. Effects of ethanol and PACAP on the resting potential and the delayed rectifier I_K current. (**A**) Ethanol (200 mM)-induced changes in membrane potential recorded in the whole-cell configuration. The fast inactivating K^+ current (I_A) was blocked by the application of 4-AP (5 mM) in the bath solution. Ethanol was ejected in the vicinity of the cell in the absence (*left*) or presence of PACAP (0.1 μM, 1 min, *right*). (**B**) Consecutive traces of I_K currents were evoked by sequential 500-msec depolarizing pulses from −50 mV to +50 mV at 15-s intervals, before (*left trace current*), during ethanol (200 mM) ejection (*middle trace current*) and after washout (*right trace current*). (**C**) Consecutive traces of I_K currents were evoked by sequential 500 m/s depolarizing pulses from −50 mV to +50 mV at 15-s intervals, before (*left trace current*) and during ethanol (200 mM) perfusion in the absence (*middle trace current*) or presence of PACAP (0.1 μM, 1 min, *right current trace*).

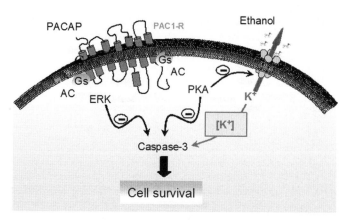

FIGURE 2. Proposed model depicting the role of delayed rectifier current channels in cerebellar granule cell survival. Ethanol rapidly activates the opening of delayed rectifier I_K current channels. The consecutive decrease in intracellular $[K^+]$ is one of the mechanisms involved in caspase-3 activation and cell death. PACAP, acting through PAC1-R, inhibits caspase-3 activity via both the adenylate cyclase (AC)/protein kinase A (PKA) and the AC/extracellular regulated kinase (ERK) signaling pathways. Activation of PKA by PACAP provokes inhibition of the I_K current and the resulting increase in intracellular $[K^+]$ causes reduction of caspase-3 activity. Thus, the inhibitory action of PACAP on I_K may antagonize the proapoptotic effect of ethanol.

DISCUSSION

The present study has demonstrated that exposure of granule cells to ethanol leads to a more negative resting membrane potential and induces a rapidly reversible stimulation of the delayed rectifier I_K current. By inhibiting the I_K amplitude, PACAP suppresses the effects of ethanol on granule cell electrical activity. In various cell types, particularly in neurons, K^+ channels are essential for governing cell volume,[14] resting membrane potential, frequency and duration of action potential,[13] and neurotransmitter release.[15] Several observations indicate that K^+ currents are involved in the regulation of neuronal cell survival/death decision[16–18] and can be responsible for intracellular K^+ loss, which may lead to apoptosis. For instance, recent studies have demonstrated that serum deprivation or staurosporine, ceramide, and β-amyloid may cause apoptotic cell death through an increase in delayed rectifier I_K currents.[16,17,19,20] PACAP has been shown to prevent caspase-3 activation through PKA- and ERK-dependent mechanisms[21–23] and to reduce apoptosis of cerebellar granule cells induced by ethanol.[11] Moreover, recent data have shown that PACAP causes inhibition of I_K current channels and inhibits programmed cell death through activation of the cAMP/PKA pathway.[12,24] The present data show that ethanol causes a marked hyperpolarization and a rapid increase of

I_K amplitude, which are both inhibited by PACAP in cerebellar granule cells. Thus, the inhibitory effect of PACAP on the I_K current amplitude in mature granule cells further supports the contention that the antiapoptotic effect of PACAP on ethanol-induced cell death may be ascribed, at least in part, to inhibition of K^+ currents, which, in turn, prevents a decrease in cytoplasmic K^+ concentration. A proposed model illustrating the signaling cascade involved in the proapoptotic effect of ethanol and the neurotrophic action of PACAP on cerebellar granule cells is shown in FIGURE 2. Ethanol, by activating the amplitude of the I_K current through an unknown mechanism, leads to a $[K^+]$ loss, which may participate in the ethanol-induced programmed cell death. PACAP, acting through PAC1-R positively coupled to AC, inhibits caspase-3 activity and promotes cell survival. Concurrently, activation of PKA by PACAP causes blockage of I_K and a resulting increase in $[K^+]$ involved in inhibition of caspase-3 activity.[12] Thus, regulation of the delayed rectifier I_K current may contribute to the proapoptotic effect of ethanol and the antiapoptotic effect of PACAP in cerebellar granule neurons.

ACKNOWLEDGMENTS

This work was supported by INSERM (U413), IREB (grant no. 2005/31), and the Conseil Régional de Haute-Normandie. T.L. is recipient of a fellowship from INSERM and the Conseil Régional de Haute-Normandie. H.V. is affiliated professor at the INRS-Institut Armand Frappier.

REFERENCES

1. ABEL, E.L. & R.F. BERMAN. 1994. Long-term behavioral effects of prenatal alcohol exposure in rats. Neurotoxicol. Teratol. **16:** 467–470.
2. OLNEY, J.W., D.F. WOZNIAK, N.B. FARBER, *et al.* 2002. The enigma of fetal alcohol neurotoxicity. Ann. Med. **34:** 109–119.
3. GOODLETT, C.R., A.D. PEARLMAN & K.R. LUNDAHL. 1998. Binge neonatal alcohol intubations induce dose-dependent loss of Purkinje cells. Neurotoxicol. Teratol. **20:** 285–292.
4. NAPPER, R.M. & J.R. WEST. 1995. Permanent neuronal cell loss in the cerebellum of rats exposed to continuous low blood alcohol levels during the brain growth spurt: a stereological investigation. J. Comp. Neurol. **362:** 283–292.
5. BASILLE, M., B.J. GONZALEZ, P. LEROUX, *et al.* 1993. Localization and characterization of PACAP receptors in the rat cerebellum during development: evidence for a stimulatory effect of PACAP on immature cerebellar granule cells. Neuroscience **57:** 329–338.
6. NIELSEN, H.S., J. HANNIBAL & J. FAHRENKRUG. 1998. Expression of pituitary adenylate cyclase activating polypeptide (PACAP) in the postnatal and adult rat cerebellar cortex. Neuroreport **9:** 2639–2642.

7. GONZALEZ, B.J., D. VAUDRY, M. BASILLE, *et al.* 2002. Function of PACAP in the central nervous system. *In* Pituitary Adenylate Cyclase-Activating Polypeptide. H. Vaudry & A. Arimura, Eds.: 125–151. Kluwer Academic Publishers. Norwell. MA.

8. BASILLE, M., D. VAUDRY, Y. COULOUARN, *et al.* 2000. Distribution of pituitary adenylate cyclase-activating polypeptide (PACAP) binding sites and PACAP receptor mRNAs in germinative neuroepithelia of the rat. J. Comp. Neurol. **425:** 495–509.

9. GONZALEZ, B.J., M. BASILLE, D. VAUDRY, *et al.* 1997. Pituitary adenylate cyclase-activating polypeptide promotes cell survival and neurite outgrowth in rat cerebellar neuroblasts. Neuroscience **78:** 419–430.

10. VAUDRY, D., B.J. GONZALEZ, M. BASILLE, *et al.* 1999. Neurotrophic activity of pituitary adenylate cyclase-activating polypeptide on rat rerebellar cortex during development. Proc. Natl. Acad. Sci. USA **96:** 9415–9420.

11. VAUDRY, D., C. ROUSSELLE, M. BASILLE, *et al.* 2002. Pituitary adenylate cyclase-activating polypeptide protects rat cerebellar granule neurons against ethanol-induced apoptotic cell death. Proc. Natl. Acad. Sci. USA **99:** 6398–6403.

12. MEI, Y.A., D. VAUDRY, M. BASILLE, *et al.* 2004. PACAP inhibits delayed rectifier potassium current via a cAMP/PKA transduction pathway: evidence for the involvement of Ik in the anti-apoptotic action of PACAP. Eur. J. Neurosci. **19:** 1446–1458.

13. SHIEH, C.C., M. COGHLAN, J.P. SULLIVAN & M. GOPALAKRISHNAN. 2000. Potassium channels: molecular defects, diseases, and therapeutic opportunities. Pharmacol. Rev. **52:** 557–594.

14. BORTNER, C.D. & J.A. CIDLOWSKI. 1999. Caspase independent/dependent regulation of K(+), cell shrinkage, and mitochondrial membrane potential during lymphocyte apoptosis. J. Biol. Chem. **274:** 21953–21962.

15. MEIR, A., S. GINSBURG, A. BUTKEVICH, *et al.* 1999. Ion channels in presynaptic nerve terminals and control of transmitter release. Physiol. Rev. **79:** 1019–1188.

16. YU, S.P., C.H. YEH, S.L. SENSI, *et al.* 1997. Mediation of neuronal apoptosis by enhancement of outward potassium current. Science **283:** 114–117.

17. COLOM, L.V., M.E. DIAZ, D.R. BEERS, *et al.* 1998. Role of potassium channels in amyloid-induced cell death. J. Neurochem. **70:** 1925–1934.

18. WANG, X., A.Y. XIAO, T. ICHINOSE & S.P. YU. 2000. Effects of tetraethylammonium analogs on apoptosis and membrane currents in cultured cortical neurons. J. Pharmacol. Exp. Ther. **295:** 524–530.

19. YU, S.P., C.H. YEH, F. GOTTRON, *et al.* 1999. Role of the outward delayed rectifier K^+ current in ceramide-induced caspase activation and apoptosis in cultured cortical neurons. J. Neurochem. **73:** 933–941.

20. BORTNER, C.D. & J.A. CIDLOWSKI. 2002. Cellular mechanisms for the repression of apoptosis. Annu. Rev. Pharmacol. Toxicol. **42:** 259–281.

21. VAUDRY, D., B.J. GONZALEZ, M. BASILLE, *et al.* 1998. Pituitary adenylate cyclase-activating polypeptide stimulates both c-fos gene expression and cell survival in rat cerebellar granule neurons through activation of the protein kinase A pathway. Neuroscience **84:** 801–812.

22. VAUDRY, D., B.J. GONZALEZ, M. BASILLE, *et al.* 2000a. Pituitary adenylate cyclase-activating polypeptide and its receptors: from structure to functions. Pharmacol. Rev. **52:** 269–324.

23. FALLUEL-MOREL, A., N. AUBERT, D. VAUDRY, *et al.* 2004. Opposite regulation of the mitochondrial apoptotic pathway by C2-ceramide and PACAP through a MAP-kinase-dependent mechanism in cerebellar granule cells. J. Neurochem. **91:** 1231–1243.
24. VAUDRY, D., B.J. GONZALEZ, M. BASILLE, *et al.* 2000b. Pituitary adenylate cyclase-activating polypeptide inhibits apoptosis of rat cerebellar granule neurons through involvement of caspase-3/CPP32. Proc. Natl. Acad. Sci. USA **97:** 13390–13395.

Spatial Approximation between the C-Terminus of VIP and the N-Terminal Ectodomain of the VPAC1 Receptor

E. CERAUDO, Y.-V. TAN, A. COUVINEAU, J.-J. LACAPERE, AND MARK LABURTHE

Unité INSERM 773, Faculté de Médecine Xavier Bichat, 75018 Paris, France

ABSTRACT: Vasoactive intestinal peptide (VIP) exerts many biological functions through interaction with the VPAC1 receptor, a class II G protein–coupled receptor. Photoaffinity labeling studies associated with receptor mapping and three-dimensional molecular modeling demonstrated that the central part of VIP (6–24) interacts with the N-terminal ectodomain of VPAC1 receptor. However, the domain of the VPAC1 receptor interacting with the C-terminus of VIP is still unknown. A photoaffinity probe, Bpa^{28}-VIP, was synthetized by substitution of amidated Asn^{28} of VIP by amidated photoreactive *para*-benzoyl-L-Phe (Bpa). Bpa^{28}-VIP was shown to be a hVPAC1 receptor agonist in CHO cells expressing the recombinant VPAC1 receptor. After obtaining a covalent ^{125}I-[Bpa^{28}-VIP]/hVPAC1 complex, it was cleaved by CNBr, PNGase F, and endopeptidase Glu-C and the cleavage products were analyzed by electrophoresis. The data demonstrated that ^{125}I-[Bpa^{28}-VIP] was covalently bonded to the 121–133 fragment within the N-terminal ectodomain of the receptor. This fragment is adjacent to those covalently attached to the central part (6–24) of VIP.

KEYWORDS: GPCR; VIP receptor; structure-activity

INTRODUCTION

The vasoactive intestinal peptide (VIP) is a prominent 28 amino acid neuropeptide with wide distribution in both central and peripheral nervous systems.[1] Consequently, VIP possesses a large spectrum of biological actions (exocrine secretions, muscle relaxation, suppression of inflammation. . .) and has potential therapeutic role in physiopathology (inflammatory diseases, such as rheumatoid arthritis or Crohn's disease, asthma or neurodegenerative diseases).[2] Its action goes through interaction with specific serpentine class II G

Address for correspondence: Marc Laburthe, INSERM U773, Faculté de Médecine Bichat, 16 rue Henri Huchard, 75018 Paris, France. Voice: 33-144856135; fax: 33-1-42288765.
e-mail: laburthe@bichat.inserm.fr

Ann. N.Y. Acad. Sci. 1070: 180–184 (2006). © 2006 New York Academy of Sciences.
doi: 10.1196/annals.1317.009

protein–coupled receptors (GPCR), named VPAC receptors.[3] Because of the importance of VIP receptors in human physiopathology, the VPAC receptors represent potential therapeutic targets. For many years, the VPAC1 receptor has been extensively characterized by site-directed mutagenesis, molecular chimerism, and molecular modeling,[4,5] suggesting the importance of its N-terminal ectodomain in VIP recognition. Moreover, the structure–activity relationship of VIP itself characterized by "Ala-scan" and molecular modeling studies[6] showed key residues for activity all along the peptide sequence. Photoaffinity labeling studies associated with receptor mapping have permitted to identify physical contacts between VIP and its receptor. In particular, amino acids at positions 6, 22, and 24 of VIP were shown to be in contact with the N-terminal ectodomain of the VPAC1 receptor, for example, the 104–108,[7] 109–120,[8] and 121–133 receptor segments, respectively. In this article, we developed a new photoaffinity probe, Bpa^{28}-VIP, in order to position the C-terminal residue of VIP within the VPAC1 receptor.

MATERIALS AND METHODS

The Bpa^{28}-VIP photoaffinity probe in which the C-terminal amidated residue (Asn^{28}) is substituted by an amidated residue carrying a photoreactive *para*-benzoyl-L-Phe residue (Bpa), was synthetized by solid-phase synthesis and tested using ligand-binding and adenylyl cyclase activity assays as described previously.[7,8] The photolabeling experiments were carried out with membranes from CHO-F7 cells, stably expressing the recombinant human VPAC1 receptor (hVPAC1R) fusioned to the green fluorescent protein (GFP) in its C-terminal end.[9] Membranes were incubated with ^{125}I-[Bpa^{28}-VIP] for 2 h in darkness and then exposed to UV (365 nm). The radiolabeled receptors were then resolved directly by SDS-NuPAGE electrophoresis under reducing conditions (DTT) or after chemical (CNBr) and/or enzymatic (PNGase F, endopeptidase Glu-C) cleavages as described previously.[7,8] Then, photolabeled bands were detected by autoradiography.

RESULTS AND DISCUSSION

In this work, we have determined the VPAC1 receptor region that is in contact with the C-terminus of VIP. Bpa^{28}-VIP, a new photoaffinity probe, also amidated at its C-terminus as Asn^{28} in VIP, was synthetized and characterized. This probe was shown to have a good affinity for hVPAC1 receptor (Ki of 9.5 nM vs. 1.0 nM for VIP) and to efficiently stimulate intracellular cyclic adenosine $3',5'$-monophosphate (cAMP) production (EC$_{50}$ of 0,1 nM for both peptides) (FIG. 1). The probe can therefore be used in photoaffinity experiments. Incubation of ^{125}I-[Bpa^{28}-VIP] with CHO-F7 membranes, followed by UV irradiation

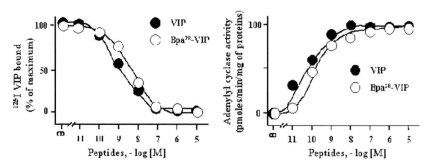

FIGURE 1. Binding and adenylyl cyclase activity assays of Bpa[28]-VIP. *Left panel,* competitive inhibition of ^{125}I-VIP binding to CHO-F7 cell membranes expressing hVPAC1 receptor by VIP and Bpa[28]-VIP. The data are expressed as a percentage of initial specific binding in the absence of competitor. *Right panel,* action of increasing concentrations of VIP and Bpa[28]-VIP on adenylyl cyclase activity in CHO-F7 membranes. The data are expressed as a percentage of maximal stimulation above basal obtained with 1 μM native VIP. The symbols are: (○) Bpa[28]-VIP and (•) VIP.

and electrophoresis analysis, showed a single photolabeled band of about 95 kDa corresponding to the labeling of the glycosylated human VPAC1 receptor (64 kDa) containing the GFP protein (25 kDa) in its C-terminal end. This 95-kDa band corresponding to [125]I-[(Bz$_2$-K[24])-VIP]/hVPAC1R complex was shifted to around 70 kDa after deglycosylation. This result is consistent with the presence of 9-kDa carbohydrate moiety on each of the three consensus N-glycosylation sites in the N-terminal domain of the hVPAC1 receptor.[10] Then, the covalent ^{125}I-probe/hVPAC1 receptor complex of 95 kDa was sequentially cleaved by cyanogen bromide (CNBr), peptide N-glycosidase F, and endopeptidase Glu-C. After CNBr cleavage of the 95-kDa complex, the SDS-NuPAGE analysis revealed a single radiolabeled fragment of 30 kDa that moved at 11 kDa after deglycosylation (FIG. 2). We can conclude that this band represents the 67–137 sequence of VPAC1 receptor. Endopeptidase Glu-C treatment of the 30-kDa fragment shifted the band to 6 kDa (FIG. 2). Under our experimental conditions, this enzyme cleaves proteins at the carboxy-terminal side of Glu residues with the notable exception of Glu-Pro sequence. Therefore, the 6-kDa band corresponds unambiguously to the 109–133 fragment. Altogether, these results support that Bpa[28]-VIP is covalently bound to the 109–133 fragment within the hVPAC1 N-terminal ectodomain. In order to better define the binding site, the mutant receptor I120M was produced by substitution of Ile[120] by a methionine residue to create a new CNBr cleavage site. This mutant had the same biological properties as the wild-type receptor, in terms of binding affinity for VIP and ability of VIP to stimulate cAMP production. Photoaffinity labeling of the mutant receptor by the [125]I-[Bpa[28]-VIP] probe followed by CNBr cleavage generated a band migrating around 5 kDa corresponding to the 121–137 sequence of the N-terminal ectodomain of the VPAC1 receptor.

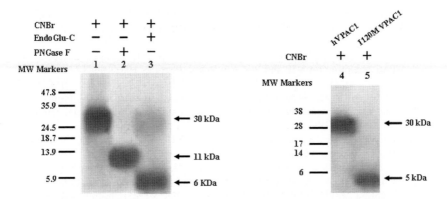

FIGURE 2. Photoaffinity labeling of the hVPAC1 receptor with the ^{125}I-[Bpa^{28}-VIP] probe. Digestion products of covalent complex ^{125}I-[Bpa^{28}-VIP] / VPAC1 receptor, after treatment with CNBr, PNGase F, or endopeptidase Glu-C for the wt (*left panel*) and the I120M mutant (*right panel*) receptors were resolved by SDS-NuPAGE gel (+DTT). See "Materials and Methods" for details. MW, molecular mass.

In conclusion, this work showed that the C-terminal end of VIP interacts with the 121–133 sequence within the N-terminal ectodomain of the receptor. Interestingly, this fragment is adjacent to other fragments previously identified to be in contact with the side chain of residues in positions 6, 22, and 24 of VIP that is, 104–108,[8] 109-120,[7] and 121–133[11] sequences. These data demonstrate that the central and C-terminal parts of VIP interact with the N-terminal ectodomain of VPAC1 receptor.

ACKNOWLEDGMENTS

E. Ceraudo and Y.-V. Tan are students supported by the Ministère de l'Enseignement Supérieur et de la Recherche. Y.-V. Tan is supported by grant FDT 20041202869 from the Fondation pour la Recherche Médicale. This research was supported by the Institut National de la Santé et de la Recherche Médicale, the Centre National de la Recherche Scientifique, the Université Paris 7, grant ACIM-2–18, 2003–5, Programmes de Microbiologie: Microbiologie Fondamentale et Appliquée, Maladies Infectieuses, Environnement et Bioterrorisme, and a grant in 2005 from Association de Recherche sur la Polyarthrite.

REFERENCES

1. SAID, S.I. 1991. Vasoactive intestinal polypeptide. Biologic role in health and disease. Trends Endocrinol. Metab. **2:** 107–122.

2. LABURTHE, M., A. COUVINEAU & T. VOISIN. 1999. Receptors for peptides of the VIP/PACAP and PYY/NPY/PP families. Gastrointestinal Endocrinol. **9:** 125–157.
3. LABURTHE, M. & A. COUVINEAU. 2002. Molecular pharmacology and structure of VPAC receptors for VIP and PACAP. Regul. Peptides **108:** 165–173.
4. LABURTHE, M., A. COUVINEAU, P. GAUDIN, *et al.* 1996. Receptors for VIP, PACAP, secretin, GRF, glucagon, GLP-1, and other members of their new family of G protein-linked receptors: structure-function relationship with special reference to the human VIP-1 receptor. Ann. N. Y. Acad. Sci. **805:** 94–111.
5. LABURTHE, M., A. COUVINEAU & J.C. MARIE. 2002. VPAC receptors for VIP and PACAP. Receptors Channels **8:** 137–153.
6. NICOLE, P., L. LINS, C. ROUYER-FESSARD, *et al.* 2000. Identification of key residues for interaction of Vasoactive Intestinal Peptide with human VPAC1 and VPAC2 Receptors And Development of a highly selective VPAC1 receptor agonist: alanine scanning and molecular modeling of the peptide. J. Biol. Chem. **275:** 24003–24012.
7. TAN, Y.V., A. COUVINEAU & M. LABURTHE. 2004. Diffuse pharmacophoric domains of Vasoactive Intestinal Peptide (VIP) and further insights into the interaction of VIP with the N-terminal Ectodomain of human VPAC1 receptor by photoaffinity labeling with [Bpa6]-VIP. J. Biol. Chem. **279:** 38889–38894.
8. TAN, Y.V., A. COUVINEAU, J. VAN RAMPELBERGH & M. LABURTHE. 2003. Photoaffinity labeling demonstrates physical contact between vasoactive intestinal peptide and the N-terminal ectodomain of the human VPAC1 receptor. J. Biol. Chem. **278:** 36531–36536.
9. GAUDIN, P., J.J. MAORET, A. COUVINEAU, *et al.* 1998. Constitutive activation of the human vasoactive intestinal peptide 1 receptor, a member of the new class II family of G protein-coupled receptors. J. Biol. Chem. **273:** 4990–4996.
10. COUVINEAU, A., C. FABRE, P. GAUDIN, *et al.* 1996. Mutagenesis of N-glycosylation sites in the human vasoactive intestinal peptide 1 receptor. Evidence that asparagine 58 or 69 is crucial for correct delivery of the receptor to plasma membrane. Biochemistry **35:** 1745–1752.
11. TAN, Y.V., A. COUVINEAU, J.J. LACAPERE & M. LABURTHE. In this book

PACAP and VIP Promote Initiation of Electrophysiological Activity in Differentiating Embryonic Stem Cells

MAGDA CHAFAI,[a] ESTELLE LOUISET,[a] MAGALI BASILLE,[a] MICHÈLE CAZILLIS,[b] DAVID VAUDRY,[a] WILLIAM ROSTÈNE,[c] PIERRE GRESSENS,[b] HUBERT VAUDRY,[a] AND BRUNO J. GONZALEZ[a]

[a]INSERM U413, IFRMP23, Université de Rouen, 76821 Mont-Saint-Aignan, France

[b]INSERM U676, Hôpital Robert-Debré, 75019 Paris, France

[c]INSERM E0350, Hôpital Saint Antoine, 75571 Paris, France

ABSTRACT: Owing to their capacity to differentiate *in vitro* into various types of neuronal cells, embryonic stem (ES) cells represent a suitable model for studying the first steps of neuronal differentiation and cerebral development. Since pituitary adenylate cyclase-activating polypeptide (PACAP) and vasoactive intestinal polypeptide (VIP) are known to control maturation of the nervous system, we have investigated the possible effects of these two neuropeptides on the differentiation of ES cells. Reverse transcription polymerase chain reaction (RT-PCR) analysis revealed that mouse ES cells express PAC1 and VPAC2 receptors. Electrophysiological recordings demonstrated that PACAP and VIP facilitate the emission of currents, suggesting that these peptides can initiate the genesis of an electrophysiological activity in differentiating ES cells.

KEYWORDS: neuropeptides; ES cells; neuronal differentiation; K^+ currents; Ca^{2+} currents

INTRODUCTION

Mouse embryonic stem (ES) cells are derived from the inner mass of the blastocyst. They can be permanently kept in cell culture as self-renewing stem cells. Their capacity of replication depends on the presence of certain factors, like the leukemia inhibitory factor that is secreted by a feeder layer of mouse embryonic fibroblasts.[1,2] It has been demonstrated that ES cells are pluripotent, since they are able to differentiate *in vitro* into endodermal, mesodermal, and ectodermal cell types.[3] ES cells cultured in suspension give rise to small

Address for correspondence: Bruno Gonzalez, INSERM U413, IFRMP 23, University of Rouen, 76821 Mont-Saint-Aignan Cedex, France. Voice: 33-235-14-6760; fax: 33-235-14-6946.
e-mail: bruno.gonzales@univ-rouen.fr

Ann. N.Y. Acad. Sci. 1070: 185–189 (2006). © 2006 New York Academy of Sciences.
doi: 10.1196/annals.1317.010

spheric embryoid bodies (EB). Specific protocols to facilitate the differentiation of EB-derived cells into neuronal cells have been established. In particular, retinoic acid enables the differentiation of these cells into various phenotypes including GABAergic, glutamatergic, cholinergic, and glycinergic neurons,[4] as well as astrocytes, oligodendrocytes, and microglial cells.[3] In addition, the association of the transcriptional factor sonic hedgehog with the growth factor fibroblast growth factor 8 (FGF 8) has been used to differentiate the ES cells into dopaminergic neurons.[5] Thus, ES cells represent a suitable model of pluripotent cells for studying the first steps of neuronal differentiation and the mechanisms of cerebral development. They also provide a potential source for cell replacement therapy in neurodegenerative diseases.

Pituitary adenylate cyclase-activating polypeptide (PACAP) and vasoactive intestinal polypeptide (VIP) act via G-protein-coupled receptors (PAC1-R, VPAC1-R, and VPAC2-R). VPAC1-R and VPAC2-R are equally recognized by VIP and PACAP, while PAC1-R only binds with high affinity to PACAP. During ontogenesis, mRNAs encoding PACAP, VIP, and their receptors are detected in the neural tube[6,7] and are expressed by differentiating neurons.[8] These observations suggest that PACAP and VIP may control cell proliferation and neuronal maturation.[9,10] Consistent with this hypothesis, it has been reported that PACAP exerts an antimitogenic effect on the developing cerebral cortex.[11] In addition, PACAP promotes cell survival by inhibiting apoptotic cell death and stimulates neurite outgrowth in cerebellar precursors.[9,10] Similarly, VIP has been found to inhibit death of mice cortical cells[12] and to favor survival of sympathetic neuroblasts.[13] Recently, *in vitro* studies have revealed that PACAP facilitates the differentiation of tumoral PC12 cells into neuroendocrine cells generating voltage-activated sodium and calcium currents.[14] These data strongly suggest that PACAP is a neurotrophic factor able to induce cellular excitability. In spite of the evidence for a crucial role of PACAP and VIP in the maturation of the nervous system, the effects of these neuropeptides on differentiation of ES cells have received little attention.

RESULTS

In order to determine whether PACAP and VIP may act on pluripotent cells, we have investigated the expression of their receptors in ES cells and EB-derived cells by reverse transcription polymerase chain reaction (RT-PCR) analysis. FIGURE 1 shows the presence PAC1-R mRNA in both ES and EB-derived cells, and VPAC2-R mRNA in ES cells. These receptors are functional and positively coupled to adenylyl cyclase (FIG. 1 B). In addition, it has been recently demonstrated that PACAP and VIP induce differentiation of EB-derived cells into GABAergic and dopaminergic neurons but do not promote the emergence of glial cells.[15] Altogether, these data led us to test whether PACAP and VIP could facilitate the functional differentiation of ES cells into excitable cells.

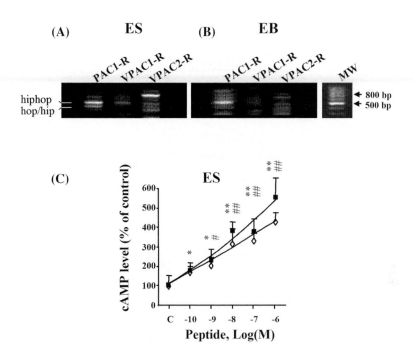

FIGURE 1. Molecular and functional characterizations of PACAP and VIP receptors in ES and EB-derived cells. (**A, B**) RT-PCR analysis revealed the expression of PAC1-R and VPAC2-R mRNAs in ES cells (**A**), and PAC1-R in EB-derived cells (**B**). Hop, hip, and hiphop indicate the positions of the respective PAC1-R variants. The molecular weights (MW) are indicated in base pairs (bp). (**C**) Effect of graded concentrations of PACAP (■) and VIP (◇) on cyclic adenosine 3′,5′-monophosphate (cAMP) production in ES cells. (From Cazillis *et al.*, 2004 with permission.)

To investigate the possible effect of PACAP and VIP on the expression of ionic channels, we have conducted an electrophysiological study on ES cells treated with PACAP and VIP (10^{-7} M each) for 2 weeks. Patch-clamp recordings in the whole cell configuration showed that PACAP increased the amplitude of outward currents generated by increasing depolarizations from -50 mV to $+30$ mV (FIG. 2 A, B). A similar effect was observed, when ES cells were treated with VIP (FIG. 2A, C). These outward currents were reduced by the potassium channel blockers TEA and Cs^{+}, indicating that PACAP and VIP increased the expression of potassium channels. PACAP and VIP also enhanced high-threshold-activated transient calcium currents evoked by depolarizations higher than -20 mV (data not shown).

CONCLUSION

In summary, RT-PCR analysis has shown that ES cells express PAC1-R and VPAC2-R. These receptors are functional suggesting that both VIP and PACAP

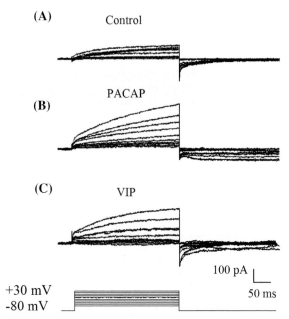

FIGURE 2. Effects of PACAP and VIP on activation of outward currents in ES cells. ES cells were treated for 2 weeks with PACAP or VIP (10^{-7} M each). The electrophysiological signals were recorded by using patch-clamp technique in the whole cell configuration. Activation of outward currents was induced by increasing depolarizing pulses from −50 mV to +30 mV by 10-mV increments from a holding potential of −80 mV.

exert biological functions at early stages of development. Electrophysiological patch-clamp recordings reveal that PACAP and VIP facilitate the generation of electrical activity in differentiating ES cells. In particular, treatments with PACAP and VIP enhance outwards currents, suggesting that both peptides can regulate potassium channels in differentiating ES cells.

ACKNOWLEDGMENTS

This work was supported by INSERM (U413) and the Conseil Régional de Haute-Normandie.

REFERENCES

1. WOBUS, A.M., H. HOLZHAUSEN, P. JAKEL & J. SCHONEICH. 1984. Characterization of pluripotent stem cell line derived from mouse embryo. Exp. Cell Res. **152:** 212–219.
2. SMITH, A.G. 2001. Embryo-derived stem cells: of mice and men. Annu. Rev. Cell Dev. Biol. **17:** 435–462.

3. GUAN, K., H. CHANG, A. ROLLETSCHEK & A.M. WOBUS. 2001. Embryonic stem cell-derived neurogenesis. Retinoic acid induction and lineage selection of neuronal cells. Cell Tissue Res. **305:** 171–176.

4. STRÜBING, C., G. AHNERT-HILGER, J. SHAN, *et al.* 1995. Differentiation of pluripotent embryonic stem cells into the neuronal lineage in vitro gives rise to mature inhibitory and excitatory neurones. Mech. Dev. **53:** 275–287.

5. KIM, J.H., J.M. AUERBACH, J.A. RODRIGUEZ-GOMEZ, *et al.* 2002. Dopamine neurons derived from embryonic stem cells function in an animal model of Parkinson's disease. Nature **418:** 50–56.

6. GRESSENS, P., J.M. HILL, I. GOZES, *et al.* 1993. Growth factor function of vasoactive intestinal polypeptide in whole cultured mouse embryos. Nature **362:** 155–158.

7. WASCHEK, J.A., R.A. CASILLIS, T.B. NGUYEN, *et al.* 1998. Neural tube expression of pituitary adenylate cyclase-activating peptide (PACAP) and receptor: potential role in patterning and neurogenesis. Proc. Natl. Acad. Sci. USA **95:** 9602–9607.

8. BASILLE, M., B.J. GONZALEZ, A. FOURNIER & H. VAUDRY. 1994. Ontogeny of pituitary adenylate cyclase-activating polypeptide (PACAP) receptors in the rat cerebellum: a quantitative autoradiographic study. Dev. Brain Res. **82:** 81–89.

9. GONZALEZ, B.J., M. BASILLE, D. VAUDRY, *et al.* 1997. Pituitary adenylate cyclase-activating polypeptide promotes cell survival and neurite outgrowth in rat cerebellar neuroblasts. Neuroscience **78:** 419–430.

10. VAUDRY, D., B.J. GONZALEZ, M. BASILLE, *et al.* 2000. The neuroprotective effect of pituitary adenylate cyclace-activating polypeptide on cerebellar granule cells is mediated through inhibition of the CED3-related cysteine protease. Proc. Natl. Acad. Sci. USA **97:** 13390–13395.

11. SUH, J., N. LU, A. NICOT, *et al.* 2001. PACAP is an anti-mitogenic signal in developing cerebral cortex. Nat. Neurosci. **4:** 123–124.

12. ZUPAN, V., A. NEHLIG, P. EVRARD & P. GRESSENS. 2000. Prenatal blockade of intestinal peptide alters cell death and synaptic equipment in the murine neocortex. Pediatr. Res. **47:** 53–63.

13. PINCUS, D.W., E. DICICCO-BLOOM & I.B. BLACK. 1994. Trophic mechanisms regulate mitotic neuronal precursors: role of vasoactive intestinal peptide (VIP). Brain Res. **663:** 51–60.

14. GRUMOLATO, L., E. LOUISET, D. ALEXANDRE, *et al.* 2003. PACAP and NGF regulate common and distinct traits of the sympathoadrenal lineage: effects on electrical properties, gene markers and transcription factors in differentiating PC12 cells. Eur. J. Neurosci. **17:** 71–82.

15. CAZILLIS, M., B.J. GONZALEZ, C. BILLARDON, *et al.* 2004 VIP and PACAP induce selective neuronal differentiation of mouse embryonic stem cells. Eur. J. Neurosci. **19:** 1–11.

Vasoactive Intestinal Peptide Generates CD4$^+$CD25$^+$ Regulatory T Cells *in vivo*

Therapeutic Applications in Autoimmunity and Transplantation

ALEJO CHORNY,[a] ELENA GONZALEZ-REY,[a] DOINA GANEA,[b] AND MARIO DELGADO[a]

[a]*Institute of Parasitology and Biomedicine, CSIC, Granada 18100, Spain*

[b]*Department of Biological Sciences, Rutgers University, Newark, New Jersey 07102, USA*

ABSTRACT: CD4$^+$ CD25$^+$ regulatory T cells (T$_{reg}$) control the immune response to a variety of antigens, including self-antigens, and several models support the idea of the peripheral generation of CD4$^+$ CD25$^+$ T$_{reg}$ from CD4$^+$ CD25$^-$ T cells. However, little is known about the endogenous factors and mechanisms controlling the peripheral expansion of CD4$^+$ CD25$^+$ T$_{reg}$. We have found that the immunosuppressive neuropeptide vasoactive intestinal peptide (VIP) induces functional T$_{reg}$ *in vivo*. The administration of VIP together with specific antigen to TCR-transgenic mice results in the expansion of the CD4$^+$ CD25$^+$, Foxp-3/neuropilin 1-expressing T cells, which inhibit responder T cell proliferation through direct cellular contact. The VIP-generated CD4$^+$ CD25$^+$ T$_{reg}$ transfer suppression, inhibiting delayed-type hypersensitivity in the hosts, prevent graft-versus-host disease in irradiated host reconstituted with allogeneic bone marrow, and significantly ameliorate the clinical score in the collagen-induced arthritis model for rheumatoid arthritis and in the experimental autoimmune encephalomyelitis model for multiple sclerosis.

KEYWORDS: regulatory T cells; immune tolerance; neuropeptide; autoimmunity; transplantation

INTRODUCTION

The immune system is faced with the daunting job of protecting the host from an array of pathogens, while maintaining tolerance to self-antigens. The

Address for correspondence: Mario Delgado, Instituto de Parasitologia y Biomedicina, CSIC, Avd. Conocimiento, PT Ciencias de la Salud, Granada 18100, Spain. Voice: 34-958-181665; fax: 34-958-181632.

e-mail: mdelgado@ipb.csic.es

Ann. N.Y. Acad. Sci. 1070: 190–195 (2006). © 2006 New York Academy of Sciences.

doi: 10.1196/annals.1317.011

induction of antigen-specific tolerance is essential to maintain immune home-ostasis, to control autoreactive T cells, and to prevent the onset of autoimmune diseases. Thymic selection prevents to a large degree the release of functional autoreactive T cells. However, potential autoreactive T cells persist in the periphery of healthy individuals and retain the capacity to initiate autoimmune disease. Thus, peripheral regulatory mechanisms are required to protect against self-directed immune responses. Active suppression by $CD4^+CD25^+$ regulatory T cells (T_{reg}) plays a key role in the control of self-reactive T cells, and in the induction of peripheral tolerance.[1] Two major types of T_{reg} cells have been characterized in the $CD4^+$ population, that is, the naturally occurring, thymus-generated T_{reg} cells, and the peripherally induced, interleukin-10 (IL-10) or Yumor growth factor-β (TGF-β)-secreting T_{reg} cells.[1] The $CD4^+CD25^+$, Foxp3-expressing, naturally occurring T_{reg} cells generated in thymus, migrate and are maintained in the periphery. Several experimental models support the idea of peripheral generation of $CD4^+CD25^+$ T_{reg} cells from $CD4^+CD25^-$ T cells,[2] in addition to the central generation. However, different mechanisms of action have been described for the peripherally induced T_{reg}. Whereas the $CD4^+CD25^+$ T_{reg} cells developing *de novo* in the periphery exert their suppressive function in a cell contact-dependent manner, the Tr1/Th3(Tr2) regulatory T cells induced by IL-10, TGF-β or through infectious tolerance exert their regulatory action through cytokines.[1–3] The endogenous factors and mechanisms controlling the peripheral expansion of $CD4^+CD25^+$ T_{reg} cells are mostly unknown. Vasoactive intestinal peptide (VIP) and pituitary adenylate cyclase-activating polypeptide (PACAP) are potent immunosuppressive agents that proved to be protective in models of autoimmune diseases, such as collagen-induced arthritis and inflammatory bowel disease.[4] Until now, VIP/PACAP were described as deactivators of macrophages, dendritic cells, and microglia, and promotors of Th2 effector differentiation and survival.[4] Since T_{reg} cells are major players in limiting the immune response, we investigated the possibility that VIP and PACAP promote T_{reg} cell development and/or activation.

VIP INDUCES CD$^+$ CD25$^+$ TREG *IN VIVO*

We compared the effects of antigen and antigen+VIP administration in TCR-transgenic mice, and concluded that VIP contributes primarily to the expansion of the $CD4^+CD25^+$ T_{reg} cell population (FIG. 1A). The increase in the numbers of $CD4^+CD25^+$ cells occurred only in the antigen-specific T cells for low doses of antigen, and was maintained long term. $CD4^+CD25^+$ T_{reg} cells have been characterized by high expression of the transcription repressor Foxp3, high surface expression of CTLA-4, GITR, neuropilin 1 (Nrp1), CD103, CD62L, and CD69, and low expression of CD45RB (FIG. 1A).[1–3] The CD4 population from antigen+VIP inoculated mice showed a decrease in CD45RB, and

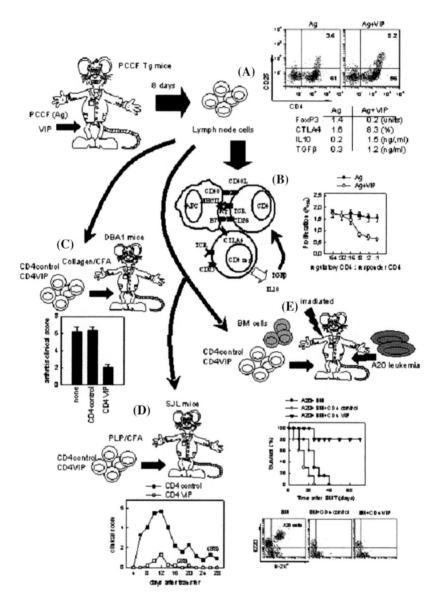

FIGURE 1. VIP augments generation of CD4$^+$CD25$^+$ T$_{reg}$ cells *in vivo*: therapeutic applications in autoimmunity and transplantation. PCCF-transgenic (Tg) mice were injected intraperitoneally (i.p.) with Ag [pigeon cytochrome C fragment (PCCF)], with or without VIP (5 nmol). **(A)** Eight days after initial Ag stimulation, lymph node cells were isolated, and analyzed for expression of CD4, CD25, CTLA4 (by flow cytometry), Foxp-3 (by RT-PCR), or IL10/TGF production (by ELISA). **(B)**. Different numbers of CD4 T cells isolated from mice inoculated with Ag or Ag plus VIP were cocultured with responder CD4. T cells isolated from naïve-Tg mice in the presence of Ag, and proliferation was determined.

increases in the expression of all the other markers, as compared to the CD4 T cells from antigen-inoculated mice. However, the purified $CD4^+CD25^+$ T cells from either antigen- or antigen+VIP inoculated mice expressed similar levels of the markers characteristic for $CD4^+CD25^+$ T_{reg} cells. These results suggest that VIP expands the $CD25^+$ population without inducing increases in Foxp3 or Nrp1 (or other markers) on a per cell basis. The *in vivo* VIP-generated $CD4^+$ T_{reg} cells appear to consist of two populations, a major $CD4^+CD25^+$ population whose suppressive mechanism is mediated through direct cellular contact and does not involve IL-10 or TGF-β, and a minor $CD4^+$ population, which produces and uses IL-10 and/or TGF-β as suppressive molecules. The major population of T_{reg} cells induced by VIP resemble indeed the recently reported $CD25^+$ cytokine-independent suppressors recruited from the peripheral $CD25^-$ population by $CD4^+CD25^+$ T cells stimulated with IL-2 and TGF-β.[5] In addition to expanding the $CD4^+CD25^+$ population, VIP also induced more efficient T_{reg} in terms of suppressive activity. On a per cell basis, the VIP-induced $CD4^+CD25^+$ T_{reg} cells are stronger suppressors of responder T cell proliferation, particularly at low T_{reg}/Tres cell ratio (FIG. 1B). The higher efficiency of VIP-induced T_{reg} cells resides in the fact that they express high levels of CTLA-4, a negative regulatory factor critical for the induction and function of T_{reg} cells.[6] Although still unconfirmed, several mechanisms have been proposed for CTLA-4 mediation in T_{reg} function. CTLA-4 binding to B7 has been shown to induce indoleamine 2,3-dioxygenase in dendritic cells, leading to tryptophan depletion and induction of proapoptotic molecules. Alternatively, CTLA-4 competes with CD28, a costimulatory signal for T cell activation, for binding to B7.

VIP-INDUCED TREG PREVENTS TRANSPLANT REJECTION AND AUTOIMMUNITY

From a therapeutic point of view, it is important to consider that the *in vivo* VIP-induced $CD4^+CD25^+$ T_{reg} cells were able to transfer antigen-specific

←

(C) CIA mice were treated with medium (control) or VIP after onset of disease (collagen injection in DBA1 mice). Spleen CD4 isolated from these mice 10 days after disease onset were transferred into arthritic mice, and arthritic clinical score was evaluated 10 days after transfer. (D) EAE mice were treated with medium (control) or VIP after onset of disease (myelin protein PLP injection in SJL mice). Spleen CD4 isolated from these mice 10 days after disease onset were transferred into EAE mice, and clinical score of paralysis was evaluated. (E) Irradiated Balb/c hosts were inoculated with allogeneic bone marrow (BM) cells (from B10 mice), and A20 leukemic cells (derived from Balb/c). CD4 T cells from PCCF-Tg mice (same haplotype than BM transplant) treated with Ag or Ag+ VIP were also injected. Survival was monitored, and tumor growth/elimination was assessed by the presence of A20 cells in blood detected by coexpression of B220 and $H-2K^d$.

suppression to naïve hosts, inhibiting both delayed-type hypersensitivity, a Th1-dependent immune reaction, and IgG1-type Ab formation, a Th2-dependent response. In addition, the spleen and lymph nodes of hosts that received the VIP-induced T_{reg} cells prior to immunization were much smaller in size compared to hosts that received T_{reg} cells from antigen-inoculated donors. The possible therapeutic use of the VIP-induced T_{reg} cells has been assessed in a model of allogeneic bone marrow transplantation (FIG. 1E). Allogeneic bone marrow transplantation is a treatment of choice in many hematopoietic malignancies. Following high-dose chemotherapy or irradiation, the host is reconstituted with bone marrow cells, and the donor T cells are responsible for the graft-versus-tumor (GVT) effects that eliminate the remaining malignant cells in the host. However, the same donor T cells initiate a graft-versus-host reaction (GVHD) that represents the major complication following allogeneic stem cell transplantation. Therefore, a desirable therapy will eliminate GVHD, without affecting the GVT response. In mice, T_{reg} cells have been shown to prevent lethal GVHD in lethally irradiated hosts reconstituted with allogeneic bone marrow. Recent studies, have also demonstrated that, while controlling GVHD, $CD4^+CD25^+$ T_{reg} cells maintain the graft-versus-leukemia (GVT) response.[7] Similar to these reports, the VIP-induced $CD4^+CD25^+$ T_{reg} cells prevented GVHD, while maintaining GVT in Gza murine model for allogeneic bone marrow transplantation (BMT). These studies suggest the possible use of VIP-induced T_{reg} cells in the survival after hematopoietic stem cell transplantation.

Recently, considerable effort has been focused on the use of antigen-specific T_{reg} cells generated *ex vivo* for the treatment of several autoimmune diseases. We have found that treatment with VIP-induced T_{reg} cells of animals with rheumatoid arthritis or experimental autoimmune encephalomyelitis (EAE) suppresses autoreactive T cells and prevents the progression of both diseases (FIG. 1C and D). The generation of highly efficient T_{reg} cells by VIP *ex vivo* could be used as an attractive therapeutic tool in the future, avoiding the administration of the peptide to the patient.

CONCLUSIONS

In this study, we characterized the functional T_{reg} cells generated *in vivo* following VIP administration. The major T_{reg} cell population induced *in vivo* following VIP and antigen administration is $CD4^+CD25^+$, Foxp-3^{hi}, neuropilin^{hi} and inhibits responder T cell proliferation through direct cellular contact. These cells exhibit increased suppressive activity compared to T_{reg} cells generated in response to antigen alone. The *in vivo* VIP-generated T_{reg} cells transfer suppression, inhibit delayed-type hypersensitivity in TCR-transgenic hosts, prevent GVHD in irradiated hosts reconstituted with allogeneic bone marrow, and ameliorate the clinical signs of autoimmune diseases, such as rheumatoid arthritis and multiple sclerosis.

ACKNOWLEDGMENTS

This work was supported by grants AI52306 and AI47325 (DG), and the Spanish Ministry of Health PI04/0674 and Ramon Areces Foundation (MD).

REFERENCES

1. WRAITH, D.C., K.S. NICOLSON & N.T. WHITLET. 2004. Regulatory CD4$^+$ T cells and the control of autoimmune disease. Curr. Opin. Immunol. **16:** 695–701.
2. AKBAR, A.N., L.S. TAAMS, M. SALMON & M. VUKMANOVIC-STEJIC. 2003. The peripheral generation of CD4$^+$CD25$^+$ regulatory T cells. Immunology **109:** 319–325.
3. THOMPSON, C. & F. POWRIE. 2004. Regulatory T cells. Curr. Opin. Pharmacol. **4:** 408–414.
4. DELGADO, M., D. POZO & D. GANEA. 2004. The significance of vasoactive intestinal peptide in immunomodulation. Pharmacol. Rev. **56:** 249–290.
5. ZHENG, S.G., J.H. WANG, J.D. GRAY, *et al*. 2004. Natural and induced CD4$^+$CD25$^+$ cells educate CD4$^+$CD25$^-$ cells to develop suppressive activity: the role of IL-2, TGF-beta, and IL-10. J. Immunol. **172:** 5213–5221.
6. READ, S., V. MALMSTROM & F. POWRIE. 2000. Cytotoxic T lymphocyte-associated antigen 4 plays an essential role in the function of CD25$^+$CD4$^+$ regulatory cells that control intestinal inflammation. J. Exp. Med. **192:** 295–302.
7. TRENADO, A., F. CHARLOTTE, S. FISSON, *et al*. 2003. Recipient-type specific CD4$^+$CD25$^+$ regulatory T cells favor immune reconstitution and control graft-versus-host disease while maintaining graft-versus-leukemia. J. Clin. Invest. **112:** 1688–1696.

Endogenous Release of Secretin From the Hypothalamus

J.Y.S. CHU,[a] W.H. YUNG,[b] AND B.K.C. CHOW[a]

[a]Department of Zoology, Kadoorie Biological Science Building, The University of Hong Kong, Pokfulam, Hong Kong

[b]Department of Physiology, Faculty of Medicine, The Chinese University of Hong Kong, Shatin, Hong Kong

ABSTRACT: Previous studies demonstrated that secretin could be released from the cerebellum, where it exerts a facilitatory action on the GABAergic inputs into the Purkinje neurons. In the present article, we provide evidence of the endogenous release of secretin in the hypothalamus and the mechanisms underlying this release. Incubation of the hypothalamic explants with KCl induces the release of secretin to 4.35 ± 0.45-fold of the basal level. This K^+-induced release was tetrodotoxin and cadmium sensitive, suggesting the involvement of voltage-gated sodium and calcium channels. The use of specific blockers further revealed the involvement of L-, N-, and P-type high voltage-activated (HVA) calcium channels. Results present in the current article provide further and more solid evidence of the role of secretin as a neuropeptide in the mammalian central nervous system.

KEYWORDS: secretin; hypothalamus; calcium channel; depolarization-evoked release

INTRODUCTION

Secretin is a 27-amino acid peptide hormone that plays a key role in gastrointestinal, pancreatic, and biliary physiology via the specific activation of its receptor in nanomolar concentrations. During the postprandial period, secretin is released from the upper small intestine into the circulation, where it is transported to the pancreas and liver. Primarily, activation of the secretin receptor in these regions leads to the elevation in cyclic adenosine $3',5'$-monophosphate (cAMP), although independent activation of Ca^{2+} signaling pathways has also been described.[1,2] Stimulation of these pathways activated various ion transporters on the pancreatic and biliary ducts, leading to the exocrine secretion of electrolytes and bicarbonate ions.[3]

Address for correspondence: Dr. Billy K.C. Chow, Department of Zoology, The University of Hong Kong, Pokfulam Road, Hong Kong, SAR, PRC. Voice: 852-22990850; fax: 852-25599114.
e-mail: bkcc@hkusua.hku.hk

Ann. N.Y. Acad. Sci. 1070: 196–200 (2006). © 2006 New York Academy of Sciences.
doi: 10.1196/annals.1317.012

Although much is known about the action of secretin in these epithelia, there is less information regarding the releasing mechanism and action of secretin elsewhere in the body. Recently, we have demonstrated that secretin could be released from the cerebellum where it modulates synaptic transmission by facilitating GABA release from the inhibitory presynaptic terminals.[4] In this article, we showed that secretin could also be released from the hypothalamus, further supporting the role of secretin as a neurotransmitter in the central nervous system (CNS).

MATERIALS AND METHODS

Immunohistochemistry

Five-micrometer coronal sections of rat brain were processed for immunofluorescence staining. The antibodies employed were rabbit anti-human secretin antibody (Phoenix Pharmaceuticals, Inc., Belmont, CA), rabbit anti-mouse secretin receptor (raised), and Alexa Fluor® 488 chicken anti-rabbit IgG (Molecular Probes, Inc., Eugene, OR). The sections were analyzed by the Quantimet 570 (Leica, Cambridge, UK) computerized image analysis system.

Peptide Release Experiment

The hypothalamic block from male Wistar rats (200–250 g) was used. Isolated tissues were equilibrated in oxygenated ACSF solution (95% O_2 and 5% CO_2) (in mM: 126 NaCl, 1.3 NaH_2PO_4, 26 $NaHCO_3$, 5 KCl, 2.4 $CaCl_2$, 1.3 $MgSO_4$, 10 glucose, 1 ascorbic acid, and 0.1mM bacitracin, pH 7.4) before experimentation. Afterward, tissues were incubated with the same medium for a 5-min preincubation period to determine the basal release of secretin and then followed by a 5-min stimulation period with 80 mM K^+ in the presence or absence of the drug being tested. The solutions containing elevated K^+ were prepared by substituting NaCl with equimolar concentrations of KCl. The incubation medium was collected and the amount of secretin released was assayed by a Rat Secretin EIA Kit (Phoenix Pharmaceuticals, Inc.).

Drugs

Sodium channel blocker TTX, Q-type calcium channel blocker ω-conotoxin MVIIC, and R-type calcium channel blocker SNX-482 were purchased from Alomone Labs (Jerusalem, Israel). The polypeptide toxin for the N-type calcium channel, ω-conotoxin GVIA, was purchased from Tocris (Ellisville, MO, USA). P-type calcium channel blocker ω-AgatoxinIVA from *Agelenopsis aperta* was purchased from Calbiochem (Darmstadt, Germany). All other chemicals and the L-type calcium channel blocker nicardipine were purchased from Sigma Co. (St. Louis, MO, USA).

RESULTS AND DISCUSSION

The study of the localization and releasing properties of a neurohormone from an isolated brain tissue represents an important step toward the identification of mechanisms involved in the modulation of such release, and ultimately, an understanding of its function. Results from this study showed that secretin and its receptor are present in high amounts within the magnocellular neurons of the paraventricular and supraoptic nuclei (FIG. 1), where the central

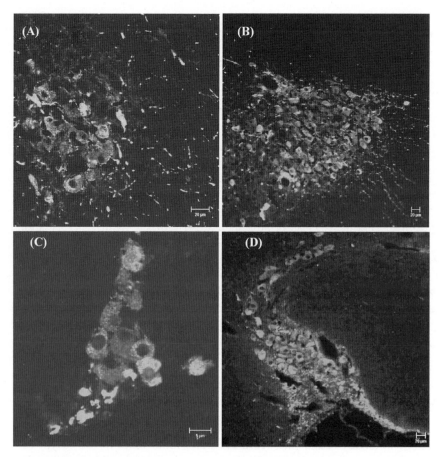

FIGURE 1. The localization of secretin and its receptor in the hypothalamic paraventricular (**A–B**) and supraoptic (**C–D**) nuclei. Both secretin receptor (**A, C**) and its nature agonist (**B, D**) were expressed in the soma of the magnocellular neurons.

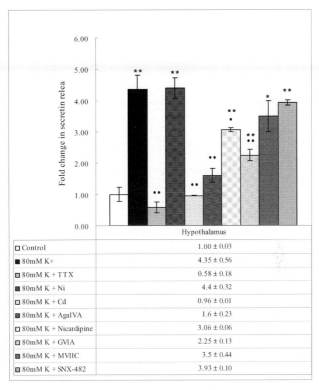

	Hypothalamus
☐ Control	1.00 ± 0.03
■ 80mM K+	4.35 ± 0.56
☐ 80mM K + TTX	0.58 ± 0.18
■ 80mM K + Ni	4.4 ± 0.32
☐ 80mM K + Cd	0.96 ± 0.01
■ 80mM K + AgaIVA	1.6 ± 0.23
☐ 80mM K + Nicardipine	3.06 ± 0.06
☐ 80mM K + GVIA	2.25 ± 0.13
■ 80mM K + MVIIC	3.5 ± 0.44
☐ 80mM K + SNX-482	3.93 ± 0.10

FIGURE 2. The releasing mechanisms of secretin from the hypothalamus. Outflow of secretin from the hypothalami explants was evoked by depolarizing with 80 mM KCl. This induced release of secretin was significantly depressed in the presence of 1 μM sodium channels blocker (TTX), 100 μM HVA channels blocker ($CdCl_2$), 90 nM ω-Agatoxin IVA (AgaIVA – P-type calcium channel blocker), 5 μM nicardipine (L-type calcium channel blocker), or 100 nM ω-conotoxin GVIA (GVIA–N-type calcium channel blocker). Various drugs were present in the incubation medium during the 5-min stimulation period with 80 mM KCl. Results were represented in mean fold change value ± SEM for the triplicate data obtained from three to five determinations. Comparison of the treated groups and controls was based on analysis of variance for multiple comparisons followed by the Student's –Newman-Keuls test. * $P < 0.05$ and ** $P < 0.01$ versus basal secretin outflow; • $P < 0.05$ and ••$P < 0.01$ versus K^+-evoked secretin outflow.

stress circuitry is clustered. Secretin could be released from the hypothalamus upon K^+-evoked depolarization (4.35 ± 0.45-fold), which is TTX sensitive. Additionally, this stimulated release is dependent on the activation of high voltage-activated (HVA) calcium channels, especially those of the L-, N-, and P-types (FIG. 2), further supporting a role of secretin as a neurotransmitter. In brief, we believed that endogenously released secretin from the hypothalamus could possibly act via an autocrine pathway to modulate the excitability of magnocellular neurons, playing a role in central stress responses.

REFERENCES

1. TRIMBLE, E.R., R. BRUZZONE, T.J. BIDEN, *et al.* 1987. Secretin stimulates cyclic AMP and inositol trisphosphate production in rat pancreatic acinar tissue by two fully independent mechanisms. Proc. Natl. Acad. Sci. USA **84:** 3146–3150.
2. WISDOM, D.M., P.J. CAMELLO, G.M. SALIDO, *et al.* 1994. Interaction between secretin and nerve-mediated amylase secretion in the isolated exocrine rat pancreas. Exp. Physiol. **79:** 851–863.
3. ISHIGURO, H., M.C. STEWARD, Y. SOHMA, *et al.* 2002. Membrane potential and bicarbonate secretion in isolated interlobular ducts from guinea-pig pancreas. J. Gen. Physiol. **120:** 617–628.
4. LEE, S.M., L. CHEN, B.K. CHOW, *et al.* 2005. Endogenous release and multiple actions of secretin in the rat cerebellum. Neuroscience **134:** 377–386.

Expression of PACAP Receptors in the Frog Brain during Development

MONICA CIARLO,[a] FEDERICA BRUZZONE,[a,b] CRISTIANO ANGELINI,[a] DAVID ALEXANDRE,[b] YOUSSEF ANOUAR,[b] MAURO VALLARINO,[a] AND HUBERT VAUDRY[b]

[a]Department of Biology, University of Genova, 16132 Genova, Italy

[b]INSERM U413, Laboratory of Cellular and Molecular Neuroendocrinology, European Institute for Peptide Research (IFRMP23), University of Rouen, 76821 Mont-Saint-Aignan, France

ABSTRACT: This study describes the expression of PAC1 and VPAC1 receptor (PAC1-R and VPAC1-R) mRNAs in the brain of frog (*Rana esculenta*) during development. PAC1-R mRNA was detected in the periventricular and subependymal layers of the thalamus and epithalamus and in the ependymal layer of the mesencephalon and rhombencephalon (stage 20), in the amygdala, in the habenular complex, in the periventricular nucleus of the hypothalamus, and in the ventral cerebellum (stage 30). VPAC1-R mRNA expression was observed only at stage 20, in the floor of the hypothalamus. These results suggest that, in amphibians, pituitary adenylate cyclase-activating polypeptide (PACAP) may play a role in brain development.

KEYWORDS: receptors; PACAP; VIP; ontogeny; frog brain; whole mount *in situ* hybridization

INTRODUCTION

Pituitary adenylate cyclase-activating polypeptide (PACAP) was initially isolated from a sheep hypothalamic extract on the basis of its ability to stimulate cAMP production in rat anterior pituitary cells.[1] Subsequently, PACAP has been characterized in various submammalian vertebrates, notably in the frog *Rana esculenta*.[2] In both amphibians and mammals, PACAP exists in two α-amidated forms with 38 (PACAP-38) and 27 (PACAP-27) amino acid residues.[1-4] The primary structure of PACAP has been remarkably well conserved during evolution from protochordates to mammals.[5] In particular, the sequence of PACAP-27 is identical in frog and mammals and the sequences of

Address for correspondence: F. Bruzzone, Department of Biology, University of Genova, V.le Benedetto XV, 5, 16132 Genova, Italy. Voice: +39-010-353-8045; fax: 39-010-353-8045.
e-mail: federicabruzzone@hotmail.com

Ann. N.Y. Acad. Sci. 1070: 201–204 (2006). © 2006 New York Academy of Sciences.
doi: 10.1196/annals.1317.014

frog PACAP-38 and mammalian PACAP-38 only differ by one substitution.[1,2] Three types of PACAP receptors have been characterized so far: PAC1 receptor (PAC1-R) that selectively binds PACAP with high affinity, and VPAC1 and VPAC2 receptors (VPAC1-R and VPAC2-R) that bind with equal affinity PACAP and vasoactive intestinal polypeptide (VIP).[6] All three types of receptors have also been cloned in amphibians that is, PAC1-R,[7,8] VPAC1-R,[9] and VPAC2-R.[10] In mammals, PACAP and its receptors are expressed early during development,[11,12] and PACAP has been found to act on immature granule cells to promote cell survival and neurite outgrowth,[13,14] suggesting that the peptide may play a key role during ontogenesis. In the present article, we have investigated the expression of PAC1-R and VPAC1-R in the frog brain during development.

RESULTS

The distribution of PAC1-R and VPAC1-R mRNAs in the brain of frog (*Rana esculenta*) tadpoles was studied by *in situ* hybridization using whole mount preparations. During early development, at stage 12 (morula stage) and stage 18, neither PAC1-R nor VPAC1-R was detected in the brain. At stage 20, PCA1-R mRNA was expressed in the periventricular and subependymal layers of the thalamus and epithalamus, and within the ependymal layer of the mesencephalon and rhombencephalon. At this stage, a low concentration of VPAC1-R mRNA was detected in the floor of the hypothalamus. At stage 25, the brain was totally devoid of PAC1-R and VPAC1-R mRNAs. At stage 30, PAC1-R mRNA was detected in the medial amygdala, the subependymal layers of the thalamus, the habenular complex, and the periventricular nucleus of the thalamus, as well as in cells lining the ventral cerebellum. Besides, no PAC1-R or VPAC1-R hybridization signal was seen in the pituitary gland during development.

CONCLUSION

The present study provides evidence for the expression of the genes encoding PAC1-R and VPAC1-R in the frog brain during development. The transient expression of PAC1-R mRNA in germinative neuroepithelia and VPAC1-R mRNA in the hypothalamus strongly suggests that, in amphibians as in mammals, PACAP may act as a neurotrophic factor promoting survival and/or differentiation of neuronal precursors.

ACKNOWLEDGMENTS

This work was supported by the University of Genova, INSERM (U413), IFRMP 23, and the Conseil Régional de Haute-Normandie.

REFERENCES

1. MIYATA, A., A. ARIMURA, R.R. DAHL, *et al.* 1989. Isolation of a novel 38 residue-hypothalamic polypeptide which stimulates adenylate cyclase in pituitary cells. Biochem. Biophys. Res. Commun. **164**: 567–574.
2. CHARTREL, N., M.C. TONON, H. VAUDRY & J.M. CONLON. 1991. Primary structure of frog pituitary adenylate cyclase-activating polypeptide (PACAP) and effects of ovine PACAP on frog pituitary. Endocrinology **129**: 3367–3371.
3. MIYATA, A., L. JIANG, R.D. DAHL, *et al.* 1990. Isolation of a neuropeptide corresponding to the N-terminal 27 residues of the pituitary adenylate cyclase-activating polypeptide with 38 residues (PACAP38). Biochem. Biophys. Res. Commun. **170**: 643–648.
4. YON, L., M. FEUILLOLEY, N. CHARTREL, *et al.* 1992. Immunohistochemical distribution and biological activity of pituitary adenylate cyclase-activating polypeptide (PACAP) in the central nervous system of the frog Rana ridibunda. J. Comp. Neurol. **324**: 485–499.
5. SHERWOOD, N.M., S.L. KRUECKL & J.E. MCRORY. 2000. The origin and function of the pituitary adenylate cyclase-activating polypeptide (PACAP)/glucagon superfamily. Endocr. Rev. **21**: 619–670.
6. VAUDRY, D., B.J. GONZALEZ, M. BASILLE, *et al.* 2000. Pituitary adenylate cyclase-activating polypeptide and its receptors: from structure to function. Pharmacol. Rev. **52**: 269–324.
7. HU, Z., V. LELIEVRE, A. CHAO, *et al.* 2000. Characterization and messenger ribonucleic acid distribution of a cloned pituitary adenylate cyclase-activating polypeptide type I receptor in the frog Xenopus laevis brain. Endocrinology **141**: 657–665.
8. ALEXANDRE, D., H. VAUDRY, V. TURQUIER, *et al.* 2002. Novel splice variants of type I pituitary adenylate cyclase-activating polypeptide receptor in frog exhibit altered adenylate cyclase stimulation and differential relative abundance. Endocrinology **143**: 2680–2692.
9. ALEXANDRE, D., Y. ANOUAR, S. JÉGOU, *et al.* 1999. A cloned frog vasoactive intestinal polypeptide/pituitary adenylate cyclase-activating polypeptide receptor exhibits pharmacological and tissue distribution characteristics of both VPAC1 and VPAC2 receptors in mammals. Endocrinology **140**: 1285–1293.
10. HOO, R.L.C., D. ALEXANDRE, S.M. CHAN, *et al.* 2001. Structural and functional identification of the pituitary adenylate cyclase-activating polypeptide receptor $VPAC_2$ receptor from the frog *Rana tigrina rugulosa*. J. Mol. Endocrinol. **27**: 229–238.
11. BASILLE, M., D. VAUDRY, Y. COULOUARN, *et al.* 2000. Comparative distribution of pituitary adenylate cyclase-activating polypeptide (PACAP) binding sites and PACAP receptor mRNAs in the rat brain during development. J. Comp. Neurol. **425**: 495–509.
12. CAZILLIS, M., B.J. GONZALEZ, C. BILLARDON, *et al.* 2004. VIP and PACAP induce selective neuronal differentiation of mouse embryonic stem cells. Eur. J. Neurosci. **19**: 798–808.
13. GONZALEZ B.J., M. BASILLE, D. VAUDRY, *et al.* 1997. Pituitary adenylate cyclase-activating polypeptide promotes cell survival and neurite outgrowth in rat cerebellar neuroblasts. Neuroscience **78**: 419–430.

14. VAUDRY, D., B.J. GONZALEZ, M. BASILLE, *et al.* 1998. Pituitary adenylate cyclase-activating polypeptide stimulates both *c-fos* gene expression and cell survival in rat cerebellar granule neurons through activation of the protein kinase A pathway. Neuroscience **84:** 801–812.

The Human VPAC1 Receptor

Identification of the N-terminal Ectodomain as a Major VIP-Binding Site by Photoaffinity Labeling and 3D Modeling

ALAIN COUVINEAU,[a] YOSSAN-VAR TAN,[a] EMILLE CERAUDO,[a]
JEAN-JACQUES LACAPÈRE,[a] SAMUEL MURAIL,[b]
JEAN-MICHEL NEUMANN,[b] AND MARC LABURTHE[a]

[a]INSERM U773, Institut National de la Santé et de la Recherche Médicale,
Faculté de Médecine Xavier Bichat, 75018 Paris, France

[b]CEA DSV/DBJC, URA CNRS 2096, Centre d'Etudes de Saclay,
91191 Gif sur Yvette Cedex, France

ABSTRACT: The human VPAC1 receptor for VIP and PACAP is a class
II Gprotein–coupled receptor (GPCR).The N-terminal ectodomain of
the VPAC1 receptor plays a crucial role in VIP binding. Photoaffinity
experiments clearly indicated that the 6–28 part of VIP physically interacts with the N-terminal ectodomain. Construction of a 3D model of the
N-terminal ectodomain of VPAC1 receptor based on the NMR structure
of the mouse CRF receptor 2 indicated the presence of short consensus repeat/Sushi domain. Docking of VIP in the N-terminal ectodomain
structural model was performed taking into account the severe constraints provided by photoaffinity. A VIP-binding site was identified
on the side of the structured core of the N-terminal ectodomain of the
receptor.

KEYWORDS: VIP; GPCR; 3D model

INTRODUCTION

The human VPAC1 receptor for vasoactive intestinal peptide (VIP) and pituitary adenylate cyclase-activating peptide (PACAP) is a 7-transmembrane
domain protein belonging to the class II family of G protein–coupled receptors (GPCR). The class II family comprises receptors for glucagon, glucagon

Address for correspondence: Alain Couvineau, INSERM U773, Faculté de Médecine X. Bichat, 16
rue Henri Huchard, 75018 Paris, France. Voice: 33-1-44-85-61-30; fax: 33-1-42-28-87-65.
e-mail: Alain.Couvineau@bichat.inserm.fr

Ann. N.Y. Acad. Sci. 1070: 205–209 (2006). © 2006 New York Academy of Sciences.
doi: 10.1196/annals.1317.015

like-peptides, secretin, VIP, PACAP, corticotropin-releasing factor (CRF), growth hormone-releasing factor (GRF), parathyroid hormone, and calcitonin.[1] These receptors display a large N-terminal domain (120–160 residues) including six highly conserved cysteine residues and share a very low sequence homology with class I receptors family.[1] The structure–function relationship analysis of VPAC1 receptor, which represents a protypical class II GPCR, has been extensively studied by mutagenesis[2] showing that the N-terminal ectodomain of the VPAC1 receptor plays a crucial role in VIP binding. However, the N-terminal ectodomain is necessary but not sufficient to ensure high affinity for VIP.[1] Assuming that VIP and its receptors play an important role in various diseases[2] and physiology[3] the molecular determination of the VIP-binding site in VPAC receptors represents a major goal. This article reviews recent data regarding VPAC1 receptor.

PHOTOAFFINITY ANALYSIS OF VPAC1 RECEPTOR

In previous studies, the use of photoaffinity probes of VIP associated with the receptor mapping and Edman degradation revealed that VIP physically interacts with the N-terminal ectodomain of VPAC1 receptor.[4,5] Indeed, the photolabile residues of VIP in position 6, 22, and 24 covalently labeled receptor amino acids (Cα or Cβ position) D^{107}, G^{116} and C^{122}, respectively (TABLE 1). Moreover, the photoaffinity probe Bpa^{28}-VIP covalently labeled the 121–133 receptor fragment in the N-terminal ectodomain (TABLE 1). This approach showed clearly that the portion 6–28 of the VIP interacts with the N-terminal ectodomain.

3D MODEL OF THE VPAC1 RECEPTOR ECTODOMAIN

On the basis of the recent determination of NMR structure of the N-terminal ectodomain of mouse CRF2 receptor,[6] a 3D molecular model was developed. Sequence alignment between the N-terminal ectodomain of VPAC1 receptor

TABLE 1. Determination of physical contacts between VIP and the VPAC1 receptor by photoaffinity experiments

Photoaffinity probes	Target residues in VPAC1 receptor
Bpa^6-VIP	D^{107}
Bpa^{22}-VIP	G^{116}
BzK^{24}-VIP	C^{122}
Bpa^{28}-VIP	Fragment A^{121}–E^{133}

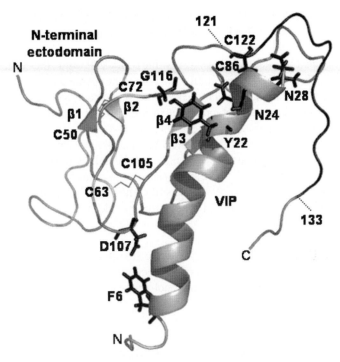

FIGURE 1. Structural model of VPAC1 receptor ectodomain and docking of VIP. Three-dimensional molecular model was based on the NMR structure of the N-terminal ectodomain of CRF2β receptor (6). NMR structure of VIP was obtained in 30% (v/v) TFE (JM Neuman, personal data). Physical contacts identified by photoaffinity were represented as: *black*, F6(VIP)–D107(receptor), Y22(VIP)–G116(receptor), N24(VIP)–C122(receptor), and N28(VIP)–(121–133 (receptor)). *Medium gray*, β sheets were β1, β2, β3, and β4. *Light gray*, disulfide bridges were C^{50}–C^{72}, C^{63}–C^{105}, C^{86}–C^{122}.

(sequence 44–137) and CRF2 receptor (sequence 39–133) revealed 42% homology. Construction of the 3D model of the N-terminal ectodomain based on sequence alignment was performed by using the Modeller software (San Francisco, CA). The fold is identified as a short consensus repeat or Sushi domain.[7] The core structure contains two antiparallel β sheets (FIG. 1) and is stabilized by three disulfide bonds between residues C^{50}–C^{72}, C^{63}–C^{105}, and C^{86}–C^{122} (FIG. 1). Putative side chain interactions probably stabilize the core structure, that is, a salt bridge involving D^{68} and R^{103} and an interaction between the two aromatic rings of W^{73} and W^{110}. Interestingly, substitution of all these residues by alanine abolishes VIP binding to its VPAC1 receptor.[1] As previously reported[1] the N-terminal ectodomain contains three N-glycosylation sites on N^{58}, N^{69} and N^{100}. The determination of the location of these asparagine residues in the model reveals that these N-glycosylation sites are accessible to aqueous environment.

DOCKING OF VIP IN THE N-TERMINAL ECTODOMAIN OF THE VPAC1 RECEPTOR

The physical contacts between residues of VIP and the N-terminal ectodomain, determined by photoaffinity, constitute severe constraints to dock VIP in its binding site. Using Haddock software, we have computed the docking calculation using the interaction constraints between residues of VIP and of VPAC1 receptor, that is, F^6, Y^{22}, N^{24} and D^{107}, G^{116}, C^{122}, respectively. As shown in FIGURE 1 the interface between VIP and the N-terminal ectodomain includes the second antiparallel β sheet (β3–β4) and the unstructured W^{110}–C^{122} segment. The VIP binds to the side of short consensus repeat/Sushi domain (FIG. 1). The C-terminal part of VIP was tilted at about 20° in order to respect contact surface between VIP and N-terminal ectodomain (FIG. 1). These results clearly demonstrate that the 6–28 part of VIP physically interacts with the N-terminal ectodomain. In contrast, the 1–5 N-terminal end of VIP is not involved in these interactions (FIG. 1) and is free to interact with the extracellular loops and transmembrane domains of the receptor. This is in good agreement with the proposed two-step activation mechanism of VPAC receptor and more generally of class II receptor.[8]

ACKNOWLEDGMENTS

Y-V Tan was supported by the grant FTD 200441202869 from the Fondation pour la Recherche Médicale. This research was supported by the Institut National de la Santé et de la Recherche Médicale and also by the grant ACIM-2–18, 2003–5, and a grant in 2005 from Association de Recherche sur la Polyarthrite.

REFERENCES

1. LABURTHE, M., A. COUVINEAU & J.C. MARIE. 2002. VPAC receptors for VIP and PACAP. Receptors Channels **8:** 137–153.
2. LABURTHE, M. & M. COUVINEAU. 2002. Molecular pharmacology and structure of VPAC receptors for VIP and PACAP. Regul. Peptides **108:** 165–173.
3. SAID, S.I. 1986. Vasoactive intestinal peptide J. Endocrinol. Investig. **9:** 191–200.
4. TAN, Y.V., A. COUVINEAU, J. VAN RAMPELBERGH & M. LABURTHE. 2003. Photoaffinity labeling demonstrates physical contact between vasoactive intestinal peptide and the N-terminal ectodomain of the human VPAC1 receptor. J. Biol. Chem. **278:** 36531–36536.
5. TAN, Y.V., A. COUVINEAU & M. LABURTHE. 2004. Diffuse pharmacophoric domains of vasoactive intestinal peptide (VIP) and further insights into the interaction of VIP with the N-terminal ectodomain of human VPAC1 receptor by photoaffinity labeling with [Bpa6]-VIP. J. Biol. Chem. **279:** 38889–38894.

6. GRACE, C.R., M.H. PERRIN, M.R. DiGRUCCIO, *et al.* 2004. NMR structure and peptide hormone binding site of the first extracellular domain of a type B1 G protein-coupled receptor. Proc. Natl. Acad. Sci. USA **101:** 12836–12841.
7. HAWROT, E., Y. XIAO, Q.L. SHI, *et al.* 1998. Demonstration of a tandem pair of complement protein modules in GABA(B) receptor 1a. FEBS Lett. **432:** 103–108.
8. HOARE, S.R.J. 2005. Mechanisms of peptide and nonpeptide ligand binding to class B G-protein-coupled receptors. Drug Discov. Today **10:** 417–427.

VPAC$_2$ Receptor Activation Mediates VIP Enhancement of Population Spikes in the CA1 Area of the Hippocampus

DIANA CUNHA-REIS, JOAQUIM ALEXANDRE RIBEIRO, AND ANA M. SEBASTIÃO

Institute of Pharmacology and Neurosciences, Faculty of Medicine and Institute of Molecular Medicine, University of Lisbon, 1649–028 Lisbon, Portugal

ABSTRACT: The receptors mediating vasoactive intestinal polypeptide (VIP) enhancement of synaptic transmission to pyramidal cell bodies were investigated. RO 25–1553 (VPAC$_2$ agonist) mimicked the excitatory effect of VIP on population spike (PS) amplitude. [K^{15}, R^{16}, L^{27}] VIP (1–7)/GRF (8–27) (VPAC$_1$ agonist) caused only a small increase in PS amplitude. The effect of VPAC$_2$ agonist (but not of the VPAC$_1$ agonist) persisted upon blockade of GABAergic transmission and was strongly attenuated upon inhibition of PKA. In conclusion, VPAC$_2$ receptor activation mediates VIP enhancement of PS amplitude in the hippocampus essentially through a PKA-dependent mechanism.

KEYWORDS: VIP; VPAC$_1$; VPAC$_2$; PKA; PKC; PSs; hippocampus

Vasoactive intestinal polypeptide (VIP), a neuropeptide expressed in the hippocampus exclusively in GABAergic interneurons,[1] modulates synaptic transmission to hippocampal pyramidal cell dendrites as well as pyramidal cell firing.[2] VIP enhancement of synaptic transmission to pyramidal cell dendrites in the CA1 area of the hippocampus is mostly dependent on VPAC$_1$ receptor activation[3] and appears to be indirectly mediated by the suppression of inhibition to pyramidal cell dendrites.[1,4] This is consistent with the preferential localization of VPAC$_1$ receptors in the stratum oriens and radiatum of the Ammon's Horn.[5,6] Some of the pathways involved in VIP modulation of hippocampal synaptic transmission directly target pyramidal cell bodies.[4,6,7] Furthermore, fine mapping of VIP receptors in the hippocampus by immunohistochemistry[5] and by autoradiography[6] indicates that the VPAC$_2$ receptor subtype is the most abundant in the pyramidal cell layer of the Ammon's Horn suggesting a major

Address for correspondence: Diana Cunha-Reis, Institute of Pharmacology and Neurosciences, Faculty of Medicine and Institute of Molecular Medicine, University of Lisbon, Av. Prof. Egas Moniz, 1649–028 Lisbon, Portugal. Voice: +351-217985183; fax: +351-217999454.
e-mail: diana-reis@fm.ul.pt

Ann. N.Y. Acad. Sci. 1070: 210–214 (2006). © 2006 New York Academy of Sciences.
doi: 10.1196/annals.1317.016

role for this receptor in the modulation of hippocampal pyramidal cell activity. We now investigated the role of VPAC$_1$ and VPAC$_2$ receptors on VIP actions at pyramidal cell bodies by studying the action of selective VPAC$_1$ and VPAC$_2$ receptor agonists on population spikes (PSs).

Electrophysiological recordings were made in rat hippocampal slices upon stimulation (0.1 ms pulses once every 15 s, 150–350 μA intensity) of the Schaffer collateral/commissural fibers, in the stratum radiatum near the CA3/CA1 border; PSs were recorded through an extracellular microelectrode (4 M NaCl, 2–4 M; resistance) placed in the *stratum piramidale* of the CA1 area (FIG. 1 A), essentially as previously performed.[4]

VIP causes a biphasic increase in PSs amplitude[4] with a maximum effect of $19.2 \pm 1.5\%$, ($n = 5$, FIG. 1) for 1 nM VIP. The VPAC$_1$ selective agonist[8] [K^{15}, R^{16}, L^{27}] VIP(1–7)/GRF(8–27) at a concentration (10 nM) reported to evoke excitatory responses in the hippocampus,[3] did not change PS amplitude ($2.5 \pm 2.6\%$, $n = 2$). A higher concentration (100 nM) of this agonist caused a small increase ($8.1 \pm 0.9\%$, $n = 3$, FIG. 1) in PS amplitude. The VPAC$_2$ selective agonist,[9] RO 251553, at a concentration (10 nM) 10 times its IC50 for VIP binding to rat VPAC$_2$ receptors[6] and that binds negligibly to rat VPAC$_1$ and rat PAC$_1$ receptors,[6,9] increased the PS amplitude by $18.9 \pm 1.7\%$ ($n = 4$, FIG. 1), an enhancement similar to that obtained with 1 nM VIP. The effect of [K^{15}, R^{16}, L^{27}] VIP(1–7)/GRF(8–27) (100 nM) was abolished ($n = 3$, FIG. 1) upon blockade of GABAergic transmission with picrotoxin (50 μM, GABA$_A$ antagonist) and CGP55845 (1 μM, GABA$_B$ antagonist), but the effect of RO 25–1553 (10 nM) was only mildly attenuated under these conditions ($n = 3$, FIG. 1). Upon inhibition of protein kinase A (PKA) with H-89 (1 μM) the effect of the VPAC$_2$ receptor agonist RO 25–1553 (10 nM) on PS amplitude was markedly attenuated to $6.4 \pm 0.6\%$ ($n = 3$, FIG. 2) whereas upon inhibition of protein kinase B (PKC) with GF 109203X (1 μM) that effect was only slightly reduced ($n = 3$, FIG. 2).

The main finding in the present work is that VIP enhancement of PS amplitude in the CA1 area of the hippocampus involves mainly VPAC$_2$ receptor activation. Blockade of GABAergic transmission occludes VIP effects on synaptic transmission to pyramidal cell dendrites.[4] In contrast, the now observed VPAC$_2$ receptor-mediated actions prevailed in the absence of GABAergic transmission, which also excludes presynaptic actions on GABA release from basket cells, the most likely source of VIP acting on pyramidal cell bodies.[1] Thus, the presently observed VPAC$_2$-mediated action appears to be mainly due to postsynaptic location VPAC$_2$ receptors on pyramidal cell bodies. These observations are consistent with the known predominance of VPAC$_2$ receptors over VPAC$_1$ receptors in the *stratum piramidale* of the hippocampus,[5,6] and with the ability of VIP to enhance excitatory post-synaptic currents (EPSCs) in the absence of GABAergic transmission.[7] The now observed VPAC$_2$ receptor-mediated action is mostly dependent on PKA and is in agreement with the known dependence on PKA of VIP effects on pyramidal cells.[1,7] The VPAC$_1$

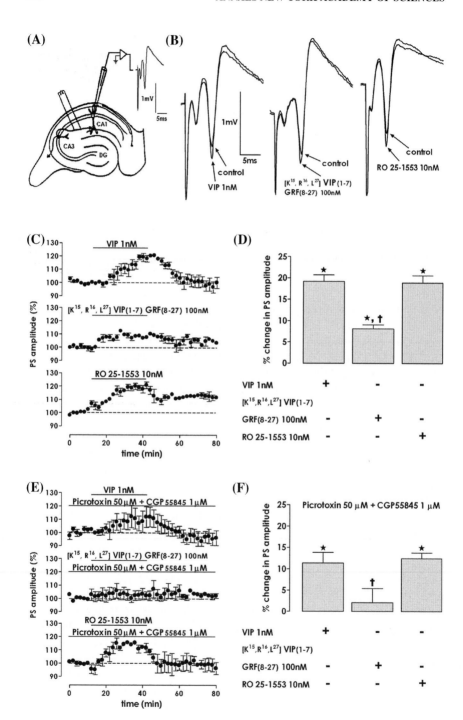

FIGURE 1. VPAC$_2$ receptor activation enhances synaptic transmission to CA1 pyramidal cell bodies. (**A**) Schematic representation of a hippocampal slice showing the location of the stimulation (left) and recording (right) electrodes. (**B**) PSs obtained in individual experiments to test the action of VIP (1 nM, left), the VPAC$_1$ receptor agonist [K^{15}, R^{16}, L^{27}] VIP(1–7)/GRF(8–27) (100 nM, middle), or the VPAC$_2$ receptor agonist RO 25–1553 (10 nM, right). (**C, E**) Time course changes in PS amplitude in the presence of VIP (1 nM, $n = 6$, top), the VPAC$_1$ receptor agonist (100 nM, $n = 3$, middle), or the VPAC$_2$ receptor agonist (10 nM, $n = 4$, bottom) when applied alone (**C**) or in the presence (**E**) of picrotoxin (50 μM) plus CGP55845 (1 μM). (**D, F**) Averaged effects of the VIP receptor agonists on PS amplitude when applied alone (**D**) or in the presence (**F**) of picrotoxin (50 μM) plus CGP55845 (1 μM) (**F**) Each column is the mean ± SEM of 3–6 experiments. $^*P < 0.05$ (Student's *t*-test) versus 0%. $†P < 0.05$ (ANOVA, followed by Dunnett's multiple comparison test) versus the effect of 1 nM VIP (first column).

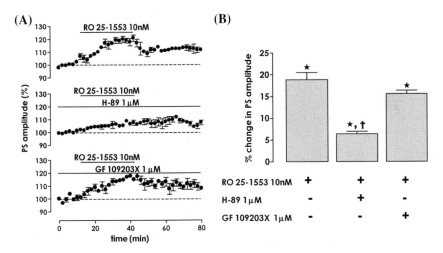

FIGURE 2. VPAC$_2$ receptor activation enhances PS activity through a PKA-dependent pathway. (**A**) Time course changes in PS amplitude when the VPAC$_2$ receptor agonist RO 251553 (10nM) was applied alone (top) or in the presence of either the PKA inhibitor H-89 (1 μM, $n = 3$, middle) or the PKC inhibitor GF 109203X (1 μM, $n = 3$, bottom). (**B**) Ability of the PKA and PKC inhibitors to modify the effects of RO 25–1553 (10 nM) on PS amplitude. Each column is the mean ± SEM of 3–4 experiments. $^*P < 0.05$ (Student's *t*-test) versus 0%. $†P < 0.05$ (ANOVA, followed by Dunnett's multiple comparison test) versus the effect of 1 nM VIP (first column).

receptor-mediated enhancement by VIP of synaptic transmission to pyramidal cell dendrites[2] results from inhibition of inhibitory circuits located at the stratum radiatum and directly impinging on pyramidal cell dendrites.[1,4] This, together with the now reported direct, probably postsynaptic, enhancement of PS amplitude by $VPAC_2$ receptor activation, suggests that different VIP receptors are involved in the two major mechanisms involved in VIP modulation of synaptic transmission to pyramidal cells, with the $VPAC_2$ receptor contributing mainly to modulation of transmission to pyramidal cell bodies and the $VPAC_1$ receptor contributing mainly to synaptic transmission to pyramidal cell dendrites.[3]

ACKNOWLEDGMENTS

We acknowledge Prof. P. Robberecht, SM, ULB, Belgium for the kind gift of VIP selective agonists. Diana Cunha-Reis and work supported by FCT.

REFERENCES

1. ACSÁDY, L., T.J. GÖRCS & T.F. FREUND. 1996. Different populations of vasoactive intestinal polypeptide-immunoreactive interneurons are specialized to control pyramidal cells or interneurons in the hippocampus. Neuroscience **73**: 317–334.
2. HAAS, H.L. & B.H. GÄHWILLER. 1992. Vasoactive intestinal polypeptide modulates neuronal excitability in hippocampal slices of the rat. Neuroscience **47**: 273–277.
3. CUNHA-REIS, D., J.A. RIBEIRO & A.M. SEBASTIÃO. 2005. VIP enhances synaptic transmission to hippocampal CA1 pyramidal cells through activation of both $VPAC_1$ and $VPAC_2$ receptors. Brain Res. **1049**: 52–60.
4. CUNHA-REIS, D., A.M. SEBASTIÃO, K. WIRKNER, et al. 2004. VIP enhances both pre- and postsynaptic GABAergic transmission to hippocampal interneurones leading to increased excitatory synaptic transmission to CA1 pyramidal cells. Br. J. Pharmacol. **143**: 733–744.
5. JOO, K.M., Y.H. CHUNG, M.K. KIM, et al. 2004. Distribution of vasoactive intestinal peptide and pituitary adenylate cyclase-activating peptide receptors ($VPAC_1$, $VPAC_2$, and PAC_1 receptor) in the rat brain. J. Comp. Neurol. **476**: 388–413.
6. VERTONGEN, P., S.N. SCHIFFMANN, P. GOURLET, et al. 1997. Autoradiographic visualization of the receptor subclasses for vasoactive intestinal polypeptide (VIP) in rat brain. Peptides **18**: 1547–1554.
7. CIRANNA, L. & S. CAVALLARO. 2003. Opposing effects by pituitary adenylate cyclase-activating polypeptide and vasoactive intestinal peptide on hippocampal synaptic transmission. Exp. Neurol. **184**: 778–784.
8. GOURLET, P., A. VANDERMEERS, P. VERTONGEN, et al. 1997. Development of high affinity selective VIP_1 receptor agonists. Peptides **18**: 1539–1545.
9. GOURLET, P., P. VERTONGEN, A. VANDERMEERS, et al. 1997. The long-acting vasoactive intestinal polypeptide agonist RO 25–1553 is highly selective of the VIP_2 receptor subclass. Peptides **18**: 403–408.

Expression and GTP Sensitivity of Peptide Histidine Isoleucine High-Affinity-Binding Sites in Rat

COLIN DEBAIGT,[a] ANNIE-CLAIRE MEUNIER,[a] STEPHANIE GOURSAUD,[a] ALICIA MONTONI,[a] NICOLAS PINEAU,[a] ALAIN COUVINEAU,[b] MARC LABURTHE,[b] JEAN-MARC MULLER,[a] AND THIERRY JANET[a]

[a]IPBC CNRS-UMR 6187 Pôle Biologie Santé, 86022 Poitiers Cedex, France

[b]INSERM U683, Faculté de Médecine X. Bichat, 75018 Paris, France

ABSTRACT: High-affinity-binding sites for the vasoactive intestinal peptide (VIP) analogs peptide histidine/isoleucine-amide (PHI)/carboxyterminal methionine instead of isoleucine (PHM) are expressed in numerous tissues in the body but the nature of their receptors remains to be elucidated. The data presented indicate that PHI discriminated a high-affinity guanosine 5′-triphosphate (GTP)-insensitive-binding subtype that represented the totality of the PHI-binding sites in newborn rat tissues but was differentially expressed in adult animals. The GTP-insensitive PHI/PHM-binding sites were also observed in CHO cells over expressing the VPAC2 but not the VPAC1 VIP receptor.

KEYWORDS: VPAC1 and VPAC2 receptors; GTP-sensitivity; PHI/PHM

INTRODUCTION

The peptide histidine/isoleucine-amide (PHI) and its human counterpart with a carboxyterminal methionine instead of isoleucine (PHM) derive from the same protein precursor as vasoactive intestinal peptide (VIP). They are considered like moderate-to-weak agonists with a modest affinity toward the well-known VPAC1 and VPAC2 forms of VIP receptors. However, utilization of ^{125}I-PHI or PHM as radiotracers, revealed that high-affinity receptors for these peptides are expressed in some tissues, such as liver, where we demonstrated that PHI discriminated a guanosine 5′-triphosphate (GTP)-insensitive subset of VIP receptors, corresponding to a 48-kDa molecular species, distinct from the 65-kDa GTP-sensitive receptor.[1] Other data also support the

Address for correspondence: Thierry Janet, IPBC CNRS-UMR 6187 Pôle Biologie Santé, 40 avenue du Recteur Pineau, 86022 Poitiers Cedex, France. Voice: 33-5-49-45-40-91; fax: 33-5-49-45-39-76. e-mail: thierry.janet@univ-poitiers.fr

Ann. N.Y. Acad. Sci. 1070: 215–219 (2006). © 2006 New York Academy of Sciences. doi: 10.1196/annals.1317.017

idea that different classes of PHI receptors could be represented, especially in liver where PHI-preferring and bivalent PHI/VIP receptors appear to be expressed.[2] Differences were observed in the dose-dependent stimulation of adrenocorticotropic hormone (ACTH) and corticosterone released in response to VIP or PHI, suggesting that these two peptides may act on different sets of receptors.[3] A VIP receptor antagonist thought to bind with equal affinities to all known VIP receptors, inhibited the effects of VIP but not those of PHI on these hormonal secretion.[4] Our group reported specific effects of PHI that suppressed neuroblastoma cell proliferation through an inhibition of MAP kinase activity, while VIP or pituitary adenylate cyclase-activating polypeptide (PACAP) inhibited cell proliferation through a PKA-dependent mechanism.[5] The question of the molecular nature of the receptor complexes corresponding to the high-affinity PHI/PHM-binding sites remains, however, unanswered: are they subsets of the VIP/PACAP receptors with distinct coupling properties that renders them, at least partially, insensitive to guanyl nucleotides or are they unknown components that remain to be characterized? It was long established in early reports on ^{125}I-VIP-binding studies that part of the VIP-binding sites expressed in the liver were sensitive to GTP.[6,7] The GTP-insensitive receptors appear to be expressed very early in the rat embryo (E9,5) and they may play a key role in the development of the central nervous system (CNS) and of the somites.[8] It was reported that a VIP derivative (stearyl-norleucine 17 VIP) could discriminate a subset of GTP-insensitive VIP-binding sites in the CNS.[9] The data presented here allow to precise further the knowledge of the expression of the high-affinity PHI/PHM-binding sites and the regulation of their GTP sensitivity.

MATERIALS AND METHODS

Membrane Preparations

Plasma membranes were obtained from freshly excised rat hearts and livers or from cultured CHO cells stably expressing VPAC1 and VPAC2 receptors[10] (generous gift from A. Couvineau and coll., Neuroendocrinology and Cell Biology, INSERM U410, Faculté de Médecine, Xavier Bichat, 75018 Paris, France). Samples were prepared following a previously described protocol.[11] Protein content was determined with a Bradford assay (BioRad Protein Assay Dye Reagent Concentrate, BioRad, Munich, Germany) and BSA as a standard. Membranes aliquots were then stored at −80°C until use.

Peptides ^{125}I-Iodination

For binding experiments, PHI or PHM was radioiodinated using a previously described chloramine-T technique[12] with slight modifications.[1]

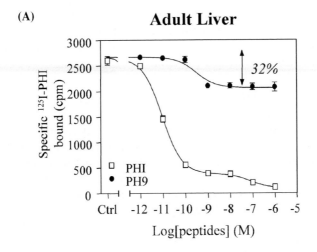

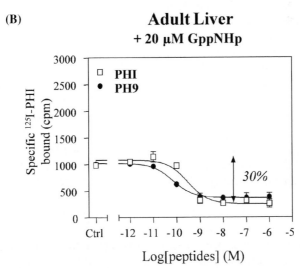

FIGURE 1. Competitive inhibition of ^{125}I-PHI binding by PHI or PH9 in the absence (**A**) or in the presence (**B**) of 20 μM GppNHp in adult liver rat tissue membranes. Membranes were incubated for 50 min at 20°C with ^{125}I-PHI (50 pM). Peptides are at indicated concentrations. Data represent the means ± SEM of three independent experiments, each performed in triplicate.

Receptor-Binding Studies

The binding reactions were conducted at 20°C for 50 min, a time at which equilibrium was reached. Each radiotracer ^{125}I-PHI or ^{125}I-PHM in the absence or in the presence of 20 μM GppNHp, a nonhydrolyzable GTP analog, was incubated according to conditions previously reported.[1,6,7] Radioactivity was measured in a γ-counter. In all experiments, specific binding corresponded to the difference between total binding, obtained in the absence of unlabeled PHI or PHM (according to the tracer used) and nonspecific binding, obtained in the presence of 10^{-6} M of PHI or PHM.

RESULTS AND DISCUSSION

The existence of a specific PHI/PHM high-affinity receptor was first suggested in rat liver[2] but not clearly demonstrated to date. Such high-affinity-binding sites were later characterized in newborn chicken liver and described as partially insensitive to GTP.[1] GTP-insensitive-binding sites expression was also highly detected during early embryos neurogenesis and their distribution changed throughout the CNS development.[8]

In adult rat liver membranes, comparison of the total ^{125}I-PHI-binding capacity, in the absence or in the presence of GppNHp (FIG. 1 A versus FIG. 1 B), reveals that the specific binding of ^{125}I-PHI in the presence of the nonhydrolyzable GTP analog (FIG. 1 B) represented only 30% of the total radiotracer binding observed in control membranes (FIG. 1 A).

This result indicated that in adult rat liver 30% of the PHI-binding sites were GTP insensitive. The GTP-insensitive-binding site is selectively discriminated by PHI/PHM and more efficiently by a peptide called PH9, isolated in our laboratory. In fact, this peptide only inhibited the specific binding of ^{125}I-PHI on GTP-insensitive-binding sites even in the absence of GppNHp (FIG. 1A).

Screening by expression cloning of a cDNA liver library with the aim to identify the PHI/PHM receptor(s) led to the isolation of a cDNA (clone6), which confers high-affinity ^{125}I-PHI-binding properties when transfected in COS7 cells.

Binding experiments performed in different newborn rat tissues indicated that only GTP-insensitive high-affinity PHI-binding sites were detected. In adult rat, representation of GTP-insensitive PHI-binding sites varies among tissues. For instance, a majority of sites were GTP insensitive in heart, while in liver, only 31% were detected (TABLE 1).

TABLE 1. Expression of GTP-insensitive-binding sites in different rat tissues and CHO cells stably tranfected with VPAC1 or VPAC2 receptors

		Binding of ^{125}I-PHI/^{125}I-PHM % of GppNHp insensitivity
Liver	new born	97%
	Adult	31%
Heart	new born	100%
	Adult	100%
Cells	CHO/VPAC1	0%
	CHO/VPAC2	80%

In CHO cells stably expressing VPAC1 and VPAC2 receptors[10] results indicated that only VPAC2 receptor presented intrinsic GTP insensitivity compared to VPAC1 receptor (TABLE 1).

In a recent review,[13] VPAC receptors interactions with potential PDZ motif-containing proteins and with receptor activity-modifying proteins (RAMPs) were highlighted. These observations led us to study the potential role of these proteins in the regulation of GTP sensitivity of specific PHI-binding sites.

REFERENCES

1. PINEAU, N., V. LELIEVRE, S. GOURSAUD, *et al*. 2001. The polypeptide PHI discriminates a GTP-insensitive form of VIP receptor in liver membranes. Neuropeptides **35:** 117–126.
2. PAUL, S., J. CHOU & E. KUBOTA. 1987. High affinity peptide histidine isoleucine-preferring receptors in rat liver. Life Sciences **41:** 2373–2380.
3. NOWAK, K.W. & L.K. MALENDOWICZ. 1998. Steroidogenic effect of peptide histidine-isoleucine (PHI) in vitro: comparison with VIP. Endocr. Res. **24:** 759–762.
4. ALEXANDER, L.D. & L.D. SANDER. 1995. VIP antagonist demonstrates differences in VIP and PHI-mediated stimulation and inhibition of ACTH and corticosterone secretion in rats. Regul. Pept. **59:** 321–333.
5. LELIÈVRE, V., N. PINEAU, J. DU, *et al.* 1998. Differential effects of peptides histidine isoleucine (PHI) and related peptides on stimulation and suppression of neuroblastoma cell proliferation: a novel VIP-independent action of PHI via MAP kinase. J. Biol. Chem. **273:** 19685–19690.
6. HILL, J.M., A. HARRIS & D.I. HILTON-CLARKE. 1992. Regional distribution of guanine nucleotide-sensitive and guanine nucleotide-insensitive vasoactive intestinal peptide receptors in rat brain. Neuroscience **48:** 925–932.
7. AMIRANOFF, B., M. LABURTHE & G. ROSSELIN. 1980. Differential effects of guanine nucleotides on the first step of VIP and glucagon action in membranes from liver cells. Biochem. Biophys. Res. Commun. **96:** 463–468.
8. GRESSENS, P., J.M. HILL, I. GOZES, *et al*. 1993. Growth factor function of vasoactive intestinal peptide in whole cultured mouse embryos. Nature **362:** 155–158.
9. HILL, J.M., S.J. LEE, D.A. DIBBERN, JR., *et al*. 1999. Pharmacologically distinct vasoactive intestinal peptide binding sites: CNS localization and role in embryonic growth. Neuroscience **93:** 783–791.
10. NICOLE, P., L. LINS, C. ROUYER-FESSARD, *et al*. 2000. Identification of key residues for interaction of vasoactive intestinal peptide with human VPAC1 and VPAC2 receptors and development of a highly selective VPAC1 receptor agonist. J. Biol. Chem. **275:** 24003–24012.
11. GOURSAUD, S., N. PINEAU, L. BECQ-GIRAUDON, *et al*. 2005. Human H9 cells proliferation is differently controlled by vasoactive intestinal peptide (VIP) or peptide histidine methionine (PHM) : implication of a GTP-insensitive form of VPAC$_1$ receptor. J. Neuroimmunol. **158:** 94–105.
12. MARTIN, J.-L., K. ROSE, G.J. HUGHES & P.J. MAGISTRETTI. 1986. [mono[^{125}I]iodo-Tyr^{10},MetO^{17}]-vasoactive intestinal polypeptide. J. Biol. Chem. **261:** 5320–5327.
13. LABURTHE, M., A. COUVINEAU, & J.-C. MARIE. 2002. VPAC receptors for VIP and PACAP. Receptors Channels **8:** 137–153.

PACAP, VIP, and PHI

Effects on AC-, PLC-, and PLD-Driven Signaling Systems in the Primary Glial Cell Cultures

AGNIESZKA DEJDA,[a] MARTA JOZWIAK-BEBENISTA,[b] AND JERZY Z. NOWAK,[a,b]

[a] Centre for Medical Biology, Polish Academy of Sciences, 93-232 Lodz, Poland

[b] Department of Pharmacology, Medical University of Lodz, 90-419 Lodz, Poland

ABSTRACT: Pituitary adenylate cyclase-activating polypeptide (PACAP), vasoactive intestinal peptide (VIP), and peptide histidine-isoleucine (PHI) are members of a superfamily of structurally related peptides widely distributed in the body and displaying pleiotropic biological activities. All these peptides are known to act via common receptors—VPAC1 and VPAC2. In addition, the effects of PACAP are mediated through its specific receptor named PAC1. The main signal transduction pathway of the mentioned receptors is adenylyl cyclase (AC)→cAMP system. PACAP and VIP may also signal through receptor-linked phospholipase C (PLC)→IP3/DAG→PKC and phospholipase D (PLD)→phosphatidic acid (PA) pathways. In the present article, we have studied the effects of PACAP, VIP, and PHI (0.001–5000 nM) on the AC-, PLC-, and PLD-driven signaling pathways in rat primary glial cell (astrocytes) cultures. All tested peptides dose-dependently and strongly stimulated cyclic adenosine 3′,5′-monophosphate (cAMP) production in this experimental model, displaying the following rank order of potency: PACAP $>>$ VIP $\geq$ PHI. Their effects on PLC-IP3/DAG were weaker, while only PACAP and VIP (0.1-5 µM) significantly stimulated PLD activity. The obtained results showed that rat cerebral cortex-derived astrocytes are responsive to PACAP, VIP and PHI/PHM and possess PAC1 and likely VPAC-type receptors linked to activation of AC-cAMP-, PLC-IP3/DAG-, and PLD-PA signaling systems.

KEYWORDS: pituitary adenylate cyclase-activating polypeptide (PACAP); vasoactive intestinal peptide (VIP); peptide histidine-isoleucine (PHI); cyclic AMP (cAMP); phospholipase C (PLC); phospholipase D (PLD); primary glial cell cultures

Address for correspondence: Agnieszka Dejda, Centre for Medical Biology, Polish Academy of Sciences, 106 Lodowa Street, 93-232 Lodz, Poland. Voice: +48-42-681-51-01; fax: +48-42-2-723-630.

e-mail: adejda@cbm.pan.pl

Ann. N.Y. Acad. Sci. 1070: 220–225 (2006). © 2006 New York Academy of Sciences.
doi: 10.1196/annals.1317.018

INTRODUCTION

Pituitary adenylate cyclase-activating polypeptide (PACAP), vasoactive intestinal peptide (VIP), and peptide histidine-isoleucine (PHI) belong to structurally related family of polypeptides embracing also, for example, peptide histidine-methionine (PHM), secretin, glucagon, and helodermin.[1] PACAP, VIP, and PHI show a widespread tissue–organ distribution. Their presence has been shown in different brain regions, adrenal glands, nerve endings of the respiratory system, gastrointestinal tract, and the reproductive system. All these peptides exert an array of biological effects. They are described as regulators of pituitary gland, pancreas, and adrenal glands, relaxants of smooth muscles in blood vessels, respiratory system, gastrointestinal tract, reproductive system, and factors affecting elements of the immune system. In the central nervous system (CNS), VIP, PACAP, and PHI act as neuroregulators, neurotransmitters, neurotrophic, or neuroprotective factors.[1–3] The biological effects of the mentioned peptides are mediated through specific membrane receptors named PAC1, VPAC1, and VPAC2, all belonging to the superfamily of receptors coupled with G proteins (GPCRs). The PAC1-type receptor preferably binds PACAP, recognizing poorly VIP and PHI/PHM, whereas VPAC-type receptors recognize similarly and with high affinity VIP and PACAP and with lower affinity PHI/PHM. The main signal transduction system linked to the PAC1 and VPAC-type receptors is the Gs-protein→ adenylyl cyclase (AC)-dependent pathway; fractions of these receptors couple also with phospholipase C (PLC)-IP3/DAG or phospholipase D (PLD)-phosphatidic acid (PA)- driven molecular cascades.[2,3]

The aim of the present article was to investigate the effects of PACAP, VIP, and PHI/PHM on three PACAP/VIP receptor-linked signaling pathways, that is, AC-cAMP, PLC-IP3/DAG, and PLD-PA in primary glial cell cultures.

MATERIALS AND METHODS

Experiments were carried out on primary astrocyte cultures derived from cerebral cortex of 1-day-old Wistar rats according to the modified method of Halonen *et al.*[4] The formation of [^{3}H]cAMP was measured in [^{3}H]adenine prelabeled glial cells with the method of Shimizu *et al.*[5] The formed [^{3}H]cAMP was isolated by sequential Dowex-alumina column chromatography according to Salomon *et al.*[6] The activity of PLC-IP3/DAG system was evaluated by measuring the accumulation of [^{3}H]IP3 in [^{3}H]*myo*-inositol prelabeled glial cells as described by Nalepa and Vetulani.[7] PLD activity was assessed by the measurement of [^{14}C]phosphatidylethanol ([^{14}C]PEt) produced in glial cells preincubated with [^{14}C]palmitic acid according to the method described by Bobeszko *et al.*[8]

All data are expressed as mean ± SEM values. For statistical evaluation of the results, analysis of variance (ANOVA) was used followed by the *post hoc* Student's –Newman-Keuls test.

RESULTS AND DISCUSSION

The following peptides were tested for their ability to stimulate the AC-cAMP, PLC-IP3/DAG, and PLD-PA signal transduction pathways in primary glial cell cultures: PACAP-38, mVIP (mammalian VIP), rPHI (rat PHI), pPHI (porcine PHI), and PHM (human PHI).

PACAP-38 (0.001–1 nM) concentration-dependently and strongly stimulated cAMP production in the examined model. The results are in line with recent data by Masmoudi *et al.*[9] obtained on primary cultures of rat cortical astrocytes. All the forms of PHI tested here (rPHI, pPHI, and PHM), as well as mVIP, used in the 0.01–5 μM concentration range, were evidently less potent than PACAP (TABLE 1). The observed rank order of potency (PACAP >> VIP ≈ PHI/PHM) supports a role of PAC1 receptors for PACAP and likely VPAC-type receptors for both VIP and PHI-related peptides.

PACAP-38, mVIP, and PHI/PHM (0.1–5 μM) also stimulated IP3 accumulation in primary glial cell cultures, indicating their ability to stimulate the PLC-IP3/DAG signaling pathway in cultured astrocytes (TABLE 1). The effects produced by the highest peptide concentration were nearly 70% for PACAP and 40–50% (of respective control value) for mVIP and PHI, suggesting that the PLC-IP3/DAG signal transduction pathway in cultured astrocytes is linked likely to VPAC-type receptors. However, our previous results obtained with the use of rat cerebral cortical slices suggested the role of PAC1 receptors in PACAP-stimulated IP3 formation as well.[10]

Concerning the PLD-PA signaling pathway, only PACAP-38 and mVIP concentration (0.1–5 μM)-dependently stimulated PLD activity, while PHI/PHM even at the highest tested concentration (5 μM) had little (rPHI, pPHI) or no (PHM) effect (FIG. 1). On the basis of the obtained data, we cannot identify the receptor type involved in the glial response to the peptides, yet the role of PAC1 as well as VPAC receptors cannot be excluded.

The results obtained in the present study provide evidence that PACAP, VIP, and PHI/PHM activate at least three (AC-cAMP, PLC-IP3/DAG, PLD-PA; PHI/PHM had no significant action on PLD activity) signaling systems in primary glial cultures originating from the newborn rat cerebral cortex. Depending on the system, the biochemical responses to the tested peptides are mediated via PAC1 and/or VPAC-type receptors, as shown in various cell/tissues.[2,11,12] It would be highly interesting to know which astrocyte function(s) is(are) regulated by particular peptide involving particular receptor/signaling pathway. In the context of the present study it is interesting to note that Ashur-Fabian *et al.*[13] recently reported the occurrence on rat astrocytes of both PACAP-preferring

TABLE 1. Effect of PACAP38, mVIP, and PHI on cAMP formation and IP$_3$ accumulation in the rat primary glial cultures

Concentration [M]	% of control cAMP formation				
	PACAP38	mVIP	pPHI	PHM	rPHI
			cAMP formation		
10^{-12}	449±25.7				
10^{-11}	975 ± 24.5				
10^{-10}	1554 ± 401.6				
3×10^{-10}	1813 ± 1				
10^{-9}	3759 ± 223.3				
10^{-8}		132 ± 10.2	271 ± 13.2	154 ± 16.5	147 ± 9.1
10^{-7}		157 ± 8.8	445 ± 25.1	157 ± 4.0	341 ± 20.2
3×10^{-7}		549 ± 58.5	526 ± 21.2	432 ± 90.9	1008 ± 211.5
10^{-6}		703 ± 24.5	766 ± 129.3	1268 ± 278	1735 ± 179
3×10^{-6}		1503 ± 207	1210 ± 164.5	2334 ± 524	2632 ± 266.8
5×10^{-6}		2368 ± 38	1365 ± 180.4	2426 ± 514	2923 ± 262.3
			IP$_3$ accumulation		
3×10^{-7}	109 ± 0.6	109 ± 12.4			
10^{-6}	117 ± 10.5	114 ± 3.16			
3×10^{-6}	125 ± 7.9	141 ± 19.6			
5×10^{-6}	166 ± 19.5	135 ± 11.6	114 ± 16.4	131 ± 20.3	148 ± 11.3

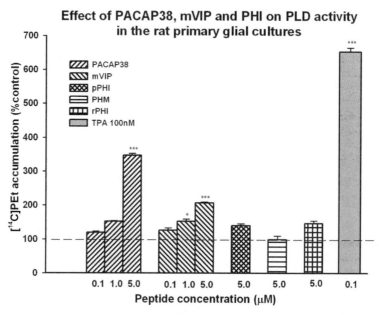

FIGURE 1. Effects of PACAP-38, mammalian VIP (mVIP), porcine PHI (pPHI), PHM, rat PHI (rPHI) on $[^{14}C]$PEt accumulation in primary glial cells cultures. Results are expressed as a percentage of control values (each taken as 100%) and represent means $\pm$ SEM from $n = 4$–24 determinations. Values statistically different from respective controls: $*P < 0.05$, $***P < 0.001$. TPA serves as a positive control of $[^{14}C]$PEt accumulation.

receptor (splice variant PACAP4 or hop2) and VPAC2 receptor, of which the former mediated the glia-dependent neuronal survival effects of VIP. Whether these hop2-like PACAP/VIP receptors present on glial cells (that mediate neurotropism) use any of the studied here signal transduction pathway (AC-cAMP, PLC-IP3, PLD-PA), or another one, remains to be established.

ACKNOWLEDGMENTS

This work was supported by grants No 2P05A-097-26 and No 3P04C-076-25 (MNiI), No 502-11-93 and No 502-11-245 (Med.Univ.), and funds from Ctr Med Biol PAS.

REFERENCES

1. SHERWOOD, N.M., S.L. KRUECKI & J.E. McRORY. 2000. The origin and function of the pituitary adenylate cyclase-activating polypeptide (PACAP)/glucagon superfamily. Endocr. Rev. **21**: 619–670.

2. VAUDRY, D., B.J. GONZALES, M. BASILLE, *et al.* 2000. Pituitary adenylate cyclase-activating polypeptide and its receptors: from structure to functions. Pharmacol. Rev. **52:** 269–324.

3. DEJDA, A., I. MATCZAK & J.Z. NOWAK. 2004. Peptide histidine-isoleucine and its human analogue peptide histidine-methionine: localization, receptors and biological function (in Polish). Postepy Hig. Med. Dosw. (on-line) **58:** 18–26.

4. HALONEN, S.K., F.C. CHIU & L.M. WEISS. 1998. Effect of cytokines on growth of toxoplasma gondii in murine astrocytes. Infect. Immun. **66:** 4989–4993.

5. SHIMIZU, H., J.W. DALY & C.R. CREVELING. 1969. A radioisotopic method for measuring the formation of adenosine 3',5'-cyclic monophosphate in incubated slices of brain. J. Neurochem. **16:** 1609–1619.

6. SALOMON, Y., C. LONDOS & M. RODBELL. 1974. A highly sensitive adenylate cyclase assay. Analyt. Biochem. **58:** 541–548.

7. NALEPA, I. & J. VETULANI. 1991. Involvement of protein kinase C in the mechanisms of in vitro effects of imipramine on generation of second messengers by noradrenaline in the cerebral cortical slices of the rat. Neuroscience **44:** 585–590.

8. BOBESZKO, M., R. CZAJKOWSKI, M. WOJCIK, *et al.* 2002. Modulation by cationic amphiphilic drugs of serine base-exchange, phospholipase D and intracellular calcium homeostasis in glioma C6 cells. Pol. J. Pharmacol. **54:** 483–493.

9. MASMOUDI O., P. GANDOLFO, J. LEPRINCE, *et al.* 2003. Pituitary adenylate cyclase-activating polypeptide (PACAP) stimulates endozepine release from cultured rat astrocytes via a PKA-dependent mechanism. FASEB J. **17:** 17–27.

10. NOWAK, J.Z., A. PIGULOWSKA, K. KUBA & J.B. ZAWILSKA. 2002. Stimulatory effects of pituitary adenylate cyclase-activating polypeptide on inositol phosphates accumulation in avian cerebral cortex and hypothalamus. Neurosci. Lett. **323:** 179–182.

11. FAVIT A., U. SCAPAGNINI & P.L. CANONICO. 1995. Pituitary adenylate cyclase-activating polypeptide activates different signal transducing mechanisms in cultured cerebellar granule cells. Neuroendocrinology **61:** 377–382.

12. MCCULLOCH, D.A., E.M. LUTZ, M.S. JOHNSON, *et al.* 2000. Differential activation of phospholipase D by VPAC and PAC1 receptors. Ann. N. Y. Acad. Sci. **921:** 175–185.

13. ASHUR-FABIAN, O., E. GILADI, D.E. BRENNEMAN & I. GOZES. 1997. Identification of VIP/PACAP receptors on rat astrocytes using antisense oligodeoxynucleotides. J. Mol. Neurosci. **9:** 211–222.

Vasoactive Intestinal Polypeptide Induces Regulatory Dendritic Cells That Prevent Acute Graft Versus Host Disease and Leukemia Relapse after Bone Marrow Transplantation

MARIO DELGADO,[a,b] ALEJO CHORNY,[a] DOINA GANEA,[b]
AND ELENA GONZALEZ-REY[a]

[a]Institute of Parasitology and Biomedicine, CSIC, Granada 18100, Spain

[b]Department of Biological Sciences, Rutgers University, Newark,
New Jersey 07102, USA

ABSTRACT: Acute graft-versus-host disease (GVHD) is a major cause of
morbidity and mortality in patients undergoing allogeneic bone marrow
transplantation (BMT) for the treatment of leukemia and other immuno-
genetic disorders. The use of tolerogenic dendritic cells (DCs) with potent
immunoregulatory properties by inducing the generation/activation of
regulatory T cells (Tr) for the treatment of acute GVHD following allogeneic BMT has been recently established. Here we report the use of the
known immunosuppressive neuropeptide, vasoactive intestinal polypeptide (VIP), as a new approach to inducing tolerogenic DCs with the
capacity to prevent acute GVHD. DCs differentiated with VIP impair
allogeneic haplotype-specific responses of donor CD4$^+$ T cells in transplanted mice by inducing the generation of Tr in the graft. Importantly,
VIP-induced tolerogenic DCs did not abrogate the graft versus leukemia
response, probably because they do not abrogate cytotoxicity of transplanted T cells against the leukemic cells. Therefore, the inclusion of
VIP-induced tolerogenic DC in future therapeutic regimens may facilitate the successful transplantation from mismatched donors, reducing
the deleterious consequences of acute GVHD, extending the applicability
of BMT.

KEYWORDS: transplantation; immune tolerance; dendritic cells; regulatory T cells; neuropeptide

Address for correspondence: Mario Delgado, Instituto de Parasitologia y Biomedicina, CSIC, Avd.
Conocimiento, PT Ciencias de la Salud, Granada 18100, Spain. Voice : 34-958-181665; fax: 34-958-
181632.
  e-mail: mdelgado@ipb.csic.es

Ann. N.Y. Acad. Sci. 1070: 226–232 (2006). © 2006 New York Academy of Sciences.
doi: 10.1196/annals.1317.019

INTRODUCTION

The induction of Ag-specific tolerance is critical for the prevention of autoimmunity and maintenance of immune tolerance. Besides their classical role as sentinels of the immune response by inducing T cell reactivity, increasing evidence now indicates that dendritic cells (DCs) can induce specific T cell tolerance. Although the mechanisms are not entirely known, the maturation–activation state of DCs might be the control point for the induction of peripheral tolerance, by promoting the generation/activation of the suppressive action of regulatory T cells (Tr). Thus, whereas mature DCs (mDCs) are potent APCs enhancing T cell immunity, immature DCs (iDCs) are involved in the induction of peripheral T cell tolerance under steady-state conditions.[1] However, the clinical use of iDCs may not be suitable for the treatment of autoimmune diseases and transplantation, because iDCs are likely to mature in inflammatory conditions.[1] In mice, tolerogenic DCs have been shown to prevent autoimmune disorders and allogeneic transplant rejection.[2] This emphasizes the need to develop tolerogenic DCs with a strong potential to induce Tr cells. Immunosuppressive therapy, which traditionally focused on lymphocytes, has been revolutionized by targeting the development and key functions of DC, and the generation of tolerogenic DCs in the laboratory has become the focus of new therapies.

Vasoactive intestinal polypeptide (VIP) is a potent immunosuppressive agent, affecting both innate and adaptive immunity.[3] Recently, we have shown that VIP affects bone marrow-derived DC (BM-DC) differently, depending on the DC maturation state. iDCs treated with VIP upregulate CD86 expression, stimulate T cell proliferation, and promote Th2-type responses, while inhibiting the Th1-type proinflammatory response.[4] In contrast, VIP downregulates CD80 and CD86 expression of mDCs, and inhibits their capacity to activate allogeneic T cells. In the present article, we show that VIP presence during the early stages of DCs differentiation from human blood monocytes or from mouse BM cells generates a distinct population of DCs, which is unable to mature following inflammatory stimuli.

MATERIALS AND METHODS

Cell Preparation

BM-DCs were generated as described.[4] Briefly, BM cells (2×10^6) obtained from Balb/c (H-2^d) mice were incubated in the complete medium (RPMI-1640 supplemented with 10% heat-inactivated fetal calf serum) containing 20 ng/mL GM-CSF in the presence or absence of VIP (10^{-8} M). At day 6, nonadherent cells were collected and stimulated for 48 h with LPS (1 μg/mL) to induce activation/maturation, and analyzed by flow cytometry and cytokine production by enzyme-linked immunosorbent assay (ELISA). Purified naïve CD4 T cells (5×10^5) were exposed to allogeneic $DC_{control}$ or DC_{VIP} (10^5). After 3 days of culture, CD4 T cells were recovered by immuodepletion of $CD11c^+$ DCs and used as potential CD4Tr cells ($Tr_{control}$ or Tr_{VIP}) in transplanted mice as indicated below.

Models for GVHD and GVT

Allogeneic transplantation was performed by a single intravenous (i.v.) injection of T cell-depleted BM cells supplemented with 1.5×10^6 spleen mononuclear cells (BMS, 1.5×10^7 cells/mouse) isolated from C57Bl/6 ($H-2^b$) into recipient Balb/c ($H-2^d$) mice lethally irradiated (10Gy TBI). Graft-versus-tumor (GVT) was induced by injecting i.v. A20 leukemic cells (derived from Balb/c mice, 10^4 cells) into irradiated Balb/c mice at the time of $H-2^b$ BMS transplantation. Recipients received single or repetitive i.v. injections of different numbers of host-matched $H-2^d$ $DC_{control}$ or DC_{VIP}, 2 and/or 5 days after transplantation. Alternatively, recipients received different numbers of $Tr_{control}$ or Tr_{VIP} ($H-2^b$) at the time of BMS transplantation. Recipients were monitored once every day from the day of transplantation until they succumbed naturally to graft-versus-host disease (GVHD) and/or tumor burden to determine survival time and body weight. Tumor growth/elimination was assessed by the presence of A20 cells in blood detected by coexpression of B220 and $H-2K^d$.

RESULTS AND DISCUSSION

BM-DCs differentiated with VIP (DC_{VIP}) exhibit a tolerogenic phenotype, characterized by low expression of costimulatory molecules (CD40, CD80, and CD86), low production of proinflammatory cytokines [tumor necrosis factor (TNF), interleukin-6 (IL-6), and IL-12], and increased production of IL-10 (FIG. 1 A). The characteristics expressed by the DC_{VIP} are quite similar to those reported for tolerogenic DCs.[1,2] In addition, tolerogenic DCs are poor stimulators of T cell proliferation and cytokine production.[1,2] DC_{VIP} do not prime T cell responses, and suppress previously primed immune responses. T cells exposed initially to $DC_{control}$ proliferated and produced a typical Th1 cytokine profile [large amounts of interferon-γ (IFN-γ) and IL-2 but no IL-4, IL-5, and IL-10] (FIG. 1 B). In contrast, T cells primed with DC_{VIP} did not proliferate following successive restimulation, and exhibited a cytokine profile characteristic of regulatory Tr1, that is, production of high amounts of IL-10 and tumor growth factor-β (TGF-β) and no or negligible synthesis of IFN-γ, IL-2, IL-4, or IL-5 (FIG. 1 C). Following TCR stimulation, Tr cells suppress the proliferation and IL-2 production of Ag-specific effector T cells. In this sense, T cells exposed to DC_{VIP} become functional Tr cells, because

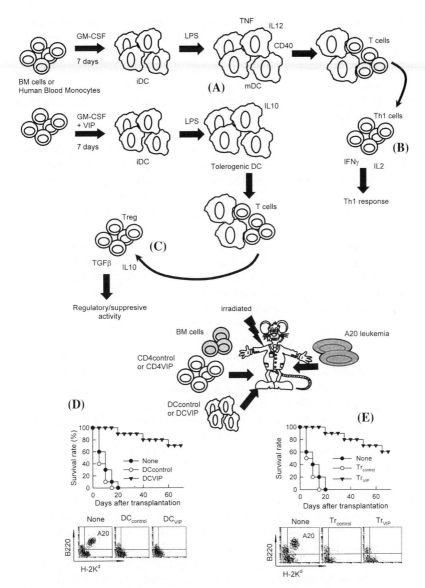

FIGURE 1. Bone marrow DCs differentiated with VIP (DC$_{VIP}$) prevent acute graft versus host disease while maintain graft versus leukemia after BMT.

they inhibit the proliferation and IL-2 production of syngeneic Th1 cells in response to allogeneic mDCs in a dose-dependent manner (FIG. 1 C).

DC$_{VIP}$ retained their T cell regulatory capacity *in vitro* and *in vivo* under inflammatory conditions. This observation is particularly relevant for conditions in which ongoing Ag presentation is associated with chronic inflammation, including autoimmune diseases and allograft rejection. Allogeneic bone marrow transplantation (BMT) is a treatment of choice in many hematopoietic malignancies. Following high-dose chemotherapy or irradiation, the host is reconstituted with bone marrow (BM) cells, and the donor T cells are responsible for the GVT effects that eliminate the remaining malignant cells in the host. However, the same donor T cells initiate a graft-versus-host reaction. In fact, acute GVHD is a major cause of morbidity and mortality in patients undergoing allogeneic BMT. Therefore, a desirable therapy will eliminate GVHD without affecting the GVT response. Most therapeutic approaches designed to reduce acute GVHD have focused on the development of immunosuppressive agents and the *ex vivo* removal of the unfractionated donor T cell proportions from the BM graft. However, removal of these T cells before grafting was shown to lead to transplant failure, sustained immunosuppression, and leukemia relapse.[5] The use of tolerogenic DCs is an innovative therapeutic strategy to limit the pathological effect of donor-alloreactive T cells.[2] Therefore, we examined whether inoculation of recipient DC_{VIP} prevents GVHD lethality. We transplanted lethally irradiated Balb/c recipients ($H-2^d$) with either T cell-depleted BM cells, or with T cell-depleted BM plus spleen mononuclear cells (BMS) from C57Bl/6 ($H-2^b$) mice. Following transplantation, we injected $DC_{control}$ or DC_{VIP} obtained from Balb/c mice. Mice given T cell-depleted BM appeared healthy, and 100% of the animals survived for at least 75 days. Mice that received BMS developed severe signs of GVHD, including weight loss, reduced mobility, hunched posture, diarrhea, and ruffled fur, and all animals died within 25 days, whereas treatment with $DC_{control}$ enhanced the lethality caused by acute GVHD, the administration of host-matched DC_{VIP} protected from lethal GVHD, and >70% of the mice survived for more than 75 days. The therapeutic effect was dose-dependent for recipients receiving a single injection with DC_{VIP}, and repetitive injections of lower doses of DC_{VIP} enhanced the survival rate (not shown).

A hallmark of acute GVHD is the expansion of alloreactive T cells in the proinflammatory environment induced by the conditioning regimen and dysregulated immune mechanisms.[5] Disease progression is characterized by the differentiation of alloreactive $CD4^+$ and $CD8^+$ T cells into effector cells leading to tissue damage, recruitment of additional inflammatory cells, and further cytokine unbalance. We therefore investigated whether infused DC_{VIP} regulate the differentiation of GVHD-causing alloreactive T effector cells in the grafted mice. We first examined the subpopulations of transplanted $I-K^b$ T cells and their ability to produce cytokines in $DC_{control}$ or DC_{VIP}-treated recipients ($H-2^d$). Inoculation of DC_{VIP} decreased the number of $CD4^+$ and $CD8^+$ $I-K^b$ donor-derived T cells, reduced the percentage of activated IL-2/IFN-γ-producing Th1 cells, and increased the number of regulatory IL-10-producing $CTLA4^+$ T cells in the $I-K^b$ $CD4^+$ T cell population (not shown). $I-K^b CD4^+$ T cells obtained from untreated or $DC_{control}$-treated transplanted mice responded vigorously to allogeneic mDC (H-2^d). In contrast, I-K^{b}CD4$^+$ T cells from DC$_{VIP}$-treated recipients were hyporesponsive, and the addition of IL-2 partially restored this response. DC$_{VIP}$ treatment reduced the levels of the proinflammatory cytokines IFN-γ, TNF-α, and IL-12 in the serum of grafted mice. These data indicate that the treatment of transplanted mice with DC$_{VIP}$ reduced the number/activation of transplanted Th1 cells, the inflammatory response against the recipient tissue, and the subsequent GVHD lethality, and induced the generation of Tr cells. This correlates with the fact that DC$_{VIP}$ induce *in vitro* the generation of IL-10/TGF-β-producing regulatory CD4$^+$CD25$^+$CTLA4$^+$ cells. *In vivo* TGF-β/IL-10 blockade and CD25$^+$ cell deletion reversed the therapeutic effect of DC$_{VIP}$ in GVHD confirming the partial involvement of newly generated Tr cells in such action (not shown).

The current preclinical immunosuppressive approaches used to suppress GVHD fail to control the balance between the anti-GVHD effect and the beneficial GVT activity of the allogeneic BM transplants. We therefore tested whether GVT effects in the recipients could be maintained when we used DC$_{VIP}$ to control GVHD. When Balb/c mice (H-2^d) received leukemic (H-2^d) A20 cells, the mice died within 20 days from leukemia, as attested by the presence of leukemic cells in the blood and by marked hepatosplenomegaly. Mice transplanted with BMS (H-2^b) together with A20 cells, and treated with DC$_{control}$ (H-2^d) died with clinical signs characteristic of GVHD, although leukemic cells could not be detected in blood, attesting to an efficient GVT effect (FIG. 1 D). In contrast, most of the leukemia-bearing transplanted mice treated with DC$_{VIP}$ were still alive at day 70. Compared to the two control groups, the presence of DC$_{VIP}$ protected mice from lethal GVHD, whereas the GVT effect was maintained (FIG. 1 D). Indeed, leukemic cells were not detectable in these mice, except for one animal that died at day 40, probably from leukemia.

In certain circumstances, successful suppression of an alloreactive response might require high numbers of Tr, and the *in vivo* administration of DC$_{VIP}$ might not be sufficient for a complete and rapid suppression. Therefore, we decided to generate *in vitro* DC$_{VIP}$-induced Tr and to subsequently determine their suppressive capacity *in vivo* both in a model of BM transplantation. We generated Tr$_{VIP}$ through stimulations of H-2^b CD4 T cells with H-2^d DC$_{VIP}$. Tr$_{control}$ were generated in the same manner with DC$_{control}$. Treatment with Tr$_{VIP}$, but not Tr$_{control}$, of H-2^d mice transplanted with H-2^b allogeneic BMS significantly increased survival by preventing GVHD (FIG. 1 E). This effect was mainly mediated through TGF-β and IL-10, since *in vivo* administration of anti-IL10 and/or anti-TGF-β Abs abrogated the protective effect (not shown). In addition, Tr$_{VIP}$ maintained GVT response in the recipient mice (FIG. 1 E). These results suggest that Tr$_{VIP}$ inhibit haplotype-matched mature T cells present in the BM graft from initiating an alloreactive response against the host.

In conclusion, the possibility of generating tolerogenic DC_{VIP} opens new therapeutic perspectives for the treatment of allogeneic transplantation. *In vitro* pulsing of tolerogenic DC_{VIP} with self- or alloantigens, followed by *in vivo* injection leads to the differentiation of antigen-specific Tr cells. Therefore, the inclusion of tolerogenic DC_{VIP} in future therapeutic regimens may minimize the dependence on nonspecific immunosuppressive drugs used currently as antirejection therapy, and may reduce the deleterious consequences of acute GVHD following allogeneic transplantation.

ACKNOWLEDGMENTS

This work was supported by grants from the Spanish Ministry of Health (PI04/0674, MD), NIH (2RO1A047325, DG and MD), and Ramon Areces Foundation (MD). MD and EG-R were funded by fellowships from Junta de Andalucia.

REFERENCES

1. STEINMAN, R.M., D. HAWIGER & M.C. NUSSENZWEIG. 2003. Tolerogenic dendritic cells. Annu. Rev. Immunol. **109:** 685–711.
2. MORELLI, A.E. & A.W. THOMSON. 2003. Dendritic cells: regulators of alloimmunity and opportunities for tolerance induction. Immunol. Rev. **196:** 125–146.
3. DELGADO, M., D. POZO & D. GANEA. 2004. The significance of vasoactive intestinal peptide in immunomodulation. Pharmacol. Rev. **56:** 249–290.
4. DELGADO, M., A. REDUTA, V. SHARMA & D. GANEA. 2004. VIP/PACAP oppositely affect immature and mature dendritic cell expression of CD80/CD86 and the stimulatory activity for CD4+ T cells. J. Leukoc. Biol. **75:** 1122–1130.
5. COOKE K.R., W. KRENGER, G. HILL, *et al.* 1998. Host reactive donor T cells are associated with lung injury after experimental allogeneic bone marrow transplantation. Blood **92:** 2571–2580.

Vasoactive Intestinal Peptide

The Dendritic Cell → Regulatory T Cell Axis

MARIO DELGADO,[a,b] ELENA GONZALEZ-REY,[b] AND DOINA GANEA[c]

[a]Department of Biological Sciences, Rutgers University, Newark, New Jersey 07102, USA

[b]Instituto de Parasitologia y Biomedicina, CSIC, 10001 Granada, Spain

[c]Temple University, 3420 N. Broad St., Philadelphia, PA 19140, USA

ABSTRACT: Tolerogenic dendritic cells (tDCs) play an important role in maintaining peripheral tolerance through the induction/activation of regulatory T cells (Treg). Endogenous factors contribute to the functional development of tDCs. In this article, we present evidence that two known immunosuppressive neuropeptides, the vasoactive intestinal peptide (VIP) and the pituitary adenylate cyclase-activating polypeptide (PACAP), contribute to the development of bone marrow-derived tDCs. The VIP/PACAP-generated DCs are CD11c^{low}CD45RB^{high}, do not upregulate CD80, CD86, and CD40 following lipopolysaccharide (LPS) stimulation, and secrete high amounts of IL-10. The VIP/PACAP-generated DCs induce functional Treg *in vitro* and *in vivo*. VIP/DCs induce antigen-specific tolerance *in vivo*, suppress delayed-type hypersensitivity (DTH), and T cells from VIP/DC-inoculated mice transfer the suppression to naïve hosts. The effect of VIP/PACAP on the DC–Treg axis represents an additional mechanism for their general anti-inflammatory role, particularly in anatomical sites that exhibit immune deviation or privilege.

KEYWORDS: regulatory T cells; tolerogenic; dendritic cells; vasoactive intestinal peptid

INTRODUCTION

Dendritic cells (DCs) are antigen-presenting cells (APCs) that contribute to innate immunity and initiate adaptive immune responses.[1] The successful initiation of the adaptive immune response requires DC maturation, following signaling through the toll-like receptors and CD40. However, in addition to their proinflammatory role, DCs also play an important role in immune homeostasis, by inducing and maintaining tolerance.[2] The neuropeptides, vasoactive intestinal peptide (VIP) and pituitary adenylate cyclase-activating polypeptide

Address for correspondence: Doina Ganea, Temple University, 3420 N. Broad St., Philadelphia, PA 19140. Voice: 215-707-9921; fax: 215-707-4003.
e-mail: doina.ganea@temple.edu

Ann. N.Y. Acad. Sci. 1070: 233–238 (2006). © 2006 New York Academy of Sciences.
doi: 10.1196/annals.1317.020

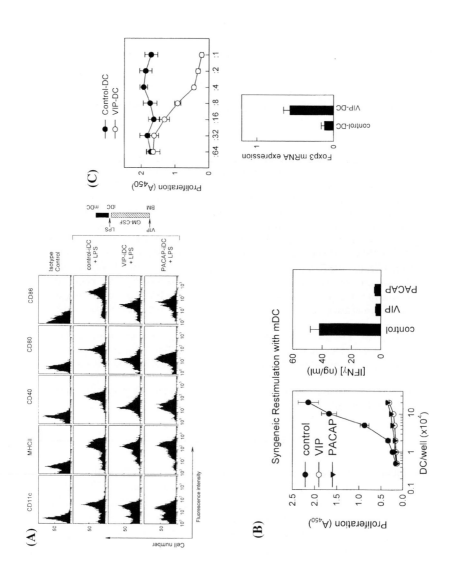

(PACAP), are potent immunosuppressive agents, affecting both innate and adaptive immunity.[3–5] In view of their immunosuppressive activity, we evaluated the VIP/PACAP potential to generate regulatory T cells (Treg) through the induction of tolerogenic DC (tDC).

RESULTS AND DISCUSSION

We generated bone marrow-derived DC in the presence or absence of VIP or PACAP. Following lipopolysaccharide (LPS) stimulation control DCs (generated in the absence of VIP/PACAP) upregulate the expression of MHCII, CD40, CD80, and CD86. In contrast, VIP/PACAP-DCs do not upregulate the expression of costimulatory molecules (CD40, 80, and 86) (FIG. 1 A). In terms of cytokine profile, control DCs secrete tumor necrosis factor (TNF), interleukin (IL)-12, and IL-6, and low levels of IL-10, whereas VIP/PACAP-DC are mostly IL-10 producers, typical of tDCs (not shown). tDCs have been shown to induce Treg that do not proliferate and do not secrete interferon-γ(IFN-γ) upon restimulation. In an *in vitro* system, we exposed TCR-transgenic pigeon cytochrome C fragment (PCCF)-specific T cells to syngeneic control DCs or VIP/PACAP-DCs pulsed with PCCF, followed by reexposure to fresh PCCF-pulsed DCs. T cells exposed initially to control DC proliferated and produced IFN-γ, whereas the T cells exposed to VIP/PACAP-DC did not proliferate and did not produce IFN-γ (FIG. 1 B). The T cells exposed to VIP-DCs function as Treg by inhibiting the proliferation and IL-2 production of fresh T cells in response to antigen (not shown). Interestingly, the Treg induced with VIP-DCs affect primarily Th1 effectors, in agreement with previous reports that VIP promotes Th2 responses to the detriment of proinflammatory Th1 immunity. *In vivo*, VIP-DCs pulsed with PCCF and injected into PCCF-transgenic mice resulted in the induction of splenic T cells that do not proliferate upon restimulation and express high levels of Foxp3, a characteristic of Treg (FIG. 1 C). The Treg generated *in vivo* upon inoculation of VIP-DCs act in an antigen-specific manner. When PCCF-loaded VIP-DCs were injected, followed by immunization and *ex vivo* restimulation with PCCF there was a significant reduction in T cell proliferation,

---

FIGURE 1. VIP and PACAP induce tDCs. (**A**) DCs grown in the presence or absence of VIP/PACAP were stimulated with LPS and analyzed for MHCII, CD40, CD80, and CD86 expression by FACS. (**B**) Control DCs or VIP/PACAP-DCs were loaded with PCCF and cultured with PCCF-specific T cells for 3 days. Repurified T cells were rested for 2 days and restimulated with fresh PCCF/APCs. Proliferation and IFN-γ secretion were determined. (**C**) Purified spleen CD4 T cells from mice injected with control DC or VIP-DC were added to responder PCCF Tg T cells in the presence of spleen APCs and PCCF and assayed for proliferation. Purified CD4 splenic T cells from mice injected with control DC or VIP-DC were analyzed for Foxp3 expression by real time RT-PCR.

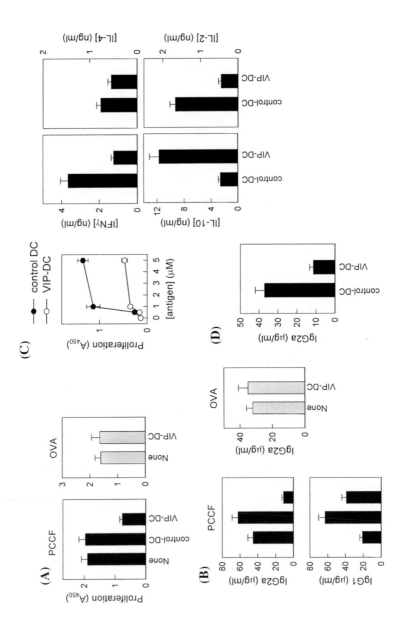

and in the anti-PCCF Ab production. In contrast, if following the administration of PCCF-loaded VIP-DCs the animals were immunized with OVA and the resulting T cells were restimulated with OVA, or injected with saline instead of DC (none) there was no effect on proliferation or anti-OVA Ab levels (FIG. 2 A, B). Since VIP-DCs seem to have a predominant effect on Th1 cells, we assessed the effect of VIP-DCs administration in a delayed-type hypersensitivity (DTH) model. Indeed, administration of VIP-DCs resulted in a significant decrease in DTH (not shown). Antigen-specific tolerance can be adoptively transferred with Treg. We investigated the potential of Treg induced by VIP-DC to transfer tolerance. Control DCs or VIP-DCs loaded with PCCF were injected into B10.A hosts that received 24 h previously PCCF-specific CD4 T cells. Splenic CD4 T cells harvested 7 days later were transferred intravenous (i.v.) into naïve B10.A mice, which were subsequently immunized with PCCF.

Two weeks after immunization, we restimulated splenic T cells *ex vivo*, and measured proliferation and the cytokine profile. Serum anti-PCCF levels were also determined. T cells from donors inoculated with VIP-DC transferred the suppressive activity, as determined by the reduction in proliferation, IL-2 and IFN-γ production, and increased IL-10 production by host splenic T cells (FIG. 2 C). In agreement with the substantial reduction in Th1 cytokines, the levels of anti-PCCF IgG2a Abs were significantly reduced (FIG. 2 D). VIP and PACAP are potent immunosuppressive agents that affect both innate and adaptive immunity.[3-5] Until now, the mechanisms described for their immunosuppressive activity included macrophage/DC/microglia deactivation, and support for Th2 effector differentiation and survival. In this study we show that bone marrow-derived DCs generated in the presence of VIP/PACAP resemble tDCs phenotypically and functionally, suggesting that VIP and PACAP provide a maturation signal similar to IL-10,[6] or to the active form of vitamin D3 1,25(OH)$_2$D$_3$[7] leading to the differentiation of tolerogenic "semi-mature" or "quiescent" DCs.[8] The effect of VIP in generating tDCs is mediated through the VPAC1 receptor and the cyclic adenosine 3',5'-monophosphate (cAMP)/PKA

FIGURE 2. VIP-DCs induce antigen-specific tolerance *in vivo* (**A** and **B**) Control and VIP-DCs loaded with PCCF or saline (none) were injected i.v. into B10.A mice transferred 24 h earlier with PCCF Tg CD4 T cells. One week later, the mice were immunized s.c. with PCCF or OVA. Splenic CD4 T cells harvested 2 weeks later were restimulated *ex vivo* with splenic APCs and PCCF or OVA and tested for proliferation (**A**). Serum PCCF-specific IgG1 and IgG2a levels were measured by enzyme-linked immunosorbent assay (ELISA) (**B**). (**C** and **D**) Splenic purified CD4 T cells from control DC or VIP-DC injected hosts. (see above) were adoptively transferred into naïve B10.A mice. One week later, the hosts were immunized with PCCF, and spleen cells harvested 2 weeks later were restimulated *ex vivo*. Proliferation and cytokine secretion was determined (**C**) and serum levels of PCCF-specific IgG2a levels were determined (**D**).

signaling pathway, and correlates with the inhibition of NF-κBp65 nuclear translocation (not shown). The connection between NF-κB transactivating activity, CD40 expression, and DC function has been established recently,[9] and the association between tDCs and lack of CD40 expression or signaling has been demonstrated.[10] Therefore, we propose that the mechanism by which VIP/PACAP induce tDCs involves the VPAC1 -> cAMP -> PKA-mediated inhibition of NF-κB nuclear translocation, leading to lack of CD40 expression. The VIP/PACAP-generated DCs induce antigen-specific tolerance upon *in vivo* administration. In addition, CD4 T cells from VIP-DC inoculated mice transfer tolerance to naïve hosts. These experiments suggest the possible use of neuropeptides *in vitro* to generate antigen-specific tDCs, followed by *in vivo* administration of these cells to patients with autoimmune diseases.

ACKNOWLEDGMENTS

This work was supported by grants AI52306 and AI47325 (DG), Johnson & Johnson Neuroimmunology Fellowships (MD), and the Spanish Ministry of Health PI04/0674 (MD).

REFERENCES

1. BANCHEREAU, J., F. BRIERE, C. CAUX, *et al.* 2000. Immunobiology of dendritic cells. Annu. Rev. Immunol. **18:** 767–811.
2. STEINMAN, R.M., D. HAWIGER & M.C. NUSSENZWEIG. 2003. Tolerogenic dendritic cells. Annu. Rev. Immunol. **109:** 685–711.
3. DELGADO, M., D. POZO & D. GANEA. 2004. The significance of vasoactive intestinal peptide in immunomodulation. Pharmacol. Rev. **56:** 249–290.
4. GANEA, D., R. RODRIGUEZ & M. DELGADO. 2003. Vasoactive intestinal peptide and pituitary adenylate cyclase-activating polypeptide: players in innate and adaptive immunity. Cell. Mol. Biol. **49:** 127–142.
5. GANEA, D. & M. DELGADO. 2003. The neuropeptides VIP/PACAP and T cells: inhibitors or activators? Curr. Pharm. Des. **9:** 639–652.
6. WAKKACH, A., N. FOURNIER, V. BRUN, *et al.* 2003. Characterization of dendritic cells that induce tolerance and T regulatory 1 cell differentiation in vivo. Immunity **18:** 605–617.
7. GRIFFIN, M.D., W.H. LUTZ, V.A. PHAN, *et al.* 2000. Potent inhibition of dendritic cell differentiation and maturation by vitamin D analogs. Biochem. Biophys. Res. Commun. **270:** 701–708.
8. LUTZ, M. & G. SCHULER. 2002. Immature, semi-mature, and fully mature dendritic cells: which signals induce tolerance or immunity. Trends Immunol. **23:** 445–449.
9. YANG, J., S.M. BERNIER, T.E. ICHIM, *et al.* 2003. LF15-0195 generates tolerogenic dendritic cells by suppression of NF-kB signaling through inhibition of IKK activity. J. Leukoc. Biol. **74:** 438–447.
10. GRACA, L., K. HONEY, E. ADAMS, *et al.* 2000. Cutting edge: anti-CD154 therapeutic antibodies induce infectious transplantation tolerance. J. Immunol. **165:** 4783–4786.

VIP and PACAP Receptor Pharmacology

A Comparison of Intracellular Signaling Pathways

LOUISE DICKSON,[a] ICHIRO ARAMORI,[b] JOHN SHARKEY,[a] AND KEITH FINLAYSON[a]

[a]*Astellas CNS in Edinburgh, University of Edinburgh, EH8 9JZ, UK*

[b]*Molecular Medicine, Astellas Pharma Inc., Ibaraki, 300–2698, Japan*

ABSTRACT: VIP/PACAP receptor activation stimulates the production of $[cAMP]_i$ and $[Ca^{2+}]_i$ by coupling to independent G-protein sub-units, although agonist potencies for the different transduction pathways appear to differ. Using CHO-K1 cells stably expressing the human VIP/PACAP receptors ($hVPAC_1R$, $hVPAC_2R$, and $hPAC_1R$), functional assays ($[cAMP]_i$ and $[Ca^{2+}]_i$) were established and the receptor pharmacology was characterized with five peptide agonists (VIP, PACAP-27, PACAP-38, $[Ala^{11,22,28}]VIP$, and R3P65). The rank order of potency (ROP) was consistent between assays for the individual receptor subtypes, however, higher agonist concentrations ($\sim$100-fold) were required for stimulating $[Ca^{2+}]_i$ when compared to $[cAMP]_i$.

KEYWORDS: $[Ala^{11,22,28}]VIP$, $[cAMP]_i$, $[Ca^{2+}]_i$; CHO-K1; GPCR; $hVPAC_1R$; $hVPAC_2R$; $hPAC_1R$; PACAP; R3P65; VIP

INTRODUCTION

Vasoactive intestinal peptide (VIP) and pituitary adenylate cyclase-activating polypeptide (PACAP) receptor activation generally results in a $G\alpha_s$-dependent increase in cAMP ($[cAMP]_i$).[1] Although less well characterized, these receptors also couple to G-proteins that regulate intracellular calcium ($[Ca^{2+}]_i$).[2] However, a shift, of varying magnitude, between agonist potencies for stimulation of these two intracellular messengers has been observed, perhaps as a consequence of the different assay methodologies.[2] We generated three individual CHO-K1 cell lines stably expressing human VIP/PACAP receptors ($hVPAC_1R$, $hVPAC_2R$, and $hPAC_1R$) and established $[cAMP]_i$ and

Address for correspondence: Dr. Keith Finlayson, Astellas CNS Research *in* Edinburgh, University of Edinburgh, The Chancellor's Building, 49 Little France Crescent, Edinburgh, EH16 4SB, UK. Voice: 44-131-242-6347; fax: 44-131-242-9371.

e-mail: Keith.Finlayson@ed.ac.uk

Ann. N.Y. Acad. Sci. 1070: 239–242 (2006). © 2006 New York Academy of Sciences.

doi: 10.1196/annals.1317.021

$[Ca^{2+}]_i$ assays in order to examine these discrepancies in receptor pharmacology. A range of agonists were examined including the endogenous, but non-selective peptides VIP, PACAP-27, and PACAP-38, and the putatively selective agonists $[Ala^{11,22,28}]VIP^3$ (VPAC1 R) and R3P65[4] (VPAC2 R).

MATERIALS AND METHODS

VIP, PACAP-27, PACAP-38, and $[Ala^{11,22,28}]VIP$ were obtained from various commercial suppliers. R3P65 was custom synthesized by Albachem (Gladsmuir, UK), with all other standard chemicals purchased from Sigma-Aldrich (Poole, UK). CHO-K1 cells stably expressing $hVPAC_1R$, $hVPAC_2R$, and $hPAC_1R$ were generated using individual cDNAs, cloned by polymerase chain reaction (PCR) amplification; (human brain cDNA; Clontech, Palo Alto, CA, USA), ligated into a eukaryotic expression vector (pMH009), and transfected into CHO-K1 cells. Clonal cell lines were isolated, and maintained at 37°C, in α-MEM containing 10% dialyzed fetal bovine serum, 2 mM L-glutamine, 100 U/mL penicillin, and 100 μg/mL streptomycin. Prior to the fluorescent $[cAMP]_i$ (CatchPoint Cyclic-AM P^{TM} ELISA; Molecular Devices [MD], Wokingham, UK) and $[Ca^{2+}]_i$ (Calcium Plus^{TM} kit; MD) assays, cells were seeded overnight in poly-D-lysine coated 96-well plates. Peptide addition and fluorescence readings were performed and monitored by the MD FlexStation. EC_{50} values were generated by non-linear, regression analysis (mean ± SEM; $n \geq 3$) using SigmaPlot 8.0 (SPSS Ltd, surrey, UK).

RESULTS

To characterize the pharmacological characteristics of the CHO-hVPAC/PAC receptor cell lines we established $[cAMP]_i$ and $[Ca^{2+}]_i$ assays. Initial studies investigated the effect of cell number on the magnitude of the response to agonists in the $[cAMP]_i$ assay. In all three cell lines, agonists induced a density (50–150,000 cells/well; 0.1 mL)-dependent increase in response size that plateaued at approximately 125,000 cells/well, with $hVPAC_1R$ shown as a representative example (FIG. 1 A). In the $[Ca^{2+}]_i$ assay, agonist-evoked responses again increased in a density-dependent manner up to 100,000 cells/well for all three hVPAC/PAC receptor cell lines, however, thereafter the signal size decreased (FIG. 1 B). As a consequence, cells were plated at 100,000 cells/well to assess agonist potencies in both the $[cAMP]_i$ and $[Ca^{2+}]_i$ assays.

VIP, PACAP-27, PACAP-38, and R3P65 all induced a concentration-dependent increase in $[cAMP]_i$ and $[Ca^{2+}]_i$ in the $hVPAC_1R$, $hVPAC_2R$, and $hPAC_1R$ cell lines, with representative data again shown for the $hVPAC_1R$ (FIG. 1 C, D). In both assays the $hPAC_1R$ was clearly PACAP preferring, with

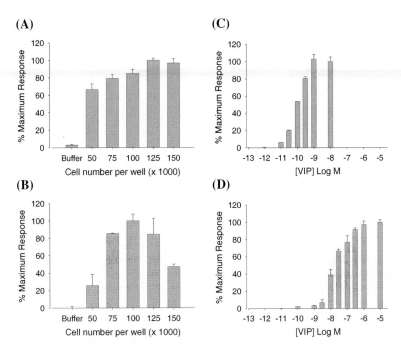

FIGURE 1. Effect of cell density (50–150,000/well) on VIP-stimulated $[cAMP]_i$ (**A**, 0.3 nM) and $[Ca^{2+}]_i$ (**B**, 30 nM) levels in $hVPAC_1R$ cells. Panels (**C**) and (**D**) show concentration response curves for VIP-stimulated $[cAMP]_i$ and $[Ca^{2+}]_i$ production in $hVPAC_1R$ cells (100,000 cells/well).

both PACAPs being 100- to 1000-fold more potent than any of the other peptides. In contrast, the VIP analog $[Ala^{11,22,28}]VIP$ was a potent agonist at the $hVPAC_1R$ in both assays but had no effect at the $hVPAC_2R$. Surprisingly, this peptide was also a full $hPAC_1R$ agonist in the $[cAMP]_i$ assay, but did not have any appreciable effect on $[Ca^{2+}]_i$. Within each assay, all the peptides (except $[Ala^{11,22,28}]VIP$) exhibited similar potencies for the $hVPAC_1R$ and $hVPAC_2R$. However, in general the agonists were approximately two orders of magnitude more potent in stimulating $[cAMP]_i$ when compared to $[Ca^{2+}]_i$, which is clearly evident when EC_{50} values from both assays are correlated (FIG. 2).

DISCUSSION

These studies clearly demonstrate that hVPAC/PAC receptor activation can result in the production of both $[cAMP]_i$ and/or $[Ca^{2+}]_i$ in CHO-K1 cells following stable receptor transfection. For the $hVPAC_1R$, $hVPAC_2R$ and $hPAC_1R$, the agonist ROPs were comparable between assays for the individual receptor subtypes, despite the fact that all five peptides were consistently less potent in

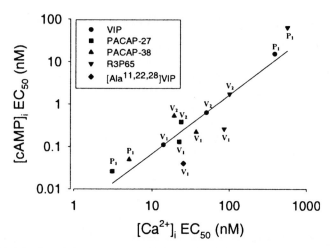

FIGURE 2. Correlation of EC_{50} values ($n \geq 3$) for VIP (●), PACAP-27 (■), PACAP-38 (▲), R3P65 (▼), and [Ala^{11,22,28}]VIP (♦) in the [cAMP]$_i$ and [Ca^{2+}]$_i$ assays using the hVPAC$_1$R, hVPAC$_2$R, and hPAC$_1$R cell lines. Linear regression analysis resulted in a correlation coefficient (r^2) of 0.83.

the [Ca^{2+}]$_i$ assay. Interestingly, [Ala^{11,22,28}]VIP[3] showed considerable selectivity for the hVPAC$_1$R in both assays, however, the agonist activity observed at the hPAC$_1$R has not been noted before. In addition, R3P65 was a potent hVPAC$_2$R agonist, but was not receptor selective as previously reported.[4]

REFERENCES

1. HARMAR, A.J. 2001. Family-B G-protein-coupled receptors. Genome Biol. **2:** 3013.1–3013.10.
2. MCCULLOCH, D.A. *et al.* 2002. Additional signals from VPAC/PAC family receptors. Biochem. Soc. Trans. **30:** 441–446.
3. NICOLE, P. *et al.* 2000. Identification of key residues for interaction of vasoactive intestinal peptide with human VPAC1 and VPAC2 receptors and development of a highly selective VPAC1 receptor agonist. Alanine scanning and molecular modeling of the peptide. J. Biol. Chem. **275:** 24003–24012.
4. YUNG, S.L. *et al.* 2003. Generation of highly selective VPAC2 receptor agonists by high throughput mutagenesis of vasoactive intestinal peptide and pituitary adenylate cyclase-activating peptide. J. Biol. Chem. **278:** 10273–10281.

Molecular Approximation between Residue 10 of Secretin and Its Receptor Demonstrated by Photoaffinity Labeling

MAOQING DONG AND LAURENCE J. MILLER

Cancer Center and Department of Molecular Pharmacology and Experimental Therapeutics, Mayo Clinic, Scottsdale, Arizona 85259, USA

ABSTRACT: Using photoaffinity labeling, we have previously explored the molecular approximations between multiple positions of secretin and its receptor. Interestingly, the amino-terminal secretin probe incorporating a photolabile residue in position 1 labels the top of the sixth transmembrane domain of the receptor, whereas other probes with photolabile residue in positions 6, 12, 13, 14, 18, 22, and 26 all label the long amino-terminal domain of the secretin receptor. Recently, we have developed a secretin probe that incorporated a radioiodinatable photolabile *p*-(4-hydroxybenzoyl)phenylalanine in position 10 and demonstrated that it efficiently labeled the secretin receptor in a saturable and specific manner. In this work, we attempted to further map its domain of labeling by cyanogen bromide (CNBr) cleavage of the wild-type and mutant receptors. Surprisingly, this position 10 probe labeled the top of the sixth transmembrane domain of the receptor, a domain labeled by the position 1 probe. These data provide an important constraint for modeling the agonist-bound G protein–coupled secretin receptor and should add substantially to our current understanding of the molecular basis of ligand binding of this important receptor.

KEYWORDS: secretin ligand; secretin receptor; G protein–coupled receptors; ligand-binding; photoaffinity labeling

INTRODUCTION

The secretin receptor is a prototypic member of the class B G protein–coupled receptors. It has an extended amino-terminal tail that contains six conserved Cys residues whose disulfide bonds have been recently mapped.[1] Like ligands for other members in the family, secretin is a moderately large peptide whose binding determinants are throughout the entire length of the

Address for correspondence: Maoqing Dong, M.D., Ph.D., Mayo Clinic Scottsdale, 13400 East Shea Blvd, Scottsdale, AZ 85259. Voice: 480-301-6830; fax: 480-301-8387.
e-mail: dongmq@mayo.edu

Ann. N.Y. Acad. Sci. 1070: 243–247 (2006). © 2006 New York Academy of Sciences.
doi: 10.1196/annals.1317.022

27 amino acid peptide. So far, substantial progress has been made in understanding how secretin binds to and activates its receptor using mutagenesis and photoaffinity labeling.[2]

Photoaffinity labeling has emerged as a powerful tool for establishing direct molecular interactions between a ligand and its receptor. To date, photoaffinity labeling studies have established eight pairs of molecular approximations between secretin and its receptor, using secretin probes incorporating photolabile residues in positions 1, 6, 12, 13, 14, 18, 22, and 26.[3–9] Of interest, the position 1 probe labels the top of the sixth transmembrane domain (TM6), whereas all other probes label residues within the amino-terminal domain of the secretin receptor. The emerging paradigm is that the high-affinity-binding determinants at the amino-terminal half, mid-region, and carboxyl-terminal half of the ligand interact primarily with the distal regions of the receptor amino-terminal domain, whereas the receptor selectivity determinant at the amino-terminal end of the ligand approaches the lipid bilayer to interact with the top of TM6 of the receptor. It should be interesting to test whether other positions of the secretin peptide (other than position 1) are adjacent to the top of TM6 or regions other than the amino terminus of the receptor.

The leucine in position 10 of secretin is interesting because it is located in the center of the amino-terminal helix within the natural secretin ligand. This is the position that we replace with a tyrosine for radioiodination and the spatial approximation between this ligand position and adjacent receptor residues has not previously been explored. Recently, a radioiodinatable photolabile secretin analog, (OH-Bpa^{10})rat secretin-27 (OH-Bpa^{10} probe) has been developed, which incorporated a photolabile p-(4-hydroxybenzoyl)phenylalanine (OH-Bpa) in position 10 of secretin.[10] Having the photolabile OH-Bpa residue and the site of radioiodination at the same site, this probe was able to covalently label the secretin receptor as did previous probes incorporating a tyrosine in position 10 of secretin.[10] Here, we explore its spatial approximation with the secretin receptor.

RESULTS AND DISCUSSION

The radioiodinatable photolabile secretin analog, the OH-Bpa probe, has been characterized to bind to the secretin receptor with high affinity and to stimulate cAMP accumulation in receptor-bearing SecR-CHO cells in a concentration-dependant manner.[10] This probe has been shown to covalently label the secretin receptor saturably and specifically, with the labeled receptor band migrating at approximate $M_r = 70,000$ that shifted to $M_r = 42,000$ after deglycosylation with endoglycosidase F (EF).[10]

To gain insight into the receptor domain of labeling by this probe, cyanogen bromide (CNBr), was used to cleave the secretin receptor labeled by the OH-Bpa^{10} probe, using the procedure we have previously described.[4] As shown

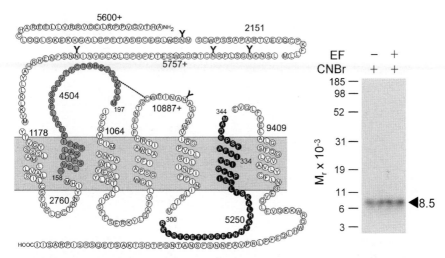

FIGURE 1. A representative autoradiograph of an SDS-polyacrylamide gel used to separate the products of CNBr cleavage of the secretin receptor labeled with the OH-Bpa[10] probe is shown. The cleavage resulted in a radioactive band migrating at approximate M_r = 8500 that did not shift after deglycosylation with EF. Considering the molecular mass of the attached probe (3251 Da) and the nonglycosylated nature, the expected receptor fragment would be in the range of ~4.5 to 5.5 kDa and the best candidates to represent these data are highlighted in gray and black circles in the diagram on the left. These data are representative of at least three independent experiments.

in FIGURE 1, CNBr cleavage of the labeled receptor yielded a labeled nonglycosylated fragment migrating at approximate M_r = 8500. Given the molecular mass of the attached OH-Bpa[10] probe (3251 Da) and clear evidence of the nonglycosylated nature of the fragment, there were two candidate fragments that best fit these data. One represents the region His[158]-Met[197] including the second transmembrane domain and the first extracellular loop and the second is within the region Arg[300]-Met[344] spanning the third intracellular loop (ICL3), TM6, and beginning of the third extracellular loop (ECL3).

Two well-characterized secretin receptor mutants, A175M and I334M, which contained an additional Met residue in each of the candidate fragments,[8] were used for further localization of the domain labeled by the OH-Bpa[10] probe. Both mutants were specifically and saturably labeled by the OH-Bpa[10] probe (FIG. 2), with the labeled receptor migrating at approximate M_r = 70,000 as the labeled wild-type receptor. Like that of the wild-type secretin receptor, CNBr cleavage of the A175M receptor yielded a radioactive band migrating at approximate M_r = 8500 (FIG. 2). However, cleavage of the I334M receptor resulted in a band migrating distinctly at approximate M_r = 4,500 (FIG. 2). This identified the Arg[300]-Met[344] fragment that spans the ICL3, TM6, and ECL3 as the domain of labeling by the OH-Bpa[10] probe. Considering the molecular

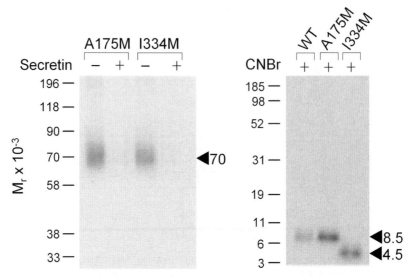

FIGURE 2. Photoaffinity labeling and CNBr cleavage of secretin receptor mutants are shown. *Left panel* shows that photoaffinity labeling of the intact A175M and I334M secretin receptor mutants with the OH-Bpa[10] probe in the presence and absence of competing secretin (1 μM). *Right panel* shows the results of CNBr cleavage of the labeled wild-type (WT), A175M, and I334M secretin receptor. As seen, CNBr cleavage of the I334M mutant yielded a band migrating at approximate $M_r = 4500$, clearly distinct from the cleavage of the WT and A175M mutant receptor. The segment Val^{335}-Met^{344} most likely contained the site of labeling of the OH-Bpa[10] probe.

mass of the attached probe (see above), the 10-residue segment Val^{335}-Met^{344} (calculated mass, 1113 Da) was felt to represent the domain of labeling.

It is interesting that the OH-Bpa[10] probe labeled the top of TM6, a domain that was labeled by the amino-terminal secretin probes.[8] This is a region that is distinct from the receptor amino terminus that was labeled by most other tested secretin probes that include probes with photolabile residues in positions 6, 12, 13, 14, 18, 22, and 26.[3–7,9] These data add another important constraint to build a peptide agonist-bound secretin receptor model, and yet suggest a more complex mechanism of receptor activation than the two domain tethering model proposed for the parathyroid hormone receptor,[11,12] another member of the class B G protein–coupled receptors. This will become clearer as we gain additional constraints.

ACKNOWLEDGMENTS

This work was supported by National Institutes of Health Grant DK46577 and the Fiterman Foundation (to LJM).

REFERENCES

1. LISENBEE, C.S., M. DONG & L.J. MILLER. 2005. Paired cysteine mutagenesis to establish the pattern of disulfide bonds in the functional intact secretin receptor. J. Biol. Chem. **280:** 12330–12338.
2. DONG, M. & L.J. MILLER. 2002. Molecular pharmacology of the secretin receptor. Receptors Channels **8:** 189–200.
3. DONG, M. *et al.* 1999. Identification of an interaction between residue 6 of the natural peptide ligand and a distinct residue within the amino-terminal tail of the secretin receptor. J. Biol. Chem. **274:** 19161–19167.
4. DONG, M. *et al.* 1999. Demonstration of a direct interaction between residue 22 in the carboxyl-terminal half of secretin and the amino-terminal tail of the secretin receptor using photoaffinity labeling. J. Biol. Chem. **274:** 903–909.
5. DONG, M. *et al.* 2000. Identification of two pairs of spatially approximated residues within the carboxyl terminus of secretin and its receptor. J. Biol. Chem. **275:** 26032–26039.
6. DONG, M. *et al.* 2002. Interaction among four residues distributed through the secretin pharmacophore and a focused region of the secretin receptor amino terminus. Mol. Endocrinol. **16:** 2490–2501.
7. DONG, M. *et al.* 2003. Spatial approximation between two residues in the mid-region of secretin and the amino terminus of its receptor. Incorporation of seven sets of such constraints into a three-dimensional model of the agonist-bound secretin receptor. J. Biol. Chem. **278:** 48300–48312.
8. DONG, M. *et al.* 2004. Spatial approximation between the amino terminus of a peptide agonist and the top of the sixth transmembrane segment of the secretin receptor. J. Biol. Chem. **279:** 2894–2903.
9. ZANG, M. *et al.* 2003. Spatial approximation between a photolabile residue in position 13 of secretin and the amino terminus of the secretin receptor. Mol. Pharmacol. **63:** 993–1001.
10. DONG, M., D.I. PINON & L.J. MILLER. 2002. Development of a biologically active secretin analogue incorporating a radioiodinatable photolabile p-(4-hydroxybenzoyl)phenylalanine in position 10. Regul. Pept. **109:** 181–187.
11. BEHAR, V. *et al.* 2000. Photoaffinity cross-linking identifies differences in the interactions of an agonist and an antagonist with the parathyroid hormone/parathyroid hormone-related protein receptor. J. Biol. Chem. **275:** 9–17.
12. BISELLO, A. *et al.* 1998. Parathyroid hormone-receptor interactions identified directly by photocross-linking and molecular modeling studies. J. Biol. Chem. **273:** 22498–22505.

Use of Photoaffinity Labeling to Understand the Molecular Basis of Ligand Binding to the Secretin Receptor

MAOQING DONG AND LAURENCE J. MILLER

Mayo Clinic, Department of Molecular Pharmacology and Experimental Therapeutics, Scottsdale, Arizona 85259, USA

ABSTRACT: The secretin receptor was the first member of the Class B family of G protein–coupled receptors that was identified in 1991, 89 years after secretin action was first recognized. That report resulted in the introduction of the term *hormone* and in the birth of the field of endocrinology. The secretin receptor has become prototypic of this receptor family, binding a moderately long linear peptide with a diffuse pharmacophoric domain. Here, we provide a detailed account of the contributions of photoaffinity labeling to establish the molecular basis of natural ligand binding to this receptor, as well as to provide insights into possible mechanisms for receptor activation and initiation of signaling. Each of the themes discussed are also relevant to other members of this physiologically and pharmacologically important receptor family.

KEYWORDS: G protein–coupled receptor; secretin receptor; ligand binding; photoaffinity labeling; mutagenesis

Secretin maintains a unique position in the history of endocrinology, gastrointestinal endocrinology, and pancreaticobiliary physiology. The first description of the ability of this peptide to stimulate secretion of a bicarbonate-rich fluid from the pancreas and biliary tree that was published in 1902 introduced the term *hormone* and launched the field of endocrinology.[1] However, our understanding of the molecular basis of the action of secretin has lagged far behind that of the classical endocrine hormones. This reflects challenges in identification of a large peptide hormone that is secreted by sparse cells scattered throughout the intestine and in the relatively late molecular identification of the secretin receptor.[2] The cDNA encoding the secretin receptor was first cloned in 1991.[2] We now understand that this receptor was the first member of a new family of guanine nucleotide-binding protein (G protein)-coupled receptors, the Class B family that includes receptors for many critically important potential

Address for correspondence: Laurence J. Miller, M.D., Director for Research, Mayo Clinic, 13400 East Shea Blvd., Scottsdale, AZ 85259. Voice: 480-301-6650; fax: 480-301-6969.
e-mail: miller@mayo.edu

Ann. N.Y. Acad. Sci. 1070: 248–264 (2006). © 2006 New York Academy of Sciences.
doi: 10.1196/annals.1317.023

drug targets, such as receptors for parathyroid hormone, calcitonin, glucagon, corticotrophin-releasing factor, and vasoactive intestinal polypeptide.[3] This family is structurally distinct from the much larger and more extensively studied Class A family of G protein–coupled receptors that includes rhodopsin and the β-adrenergic receptor.

Members of the Class B family of G protein–coupled receptors share a long amino-terminal tail that contains three conserved pairs of Cys residues that define three intra-domain disulfide bonds.[4–8] It is also noteworthy that the transmembrane segments of this group of G protein–coupled receptors do not include the signature residues that are highly conserved in the Class A family. The residues that are conserved within the Class B family are distinct and suggest a pattern of helical confluence that is also distinct from the Class A family.[9,10] Thus, while the members of the Class B receptor family share key structural features with each other, these are quite distinct from those of the Class A G protein–coupled receptors. Of interest, the natural ligands of the Class B receptors also share a number of structural features. All are moderately long peptides (in excess of 25 residues) that possess diffuse pharmacophoric domains[3,11] and have conformational features in common.[12] The amino-terminal regions of these ligands seem to be critical for the selectivity of binding and activating receptors, while the carboxyl-terminal regions contribute to high-affinity binding.[13,14]

In this article, we have focused on the secretin receptor as prototypic of the Class B family of G protein–coupled receptors. We are particularly interested in the structure of this receptor and the molecular basis of ligand binding. It also provides insights into the molecular basis of receptor activation. A detailed understanding of these events provides important insights that can theoretically be quite useful in the design, optimization, and development of drugs acting at this and structurally related receptors.

SECRETIN PHYSIOLOGY

After many years of awareness of a secretin effect, the linear 27-residue amidated polypeptide responsible for mediating this was finally isolated and characterized in 1970.[15] This provided the basis for synthesizing adequate amounts of this hormone to explore its physiology and to raise antisera and develop an assay useful for its quantitation. It soon became clear that secretin was produced in single endocrine cells that were scattered throughout the small intestine, with a gradient from the area of highest density in the proximal duodenum that falls off more distally in the aboral direction. Secretin is secreted in response to a meal, with the most stimulatory fractions of chyme representing endogenous acid, bile acids, and fatty acids introduced in the meal.[16]

Major targets for this hormone include pancreatic and biliary mucosa, where secretin stimulates a bicarbonate-rich fluid secretion.[16] Other physiologic effects include inhibition of gastric emptying, stimulation of Brunner's gland secretion, and inhibition of gastric secretion. It is noteworthy that the pharmacology of these effects is consistent with a single molecular form of the secretin receptor. Potential roles for secretin include effects on islet cell function, pancreatic exocrine cell growth, renal function, and cardiac function, although the receptors mediating these effects have not been well defined.

SECRETIN RECEPTOR CHARACTERIZATION

Before the molecular nature of the secretin receptor was known, it was characterized pharmacologically. Of interest, binding studies demonstrated both high- and low-affinity binding sites on various cells. The pancreatic acinar cell was particularly well studied. On that cell, there also were high- and low-affinity vasoactive intestinal polypeptide-binding sites that were demonstrated.[17,18] It ultimately became clear that the high-affinity secretin-binding site corresponded to the low-affinity vasoactive intestinal polypeptide-binding site. Similarly, the high-affinity vasoactive intestinal polypeptide-binding site corresponded to the low-affinity secretin-binding site. With the ultimate molecular characterization of these binding sites, the former was shown to represent the secretin receptor, while the latter represents a vasoactive intestinal polypeptide receptor (VPAC1 receptor). Both of these receptors are members of the Class B family of G protein–coupled receptors.

The molecular identity of the secretin receptor was initially determined using affinity labeling with *p*-azidophenylglyoxal cross-linking.[19] Because no particularly rich source of this receptor was known or available, most of our understanding has followed the cloning of the receptor cDNA. This was finally achieved when Ishihara *et al.* used the approach of expression cloning.[2] The predicted primary sequence and sequence motifs for potential sites of glycosylation, phosphorylation, and disulfide bond formation provided powerful insights. This clone also provided the ability to engineer high levels of receptor expression, permitting biochemical studies of this receptor and careful structure-activity evaluations using mutagenesis techniques. The gene encoding the secretin receptor is located on human chromosome 2 at locus 2q14.1.[20] This gene contains 13 exonic domains.

The cDNAs for the secretin receptor from multiple species are now available.[2,21–24] These reflect strong similarities, with a signal sequence and predicted mature receptors ranging from 418 to 427 residues. Across the species the sequences are approximately 80% identical, with the most marked differences residing in the signal sequences and in the carboxyl-terminal tail regions. The highest degree of similarity resides in the transmembrane segments that likely determine the conserved architecture for this family and the amino-terminal tail region that has been determined to be critical for natural secretin binding and activity.[13,25,26] Similar to all receptors in this superfamily, the secretin receptor has seven hydrophobic segments that are felt to correspond with transmembrane helical domains. The heptahelical topology has, indeed, been confirmed by the accessibility of epitope tags in extracellular loop and tail domains.[27,28] As noted above, the positions of charged residues and conserved residues within these helical domains have predicted a distinct helical bundle pattern from that which has been best delineated for the Class A family member, rhodopsin.[9,10] It is now clear that the rhodopsin template has little to offer for the Class B family members.

Disulfide bonds in the amino-terminal tail domain are quite important for normal function, as demonstrated by studies resulting in loss of function after treating the receptor with chemical reduction and alkylation.[29] The three specific disulfide bonds that are conserved across this receptor family have now been well established, using biochemical, nuclear magnetic resonance, and paired mutagenesis techniques.[4–8] These represent the bonding of the first to the third, second to the fifth, and fourth to the sixth conserved Cys residues.

The amino-terminal tail is also a glycosylated domain. While the number of carbohydrate chains are somewhat varied, it is clear that this facilitates efficient biosynthesis and trafficking to the cell surface. In the human receptor, the first and fourth positions of glycosylation seem to be particularly functionally important.[30] Glycosylation also likely protects the receptor from proteolysis.

MOLECULAR BASIS OF SECRETIN BINDING

The diversity in molecular mechanisms for binding of natural ligands of G protein–coupled receptors is as remarkable as the structural diversity in these ligands. These range from photons and small odorants and biogenic amines to larger peptides and very large glycoproteins and even viral particles. The themes for binding correspond to these sizes and chemical characteristics. The smallest ligands tend to bind within the confluence of helical domains within the lipid bilayer, while the larger ligands tend to interact, at least partially, with extracellular loop and tail domains.[31] As might be predicted from the relatively large size and hydrophilicity of the natural peptide ligands for the Class B family of G protein–coupled receptors, extracellular regions have been targeted as likely sites of binding. This is further emphasized by the diffuse nature of the pharmacophoric domains of these peptides.

Truncation, site-directed mutagenesis, and chimeric receptor studies of the secretin receptor all have focused attention on the unique role played by the structurally unique amino-terminal tail domain for natural ligand binding and biological activity (key studies summarized in TABLE 1). If the amino terminus is truncated or extensively modified by segmental deletion or mutagenesis, binding and signaling in response to secretin are eliminated. This theme is consistent through other members of the Class B family of G protein–coupled receptors.[32–34] The specificity of the receptor is carried largely by the amino

TABLE 1. Critical residues within the amino terminus of the secretin receptor identified by site-directed mutagenesis and chimeric receptor studies[a]

Secretin receptor constructs	Critical receptor residues	References
Secretin-VIP receptor chimera	Ala[1]-Met[123]	13,25,26,52–55
Secretin-VIP receptor chimera	Ala[1]-Leu[10]	52
Secretin-VIP receptor chimera	Val[103]-Asn[110], Tyr[116]-Leu[120]	56
D98N	Asp[98]	36
D48H, D49R, R83D, R83L	Asp[49], Arg[83]	37
Cys to Ser or Ala	Cys[24], Cys[44], Cys[53], Cys[67], Cys[85], Cys[101]	28,29,38
N50L, N106L	Asn[50], Asn[106]	30

[a]VIP, vasoactive intestinal polypeptide.

terminus in chimeric forms of two members of this family.[13,35] Site-directed mutagenesis studies have focused interest on unique residues within this region of the secretin receptor.[28–30,36–38]

PHOTOAFFINITY LABELING

Photoaffinity labeling represents a complementary approach to mutagenesis experiments for the mapping of receptor ligand-binding domains. This technique provides a mechanism to identify not only a receptor molecule, but also domains and even specific residues within that molecule. The experimental approach depends on spatial approximation between the position of a photolabile residue within a probe and the adjacent residues within the receptor as it is docked. Constraints obtained with affinity labeling are particularly powerful, since they are direct and clearly establish spatial approximations between distinct residues within a ligand and a receptor that are extremely useful for molecular modeling.

At this time, there have been successful photoaffinity labeling studies of the secretin receptor that have used probes with sites of covalent attachment spanning the entire pharmacophoric domain.[27,39–45] Indeed, these are the major focus of the current review.

Intrinsic photoaffinity labeling, in which a photoreactive benzophenone moiety has been inserted within the pharmacophoric domain of secretin analogues, has been most successful and informative. Recently, we have successfully generated a series of secretin analogues with photolabile residues situated at various positions within the secretin pharmacophore.[27,39–45] With these, a series of residue–residue approximations between the secretin ligand and receptor have been obtained (TABLE 2). These data have provided key constraints for developing and refining molecular models of the natural agonist-bound secretin receptor.

TABLE 2. Molecular approximations between secretin and its receptor demonstrated by photoaffinity labeling studies[a]

Photoprobes	Peptide position of covalent labeling	Receptor domain labeled	Receptor residue labeled	References
[Bpa^{22}]secretin	22	N-ECD	Leu^{17}	27,39
[Bpa^{6}]secretin	6	N-ECD	Val^{4}	40
[Bpa^{26}]secretin	26	N-ECD	Leu^{36}	39
[(BzBz)Lys^{18}]secretin	18	N-ECD	Arg^{14}	41
[Bpa^{13}]secretin	13	N-ECD	Val^{103}	44
[(BzBz)Lys^{12}]secretin	12	N-ECD	Val^{6}	45
[(BzBz)Lys^{14}]secretin	14	N-ECD	Pro^{38}	45
[Bpa^{1}]secretin	1	TM6	Phe^{338}	42
[Bpa^{-1}]secretin	−1	TM6	Tyr^{333}	42
[Bpa^{-2},Gly^{-1}]secretin	−2	TM6	Phe^{336}	42
[Ac-Bpa^{1},]secretin	1	TM6	Val^{335}	43
[Ac-Bpa^{-1}]secretin	−1	N-ECD	Thr^{3}	43
[Ac-Bpa^{-2},Gly^{-1}]secretin	−2	N-ECD	Thr^{3}	43
[Ac-Lys^{-2},Bpa^{-1}]secretin	−1	N-ECD	Val^{4}	43
[Ac-Arg^{-2},Bpa^{-1}]secretin.	−1	N-ECD and TM6	Val^{4} andVal^{335}	43

[a]Bpa, p-benzoyl-L-phenylalanine; (BzBz)Lys, p-benzoylbenzoyl-L-lysine; Ac, acetyl; N-ECD, amino-terminal extracellular domain; TM6, the sixth transmembrane domain.

The first application in this series of studies was the use of a photoreactive secretin analogue incorporating a p-benzoyl-L-phenylalanine (Bpa) in the amino-terminal half of the 27-amino acid secretin ligand, in position 22.[27] This probe was fully characterized to bind to the secretin receptor with similar affinity to natural secretin and to elicit a full biological response in CHO-SecR cells stably expressing this receptor. This probe was also demonstrated to covalently label the secretin receptor in a saturable and specific manner, with the labeled receptor migrating at approximate $M_r = 70,000$ and shifting to $M_r = 42,000$ after deglycosylation with endoglycosidase F. Peptide mapping by cyanogen bromide (CNBr) cleavage of the labeled receptor established the first (Ala^{1}-Met^{51}) and third (Leu^{74}-Met^{123}) CNBr fragments were the best candidates to represent the region of labeling by the [Bpa]^{22} secretin probe (FIG. 1A). To determine which of these regions included the site of covalent attachment, two secretin receptor constructs (SecR-HA37 and SecR-HA79) were prepared that incorporated hemagluttinin (HA) epitope tags within each of these CNBr fragments. Immunoprecipitation with anti-HA monoclonal antibody of these HA-tagged receptor constructs stably expressed in CHO cells identified the Ala^{1}-Met^{51} fragment at the distal end of the amino terminus of the secretin receptor contained the site of labeling (FIG. 1B). This was narrowed down to the Ala^{1}-Lys^{30} fragment of the receptor by endoproteinase Lys-C cleavage of the labeled intact receptor, as well as of the CNBr fragment (FIG. 1C).[27] This was still further localized to segment Leu[17]-Lys[30] using CNBr cleavage of V13M and V16M mutant secretin receptors.[39] Leu[17] of the secretin receptor was subsequently identified as the specific site of labeling for the Bpa[22] probe using radiochemical Edman degradation sequencing of the radiochemically pure CNBr fragments coming from affinity labeled V13M and V16M secretin receptor mutants (FIG. 1D).[39]

A second receptor site of photoaffinity labeling has been identified using a secretin analogue having the photolabile Bpa in the amino-terminal half, in position 6, of the ligand.[40] This probe has relatively lower affinity to bind the secretin receptor and was less potent to elicit a cAMP response in receptor bearing SecR-CHO cells than natural secretin. It also covalently labeled the receptor in a saturable and specific manner. The site of labeling with this probe was identified within the first 30 receptor residues by CNBr and endoproteinase Lys-C cleavage of wild-type and V16M mutant secretin receptors, as well as by immunoprecipitation of HA-tagged secretin receptor fragments. Radiochemical sequencing of the CNBr fragment representing Ala[1]-Met[51] identified receptor residue Val[4] within the distal amino terminus of the receptor as the site of labeling by the Bpa[6] probe.

It is interesting that both the amino- and carboxyl-terminal probes having photolabile Bpa in positions 22 and 6, which are 16 amino acids away from each other, labeled the receptor residues within a much focused region at the distal amino-terminal tail of the receptor. We next moved our focus toward the carboxyl terminus of the ligand by placing the Bpa in position 26.[39] This probe bound the secretin receptor with affinity similar to that of natural secretin and was also a full agonist that was as potent as natural secretin in stimulating a cAMP response in receptor-bearing cells. The domain of labeling by this probe was localized to the segment between residues Gly[34] and Ala[41] using CNBr and endoproteinase Lys-C cleavage of labeled wild-type and A41M mutant receptor constructs and immunoprecipitation of epitope-tagged receptor

---

FIGURE 1. Example of identification of a site of covalent attachment within the receptor amino terminus. (**A**) CNBr cleavage of labeled native and deglycosylated secretin receptor identified the Ala[1]-Met[51] and Leu[74]-Met[123] fragments at the distal amino terminus of the receptor as the two candidates to represent the region of labeling by the [Bpa[22]]secretin probe. (**B**) Immunoprecipitation of HA-tagged secretin receptor using anti-HA monoclonal antibody established Ala[1]-Met[51] as the domain of labeling. (**C**) The Ala[1]-Lys[30] segment was further identified to contain the site of labeling by endoprotease Lys-C cleavage of the labeled intact receptor or its CNBr fragment. (**D**) Radiochemical Edman degradation sequencing of CNBr fragments from labeled V13M and V16M secretin receptor mutants identified receptor residue Leu[17] as the site of labeling. (Reproduced from Dong et al.[27,39] by permission of the American Society for Biochemistry and Molecular Biology.) EF, endoglycosidase F.

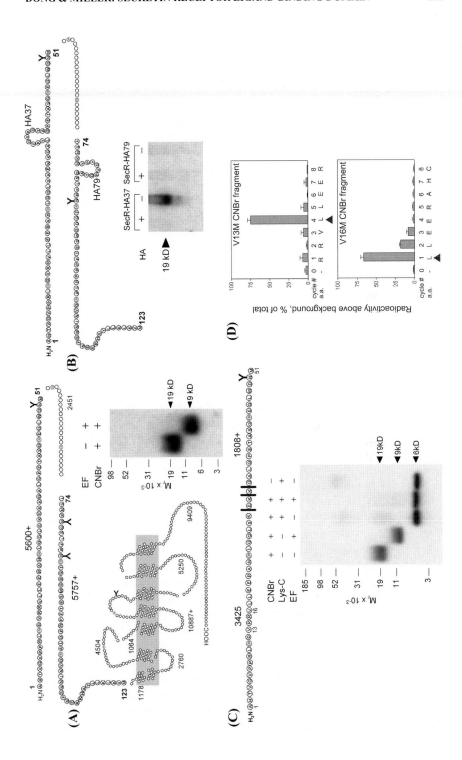

fragments. The receptor residue Leu[36] was identified as the site of covalent attachment using radiochemical sequencing of the Gly[34]-Lys[69] fragment resulting from endoproteinase Lys-C cleavage of the wild-type receptor. This residue was again within the distal amino terminus of the secretin receptor, a domain labeled by the Bpa[22] and Bpa[6] probes. This further strengthened the key role of the distal amino terminus of the secretin receptor for ligand binding and activation, as had been suggested by mutagenesis studies.

This series of studies has been extended with studies using an additional probe having its site of covalent attachment in a distinct region of the peptide, between amino- and carboxyl-terminal helical domains and closer to the mid-region of the ligand. This probe incorporated a photolabile p-benzoylbenzoyl-L-lysine [(BzBz)Lys] in position 18.[41] Although it bound the secretin receptor with lower affinity and was less potent to stimulate a biological response in receptor-bearing cell line than natural secretin, this probe was a full agonist and covalently labeled the secretin receptor efficiently. The labeled domain was demonstrated to be within the tiny segment between Arg[14] and Val[16] by CNBr cleavage of the wild-type and V13M and V16M mutant receptors. The specific site of labeling was identified by radiochemical sequencing to represent receptor residue Arg[14]. Remarkably, this residue also resides within the same receptor subdomain as the residues labeled with the aforementioned probes.

It is of great interest that a large number of photolabile residues distributed throughout the ligand have been used to establish spatial approximation with residues within a focused subdomain (within the first 40 amino acids) of this receptor. These sites of covalent attachment are spread throughout both the amino-terminal (position 6) and carboxyl-terminal (position 22) helical domains and the intermediate turn (position 18) domain of secretin. Without having had an accepted, meaningful conformational model for the entire amino-terminal tail domain of the secretin receptor to provide the setting for docking this ligand, these four pairs of extant residue–residue approximations have been accommodated in a credible model of secretin bound to the first 40 residues of the receptor amino-terminal tail.[41]

It is of particular interest that all four photoaffinity labeling studies of the secretin receptor described above have established spatial approximations not only with the amino-terminal receptor tail domain, but also with the same small portion of this domain. Although this raises the interesting possibility that this region makes an important contribution to a ligand-binding motif, there are not yet adequate data to understand its conformation or its relation to the rest of the amino-terminal domain in an intact receptor. Position 13 within secretin is very interesting because it is located in the mid-region and near the turn between two helical domains of the solution structure of this hormone. Therefore, a probe incorporating a Bpa in position 13 has been designed and used for photoaffinity labeling studies.[44] A series of experimental strategies were used to ultimately define the receptor residue that is covalently labeled by this probe.

CNBr cleavage and immunoprecipitation of HA-tagged fragments identified that the third CNBr fragment (Leu^{74}-Met^{123}, refer to diagram in FIG. 1) within the extracellular amino terminus of the secretin receptor contained the site of labeling for the Bpa^{13} probe. This was the first photolabile probe that labeled a CNBr fragment other than the first one (Ala^1-Met^{51}, FIG. 1). The labeled domain was further localized to a smaller fragment, Asp^{98}-Met^{123}, by sequential CNBr, BNPS-skatole, and endoproteinase Asp-N cleavage of the labeled receptor. Receptor residue Val^{103} was identified as the specific site of labeling by radiochemical sequencing of the fragments resulting from CNBr and endoproteinase Asp-N cleavage and from CNBr, BNPS-skatole, and Asp-N cleavage of the deglycosylated secretin receptor that had been labeled with the Bpa^{13} probe. Although still within the amino-terminal tail domain of the secretin receptor, this residue is in a region of the tail that is close to the first transmembrane domain and, therefore, is distinct from that which had previously been labeled.

Using a 93-residue fragment from an intact structural domain of S-adenosylmethionine synthetase as a template, a physically plausible three-dimensional model for the amino-terminal domain of the secretin receptor was generated that satisfied each of these five spatial constraints derived from existing photoaffinity labeling experiments.[44] This conformational model of the secretin receptor amino terminus accommodated the peptide hormone in a conformation quite similar to the secretin NMR structure and enabled the formation of disulfide bonds that have subsequently been reported for the extracellular domain of the secretin receptor.[38]

The Bpa^{13} probe was the first probe that labeled a receptor residue other than within the distal region of the amino terminus of the secretin receptor. Two more probes have thus been designed to incorporate a photolabile residue [(BzBz)Lys] straddling this informative position 13, in positions 12 and 14.[45] Both probes represented high-affinity and potent full agonists and covalently labeled the secretin receptor efficiently. Surprisingly, the domains of labeling were localized to the Ala^1-Pro^8 segment for the position 12 probe and to the Gly^{34}-Met^{51} segment for the position 14 probe, respectively. This was achieved using sequential CNBr and endoproteinase Lys-C cleavage of wild-type and mutant receptors, and immunoprecipitation of HA-tagged fragments. The specific site of labeling was further identified as single receptor residue Val^6 for the position 12 probe and Pro^{38} for the position 14 probe, respectively. This was accomplished by radiochemical sequencing of relevant labeled fragments. Both of these sites reside within the distal amino terminus, in the first 40 amino acid residues of the secretin receptor.

These seven photoaffinity labeling constraints combined with the three disulfide bonds in the receptor amino-terminal domain provided a broad array of constraints that were distributed throughout the peptide and the amino-terminal domain of the receptor that allowed the generation of a new, more refined molecular model of the ligand-occupied secretin receptor.[45] The three-dimensional conformation of secretin receptor residues 9 through 108 was modeled by direct homology with the amino-terminal domain of ribonuclease Mc1. The amino-terminal domain of the secretin receptor presented a stable platform for peptide ligand interaction, with the amino terminus of the peptide hormone extended toward the transmembrane helix domain of the receptor. This provided new insights into the molecular basis of natural ligand binding and supplied new, testable hypotheses for the molecular basis of activation of this receptor.

To test whether the amino terminus of secretin approaches the body of the secretin receptor that includes the transmembrane helical bundle, as predicted by the model described above, we developed probes having photolabile residues at the amino terminus of the secretin-like probes.[42] The first probe incorporated a photolabile Bpa in the position of His^1 of secretin. Because His^1 is critical for function,[46] we also positioned a photolabile Bpa as an amino-terminal extension of intact peptides, in positions -1 and -2. Each probe was shown to represent a full agonist that bound to the secretin receptor specifically and saturably, although the position 1 analogue had lowest affinity, reflecting the modification of this critical residue. The probes were all able to efficiently label the secretin receptor. CNBr cleavage of the secretin receptor labeled by each of these probes yielded a non-glycosylated band migrating at M_r = 8500 (FIG. 2A). This pattern of migration was distinct from that of the regions labeled by probes with sites of covalent attachment within the amino-terminal half (Bpa^6), mid-region [$(BzBz)Lys^{12}$, Bpa^{13}, and $(BzBz)Lys^{14}$], and carboxyl-terminal regions [Bpa^{26}, Bpa^{22}, and $(BzBz)Lys^{18}$] of secretin that all labeled the glycosylated secretin receptor amino terminus. Based on the apparent size of the labeled band, there were two strong candidates to represent the labeled region, His^{158}-Met^{197} including the second transmembrane domain and the first extracellular loop, and Arg^{300}-Met^{344} spanning the third intracellular loop, the sixth transmembrane domain (TM6), and the beginning of the third extracellular loop domain (FIG. 2A). CNBr cleavage of two mutant receptors that introduced an additional Met into each of these two candidate fragments, A175M and I334M, helped to establish that the sites of covalent attachment for each of these probes were at the region of the top of TM6 (FIG. 2B). Radiochemical sequencing of the labeled His^{332}-Ile^{427} fragment (from CNBr cleavage of labeled I331M/M344I mutant receptor) identified Tyr^{333} as the site of labeling by the Bpa^{-1} probe. Phe^{336} and Phe^{338} were identified as the receptor residues labeled by the Bpa^{-2} and Bpa^1 probes, respectively, by radiochemical sequencing of the labeled Val^{335}-Ile^{427} fragment (from CNBr cleavage of labeled I334M/M344I mutant receptor).

These three new pairs of approximations became important constraints for refining our previous model.[45] The refined ligand-receptor complex satisfied each of the original cross-linking and disulfide bond constraints, as well as these new photoaffinity labeling constraints.[42] These include the covalent labeling of receptor residues located in both the amino terminus and TM6. This

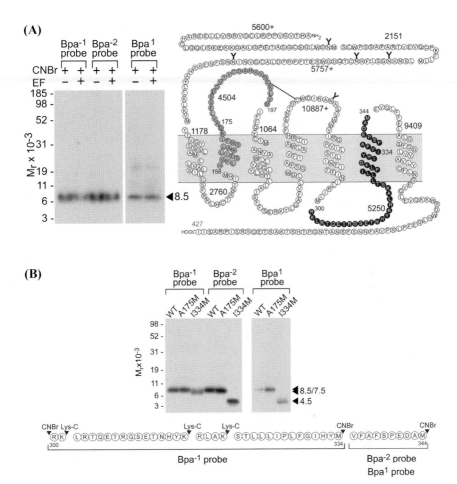

FIGURE 2. Example of identification of a site of covalent attachment within the receptor core using [Bpa^1]secretin, [Bpa^{-1}]secretin, and [Bpa^{-2}]secretin probes. (**A**) CNBr cleavage of the secretin receptor labeled by each probe yielded a non-glycosylated fragment migrating at $M_r = 8500$. The His^{158}-Met^{197} and Arg^{300}-Met^{344} fragments highlighted in bold and grey in the receptor diagram are the best candidates to represent the regions of labeling by each probe. (**B**) CNBr cleavage of A175M and I334M secretin receptor mutants established the Arg^{300}-Met^{344} fragment spanning TM6 as the region of labeling by each probe. (Reproduced from Dong *et al.*[42] by permission of the American Society for Biochemistry and Molecular Biology.) EF, endoglycosidase F.

was interpreted as suggesting that secretin may exert tension between the receptor amino terminus and body regions to elicit a conformational change effecting receptor activation, a signal transduction mechanism similar to that proposed for parathyroid hormone receptors.[47–49] This was felt to possibly represent a common theme for the Class B G protein–coupled receptors, particularly since analogous observations have subsequently been made for the calcitonin receptor, another member of this family.[50,51]

To closely examine how critical the spatial approximation between the amino terminus of secretin and its receptor may be, we investigated whether the charge of the amino terminus of the ligand might affect the spatial approximations between secretin and its receptor.[43] Another series of amino-terminal probes were prepared that included the blocking of their α-amino groups by acetylation to eliminate that single positive charge. In some of these, a positive charge was added back with an additional basic residue, while maintaining the blocked amino terminus. It is interesting that high-affinity amino-terminally blocked probes labeled the distal amino terminus of the receptor, rather than TM6, while adding a basic residue, again resulted in labeling of TM6. This supported the critical nature of charge for the positioning of the peptide amino terminus adjacent to TM6 of the receptor, and suggested that the tethering role of the peptide ligand may be less critical for biological activity than previously suggested (see above).

To explore possible charge–charge interactions between the secretin amino terminus and an acidic residue within the body of the receptor in the region of TM6 or the adjacent third extracellular loop, candidate acidic residues were mutated to Ala (E341A, D342A, E345A, E351A).[43] Among these, the E351A mutant was demonstrated to markedly interfere with binding and biological activity, supporting the possibility that there was a charge–charge interaction between this residue and the amino terminus of secretin that might contribute to normal ligand docking. In brief, the amino-terminal half of secretin apparently resides in a groove between the distal and proximal ends of the receptor amino-terminal tail, with the amino terminus residing just above the TM6 region of the receptor, possibly interacting with acidic residue E351. Structural variation in the peptide amino terminus retains biological activity, while changing spatial approximation with the receptor body. We postulate that the base of the stable amino-terminal architecture, rather than the peptide ligand exerts the tension to change the conformation of the receptor body that mediates initiation of the intracellular signaling cascade.

ACKNOWLEDGMENTS

This work was supported by National Institutes of Health Grant DK46577 and the Fiterman Foundation.

REFERENCES

1. BAYLISS, W.M. & E.H. STARLING. 1902. On the causation of the so-called 'peripheral reflex secretion' of the pancreas. Proc. R. Soc. Lond. [Biol] **69:** 352–353.

2. ISHIHARA, T., S. NAKAMURA, Y. KAZIRO, et al. 1991. Molecular cloning and expression of a cDNA encoding the secretin receptor. EMBO J. **10:** 1635–1641.

3. ULRICH, C.D., M. HOLTMANN & L.J. MILLER. 1998. Secretin and vasoactive intestinal peptide receptors: members of a unique family of G protein–coupled receptors. Gastroenterology **114:** 382–397.

4. GRAUSCHOPF, U., H. LILIE, K. HONOLD, et al. 2000. The N-terminal fragment of human parathyroid hormone receptor 1 constitutes a hormone binding domain and reveals a distinct disulfide pattern. Biochemistry **39:** 8878–8887.

5. BAZARSUREN, A., U. GRAUSCHOPF, M. WOZNY, et al. 2002. In vitro folding, functional characterization, and disulfide pattern of the extracellular domain of human GLP-1 receptor. Biophys. Chem. **96:** 305–318.

6. GRACE, C.R., M.H. PERRIN, M.R. DIGRUCCIO, et al. 2004. NMR structure and peptide hormone binding site of the first extracellular domain of a type B1 G protein–coupled receptor. Proc. Natl. Acad. Sci. USA **101:** 12836–12841.

7. PERRIN, M.H., W.H. FISCHER, K.S. KUNITAKE, et al. 2001. Expression, purification, and characterization of a soluble form of the first extracellular domain of the human type 1 corticotropin releasing factor receptor. J. Biol. Chem. **276:** 31528–31534.

8. PERRIN, M.H., M.R. DIGRUCCIO, S.C. KOERBER, et al. 2003. A soluble form of the first extracellular domain of mouse type 2beta corticotropin-releasing factor receptor reveals differential ligand specificity. J. Biol. Chem. **278:** 15595–15600.

9. DONNELLY, D. 1997. The arrangement of the transmembrane helices in the secretin receptor family of G-protein-coupled receptors. FEBS Lett. **409:** 431–436.

10. TAMS, J.W., S.M. KNUDSEN & J. FAHRENKRUG. 1998. Proposed arrangement of the seven transmembrane helices in the secretin receptor family. Receptors Channels **5:** 79–90.

11. SEGRE, G.V. & S.R. GOLDRING. 1993. Receptors for secretin, calcitonin, parathyroid hormone (PTH)/PTH-related peptide, vasoactive intestinal peptide, glucagonlike peptide 1, growth hormone-releasing hormone, and glucagon belong to a newly discovered G-protein-linked receptor family. Trends Endocrinol. Metab. **4:** 309–314.

12. BODANSZKY, M. & A. BODANSZKY. 1986. Conformation of peptides of the secretin-VIP-glucagon family in solution. Peptides **7:** 43–48.

13. HOLTMANN, M.H., E.M. HADAC & L.J. MILLER. 1995. Critical contributions of amino-terminal extracellular domains in agonist binding and activation of secretin and vasoactive intestinal polypeptide receptors. Studies of chimeric receptors. J. Biol. Chem. **270:** 14394–14398.

14. WULFF, B., S.M. KNUDSEN, K. ADELHORST & J. FAHRENKRUG. 1997. The C-terminal part of VIP is important for receptor binding and activation, as evidenced by chimeric constructs of VIP/secretin. FEBS Lett. **413:** 405–408.

15. MUTT, V., J.E. JORPES & S. MAGNUSSON. 1970. Structure of porcine secretin: the amino acid sequence. Biochem. J. **15:** 513–519.

16. MUTT, V. 1980. Secretin: isolation, structure, and functions. *In* Gastrointestinal Hormones. G.B.J. Glass, Ed.: 85–126. Raven Press. New York.

17. BISSONNETTE, B.M., M.J. COLLEN, H. ADACHI, et al. 1984. Receptors for vasoactive intestinal peptide and secretin on rat pancreatic acini. Am. J. Physiol. **246:** G710–G717.

18. ZHOU, Z.C., J.D. GARDNER & R.T. JENSEN. 1987. Receptors for vasoactive intestinal peptide and secretin on guinea pig pancreatic acini. Peptides **8:** 633–637.

19. GOSSEN, D., P. POLOCZEK, M. SVOBODA & J. CHRISTOPHE. 1989. Molecular architecture of secretin receptors: the specific covalent labelling of a 51 kDa peptide after cross-linking of iodosecretin to intact rat pancreatic acini. FEBS Lett. **243:** 205–208.

20. HO, P.K., R.S.M. FONG, H.S.T. KAI, *et al.* 1999. The human secretin receptor gene: genomic organization and promoter characterization. FEBS Lett. **455:** 209–214.

21. ULRICH, C.D., D.I. PINON, E.M. HADAC, *et al.* 1993. Intrinsic photoaffinity labeling of native and recombinant rat pancreatic secretin receptors. Gastroenterology **105:** 1534–1543.

22. JIANG, S. & C.D. ULRICH. 1995. Molecular cloning and functional expression of a human pancreatic secretin receptor. Biochem. Biophys. Res. Commun. **207:** 883–890.

23. CHOW, B.K.C. 1995. Molecular cloning and functional characterization of a human secretin receptor. Biochem. Biophys. Res. Commun. **212:** 204–211.

24. SVOBODA, M., M. TASTENOY, P. DE NEEF, *et al.* 1998. Molecular cloning and in vitro properties of the recombinant rabbit secretin receptor. Peptides **19:** 1055–1062.

25. GOURLET, P., J.P. VILARDAGA, P. DE NEEF, *et al.* 1996. Interaction of amino acid residues at positions 8-15 of secretin with the N-terminal domain of the secretin receptor. Eur. J. Biochem. **239:** 349–355.

26. GOURLET, P., J.P. VILARDAGA, P. DE NEEF, *et al.* 1996. The C-terminus ends of secretin and VIP interact with the N- terminal domains of their receptors. Peptides **17:** 825–829.

27. DONG, M., Y. WANG, D.I. PINON, *et al.* 1999. Demonstration of a direct interaction between residue 22 in the carboxyl-terminal half of secretin and the amino-terminal tail of the secretin receptor using photoaffinity labeling. J. Biol. Chem. **274:** 903–909.

28. ASMANN, Y.W., M. DONG, S. GANGULI, *et al.* 2000. Structural insights into the amino-terminus of the secretin receptor: I. Status of cysteine and cystine residues. Mol. Pharmacol. **58:** 911–919.

29. VILARDAGA, J.P., E. DI PAOLO, C. BIALEK, *et al.* 1997. Mutational analysis of extracellular cysteine residues of rat secretin receptor shows that disulfide bridges are essential for receptor function. Eur. J. Biochem. **246:** 173–180.

30. PANG, R.T.K., S.S.M. NG, C.H.K. CHENG, *et al.* 1999. Role of N-linked glycosylation on the function and expression of the human secretin receptor. Endocrinology **140:** 5102–5111.

31. SCHWARTZ, T.W. 1994. Locating ligand-binding sites in 7TM receptors by protein engineering. Curr. Opin. Biotech. **5:** 434–444.

32. CAO, Y.-J., G. GIMPL & F. FAHRENHOLZ. 1995. The amino-terminal fragment of the adenylate cyclase activating polypeptide (PACAP) receptor functions as a high affinity PACAP binding domain. Biochem. Biophys. Res. Commun. **212:** 673–680.

33. ADAMS, A.E., A. BISELLO, M. CHOREV, *et al.* 1998. Arginine 186 in the extracellular n-terminal region of the human parathyroid hormone 1 receptor is essential for contact with position 13 of the hormone. Mol. Endocrinol. **12:** 1673–1683.

34. STROOP, S.D., H. NAKAMUTA, R.E. KUESTNER, *et al.* 1996. Determinants for calcitonin analog interaction with the calcitonin receptor N-terminus and transmembrane-loop regions. Endocrinology **137:** 4752–4756.

35. STROOP, S.D., R.E. KUESTNER, T.F. SERWOLD, *et al.* 1995. Chimeric human calcitonin and glucagon receptors reveal two dissociable calcitonin interaction sites. Biochemistry **34:** 1050–1057.

36. HOLTMANN, M.H., E.M. HADAC, C.D. ULRICH & L.J. MILLER. 1996. Molecular basis and species specificity of high affinity binding of vasoactive intestinal polypeptide by the rat secretin receptor. J. Pharmacol. Exp. Ther. **279:** 555–560.
37. DI PAOLO, E., J.P. VILARDAGA, H. PETRY, et al. 1999. Role of charged amino acids conserved in the vasoactive intestinal polypeptide/secretin family of receptors on the secretin receptor functionality. Peptides **20:** 1187–1193.
38. LISENBEE, C.S., M. DONG & L.J. MILLER. 2005. Paired cysteine mutagenesis to establish the pattern of disulfide bonds in the functional intact secretin receptor. J. Biol. Chem. **280:** 12330–12338.
39. DONG, M., Y.W. ASMANN, M. ZANG, et al. 2000. Identification of two pairs of spatially approximated residues within the carboxyl terminus of secretin and its receptor. J. Biol. Chem. **275:** 26032–26039.
40. DONG, M., Y. WANG, E.M. HADAC, et al. 1999. Identification of an interaction between residue 6 of the natural peptide ligand and a distinct residue within the amino-terminal tail of the secretin receptor. J. Biol. Chem. **274:** 19161–19167.
41. DONG, M., M.W. ZANG, D.I. PINON, et al. 2002. Interaction among four residues distributed through the secretin pharmacophore and a focused region of the secretin receptor amino terminus. Mol. Endocrinol. **16:** 2490–2501.
42. DONG, M., Z.J. LI, D.I. PINON, et al. 2004. Spatial approximation between the amino terminus of a peptide agonist and the top of the sixth transmembrane segment of the secretin receptor. J. Biol. Chem. **279:** 2894–2903.
43. DONG, M.Q., D.I. PINON & L.J. MILLER. 2005. Insights into the structure and molecular basis of ligand docking to the G protein–coupled secretin receptor using charge-modified amino-terminal agonist probes. Mol. Endocrinol. **19:** 1821–1836.
44. ZANG, M.W., M. DONG, D.I. PINON, et al. 2003. Spatial approximation between a photolabile residue in position 13 of secretin and the amino terminus of the secretin receptor. Mol. Pharmacol. **63:** 993–1001.
45. DONG, M., Z.J. LI, M.Q. ZANG, et al. 2003. Spatial approximation between two residues in the mid-region of secretin and the amino terminus of its receptor: incorporation of seven sets of such constraints into a three-dimensional model of the agonist-bound secretin receptor. J. Biol. Chem. **278:** 48300–48312.
46. HEFFORD, M.A. & H. KAPLAN. 1989. Chemical properties of the histidine residue of secretin: evidence for a specific intramolecular interaction. Biochim. Biophys. Acta **998:** 267–270.
47. GARDELLA, T.J., H. JÜPPNER, A.K. WILSON, et al. 1994. Determinants of [Arg^2]PTH-(1-34) binding and signaling in the transmembrane region of the parathyroid hormone receptor. Endocrinology **135:** 1186–1194.
48. GARDELLA, T.J., M.D. LUCK, M.H. FAN & C.W. LEE. 1996. Transmembrane residues of the parathyroid hormone (PTH)/PTH- related peptide receptor that specifically affect binding and signaling by agonist ligands. J. Biol. Chem. **271:** 12820–12825.
49. BERGWITZ, C., S.A. JUSSEAUME, M.D. LUCK, et al. 1997. Residues in the membrane-spanning and extracellular loop regions of the parathyroid hormone (PTH)-2 receptor determine signaling selectivity for PTH and PTH-related peptide. J. Biol. Chem. **272:** 28861–28868.
50. DONG, M., D.I. PINON, R.F. COX & L.J. MILLER. 2004. Importance of the amino terminus in secretin family G protein–coupled receptors: intrinsic photoaffinity labeling establishes initial docking constraints for the calcitonin receptor. J. Biol. Chem. **279:** 1167–1175.

51. DONG, M., D.I. PINON, R.F. COX & L.J. MILLER. 2004. Molecular approximation between a residue in the amino-terminal region of calcitonin and the third extracellular loop of the class B G protein–coupled calcitonin receptor. J. Biol. Chem. **279:** 31177–31182.
52. HOLTMANN, M.H., S. GANGULI, E.M. HADAC, *et al.* 1996. Multiple extracellular loop domains contribute critical determinants for agonist binding and activation of the secretin receptor. J. Biol. Chem. **271:** 14944–14949.
53. GOURLET, P., A. VANDERMEERS, M.C. VANDERMEERS-PIRET, *et al.* 1996. Effect of introduction of an arginine[16] in VIP, PACAP and secretin on ligand affinity for the receptors. Biochim. Biophys. Acta **1314:** 267–273.
54. OLDE, B., A. SABIRSH & C. OWMAN. 1998. Molecular mapping of epitopes involved in ligand activation of the human receptor for the neuropeptide, VIP, based on hybrids with the human secretin receptor. J. Mol. Neurosci. **11:** 127–134.
55. VILARDAGA, J.-P., P. DE NEEF, E. DI PAOLO, *et al.* 1995. Properties of chimeric secretin and VIP receptor proteins indicate the importance of the N-terminal domain for ligand discrimination. Biochem. Biophys. Res. Commun. **211:** 885–891.
56. ROBBERECHT, P., E. DI PAOLO, N. MOGUILEVSKY, *et al.* 2000. Sequences (103–110) and (116–120) of the rat secretin receptor are implicated in secretin and VIP recognition. Ann. N. Y. Acad. Sci. **921:** 362–365.

PACAP and Ceramides Exert Opposite Effects on Migration, Neurite Outgrowth, and Cytoskeleton Remodeling

ANTHONY FALLUEL-MOREL,[a] DAVID VAUDRY,[a] NICOLAS AUBERT,[a] LUDOVIC GALAS,[a] MAGALIE BENARD,[a] MAGALI BASILLE,[a] MARC FONTAINE,[a] ALAIN FOURNIER,[b] HUBERT VAUDRY,[a] AND BRUNO J. GONZALEZ[a]

[a]INSERM U413, IFRMP 23, University of Rouen, 76821 Mont-Saint-Aignan, France

[b]INRS-Institut Armand Frappier, University of Québec, H9R 1G6 Pointe-Claire, Canada

ABSTRACT: During brain development, cells that fail to reach their final destination or to establish proper connections are eliminated. It has been shown that the proinflammatory cytokine second messenger ceramides and the neuropeptide pituitary adenylate cyclase-activating polypeptide (PACAP) play pivotal roles in the histogenesis of the cerebellum. However, little is known regarding the effects of these two factors on cerebellar granule cell migration. We have found that PACAP prevents the effects of C2-ceramide on granule cell motility and neurite outgrowth. These actions are attributable to opposite effects on actin distribution, tubulin polymerization, and Tau phosphorylation. These data suggest that PACAP and factors inducing ceramide production may control granule cell migration during cerebellar development.

KEYWORDS: sphingolipids; cerebellar corticogenesis; cell migration; differentiation

INTRODUCTION

During neurodevelopment, newly generated neurons migrate to reach their final destination and undergo neurite outgrowth to get integrated into neuronal networks.[1] Neurons that have not established proper connections will be eliminated through apoptosis, and defects in the differentiation processes can lead to severe neuronal malformations.[2] The neuropeptide pituitary adenylate cyclase-activating polypeptide (PACAP) and its receptors are actively expressed in the

Address for correspondence: Dr. Hubert Vaudry, INSERM U413, IFRMP 23, University of Rouen, 76821 Mont-Saint-Aignan, France. Voice: 33-235-14-6624; fax: 33-235-14-6946.
e-mail: hubert.vaudry@univ-rouen.fr

Ann. N.Y. Acad. Sci. 1070: 265–270 (2006). © 2006 New York Academy of Sciences.
doi: 10.1196/annals.1317.024

central nervous system during development, notably in the germinative neu-roepithelia and migratory zones.[3,4] PACAP exerts neurotrophic and prodiffer-entiating effects especially on cerebellar granule cells,[5] and increases *in vivo* the number of granule neurons in the internal granule cell layer.[6] While these observations suggest a role for PACAP in brain plasticity, the possible effect of PACAP on neuronal migration has not yet been investigated. The proinflam-matory cytokines, TNF-α and FasL, are thought to play a pivotal role in the histogenesis of the cerebellar cortex. TNF-α and FasL stimulate the biosynthe-sis of ceramides that cause the death of cerebellar granule neurons,[7] and the proapoptotic effect of ceramides on granule cells can be prevented by PACAP.[8] In the present article, we describe interactions between ceramides and PACAP on granule cell migration, neurite outgrowth, and cytoskeleton remodeling.

EFFECTS OF C2-CERAMIDE AND PACAP ON GRANULE CELL MOTILITY AND NEURITE OUTGROWTH

The effects of ceramide and PACAP on cell motility have been studied by using time-lapse microscopy.[9] In control conditions, granule neurons move regularly from their origin during the whole recording period (FIG. 1 A). Ex-posure of granule cells to C2-ceramide (20 μM) induces an immediate and transient increase of cell motility (FIG. 1 A). After 10 to 30 min of C2-ceramide treatment, the distance to origin is significantly higher ($P < 0.001$) than in control conditions. In contrast, after 3 h of treatment, C2-ceramide inhibits cell motility ($P < 0.001$). Treatment with PACAP (10^{-7} M) does not affect cell motility during the first 6 h but inhibits cell displacement during the next 6 h (FIG. 1 A). Coincubation of granule cells with C2-ceramide and PACAP abolishes the early stimulatory effect of C2-ceramide on cell motility (FIG. 1 A; $P < 0.001$). In contrast, PACAP does not affect C2-ceramide-induced inhibi-tion of cell movements during a prolonged incubation.

In control conditions, *in vitro,* differentiating cerebellar granule cells extend long neurites (FIG. 1 B) with a mean length of 17 μm (FIG. 1 C). Treatment of cells with C2-ceramide (20 μM) for 12 h reduces the proportion of cells bearing neurites (FIG. 1 B). Cell morphology is characterized by a single neurite with a mean length of 10 μm (FIG. 1 C). Addition of graded concentrations of PACAP significantly enhances the number of cells bearing neurites (FIG. 1 B) and induces a dose-dependent increase in the mean neurite length (FIG. 1 C).

EFFECTS OF C2-CERAMIDE AND PACAP ON CYTOSKELETON ORGANIZATION

Cytoskeleton components play a pivotal role in cell migration and neu-rite outgrowth. For instance, leading edge extension largely depends on actin

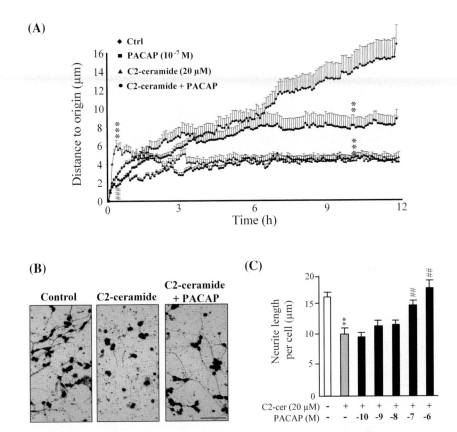

FIGURE 1. Effects of C2-ceramide and PACAP on granule cell motility and neurite outgrowth. (**A**) Distance to origin covered by cerebellar granule cells cultured in control conditions ($\blacklozenge$; $n = 100$), in the presence of 20 μM C2-ceramide ($\blacktriangle$; $n = 90$), 10^{-7} M PACAP ($\blacksquare$; $n = 80$), and C2-ceramide plus PACAP ($\bullet$; $n = 85$). (**B**) Microphotographs illustrating the growth of neuritic processes after 12 h of culture in control conditions, in the presence of 20 μM C2-ceramide, and C2-ceramide plus 10^{-7} M PACAP. Scale bar: 50 μm. (**C**) Effect of C2-ceramide (20 μM) in the absence or presence of graded concentrations of PACAP (10^{-10}–10^{-6} M) on the mean length of neurites. **, $P < 0.01$; ***, $P < 0.001$ versus control; #, $P < 0.05$; ##, $P < 0.01$; ###, $P < 0.001$ versus C2-ceramide. (From Reference 9, with permission.)

rich protrusions while nucleokinesis requires microtubule elongation.[1] Neither C2-ceramide nor PACAP has any effect on *de novo* synthesis of cytoskeletal proteins but the two compounds markedly alter the distribution of tubulin in cerebellar granule cell[9]: exposure of granule cells to C2-ceramide provokes depolymerization of microtubules, and the organization of the microtubular system is largely restored by coadministration of PACAP (FIG. 2 A). As nucleokinesis is critically dependent on tubulin polymerization,[10] these results

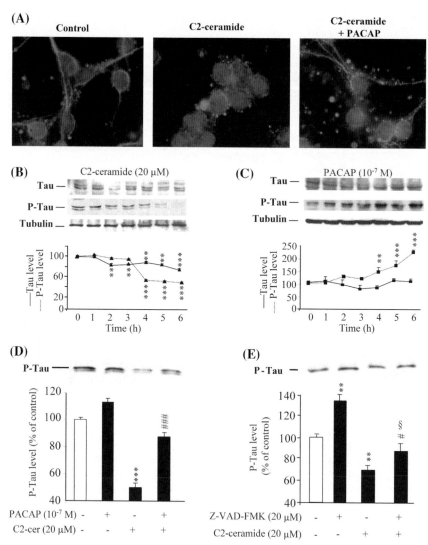

FIGURE 2. Effects of C2-ceramide and PACAP on cytoskeleton remodeling. **(A)** Immunohistochemical micrographs illustrating the distribution of tubulin in control conditions, in the presence of 20 μM C2-ceramide and C2-ceramide plus 10^{-7} M PACAP. **(B–C)** Time course of the effect of 20 μM C2-ceramide **(B)** and 10^{-7} M PACAP **(C)** on Tau and phospho-Tau levels. Tubulin was used as a loading control. **(D)** Effect of a 6-h incubation of cells in the absence or presence of 10^{-7} M PACAP, 20 μM C2-ceramide, and C2-ceramide plus PACAP on phospho-Tau level. **(E)** Western blot analysis showing the effect of 20 μM C2-ceramide, 20 μM Z-VAD-FMK, and C2-ceramide plus Z-VAD-FMK on phospho-Tau level. **, $P < 0.01$; ***, $P < 0.001$ versus control; #, $P < 0.05$; ##, $P < 0.01$; ###, $P < 0.001$ versus C2-ceramide; §, not statistically different from control. (From Reference 9, with permission.)

strongly suggest that inhibition of granule cell nucleokinesis induced by long-term (3 to 12 h) C2-ceramide treatment results from massive disorganization of tubulin.

Tau, a major microtubule-associated protein found in neurons, stimulates tubulin polymerization and thus maintains microtubule structure. Tau possesses many phosphorylation sites and the biological activity of Tau is regulated by its degree of phosphorylation. C2-ceramide produces a marked inhibition of tau phosphorylation after 4 h of treatment and a moderate reduction of the total level of Tau after 2 h (FIG. 2 B). In contrast, PACAP increases Tau phosphorylation and has no effect on the total amount of Tau (FIG. 2 C). In addition, PACAP prevents the effect of C2-ceramide on phosphorylated Tau (FIG. 2 D). Tau can be phosphorylated by various kinases,[11] dephosphorylated by several protein phosphatases,[12] and can also be cleaved by caspase-3.[13] The fact that ceramides activate caspase-3 while PACAP stimulates PKA and MAP-kinase and inhibits caspase-3 activity strongly suggests that C2-ceramide and PACAP can regulate phosphorylation and proteolysis of Tau.[8] Indeed, the cell permeable caspase inhibitor Z-VAD-FMK suppresses the effect of C2-ceramide on the level of phospho-Tau (FIG. 2 E), indicating that the early action of C2-ceramide on "granule cell dance" can be accounted for by a caspase-dependent degradation of the cytoskeleton.[9] Consistent with this notion, it has been recently shown that Tau can be hydrolyzed by caspase-3 at the Asp[421] residue.[13]

CONCLUSION

C2-ceramide and PACAP exert opposite effects on granule cell motility and neurite outgrowth: C2-ceramide provokes an immediate and transient activation of cell movements, inhibits neuritogenesis, and induces apoptosis, whereas PACAP inhibits cell migration and promotes neurite outgrowth.[9] These opposite effects of C2-ceramide and PACAP on cell motility and neurite extension can be ascribed to their reverse actions on the cytoskeletal architecture and on Tau integrity, through phosphorylation and caspase-dependent mechanisms. Taken together, these data strongly suggest that cytokines and PACAP are involved in the control of granule cell migration and synaptogenesis during the histogenesis of the cerebellar cortex.

ACKNOWLEDGMENTS

This work was supported by INSERM (U413), an exchange program from INSERM and FRSQ, and the Conseil Régional de Haute-Normandie. A.F.M. and M. Benard are recipients of a fellowship from the LARC-Neuroscience network. N.A. is the recipient of a fellowship from CIT and the Conseil

Régional de Haute-Normandie. H.V. is an affiliated professor at the INRS-Institut Armand Frappier (Montreal).

REFERENCES

1. Tojima, T. & E. Ito. 2004. Signal transduction cascades underlying de novo protein synthesis required for neuronal morphogenesis in differentiating neurons. Prog. Neurobiol. **72:** 183–193.
2. Vaillant, C., C. Meissirel, M. Mutin, et al. 2003. MMP-9 deficiency affects axonal outgrowth, migration, and apoptosis in the developing cerebellum. Mol. Cell. Neurosci. **24:** 395–408.
3. Sheward, W.J., E.M. Lutz, A.J. Copp & A.J. Harmar. 1998. Expression of PACAP, and PACAP type 1 (PAC1) receptor mRNA during development of the mouse embryo. Dev. Brain Res. **109:** 245–253.
4. Basille, M., B.J. Gonzalez, D. Vaudry, et al. 2000. Distribution of pituitary adenylate cyclase-activating polypeptide (PACAP) receptor mRNAs and PACAP binding sites in the rat brain germinative neuroepithelia. J. Comp. Neurol. **425:** 495–509.
5. Vaudry, D., A. Falluel-Morel, S. Leuillet, et al. 2003. Regulators of cerebellar granule cell development act through specific signaling pathways. Science **300:** 1532–1534.
6. Vaudry, D., B.J. Gonzalez, M. Basille, et al. 1999. Neurotrophic activity of pituitary adenylate cyclase-activating polypeptide on rat cerebellar cortex during development. Proc. Natl. Acad. Sci. USA **96:** 9415–9420.
7. Birbes, H., C. Luberto, Y.T. Hsu, et al. 2005. A mitochondrial pool of sphingomyelin is involved in TNFalpha-induced Bax translocation to mitochondria. Biochem. J. **386:** 445–451.
8. Falluel-Morel, A., N. Aubert, D. Vaudry, et al. 2004. Opposite regulation of the mitochondrial apoptotic pathway by C2-ceramide and PACAP through a MAP-kinase-dependent mechanism in cerebellar granule cells. J. Neurochem. **91:** 1231–1243.
9. Falluel-Morel, A., D. Vaudry, N. Aubert, et al. 2005. Pituitary adenylate cyclase-activating polypeptide prevents the effects of ceramides on migration, neurite outgrowth, and cytoskeleton remodeling. Proc. Natl. Acad. Sci. USA **102:** 2637–2642.
10. Lambert de Rouvroit, C. & A.M. Goffinet. 2001. Neuronal migration. Mech. Dev. **105:** 47–56.
11. Raghunandan, R. & V.M. Ingram. 1995. Hyperphosphorylation of the cytoskeletal protein Tau by the MAP-kinase PK40erk2: regulation by prior phosphorylation with cAMP-dependent protein kinase A. Biochem. Biophys. Res. Commun. **215:** 1056–1066.
12. Sun, L., S.Y. Liu, X.W. Zhou, et al. 2003. Inhibition of protein phosphatase 2A- and protein phosphatase 1-induced tau hyperphosphorylation and impairment of spatial memory retention in rats. Neuroscience **118:** 1175–1182.
13. Gamblin, T.C., F. Chen, A. Zambrano, et al. 2003. Caspase cleavage of tau: linking amyloid and neurofibrillary tangles in Alzheimer's disease. Proc. Natl. Acad. Sci. USA **100:** 10032–10037.

The Effects of PACAP and VIP on the *in Vitro* Melatonin Secretion from the Embryonic Chicken Pineal Gland

N. FALUHELYI, D. REGLŐDI, AND V. CSERNUS

Department of Anatomy, Pécs University, Medical School, and Neurohumoral Regulations Research Group of the Hungarian Academy of Sciences, 7624 Pécs, Hungary

ABSTRACT: The effects of *in vitro* VIP administration and those of *in ovo* pretreatment with PACAP antagonist (PACAP6-38) on the development of the embryonic melatonin (MT) secretion were investigated. With dynamic *in vitro* bioassay we showed that (1) the development of the circadian MT secretion seems to be unaffected by VIP administrations or by PACAP6-38 pretreatments; (2) exposure of the embryonic chicken pineal gland to VIP induces transitory increase in MT secretion at or before the 14th embryonic day *in vitro*.

KEYWORDS: avian; *in vitro*; perifusion; melatonin; phaseshift

INTRODUCTION

Pituitary adenylate cyclase-activating polypeptide (PACAP) and vasoactive intestinal peptide (VIP) have been shown to participate in the modulation of the circadian rhythm of chicken pineal glands. It has been reported that both PACAP and VIP stimulate melatonin (MT) release from the adult pineal gland in chicken, without influencing the intrinsic circadian clock.[1–3] Recently, we have demonstrated that *in vitro* PACAP administration transiently elevates the embryonic MT secretion.[4,5] However, no data are available on the responsiveness of embryonic chicken pineals to VIP. The simultaneous presence of the specific PAC1 receptor and the common VIP/PACAP receptors, VPAC1 and VPAC2, has been demonstrated in the chicken pineal gland,[6] and furthermore, VIP and PACAP differentially regulate melatonin synthesis in pinealocytes.[7] These observations prompted us to investigate the effects of VIP on the embryonic melatonin secretion. Embryonic treatment with the PACAP antagonist PACAP6-38 has been shown to cause transient changes in motor behavior,

Address for correspondence: Valér Csernus, Department of Anatomy, Pécs University, 7624 Pécs, Szigeti u 12. Hungary. Voice: +36-72-536392; fax: +36-72-536393.
e-mail: valer.csernus@aok.pte.hu

Ann. N.Y. Acad. Sci. 1070: 271–275 (2006). © 2006 New York Academy of Sciences.
doi: 10.1196/annals.1317.025

long-lasting alteration in social behavior, and inhibition of olfactory memory formation in chicken.[8,9] The possible effects of similar treatment on the development of the circadian MT rhythm have not been investigated yet. We planned to monitor the MT secretion from embryonic chicken pineal glands after *in ovo* PACAP6-38, and *in vitro* VIP treatment.

METHODS

Fertilized eggs of domestic chicken were obtained from a local hatchery (Mohacs, Hungary) and were incubated at 37.5°C, with a relative humidity of 60%. Chicken embryos were injected *in ovo* with 50-µg PACAP6-38 dissolved in 25-µL physiological saline or only with saline in the same volume (controls) once either on the 8th or on the 15th day of embryonic life (E8 and E15, respectively), in a way described elsewhere.[6] All procedures were performed in accordance with the ethical guidelines approved by the University of Pécs (No. BA02/2000-31/2001).

Pineal glands of the chicken embryos at ages of E13 and E18 were collected. MT release of pineal glands (3 in each column) were analyzed in dynamic, *in vitro* bioassays (perifusion system) as described earlier.[3] All experiments were carried out under continuous darkness. In some of our experiments, the pineals were exposed to VIP in 100-nM concentration for 1 h. MT content from the collected perifusion fluid was assayed with radioimmunoassay (RIA).[3] The assay had a sensitivity of 80 pmol/L. The intra- and interassay coefficients of variation were 5–6% and 7–8%, respectively. Each figure shows results of one of 3–4 similarly designed experiments, which resulted in visibly identical graphs.

RESULTS AND DISCUSSION

MT secretion from E13-16 pineals showed an irregular pattern without apparent rhythmicity (FIG. 1). E18-21 pineals, analyzed under identical conditions, showed similar result (data not shown). These are in accordance with our previous results.[5] Repeated VIP administration caused only transient increase in the MT secretion but did not initiate a circadian pattern (FIG. 1). These results indicate that the VIP is not directly involved in the development of the circadian MT secretion. In contrast, when egg turnings were carried out periodically, the *in vitro* MT secretion showed circadian rhythm from the 18th embryonic day.[4,5]

In another experiment, pineals, explanted from E18 embryos after PACAP6-38 pretreatment on E15 were investigated. The MT secretion did not show regular circadian rhythm, but the secretion pattern was parallel in pretreated and control pineals (FIG. 2). MT secretion from E13 pineals pretreated

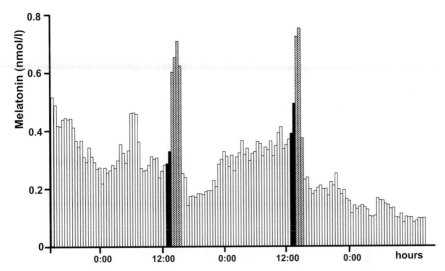

FIGURE 1. *In vitro* melatonin release from pineals of 13-day-old chicken embryos. The experiment was carried out under continuous darkness. Melatonin contents of consecutive 30-min fractions, collected during an experiment, are shown. On the horizontal axis, the real time (0:00 = midnight) is indicated. VIP was applied in 100-nM concentration for 60 min at 13:00 each day (black bars). Hatched bars indicate the period of a transient increase of MT release. The melatonin secretion did not show regular, circadian rhythm.

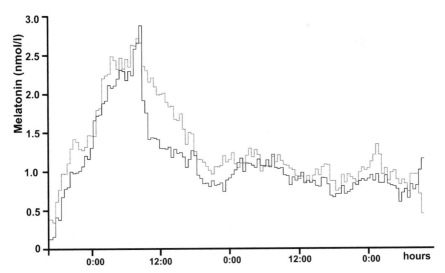

FIGURE 2. *In vitro* MT secretion of pineals explanted from 18-day-old chicken embryos, after PACAP6-38 pretreatment on the 15th embryonic day (continuous line) and that of control pineals (dotted line). The pattern of melatonin secretion is irregular but nearly parallel in the two simultaneously performed experiments.

on 8th embryonic day was also similar to that of controls (data not shown). It is concluded that under the observed experimental parameters PACAP antagonist did not cause changes in the embryonic MT secretion. Compared with our previous results we suggest that PACAP is not involved in the maturation of the MT secretion or the circadian clock. To the best of our knowledge, the effects of other neurotransmitters on these developmental procedures have not been studied in birds. We plan further investigations to collect more details on this area.

In summary, we found no changes in the embryonic MT secretion after periodic administration of VIP or after *in ovo* PACAP antagonist pretreatments at different ages. In accordance with our previous studies, we suggest that the examined neuropeptides influence embryonic MT secretion but have no direct effect on the maturation of chicken embryonic MT secretion.

ACKNOWLEDGMENTS

This work was supported by the Hungarian Medical Research Council (ETT – 635/2003), the Hungarian Scientific Research Fund (OTKA T-046256 and T-046589), and the Hungarian Academy of Sciences.

REFERENCES

1. ZATZ, M., G. KASPER & C.R. MARQUEZ. 1990. Vasoactive intestinal peptide stimulates chick pineal melatonin production and interacts with other stimulatory and inhibitory agents but does not show alpha 1-adrenergic potentiation. J. Neurochem. **55:** 1149–1153.
2. NAKAHARA, K., Y. ABE, T. MURAKAMI, *et al.* 2002. Pituitary adenylate cyclase-activating polypeptide (PACAP) is involved in melatonin release via the specific receptor PACAP-r1, but not in the circadian oscillator, in chick pineal cells. Brain Res. **939:** 19–25.
3. CSERNUS, V., R. JOZSA, D. REGLODI, *et al.* 2004. The effect of PACAP on rhythmic melatonin release of avian pineals. Gen. Comp. Endocrinol. **135:** 62–69.
4. FALUHELYI, N., D. REGLODI, I. LENGVARI, *et al.* 2004. Development of the circadian melatonin rhythm and the effect of PACAP on melatonin release in the embryonic chicken pineal gland. An in vitro study. Regul. Pept. **123:** 23–28.
5. FALUHELYI, N., D. REGLODI & V. CSERNUS. 2005. Development of the circadian melatonin rhythm and its responsiveness to PACAP in the embryonic chicken pineal gland. Ann. N. Y. Acad. Sci. **1040:** 305–309.
6. NOWAK, J.Z. & J.B. ZAWILSKA. 2003. PACAP in avians: origin, occurrence, and receptors—pharmacological and functional considerations. Curr. Pharm. Des. **9:** 467–481.
7. REKASI, Z. & T. CZOMPOLY. 2002. Accumulation of rat pineal serotonin N-acetyltransferase mRNA induced by pituitary adenylate cyclase activating polypeptide and vasoactive intestinal peptide in vitro. J. Mol. Endocrinol. **28:** 19–31.

8. HOLLOSY, T., R. JOZSA, B. JAKAB, *et al.* 2004. Effects of in ovo treatment with PACAP antagonist on general activity, motor and social behavior of chickens. Regul. Pept. **123:** 99–106.
9. JOZSA, R., T. HOLLOSY, A. TAMAS, *et al.* 2005. Pituitary adenylate cyclase activating polypeptide plays a role in olfactory memory formation in chicken. Peptides **26:** 2344–2350.

VIP Prevents Experimental Multiple Sclerosis by Downregulating Both Inflammatory and Autoimmune Components of the Disease

AMELIA FERNANDEZ-MARTIN,[a] ELENA GONZALEZ-REY,[a]
ALEJO CHORNY,[a] JAVIER MARTIN,[a] DAVID POZO,[b]
DOINA GANEA,[c] AND MARIO DELGADO[a]

[a] Institute of Parasitology and Biomedicine, CSIC, Granada 18100, Spain

[b] Department of Biochemistry and Molecular Biology, Medical School of Seville, Seville, Spain

[c] Department of Biological Sciences, Rutgers University, Newark, New Jersey 07102, USA

ABSTRACT: Multiple sclerosis (MS) is a disabling inflammatory, autoimmune demyelinating disease of the central nervous system (CNS). Despite intensive investigation, the mechanisms of disease pathogenesis remain unclear, and curative therapies are unavailable for MS. The current study describes a new possible strategy for the treatment of MS, based on the administration of the vasoactive intestinal peptide (VIP). Treatment with VIP significantly reduced incidence and severity of experimental autoimmune encephalomyelitis (EAE), an MS-related rodent model. VIP suppressed EAE neuropathology by reducing CNS inflammation and by selective blocking encephalitogenic T-cell reactivity, emerging as an attractive candidate for the treatment of human MS.

KEYWORDS: inflammation; autoimmunity; multiple sclerosis; neuroimmunology; neuropeptide

Experimental autoimmune encephalomyelitis (EAE) is an inflammatory, autoimmune demyelinating disease of the central nervous system (CNS), which shows pathologic and clinical similarities to human multiple sclerosis (MS) and is used as a model system to test potential therapeutic agents.[1] Both EAE and MS are considered archetypal CD4 Th1 cell-mediated autoimmune diseases. The initial stages of EAE involve multiple steps, which can be divided into two

Address for correspondence: Mario Delgado, Instituto de Parasitologia y Biomedicina, CSIC, Avd. Conocimiento, PT Ciencias de la Salud, Granada 18100, Spain. Voice: 34-958-181665; fax: 34-958-181632.
e-mail: mdelgado@ipb.csic.es

Ann. N.Y. Acad. Sci. 1070: 276–281 (2006). © 2006 New York Academy of Sciences.
doi: 10.1196/annals.1317.026

main phases: initiation and establishment of autoimmunity, and later events associated with the evolving immune and inflammatory responses. The crucial process underlying disease initiation is the induction of autoimmunity to myelin sheath components; later events involve a destructive inflammatory process. Progression of the autoimmune response involves the development of reactive Th1 cells with encephalitogenic potential, their entry into the CNS, and future recruitment of inflammatory cells through multiple mediators.[1–3] Inflammatory mediators, such as cytokines [i.e., interleukin-12 (IL-12), inteferon-γ (IFNγ), and tumor necrosis factor-α (TNFα)] and nitric oxide (NO), produced by infiltrating cells and resident microglia, play a critical role in demyelination, contributing to oligodendrocyte loss and degenerative axonal pathology.[2,3] Although available therapies based on immunosuppressive agents inhibit the inflammatory component of MS and either reduce the relapse rate or delay disease onset, they do not suppress progressive clinical disability. This illustrates the need for novel therapeutic approaches to prevent the inflammatory and autoimmune components of the disease and to promote repair and regeneration mechanisms.

Vasoactive intestinal peptide (VIP) is a potent Th2-produced immunosuppressive agent that proved to be protective in several models of autoimmune diseases, such as collagen-induced arthritis, inflammatory bowel disease, and uveoretinitis.[4–6] Until now, the mechanisms described for the VIP immunosuppressive activity included the deactivation of macrophages, dendritic cells, and microglia, and promoting Th2 effector differentiation and survival, and inhibiting Th1 responses.[7] In this article, we investigated the effect of the administration of VIP in an EAE model that mirrors different clinical characteristics of MS. In the majority of MS patients, clinical disease follows a relapsing–remitting course. Therefore, we tested the effect of VIP on the relapsing–remitting EAE (RR-EAE) model induced by proteolipid protein (PLP$_{139-151}$) in SJL/J mice. Disease severity was substantially reduced in mice receiving VIP treatment during the efferent phase of disease, as reflected by a delay in disease onset, a decrease in the mean clinical score, a decrease in the rate of relapse, and a reduction in the cumulative disease index (CDI) (TABLE 1). The therapeutic properties of VIP were also apparent when administered at the onset of clinical symptoms (day 10), at the peak of the disease (day 15, acute phase), or after initial remission (day 20, relapsing phase). The VIP effects were dose-dependent, and as previously shown for other autoimmune models, 2 nmol was the most efficient therapeutic dose. In addition, the protective effect of VIP was long lasting, with no clinical symptoms up to 60 days after VIP administration. The VPAC1 agonist mimicked the VIP effects on RR-EAE, whereas the VPAC2 agonist showed only a weak effect (TABLE 1). VIP specificity was confirmed as glucagon, secretin, and the two fragments VIP$_{1-12}$ and VIP$_{10-28}$ did not show any effect. In agreement with a recent report,[8] the structurally related pituitary adenylate cyclase-activating polypeptide (PACAP) reduced EAE progression, severity, and incidence.

TABLE 1. Effect of VIP and VIP-related peptides in EAE

	Incidence[a]			Onset	Peak[b]	Relapse	CDI[c]
	Severe	Mild	None (%)	(days)		(%)	
Control	19/22(86)	3/22(14)	0/22(0)	11.3 ± 0.8	5.4 ± 0.5	14/22(64)	56.4 ± 6.6
VIP	0/26(0)	7/26(27)	19/26(73)	21.3 ± 1.3	2.4 ± 0.3	4/26(15)	21.3 ± 2.7
PACAP	0/12(0)	4/12(33)	8/12(67)	19.4 ± 1.6	2.1 ± 0.4	3/12(25)	22.4 ± 3.6
VIP$_{1-12}$	10/12(83)	2/12(17)	0/12(0)	12.1 ± 1.1	5.5 ± 0.7	7/12(58)	54.3 ± 7.1
VIP$_{10-28}$	9/12(75)	2/12(17)	1/12(8)	10.8 ± 1.2	5.4 ± 0.3	7/12(58)	55.2 ± 3.6
VPAC1	0/12(0)	3/12(25)	9/12(75)	22.6 ± 2.7	2.0 ± 0.3	2/12(17)	19.7 ± 2.1
VPAC2	3/12(25)	6/12(50)	3/12(25)	15.3 ± 1.6	3.9 ± 0.6	5/12(42)	39.3 ± 6.1
Secretin	8/10(80)	2/10(20)	0/10(0)	10.7 ± 1.1	5.6 ± 0.7	6/10(60)	54.3 ± 5.5
Glucagon	7/10(70)	3/10(30)	0/10(0)	11.4 ± 1.2	5.3 ± 0.4	5/10(50)	53.4 ± 5.6

RR-EAE was induced in SJL/J mice by s.c. immunization with 150 µg of PLP$_{139-151}$ emulsified in CFA containing 500 µg of *M. tuberculosis* H37 RA. EAE mice were treated i.p. for 3 days with PBS (control) with 2 nmol/day of VIP, PACAP, VIP fragments (VIP$_{1-12}$, VIP$_{10-28}$), VIP receptor agonists (VPAC1, VPAC2), or other peptides of the family (secretin, glucagon) starting 2 days after onset of disease (approximately at day 11 post immunization).
[a]Disease incidence is graded as severe (clinical score: 4–6), mild (clinical score: 0.5–3) or none (no clinical signs). Numbers in parentheses represent percentages.
[b]Peak defines the mean of maximal clinical score.
[c]CDI is the mean of the sum of the daily disease scores.

These results indicate that VPAC1 plays a major role in the protective effect of VIP in EAE.

We next investigated the mechanisms underlying the amelioration of EAE after VIP treatment. The pathology of MS and EAE features focal areas of inflammatory infiltration and demyelination with oligodendrocyte depletion.[1-3] Histopathologic examination of spinal cords confirmed that the beneficial actions of VIP were due to a decrease in inflammatory infiltrates (mainly CD4 T cells and macrophages), and subsequent demyelination, oligodendrocyte loss, and axonal damage, suggesting that VIP prevents the entry or retention of inflammatory autoreactive cells into the CNS. This effect was accompanied by a reduction of brain and spinal cord inflammatory mediators, including cytokines (TNF-α, IL-6, IL-1β, IL-18, and IL-12), enzymes (iNOS), chemokines (Rantes, MIP-1, MCP-1, IP-10, and MIP-2), and chemokine receptors (CCR-1, CCR-2, and CCR-5) associated with EAE pathology. In addition, VIP-treated mice showed increased levels of the anti-inflammatory cytokines IL-10, IL-1Ra, and TGF-β in spinal cords.

EAE is a Th1-type cell-mediated autoimmune disease. High levels of Th1-type cytokines (e.g., IFN-γ) are detected in the CNS in both MS and EAE, and conversely, neutralizing Th1 cytokine antibodies ameliorate disease progression in the murine model.[9] In addition, Th2-type cytokines are predominantly present in murine brain during recovery, and treatment based on Th2-type cytokines suppressed the disease, suggesting a switch from Th1- to Th2-type responses during remission.[9] VIP ameliorates EAE by reducing encephalitogenic T cell responses and/or migration to the CNS. VIP administration during EAE progression partially inhibits autoreactive T cell clonal expansion. Analysis of the effector Th phenotype by its cytokine profile showed that peripheral (draining lymph nodes) and infiltrating (CNS) T cells from VIP-treated EAE mice produced lower levels of Th1-type cytokines (IFN-γ, IL-2, and TNF-γ), and higher levels of Th2-type cytokines (IL-4, IL-5, and IL-10) than control EAE mice, indicating a shift in the Th1/Th2 balance. High levels of circulating antibodies directed against myelin antigens invariably accompany the development of MS, and their production is a major factor in determining susceptibility to the disease.[1-3] VIP administration resulted in reduced serum levels of PLP-specific IgG, particularly autoreactive IgG2a antibodies, generally reflective of Th1 activity. These data provide further evidence that VIP administration in EAE reduces the Th1 autoreactive response and promotes Th2 responses both in the CNS and the periphery.

In summary, we show that the neuropeptide VIP provides a highly effective therapy for EAE. The therapeutic effect of VIP is associated with a striking reduction of the two deleterious components of the disease, that is, the autoimmune and inflammatory response (FIG. 1). VIP treatment decreases the presence of encephalitogenic Th1 cells in the periphery and the CNS. In addition, VIP strongly reduced the inflammatory response during EAE progression by downregulating the production of several inflammatory mediators, various

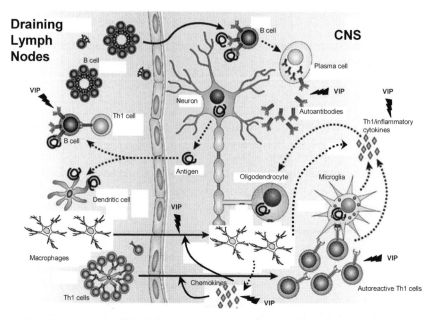

FIGURE 1. VIP and PACAP protects against EAE by downregulating the two components of the disease: inflammation and Th1-driven autoreactive response.

chemokines, and their receptors in both spinal cord and brain parenchyma. At the same time, VIP induced the production of the anti-inflammatory cytokines, which ameliorate the disease. As a consequence, VIP reduced the appearance of inflammatory infiltrates in the CNS, and the subsequent demyelination and axonal damage typical of EAE. In addition, the infiltrating cells found in VIP-treated mice were not able to produce Th1 cytokines, suggesting that VIP directly suppresses the inflammatory/autoreactive cells. Our results serve as a basis for proposing a novel strategy to treat MS, targeted to the inhibition of the different neuropathologic components of the disease. The fact that delayed VIP administration ameliorates ongoing disease also fulfills an essential prerequisite for a therapeutic agent. These observations provide a powerful rationale for the assessment of the efficacy of VIP as a novel therapeutic approach to the treatment of MS.

ACKNOWLEDGMENTS

This work was supported by the following grants: NIH (2RO1A047325, DG, and MD), Spanish Ministry of Health (PI03/0526, MD), and La Caixa Foundation (MD).

REFERENCES

1. STEINMAN, L. 1999. Assessment of animal models for MS and demyelinating disease in the design of rational therapy. Neuron **24:** 511–514.
2. BAUER, J., H. RAUSCHKA & H. LASSMANN. 2001. Inflammation in the nervous system: the human perspective. Glia **36:** 235–243.
3. OWENS, T. 2003. The enigma of multiple sclerosis: inflammation and neurodegeneration cause heterogeneous dysfunction and damage. Curr. Opin. Neurol. **16:** 259–265.
4. DELGADO, M., C. ABAD, C. MARTINEZ, *et al.* 2001. Vasoactive intestinal peptide prevents experimental arthritis by downregulating both autoimmune and inflammatory components of the disease. Nat. Med. **7:** 563–568.
5. ABAD, C., C. MARTINEZ, M.G. JUARRANZ, *et al.* 2003. Therapeutic effects of vasoactive intestinal peptide in the trinitrobenzene sulfonic acid mice model of Crohn's disease. Gastroenterology **124:** 961–971.
6. KEINO, H., T. KEZUKA, M. TAKEUCHI, *et al.* 2004. Prevention of experimental autoimmune uveoretinitis by vasoactive intestinal peptide. Arch. Ophthalmol. **122:** 1179–1184.
7. DELGADO, M., D. POZO & D. GANEA. 2004. The significance of the vasoactive intestinal peptide in immunomodulation. Pharmacol Rev. **56:** 249–290.
8. KATO, H., A. ITO, J. KAWANOKUCHI, *et al.* 2004. Pituitary adenylate cyclase-activating polypeptide (PACAP) ameliorates experimental autoimmune encephalomyelitis by suppressing the functions of antigen presenting cells. Mult. Scler. **10:** 651–659.
9. MILLER, S.D. & E.M. SHEVACH. 1998. Immunoregulation of experimental autoimmune encephalomyelitis: editorial overview. Res. Immunol. **149:** 753–759.

C-Type Natriuretic Peptide Is Specifically Augmented by Pituitary Adenylate Cyclase-Activating Polypeptide in Rat Astrocytes

KOHSHO FUJIKAWA,[a,b] TETSUYA NAGAYAMA,[a] KAZUHIKO INOUE,[b]
NAOTO MINAMINO,[c] KENJI KANGAWA,[c] MASAKI NIIRO,[a] AND
ATSURO MIYATA[b]

[a]Department of Pharmacology, Graduate School of Medical and Dental
Sciences, Kagoshima University, Kagoshima 890-8544, Japan

[b]Department of Neurosurgery, Graduate School of Medical and Dental Sciences,
Kagoshima University, Kagoshima 890-8544, Japan

[c]Research Institute, National Cardiovascular Center, Suita Osaka 565-8565,
Japan

ABSTRACT: In rat-cultured astrocytes, pituitary adenylate cyclase-
activating polypeptide (PACAP) activates gene expression and secretion
of C-type natriuretic peptide (CNP) in a dose- and time-dependent manner. These results suggest that PACAP might be involved in the regulation
of CNP biosynthesis in astrocytes.

KEYWORDS: PACAP; astrocytes; RIA; RT-PCR

INTRODUCTION

C-type natriuretic peptide (CNP) belongs to the natriuretic peptide family
and has a ring structure formed by an intramolecular disulfide bond, which is
conserved among atrial natriuretic peptide (ANP) and brain natriuretic peptide (BNP), and CNP. ANP and BNP are mainly secreted from heart as cardiac
hormones, whereas CNP is mainly localized in the central nervous system
(CNS),[1,2] but the function of CNP in the CNS has not been clarified so far.
Receptors for natriuretic peptides are mainly guanylate cyclase-A (GC-A) and
guanylate cyclase-B (GC-B).[3] ANP and BNP show high affinity to GC-A,
whereas, CNP preferentially binds to GC-B. Both GC-A and GC-B elicit an

Address for correspondence: Atsuro Miyata, Pharmacology, Graduate School of Medical and Dental
Science, Kagoshima University, Kagoshima, 890-8544, Japan. Voice: +81-99-275-5256; fax: +81-99-
265-8567.
e-mail: amiyata@m3.kufm.kagoshima-u.ac.jp

Ann. N.Y. Acad. Sci. 1070: 282–285 (2006). © 2006 New York Academy of Sciences.
doi: 10.1196/annals.1317.027

increase in intracellular cGMP that mediates most of the biological actions of natriuretic peptides.[4] It has been reported that the neurotrophic peptide pituitary adenylate cyclase-activating polypeptide (PACAP)[5] stimulates adenylate cyclase activity in cultured astroytes.[6] During the course of a study regarding the effect of PACAP on astrocytes with DNA microarray, we found that CNP gene expression was markedly increased in astrocytes stimulated by PACAP. Then, we tried to clarify how PACAP regulates the biosynthesis and secretion of CNP in the astrocytes.

MATERIALS AND METHODS

Rat astrocytes in primary culture were prepared from the cerebral cortex of newborn rat, and maintained in Dulbecco's Modified Eagle Medium (DMEM) with 10% fetal bovine serum, penicillin, and streptomycin with Poly–L–Lysin-coated dish. In the second or third passages, after overnight serum starvation, cells were treated with different concentrations of PACAP for 24 h, or with PACAP (10^{-8} M) for different incubation periods.

Total RNA was prepared from astrocytes by TRIZOL LS (Invitrogen, Carlsbad, CA). The first stand cDNA was synthesized from total RNA (2 μg), with random primers and reverse transcriptase (SuperScript^{TM} RT, Invitrogen). The resulting cDNA was amplified with AmpliTaq polymerase, and the PCR products were electrophorezed.

CNP was prepurified using a cartridge column (Sep-pak C_{18}, Waters, Milford, MA) and assessed by means of a specific radioimmunoassay (RIA) for CNP as reported previously.[1,7] CNP-immunoreactivity was characterized by reversed phase high-performance liquid chromatography (RP-HPLC) combined with CNP RIA.

RESULTS AND DISCUSSION

RT-PCR analysis showed that CNP gene expression was observed in the neuronal cells but not in astrocytes under basal conditions (FIG. 1 A). PACAP treatment provoked the expression of the CNP gene in astrocytes in a dose- and time-dependent manner (FIG. 1 B, C). Expression of GC-B, a specific receptor for CNP, was also observed in astrocytes and neuronal cells, but GC-B mRNA levels were not changed by PACAP treatment (FIG. 1 A, C).

Measurement of CNP in the culture medium revealed that PACAP stimulated the secretion of CNP-immunoreactivity in the time- and dose-dependent manner (FIG. 2). Thus, the basal secretion from astrocytes was 15.9 pg/10^6 cells and increased to 796 pg/10^6 cells in the presence of 10 nM PACAP for 24 h. These results suggest that PACAP could be involved in the regulation of CNP biosynthesis and secretion in the CNS.

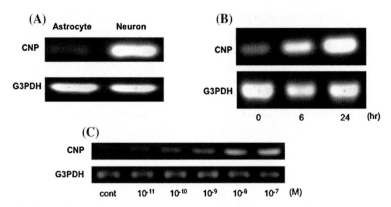

FIGURE 1. The effect of PACAP treatment on CNP gene expression in astrocytes. (**A**) Comparison of CNP expression between neuronal cells and astrocytes under basal condition. (**B**) Time-dependent increase of CNP gene expression in astrocytes by PACAP (10^{-8} M) treatment for various periods. (**C**) Dose-dependent increase of CNP gene expression in astrocytes.

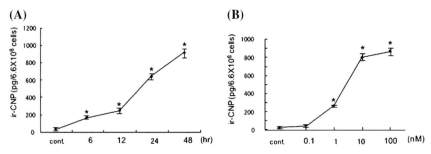

FIGURE 2. The effect of PACAP treatment on CNP secretion from astrocytes. PACAP treatment increased CNP secretion from rat astrocytes in a dose (**A**)- and time (**B**)-dependent manner. $^*P < 0.005$ as compared to control.

REFERENCES

1. MINAMINO, N., Y. MAKINO, et al. 1991. Characterization of immunoreactive human C-type natriuretic peptide in brain and heart. Biochem. Biophys. Res. Commun. **179:** 535–542.
2. KANEKO, T., G. SHIRAKAMI, K. NAKAO, et al. 1993. C-type natriuretic peptide (CNP) is the major natriuretic peptide in human cerebrospinal fluid. Brain Res. **612:** 104–109.
3. KOBAYASHI, H., T. MIZUKI, M. TSUTSUI, et al. 1993. Receptors for C-type natriuretic peptide in cultured rat glial cells. Brain Res. **617:** 163–166.
4. YEUNG, V.T., S.K. HO, M.G. NICHOLLS, et al. 1996. Binding of CNP-22 and CNP-53 to cultured mouse astrocytes and effects on cyclic GMP. Peptides **17:** 101–106.

5. VAUDRY, D., B.J. GONZALEZ, M. BASILLE, *et al.* 2000. Pituitary adenylate cyclase-activating polypeptide and its receptors: from structure to function. Pharmacol. Rev. **52:** 269–324.
6. MASMOUDI, O., P. GANDOLFO, J. LEPRINCE, *et al.* 2003. Pituitary adenylate cyclase-activating polypeptide (PACAP) stimulates endozepine release from cultured rat astrocytes via a PKA-dependent mechanism. FASEB J. **17:** 17–27.
7. UEDA, S., N. MINAMINO, M. ABURAYA, *et al.* 1991. Distribution and characterization of immunoreactive porcine C-type natriuretic peptide. Biochem. Biophys. Res. Commun. **175:** 759–767.

Aromatase Gene Expression and Regulation in the Female Rat Pituitary

GUILLAUME GALMICHE,[a] SOPHIE CORVAISIER,[b] AND MARIE-LAURE KOTTLER[a,b]

[a]Laboratoire "Estrogènes et Reproduction" EA 2608 USC 2006 INRA, Université de Caen Basse-Normandie, 14032 Caen, France

[b]Département de Génétique et Reproduction, UFR de médecine CHU, 14032 Caen, France

ABSTRACT: Aromatase cytochrome P450, the key enzyme of estrogen biosynthesis from androgens, is encoded by *CYP19*. Its structure shows some peculiarities: exons II to X encode the protein, while multiple alternative exons I encode unique 5'-untranslated regions of the aromatase mRNA transcripts. Immunohistochemistry studies in the rat have shown that pituitary aromatase expression is sex-dependent and varies across the estrous cycle, suggesting that estrogens might be involved in the regulation of aromatase activity and might act locally as a paracrine or autocrine factor in the pituitary. In the present study, we used RT-PCR to characterize aromatase transcripts and real-time PCR to quantify the expression of the total aromatase mRNA at the different stages of the estrous cycle and from an ovariectomy and estradiol replacement model. We identified the two previously described aromatase transcripts with a specific 5'untranslated region of the brain 1f and the gonadal PII transcripts. Total aromatase mRNA expression in the pituitary varied significantly during the estrous cycle, with the highest level occurring on the day of metestrus. After ovariectomy, we observed an increase of aromatase mRNA levels, and this effect was completely prevented by estradiol administration. These results suggest that pituitary aromatase mRNA expression is downregulated by estrogens.

KEYWORDS: aromatase; estrogen; pituitary; reproduction

INTRODUCTION

It is established that estrogens are involved in the modulation of the hypothalamic-pituitary-gonadal axis, ovaries, hypothalamus, and pituitary. However, it is difficult to delineate the overall contribution of each site of

Address for correspondence: Marie-Laure Kottler, Département de Génétique et Reproduction, UFR de médecine CHU Caen, France. Voice: 33-231-27-2417; fax: 33-231-56-5120.
e-mail: kottler-ml@chu-caen.fr

Ann. N.Y. Acad. Sci. 1070: 286–292 (2006). © 2006 New York Academy of Sciences.
doi: 10.1196/annals.1317.028

action in reproductive function, particularly at the level of the pituitary gland. While a negative action on gonadotropin release is well documented, the mechanisms involved in the stimulating action in driving the preovulatory LH surge are not completely clarified. An increase of the frequency and the amplitude of GnRH pulses in the preovulatory period is admitted but the deficient GnRH patients receiving GnRH pulses can ovulate, although the frequency of pulses and the quantity delivered remain identical all along the cycle.[1] Therefore, other mechanisms are involved in the regulation of pituitary functions. Regarding the subcellular mechanisms for GnRH action in gonadotrophs, there is little information on the means by which estrogens act on these cells. We hypothesized that local estrogen production, in pituitary, could act in an auto- or paracrine manner to play a role in gonadotropin release. Indeed, studies using *in vitro* and *in vivo* models have demonstrated that *in situ* aromatase, which is responsible for the formation of estrogens from androgens, was involved in local estrogen production in both bone[2] and breast cancer.[3–5]

The aromatase complex is composed of a specific glycoprotein, cytochrome P450 (P450arom), and a ubiquitous flavoprotein, NADPH-cytochrome P450 reductase.[6] The enzyme is encoded by a single copy gene, *cyp19*, and its structure shows some peculiarities: exons II to X encode the protein, while multiple alternative exons I encode unique 5'-untranslated regions of the aromatase mRNA transcripts. Aromatase gene transcription is therefore distinct in different estrogen producing tissues and is under the control of tissue-specific promotors.[7] In humans and higher primates, aromatase is expressed in many tissues, including the gonads, brain, placenta, and bone[7] while in rodents, it is only expressed in the gonads and in the brain[8,9]; these transcripts carry the gonadal-specific first exon II, under the control of promotor II and the brain-specific exon If, under the control of promotor 1f (P1f), respectively. Exon If is the major transcript present in the thalamic–hypothalamic areas of the mouse[10] and in the hypothalamus and amygdala of adult rats,[9] although in both cases, low amounts of transcripts, containing the gonadal subtype, may also be present in these regions.[10,11]

Previous studies have shown that key genes of estrogen signaling and biosynthesis ERα, ERβ, and *cyp19* are expressed in the rat pituitary.[12,13] Carretero *et al.* have shown by immunohistochemistry that aromatase expression is sex-dependent and changes across the estrous cycle,[12] suggesting that estrogens might be involved in the regulation of aromatase activity in the pituitary and thus may act locally as a paracrine or autocrine factor. However, the mechanism responsible for the regulation of aromatase expression and its physiological role in the pituitary remains unclear. In the present article, the occurrence of aromatase transcripts expressed in the female rat pituitary was examined and the potential role of estrogen in aromatase regulation in the rat pituitary was investigated.

MATERIALS AND METHODS

Animals

Mature female Sprague-Dawlay rats were maintained under standard laboratory conditions (12-h light, 12-h dark cycle) and housed individually with water available *ad libitum*. All procedures were approved by the Ministère de l'Agriculture et de la Pêche-Service Santé Animale (France).

Experimental Design

Experiment 1 : Rats were distributed according to three phases of the estrous cycle by examining vaginal smears between 09:00 and 10:00 h ($n = 6$ per group): metestrus, proestrus, and estrous. Only animals exhibiting two consecutive normal estrous cycles were included.

Experiment 2 : We used a treatment paradigm such that the measurements were taken at the time when negative feedback action of estrogens is operative. Rats ($n = 7$ per group) were divided into two groups: ovariectomized (OVX), and ovariectomized plus 17β-estradiol replacement (OVX + E2). OVX rats were injected subcutaneously (s.c.) every 2 days for 3 weeks with 10 μg of 17β-estradiol. The effectiveness of the estrogenic treatment was checked daily by examining vaginal smears with saline serum, between 09:00 and 10:00 h.

Tissue Collection

After 3 weeks, the rats were killed and blood samples (3–5 mL) were obtained from each animal through a cardiac catheter, for serum estradiol evaluation. The blood was immediately centrifuged in dry tubes, and the serums stored at $-20°C$. Pituitaries were dissected out immediately after death, flash frozen in liquid nitrogen, and stored at $-80°C$ for subsequent RNA extraction.

Measurement of Pituitary Aromatase mRNA

Total RNA was extracted using the TRIZOL reagents kit (Life Technologies, Inc., Cergy Pontoise, France), and then converted into cDNA by reverse transcription, according to the manufacturer's directions.

Quantitative Real-Time PCR

Relative levels of total aromatase mRNA (TaqMan, Applied Biosystems, Courtaboeuf, France) were determined by real-time PCR, using an ABI Prism

7,000 Sequence Detector System (Applied Biosystems). A specific standard curve was constructed by RT-PCR. The sequences of the primers selected for total aromatase analyses were located on exons IX and X. This allowed for evaluation of all aromatase variants, independent of the tissue-specific exon I inserted in the mRNA. Standard curves were generated by 10-fold serial dilution. Because these experimental manipulations are known to induce dramatic changes in the functioning of the gonadotropic axis, LHβ mRNA expression (SYBR Green, Applied Biosystems) was also monitored. For all samples, the quantification of β-actin gene (SYBR Green, Applied Biosystems) was used as the endogenous control for normalization of initial RNA levels. PCR reactions were performed in duplicate and a reagent blank prepared using the RT blank was included with each plate to detect contamination by genomic DNA.

Estradiol Production

Serum estradiol levels were measured by using a radioimmunoassay kit (ESTRADIOL, DiaSorin, Antony, France) validated for use in the rat. The limit of detection was 5 pg/mL of serum. The intra- and interassay coefficients of variation were 4% and 11.2%, respectively.

Statistical Analysis

Data are expressed as the mean $\pm$ SEM for each group and compared to the metestrus group. Statistical analysis was performed using one-way ANOVA, and differences between groups were determined by the Fisher-protected least significant difference test (SigmaStat for Windows, Version 3.1; SPSS Inc., Chicago, IL).

RESULTS

Two Aromatase Transcripts are Expressed in Rat Pituitary

RT-PCR analysis revealed that brain and gonadal-specific transcripts were expressed in the pituitary gland (data not shown).

Aromatase Gene Expression is Regulated Across the Rat Estrous Cycle

For high levels of estrogen at proestrus, aromatase transcripts were significantly decreased ($P < 0.001$) compared to the metestrus period (FIG. 1). We have also observed a significant decrease ($P < 0.05$) of aromatase transcripts, during the estrus stage (FIG. 1).

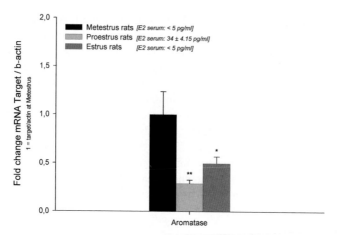

FIGURE 1. Pituitary profile of expression of total aromatase in adult (90 days old) rats across the estrous cycle. Aromatase mRNA levels were measured by real-time RT-PCR using appropriate primers and normalized to β-actin mRNA levels. Data are expressed as fold-change versus metestrus group. Each column with a bar shows the mean ± SEM. $^*P < 0.05$ versus metestrus; $^{**}P < 0.01$ versus metestrus. E2 serum levels are represented *in italics*.

Estradiol Decreases Aromatase Gene Expression in Pituitary

OVX resulted in a significant increase ($P < 0.001$) in the pituitary expression of aromatase mRNA (FIG. 2). E2 replacement completely reversed the effects of OVX (FIG. 2). Similarly, and in agreement with previous studies, we found that LHβ mRNA increase ($P < 0.001$) by OVX, while E2 replacement totally prevented this effect (FIG. 2). The predictable changes in plasma estradiol levels were observed.

DISCUSSION

Using RT-PCR, we found in the rat pituitary the two species of aromatase RNA, characterized by their specific exon 1 driven by their cognate promotor, namely PII and P1f. Quantification of total aromatase gene expression in normal cycling females demonstrated that the expression is higher during the metestrus stage. Several factors may account for the differential regulation of aromatase in the pituitary across the estrous cycle including ovarian steroids, hypothalamic peptides, and intrapituitary factors. It is well known that the reproductive hormones LH and FSH are under positive and negative regulation by estrogens. Thus, we have examined the effects of castration and E2 replacement. We have shown that ovariectomy increases the expression of aromatase and that this effect is completely prevented by estradiol

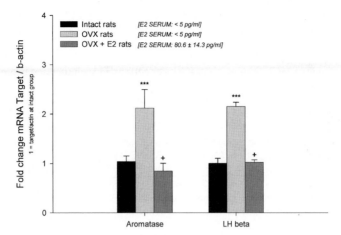

FIGURE 2. Profile of expression of total aromatase in pituitary of adult (90 days old) rats, 3 weeks after OVX with or without estrogen replacement (40 μg/kg). Aromatase mRNA levels were measured by real-time RT-PCR using appropriate primers and normalized to β-actin mRNA levels. The LHβ quantification was used as a control. Data are expressed as fold-change versus metestrus. Each column with a bar shows the mean ± SEM. ***$P < 0.001$ versus metestrus; $^+P < 0.001$ versus OVX. E2 serum levels are represented *in italics*.

administration. Similar results were found for the expression of LHβ.[14–16] There is compelling evidence showing that gonadotropin secretion is regulated by endogenous estradiol at the sites of both hypothalamus and pituitary. Alpha estrogen receptors appeared to play a critical role in the control of pituitary gland and gonadotropin secretion. It was also shown that they can be transcriptionally activated in gonadotrophs.[17] In young men, inhibition of P450 aromatase enhanced gonadotropin secretion at the site of the pituitary.[18] Thus, the mechanisms responsible for pituitary aromatase regulation by estrogens remain to be clarified. Further studies to delineate the respective role of GnRH and estrogen action at the pituitary are in progress.

ACKNOWLEDGMENT

This work was supported by grant from FARO, Oganon, France.

REFERENCES

1. LETTERIE, G.S. *et al.* 1996. Ovulation induction using s.c. pulsatile gonadotrophin-releasing hormone: effectiveness of different pulse frequencies. Hum. Reprod. **11:** 19–22.

2. EYRE, L.J. *et al.* 1998. Characterization of aromatase and 17 beta-hydroxysteroid dehydrogenase expression in rat osteoblastic cells. J. Bone Miner. Res. **13:** 996–1004.

3. KITAWAKI, J. *et al.* 1992. Contribution of aromatase to the deoxyribonucleic acid synthesis of MCF-7 human breast cancer cells and its suppression by aromatase inhibitors. J. Steroid Biochem. Mol. Biol. **42:** 267–277.

4. SANTNER, S.J. *et al.* 1993. Effect of androstenedione on growth of untransfected and aromatase-transfected MCF-7 cells in culture. J. Steroid Biochem. Mol. Biol. **44:** 611–616.

5. YUE, W. *et al.* 1998. In situ aromatization enhances breast tumor estradiol levels and cellular proliferation. Cancer Res. **58:** 927–932.

6. SIMPSON, E.R. *et al.* 1994. Aromatase cytochrome P450, the enzyme responsible for estrogen biosynthesis. Endocr. Rev. **15:** 342–355.

7. SIMPSON, E.R. *et al.* 1997. Cytochromes P450 11: expression of the CYP19 (aromatase) gene: an unusual case of alternative promoter usage. FASEB J. **11:** 29–36.

8. HONDA, S., N. HARADA & Y. TAKAGI. 1996. The alternative exons 1 of the mouse aromatase cytochrome P-450 gene. Biochim. Biophys. Acta **1305:** 145–150.

9. YAMADA-MOURI, N., S. HIRATA & J. KATO. 1996. Existence and expression of the untranslated first exon of aromatase mRNA in the rat brain. J. Steroid Biochem. Mol. Biol. **58:** 163–166.

10. GOLOVINE, K., M. SCHWERIN & J. VANSELOW. 2003. Three different promoters control expression of the aromatase cytochrome p450 gene (cyp19) in mouse gonads and brain. Biol. Reprod. **68:** 978–984.

11. KATO, J., N. YAMADA-MOURI & S. HIRATA. 1997. Structure of aromatase mRNA in the rat brain. J. Steroid Biochem. Mol. Biol. **61:** 381–385.

12. CARRETERO, J. *et al.* 1999. Immunohistochemical evidence of the presence of aromatase P450 in the rat hypophysis. Cell Tissue Res. **295:** 419–423.

13. SCHREIHOFER, D.A., M.H. STOLER & M.A. SHUPNIK. 2000. Differential expression and regulation of estrogen receptors (ERs) in rat pituitary and cell lines: estrogen decreases ERalpha protein and estrogen responsiveness. Endocrinology **141:** 2174–2184.

14. COUNIS, R., M. CORBANI & M. JUTISZ. 1983. Estradiol regulates mRNAs encoding precursors to rat lutropin (LH) and follitropin (FSH) subunits. Biochem. Biophys. Res. Commun. **114:** 65–72.

15. PAPAVASILIOU, S.S. *et al.* 1986. Alpha and luteinizing hormone beta messenger ribonucleic acid (RNA) of male and female rats after castration: quantitation using an optimized RNA dot blot hybridization assay. Endocrinology **119:** 691–698.

16. SHUPNIK, M.A., S.D. GHARIB & W.W. CHIN. 1989. Divergent effects of estradiol on gonadotropin gene transcription in pituitary fragments. Mol. Endocrinol. **3:** 474–480.

17. DEMAY, F. *et al.* 2001. Steroid-independent activation of ER by GnRH in gonadotrope pituitary cells. Endocrinology **142:** 3340–3347.

18. WICKMAN, S. & L. DUNKEL. 2001. Inhibition of P450 aromatase enhances gonadotropin secretion in early and midpubertal boys: evidence for a pituitary site of action of endogenous E. J. Clin. Endocrinol. Metab. **86:** 4887–4894.

PACAP Inhibits Oxidative Stress-Induced Activation of MAP Kinase-Dependent Apoptotic Pathway in Cultured Cardiomyocytes

BALAZS GASZ,[a] BOGLARKA RÁCZ,[a] ERZSEBET RÖTH,[a] BALAZS BORSICZKY,[a] ANDREA TAMÁS,[b] ARPAD BORONKAI,[c] FERENC GALLYAS JR.,[c] GABOR TÓTH,[d] AND DORA REGLŐDI[b]

[a] *Departments of Surgical Research and Techniques, Neurohumoral Regulations Research Group of the Hungarian Academy of Sciences, University of Pécs, 7624 Pécs, Hungary*

[b] *Department of Anatomy, Neurohumoral Regulations Research Group of the Hungarian Academy of Sciences, University of Pécs, 7624 Pécs, Hungary*

[c] *Departments of Biochemistry and Medical Chemistry, University of Pécs, 7624 Pécs, Hungary*

[d] *Department of Medical Chemistry, University of Szeged, Szeged 6701, Hungary*

ABSTRACT: The present article investigated the effect of pituitary adenylate cyclase-activating polypeptide (PACAP) on oxidative stress-induced apoptosis in neonatal rat cardiomyocytes. Our results show that PACAP decreased the ratio of apoptotic cells following H_2O_2 treatment. PACAP also diminished the activity of apoptosis signal-regulating kinase. These effects of PACAP were counteracted by the PACAP antagonist PACAP6-38. In summary, our results show that PACAP is able to attenuate oxidative stress-induced cardiomyocyte apoptosis and suggest that its cardioprotective effect is mediated through inhibition of the MAP kinase-dependent apoptotic pathway.

KEYWORDS: PACAP; apoptosis; cardiomyocyte; apoptosis signal-regulating kinase

INTRODUCTION

It is well known that pituitary adenylate cyclase-activating polypeptide (PACAP) has a protective effect on various cell types against oxidative injury.[1,2] PACAP and its receptors are present in cardiac neurons, cardiac myocytes, and

Address for correspondence: Dora Reglodi, M.D., Ph.D., Department of Anatomy, University of Pécs, 7624 Pécs, Szigeti u 12. Hungary. Voice: +36-72-536001; ext.: 5398; fax: +36-72-536-393.
e-mail: dora.reglodi@aok.pte.hu

Ann. N.Y. Acad. Sci. 1070: 293–297 (2006). © 2006 New York Academy of Sciences.
doi: 10.1196/annals.1317.029

coronary blood vessels.[3] PACAP inhibits protein and DNA synthesis in cardiac fibroblasts, which implies that it may act as a protective mediator against myocardial fibrosis,[4] but direct cardioprotective effect of the peptide has not been investigated yet. Our recent study has shown the protective role of PACAP in cultured cardiomyocytes, where we demonstrated the antiapoptotic effects of PACAP in cardiac cells exposed to oxidative stress.[5] Factors associated with execution of apoptosis, such as caspase-3 and bcl-2 family members were evaluated.[5] Recent studies emphasize the pivotal role of apoptosis signal-regulating kinase 1 (ASK1) in various stress-induced apoptosis.[6] ASK1 appears to be a crucial and early contributor in life and apoptotic death decision.[7] Therefore, we investigated the ability of PACAP to inhibit ASK1-mediated apoptotic pathways.

MATERIALS AND METHODS

Primary culture of neonatal rat cardiomyocyte was prepared as described previously.[5] Briefly, cardiomyocytes were acquired from ventricular slices of 2- to 4-day-old Wistar rats and plated on collagen-coated plates. Cells were incubated in a serum-free medium for 1 day and divided into six groups. The control group was further incubated with medium. Other two groups were treated with 20 nM PACAP-38 or 250 nM PACAP6-38 alone for 4 h. Three additional groups of cells were randomized to receive administration of either 1 mM H_2O_2 alone or H_2O_2 and PACAP-38 (20 nM) or H_2O_2 and PACAP-38 (20 nM) and PACAP6-38 (250 nM) for 4 h. Cells were harvested by trypsin-EDTA and the ratio of apoptotic and intact living cells was evaluated by double staining with fluorescein isothicyanate (FITC)-labeled annexin V and propidium iodide using flow cytometry.[5] Moreover, appearance of phosphorylated ASK1 was quantified using intracellular staining for flow cytometry.[5,8] Samples were measured with a flow cytometer and differences between groups were assessed with analysis of variance (ANOVA).

RESULTS AND DISCUSSION

The distribution of living, apoptotic, and necrotic cells are summarized in TABLE 1. Briefly, administration of PACAP-38 and PACAP6-38 alone did not cause changes in the percentage of living and apoptotic cells compared to control values. H_2O_2 treatment led to a marked increase in the ratio of early apoptotic cells and a reduction in the amount of living cells compared to the untreated group. The data indicate that PACAP-38 reduced the ratio of apoptotic cells and raised that of living cells during H_2O_2 exposure ($P < 0.05$ versus treatment with H_2O_2). The effect of PACAP-38 could be antagonized by coincubation with PACAP6-38, provoking an increased ratio of apoptosis ($P < 0.05$ versus cotreatment with H_2O_2 and PACAP-38). Considerable upregulation of

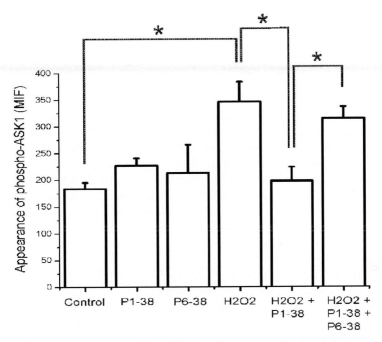

FIGURE 1. Appearance of active ASK1 in cardiomyocytes. Data are presented as mean ± SEM ($n = 6$). P1-38: PACAP-38, P6-38: PACAP6-38. $*P < 0.05$.

ASK1 was observed upon H_2O_2 administration (FIG. 1). PACAP-38 significantly inhibited the H_2O_2-related activation of ASK1 that could be diminished by simultaneous administration of PACAP6-38 (FIG. 1).

The present study showed that PACAP treatment markedly attenuated oxidative stress-induced apoptosis and inhibited ASK1 activation in cardiomyocytes. Apoptosis by oxidative stress has been implicated in several pathological processes in the cardiovascular system: heart failure, ischemia reperfusion injury,

TABLE 1. Percentage of living, apoptotic, and necrotic cells. Data are presented as mean ± SEM ($n = 6$)

	Living cells (%)	Early apoptotic cells (%)	Late apoptotic cells (%)	Necrotic cells (%)
Control	90.49 ± 1.34	5.44 ± 1.65	3.86 ± 0.35	0.36 ± 0.14
PACAP-38	93.88 ± 1.71	3.22 ± 2.06	2.47 ± 0.64	0.355 ± 0.12
PACAP6-38	91.37 ± 3.4	5.69 ± 3.44	2.45 ± 0.23	0.48 ± 0.39
H_2O_2	76.76 ± 3.66	20.61 ± 3.8	2.12 ± 0.34	0.49 ± 0.2
H_2O_2+ PACAP-38	91.06 ± 0.99	6.77 ± 1.17	1.83 ± 0.255	0.28 ± 0.09
H_2O_2+ PACAP-38+ PACAP6-38	81.85 ± 1.29	14.1 ± 1.82	3.1 ± 0.87	1.14 ± 0.42

and aortic cross-clamping during cardiac surgery.[9,10] As an initial event of apoptosis, H_2O_2 dissociates thioredoxin from ASK1, which is identified as an interaction partner of the molecule inhibiting its kinase activity.[6] Thus, activated ASK1 triggers apoptogenic kinase cascade leading to phosphorylation of c-Jun N-terminal kinase (JNK) and p38-MAP kinase, which are responsible for apoptotic cell death. The overexpression of constitutively active ASK1 induces apoptosis and apoptosis induced by various stimulations can be impaired by expression of kinase inactive mutant of Ask and is significantly inhibited in ASK1-deficient mice.[6] These findings suggest that ASK1 plays a pivotal role in stress-mediated and especially oxidative injury-induced apoptosis.[6] Accumulating evidence indicates that sustained, but not transient, activation of JNK and p38 kinase is sufficient for induction of apoptosis.[6] Duration and severity of oxidative stimulus are responsible for cell fate decision of apoptotic death that is mediated through activation state of ASK1 and MAP kinases.[11] Therefore, activation of ASK1 appears to be an initial and decisive step in the execution of apoptosis.

We have previously shown that following H_2O_2 administration, PACAP reduces the activation of caspase-3 and increases the activation of Bcl-2 and phospho-Bad proteins.[5] Activation of proapoptotic members of bcl-2 family proteins indicates stress- or mitochondria-mediated pathway of apoptosis resulting in cytochrome c release and activation of caspase-3 further leading to nuclear degradation and cellular morphological changes.[6] Here, we demonstrated that PACAP can reduce the activation of ASK1 in H_2O_2-treated cardiomyocytes. ASK1 activation is a preceding event of the apoptotic process leading to proapoptotic activation of bcl-2 family proteins and forced release of cytochrome c.[11] The effect of PACAP on ASK1 activation has not yet been reported. It has been shown that PACAP inhibits the activation of JNK and p38 signal transduction pathways.[12] On the other hand, it is known that PACAP treatment can induce activation of survival pathways, such as protein kinase B (Akt) or extracellular signal-regulated kinase (ERK).[2] Through inhibition of ASK1, PACAP may advantageously regulate the dynamic balance between apoptotic (ASK1/JNK, p38) and survival (AKT, ERK) pathways. Further studies are required to clarify the signal pathways on the protective effects of PACAP in cardiomyocytes.

In conclusion, it was observed that cytoprotective, antiapoptotic effects of PACAP is also present in cardiomyocytes. Moreover, the present data suggest that inhibition of ASK1 is involved in PACAP-mediated protection of cardiomyocyte apoptosis.

ACKNOWLEDGMENTS

This work was supported by the Grants OTKA T046589, T048848, T048851, F046504, ETT 596/2003 and the Hungarian Academy of Sciences.

REFERENCES

1. VAUDRY, D. *et al.* 2002. PACAP protects cerebellar granule neurons against oxidative stress-induced apoptosis. Eur. J. Neurosci. **15:** 1451–1560.
2. SOMOGYVARI-VIGH, A. & D. REGLODI. 2004. Pituitary adenylate cyclase activating polypeptide: a potential neuroprotective peptide. Review. Curr. Pharm. Des. **10:** 2861–2889.
3. DEHAVEN, W.I. & J. CUEVAS. 2002. Heterogenity of pituitary adenylate cyclase activating polypeptide and vasoactive intestinal polypeptide receptors in rat intrinsic cardiac neurons. Neurosci. Lett. **328:** 45–49.
4. SANO, H. *et al.* 2002. The effect of pituitary adenylate cyclase activating polypeptide on cultured rat cardiocytes as a cardioprotective factor. Regul. Pept. **109:** 107–113.
5. GASZ, B. *et al.* 2006. Pituitary adenylate cyclase activating polypeptide protects cardiomyocytes against oxidative stress-induced apoptosis. Peptides **27:** 87–94.
6. TAKEDA, K. *et al.* 2003. Roles of MAPKKK ASK1 in stress induced cell death. Cell Struct. Funct. **28:** 23–29.
7. MATSUZAWA, A. *et al.* 2002. Physiological roles of ASK1-mediated signal transduction in oxidative stress- and endoplasmic reticulum stress-induced apoptosis: advanced findings from ASK1 knockout mice. Antioxid. Redox. Signal. **3:** 415–425.
8. SCHROFF, R.W. *et al.* 1984. Detection of intracytoplasmic antigens by flow cytometry. J. Immunol. Methods **70:** 167–177.
9. KUMAR, D. & B.I. JUGDUTT. 2003. Apoptosis and oxidants in the heart. J. Lab. Clin. Med. **142:** 288–297.
10. THATTE, H.S. *et al.* 2004. Acidosis-induced apoptosis in human and porcine heart. Ann. Thorac. Surg. **77:** 1376–1383.
11. YOON, S., C.H. YUN & A.S. CHUNG. 2002. Dose effect of oxidative stress on signal transduction in aging. Mech. Ageing Dev. **123:** 1597–1604.
12. DOHI, K. *et al.* 2002. Pituitary adenylate cyclase activating polypeptide (PACAP) prevents hippocampal neurons from apoptosis by inhibiting JNK/SAPK and p38 signal transduction pathways. Regul. Pept. **109:** 83–88.

Modulation of Pituitary Adenylate Cyclase-Activating Polypeptide (PACAP) Expression in Explant-Cultured Guinea Pig Cardiac Neurons

BEATRICE M. GIRARD, BETH A. YOUNG, THOMAS R. BUTTOLPH, SHERYL L. WHITE, AND RODNEY L. PARSONS

Department of Anatomy & Neurobiology, University of Vermont College of Medicine, Burlington, Vermont 05405, USA

ABSTRACT: Pituitary adenylate cyclase-activating polypeptide (PACAP) expression was quantified in explant-cultured guinea pig cardiac ganglia neurons. In explant culture, both the percentage of PACAP-immunoreactive neurons and pro-PACAP transcript levels increased significantly. Treatment with neurturin or glial-derived neurotrophic factor significantly suppressed the percentage of PACAP-IR neurons, but not pro-PACAP transcript levels.

KEYWORDS: pituitary adenylate cyclase-activating polypeptide; cardiac neurons; trophic factors

INTRODUCTION

In freshly fixed guinea pig cardiac ganglia, few cholinergic neurons are pituitary adenylate cyclase-activating polypeptide (PACAP) immunoreactive, although all the cardiac neurons are innervated by extrinsic PACAP-IR fibers and express PAC1 receptors.[1,2] Following explant-culture, the percentage of cardiac neurons-expressing PACAP increases significantly.[2] The mechanisms regulating PACAP expression in explant cultured cardiac neurons have not been investigated; thus, we quantified PACAP expression and investigated whether trophic factors could modulate PACAP expression in explant-cultured guinea pig cardiac ganglia neurons.

Address for correspondence: Beatrice M. Girard, Ph.D., University of Vermont College of Medicine, Department of Anatomy & Neurobiology, 149 Beaumont Avenue, HSRF 416A, Burlington, VT 05405. Voice: 802-656-0398; fax: 802-656-4674.

e-mail: Beatrice.Girard@uvm.edu

Ann. N.Y. Acad. Sci. 1070: 298–302 (2006). © 2006 New York Academy of Sciences.

doi: 10.1196/annals.1317.030

PREPARATION AND METHODS

Experiments were performed *in vitro* on freshly dissected and explant-cultured cardiac ganglia whole mount preparations of Hartley guinea pigs (mixed sex; 250–400 g) following protocols described previously.[1–3]

For immunocytochemistry, the control and the cultured explanted preparations were fixed in 2% paraformaldehyde containing 0.2% picric acid, immunolabeled and examined with an Olympus AX70 fluorescence photomicroscope (Olympus America Inc., Melville, NY).[1–3] Quantification of PACAP-IR parasympathetic neurons was made from the ratio: ([PACAP-IR neurons/choline acetyl transferase-IR neurons] $\times$ 100 = %).[3]

Standard reverse transcription-polymerase chain reaction (PCR) and quantitative PCR for pro-PACAP transcript levels were performed using total RNA extracted from whole mount ganglia preparations as described previously.[1,3] PCR was conducted using the primer templates for guinea pig pro-PACAP transcripts (FIG. 1A) designed according to the sequence previously published.[1]

RESULTS AND DISCUSSION

In freshly fixed preparations only $1.3 \pm 0.4\%$ of the cholinergic cardiac neurons were PACAP immunoreactive (FIG. 1C). However, after 48–72 h in explant culture, the percentage of PACAP-IR neurons increased $\sim$20-fold (FIG. 1A, C). In two whole mount preparations maintained in explant culture for 72 h without added horse serum, the percentage of PACAP-IR cells was 24% and 22%. These values were similar to the average percentage value for cardiac ganglia whole mounts maintained for 72 h in culture media with added serum ($24.5 \pm 4.7\%$, $n = 3$ preparations). Thus, the increase in PACAP expression was not dependent on the presence of serum-derived factors. In explant cultured cardiac neurons, PACAP immunoreactivity was not restricted to the cell body, but was evident in fibers within nerve bundles and in regenerating axons extending out from cut ends of nerve trunks (FIG. 2B).

Transcript levels were determined in total RNA extracts from freshly dissected and from preparations explant cultured for 72 h. The number of cardiac neurons within each whole mount can vary and total extracted RNA reflects RNA from multiple cell types in the whole mount preparation. Thus, to compare results between preparations, the pro-somatostatin transcript level was used to determine relative numbers of cardiac neurons in each cardiac ganglia whole mount preparation extracted for PCR analysis.[3] The primers to quantify guinea pig pro-somatostatin transcript levels were those used in our recent study.[3] Results from both semiquantitative and quantitative PCR demonstrated that pro-PACAP transcript levels were present at low levels in freshly dissected cardiac ganglia, but increased over time in cardiac ganglia preparations maintained in explant culture (FIG. 1C, D). In contrast, the expression

(A)

Oligo	Sequence	Position	T$_a$	Size
gp PACAP	5'-TCGGACGGAATCTTCACA-3'	4U18	60°C	93 bp
gp PACAP	5'-TTTATACCTTTTCCCCAGGA-3'	77L20		

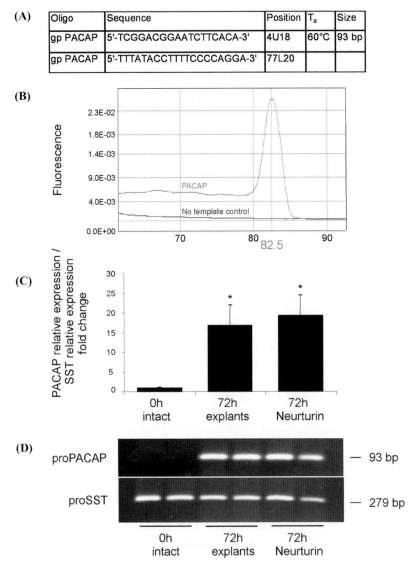

FIGURE 1. PACAP mRNA expression in guinea pig cardiac ganglia preparations. (**A**) Table of the PACAP primers specific for guinea pigs. (**B**) SYBR Green I melting analysis of the amplified product demonstrated a single unique dissociation curve to validate the amplification parameters. (**C**) Explants cultured for 72 h increased in PACAP mRNA expression compared to intact ganglia while neurturin treatment for 72 h did not significantly alter PACAP gene expression. RNA extracted from each cardiac ganglia explant was reverse transcribed for quantitative and semiquantitative RT-PCR. Amplification of the same templates for SST is shown to demonstrate similar levels of neuronal cDNA in the reaction mixtures. Data for each condition represent the mean of five individual explants ± SEM ($P < 0.001$).

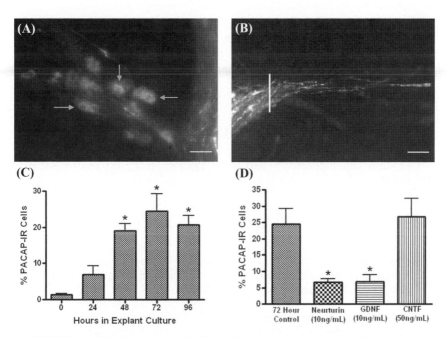

FIGURE 2. PACAP expression in 72-h explant-cultured cardiac ganglia preparations. (**A**) Many cardiac neurons are immunoreactive for PACAP (examples indicated by arrows). (**B**) Regenerating PACAP-IR fibers extend out of a cut (indicated by vertical line) nerve bundle. (**C**) The increase in the percentage of PACAP-IR neurons increases significantly (indicated by asterisks) over time in explant culture. (**D**) Neurturin and GDNF, but not CNTF, treatment significantly decreased (indicated by asterisks) the percentage of PACAP-IR cells in 72-h explant-cultured cardiac ganglia. The calibration bar equals 50 μm.

of pro-somatostatin transcript levels was not different in extracts from freshly dissected or cultured cardiac ganglia preparations (FIG. 2D).

Treatment of 72-h explant-cultured cardiac ganglia preparations with human neurturin (10 ng/mL) or human glial-derived neutrophic factor (GDNF) (10 ng/mL) significantly diminished the percentage of PACAP-IR neurons (FIG. 1D). This was not a nonspecific effect as treatment with recombinant rat ciliary neurotrophic factor (CNTF, 50 ng/mL) had no affect on the percentage of PACAP-IR cells (FIG. 1D). Although the percentage of PACAP-IR cells decreased after 72 h in explant culture, PCR analysis indicated that treatment with 10 ng/mL neurturin did not significantly change pro-PACAP transcript levels (FIG. 2).

PACAP peptides are potent trophic and intercellular signaling molecules. It is commonly considered that increased PACAP expression following injury may be important in neuronal survival and regeneration.[4,5] An increase in PACAP expression during explant culture may be a protective mechanism generated in response to cardiac neuron injury. The decrease in explant-cultured PACAP-IR cardiac neurons cells by neurturin or GDNF without an accompanying decrease in pro-PACAP transcript levels suggests that these trophic molecules either alter translational mechanisms or enhance peptide secretion and thereby diminish cellular peptide levels.[3]

ACKNOWLEDGMENTS

This study was supported in part by NIH grants HL-65481 and NCRR P20 RR-16435.

REFERENCES

1. BRAAS K.M., V. MAY, S.A. HARAKALL, et al. 1998. Pituitary adenylate cyclase-activating polypeptide expression and modulation of neuronal excitability in guinea pig cardiac ganglia. J. Neurosci. **18:** 9766–9779.
2. CALUPCA M.A., M.A. VIZZARD & R.L. PARSONS. 2000. Origin of pituitary adenylate cyclase-activating polypeptide (PACAP)-immunoreactive fibers innervating guinea pig parasympathetic cardiac ganglia. J. Comp. Neurol. **423:** 26–39.
3. BRAAS K.M., T.M. ROSSIGNOL, B.M. GIRARD, et al.2004. Pituitary adenylate cyclase activating polypeptide (PACAP) decreases neuronal somatostatin immunoreactivity in cultured guinea-pig parasympathetic cardiac ganglia. Neuroscience **126:** 335–346.
4. ARIMURA A. 1998. Perspectives on pituitary adenylate cyclase activating polypeptide (PACAP) in the neuroendocrine, endocrine, and systems. Jpn J. Physiol. **48:** 301–331.
5. VAUDRY D., B.J. GONZALEZ, M. BASILLE, et al. 2000. Pituitary adenylate cyclase-activating polypeptide and its receptors: from structure to functions. Pharmacol. Rev. **52:** 269–324.

VIP: An Agent with License to Kill Infective Parasites

ELENA GONZALEZ-REY, ALEJO CHORNY, AND MARIO DELGADO

Instituto de Parasitologia y Biomedicina, CSIC, Granada 18100, Spain

ABSTRACT: Antimicrobial peptides are small, cationic, and amphipathic peptides of variable length, sequence, and structure. They are effector molecules of innate immunity with microbicidal and both pro- or anti-inflammatory activities. Vasoactive intestinal polypeptide (VIP) and the structurally related pituitary adenylate cyclase-activating polypeptide (PACAP) are well-known immunomodulators. On the basis of their cationic and amphipathic structures, resembling antimicrobial peptides, we propose that their immune role could also include a direct lethal effect against pathogens. We thus investigated the potential antiparasitic activities of VIP and PACAP against the African trypanosome *Trypanosoma brucei (T. brucei)*. Both peptides killed the bloodstream (infective) form but not the insect (noninfective) form of the parasite. VIP and PACAP caused complete destruction of the parasite integrity through a mechanism involving their entry and accumulation into the cytosol. These results provide the basis for further studies of these and other structurally related peptides as alternative treatments for parasitic diseases mainly with associated drug resistances.

KEYWORDS: neuropeptides; parasitic diseases; trypanosome; antiparasitic agent

INTRODUCTION

Membrane-disrupting peptides are found in a wide range of eukaryotic organisms, where they act as important components of the innate immune system during the early host defenses against invading pathogens. Despite their broad structural diversity, these peptides share common properties, including small size (< 10 kDa), high positive charge, and amphipathic α-helix and/or β-sheet structures adopted upon interaction with membranes.[1] A direct antimicrobial action of these peptides has been increasingly recognized lately.[2] Although the exact mechanism of action of these peptides is a matter of controversy, most of

Address for correspondence: Elena González-Rey, Instituto de Parasitologia y Biomedicina, CSIC, Avd. Conocimiento, PT Ciencias de la Salud, Granada 18100, Spain. Voice: 34-958-181665; fax: 34-958-181632.

e-mail: elenag@ipb.csic.es

Ann. N.Y. Acad. Sci. 1070: 303–308 (2006). © 2006 New York Academy of Sciences.
doi: 10.1196/annals.1317.032

them lyse the cell by disrupting the plasma membrane. Cationic amphipathic peptides target specific membranes based on differences in composition, fluidity, and membrane potential. These peptides seem to preferentially recognize and interact with the anionic phospholipids exposed on the outer surface of the bacterial membrane, but not on the mammalian cell membrane.[3,4] Antimicrobial peptides are produced by tissues and cells primarily involved in contact with microorganisms, such as mucosal, epithelial, or phagocytic cells.[5] Their role as modulators of the inflammatory process has received special attention lately.[5] In contrast to the increasing evidences about the antibacterial and antifungal activities of these cationic peptides, few reports describe their activities and mode of action against protozoan and metazoan parasites.

Trypanosoma brucei (*T. brucei*) is a protozoan parasite responsible for significant morbidity and mortality in both humans (sleeping sickness, with an estimated risk of infection for 60 million people) and animals (nagana).[6] The course of infection is characterized by parasitic waves corresponding with the antigenic variation of the parasite, and the untreated infection is lethal. However, most drugs used for its treatment are antiquated, ineffective, toxic, and susceptible to resistance by the parasite.[7] *T. brucei* alternately infects tsetse fly vectors (procyclic form) and mammalian hosts (bloodstream form), showing the two life-cycle stages of the parasite as unique properties of their surface membranes. Both parasite forms present a high concentration of membrane proteins anchored through glycosylphosphatidylinositol (GPI), although their relative abundance and structure differ in the two forms.[8-10] The bloodstream form is abundantly covered by a layer of the GPI-variant surface glycoprotein (VSG) that confers it a highly negative glycocalix.[11] In contrast, the procyclic form expresses a different surface coat, consisting of an abundant GPI-anchored protein called "procyclin."[12]

To date, vasoactive intestinal polypeptide (VIP) and the structurally related pituitary adenylate cyclase-activating polypeptide (PACAP) have been described as immunomodulatory factors involved in the maintenance of the immune system homeostasis. On the basis of their characteristics as amphipathic α-helix cationic peptides resembling antimicrobial peptides,[13] we investigated the role of these neuropeptides as antiparasitic agents.

RESULTS AND DISCUSSION

The susceptibility of *T. brucei* to VIP/PACAP was examined by the incubation of different numbers of parasites with the neuropeptides for 48 h at either 28°C for the procyclic forms or 37°C for the bloodstream forms. The effects of the peptides in parasite viability were assessed by MTT reduction assay and by microscopic examination. Controls consisted of parasites incubated with medium alone. We observed that a single administration of VIP or PACAP (3 μM) induced a mortality of 50% in the bloodstream forms, in comparison with controls without peptides. The parasiticidal effect of VIP and PACAP

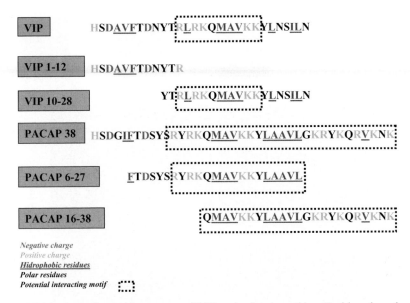

FIGURE 1. Amino acid sequence of VIP and related peptides. Residues have been labeled according to their hidrophobicity and charge as shown in the legend.

was markedly increased after a second administration of the neuropeptides 24 h later (approximately 82% mortality). Interestingly, the noninfective procyclic forms were resistant to VIP and PACAP treatment. The different susceptibility between the mammalian and the insect forms to VIP and PACAP could be on account of their very different surface properties, mainly based on the different amounts and distribution of their membrane-negative charges.

In order to determine the exact region of these peptides responsible of the antiparasitic effect, we tested different fragments of VIP and PACAP in the same conditions as described above. Parasites incubated with PACAP6–27, PACAP16–38, or VIP10–28 showed a mortality of 73%, 94%, and 95%, respectively. In contrast, VIP1–12 had no parisiticidal effect. Based on the amino acid sequence of these peptides shown in FIGURE 1, we hypothesized that only the region of the neuropeptides with a rich combination of positively charged amino acids and hydrophobic residues are determinants in their antiparasitic effect. Similarly, other neuropeptides with antiparasitic capacity, including adrenomedullin, urocortin, alpha-melanocyte stimulating-hormone, and ghrelin, have also shown this dependence on cationic–hydrophobic residue combination.

We next investigated the morphological changes in the structural integrity of the infective forms of the parasites caused by VIP and PACAP. Microscopic analysis showed that parasites treated with VIP or PACAP loss their normal shape, swelling, and exhibited a more-rounded morphology, appearing as "ghost-like" forms (FIG. 2 A). This is in agreement with the hypothesis that

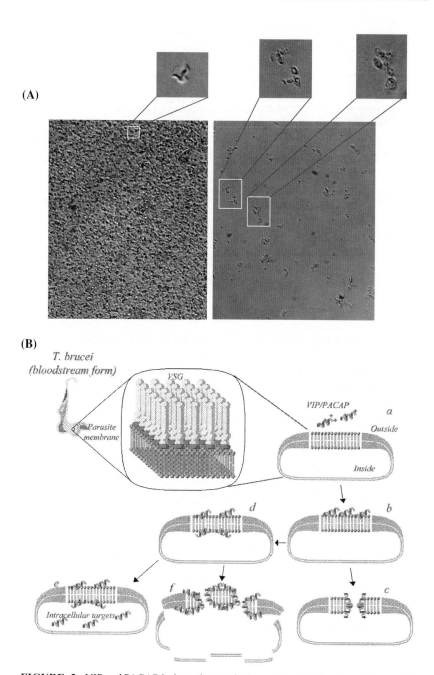

FIGURE 2. VIP and PACAP induce changes in the membrane integrity of the parasites. (**A**) Micrographs of untreated (left) or VIP-treated (3 μM, right) *T. brucei* after 48 h of culture. (**B**) Diagram showing the potential mechanisms involved in the antiparasitic action of VIP/PACAP.

these peptides could be acting by disrupting the membrane integrity. Different models have been proposed to explain the lytic peptide–membrane interactions. The Shai-Matsuzaki-Huang model (FIG. 2 B) shows that, after the first interaction between the peptide and the membrane in a carpet-like manner (FIG. 2 B-b), the peptide can integrate into the membrane expanding the outer leaflet. This can be derived in the formation of transient pores (FIG. 2 B-C). Alternatively, interaction of the peptide with the membrane can cause displacement of lipids and subsequent alteration of the membrane. This can provoke the collapse of the membrane into fragments (FIG. 2 B-d, B-f). In addition, the peptides can diffuse and react with potential intracellular targets (FIG. 2 B-e). Thereafter, disruption of the membrane integrity probably leads to osmotic instability, causing cell swelling and eventual lysis. In this sense, by using a fluorescein-labeled VIP, we have observed the accumulation of the peptide in intracellular vesicles in the infective forms. The neuropeptide accumulation in the parasitic cytosol could be subsequent to the peptide-mediated membrane integrity alteration and/or direct diffusion of the peptide from the inner leaflet of the membrane.

Therefore, we can hypothesize that treatment with VIP or PACAP of the infective bloodstream forms of *T. brucei* results in the killing of the parasites, by disrupting the parasite membrane. However, the fact that VIP accumulates into the cytosol of the parasite in round vesicles suggests that the antiparasitic effects could be also mediated by additional mechanisms, including autophagy and apoptosis-like programmed cell death, although these possibilities have to be adequately addressed. The membrane-lytic effect of VIP/PACAP makes the development of resistance by the parasite difficult, and points them as potential candidates for future design of antiparasitic drugs with a new mode of action. The new role given in this study to VIP and PACAP as ancient elements of the innate host defenses, in combination with their immunomodulatory effects, could be extremely useful in disorders in which inflammation and infection coexist.

ACKNOWLEDGMENTS

This work was supported by the Spanish Ministry of Health PI04/0674 and Ramon Areces Foundation. We would like to thank Francisco Jose Sánchez-Luque for his help in preparation of the figure.

REFERENCES

1. ZASLOFF, M. 2002. Antimicrobial peptides of multicellular organisms. Nature **415:** 389–395.
2. REDDY, K.V.R., R.D. YEDERY & C. ARANHA. 2004. Antimicrobial peptides: premises and promises. Int. J. Antimicrob. Agents **24:** 536–547.

3. TYTLER, E.M., G.M. ANANTHARAMAIAH, D.E. WALKER, *et al.* 1995. Molecular basis for prokaryotic specificity of magainin induced lysis. Biochemistry **34:** 4393–4401.
4. PAPO, N., M. SHAHAR, L. EISENBACH & Y. SHAI. 2003. A novel lytic peptide composed of DL aminoacids selectively kills cancer cells in culture and in mice. J. Biol. Chem. **278:** 21018–21023.
5. HANCOCK, R.E.W. & G. DIAMOND. 2000. The role of cationic antimicrobial peptides in innate host defences. Trends Microbiol. **8:** 402–409.
6. WORLD HEALTH ORGANIZATION (WHO). Control and surveillance of African trypanosomiasis: report of a WHO expert committee. 2001. World Health Organ Fact Sheet n° 259.
7. KEISER, J., A. STICH & C. BURRI. 2001. New drugs for the treatment of human African trypanosomiasis: research and development. Trends Parasitol. **17:** 42–49.
8. FIELD, M.C., A.K. MENON & G.A. CROSS. 1992. Developmental variation of glycosylphosphatidylinositol membrane anchors in *Trypanosoma brucei*: in vitro biosynthesis of intermediates in the construction of the GPI anchor of the major procyclic surface glycoproteins. J. Biol. Chem. **267:** 5324–5329.
9. HOMANS, S.W., C.J. EDGE, M.A. FERGUSON, *et al.* 1989. Solution structure of the glycosylphosphatidylinositol membrane anchor glycan of *Trypanosoma brucei* variant surface glicoprotein. Biochemistry **28:** 2881–2887.
10. PATURIAUX-HANOCQ, F., N. ZITZMANN, J. HANOCQ-QUERTIER, *et al.* 1997. Expression of a variant surface glycoprotein of *Trypanosoma gambiense* in procyclic forms of *Trypanosoma brucei* shows that the cell type dictates the nature of the glycosylphosphatidylinositol membrane anchor attached to the glycoprotein. Biochem. J. **324:** 885–895.
11. CARDOSO DE ALMEIDA, M.L. & M.J. TURNER. 1983. The membrane form of variant surface glicoproteins of *Trypanosoma brucei*. Nature **302:** 349–352.
12. ACOSTA-SERRANO, A., R.N. COLE, A. MEHLERT, *et al.* 1999. The procyclin repertoire of *Trypanosoma brucei*: identification and structural characterization of the Glu-Pro-rich polypeptides. J. Biol. Chem. **274:** 29763–29771.
13. VAUDRY, D., B.J. GONZALEZ, M. BASILLE, *et al.* 2000. Pituitary adenylate cyclase-activating polypeptide and its receptors: from structure to function. Pharmacol. Rev. **52:** 269–324.

PACAP Stimulates the Release of the Secretogranin II-Derived Peptide EM66 from Chromaffin Cells

JOHANN GUILLEMOT,[a] DJIDA AIT-ALI,[a] VALERIE TURQUIER,[a] MAITE MONTERO-HADJADJE,[a] ALAIN FOURNIER,[b] HUBERT VAUDRY,[a] YOUSSEF ANOUAR,[a] AND LAURENT YON[a]

[a]INSERM U413, Laboratory of Cellular and Molecular Neuroendocrinology, European Institute for Peptide Research (IFRMP 23), University of Rouen, 76821 Mont-Saint-Aignan, France

[b]INRS–Institut Armand Frappier, University of Quebec, Pointe-Claire, Canada

ABSTRACT: The aim of the present article was to examine the effect of PACAP on the release of the SgII-derived peptide EM66 from primary cultures of bovine chromaffin cells. PACAP dose dependently stimulated EM66 release from cultured chromaffin cells. A significant response was observed after 6 h of treatment with PACAP and increased to reach a 3.6-fold stimulation at 72 h. The stimulatory effect of PACAP was mediated through multiple signaling pathways, including calcium influx through L-type channels, PKA, PKC, and MAP-kinase cascades, to regulate EM66 release from chromaffin cells. These data suggest that EM66 may act downstream of the trans-synaptic stimulation of the adrenal medulla by neurocrine factors.

KEYWORDS: bovine chromaffin cells; EM66; PACAP; secretogranin II; transduction pathways

INTRODUCTION

Chromogranins (Cgs) are acidic proteins present in dense core vesicles of adrenal medullary cells, where they are co-stored and co-released with catecholamines. The presence in their sequences of multiple pairs of basic amino acids that are potential cleavage motifs for prohormone convertases, suggests that Cgs may serve as precursors to generate biologically active peptides. Consistent with this notion, the secretogranin II (SgII)-derived peptide secretoneurin (SN)[1] has been shown to induce dopamine release from the rat

Address for correspondence: Laurent Yon, INSERM U413, Laboratory of Cellular and Molecular Neuroendocrinology, European Institute for Peptide Research (IFRMP23), University of Rouen, 76821 Mont-Saint-Aignan, France. Voice: (33) 235-14-6945; fax: (33) 235-14-6946.
e-mail: laurent.yon@univ-rouen.fr

Ann. N.Y. Acad. Sci. 1070: 309–312 (2006). © 2006 New York Academy of Sciences.
doi: 10.1196/annals.1317.033

striatum and to exert chemotactic activity for monocytes.[2] SgII is also the precursor of a SN-flanking peptide termed EM66 that has been highly conserved during evolution.[3] EM66 is present during early development in the fetal human adrenal gland[4] and, in the rat adrenal medulla, EM66 is localized in secretory granules of adrenergic cells.[5] These observations suggest that EM66 may be released upon stimulation of endocrine cells to act as an autocrine, paracrine, or endocrine factor. Recently, it has been shown that pituitary adenylate cyclase-activating polypeptide (PACAP), a potent regulator of chromaffin cell activity, regulates SgII gene transcription and SN release in adrenochromaffin cells.[6] The aim of the present article was to examine the effect of PACAP on EM66 release from cultured bovine chromaffin cells.

MATERIALS AND METHODS

Primary cultures of bovine adrenochromaffin cells were obtained after retrograde perfusion of bovine adrenal glands with 0.1% collagenase and 30 U/mL DNase I, followed by mechanical dissociation of the digested adrenal medulla. The isolated cells were cultured in DMEM supplemented with 5% calf serum, 100 U/mL penicillin, 100 μg/mL streptomycin, and 0.25 μg/mL fungizone. Chromaffin cells were purified by differential plating to remove adherent contaminating cells, and then plated at a density of 10^6 cells/mL in poly-L-lysine-coated 24-well plates. After a resting period of 1 day, chromaffin cells were incubated for 24 h with PACAP-38 in the absence or presence of the nonspecific protein kinase inhibitor H7 (100 μM), the protein kinase A (PKA) inhibitor H89 (20 μM), the protein kinase C (PKC) inhibitor chelerythrine (5 μM), the nonselective calcium channel blocker $NiCl_2$ (3 mM), the cytosolic calcium chelator BAPTA-AM (50 μM), the L-type calcium channel blocker nimodipine (10 μM), the p42/44 ERK 1/2 MAPK inhibitor U0126 (10 μM), and the p38 MAPK inhibitor SB203580 (10 μM). At the end of the incubation period, aliquots of culture medium were taken and immediately frozen at $-20°C$ until RIA determination of EM66.[4]

RESULTS AND DISCUSSION

PACAP-38 induced a dose-dependent stimulation of EM66 release (up to 2.5-fold) with a half-maximal effect at a concentration of 4.8 nM. Kinetic experiments performed with a dose of 50 nM of PACAP-38 showed that the peptide significantly stimulated EM66 release after 6 h of treatment (FIG. 1A). The effect of PACAP-38 increased gradually to reach a 3.6-fold stimulation after 72 h and then markedly diminished (FIG. 1A). We have recently shown that PACAP stimulates SN release within 5 min and that the effect peaks at 16 h.[6] Altogether, these data indicate that PACAP exerts long-lasting effects on

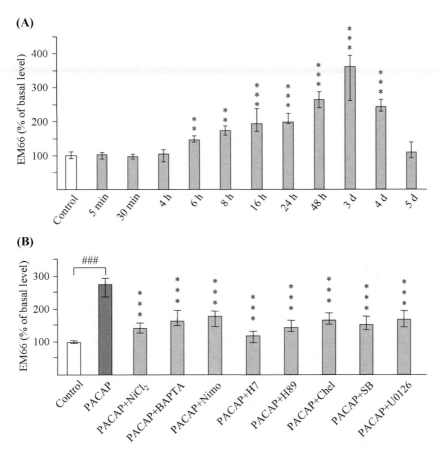

FIGURE 1. (**A**) Time course of PACAP-induced EM66 release. Chromaffin cells were incubated for 24 h in the absence (control) or presence of 50-nM PACAP-38 for various durations. Values represent the median and 25th to 75th percentile of the EM66 concentration in the culture medium, expressed as percentages of basal level (corresponding controls). $**P < 0.01$, $***P < 0.001$ versus the corresponding control. (**B**) Effect of various inhibitors on PACAP-induced EM66 release. Chromaffin cells were incubated for 24 h in control conditions or with 50-nM PACAP, in the absence or presence of 3-mM $NiCl_2$, 50-μM BAPTA, 10-μM nimodipine (Nimo), 100-μM H7, 20-μM H89, 5-μM chelerythrine (Chel), 10-μM SB203580 (SB), or 10-μM U0126. Values represent the median and 25th to 75th percentile of EM66 concentration in the culture medium, expressed as percentages of basal level (corresponding controls). $###P < 0.001$, PACAP versus control; $***P < 0.001$, inhibitor versus PACAP alone.

the release of sg II-derived peptides in chromaffin cells, suggesting that, under PACAP action, adrenomedullary cells may continuously release several neuropeptides that can exert various regulatory effects. To determine the transduction pathways involved in the effect of PACAP on EM66 secretion, we applied calcium and kinase blockers. The cytosolic calcium chelator BAPTA-AM, the nonselective calcium channel blocker, $NiCl_2$, and the L-type calcium channel blocker, nimodipine, significantly reduced the stimulatory effect of PACAP on EM66 release, suggesting that calcium mobilization is required for PACAP-induced EM66 secretion (FIG. 1B). The nonselective protein kinase inhibitor H7 reduced by 90% the effect of PACAP on EM66 release (FIG. 1B). The action of PACAP was also reduced significantly by H89 and by chelerythrine (FIG. 1B), indicating that both PKA and PKC are involved in PACAP-evoked stimulation of EM66 secretion. Finally, the effect of PACAP was reduced by the MAP-kinase inhibitors SB203580 and U0126 (FIG. 1B), indicating that ERK 1/2 and p38 are also implicated in PACAP stimulation of EM66 release.

Collectively, these data indicate that PACAP activates multiple signaling pathways to regulate EM66 release from medullary chromaffin cells, suggesting that EM66 may act downstream of the trans-synaptic stimulation of the adrenal medulla by neurocrine factors.

ACKNOWLEDGMENTS

This work was supported by INSERM (U 413), the European Institute for Peptide Research (IFRMP 23), and the Conseil Régional de Haute-Normandie. JG is the recipient of a doctoral fellowship from the Conseil Régional de Haute-Normandie. HV is affiliated professor at INRS—Institut Armand Frappier.

REFERENCES

1. VAUDRY, H. & J.M. CONLON. 1991. Identification of a peptide arising from the specific post-translation processing of secretogranin II. FEBS Lett. **284:** 31–33.
2. WIEDERMANN, C.J. 2000. Secretoneurin: a functional neuropeptide in health and disease. Peptides **21:** 1289–1298.
3. ANOUAR, Y., S. JEGOU, D. ALEXANDRE, *et al.* 1996. Molecular cloning of frog secretogranin II reveals the occurrence of several highly conserved potential regulatory peptides. FEBS Lett. **394:** 295–299.
4. ANOUAR, Y., C. DESMOUCELLES, L. YON, *et al.* 1998. Identification of a novel secretogranin II-derived peptide [SgII(187–252)] in adult and fetal human adrenal glands using antibodies raised against the human recombinant peptide. J. Clin. Endocrinol. Metab. **83:** 2944–2951.
5. MONTERO-HADJADJE, M., G. PELLETIER, L. YON, *et al.* 2003. Biochemical characterization and immunocytochemical localization of EM66, a novel peptide derived from secretogranin II, in the rat pituitary and adrenal glands. J. Histochem. Cytochem. **51:** 1083–1095.
6. TURQUIER, V., L. YON, L. GRUMOLATO, *et al.* 2001. Pituitary adenylate cyclase-activating polypeptide stimulates secretoneurin release and secretogranin II gene transcription in bovine adrenochromaffin cells through multiple signaling pathways and increased binding of pre-existing activator protein-1-like transcription factors. Mol. Pharmacol. **60:** 42–52.

New Nonradioactive Technique for Vasoactive Intestinal Peptide-Receptor-Ligand-Binding Studies

INES HABERL,[a] DANIELA SUSANNE HABRINGER,[a] FRITZ ANDREAE,[b] ANDREAS ARTL,[b] AND WILHELM MOSGOELLER[a]

[a]*Department of Internal Medicine I, Division: Institute of Cancer Research, Medical University of Vienna, Borschkegasse 8a, A-1090 Vienna, Austria*

[b]*piCHEM Research & Development, Kahngasse 20, A-8045 Graz, Austria*

ABSTRACT: We describe fluorescent-labeled peptide (FLP) studies on living cells. The new technique is nonradioactive and it allows monitoring of the binding and internalization of Vasoactive Intestinal Peptide (VIP) in VIP receptor-expressing cells. The technique is easy to perform and the observed reaction is peptide sequence specific.

KEYWORDS: VIP; peptide internalization; nonradioactive; fluorescence; FLP

INTRODUCTION

The intracellular tracing of peptides like Vasoactive Intestinal Peptide (VIP) has been investigated by several groups using radiolabeled ^{125}I-VIP.[1,2] We now present a new and easy method that uses fluorescence-labeled peptides (FLP) to observe VIP binding and internalization into living cells, and provides an attractive alternative that is less sensitive but environmentally safe, and fast. Using FLP allows for real-time imaging of living cells.[3] It eliminates fixation artifacts.[4] Confocal laser scanning microscopy is superior to flow cytometry analysis, and it may discriminate labeled ligands on the cell surface from cytoplasmatic (internalized) peptides. We used pulmonary artery smooth muscle cells (PASMCs) that natively express both types of VIP receptors.[5]

Address for correspondence: Wilhelm Mosgoeller, Department of Internal Medicine I, Division: Institute of Cancer Research, Medical University of Vienna, Borschkegasse 8a, A-1090 Vienna, Austria. Voice: +43-1-4277-65260; fax: +43-1-4277-65264.
e-mail: wilhelm.mosgoeller@meduniwien.ac.at

Ann. N.Y. Acad. Sci. 1070: 313–316 (2006). © 2006 New York Academy of Sciences.
doi: 10.1196/annals.1317.034

MATERIALS AND METHODS

VIP was labeled with Cy3 on the amino terminus (Cy3VIP) or on the carboxy-terminus (VIP-Cy3, piCHEM, Graz, Austria). PASMCs (Cell Systems, St. Katharinen, Germany) were grown on coverslips. Following PBS rinses, unspecific binding was blocked by fetal calf serum (10%). The incubation with FLP varied in concentration, time, and temperature as indicated. Cell nuclear counterstaining was done simultaneously with TO-PRO®-3 iodide (Molecular Probes, Leiden, the Netherlands; 1:1000) under dimmed light conditions. The washed coverslips were mounted in antifade medium (Citifluor; Gröpel, Tulln, Austria) and viewed immediately on a confocal laser scanning microscope (Leica Microsystems, Wetzlar, Germany). For specificity controls, cells were incubated with unlabeled VIP or a labeled peptide termed scrambled VIP (ScrVIP-Cy3). Scr-VIP contains all amino acids of VIP in equimolar amounts but a randomized sequence. To control for peptide unrelated staining we incubated the cells with Cy3 alone.

RESULTS AND DISCUSSION

This may be the first report that VIP internalization is directly observable in living cells by FLP.

Internalization of Labeled VIP is Peptide Specific

The labeling, either at the amino-terminus or the carboxy-terminus appeared to play a role for the signal- and ligand-binding intensity (compare FIG. 1A, B), emphasizing the importance of the free amino-terminus.

The signal distribution and internalization process is concentration dependent (compare FIG. 1B, C). The observable nuclear signal in FIGURE 1C most

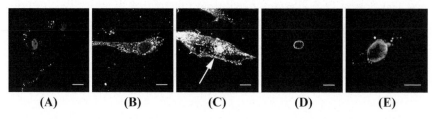

(A)   (B)   (C)   (D)   (E)

FIGURE 1. FLP studies with Cy3-labeled VIP. Cells were incubated at 37°C with 1 μM Cy3VIP (**A**), 1μM VIP-Cy3 (**B**), or 5μM VIP-Cy3 (**C**), for 30 min. For controls, cells were incubated with unlabeled VIP (**D**) or ScrVIP-Cy3 (**E**). Incubation with 5μM VIP-Cy3 (**C**) overloads the cell nucleus (arrow). Note that ScrVIP-Cy3 is not internalized and the binding and internalization of VIP-Cy3 is peptide specific (Bar = 8 μM).

likely corresponds to a peptide overload. It may be a technology-inherent artifact, not observable with radiolabeled peptides.[1,2] We cannot exclude that degraded fluorescent probe reaches the nucleus. In our experimental series with VIP-Cy3 and PASMCs, 1 μM of peptide concentration appeared as optimal condition. No signal was observed with unlabeled VIP (FIG. 1D), or with co-incubation with a 10-fold access of unlabeled VIP (not shown). Occasionally, insignificant labeling was observed after incubation with ScrVIP-Cy3 (FIG. 1E). Cy3 alone (data not shown) revealed no signal. The functional properties of VIP were conserved when using VIP-Cy3, as observed by the known vasorelaxing effect on arteries (data not shown, personal communication: A. Shahbazian). We conclude that the observed fluorescence signal and the translocation to the cytoplasm is peptide specific.

Internalization in Dependence on Time and Temperature

FIGURE 2 shows cells incubated with 1 μM VIP-Cy3 at 4°C, 20°C, or 37°C for times between 15 and 60 min. At 4°C, the ligand remains mainly at the cell surface, indicating ligand membrane binding without internalization. At 20°C, the

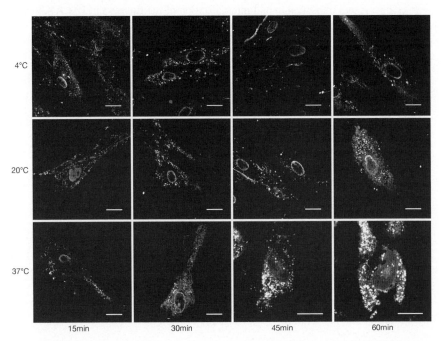

FIGURE 2. Internalization in a time course at 4°C, 20°C, and 37°C. PASMCs were incubated for 15, 30, 45, and 60 min with 1μM VIP-Cy3 and TO-PRO®-3 at indicated temperatures (Bar = 8 μM).

fluorescence inside the cytoplasm increased with time. Rapid internalization occurred within 15 min at 37°C. The end of the time course is characterized by a punctate fluorescense accumulation within the cytoplasm and also the nuclear region. These results are in agreement with previous data using ^{125}I-VIP for internalization studies in HT-29 colon carcinoma cell lines.[6]

In conclusion, FLP studies allow for a very fast and elegant screening method for binding and internalization (assisting tests for biological activity) of VIP and analogues. FLPs may therefore assist the function screening of other peptides.

REFERENCES

1. COUVINEAU, A., C. ROUYER-FESSARD, A. FOURNIER, et al. 1984. Structural requirements for VIP interaction with specific receptors in human and rat intestinal membranes: effect of nine partial sequences. Biochem. Biophys. Res. Commun. **121:** 493–498.
2. ROBBERECHT, P., D.H. COY, P. DE NEEF, et al. 1986. Interaction of vasoactive intestinal peptide (VIP) and N-terminally modified VIP analogs with rat pancreatic, hepatic and pituitary membranes. Eur. J. Biochem. **159:** 45–49.
3. GERMANO, P.M., J. STALTER, S.V. LE, et al. 2001. Characterization of the pharmacology, signal transduction and internalization of the fluorescent PACAP ligand, fluor-PACAP, on NIH/3T3 cells expressing PAC1. Peptides **22:** 861–866.
4. LUNDBERG, M. & M. JOHANSSON. 2002. Positively charged DNA-binding proteins cause apparent cell membrane translocation. Biochem. Biophys. Res. Commun. **291:** 367–371.
5. PETKOV, V., W. MOSGOELLER, R. ZIESCHE, et al. 2003. Vasoactive intestinal peptide as a new drug for treatment of primary pulmonary hypertension. J. Clin. Invest. **111:** 1339–1346.
6. HEJBLUM, G., P. GALI, C. BOISSARD, et al. 1988. Combined ultrastructural and biochemical study of cellular processing of vasoactive intestinal peptide and its receptors in human colonic carcinoma cells in culture. Cancer Res. **48:** 6201–6210.

Calcium Influx through Channels Other than Voltage-Dependent Calcium Channels Is Critical to the Pituitary Adenylate Cyclase-Activating Polypeptide-Induced Increase in Excitability in Guinea Pig Cardiac Neurons

JEAN C. HARDWICK,[a] JOHN D. TOMPKINS,[b] SARAH A. LOCKNAR,[b]
LAURA A. MERRIAM,[b] BETH A. YOUNG,[b] AND RODNEY L. PARSONS[b]

[a]Department of Biology, Ithaca College, Ithaca, New York 14850, USA

[b]Department of Anatomy and Neurobiology, University of Vermont, Burlington, Vermont 05405, USA

ABSTRACT: Pituitary adenylate cyclase-activating polypeptide (PACAP) effects on intracellular calcium ($[Ca^{2+}]_i$) and excitability have been studied in adult guinea pig intracardiac neurons. PACAP increased excitability, but did not elicit Ca^{2+} release from intracellular stores. Exposure to a Ca^{2+}-deficient solution did not deplete $[Ca^{2+}]_i$ stores but did eliminate the PACAP-induced increase in excitability. We postulate that Ca^{2+} influx is required for the PACAP-induced increase in excitability.

KEYWORDS: PACAP; calcium influx; intracellular calcium release; action potential generation; voltage-dependent calcium channels

INTRODUCTION

Pituitary adenylate cyclase-activating polypeptide (PACAP) is present in the cholinergic parasympathetic preganglionic fibers innervating neurons in the guinea pig intracardiac ganglia.[1] Braas et al.,[2] determined that the intracardiac neurons express the PACAP-selective PAC1 receptor and that exogenous PACAP markedly increases neuronal excitability. DeHaven and Cuevas[3]

Address for correspondence: Rodney L. Parsons, Ph.D., University of Vermont College of Medicine, Department of Anatomy and Neurobiology, 89 Beaumont Ave., Given C427, Burlington, VT 05405. Voice: 802-656-2230; fax: 802-656-8704.
e-mail: Rodney.Parsons@uvm.edu

Ann. N.Y. Acad. Sci. 1070: 317–321 (2006). © 2006 New York Academy of Sciences.
doi: 10.1196/annals.1317.036

reported recently that PACAP also increased excitability of rat neonatal intracardiac neurons, which can express both PAC1 and VPAC receptors.[4] It is postulated for rat neonatal neurons that the PACAP-induced increase in excitability requires release of Ca^{2+} from internal stores.[3] The present experiments tested if a PACAP-induced release of Ca^{2+} from intracellular stores is required for the PACAP-induced increase in excitability in adult guinea pig cardiac neurons.

PREPARATION AND METHODS

Experiments were performed *in vitro* on guinea pig intracardiac neurons either in ganglion whole mount preparations or dissociated from the ganglia following protocols described previously.[2,5] Two bath solutions were used, a control solution containing 2.5 mM calcium and a Ca^{2+}-deficient solution in which Ca^{2+} was replaced by magnesium (Mg^{2+}). No Ca^{2+} chelator was included in the Ca^{2+}-deficient solution. Measurements of intracellular Ca^{2+} transients used fluorescence measurements of the Ca^{2+}-sensitive dye fluo-3.[5] Standard intracellular recording techniques were used to determine the effect of PACAP on excitability.[2] PACAP effects on voltage-dependent calcium currents were tested using barium (Ba^{2+}) as a charge carrier.[6]

RESULTS AND DISCUSSION

First, we tested whether PACAP alone initiated a rise in $[Ca^{2+}]_i$ in guinea pig intracardiac neurons. No change in the fluo-3 fluorescence ratio (F/F_o) was observed in 13 cells following 100 nM PACAP application for 1–3 min (FIG. 1). We also challenged the same cells with 10 mM caffeine to initiate Ca^{2+} release from internal stores to insure that the cells were adequately loaded with the Ca^{2+}-sensitive dye and changes in $[Ca^{2+}]_i$ could be observed.[5] Caffeine elicited a transient rise in $[Ca^{2+}]_i$ even though no rise in $[Ca^{2+}]_i$ was observed during PACAP application ($n = 13$).

Using intracellular recordings, we analyzed the effect of PACAP on excitability in neurons maintained in intracardiac ganglia whole mount preparations.[2] Following PACAP application, a short duration (<2 min) variable depolarization (4–15 mV) was produced by PACAP. Once the peptide-induced depolarization subsided, the prolonged effect of PACAP on excitability was determined by applying long suprathreshold current pulses of increasing intensity to elicit action potentials.[2] An excitability curve was generated by plotting the number of action potentials initiated as a function of the current intensity.[2] The effect of PACAP on excitability was quantified in 6 intracardiac neurons (5 phasic cells and 1 tonic cell) bathed in the control Ca^{2+}-containing solution. PACAP significantly increased excitability. Example recordings from 1 phasic cell are

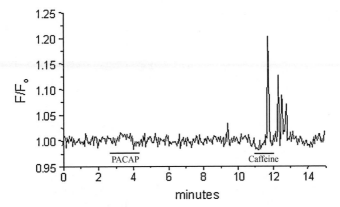

FIGURE 1. Cardiac neurons do not show an increase in cytosolic calcium concentration due to PACAP application, but do respond to caffeine. Cell was exposed to 100 nM PACAP via bath application during the first bar and to 10 mM caffeine during the second bar. An increase in the fluo-3 fluorescence ratio (F/Fo) indicates an increase in intracellular calcium.

presented in FIGURE 2 A and the average results from the 5 phasic cells exposed to PACAP are shown in FIGURE 2 B. PACAP also increased excitability in the tonic cell; the number of action potentials produced by a 500 ms, 0.3 nA depolarizing current pulse doubled following peptide application. In cells bathed in the Ca^{2+}-deficient solution, PACAP did not increase excitability (FIG. 2 C).

In additional experiments, we determined in dissociated neurons voltage clamped with the perforated patch recording mode whether PACAP affected voltage-dependent calcium currents.[6] PACAP (100 nM) decreased the peak barium current by $46 \pm 6\%$ ($n = 7$ cells).

We then tested whether exposure to the Ca^{2+}-deficient solution depleted intracellular Ca^{2+} stores using fluorescence measurements of the Ca^{2+}-sensitive dye fluo-3 in dissociated intracardiac neurons. Dissociated neurons were kept either in the Ca^{2+}-containing bath solution or the Ca^{2+}-deficient bath solution for 15 min and then exposed to 20 mM caffeine. Caffeine (20 mM for 1–2 min) elicited a comparable rise in intracellular Ca^{2+} in cells maintained either in the Ca^{2+} containing (fluo-3 fluorescence ratio $F/F_o = 1.24 \pm 0.05$, $n = 15$ cells) or Ca^{2+}-deficient solution (fluo-3 fluorescence ratio $F/F_o = 1.24 \pm 0.05$, $n = 9$ cells). These results suggested that Ca^{2+} stores were not depleted following a short exposure to a Ca^{2+}-deficient solution.

Taken together, these experiments suggest that in the case of adult guinea pig cardiac neurons, a PACAP-induced release of Ca^{2+} from internal stores is not required for the peptide-induced increase in excitability. Rather, it appears that a PACAP-induced influx of Ca^{2+} through plasma membrane Ca^{2+}-permeable channels is required. Our results also suggest that the PACAP-activated Ca^{2+} influx pathway is not a voltage-dependent calcium channel, but rather another

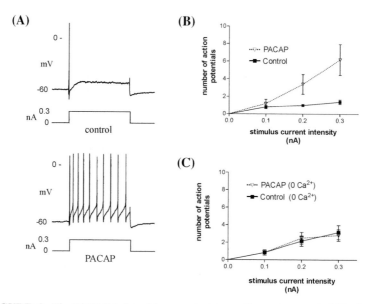

FIGURE 2. The PACAP-induced increase in intracardiac neuron excitability is eliminated by replacing calcium with magnesium. (**A**) an intracellular recording illustrating the increase in action potential firing following a 500 ms "puffer application" of PACAP (50 μM in puffer pipette). The neuron was bathed in calcium-containing solution. (**B**) an excitability curve showing the average number of action potentials produced by 500 msec depolarizing stimuli of increasing intensity. Before PACAP application, only 1–2 action potentials were elicited. After PACAP application, a much greater number of action potentials were elicited by the same stimuli. Data give mean ± SEM from 5 phasic cells. (**C**) an excitability curve for 6 other phasic cells maintained in a calcium-deficient solution. Note that following PACAP application, there was no increased action potential firing.

type of Ca^{2+} permeable membrane channel, very likely a receptor operated nonselective cation channel.

ACKNOWLEDGMENTS

This study was supported in part by NIH grants HL-65481 and NCRR P20 RR-16435. Dr. John D. Tompkins was supported by NIH training grant HL07594.

REFERENCES

1. CALUPCA, M.A., M.A. VIZZARD & R.L. PARSONS. 2000. Origin of pituitary adenylate cyclase-activating polypeptide (PACAP)-immunoreactive fibers innervating guinea pig parasympathetic cardiac ganglia. J. Comp. Neurol. **423:** 26–39.

2. BRAAS, K.M., V. MAY, S.A. HARAKALL, *et al.* 1998. Pituitary adenylate cyclase-activating polypeptide expression and modulation of neuronal excitability in guinea pig cardiac ganglia. J. Neurosci. **18:** 9766–9779.
3. DEHAVEN, W.I. & J. CUEVAS. 2004. VPAC receptor modulation of neuroexcitability in intracardiac neurons: dependence on intracellular calcium mobilization and synergistic enhancement by PAC$_1$ receptor activation. J. Biol. Chem. **279:** 40609–40621.
4. DEHAVEN, W.I. & J. CUEVAS. 2002. Heterogeneity of pituitary adenylate cyclase-activating polypeptide and vasoactive intestinal polypeptide receptors in rat intrinsic cardiac neurons. Neurosci. Lett. **328:** 45–49.
5. LOCKNAR, S.A., K.L. BARSTOW, J.D. TOMPKINS, *et al.* 2004. Calcium-induced calcium release regulates action potential generation in guinea-pig sympathetic neurones. J. Physiol. **555.3:** 627–635.
6. MERRIAM, L.A. & R.L. PARSONS. 1995. Neuropeptide galanin inhibits ω-conotoxin GVIV-sensitive calcium channels in parasympathetic neurons. J. Neurophysiol. **73:** 1374–1382.

Mechanisms and Modulation of Pituitary Adenylate Cyclase-Activating Protein-Induced Calcium Mobilization in Human Neutrophils

ISSAM HARFI AND ERIC SARIBAN

Hemato-Oncology Unit and Laboratory of Pediatric Oncology, Hôpital Universitaire des Enfants, 1020 Brussels, Belgium

ABSTRACT: The neuropeptide pituitary adenylate cyclase-activating protein (PACAP) acts via the G protein–coupled receptor vasoactive intestinal peptide (VIP)/PACAP receptor-1 to induce phospholipase C (PLC)/calcium and mitogen-activated protein kinase (MAPK)-dependent proinflammatory activities in human polymorphonuclear neutrophils (PMNs). In this article, we evaluate other mechanisms that regulate PACAP-evoked calcium transients, the nature of the calcium sources, and the role of calcium in proinflammatory activities. Reduction in the activity of PMNs to respond to PACAP was observed after cell exposure to inhibitors of the cAMP/protein kinase A (PKA), protein kinase C (PKC), and PI3K pathways, to pertussis toxin (PTX), genistein, and after chelation of intracellular calcium or after extracellular calcium depletion. Mobilization of intracellular calcium stores was based on the fact that PACAP-associated calcium transient was decreased after exposure to (*a*) thapsigargin (Tg), (*b*) xestospongin C (XeC), and (*c*) the protonophore carbonyl cyanide 4-(trifluoromethoxy)phenylhydrazone; inhibition of calcium increase by calcium channel blockers, by nifedipine and verapamil, indicated that PACAP was also acting on calcium influx. Such mobilization was not dependent on a functional actin cytoskeleton. Homologous desensitization with nanomoles of PACAP concentration and heterologous receptors desensibilization by G protein–coupled receptor agonists were observed. Intracellular calcium depletion modulated PACAP-associated ERK but not p38 phosphorylation; in contrast, extracellular calcium depletion modulated PACAP-associated p38 but not ERK phosphorylation. In PACAP-treated PMNs, reactive oxygen species production and CD11b membrane upregulation in contrast to lactoferrin release were dependent on both intra- and extracellular calcium, whereas matrix metalloproteinase-9 release was unaffected by

Address for correspondence: Issam Harfi, Hemato-Oncology Unit and Laboratory of Pediatric Oncology, Hôpital Universitaire des Enfants, 1020 Brussels, Belgium. Voice: 003224772678; fax: 003224772678.

e-mail: issam.harfi@huderf.be

Ann. N.Y. Acad. Sci. 1070: 322–329 (2006). © 2006 New York Academy of Sciences.
doi: 10.1196/annals.1317.037

extracellular calcium depletion. These data indicate that both extracellular and intracellular calcium play key roles in PACAP proinflammatory activities.

KEYWORDS: PACAP; calcium; neutrophils

INTRODUCTION

Calcium ions are universal secondary messengers that play a key role in many cellular signal transduction pathways regulating diverse functions, such as secretion, cell motility, proliferation, and cell death. Cytosolic calcium concentration is tightly regulated. Ligand-stimulated increases in intracellular calcium concentration come mainly from two sources: internal stores that release calcium and plasma membrane channels that open to allow external calcium to flow into the cell. In a number of cells the release of intracellular stored calcium following receptor activation acts as a trigger for longer calcium signals derived from calcium influx through membrane calcium channel.[1,2] In these cells, calcium signaling has been implicated in oxidase activation, cell degranulation, and priming response to a wide variety of proinflammatory molecules. In contrast, other cellular events, such as cell shape change, chemotaxis, actin polymerization, and phagocytosis do not initially require an increase in cytosolic-free calcium concentration. In human polymorphonuclear neutrophils (PMNs), G protein–coupled receptors (GPCRs) constitute the major group of cell surface receptors. Receptors to fMLP, C5a, platelet-activating factor, and IL-8 are all GPCRs and convey through cytosolic-free calcium signaling a number of proinflammatory activities, such as chemoattraction, oxidase activation, and degranulation.

Pituitary adenylate cyclase-activating polypeptide (PACAP), a member of the vasoactive intestinal polypeptide (VIP)/secretin/glucagon/growth hormone-releasing factor family acts through the activation of 3 PACAP/VIP GPCRs that have been cloned: VPAC1 and VPAC2 receptors, which bind VIP and PACAP with equal affinity and PAC1 receptor, which is PACAP selective. All three VPAC receptors are coupled to stimulation of adenylyl cyclase via the heterotrimeric Gαs protein, triggering a cAMP/protein kinase A (PKA) transduction pathway, and to activation of phospholipase C (PLC) via the heterotrimeric Gαi/αq protein with subsequent increase in inositol 1,4,5-triphosphate (IP3) and calcium.[3] In addition, VPAC1-mediated calcium increase is partially blocked by pertussis toxin (PTX) indicating that this receptor is coupled to Gαi.[4] In nonhematopoietic cells, PACAP-evoked calcium transient has been reported to be dependent on the activation of cAMP/PKA, PLC-β, protein kinase C (PKC), or to be PKA- and PKC-independent.[4] These mechanisms act differentially on the mobilization of intracytoplasmic calcium from IP3-sensitive and -insensitive stores and on the modulation of multiple calcium channels or specific L-type calcium channel present on the plasma membrane.[3]

We have recently found that PACAP, through the GPCR VPAC1, acts as a proinflammatory molecule in PMNs.[5] We have shown that secondary messengers activated by PACAP in neutrophils include c-AMP, IP3, and calcium with downstream activation of the mitogen-activated protein kinase (MAPK) ERK and p38. The inflammatory activities stimulated by PACAP in these cells include the production of reactive oxygen species (ROS), upregulation in the membrane expression of the integrin CD11b, and release of secondary (lactoferrin) and tertiary (matrix metalloproteinase-9 [MMP-9]) granules.[5]

In the present article, we investigated the intracellular mechanisms following PACAP receptor activation that leads to calcium release in human neutrophils. We evaluated the contribution of the extracellular and the intracellular calcium as well as the different sources of the intracellular calcium stores sensitive to PACAP exposure. We investigated whether the calcium signal is consistent with G protein mediation by manifesting the phenomenon of homologous desensitization.

RESULTS

G Protein-Dependent Pathways and Calcium Response in PACAP-Activated PMN'S

The VPAC1 receptor-mediated IP3 and calcium increases were partially blocked by PTX treatment, which uncouples $G\alpha i$ proteins from G protein-linked GPCRs, attesting to a contribution of $G\alpha i$ protein.[4,6] We tested the effect of increasing concentration of PTX on the calcium response in PACAP-treated PMNs. As shown in FIGURE 1 A, stimulation with PACAP caused a biphasic calcium response including an immediate peak followed by a plateau phase characterized by a sustained elevation of calcium. Both phases were dose-dependently inhibited by PTX, with partial inhibition after 50 ng/mL PTX and complete inhibition after 500 ng/mL PTX exposure (FIG. 1 A).

Intra- and Extracellular Calcium Mobilization by PACAP

In PMNs, most of the intracellular releasable calcium is stored in the IP3-sensitive endoplasmic reticulum. Entry of calcium in IP3-sensitive organelles is regulated by a calcium-dependent ATPase. Thapsigargin (Tg), by inhibiting the reuptake of calcium through calcium-dependent ATPase, depletes intracellular calcium in neutrophils. PACAP applied after Tg (2 μM) exposure was unable to induce a calcium signal indicating that PACAP-induced calcium signals were associated with calcium release from the endoplasmic reticulum stores (FIG. 1 B).

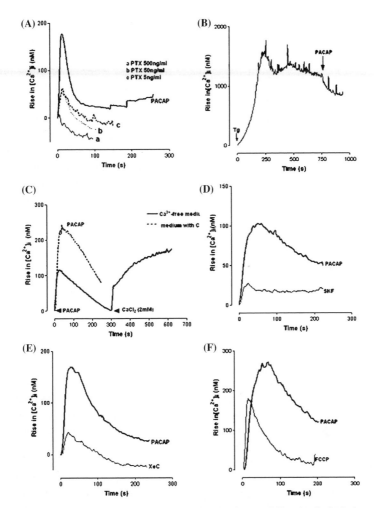

FIGURE 1. (**A**) PACAP-induced intracellular calcium mobilization is Gαi -dependent. Dose-response inhibitory effect of PTX on the mobilization of intracellular calcium in PMNs exposed to PACAP. PACAP did not induce calcium signal in PMNs preincubated with Tg (**B**). Twelve minutes after addition of 2 μM Tg, 1 μM PACAP was added in fluo-3-loaded cell suspension (1×10^6 cells/mL, RPMI containing 5 mM calcium, pH 7.2). (**C**) PACAP stimulated a transient calcium signal in PMNs that depended on the presence of extracellular calcium. PMNs were stimulated with 1 μM PACAP in calcium-free PBS containing 2 mM EGTA; at the time indicated, 2 mM $CaCl_2$ was added to calcium-depleted PMNs. (**D**) Inhibitory effect of SKF96365 (SKF) on the mobilization of intracellular calcium in PMNs exposed to PACAP. PACAP-induced intracellular calcium mobilization was IP3- and mitochondria-dependent. Inhibitory effect of XeC (**E**), and FCCP (**F**) on the mobilization of intracellular calcium in PMNs exposed to PACAP. Preincubation time before PACAP exposure was 2 h for PTX, 10 min for SKF96365, 10 min for XeC, and 20 min for FCCP. Data from a single experiment representative of a minimum of four independent experiments are shown.

According to the nature of the stimulus, this initial rise in calcium due to the mobilization of intracellular stores may or may not activate calcium influx across the plasma membrane to amplify and sustain the overall increase in intracytoplasmic calcium. To evaluate the role of external calcium influx in PACAP signaling, experiments were performed in calcium-free medium (0 mM calcium, 2 mM EGTA). Application of PACAP in such a medium induced an increase in calcium (FIG. 1 C) significantly lower from the one recorded in the presence of calcium (FIG. 1 C). Further confirmation of the role of extracellular calcium was provided by experiments where the addition of calcium (2 mM) back to the calcium-free medium resulted in an increase in intracytoplasmic calcium in PACAP-treated PMNs (FIG. 1 C). This supports that PACAP also causes mobilization of extracellular stores. Indeed, the application of SKF96365, a specific calcium channel blocker, induced a lower calcium increase in PACAP-treated cells: SKF96365 (10 μM) reduced by $73 \pm 7\%$ the calcium response to PACAP (FIG. 1 D).

Nature of the Intracellular Calcium Pool

Activation of PLC leads to the generation of IP3 and diacylglycerol. In neutrophils, calcium release from the endoplasmic reticulum into the cytosolic compartment occurs through the opening of the universal IP3 receptor channel that is modulated by IP3. PMNs exposure to xestospongin C (XeC,10 μM), a specific blocker of the IP3 receptor, resulted in a decrease by $92 \pm 7\%$ of the calcium response to PACAP indicating that IP3-sensitive calcium pools are involved in PACAP-evoked calcium increase (FIG. 1 E). IP3-insensitive stores, such as the one located in mitochondria have been described in neutrophils. The use of the protonophore carbonyl cyanide 4-(trifluoromethoxy)phenylhydrazone (FCCP) (5 μM), which causes depletion of mitochondrial calcium, significantly decreased by $59 \pm 14\%$ the PACAP-associated calcium increase indicating that mitochondria also contribute to the PACAP-associated calcium signal (FIG. 1 F).

Homologous Desensitization

Prolonged stimulation of GPCRs results in desensitization that can be homologous and/or heterologous. Homologous desensitization is specific for receptor and its agonist. We then examined whether the response of PMNs to PACAP manifests homologous desensitization. As shown in FIGURE 2 A, a 5-min preexposure of PMNs to PACAP decreased by $56 \pm 10\%$ the capacity of the neuropeptide to mobilize intracellular-free calcium. Homologous desensitization was less pronounced after 15 min (FIG. 2 B) and vanished after 30 min (data not shown); it was not associated with a cAMP/PKA-mediated effect since pretreatment with forskolin, a stimulator of adenylyl cyclase, had

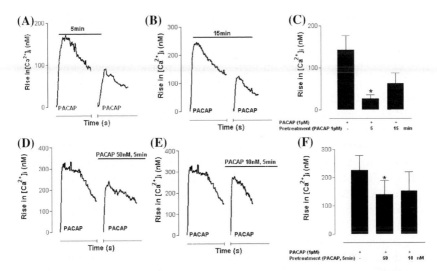

FIGURE 2. PACAP elicits a calcium signal in PMNs that undergoes homologous desensitization. PMNs loaded with fluo-3 were sequentially stimulated with PACAP. PACAP (1 μM) was added at time 0 and again at 5 min (**A**) or after 15 min (**B**). PMNs were stimulated with PACAP (1 μM) at time 0 (control peak) while an aliquot was preincubated with 50 nM PACAP (**D**) or 10 nM PACAP (**E**) for 5 min before 1 μM PACAP exposure. Data from a single experiment representative of a minimum of four independent experiments are shown. Peak values are recorded after PACAP treatment and are reported as the mean ± SEM of a minimum of four different experiments (**C** and **F**). *$P < 0.05$ versus PACAP alone.

no effect on the PACAP-mediated calcium peak (data not shown). We then took advantage of this biological effect of PACAP to evaluate the minimal PACAP concentration necessary to desensitize the VPAC1 receptor for calcium mobilization. As shown in FIGURE 2 D and E, desensitization was still observed after exposure to 50 nM PACAP and disappeared at 10 nM of peptide treatment.[6]

DISCUSSION

PACAP is a neurotrophic, neuroprotective peptide that displays numerous functions in both cells of neuronal and nonneuronal origin.[3] After interaction with one of its three receptors, PACAP induces the formation of second messengers including cAMP, IP3, and calcium. Inflammatory processes expose PMNs to a wide variety of agonists that can attract, prime, or activate PMNs via GPCRs, which typically signal by increasing calcium in addition to activating other second messengers. We have recently reported that PACAP in human neutrophils activates multiple proinflammatory functions through PLC/ERK and p38 MAPK activation. In these cells, PACAP also increases intracytoplasmic calcium through mechanisms that are PLC/ERK-dependent, but p38-independent.[5]

In nonhematopoietic cells, PACAP mobilizes calcium from both intra- and extracellular sources through multiple pathways according to cell types and nature of the VPAC receptors present on the cell surface. In these cells, calcium mobilization has been linked to mechanisms that are cAMP/PKA- and PLC/PKC-dependent, cAMP/PKA- and PLC-independent, and ryanodine/caffeine-dependent.[2] Calcium influx through L-, N-, and Q-type calcium channels has also been reported.[7] In hematopoietic cells, PACAP has been reported to have both pro- and anti-inflammatory properties. Although there are numerous reports on PACAP-associated calcium mobilization in non-hematopoietic cells,[2] very few studies have been conducted on such effect in hematopoietic cells.[8,9] Similar to what has been described with other GPCRs, we found that PACAP-mediated calcium signals in PMNs is PLC/IP3- and PTX-dependent as well as cAMP/PKA-dependent indicating the participation of Gαi/αq and Gαs proteins. In these cells, IP3-containing calciosomes and mitochondria are sources of intracellular calcium. The role of extracellular calcium is suggested by the fact that PACAP-mediated calcium signal is abolished in cells cultured in calcium-free medium and that the addition of calcium in the medium restores the calcium signal. In addition, PMNs preincubated with the cation channel entry inhibitor SKF96365 demonstrate a markedly diminished calcium response to PACAP. Altogether, these results indicate PACAP induces calcium influx that is most likely activated by depletion of intracellular calcium stores.

Chemoattractant-mediated inflammatory responses are regulated by desensitization whereby prolonged stimulation with appropriate agonist results in the decrease in chemoattractant receptors mediated signals. Human neutrophils are known to undergo rapid homologous chemoattractant receptor desensitization while cross-desensitization among these receptors also occurs.[10] We also found that PACAP-treated PMNs display homologous desensitization of calcium mobilization. Of interest, downregulation of VPAC1 receptor was also observed in stimulated human T lymphocytes and might represent a mechanism for restricting bystander activation of T cells.[11] Affinity of the VPAC1 receptor is in the nM range. Investigating PACAP homologous desensitization mechanism we demonstrated that PACAP at nM concentration was effective to desensitize VPAC1 receptor for calcium mobilization and experiments are in progress in our laboratory to evaluate how this desensitization translates into decrease in PMN activity.

ACKNOWLEDGMENTS

This work was supported by Grants 3.4553.00, 7.4562.01, and 7.4593.03 from the Fonds National de la Recherche Scientifique, by the Lambeau-Marteaux and Wajnman-Mandelbaum Foundations, by the Fondation Aide aux Enfants Atteints de Cancer du Luxembourg, and by the Hellef für krisbskrankkanner Foundation of Luxembourg.

REFERENCES

1. BERRIDGE, M.J., M.D. BOOTMAN & H.L. RODERICK. 2003. Calcium signalling: dynamics, homeostasis and remodelling. Nat. Rev. Mol. Cell. Biol. **4:** 517–529.
2. SPASSOVA, M.A., J. SOBOLOFF, L.P. HE, et al. 2004. Calcium entry mediated by SOCs and TRP channels: variations and enigma. Biochim. Biophys. Acta **1742:** 9–20.
3. VAUDRY, D., B.J. GONZALEZ, M. BASILLE, et al. 2000. Pituitary adenylate cyclase-activating polypeptide and its receptors: from structure to functions. Pharmacol. Rev. **52:** 269–324.
4. LANGER, I. & P. ROBBERECHT. 2005. Mutations in the carboxy-terminus of the third intracellular loop of the human recombinant VPAC1 receptor impair VIP-stimulated [Ca2+]i increase but not adenylate cyclase stimulation. Cell. Signal. **17:** 17–24.
5. HARFI, I., S. D'HONDT, F. CORAZZA & E. SARIBAN. 2004. Regulation of human polymorphonuclear leukocytes functions by the neuropeptide pituitary adenylate cyclase-activating polypeptide after activation of MAPKs. J. Immunol. **173:** 4154–4163.
6. HARFI, I., F. CORAZZA, S. D'HONDT & E. SARIBAN. 2005. Differential calcium regulation of proinflammatory activities in human neutrophils exposed to the neuropeptide pituitary adenylate cyclase activating protein. J. Immunol. **175:** 4091–4102.
7. O'FARRELL, M. & P.D. MARLEY. 1997. Multiple calcium channels are required for pituitary adenylate cyclase-activating polypeptide-induced catecholamine secretion from bovine cultured adrenal chromaffin cells. Naunyn Schmiedebergs Arch. Pharmacol. **356:** 536–542.
8. VAN RAMPELBERGH, J., P. POLOCZEK, I. FRANCOYS, et al. 1997. The pituitary adenylate cyclase activating polypeptide (PACAP I) and VIP (PACAP II VIP1) receptors stimulate inositol phosphate synthesis in transfected CHO cells through interaction with different G proteins. Biochim. Biophys. Acta **1357:** 249–255.
9. HAYEZ, N., I. HARFI, R. LEMA-KISOKA, et al. 2004. The neuropeptides vasoactive intestinal peptide (VIP) and pituitary adenylate cyclase activating polypeptide (PACAP) modulate several biochemical pathways in human leukemic myeloid cells. J. Neuroimmunol. **149:** 167–181.
10. RICHARDSON, R.M., H. ALI, E.D. TOMHAVE, et al. 1995. Cross-desensitization of chemoattractant receptors occurs at multiple levels. Evidence for a role for inhibition of phospholipase C activity. J. Biol. Chem. **270:** 27829–27833.
11. LARA-MARQUEZ, M., M. O'DORISIO, T. O'DORISIO, et al. 2001. Selective gene expression and activation-dependent regulation of vasoactive intestinal peptide receptor type 1 and type 2 in human T cells. J. Immunol. **166:** 2522–2530.

PACAP Enhances Mouse Urinary Bladder Contractility and Is Upregulated in Micturition Reflex Pathways after Cystitis

GERALD M. HERRERA,[a,b] KAREN M. BRAAS,[c] VICTOR MAY,[a,c] AND MARGARET A. VIZZARD[c,d]

[a]University of Vermont College of Medicine, Department of Pharmacology, Burlington, Vermont 05405, USA

[b]Med Associates, Inc., St. Albans, Vermont, 05478, USA

[c]University of Vermont College of Medicine, Department of Anatomy and Neurobiology Burlington, Vermont 05405, USA

[d]University of Vermont College of Medicine, Department of Neurology, Burlington, Vermont 05405, USA

ABSTRACT: Pituitary adenylate cyclase-activating polypeptide (PACAP) elicits a transient contraction, sustained increase in the amplitude of spontaneous phasic contractions, and significantly increases the amplitude of nerve-mediated contractions in mouse urinary bladder smooth muscle (UBSM) strips. PACAP immunoreactivity (IR) is increased in micturition reflex pathways following cystitis. PACAP may contribute to altered sensation and bladder overactivity in the chronic bladder inflammatory syndrome, interstitial cystitis.

KEYWORDS: urinary bladder; bladder overactivity; cyclophosphamide; myograph; immunohistochemistry; lumbosacral spinal cord

INTRODUCTION

Pituitary adenylate cyclase-activating polypeptide (PACAP), a member of the vasoactive intestinal peptide (VIP)/secretin/glucagon superfamily,[1] is widely expressed and has diverse functions in the endocrine, nervous, cardiovascular, and gastrointestinal systems.[2,3] PACAP exists as two alternatively processed forms, PACAP-27 and PACAP-38.[4] High levels of PACAP and VIP

Address for correspondence: Margaret A. Vizzard, Ph.D., University of Vermont College of Medicine, Department of Neurology, D411 Given Building, Burlington, VT 05405. Voice: 802-656-3209; fax: 802-656-8704.

e-mail: margaret.vizzard@uvm.edu

Ann. N.Y. Acad. Sci. 1070: 330–336 (2006). © 2006 New York Academy of Sciences.
doi: 10.1196/annals.1317.040

expression have been identified in CNS neurons and in sensory and auto-nomic ganglia.[2,4-7] PACAP may have prevalent roles in the lower urinary tract (LUT).[8-10] Widespread PACAP-immunoreactivity (IR) exists in nerve fibers along the rat urinary tract, including urinary bladder smooth muscle (UBSM), suburothelial plexus, and blood vessels.[10] PACAP-IR fibers appear to be in sensory neurons, as neonatal capsaicin (C-fiber neurotoxin) treatment signifi-cantly reduced PACAP-IR,[10] and PACAP-IR nerve fibers in the bladder express vanilloid receptors.[11] Few studies have examined PACAP expression after in-duction of inflammatory states.[12,13] We have previously demonstrated an up-regulation of PACAP levels in micturition pathways after cyclophosphamide (CYP)-induced cystitis[8] in rats. Intrathecal or intra-arterial administration of PACAP-27 near the urinary bladder facilitates micturition in conscious normal rats, but PACAP-27 has small direct effects on isolated bladder strips.[9] The present article in mouse determined if PACAP peptides facilitate LUT function and PACAP expression is upregulated in LUT tissues with cystitis.

CYP-INDUCED CYSTITIS

Chemical cystitis was induced in adult female C57Bl/6 mice ($n = 6$; 25–30 g) by CYP (150 mg/kg; i.p.) treatment 48 h prior to euthanasia.[8,14] Control mice ($n = 6$) were injected with distilled water. Animal use was approved by The University of Vermont IACUC.

ISOMETRIC TENSION RECORDING

Isometric tension was recorded from mouse UBSM strips using MyoMED Myograph System (MED Associates Inc., St. Albans, VT).[15] Bladder contrac-tions were elicited by electric field stimulation (EFS) from 0.5 Hz to 5 Hz every 3 min (20 V amplitude, alternating polarity, 0.2 msc pulse width). These conditions evoke UBSM contractions that are eliminated by blocking neuronal sodium channels with tetrodotoxin (TTX).[15] Peptides (PACAP-27, PACAP-38; 80 nM; BACHEM, Torrance, CA) were added to the bath. Time control ex-periments were performed for EFS providing the basis for which effects of peptides were measured. All reagents were from Sigma ImmunoChemicals (St. Louis, MO) unless stated.

IMMUNOHISTOCHEMISTRY

Control and CYP-treated mice were euthanized.[8] Lumbosacral (L6-S1) spinal cord sections (20 μm) were incubated with antiserum (rabbit anti-PACAP-38; 1:3000; Phoenix Pharmaceuticals, Inc., CA).[8] Density of PACAP-IR in the L6-S1 spinal cord was determined.[8]

STATISTICS

All values are means ± SEM. Data were compared using Student's t-test, one- or two-way analysis of variance (ANOVA), where appropriate. $P \leq 0.05$ was considered significant.

RESULTS

Application of PACAP-27 (80 nM; FIG. 1 A) or PACAP-38 (80 nM; FIG. 1 B) increased UBSM tone and increased the amplitude of spontaneous phasic UBSM contractions. Blocking nerve-evoked contractions with TTX (1 μM) did not alter PACAP-38 (absence versus presence of TTX: 0.26 ± 0.07 mN versus 0.18 ± 0.80 mN) or PACAP-27 (absence versus presence of TTX: 0.64 ± 0.16 mN versus 0.52 ± 0.12 mN) induced increases in UBSM tone, suggesting a direct effect of PACAP on UBSM. Application of PACAP-27 (80 nM) (FIG. 1 C) or PACAP-38 (80 nM) (FIG. 1 D) increased the amplitude of nerve-evoked contractions (red lines) of detrusor smooth muscle at the lower stimulation frequencies tested compared to control (no peptide, black lines; FIG. 1 C, D) (FIG. 1 E).

After CYP-induced cystitis (FIG. 2 B), the intensity of PACAP-IR was significantly increased in specific regions of the L6-S1 spinal cord compared to control (FIG. 2 A). After cystitis, the density of PACAP-IR increased in the superficial laminae (I–II), medial (1.8-fold increase) to lateral (1.5-fold increase) dorsal horn (DH) (FIG. 2 B, C). PACAP-IR was also increased after cystitis in the lateral collateral pathway of Lissauer (LCP; 1.5-fold increase) and in the sacral parasympathetic nucleus (SPN) (FIG. 2 B, C). Increases in PACAP-IR were similarly observed in the S1 spinal segment after cystitis (FIG. 2 D).

DISCUSSION

These studies demonstrate that (*a*) PACAP elicits a sustained contraction of UBSM that is not blocked by TTX, suggesting a direct PACAP action on UBSM; (*b*) PACAP significantly increases the amplitude of nerve-mediated EFS UBSM contractions; and (*c*) CYP-induced cystitis increases PACAP-IR in specific regions of the L6-S1 spinal cord involved in micturition reflexes. The PACAP-induced increase in UBSM contractility demonstrated in the present study is consistent with a previous study where *in vivo* administration of PACAP-27 facilitated micturition in conscious rats.[9] However, in the present study, robust direct effects of PACAP on detrusor smooth muscle were also observed and PACAP also facilitated the amplitude of nerve-mediated bladder contractions. The responses in these experiments appeared greater than those in previous work[9] possibly due to the removal of urothelium that may have mitigating effects. The facilitatory effects of PACAP are consistent

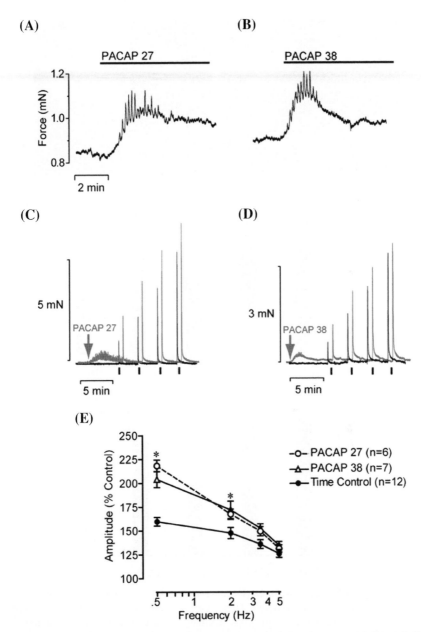

FIGURE 1. Facilitatory effects of PACAP-27 (80 nM) (**A**) or PACAP-38 (80 nM) (**B**) on resting tone and on nerve-evoked contractions (**C, D**) in UBSM strips. (**C, D**) Tick marks indicate the approximate start of nerve-evoked contractions at 0.5, 2, 3.5, and 5 Hz. (**E**) Summary data from nerve-evoked contraction experiments with contraction amplitude in the presence of peptide normalized to control contraction amplitude (no peptide) from same detrusor strips. *$P \le 0.05$.

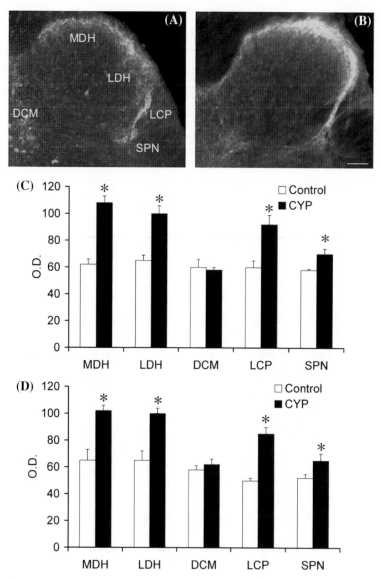

FIGURE 2. CYP-induced cystitis (**B**) increases PACAP expression in the medial dorsal horn (MDH) and lateral dorsal horn (LDH), LCP, SPN but not in the DCM of the L6-S1 spinal segments compared to control (**A**). Summary of the density of PACAP staining in specific spinal cord regions in the L6 (**C**) and S1 (**D**) spinal segments. *$P \leq 0.05$.

with the action of other neuropeptides on LUT tissues including substance P,[16] VIP,[17] and cocaine and amphetamine-regulated transcript peptide (CARTp)[18] but contrast with the inhibitory effects of calcitonin gene-related peptide.[19]

CYP-induced cystitis increased PACAP expression in a number of regions involved in micturition reflexes including the DH, SPN, and LCP. These results are consistent with our previous studies in rats.[8] Tissue inflammation or irritation can induce changes in the properties of somatic sensory pathways, leading to hyperalgesia and allodynia.[20] Peripheral sensitization of primary afferents or changes in central synapses can contribute to the increased pain sensation.[21,22] Increased PACAP expression in micturition pathways may contribute to altered sensory processing and bladder overactivity after CYP-induced cystitis and may have implications for the chronic bladder inflammation syndrome, interstitial cystitis.[23]

ACKNOWLEDGMENTS

This work was supported by NIH DK051369, DK060481, DK065989, and NS040796.

REFERENCES

1. MIYATA, A. *et al.* 1990. Isolation of a neuropeptide corresponding to the N-terminal 27 residues of the pituitary adenylate cyclase activating polypeptide with 38 residues (PACAP38). Biochem. Biophys. Res. Comm. **170:** 643–648.
2. ARIMURA, A. *et al.* 1991. Tissue distribution of PACAP as determined by RIA: highly abundant in the rat brain and testes. Endocrinology **129:** 2787–2789.
3. HANNIBAL, J. 2002. Pituitary adenylate cyclase-activating peptide in the rat central nervous system: an immunohistochemical and in situ hybridization study. J. Comp. Neurol. **453:** 389–417.
4. ARIMURA, A. 1998. Perspectives on pituitary adenylate cyclase activating polypeptide (PACAP) in the neuroendocrine, endocrine, and nervous systems. Jpn. J. Physiol. **48:** 301–331.
5. BRAAS, K.M. *et al.* 1998. Pituitary adenylate cyclase-activating polypeptide expression and modulation of neuronal excitability in guinea pig cardiac ganglia. J. Neurosci. **18:** 9766–9779.
6. BRANDENBURG, C.A., V. MAY & K.M. BRAAS. 1995. Expression, secretion and plasticity of endogenous pituitary adenylate cyclase activating polypeptides (PACAP) in rat superior cervical ganglion neurons. Soc. Neurosci. [Abstr.] **21:** 1598.
7. MOLLER, K. *et al.* 1993. Pituitary adenylate cyclase activating peptide is a sensory neuropeptide: immunocytochemical and immunochemical evidence. Neuroscience **57:** 725–732.
8. VIZZARD, M.A. 2000. Up-regulation of pituitary adenylate cyclase-activating polypeptide in urinary bladder pathways after chronic cystitis. J. Comp. Neurol. **420:** 335–348.
9. ISHIZUKA, O. *et al.* 1995. Facilitatory effect of pituitary adenylate cyclase-activating polypeptide on micturition in normal, conscious rats. Neuroscience **66:** 1009–1014.

10. FAHRENKRUG, J. & J. HANNIBAL. 1998. Pituitary adenylate cyclase activating polypeptide immunoreactivity in capsaicin-sensitive nerve fibres supplying the rat urinary tract. Neuroscience **83:** 1261–1272.
11. ZVAROVA, K., J.D. DUNLEAVY & M.A. VIZZARD. 2005. Changes in pituitary adenylate cyclase activating polypeptide expression in urinary bladder pathways after spinal cord injury. Exp. Neurol. **192:** 46–59.
12. WANG, Z.-Y., P. ALM & R. HAKANSON. 1996. PACAP occurs in sensory nerve fibers and participates in ocular inflammation in the rabbit. Ann. N. Y. Acad. Sci. **805:** 779–783.
13. ZHANG, Y.Z. *et al.* 1998. Pituitary adenylate cyclase-activating peptide is upregulated in sensory neurons by inflammation. Neuroreport **9:** 2833–2836.
14. COX, P.J. 1979. Cyclophosphamide cystitis—identification of acrolein as the causative agent. Biochem. Pharmacol. **28:** 2045–2049.
15. HERRERA, G.M., T.J. HEPPNER & M.T. NELSON. 2000. Regulation of urinary bladder smooth muscle contractions by ryanodine receptors and BK and SK channels. Am. J. Physiol. **279:** R60–R68.
16. CHIEN, C.T. *et al.* 2003. Substance P via NK1 receptor facilitates hyperactive bladder afferent signaling via action of ROS. Am. J. Physiol. **284:** F840–F851.
17. IGAWA, Y. *et al.* 1993. Facilitatory effect of vasoactive intestinal polypeptide on spinal and peripheral micturition reflex pathways in conscious rats with and without detrusor instability. J. Urol. **149:** 884–889.
18. ZVAROVA, K. & M.A. VIZZARD. 2005. Ontogeny of cocaine-and amphetamine-regulated transcript peptide (CARTp) in urinary bladder and lumbosacral spinal cord of neonatal rat. J. Comp. Neurol. **489:** 501–517.
19. GILLESPIE, J.I. 2005. Inhibitory actions of calcitonin gene-related peptide and capsaicin: evidence for local axonal reflexes in the bladder wall. BJU Int. **95:** 149–156.
20. SENGUPTA, J.N. & G.F. GEBHART. 1994. Mechanosensitive properties of pelvic nerve afferent fibers innervating the urinary bladder of the rat. J. Neurophysiol. **72:** 2420–2430.
21. MCMAHON, S.B. 1996. NGF as a mediator of inflammatory pain. Philos. Trans. R. Soc. Lond. B. Biol . Sci. **351:** 431–440.
22. DMITRIEVA, N. *et al.* 1997. The role of nerve growth factor in a model of visceral inflammation. Neuroscience **78:** 449–459.
23. PETRONE, R.L. *et al.* 1995. Urodynamic findings in patients with interstitial cystitis. J. Urol. **153:** 290A.

Protective Role for Plasmid DNA-Mediated VIP Gene Transfer in Non-Obese Diabetic Mice

JUAN LUIS HERRERA,[a] RAFAEL FERNÁNDEZ-MONTESINOS,[a]
ELENA GONZÁLEZ-REY,[b] MARIO DELGADO,[b]
AND DAVID POZO[a]

[a]Department of Medical Biochemistry and Molecular Biology, University of
Seville Medical School, 14009 Sevilla, Spain

[b]Institute of Biomedicine and Parasitology, CSIC, Granada, Spain

ABSTRACT: Studies focused on the development of diabetes in NOD
mice—a model for human type 1 diabetes—have revealed that an autoimmune inflammatory process is produced by the effect of Th1 cells and
their secreted cytokines. DNA vaccination has been shown to be an effective method for modulating immunity in viral infections and experimental autoimmune diseases, including diabetes. VIP's immunomodulatory
properties are partly mediated by skewing the pattern of cytokines from a
proinflammatory response to an anti-inflammatory response. Using gene
delivery to express VIP, we interfered in the immune process leading to
diabetes in prone, cyclophosphamide-treated NOD mice. Our results extend the role of VIP in the control of immunoregulatory networks and
open new perspectives for immunointervention through VIP-based gene
therapy.

KEYWORDS: vasoactive intestinal peptide; gene delivery; diabetes;
neuroimmunology

INTRODUCTION

Insulin-dependent diabetes mellitus (IDDM) is a metabolic disorder caused
by the autoimmune destruction of the insulin-producing β cells of the pancreas. The inflammatory process in early diabetes is thought to be initiated
and propagated by the effect of Th1-secreted cytokines alongside malfunction
of key innate effector cells.[1–3] The non-obese diabetic (NOD) mice develop
an autoimmune syndrome similar in several aspects to human IDDM.[4,5] NOD

Address for correspondence: Dr. David Pozo, Department of Medical Biochemistry and Molecular
Biology, the University of Seville School of Medicine, Avda. Sanchez Pizjuan, 4, 41009 Sevilla, Spain.
Voice: +34-95-4559852; fax: +34-95-4907048.
e-mail: dpozo@us.es

Ann. N.Y. Acad. Sci. 1070: 337–341 (2006). © 2006 New York Academy of Sciences.
doi: 10.1196/annals.1317.041

mice can develop two types of autoimmune diabetes, spontaneous diabetes and cyclophosphamide-accelerated diabetes (CAD). Cyclophosphamide induces synchronization and acceleration of IDDM in NOD mice involving a cytokine shift from Th2 to Th1 and the alteration of various immunocompetent cell populations.[4,5] When compared with spontaneous NOD diabetes, CAD stands as a stronger variant of the autoimmune diabetes. Moreover, since the spontaneous autoimmune diabetes onset in NOD mice is quite variable, cyclophosphamide treatment synchronizes the pathogenic events prior to the onset of hyperglycemia and to some extent increases the robustness of animal model.[4]

After pioneering work that disclosed functional vasoactive intestinal peptide (VIP) receptors[6,7] and the presence of VIP[8] in the immune system, a compelling experimental evidence has moved VIP from the classical role of a neurohumoral hormone into a cytokine-like molecule.[9–12] Mindful that VIP acts as a potent, endogenous anti-inflammatory molecule and that promotes Th2-type responses *in vivo* and *in vitro*, the current study was performed to examine the impact of somatic VIP DNA gene delivery on CAD. By using gene delivery to express the targeted protein, we can overcome some of the conventional problems of neuropeptide administration: complexity of purification or treatment costs, need for frequent administration due to a short half-life, and unstable biological activity.

MATERIALS AND METHODS

Mice

Female NOD mice (8- to 10-weeks old) were purchased from Charles River Laboratories (Wilmington, MA) and maintained under pathogen-free conditions in the University of Seville Animal Breeding Center. Experiments were conducted under the supervision and guidelines of the Ethic Experimental Committee according to EU regulations (86/609/CEE).

Plasmid DNA Preparation and Induction/Diagnosis of Diabetes

The VIP cDNA was produced by RT-PCR using *Pfu* DNA polymerase (Stratagene, La Jolla, CA) from mouse brain and cloned into compatible enzyme restriction sites of pcDNA3 vector (Invitrogen, Carlsbad, CA). The integrity of the construct was confirmed by sequencing. Plasmid DNA was prepared in large scale using the low-free alkaline lysis method of Wizard Megaprep (Promega, Madison, WI). Preps were endotoxin-free determined by E-toxate *Limulus polyphemus* assay (Sigma-Aldrich, St. Louis, MO). Plasmid DNA was precipitated with ethanol and resuspended in sterile PBS.

Spectrophotometric analysis revealed 260:280 ratios nm ≤1.80. Three days before the first DNA treatment, the tibialis anterior (TA) muscles had been injected with 100 μL of 0.5% bupivacaine hydrochloride (Sigma-Aldrich) in isotonic NaCl to induce muscle regeneration, which increases the efficiency of gene transfer by direct DNA injection. Plasmid DNA intake and sustained expression by treated muscle tissue was verified by PCR, RT-PCR, and Western blot (data not shown).

To accelerated diabetes, female NOD mice were treated intraperitoneally (i.p.) with 200 mg/kg cyclophosphamide in PBS (Sigma-Aldrich) twice, 10 days apart, at the age of 8 weeks. Intramuscular TA injections of plasmid DNA (300 μg total, 150 μg in each TA muscle, 1.5 μg/μL) were done using a sterile 29G insulin syringe fitted with a plastic collar to limit needle penetration to 2–3 mm 2 days before cyclophosphamide treatment. For the assessment of diabetes, a commercial glucose analyzer was used. Animals that scored positively (>250 mg/dL) on two consecutive tests were classified as diabetic. Statistical analysis was performed with Student's *t*-test.

RESULTS AND DISCUSSION

The NOD mouse is an animal model system that is often used to study type 1 diabetes. Type 1 diabetes is a complex chronic disease characterized by an autoimmune/inflammatory process targeted to β cells of the pancreas. Here we report for the first time on immune intervention by VIP gene delivered with plasmid DNA. Administration of pcDNA3-*VIP* significantly reduced the incidence of diabetes in NOD female mice (FIG. 1). The incidence of diabetes

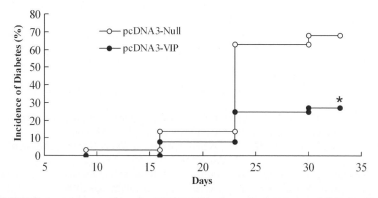

FIGURE 1. Administration of pcDNA3-*VIP* reduces the incidence of diabetes. Female NOD mice were allocated to groups of 15 mice. Diabetes was induced in 8- to 10-week-old female NOD mice by i.p administration of cyclophosphamide (200 mg/kg). The pcDNA3-*VIP*-treated mice developed a significantly lower incidence of diabetes compared with pcDNA3 null-treated mice (*$P<0.001$).

was around two times higher in pCDNA3 null-treated mice compared with mice receiving pcDNA3-*VIP*. The mechanisms of action of VIP in the onset of diabetes in NOD mice are under current investigation in our lab and include modulation of the Th1–Th2 balance and the analysis of regulatory T cell populations.

It has previously been shown using various delivery methods that diabetes can be prevented by Th2-secreted cytokines.[4,5] The effects of these cytokines in the treatment of type 1 diabetes in humans have not yet been tested. Besides, the NOD mouse might represent a less complex system compared with human. Thus, immune modulation using VIP or a combination of VIP with others immunomodulators in averting insulitis and postponing or preventing diabetes requires further work.

ACKNOWLEDGMENTS

This study was supported by extramural grants from the Instituto de Salud Carlos III, Fondo de Investigacion Sanitaria, Spanish Ministry of Health (PI 030359 to D.P and PI 030526 to M.D) and from the European Union 6th Framework Grants Program (MERG-CT-2004–006380 to D.P).

REFERENCES

1. RAZ, I., R. ELDOR & Y. NAPARSTEK. 2005. Immune modulation for prevention of type 1 diabetes mellitus. Trends Biotechnol. **23:** 128–134.
2. BEYAN, H. *et al*. 2003. A role for innate immunity in type 1 diabetes? Diabetes Metab. Res. Rev. **19:** 89–100.
3. KUKREJA, A. *et al*. 2002. Multiple immuno-regulatory defects in type-1 diabetes. J. Clin. Invest. **109:** 131–140.
4. ATKINSON, M.A. & E.H. LEITER. 1999. The NOD mouse model of type 1 diabetes: as good as it gets? Nat. Med. **5:** 601–604.
5. GALLEGOS, A.M. & M.J. BEVAN. 2004. Driven to autoimmunity: the nod mouse. Cell **117:** 149–151.
6. GUERRERO, J. *et al*. 1981. Interaction of vasoactive intestinal peptide with human blood mononuclear cells. Mol. Cell Endocrinol. **21:** 151–160.
7. O'DORISIO, M.S. *et al*. 1981. Vasoactive intestinal polypeptide modulation of lymphocyte adenylate cyclase. J. Immunol. **127:** 2551–2554.
8. GOMARIZ, R. *et al*. 1990. Demonstration of immunoreactivity vasoactive intestinal peptide (IR-VIP) and somatostatin (IR-SOM) in rat thymus. Brain Behav. Immun. **4:** 151–161.
9. DELGADO, M. *et al*. 2002. Vasoactive intestinal peptide in the immune system: potential therapeutic role in inflammatory and autoimmune diseases. J. Mol. Med. **80:** 16–24.

10. GANEA, D. & M. DELGADO. 2002. Vasoactive intestinal peptide (VIP) and pituitary adenylate cyclase-activating polypeptide (PACAP) as modulators of both innate and adaptive immunity. Crit. Rev. Oral Biol. Med. **13:** 229–237.
11. DELGADO, M., D. POZO & D. GANEA. 2004. The significance of vasoactive intestinal peptide in immunomodulation. Pharmacol. Rev. **56:** 249–290.
12. POZO, D. & M. DELGADO. 2004. The many faces of VIP in neuroimmunology: a cytokine rather a neuropeptide? FASEB J. **18:** 1325–1334.

Inhibition of Self-Renewal and Induction of Neural Differentiation by PACAP in Neural Progenitor Cells

MEGUMI HIROSE,[a,b] HITOSHI HASHIMOTO,[a] JUNKO IGA,[a] NORIHITO SHINTANI,[a] MEGUMI NAKANISHI,[a] NAOHISA ARAKAWA,[a] TAKESHI SHIMADA,[a] AND AKEMICHI BABA[a]

[a]Laboratory of Molecular Neuropharmacology, Graduate School of Pharmaceutical Sciences, Osaka University, Suita, Osaka 565-0871, Japan

[b]The Japan Society for the Promotion of Science (JSPS) Research Fellow, Japan

ABSTRACT: Several lines of evidence have suggested roles for pituitary adenylate cyclase-activating polypeptide (PACAP) in the developing nervous system. Previously, we showed that mRNA for PACAP, vasoactive intestinal peptide (VIP), and their three receptor subtypes, is differentially expressed in embryonic stem (ES) cells, ES cell-derived, neural stem cell-enriched cultures, and differentiated neurons, by using the five steps of the *in vitro* neuronal culture model of ES cell differentiation. Here, we examined the effects of PACAP on self-renewal and cell lineage determination of neural progenitor/stem cells. PACAP inhibited the basic fibroblast growth factor-induced proliferation (self-renewal), as assessed by neurosphere formation. PACAP increased microtubule-associated protein 2-positive neurons without affecting the number of cells positive for the neural stem cell marker nestin, astrocyte marker glial fibrillary acidic protein, and oligodendrocyte marker CNPase. These results suggest that PACAP inhibits self-renewal but, instead, induces early neuronal differentiation of neural progenitor cells.

KEYWORDS: differentiation; neural progenitor cells; neurons; PACAP; self-renewal

INTRODUCTION

Pituitary adenylate cyclase-activating polypeptide (PACAP) has been implicated in a variety of central nervous system functions, including hypophysiotropic function, memory formation, psychomotor function, and neuronal degeneration.[1,2] In addition, convergent evidence has suggested that PACAP

Address for correspondence: Hitoshi Hashimoto, Laboratory of Molecular Neuropharmacology, Graduate School of Pharmaceutical Sciences, Osaka University, 1-6 Yamadaoka, Suita, Osaka 565-0871, Japan. Voice: 81-6-6879-8181; fax: 81-6-6879-8184.

e-mail: hasimoto@phs.osaka-u.ac.jp

Ann. N.Y. Acad. Sci. 1070: 342–347 (2006). © 2006 New York Academy of Sciences.
doi: 10.1196/annals.1317.042

plays diverse roles in mammalian neurogenesis. PAC_1 receptor mRNA is expressed at very high levels in ventricular zones throughout the neuraxis, while PACAP mRNA is primarily expressed in postmitotic parenchymal tissue.[3] Cultures of rat cortical precursor cells produce PACAP as an autocrine signal to elicit cell cycle withdrawal,[4] and intracerebroventricular injection of PACAP into embryonic day 15.5 (E15.5) rats inhibits precursor mitosis and neurogenesis.[5] In cerebellar granule neuron precursors, PACAP inhibits Sonic Hedgehog (Shh)-induced proliferation, while stimulating mitosis in the absence of Shh, suggesting a role for PAC_1 receptor signaling as a sensor of environmental cues.[6]

Cazillis *et al.* have recently shown that PAC_1 and VPAC receptors are functionally expressed in mouse embryonic stem (ES) cells and embryoid body-derived cells and that PACAP and vasoactive intestinal peptide (VIP) induce differentiation of ES cells into a neuronal phenotype.[7] In addition, we have shown that PACAP, VIP, and their receptor mRNAs are differentially expressed in ES cells, ES-cell-derived neural stem cell-enriched cultures, and differentiated neurons.[8]

In this article, we aimed to elucidate the effects of PACAP on proliferation and cell lineage determination of neural progenitor/stem cells. Using cultured rat neurospheres and multipotential neural stem cells, we examined the effects of PACAP-38 on self-renewal (proliferation to form secondary spheres after dissociation) and differentiation of neural progenitor cells.

MATERIAL AND METHODS

Neurosphere Cultures and Secondary Sphere Formation

Neurosphere cultures were prepared from cerebral cortices of E14.5 Wistar rats as described previously.[9] After 7 days culture *in vitro*, primary spheres were collected by centrifugation, mechanically dissociated, and re-plated into 96-well plates at 10,000 cells per well. The cells were further cultured for 7 days in the presence of basic fibroblast growth factor (bFGF) (10 ng/mL) with different concentrations of PACAP-38.

Immunocytochemistry

Another type of multipotential neural stem cells was also prepared by the method of Johe *et al.*[10] The cells were cultured for 5 days on poly-L-ornithine- and fibronectin-coated culture slides in the presence of different concentrations of PACAP-38. After fixation with 4% paraformaldehyde, cells were stained with the following primary antibodies: anti-nestin (1:250; Pharmingen, San Diego, CA), anti-microtubule-associated protein 2 (MAP2) (1:250; Chemicon,

Temecula, CA), anti-glial fibrillary acidic protein (GFAP) (1:250; Chemicon), and anti-CNPase (1:500; Chemicon). Primary antibodies were visualized by fluorescence microscopy using rhodamine- or FITC-conjugated secondary antibodies (1:1000; Cappel, Durham, NC).

RESULTS

PACAP-Induced Inhibition of Self-Renewal of Neural Progenitor Cells

Self-renewal of the primary neurosphere-forming cells was assessed by bFGF-induced formation of secondary neurospheres in the presence or absence of PACAP-38. After 7 days culture *in vitro*, secondary spheres contained a significant number of nestin-positive cells, indicating the presence of neural stem cells (data not shown). In the presence of 1 and 100 nM PACAP-38,

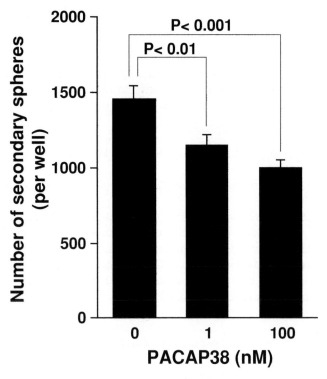

FIGURE 1. PACAP-38-induced inhibition of neural progenitor cell self-renewal. Primary neurosphere-forming cells were dissociated and cultured for 7 days with indicated concentrations of PACAP-38 in the presence of bFGF (10 ng/mL), and the number of the resultant secondary spheres was counted. Data are expressed as mean $\pm$ SE ($n = 16$–23) from three independent experiments. Statistically significant differences were assessed by ANOVA with Fisher's protected least square difference (PLSD) test.

TABLE 1. Cell type-specific differentiation of multipotential neural stem cells by PACAP-38

PACAP-38 (nM)	Marker-positive cells (% of total cells)		
	Nestin	MAP2	GFAP
0	31.0 ± 8.7	6.4 ± 2.2	2.77 ± 0.02
1	35.9 ± 2.0	9.4 ± 1.5	2.5 ± 1.2
10	30.3 ± 1.8	10.0 ± 2.0	1.5 ± 0.2
100	31.4 ± 0.1	15.2 ± 1.8*	2.0 ± 0.7

Multipotential neural stem cells were cultured for 5 days with indicated concentrations of PACAP-38, and the cell-type composition was analyzed by immunostaining. Data are expressed as mean ± SE from two or three independent experiments performed in duplicate. *$P < 0.05$ compared with 0 nM PACAP-38 by ANOVA with Fisher's PLSD test.

the number of secondary spheres decreased significantly by 21% and 31%, respectively (FIG. 1).

PACAP-Induced Differentiation into MAP2-Positive Neurons

Possible effects of PACAP on cell type-specific differentiation were assessed using cultured multipotential neural stem cells. After 5 days culture, PACAP-38 increased the number of MAP2-positive cells in a dose-dependent manner. By contrast, PACAP-38 did not affect the number of cells positive for nestin or GFAP (TABLE 1). Only a few cells were positive for the oligodendrocyte marker CNPase, irrespective of the presence or absence of PACAP-38 (data not shown).

DISCUSSION

In the present article, we showed that, in neural progenitor/stem cells, PACAP-38 inhibited bFGF-induced self-renewal, while it induced differentiation into neurons but not astrocytes. Although the underlying mechanisms remain to be investigated, anti-mitogenic actions of PACAP were likely to be involved in both processes.[4–6,11]

Previous studies show that PACAP and its receptors are expressed in mouse ES cell-derived neuronal progenitors and neurons, as well as in ES cells themselves.[7,8] These findings raise the possibility that PACAP is involved in neural development, even at a very early stage, and its effects are at least partly mediated via autocrine or paracrine mechanisms.

Recently developed PACAP- or PACAP receptor-deficient mice show a plethora of knockout phenotypes relevant to human diseases.[12] Future studies are necessary to investigate putative developmental and/or neuroplastic abnormalities in these mutant mice.

The present results support and extend previous findings concerning the role of PACAP receptor signaling in neural development, particularly in processes leading to neurogenesis.

ACKNOWLEDGMENTS

This research was supported, in part, by Grants-in-Aid for Scientific Research (A) and (B), and for Young Scientists (B) from Japan Society for the Promotion of Science; by a grant from Taisho Pharmaceutical Co. Ltd.; and by a Grant-in-Aid from the Japan Society for the Promotion of Science Research Fellows to M.H. She is also supported by Research Fellowships of the Japan Society for the Promotion of Science for Young Scientists.

REFERENCES

1. ARIMURA, A. 1998. Perspectives on pituitary adenylate cyclase activating polypeptide (PACAP) in the neuroendocrine, endocrine, and nervous systems. Jpn. J. Physiol. **48:** 301–331.
2. VAUDRY, D., B.J. GONZALEZ, M. BASILLE, et al. 2000. Pituitary adenylate cyclase-activating polypeptide and its receptors: from structure to functions. Pharmacol. Rev. **52:** 269–324.
3. JAWORSKI, D.M. & M.D. PROCTOR. 2000. Developmental regulation of pituitary adenylate cyclase-activating polypeptide and PAC_1 receptor mRNA expression in the rat central nervous system. Brain Res. Dev. Brain Res. **120:** 27–39.
4. LU, N. & E. DICICCO-BLOOM. 1997. Pituitary adenylate cyclase-activating polypeptide is an autocrine inhibitor of mitosis in cultured cortical precursor cells. Proc. Natl. Acad. Sci. USA. **94:** 3357–3362.
5. SUH, J., N. LU, A. NICOT, et al. 2001. PACAP is an anti-mitogenic signal in developing cerebral cortex. Nat. Neurosci. **4:** 123–124.
6. NICOT, A., V. LELIEVRE, J. TAM, et al. 2002. Pituitary adenylate cyclase-activating polypeptide and sonic hedgehog interact to control cerebellar granule precursor cell proliferation. J. Neurosci. **22:** 9244–9254.
7. CAZILLIS, M., B.J. GONZALEZ, C. BILLARDON, et al. 2004. VIP and PACAP induce selective neuronal differentiation of mouse embryonic stem cells. Eur. J. Neurosci. **19:** 798–808.
8. HIROSE, M., H. HASHIMOTO, N. SHINTANI, et al. 2005. Differential expression of mRNAs for PACAP and its receptors during neural differentiation of embryonic stem cells. Regul. Pept. **126:** 109–113.
9. REYNOLDS, B.A., W. TETZLAFF & S. WEISS. 1992. A multipotent EGF-responsive striatal embryonic progenitor cell produces neurons and astrocytes. J. Neurosci. **12:** 4565–4574.
10. JOHE, K.K., T.G. HAZEL, T. MULLER, et al. 1996. Single factors direct the differentiation of stem cells from the fetal and adult central nervous system. Genes Dev. **10:** 3129–3140.

11. CAREY, R.G., B. LI & E. DICICCO-BLOOM. 2002. Pituitary adenylate cyclase activating polypeptide anti-mitogenic signaling in cerebral cortical progenitors is regulated by p57^{Kip2}-dependent CDK2 activity. J. Neurosci. **22:** 1583–1591.
12. HASHIMOTO, H., N. SHINTANI & A. BABA. 2006. New insights into the central PACAPergic system from the phenotypes in PACAP- and PACAP receptor-knockout mice. Ann. N. Y. Acad. Sci. **1070:** 75–89.

Presence of PACAP and VIP in Embryonic Chicken Brain

RITA JOZSA,[a] TIBOR HOLLOSY,[a] JOZSEF NEMETH,[b]
ANDREA TAMÁS,[a] ANDREA LUBICS,[a] BALAZS JAKAB,[b]
ANDRAS OLAH,[a] AKIRA ARIMURA,[c] AND DORA REGLÖDI[a]

[a]Department of Anatomy, Neurohumoral Regulations Research Group of the
Hungarian Academy of Sciences, University of Pécs, 7624 Pécs, Hungary

[b]Departments of Pharmacology and Pharmacotherapy Neuropharmacology
Research Group of the Hungarian Academy of Sciences, University of Pécs,
7624 Pécs, Hungary

[c]US-Japan Biomedical Research Laboratories, Tulane University,
New Orleans, USA

ABSTRACT: The aim of the present article was to investigate the occurrence and temporary changes of pituitary adenylate cyclase-activating polypeptide (PACAP)-38 and vasoactive intestinal peptide (VIP) in various brain areas of chicken embryos by means of radioimmunoassay. The highest concentrations of PACAP-38 were measured in the brain stem followed by the hypothalamus, cerebellum, and telencephalon. PACAP-38 levels were significantly higher than those of VIP in all examined brain areas. The levels of both PACAP-38 and VIP showed a tendency to decrease until hatching during embryonic development of the chicken.

KEYWORDS: development; hypothalamus; telencephalon; cerebellum; brain stem

INTRODUCTION

The presence of pituitary adenylate cyclase-activating polypeptide (PACAP) has been shown in several vertebrate species, among which the sequence varies only by 1–6 amino acids.[1] In avians, the structure of PACAP-38 differs only by one amino acid from the mammalian peptide.[2,3] The presence and distribution of PACAP and its receptors and the gene encoding PACAP in the chicken are well known.[3] Several studies have been carried out on receptor binding and the effects of PACAP on signal transduction pathways in the chicken have also been shown.[3–5] However, less is known about the expression and functions of

Address for correspondence: Rita Jozsa, Department of Anatomy, University of Pécs, 7624 Pécs, Szigeti u 12. Hungary. Voice: +36-72-536001; ext.: 5398; fax: +36-72-536-393.
e-mail: rita.jozsa@aok.pte.hu

Ann. N.Y. Acad. Sci. 1070: 348–353 (2006). © 2006 New York Academy of Sciences.
doi: 10.1196/annals.1317.039

PACAP in the embryonic chicken brain. *In vitro*, the presence of PACAP has been shown in 3.5-embryonic day (ED)-old chick neuroblasts[6] and PACAP can be detected in nerves of the chicken gut as early as ED 4–5.[7] Early expression of PACAP is also known in other species, and is thought to play an important role during the development of the nervous system as shown by numerous *in vitro* and a few *in vivo* studies.[8,9] In chicken, *in vivo* studies have shown that PACAP plays a role in regulating apoptosis during development[10] and is involved in the development of social behavior, locomotion, and olfactory memory formation.[11,12] The primary aim of the present article was to investigate the presence of PACAP-38 and changes in PACAP-38 levels in different brain areas of the chicken embryo by means of radioimmunoassay (RIA).

PACAP shows closest structural homology to vasoactive intestinal peptide (VIP). Chicken VIP differs from its mammalian counterpart by only four amino acid residues.[13] The distribution of VIP-containing neuronal structures and sites of gene expression for VIP has been mapped in the chicken brain.[14] In the chicken embryonic nervous system, VIP immunoreactive elements have been described in the spinal cord,[15] sympathetic ganglia,[16] and in the mucosal plexus,[17] furthermore, pituitary cells have been reported to be responsive to VIP from ED15.[18] However, less is known about the embryonic appearance of VIP in the chicken brain. In the present article, we have compared PACAP-38 levels in the chicken embryonic brain with those of VIP, starting from ED15, the time when the concentration of these peptides can already be accurately measured in different parts of the chicken embryonic brain by RIA according to our preliminary observations.

MATERIALS AND METHODS

Fertilized broiler eggs were obtained from a commercial hatchery (Mohacs, Hungary). Eggs were incubated at 37.5°C with a relative humidity of 60%. Eggs were turned automatically every hour. Brains of chicken embryos at different developmental stages were removed daily (from ED15), and the brain areas, showing high concentrations of PACAP in posthatched chicken in a previous study,[11] were collected: brain stem (BS), hypothalamus (HTH), cerebellum (CER), and telencephalon (TEL). After weighing, tissues were homogenized in ice-cold distilled water. The homogenate was centrifuged (12,000 rpm, 4°C, 30 min) and the supernatant was transferred for PACAP-38 and VIP RIA.

PACAP-38 RIA procedure was performed as previously described.[19] Briefly, the antiserum "88111-3" was raised against a conjugate of Cys23-PACAP24–38 and bovine thyroglobulin coupled by carbodiimide in rabbit.[20] The tracer, mono-^{125}I-labeled ovine PACAP24–38 C-terminal fragment, was prepared in our laboratory. Ovine PACAP-38 was used as a RIA standard ranging from 0 to 1000 fmol/mL. Assays were prepared in 1 mL phosphate buffer (0.05 mol/L, pH 7.4) containing 0.1 mol/L sodium chloride, 0.25% (w/v) bovine serum albumine (BSA), and 0.05% (w/v) sodium azide. One hundred microliter anti-serum at a working dilution of 1:10,000, 100 μL RIA tracer (5000 cpm/tube), and 100 μL PACAP-38 standard or unknown samples were measured into polypropylene tubes with the assay buffer.

VIP RIA procedure was also performed according to previous descrip-tions.[21,22] Briefly, the antiserum "85/24" was raised against a conjugate of porcine VIP and bovine thyroglobulin coupled by glutaraldehyde in rabbit. The tracer, mono-^{125}I-labeled porcine VIP on Tyr[22] residue was prepared in our lab-oratory. Porcine VIP was used as RIA standard ranging from 0 to 100 fmol/mL. Assays were prepared in 1 mL phosphate buffer (0.02 mol/L, pH 6.5) contain-ing 0.05% (w/v) sodium azide, 0.1% (w/v) EDTA, 0.05% (w/v) polybrene, and 0.25% (w/v) BSA. One hundred microliter antiserum (working dilution 1:21,000), 100 μL RIA tracer (5000 cpm/tube), and 100 μL standard or un-known samples were measured into polypropylene tubes with assay buffer. After 48- to 72-h incubation at 4°C, the antibody-bound PACAP-38 or VIP was separated from the free one by addition of 100 μL separating solution containing 10 g charcoal, 1 g dextran, and 0.5 g commercial fat-free milk powder in 100 mL distilled water. Following centrifugation (3000 rpm, 4°C, 15 min) the tubes were gently decanted and the radioactivity of the precipitates was measured in a gamma counter. PACAP-38 and VIP concentrations of the unknown samples were read from a calibration curve and were expressed as mean (±SEM) fmol/mg wet tissue weight. Peptide concentrations were com-pared using analysis of variance (ANOVA) test, where statistical significance was considered at $P < 0.05$.

RESULTS AND DISCUSSION

The results of the RIA measurements show that both PACAP-38 and VIP are present at high concentrations during the second half of the embryonic devel-opmental period (FIGS. 1 and 2). PACAP-38 levels were significantly higher than those of VIP in all examined brain areas. Highest PACAP-38 concentra-tions were measured in the BS, followed by the HTH and CER, while TEL showed the lowest PACAP-38 levels among the investigated brain areas. Levels of PACAP-38 showed a tendency to decrease during the second half of embry-onic development in the chicken in almost every observed brain area, reaching statistical significance in the BS and HTH between ED15 and ED20. VIP levels also showed a slight decrease during development, which was significant only in the BS. Compared with PACAP-38, VIP levels were significantly lower in all examined brain areas, which might suggest that regulatory mechanisms me-diated by PACAP-38 are more dominant than those of VIP during the second half of embryonic development. Thus, the high concentration of PACAP-38 and lower levels of VIP during the second half of embryonic development may indicate important region- and stage-specific roles of these peptides in brain

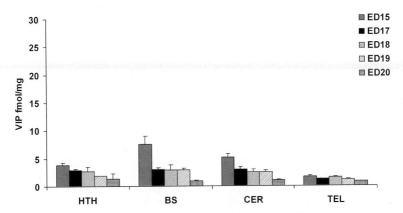

FIGURE 1. PACAP concentration in the HTH, BS, CER, and TEL in chicken embryos from ED15 to ED20. Results are given as mean ± SEM.

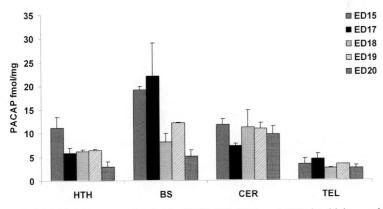

FIGURE 2. VIP concentration in the HTH, BS, CER, and TEL in chicken embryos from ED15 to ED20. Results are given as mean ± SEM.

development, the discussion of which is beyond the scope of the present article. In summary, the present study shows that both PACAP-38 and VIP are present in various brain areas in the chicken embryo and that PACAP-38 occurs at significantly higher concentrations than VIP.

ACKNOWLEDGMENTS

This work was supported by OTKA T046589, 043467, F048908, ETT 82/2003, 03597/2003, and the Hungarian Academy of Sciences. The authors thank Dora Omboli for her help.

REFERENCES

1. VAUDRY, D. *et al*. 2000. Pituitary adenylate cyclase activating polypeptide and its receptors: from structure to functions. Pharmacol. Rev. **52:** 269–324.
2. ARIMURA, A. 1998. Perspectives on pituitary adenylate cyclase activating polypeptide (PACAP) in the neuroendocrine, endocrine, and nervous systems. Jpn. J. Physiol. **48:** 301–331.
3. NOWAK, J.Z. & J.B. ZAWILSKA. 2003. PACAP in avians: origin, occurrence, and receptors—pharmacological and functional considerations. Curr. Pharm. Des. **9:** 467–481.
4. ZAWILSKA, J.B., P. NIEWIADOMSKI & J.Z. NOWAK. 2003. Characterization of vasoactive intestinal peptide/pituitary adenylate cyclase activating polypeptide receptors in chick cerebral cortex. J. Mol. Neurosci. **20:** 153–162.
5. NOWAK, J.Z. *et al*. 2002. Stimulatory effects of pituitary adenylate cyclase activating polypeptide on inositol phosphates accumulation in avian cerebral cortex and hypothalamus. Neurosci. Lett. **323:** 179–182.
6. ERHARDT, N.M. *et al*. 2001. Early expression of pituitary adenylate cyclase activating polypeptide and activation of its receptor in chick neuroblasts. Endocrinology **142:** 1616–1625.
7. SALVI, E.P., R. VACCARO & T.G. RENDA. 2000. Ontogeny of PACAP immunoreactivity in extrinsic and intrinsic innervation of chicken guts. Peptides **21:** 1703–1709.
8. WASHEK, J.A. 2002. Multiple actions of pituitary adenylyl cyclase activating peptide in nervous system development and regeneration. Dev. Neurosci. **24:** 14–23.
9. SOMOGYVÁRI-VIGH, A. & D. REGLODI. 2004. Pituitary adenylate cyclase activating polypeptide: a potential neuroprotective peptide. Review. Curr. Pharm. Des. **10:** 2861–2889.
10. ARIMURA, A. *et al*. 1994. PACAP functions as a neurotrophic factor. Ann. N. Y. Acad. Sci. **739:** 228–243.
11. HOLLOSY, T. *et al*. 2004. Effects of in ovo treatment with PACAP antagonist on general activity, motor and social behavior in chickens. Regul. Pept. **123:** 99–106.
12. JOZSA, R. *et al*. 2005. Pituitary adenylate cyclase activating polypeptide plays a role in olfactory memory formation in chicken. Peptides **26:** 2344–2350.
13. NOWAK, J.Z. *et al*. 2001. Vasoactive intestinal peptide-stimulated adenosine 3′,5′-cyclic monophosphate formation in cerebral cortex and hypothalamus of chick and rat: comparison of the chicken and mammalian peptide. Neurosci. Lett. **297:** 93–96.
14. KUENZEL, W. *et al*. 1997. Sites of gene expression for vasoactive intestinal polypeptide throughout the brain of the chick (Gallus domesticus). J. Comp. Neurol. **381:** 101–118.
15. DU, F., J.A. CHAYVIALLE & P. DUBOIS. 1988. Distribution and development of VIP immunoreactive neurons in the spinal cord of the embryonic and newly hatched chick. J. Comp. Neurol. **268:** 600–614.
16. NEW, H.V. & A.W. MUDGE, 1986. Distribution and ontogeny of SP, CGRP, SOM, and VIP in chick sensory and sympathetic ganglia. Dev. Biol. **116:** 337–346.
17. EPSTEIN, M.L., J. HUDIS & J.L. DAHL. 1983. The development of peptidergic neurons in the foregut of the chick. J. Neurosci. **3:** 2431–2447.
18. WOODS, K.L. & T.E. PORTER. 1998. Ontogeny of prolactin-secreting cells during chick embryonic development: effect of vasoactive intestinal peptide. Gen. Comp. Endocrinol. **112:** 240–246.

19. JAKAB, B. *et al.* 2004. Distribution of PACAP-38 in the central nervous system of various species determined by a novel radioimmunoassay. J. Biochem. Biophys. Meth. **61:** 189–198.
20. ARIMURA, A. *et al.* 1991. Tissue distribution of PACAP as determined by RIA: highly abundant in the rat brain and testes. Endocrinology **129:** 2787–2789.
21. NEMETH, J. *et al.* 2002. Comparative distribution of VIP in the central nervous system of various species measured by a new radioimmunoassay. Regul. Pept. **109:** 3–7.
22. NÉMETH, J. *et al.* 2002. ^{125}I-labelling and purification of peptide hormones and bovine serum albumin. J. Radioanal. Nucl. Chem. **251:** 129–133.

Short-Term Fasting Differentially Alters PACAP and VIP Levels in the Brains of Rat and Chicken

RITA JOZSA,[a] JOZSEF NEMETH,[b] ANDREA TAMAS,[a]
TIBOR HOLLOSY,[a] ANDREA LUBICS,[a] BALAZS JAKAB,[b]
ANDRAS OLAH,[a] ISTVAN LENGVARI,[a] AKIRA ARIMURA,[c]
AND DORA REGLÖDI[a]

[a]Department of Anatomy, Neurohumoral Regulations Research Group of the Hungarian Academy of Sciences, University of Pécs, 7624 Pécs, Hungary

[b]Department of Pharmacology and Pharmacotherapy Neuropharmacology Research Group of the Hungarian Academy of Sciences, University of Pécs, 7624 Pécs, Hungary

[c]US-Japan Biomedical Research Laboratories, Tulane University, New Orleans, Louisiana 70037, USA

ABSTRACT: The present article investigated the levels of pituitary adenylate cyclase-activating polypeptide (PACAP) and vasoactive intestinal polypeptide (VIP) in the brains of rats and chickens 12, 36, and 84 h after starvation. PACAP levels increased in both species, 12 h after food deprivation in rats, and with a 24-h delay in chickens. VIP levels showed a more complex pattern: a gradual increase in the hypothalamus and telencephalon, and a significant decrease in the brain stem of rats. In chickens, a decrease was observed in every brain area after 36 h of starvation. These data show that PACAP and VIP are differentially regulated and are involved in the regulatory processes under a food-restricted regimen, and are differentially altered in nocturnal and diurnal species.

KEYWORDS: starvation; radioimmunoassay; hypothalamus; brain stem; telencephalon

INTRODUCTION

Both vasoactive intestinal polypeptide (VIP) and pituitary adenylate cyclase-activating polypeptide (PACAP), and their receptors are expressed in the areas involved in the regulation of feeding in rats and chickens.[1–5] Several lines of

Address for correspondence: Dora Reglodi, M.D., Ph.D., Department of Anatomy, University of Pécs, 7624 Pécs, Szigeti u 12. Hungary. Voice: +36-72-536001; ext.: 5398; fax: +36-72-536 393.
e-mail: dora.reglodi@aok.pte.hu

Ann. N.Y. Acad. Sci. 1070: 354–358 (2006). © 2006 New York Academy of Sciences.
doi: 10.1196/annals.1317.044

evidence suggest that both peptides are involved in regulation of feeding mechanisms in various species.[5-9] In addition to the central role of the hypothalamus in regulation of feeding, brain stem and telencephalon are also involved in the integration of feeding behavior.[10] Therefore, the aim of the present article was to investigate by means of radioimmunoassay (RIA), whether PACAP and VIP levels change in the hypothalamus, brain stem, and telencephalon of two species: rats and chickens, after different times of food deprivation.

MATERIALS AND METHODS

Adult male Wistar rats and broiler chickens were used. Food deprivation was started at 8 PM, and animals were sacrificed 12, 36, and 84 h after the beginning of starvation ($n = 6$ in each group). Water was available *ad libitum*. Control animals were sacrificed at each time point (both rats and chicken $n = 6$). Hypothalamus, brain stem, and telencephalon were removed, weighed, and further processed for RIA analysis of PACAP and VIP content. RIA procedure was performed as previously described.[11,12] Briefly, PACAP and VIP antisera (88111–3, 1:10,000 and 85/24, 1:21,0000, respectively) raised in rabbit were used.[12,13] *Tracer:* mono-^{125}I-labeled ovine PACAP24–38 and porcine VIP were prepared in our laboratory (5000 cpm/tube).[14] *Standard:* ovine PACAP-38 and porcine VIP were used as a RIA standard ranging from 0 to 1000 and 0 to 100 fmol/mL, respectively. *Buffer:* PACAP assay was prepared in 1 mL 0.05 mol/L (pH 7.4), VIP assay in 1 mL 0.02 mol/L (pH 6.5) phosphate buffer-containing sodium chloride, BSA, EDTA, and sodium azide. *Incubation time:* 48–72 h incubation at 4°C. *Separation solution:* charcoal/dextran/milk powder (10:1:0.5 g in 100 mL distilled water). Detection limit for PACAP-38 is

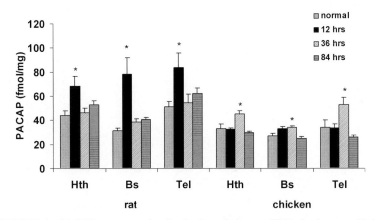

FIGURE 1. PACAP concentration in the hypothalamus (Hth), brain stem (Bs), and telencephalon (Tel) of rats and chickens after different times of starvation. Data are expressed as mean ± SEM. *$P < 0.05$ versus *ad libitum* fed, normal state.

2 fmol/mL, for VIP 0.1 fmol/mL. Statistical comparisons were made using the analysis of variance (ANOVA) test.

RESULTS AND DISCUSSION

PACAP and VIP contents of the investigated brain areas are summarized in FIGURE 1 (PACAP) and FIGURE 2 (VIP). In rats, a significant elevation of PACAP was observed after 12 h of starvation in the hypothalamus, brain stem, and telencephalon. At later time points, levels returned to those measured in normal control animals. In the chicken, a pattern similar to that observed in rats was found in the PACAP levels. However, the significant elevation was observed 36 h after food deprivation, so there was a 24-h delay compared to rats (FIG. 1).

In contrast to the one peak observed in the PACAP levels, VIP concentrations showed a gradual increase in the rat hypothalamus and telencephalon, then levels returned to nearly *ad libitum* fed state. However, these changes were not significant. In the brain stem, a significant decrease was observed after 84 h of starvation in rats. VIP levels in the chicken showed an opposite pattern to that found with PACAP: levels gradually decreased, and it was significant 36 h after food deprivation. By 84 h, concentrations returned to normal, except in the telencephalon (FIG. 2).

In summary, our results show that PACAP levels are increased in both species, 12 h after food deprivation in rats, and 24 h later in chickens. VIP levels show a more complex pattern: a gradual increase in the hypothalamus and telencephalon, and a significant decrease in the brain stem in rats. In chickens, a significant decrease was observed in every brain area after 36 h of starvation.

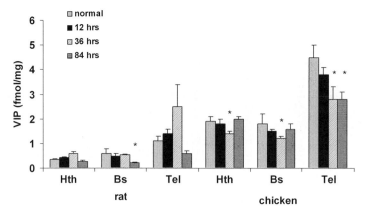

FIGURE 2. VIP concentration in the hypothalamus (Hth), brain stem (Bs), and telencephalon (Tel) of rats and chickens after different times of starvation. Data are expressed as mean ± SEM. *$P < 0.05$ versus *ad libitum* fed, normal state.

In addition to the main neuropeptides (CCK and NPY) involved in the control of food intake and regulation of complex feeding behaviors, many other neuropeptides have been shown to participate in these processes. Several lines of evidence suggest that both PACAP and VIP are involved in the regulation of feeding mechanisms in mammals, chicken, and goldfish.[5–9] In the chicken, both peptides inhibit feeding when injected intracerebroventricularly (i.c.v.),[5] although by different mechanisms.[15,16] In mammals, results are more inconsistent: VIP has been shown to both reduce and increase food intake,[5,17] while PACAP reduces food intake both in rats and mice.[6,7] As far as the levels of these peptides in different feeding states are concerned, relatively few data are available. In a recent study, PACAP and PAC1 receptor mRNA has been shown to increase after 7 days of excessive feeding, while there was no change after 7 days of starvation in goldfish.[9] VIP levels have been found to be unchanged in the rat brain after 4 or 5 days of fasting.[18,19]

The exact mechanisms by which PACAP and VIP participate in regulating feeding are still not clear. However, our present results provide further evidence that these peptides are involved in regulatory processes under a food-restricted regimen, and their levels respond to feeding states. In accordance with other findings,[5] the function of PACAP seems to be more conserved between species, since the pattern of change was the same in rats and chicken. The nocturnal and diurnal feeding behavior may be responsible for the 24-h delay observed in chicken. However, the pattern of changes in VIP levels was different from that of PACAP and was different between the two species. This indicates, in accordance with other findings,[15] that the two peptides are differentially involved in regulatory processes of feeding.

ACKNOWLEDGMENTS

This work was supported by the National Science Research Fund (OTKA T046589, 043467, F048908, ETT 82/2003, 03597/2003), the Hungarian Academy of Sciences, and Szechenyi Scholarship (J. Nemeth). The authors thank Brian K. Lucas and Dora Omboli for their help.

REFERENCES

1. ZHOU, C.J. *et al.* 2002. PACAP and its receptors exert pleiotropic effects in the nervous system by activating multiple signaling pathways. Curr. Prot. Pept. Sci. **3:** 423–439.
2. NOWAK, J.Z. & J.B. ZAWILSKA. 2003. PACAP in avians: origin, occurrence, and receptors—pharmacological and functional considerations. Curr. Pharm. Des. **9:** 467–481.
3. PEETERS, K., L.R. BERGHMAN & F. VANDESANDE. 1998. Comparative distribution of pituitary adenylate cyclase activating polypeptide and vasoactive intestinal

polypeptide immunoreactivity in the chicken forebrain. Ann. N. Y. Acad. Sci. **839:** 417–419.

4. KOVES, K. *et al.* 1991. Comparative distribution of immunoreactive pituitary adenylate cyclase activating polypeptide and vasoactive intestinal polypeptide in rat forebrain. Neuroendocrinology **54:** 159–169.

5. TACHIBANA, T. *et al.* 2003. Intracerebroventricular injection of vasoactive intestinal peptide and pituitary adenylate cyclase activating polypeptide inhibits feeding in chicks. Neurosci. Lett. **339:** 203–206.

6. CHANCE, W.T. *et al.* 1995. Anorectic and neurochemical effects of pituitary adenylate cyclase activating polypeptide in rats. Peptides **16:** 1511–1516.

7. MORLEY, J.E. *et al.* 1992. Pituitary adenylate cyclase activating polypeptide (PACAP) reduces food intake in mice. Peptides **13:** 1133–1135.

8. MATSUDA, K. *et al.* 2005. Inhibitory effects of pituitary adenylate cyclase activating polypeptide (PACAP) and vasoactive intestinal peptide (VIP) on food intake in the goldfish, Carassius auratus. Peptides **26:** 1611–1616.

9. MATSUDA, K. *et al.* 2005. Anorexigenic action of pituitary adenylate cyclase activating polypeptide (PACAP) in the goldfish: feeding-induced changes in the expression of mRNAs for PACAP and its receptors in the brain, and locomotor response to central injection. Neurosci. Lett. **386:** 9–13.

10. GRILL, H.J. & J.M. KAPLAN. 2002. The neuroanatomical axis for control of energy balance. Front Neuroendocrinol. **23:** 2–40.

11. JAKAB, B. *et al.* 2004. Distribution of PACAP-38 in the central nervous system of various species determined by a novel radioimmunoassay. J. Biochem. Biophys. Meth. **61:** 189–198.

12. NÉMETH, J. *et al.* 2002. Comparative distribution of VIP in the central nervous system of various species measured by a new radioimmunoassay. Regul. Pept. **109:** 3–7.

13. ARIMURA, A. *et al.* 1991. Tissue distribution of PACAP as determined by RIA: highly abundant in the rat brain and testes. Endocrinology **129:** 2787–2789.

14. NÉMETH, J. *et al.* 2002. ^{125}I-labelling and purification of peptide hormones and bovine serum albumin. J. Radioanal. Nucl. Chem. **251:** 129–133.

15. TACHIBANA, T. *et al.* 2003. Pituitary adenylate cyclase activating polypeptide and vasoactive intestinal peptide inhibit feeding in the chick brain by different mechanisms. Neurosci. Lett. **348:** 25–28.

16. TACHIBANA, T. *et al.* 2004. Anorexigenic effects of pituitary adenylate cyclase activating polypeptide and vasoactive intestinal peptide in the chick brain are mediated by corticotrophin-releasing factor. Regul. Pept. **120:** 99–105.

17. KULKOSKY, P.J. *et al.* 1989. Vasoactive intestinal peptide: behavioral effects in the rat and hamster. Pharmacol. Biochem. Behav. **34:** 387–393.

18. ZHENG, B. *et al.* 1987. Brain/gut peptides in fed and fasted rats. Endocrinology **120:** 714–717.

19. SHULKES, A. *et al.* 1983. Starvation in the rat: effect on peptides of the gut and brain. Aust. J. Exp. Biol. Med. Sci. **61:** 581–587.

VIP Decreases TLR4 Expression Induced by LPS and TNF-α Treatment in Human Synovial Fibroblasts

Y. JUARRANZ,[a] I. GUTIÉRREZ-CAÑAS,[b] A. ARRANZ,[a] C. MARTÍNEZ,[c] C. ABAD,[a] J. LECETA,[a] J.L. PABLOS,[b] AND R.P. GOMARIZ[a]

[a]Departamento de Biología Celular, Facultad de Biología, Universidad Complutense de Madrid, 28040 Madrid, Spain

[b]Servicio de Reumatología y Unidad de Investigación, Hospital 12 de Octubre, 28040 Madrid, Spain

[c]Departamento de Biología Celular, Facultad de Medicina, Universidad Complutense de Madrid, 28040 Madrid, Spain

ABSTRACT: It has been demonstrated that VIP produces beneficial effects both in a murine model of rheumatoid arthritis and in human rheumatoid synovial fibroblasts through the modulation of proinflammatory mediators. Toll-like receptors (TLRs) play a key role in the immediate recognition of microbial surface components by immune cells prior to the development of adaptative microbe-specific immune responses. In this study, we demonstrate that VIP decreases lipopolysaccharide (LPS) and TNF-α-induced expression of TLR4 and its correlation with the production of CCL2 and CXCL8 chemokines in human synovial fibroblasts from patients with rheumatoid arthritis and osteoarthritis. Our results add a new step for the use of VIP, as a promising candidate, for the treatment of rheumatoid arthritis.

KEYWORDS: VIP; TLR4; TLR2; rheumatoid arthritis; CCL2; CXCL8

INTRODUCTION

Toll-like receptors (TLRs) belong to a family of pattern recognition units that allow the innate immune system to distinguish microbial and other foreign structures from self.[1] TLRs mediate the recruitment of immunological signaling cascades that activate nuclear factor (NF)-κB and other transcription factors capable of inducing proinflammatory immune responses.[2] The TLR family comprises at least 11 members, of which TLR4 was the first described

Address for correspondence: Yasmina Juarranz, Departamento de Biología Celular, Facultad de Biología, UCM, 28040 Madrid, Spain. Voice: 34-91-394-4971; fax: 34-91-394-4981.
e-mail: yashina@bio.ucm.es

Ann. N.Y. Acad. Sci. 1070: 359–364 (2006). © 2006 New York Academy of Sciences.
doi: 10.1196/annals.1317.045

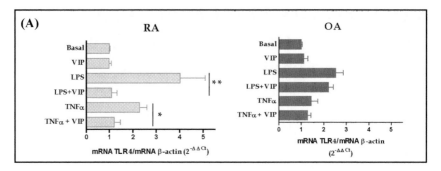

(A)

RA

OA

mRNA TLR4/mRNA β-actin $(2^{-\Delta\Delta Ct})$

mRNA TLR4/mRNA β-actin $(2^{-\Delta\Delta Ct})$

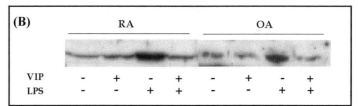

(B)

RA   OA

| VIP | – | + | – | + | – | + | – | + |
| LPS | – | – | + | + | – | – | + | + |

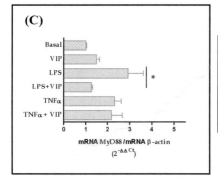

(C)

mRNA MyD88 /mRNA β-actin $(2^{-\Delta\Delta Ct})$

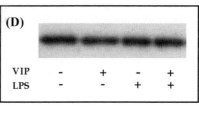

(D)

| VIP | – | + | – | + |
| LPS | – | – | + | + |

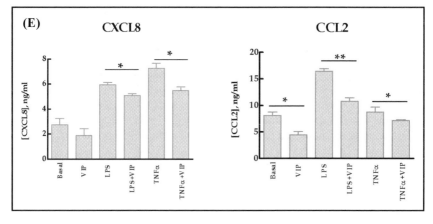

(E) CXCL8   CCL2

[CXCL8], ng/ml

[CCL2], ng/ml

in humans.[3] This receptor is essential for lipopolysaccharide (LPS) signaling, which also requires an additional molecule, MD-2, which associates with the extracellular portion of TLR4.[2] There is growing evidence of the involvement of TLRs in the pathogenesis of autoimmune disorders, such as Crohn's disease or rheumatoid arthritis (RA). RA is a chronic inflammatory, systemic, and autoimmune disorder that mainly affects the synovial tissues in multiple joints.[4] Activation and proliferation of fibroblasts-like synoviocytes (FLS) contribute to the pannus formation and lead to joint destruction in patients with RA.[4,5] The pathogenesis of RA is still largely unknown; however, on the basis of the fact that bacterial DNA containing CpG motifs and bacterial cell wall fragments have been detected in the synovial fluid of patients with RA, TLR signaling could have an important role in the pathogenesis of this disease.[6] Recently, a close relation between the expression of TLR2 and TLR4 in human rheumatoid synovial tissues and proinflammatory cytokines production has been described.[7]

Vasoactive intestinal polypeptide (VIP) is a pleiotropic peptide, produced by immune cells, showing anti-inflammatory and immunoregulatory properties.[8] The effects of VIP lead to the amelioration or prevention of several inflammatory and autoimmune disorders in animal models of septic shock, Crohn's disease, or RA.[8–11] Related to RA, VIP prevents bone destruction in a collagen-induced arthritis murine model.[12] Our previous data have also shown that VIP modulates the synthesis of proinflammatory mediators in human rheumatoid synovial fibroblasts.[13] Recently, it has been described for the first time a relation between VIP treatment and the decrease in TLR2 and TLR4 expression in TNBS-induced colitis, a model of Crohn's disease.[14]

To examine the anti-inflammatory role of VIP in RA, we study the effect of VIP in TLR4 expression and in its signaling in human RA synovial fibroblasts, as important mediators of the inflammatory response in RA.

FIGURE 1. (**A**) The effect of VIP on TLR4 mRNA expression was studied by quantitative real-time PCR in cultured FLS from patients with RA or OA under basal (nonstimulated) conditions and after stimulation with 20 μg/mL LPS or 10 nM TNF-α. (**B**) Western blot analysis of TLR4 protein expression in cultured FLS unstimulated or stimulated with LPS at the indicated concentrations in RA-FLS and OA-FLS. A representative experiment is shown. (**C**) Quantitative real-time PCR for MyD88 expression in cultured FLS from patients with RA under the same conditions described in A. (**D**) Western blot analysis of MyD88 protein expression in cultured RA-FLS. A representative experiment is shown. (**E**) Effect of VIP on CCL2 and CXCL8 production in cultured RA-FLS determined by ELISA method in the same conditions described previously in (**A**). Results from quantitative real-time PCR were corrected in all case by β-actin mRNA expression within each sample. Results are the mean $\pm$ SEM of two experiments performed in duplicate, including three FLS lines from different patients with RA or OA.

RESULTS AND DISCUSSION

Firstly, we studied the constitutive expression of TLR4 by quantitative real-time polymerase chain reaction and Western blot in FLS from patients with RA or osteoarthrosis (OA). mRNA levels and protein levels are higher in RA-FLS than OA-FLS (data not shown), pointing out an important role of this receptor in the pathogenesis of arthritis and being in agreement with previous results.[7] In both diseases, LPS was able to increase significantly TLR4 expression, with a higher degree in RA-FLS (FIG. 1A). After stimulation with the proinflammatory cytokine tumor necrosis factor-α (TNF-α), TLR4 mRNA and protein levels were lightly increased in RA-FLS and OA-FLS, compared with LPS stimulation (FIG. 1A, B). The addition of VIP resulted in a decrease of TLR4 at mRNA and protein levels in both OA and RA samples showing a higher effect in RA-FLS (FIG. 1A, B). These results confirm the anti-inflammatory role of VIP in human RA[10,12,13] with the involvement of TLR pathway, confirming previous results in a murine model of Crohn's disease.[14] MyD88 was the first adaptor protein described as being involved in TLRs signaling of the four adaptors' proteins described to date.[2,3] This cytosolic molecule is required for signaling by all TLRs, with probably the exception of TLR3[3] and is essential for the production of inflammatory molecules. As MyD88 is involved in the streptococcal model of arthritis,[15] we have studied its possible involvement in OA and RA as well as the effect of VIP. As no significant differences between RA-FLS and OA-FLS were detected, we studied the LPS, TNF-α, and VIP effect in RA-FLS. Only the presence of LPS in RA-FLS cultures increased in a significant manner the mRNA expression for MyD88 whereas VIP also decreased this stimulated mRNA expression (FIG. 1C). In contrast, this fact was not confirmed at protein level (FIG. 1D). Further studies are needed to explain this discrepancy.

Proinflammatory chemokines represent an important group of mediators that can be induced in RA-FLS in response to cytokines and TLR ligands. To analyze whether the observed changes in TLR4 expression were correlated with functional changes, we studied by ELISA assays the production of two important chemokines in response to LPS- and TNF-α stimulation, CCL2 (MCP-1) and CXCL8 (IL-8). The release of both chemokines by FLS cultures after LPS- and TNF-α stimulation was induced significantly (FIG. 1E). Addition of VIP produced a significant decrease of LPS- and TNF-α-induced chemokines levels. Moreover, VIP had also effect on basal release of CCL2 and CXCL8 (FIG. 1E). The transcription of both chemokines is under NF-κB control.[16] TLR and cytokines receptors share common intracellular pathways where NF-κB plays an important role.[2,16] Therefore, a possible explanation of the VIP-mediated effect on TLR4 (FIG. 2) could be its *in vivo* and *in vitro* known negative effect on NF-κB activation[8,12] previously described in an animal model of RA.[12] In conclusion, the VIP effect in TLR4 signaling in humans adds a new step for the use of VIP, as promising candidate, for the treatment of RA.

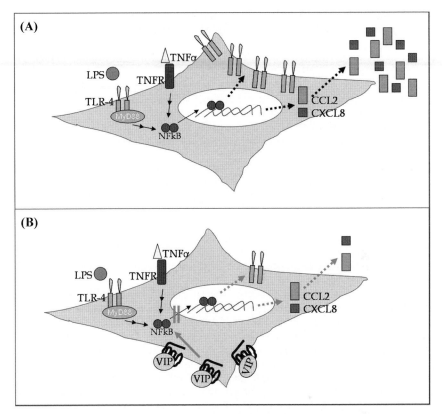

FIGURE 2. Hypothesis of VIP effect on TLR4 expression and TLR4-mediated chemokines production in human rheumatoid synovial fibroblast. (**A**) LPS or TNF-α stimulation of FLS, results in an increase of TLR4 expression and TLR4-mediated CXCL8 and CCL2 production. (**B**) The addition of VIP to culture medium in the presence of both stimulus decreases TLR4 expression and TLR4-mediated chemokines.

ACKNOWLEDGMENTS

This work was supported by grants BFI 2002-03489 from Ministerio de Ciencia y Tecnología (Spain), G03/152 from Fondo de Investigación Sanitaria (Spain), a predoctoral fellowship from Ministerio de Ciencia y Tecnología (to AA), and a postdoctoral fellowship from Comunidad de Castilla la Mancha (to IGC).

REFERENCES

1. SEIBL, R., D. KYBURZ, R.P. LAUENER, *et al.* 2004. Pattern recognition receptors and their involvement in the pathogenesis of arthritis. Curr. Opin. Rheumatol. **16:** 411–418.

2. AKIRA, S. & K. TAKEDA. 2004. Toll-like receptor signalling. Nat. Rev. **4:** 499–511.
3. O'NEILL, L.A.J. 2004. TLRs: Professor Mechnikov, sit on your hat. Trends Immunol. **25:** 687–693.
4. FELDMANN, M., F.M. BRENNAN & R.N. MAINI. 1996. Role of cytokines in rheumatoid arthritis. Ann. Rev. Immunol. **14:** 397–440.
5. MOR, A., S.B. ABRAMSON & M.H. PILLINGER. 2005. The fibroblast-like synovial cell in rheumatoid arthritis: a key player in inflammation and joint destruction. Clin. Immunol. **115:** 118–128.
6. OSPELT, C., D. KYBURZ, M. PIERER, et al. 2004. Toll-like receptors in rheumatoid arthritis joint destruction mediated by two different pathways. Ann. Rheum. Dis. **63:** ii90–ii91.
7. RADSTAKE, T.R.D.J., M.F. ROELOFS, Y.M. JENNISKENS, et al. 2004. Expression of toll-like receptors 2 and 4 in rheumatoid synovial tissue and regulation by proinflammatory cytokines interleukin-12 and -18 via interferon-γ. Arthritis Rheum. **50:** 3856–3865.
8. GOMARIZ, R.P., C. MARTINEZ, C. ABAD, et al. 2001. Immunology of VIP: a review and therapeutical perspectives. Curr. Pharm. Des. **7:** 89–111.
9. MARTINEZ, C., C. ABAD, M. DELGADO, et al. 2002. Anti-inflammatory role in septic shock of pituitary adenylate cyclase-activating polypeptide receptor. Proc. Natl. Acad. Sci. USA **99:** 1053–1058.
10. DELGADO, M., C. ABAD, C. MARTINEZ, et al. 2001. Vasoactive intestinal peptide prevents experimental arthritis by downregulating both autoimmune and inflammatory components of the disease. Nat. Med. **7:** 563–568.
11. ABAD, C., C. MARTINEZ, M.G. JUARRANZ, et al. 2003. Therapeutic effects of vasoactive intestinal peptide in the trinitrobenzene sulfonic acid mice model of Crohn's disease. Gastroenterology **124:** 961–971.
12. JUARRANZ, Y., C. ABAD, C. MARTINEZ, et al. 2005. Protective effect of vasoactive intestinal peptide on bone destruction in the collagen-induced arthritis model of rheumatoid arthritis. Arthritis Res. Ther. **7:** R1034–R1045.
13. JUARRANZ, M.G., B. SANTIAGO, M. TORROBA, et al. 2004. Vasoactive intestinal peptide modulates proinflammatory mediator synthesis in osteoarthritic and rheumatoid synovial cells. Rheumatology **43:** 416–422.
14. GOMARIZ, R.P., A. ARRANZ, C. ABAD, et al. 2005. Time-course expression of Toll-like receptors 2 and 4 in inflammatory bowel disease and homeostatic effect of VIP. J. Leukoc. Biol. Apr 27 **78:** 491–502.
15. JOOSTEN, L.A., M.I. KOENDERS, R.L. SMEETS, et al. 2003. Toll-like receptor 2 pathway drives streptococcal cell wall-induced joint inflammation: critical role of myeloid differentiation factor 88. J. Immunol. **171:** 6145–6153.
16. BONIZZI, G. & M. KARIN. 2004. The two NF-κB activation pathways and their role in innate and adaptive immunity. TRENDS Immunol. **25:** 280–288.

Effects of Systemic PACAP Treatment in Monosodium Glutamate-Induced Behavioral Changes and Retinal Degeneration

P. KISS,[a] A. TAMÁS,[a] A. LUBICS,[a] I. LENGVÁRI,[a] M. SZALAI,[a] D. HAUSER,[a] Z.S. HORVATH,[a] B. RACZ,[b] R. GABRIEL,[c] N. BABAI,[c] G. TOTH,[d] AND D. REGLÓDI[a]

[a]Departments of Anatomy (Neurohumoral Regulations Research Group of the Hungarian Academy of Sciences), Pécs University, 7624 Pécs, Hungary
[b]Departments of Surgical Research and Techniques, Pécs University, 7624 Pécs, Hungary
[c]Departments of General Zoology and Neurobiology (MTA-PTE Adaptational Biology), Pécs University, 7624 Pécs, Hungary
[d]Department of Medical Chemistry, University of Szeged, Szeged Hungary

ABSTRACT: The present article investigated effects of systemic pituitary adenylate cyclase-activating polypeptide (PACAP) treatment in monosodium glutamate (MSG)-induced retinal degeneration and neurobehavioral alterations in neonatal rats. It was found that the dose of PACAP that effectively enhances neurobehavioral development in normal rats was able to counteract the retarding effect of MSG on righting, forelimb placing, and grasp reflexes and caused a significant amelioration of the righting and gait reflex performance and motor coordination at 2 weeks of age. In the retina, significant amelioration of neuronal loss in the inner retinal layers was achieved, but it was much less than that observed by local administration.

KEYWORDS: neonatal rat; retinotoxicity; neurobehavioral development

INTRODUCTION

Pituitary adenylate cyclase-activating polypeptide (PACAP) plays an important role in the development of the nervous system, exerting a variety of growth factor-like actions *in vitro*.[1,2] *In vivo*, local administration of PACAP increases the volume of the cerebellar cortex in 8-day-old rats,[3] attenuates

Address for correspondence: Dora Reglódi, M.D., Ph.D., Department of Anatomy, University of Pécs, 7624 Pécs, Szigeti u 12. Hungary. Voice: +36-72-536001; fax: +36-72-536393.
e-mail: dora.reglodi@aok.pte.hu

Ann. N.Y. Acad. Sci. 1070: 365–370 (2006). © 2006 New York Academy of Sciences.
doi: 10.1196/annals.1317.046

excitotoxic injury during brain development,[4] and reduces the number of degenerating cells in chicken dorsal root ganglia and spinal cord when given *in ovo*.[5] We have shown that treatment of neonatal rats with PACAP during the first 2 weeks of postnatal life enhances neurobehavioral development.[6] Treatment with PACAP- or vasoactive intestinal polypeptide (VIP) antagonists has been found to retard the development of various neural reflexes.[6,7] Monoseodium glutamate (MSG) is used as a food additive that leads to several profound changes when given to neonatal rats. It causes degeneration in various brain areas and the retina, and leads to neurochemical and behavioral alterations.[8] Recently, we have demonstrated that PACAP significantly attenuates the severe MSG-induced degeneration of the retina when administered locally, into the vitreous body.[9,10] Given that PACAP effectively crosses the blood-brain barrier, a few studies have shown neuroprotective effects of PACAP in adult rats when given systemically.[11,12] The aim of the present article was to investigate whether systemic treatment with PACAP is able to attenuate MSG-induced retinal degeneration and neurobehavioral alterations in neonatal rats.

MATERIALS AND METHODS

Litters of Wistar rats were used from postnatal day (PD) 1. Procedures were performed in accordance with approved ethical guidelines. Based on previous observations that there is no gender difference in neurobehavioral development during the first 3 weeks,[8] we used both male and female pups. Drug administration was performed as previously described.[8] Briefly, pups received 4 mg/g body weight MSG dissolved in saline subcutaneously on PD 1, 3, 5, 7, and 9. PACAP treatment was performed during the first 14 days, a period during which the blood-brain barrier is not yet complete and which is a critical phase of neurobehavioral maturation.[13] Rats ($n = 15$ in all groups, from mixed dams) received 1-μg PACAP-38 subcutaneously, a dose that was shown to enhance neurobehavioral development,[6] while control group received only daily saline injections. In order to compare development to normal pups, a group of rats received only saline injection without MSG/or PACAP treatment. Physical signs, body weight, and neural reflexes were tested daily during the first 3 weeks. The day of appearance of the following signs were examined: eye opening, incisor eruption, ear unfolding, negative geotaxis and ear twitch, eyelid, limb placing and grasp, gait, auditory startle, and air righting reflexes. In cases of negative geotaxis, air righting and gait reflexes, also the time to perform the tasks was measured. As a measure of motor coordination, foot-fault test was performed at 2 and 3 weeks of age. Detailed behavioral testing has been described elsewhere.[6–8,13,14]

The most well-established morphological effect of neonatal MSG treatment is the destruction of neurons in the arcuate nucleus and the retina. Therefore, histological analysis of these areas was performed as previously described.[8,9]

After completing the behavioral testing, animals were sacrificed at PD 21. Half of the pups from each group were perfused with 4% paraformaldehyde and sections from the arcuate nucleus were further processed for histological staining with toluidine blue and for immunohistological staining with tyrosine-hydroxylase antiserum. The retinas of the other half of the animals were removed and were processed for histological analysis. Semithin sections were stained with toluidine blue.

RESULTS AND DISCUSSION

Appearance of physical characteristics and neural reflexes are summarized in TABLE 1. MSG treatment led to retardation of forelimb placing, forelimb grasp, and air righting reflexes, while in PACAP-treated animals, these reflexes appeared significantly earlier, similar to normal rats. MSG is also known to cause retardation of somatic development, as indicated by the reduced weight of the animals (TABLE 2). By the end of the observation period, MSG-treated rats weighed by almost one third less than normal rats. The average body weight of PACAP-treated animals was about 1 g more than MSG-treated controls every observed day starting from day 6, but differences were not significant. MSG-treated pups performed worse in righting and gait reflexes that were ameliorated by PACAP treatment at 2 weeks of age (TABLE 2). Foot-fault test has been shown to be a good indicator of MSG-induced retardation of motor coordination development[8] and the number of foot faults was less in the PACAP-treated group than in the MSG-treated pups, but only at 2 weeks of

TABLE 1. Days of appearance of physical and neurological signs in normal, untreated, and MSG-treated rats

	Days of appearance		
Signs	Untreated	MSG	MSG+PACAP
Eye opening	14.53 ± 0.18	13.94 ± 0.17	13.86 ± 0.20
Incisor eruption	9.84 ± 0.18	9.64 ± 0.18	9.78 ± 0.16
Ear unfolding	13.23 ± 0.16	13.35 ± 0.11	13.21 ± 0.11
Negative geotaxis	11.07 ± 0.32	11.76 ± 0.34	10.57 ± 0.28
Ear twitch reflex	15.23 ± 0.50	15.00 ± 0.51	14.71 ± 0.59
Eyelid reflex	12.92 ± 0.43	13.11 ± 0.20	13.21 ± 0.23
Forelimb placing	10.46 ± 0.46	12.17 ± 0.32**	11.93 ± 0.26*#
Hindlimb placing	20.15 ± 0.31	20.64 ± 0.20	20.00 ± 0.31
Forelimb grasp	6.15 ± 0.31	7.05 ± 0.32*	6.07 ± 0.23#
Hindlimb grasp	8.31 ± 0.26	8.76 ± 0.32	8.35 ± 0.19
Gait	10.15 ± 0.63	10.12 ± 0.50	9.35 ± 0.28
Auditory startle	13.61 ± 0.21	13.94 ± 0.30	13.71 ± 0.19
Air righting	4.76 ± 0.30	6.35 ± 0.41**	5.71 ± 0.19#

Data are expressed as mean $\pm$ SEM. *$P < 0.05$, **$P < 0.01$ vs. untreated rats, #$P < 0.05$ vs. MSG-treated rats.

TABLE 2. Body weight, reflex performances, and foot-fault test in 2- and 3-week-old normal (untreated) and MSG-treated pups

	Untreated	MSG	MSG+PACAP
Body weight (g)			
2 weeks	22.77 ± 0.83	16.97 ± 0.64***	17.54 ± 0.42***
3 weeks	33.65 ± 1.47	24.26 ± 0.90***	25.19 ± 0.66***
Righting (sec)			
2 weeks	0.61 ± 0.02	1.32 ± 0.14*	0.75 ± 0.07
3 weeks	0.10 ± 0.00	0.15 ± 0.03	0.13 ± 0.01
Geotaxis (sec)			
2 weeks	21.96 ± 1.6	22.03 ± 1.94	19.70 ± 1.77
3 weeks	6.76 ± 1.20	7.98 ± 1.17	10.58 ± 1.68
Gait (sec)			
2 weeks	13.84 ± 2.58	16.64 ± 2.10*	15.85 ± 1.74
3 weeks	5.20 ± 1.40	7.69 ± 1.10	6.17 ± 1.16
Foot fault (No)			
2 weeks	8.4 ± 1.0	11.3 ± 1.9**	8.1 ± 1.7
3 weeks	3.5 ± 0.3	6.5 ± 1.0*	6.1 ± 0.9*

Data are expressed as mean ± S.E.M. *$P < 0.05$, **$P < 0.01$, ***$P < 0.001$ vs. untreated rats.

age (TABLE 2). At this time point, significant differences were observed in the number of foot faults with both the fore- and hindlimbs.

Histological analysis of the arcuate nucleus revealed severe loss of tyrosine-hydroxylase immunopositive neurons after MSG treatment. When compared to the number of immunopositive neurons of normal animals (100%), the remaining neurons in the MSG-treated group was only 12.7 ± 0.1% ($P < 0.001$). In the PACAP-treated group, the number of neurons was higher, 35.2 ± 8.8% ($P < 0.01$ vs. normal group), but it was not significantly different from the MSG-treated control pups. In the retina, the total thickness was reduced to approximately half of the normal in both MSG- and PACAP-treated groups. While the photoreceptor layer was unchanged, the ganglionic, inner nuclear, and inner plexiform layers seemed fused after MSG treatment (to 2.9% ± 0.6 of normal), similar to earlier observations.[9,10] The inner retinal layers remained distinguishable in the PACAP-treated group (8.8% ± 0.5 of normal) and were significantly thicker than those in the MSG-treated controls ($P < 0.001$), but did not reach the thickness after local PACAP treatment, where the inner retinal layers were approximately half of the normal.[9,10]

In summary, our results show that the dose of PACAP that effectively enhances neurobehavioral development in normal rats was able to counteract the retarding effect of MSG on righting, forelimb placing, and grasp reflexes and caused a significant amelioration of the righting and gait reflex performance and motor coordination at 2 weeks of age. However, the assigned dose was not able to counteract the effects of MSG on the body weight and the neuronal loss in the arcuate nucleus. Previously, we have shown that the same treatment paradigm enhanced neurobehavioral development in normal pups, and local treatment ameliorated the effects of MSG in the retina.[9,10] In the retina, the present study also showed significant amelioration of neuronal loss in the inner retinal layers, but it was much less than that observed by local administration.[9,10] Since PACAP has been shown to effectively cross the blood-brain barrier,[11] most probably the concentration of PACAP did not reach the effective concentration in the examined areas. This might be the explanation for findings of others that have shown *in vivo* neurotrophic effects of PACAP only by local treatments.[3–5] However, the slight protective effect of PACAP found in the present study calls for other investigations to find a possible optimal treatment paradigm with PACAP for neonatal nervous injuries.

ACKNOWLEDGMENTS

This work was supported by OTKA T046589/061766, T048848, F048908, and the Hungarian Academy of Sciences.

REFERENCES

1. WASCHEK, J.A. 2002. Multiple actions of pituitary adenylyl cyclase activating peptide in nervous system development and regeneration. Dev. Neurosci. **24:** 14–23.
2. SOMOGYVARI-VIGH, A. & D. REGLODI. 2004. Pituitary adenylate cyclase activating polypeptide: a potential neuroprotective peptide. Review. Curr. Pharm. Des. **10:** 2861–2889.
3. VAUDRY, D. *et al.* 2000. PACAP acts as a neurotrophic factor during histogenesis of the rat cerebellar cortex. Ann. N. Y. Acad. Sci. **921:** 293–299.
4. GRESSENS, P. *et al.* 2000. VIP and PACAP38 modulate ibotenate-induced neuronal heteroropias in the newborn hamster neocortex. J. Neuropathol. Exp. Neurol. **59:** 1051–1062.
5. ARIMURA, A. *et al.* 1994. PACAP functions as a neurotrophic factor. Ann. N. Y. Acad. Sci. **739:** 228–243.
6. REGLODI, D. *et al.* 2003. The effects of PACAP and PACAP antagonist on the neurobehavioral development of newborn rats. Behav. Brain Res. **140:** 131–139.
7. HILL, J.M. *et al.* 1991. Vasoactive intestinal peptide antagonist retards the development of neonatal behaviors in the rat. Peptides **12:** 187–192.
8. KISS, P. *et al.* 2005. Development of neurological reflexes and motor coordination in rats neonatally treated with monosodium glutamate. Neurotox. Res. **8:** 235–244.
9. TAMAS, A. *et al.* 2004. Effects of pituitary adenylate cyclase activating polypeptide in retinal degeneration induced by monosodium-glutamate. Neurosci. Lett. **372:** 110–113.
10. BABAI, N. *et al.* 2005. Degree of damage compensation by various PACAP treatments in monosodium glutamate-induced retina degeneration. Neurotox. Res. **8:** 227–233.

11. BANKS, W.A. *et al*. 1996. Transport of pituitary adenylate cyclase activating polypeptide across the blood-brain barrier and the prevention of ischemia-induced death of hippocampal neurons. Ann. N. Y. Acad. Sci. **805:** 270–277.
12. REGLODI, D. *et al*. 2000. Neuroprotective effects of PACAP38 in a rat model of transient focal ischemia under various experimental conditions. Ann. N. Y. Acad. Sci. **921:** 119–128.
13. ALTMAN, J. & K. SUDARSHAN. 1975. Postnatal development of locomotion in the laboratory rat. Anim. Behav. **23:** 896–920.
14. SMART, J.L. & J. DOBBING. 1971. Vulnerability of developing brain. II. Effects of early nutritional deprivation on reflex ontogeny and development on behavior in the rat. Brain Res. **28:** 85–95.

Localization of Small Heterodimer Partner (SHP) and Secretin in Mouse Duodenal Cells

IAN P.Y. LAM,[a] LEO T.O. LEE,[a] H.S. CHOI,[b] AND BILLY K.C. CHOW[a]

[a]Department of Zoology, The University of Hong Kong, Hong Kong, China

[b]Hormone Research Center, School of Biological Sciences and Technology, Chonnam National University, Republic of Korea

ABSTRACT: Previous studies have demonstrated the transcriptional repressive property of the atypical nuclear receptor, small heterodimer partner (SHP), on NeuroD. NeuroD is a basic helix-loop-helix transcription factor that has also been shown to be important in modulating secretin gene expression. The present study revealed the activation of the human secretin core promotor by overexpressing NeuroD, and the localization of SHP and secretin-producing cells in mouse duodenal epithelium by immunohistochemical stainings. These results indicated that SHP and secretin are potentially co-expressed and lead us to propose a novel regulatory pathway, in which SHP represses NeuroD's positive regulatory activity on secretin gene.

KEYWORDS: secretin; small heterodimer partner; NeuroD

INTRODUCTION

Secretin is mainly secreted from the enteroendocrine S-cells located in the proximal small intestine.[1] The cell-specific expression of secretin is directed by an E-box motif (CANNTG),[2,3] which is the binding site of basic helix-loop-helix (bHLH) family of transcription factors. The bHLH factors are classified into class A and B according to their DNA-binding properties and tissue distributions.[4] Class A proteins include E2A (E12 and E47), are ubiquitously expressed, and able to form homodimer or heterodimer with a class B protein, while class B proteins, such as NeuroD, are expressed in a cell-specific manner and they only form heterodimers with class A proteins upon binding to the E-box sequence.[4] Recently, we have shown that the interaction of the heterodimer NeuroD and E2A with the E-box sequence is important in activating

Address for correspondence: Billy K.C. Chow, Department of Zoology, The University of Hong Kong, Pokfulam Road, Hong Kong, PRC. Voice: 852-2299-0850; fax: 852-2559-9114.
e-mail: bkcc@hkusua.hku.hk

Ann. N.Y. Acad. Sci. 1070: 371–375 (2006). © 2006 New York Academy of Sciences.
doi: 10.1196/annals.1317.047

human secretin gene expression.[3] Other studies have shown that activation of E-box-dependent transcription by NeuroD is potentiated by the co-activator p300 (also known as cAMP response element binding protein-binding protein, CBP).[5]

On the other hand, the small heterodimer partner (SHP, NR0B2) has previously been identified to physically interact with NeuroD and repress its transcriptional activity.[6] SHP is an atypical nuclear receptor that lacks a DNA-binding domain and the ligands that bind to SHP are still unknown. One of the proposed repressing mechanisms of SHP is by interfering the interactions of co-activator p300 with NeuroD. This novel repressive property of SHP prompted us to investigate the regulation of secretin gene expression by the SHP, p300, and NeuroD axis. For this reason, in this article, we showed the activation of the human secretin core promotor by overexpressing NeuroD, and we also suggested that SHP and secretin-producing cells are co-localized in duodenal epithelial cells by immunohistochemical stainings.

METHODS AND MATERIALS

Cell Culture and Transient Transfection and Luciferase Assay

The human duodenal adenocarcinoma cells, HuTu-80 (purchased from American Type Culture Collection, Manassas, VA) and the neuroblastoma cell line, SH-SY5Y (kindly provided by Prof. J. Hugon, The University of Hong Kong), were cultured in MEM (Invitrogen, Carlsbad, CA) with nonessential amino acids, 10% FBS, 100 U/mL penicillin, and 100 μg/mL streptomycin at 37°C and 5% CO_2.

SH-SY5Y cells and HuTu-80 cells were plated at a density of 1.5×10^5 (HuTu-80) or 2.0×10^5 (SH-SY5Y) cells/ 35-mm well (6-well plate, Costar, San Diego, CA). After 2 days of incubation, 2-μg promotor-luciferase construct, 0.5-μg pCMV-β-gal, and various amounts of NeuroD expression vectors (NeuroD/CMV) in a total of 5-μg DNA (adjusted with pcDNA 3.1) were co-transfected into cells using 10 μL GeneJuice reagent (Novagen, Darmstadt, Germany) according to the manufacturer's protocol. Cells were harvested 48 h after the transfection. The cell extracts were assayed for luciferase and β-galactosidase activities as described previously.[7]

Immunohistochemical Staining

The mouse duodenum sections were dewaxed by xylene and rehydrated in decreasing concentrations of ethanol. The endogenous peroxidase activity was blocked by 4% hydrogen peroxide in methanol. Microwave antigen retrieval with 0.01-M citric acid buffer at pH 6.0 was performed and followed by incubation in FBS. The SHP and secretin present in the sections were detected by goat anti-SHP IgG (Santa Cruz Biotechnology, Santa Cruz, CA) and goat anti-secretin IgG (Phoenix Pharmaceuticals, Belmont, CA), respectively, at 1:250 dilution. Afterward, the sections were incubated with HRP-conjugated anti-goat-IgG (1:1000 dilution, Santa Cruz Biotechnology), followed by the substrate DAB (Zymed, San Francisco, CA) and finally counterstained by hematoxylin.

RESULTS AND DISCUSSION

In attempts to study the functional role of NeuroD in regulating the human secretin gene, it was transiently overexpressed in duodenal (HuTu-80) and neuronal cells (SH-SY5Y). Overexpression of NeuroD upregulated the human secretin core promotor (p341) activities to 3.3 ± 0.4-fold in HuTu-80 and 3.1 ± 0.2-fold in SH-SY5Y (2.0 μg of NeuroD/CMV) cells, respectively (FIG.1), clearly indicating an important regulator role of NeuroD in activating human secretin gene. In previous studies, using the rat insulin promotor, element 3 (RIPE3), it was demonstrated that overexpression of SHP downregulated NeuroD-E47-dependent RIPE3 transcription in a dose-dependent manner.[8] However, overexpression of SHP did not significantly affect the activity of a mutant E-box construct (RIPE3 Em). Together with the GST-pull down assay, it was believed that SHP downregulated NeuroD-dependent expressions by inhibiting its interactions with p300. We therefore propose here that SHP can downregulate the secretin gene via a similar mechanism.

To investigate this, we tested the possible co-expression of SHP and secretin in mouse duodenum by immunohistochemical stainings. Most of the

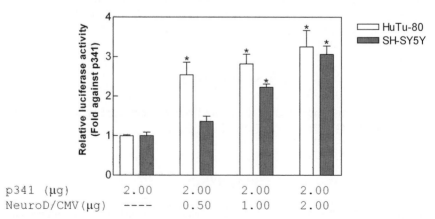

FIGURE 1. Effects of overexpression of NeuroD on the human secretin core promotor in HuTu-80 and SH-SY5Y cells. Various amounts (from 0 to 2.0 μg of Neuro/CMV) were co-transfected with p341 into HuTu-80 or SH-SY5Y cells. *$P < 0.001$ versus the control, p341 alone.

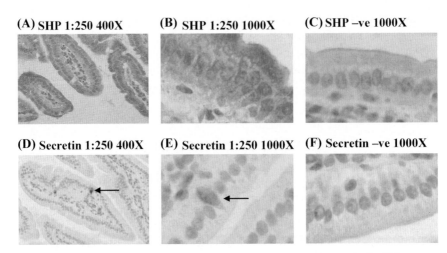

FIGURE 2. Immunohistochemical staining of SHP (**A, B**) and secretin (**D, E**) in mouse duodenal sections. The secretin-immunoreactive cells were indicated by arrows. Negative controls (**C, F**) were performed by omitting the primary antibody and no signals were observed. The pictures were captured with 400× (**A, D**) and 1000× (**B, C, E** and **F**) magnifications.

epithelial cells in the duodenal villi were SHP-immunoreactive while other cell layers further away from the lumen, including mucosa, submucosa, muscle, and serosa were negatively stained (FIG. 2A, B). On the other hand, distinct and isolated secretin-immunoreactive cells were found only in the epithelial cell layer (FIG. 2D, E). Taken together, it is likely that secretin-producing cells would also express SHP, and these data are consistent with our hypothesis that SHP downregulates secretin via NeuroD in the duodenal cells. We are in the process of testing the cellular mechanisms involved.

ACKNOWLEDGMENTS

This work was supported by the Hong Kong Government RGC HKU7219/02M and CRCG 10205783 to Billy K.C. Chow.

REFERENCES

1. BRYANT, M.G. & S.R. BLOOM. 1979. Distribution of the gut hormones in the primate intestinal tract. Gut **20**: 653–659.
2. WHEELER, M.B., J. NISHITANI, A.M.J. BUCHAN, et al. 1992. Identification of a transcriptional enhancer important for enteroendocrine and pancreatic islet cell-specific expression of the secretin gene. Mol. Cell. Biol. **12**: 3531–3539.

3. LEE, L.T., K.C. TAN-UN, R.T. PANG, *et al.* 2004. Regulation of the human secretin gene is controlled by the combined effect of CpG methylation, Sp1/Sp3 ratio, and the E-box element. Mol. Endocrinol. **18:** 1740–1755.
4. JONES, S. 2004. An overview of the basic helix-loop-helix proteins. Genome Biol. **5:** 226.
5. MUTOH, H., F.J. NAYA, M.J. TSAI & A.B. LEITER. 1998. The basic helix-loop-helix protein BETA2 interacts with p300 to coordinate differentiation of secretin-expressing enteroendocrine cells. Genes Dev. **12:** 820–830.
6. KIM, J.Y., K. CHU, H.J. KIM, *et al.* 2004. Orphan nuclear receptor small heterodimer partner, a novel corepressor for a basic helix-loop-helix transcription factor BETA2/NeuroD. Mol. Endocrinol. **18:** 776–790.
7. NGAN, E.S., P.K. CHENG, P.C. LEUNG & B.K. CHOW. 1999. Steroidogenic factor-1 interacts with a gonadotrope-specific element within the first exon of the human gonadotropin-releasing hormone receptor gene to mediate gonadotrope-specific expression. Enodocrinology **140:** 2452–2462.
8. KIM, J.W., V. SEGHERS, V. CHO, *et al.* 2002. Transactivation of the mouse sulfonylurea receptor 1 gene by BETA2/NeuroD. Mol. Enodocrinol. **16:** 1097–1107.

Differential Mechanisms for PACAP and GnRH cAMP Induction Contribute to Cross-talk between both Hormones in the Gonadotrope LβT2 Cell Line

SIGOLÈNE LARIVIÈRE, GHISLAINE GARREL,
MARIE-THÉRÈSE ROBIN, RAYMOND COUNIS,
AND JOËLLE COHEN-TANNOUDJI

*UMR CNRS 7079, Physiologie & Physiopathologie, Université Pierre &
Marie Curie-Paris 6, 75252 Paris, France*

ABSTRACT: The effects and respective influence of pituitary adenylate cyclase-activating polypeptide (PACAP) and gonadotropin-releasing
hormone (GnRH) on cyclic AMP (cAMP) production in pituitary gonadotropes were analyzed using the LβT2 cell line. Both hormones induced cAMP with, however, different intensity and time course. In addition, the GnRH effect was markedly reduced by PKC inhibitors. Despite
its positive coupling to cAMP pathway, GnRH counteracted PACAP induction of cAMP and this effect was mimicked by the PKC activator phorbol 12-myristate 13-acetate (PMA). The data reveal major differences in
the mechanisms by which PACAP and GnRH activate cAMP/PKA pathway in LβT2 cells and suggest that PKC activation serves GnRH not only
to increase cAMP but also to counteract the PACAP stimulation of this
signaling pathway.

KEYWORDS: PACAP; GnRH; cross-talk; cell signaling; LβT2
gonadotrope cell line

INTRODUCTION

Pituitary adenylate cyclase-activating polypeptide (PACAP) and gonadotropin-releasing hormone (GnRH), the primary regulator of the pituitary
gonadotropes, act separately or synergistically to regulate exocytotic gonadotropin (LH, FSH) release, gonadotropin subunit gene expression, and

Address for correspondence: Dr. J. Cohen-Tannoudji, UMR CNRS 7079, Université P. & M. Curie-
Paris 6, Case 256, 4 Place Jussieu, 75252 Paris cedex 05, France. Voice: 331-44.7.26.55; fax: 331-
44.27.265.

e-mail: joelle.cohen-tannoudji@snv.jussieu.fr

Ann. N.Y. Acad. Sci. 1070: 376–379 (2006). © 2006 New York Academy of Sciences.
doi: 10.1196/annals.1317.048

gonadotrope responsiveness to GnRH.[1-5] Altogether, data clearly demonstrate the ability of PACAP to modulate GnRH signaling. The reciprocal interaction is much less documented. Both PACAP and GnRH interact with specific G protein–coupled receptors and activate a common network of signaling pathways, however, with distinct efficacy. In particular, GnRH preferentially stimulates the inositol triphosphate (IP3)/protein kinase C (PKC) cascade whereas PACAP preferentially stimulates the cyclic AMP (cAMP)/protein kinase (PKA) cascade.[6]

To examine the respective coupling of PACAP and GnRH with the IP3/PKC and cAMP/PKA pathways we took advantage of the gonadotrope cell line, LβT2. This cell line developed by Dr. P. Mellon at the University of San Diego[7] expresses not only the GnRH receptor and α subunit but also gonadotropin β subunits and is thus more differentiated than the αT3-1 cell line. Further, and in contrast with αT3-1 cells, LβT2 cells appeared to produce cAMP in response to GnRH whereas responsiveness to PACAP remains controversial.[8]

RESULTS

We observed that PACAP-38 efficiently stimulated cAMP generation in LβT2 cell line in a time- and concentration-dependent fashion, implying that these cells express PACAP receptors coupled with functional signal transduction pathways. PACAP-38 rapidly increased intracellular cAMP (significant after 5 min) and its effect was maximal, 120-fold over basal, within 30 min. Both PACAP-27 and -38 induced concentration-dependent increase in cAMP production with an equivalent potency whereas concentrations of VIP up to 10^{-3} M were ineffective. This rank order of potency suggests the involvement of PAC1 receptor and this was supported by the identification, by RT-PCR, of this receptor subtype in LβT2 cell line.

GnRH was also able to stimulate cAMP generation, however, with an amplitude and time course different from PACAP as significant cAMP accumulation was detected after 1 h stimulation and maximal effect (2.7-fold over basal only) was achieved after 4 h. Interestingly, the effects of GnRH on cAMP production were mimicked by the phorbol ester PMA and abolished by the PKC inhibitor GF109203X thus in favor of a PKC-mediated adenylate cyclase (AC) activation.

Although GnRH is able to stimulate AC/cAMP pathway, its coincubation with PACAP (FIG. 1 A) did not synergize but instead, profoundly inhibited PACAP-induced cAMP production. Whether this inhibitory effect was mediated by PKC was investigated. As shown in FIGURE 1 B, both GnRH and PKC activation with PMA caused a 70% inhibition of PACAP-induced cAMP accumulation, and the inhibitory effect of GnRH was markedly reduced in the presence of PKC inhibitor.

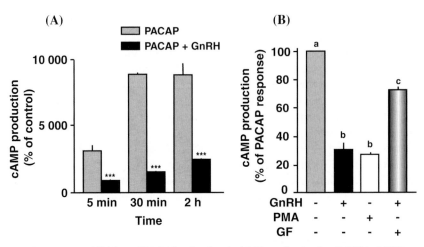

FIGURE 1. Inhibition of PACAP-stimulated cAMP production by GnRH in LβT2 gonadotrope cells. LβT2 cells were loaded overnight with [2,8-^{3}H]adenine and then stimulated with various drugs in the presence of the phosphodiesterase inhibitor IBMX. Reactions were stopped with cold lysis buffer and intracellular [^{3}H]cAMP was assayed.[10] Basal cAMP production ranged from 20 to 40 fmol/well. (**A**) *Time course of inhibition:* cells were treated with 20 nM PACAP-38 alone (shaded columns) or in the presence (black columns) of the GnRH agonist D[Trp^6]GnRH (100 nM) for the indicated periods of time. (**B**) *Involvement of PKC in the GnRH inhibitory effect:* cells were incubated for 30 min with 20 nM PACAP-38 in the absence or presence of D[Trp^6]GnRH (100 nM), the phorbol ester PMA (50 nM) or a combination of D[Trp^6]GnRH and the PKC inhibitor GF109203X (GF, 2 μM). Data are the mean ± SEM of 3–5 experiments performed in duplicate and expressed as a percentage of control responses. ***$P \leq 0.001$ as compared with PACAP (**A**). Columns labeled with the same letter denote nonstatistically different values (**B**).

DISCUSSION

Our study demonstrates that PACAP readily stimulates cAMP production in LβT2 cells. In addition, the capacity of these cells to generate cAMP in response to GnRH has been further established. Interestingly, the stimulatory effect of GnRH clearly appears mediated by PKC suggesting that cross-talk occurs via phosphorylation of element(s) involved in AC activation. In this regard, it was surprising that PACAP and GnRH did not cooperate at increasing intracellular cAMP level. Moreover, the observed inhibitory action of GnRH again seems mediated through PKC. Such an inhibition was described previously in αT3-1 cells, however, in these cells GnRH is incapable of inducing cAMP.[9] The fact that GnRH acts via PKC both to induce a modest cAMP production and antagonize the substantial cAMP production by PACAP denotes co-ordinate control of intracellular cAMP pathways. It also suggests that GnRH would require low intracellular concentrations of cAMP for optimal function. Conversely, one can expect that the cAMP-mediated PACAP regulation in pituitary gonadotropes[2-5] would be maximal under conditions of weak GnRH stimulation only. The mechanisms by which PACAP and GnRH dialogue within gonadotropes and the relevance of such interactions in physiological contexts are subjects of current investigation.

REFERENCES

1. COUNIS, R. 1999. Gonadotropin biosynthesis. *In* Encyclopedia of Reproduction, Vol. 4. J. Neill & E. Knobil, Eds.: 507–520. Academic Press. New York & London.
2. GARREL, G. *et al.* 2002. Pituitary adenylate cyclase-activating polypeptide stimulates nitric oxide synthase type I expression and potentiates the cGMP response to gonadotropin-releasing hormone of rat pituitary gonadotrophs. J. Biol. Chem. **277:** 46391–46401.
3. TSUJI, T. *et al.* 1994. Effects of pituitary adenylate cyclase-activating polypeptide on gonadotropin secretion and subunit messenger ribonucleic acids in perifused rat pituitary cells. Endocrinology **135:** 826–833.
4. CULLER, M.D. & C.S. PASCHALL 1991. Pituitary adenylate cyclase-activating polypeptide (PACAP) potentiates the gonadotropin-releasing activity of luteinizing hormone-releasing hormone. Endocrinology **129:** 2260–2262.
5. PINCAS, H. *et al.* 2001. Pituitary adenylate cyclase-activating polypeptide and cyclic adenosine 3', 5'-monophosphate stimulate the promoter activity of the rat GnRH receptor gene via a bipartite response element in gonadotrope-derived cells. J. Biol. Chem. **276:** 23562–23571.
6. SCHOMERUS, E. *et al.* 1994. Effects of pituitary adenylate cyclase-activating polypeptide in the pituitary: activation of two signal transduction pathways in the gonadotrope-derived alpha T3-1 cell line. Endocrinology **134:** 315–323.
7. ALARID, E.T. *et al.* 1996. Immortalization of pituitary cells at discrete stages of development by directed oncogenesis in transgenic mice. Development **122:** 3319–3329.
8. FOWKES, R.C. *et al.* 2003. Absence of pituitary adenylate cyclase-activating polypeptide-stimulated transcription of the human glycoprotein alpha-subunit gene in LβT2 gonadotrophs reveals disrupted cAMP-mediated gene transcription. J. Mol. Endocrinol. **31:** 263–278.
9. MCARDLE, C.A. *et al.* 1994. Pituitary adenylate cyclase-activating polypeptide effects in pituitary cells: modulation by gonadotropin-releasing hormone in alpha T3-1 cells. Endocrinology **134:** 2599–2605.
10. JOHNSON, R.A. *et al.* 1994. Determination of adenylyl cyclase catalytic activity using single and double column procedures. Methods Enzymol. **238:** 31–56.

Identification of Proteins Regulated by PACAP in PC12 Cells by 2D Gel Electrophoresis Coupled to Mass Spectrometry

ALEXIS LEBON,[a,b] DAMIEN SEYER,[a,c] PASCAL COSETTE,[a,c]
LAURENT COQUET,[a,c] THIERRY JOUENNE,[a,c] PHILIPPE CHAN,[a,b]
JEROME LEPRINCE,[a,b] ALAIN FOURNIER,[d] HUBERT VAUDRY,[a,b]
BRUNO J. GONZALEZ,[a,b] AND DAVID VAUDRY[a,b]

[a]European Institute for Peptide Research (IFRMP23), University of Rouen,
76821 Mont-Saint-Aignan, France

[b]INSERM U413, Laboratory of Cellular and Molecular Neuroendocrinology,
University of Rouen, 76821 Mont-Saint-Aignan, France

[c]IBBR Group, CNRS UMR 6522, Proteomic Platform of the IFRMP23,
University of Rouen, 76821 Mont-Saint-Aignan, France

[d]INRS-Institut Armand Frappier, University of Québec, Pointe-Claire,
Canada H9R 1G6

ABSTRACT: The rat pheochromocytoma PC12 cell line has been widely
used as a model to study neuronal differentiation. In particular, after
serum depletion, PC12 cells stop to proliferate and undergo apoptosis. Under such conditions, treatment with pituitary adenylate cyclase-
activating polypeptide (PACAP) promotes cell survival and induces neurite outgrowth. The identification of the proteins regulated by PACAP in
PC12 cells under apoptotic conditions should provide valuable information concerning the mechanisms controlling neuronal cell survival and
differentiation. To this aim, PC12 cells cultured in serum-free medium
were treated with PACAP (10^{-7} M), proteins were extracted, separated by
two-dimensional gel electrophoresis (2-DE), and identified by MALDI-
ToF mass spectrometry. The comparison between 16 2-DE maps led to
the characterization of 110 proteins regulated by PACAP among which
22 have been identified by automatic query of the Mascot, Aldente, and
Profound servers with the ProGeR-CDD database. Seventy-six percent of
these proteins, including the p17 subunit of caspase-3, the heat shock protein hsp60, and the GTPase ran were found to be repressed whereas the
others notably hsp27, tubulin β-5, and calmodulin were overexpressed.
Investigation of the putative functions indicated that some of the proteins

Address for correspondence: Dr. Hubert Vaudry, INSERM U413, Laboratory of Cellular and Molecular Neuroendocrinology, European Institute for Peptide Research (IFRMP 23), University of Rouen,
76821 Mont-Saint-Aignan, France. Voice: +33 235-14-6624; fax: +33 235-14-6946.
e-mail: hubert.vaudry@univ-rouen.fr

Ann. N.Y. Acad. Sci. 1070: 380–387 (2006). © 2006 New York Academy of Sciences.
doi: 10.1196/annals.1317.049

regulated by PACAP and identified in the present article could control cell survival or differentiation.

KEYWORDS: proteomics; PC12 cells; PACAP

INTRODUCTION

Pituitary adenylate cyclase-activating polypeptide (PACAP) was isolated from the sheep hypothalamus on the basis of its capacity to stimulate the production of cAMP in rat anterior pituitary cells.[1] PACAP and its receptors are widely expressed in the brain and peripheral organs where they exert many biological activities.[2] In particular, PACAP has been shown to act as a neurotrophic factor promoting cell differentiation during development and inhibiting apoptosis induced by neurotoxic agents.[3–5]

The rat pheochromocytoma PC12 cell line can easily be manipulated to study neuronal differentiation and apoptosis.[6,7] When cells are cultured with serum, addition of PACAP inhibits cell proliferation and induces neurite outgrowth.[8,9] When cells are cultured in serum-free medium, PACAP promotes PC12 survival and neuritogenesis.[10] The effect of PACAP on neurite outgrowth requires the phosphorylation of the MAP kinases ERK1/2[11] but the downstream targets controlling cell differentiation remain largely unknown. The aim of the present article was thus to identify new proteins regulated by PACAP in PC12 cells using two-dimensional gel electrophoresis (2-DE) analysis coupled to mass spectrometry identification.

MATERIALS AND METHODS

PC12 cells were grown as previously reported.[7] When cell density reached 200,000 cells/mL, cultures were shifted to serum-free medium for 12 h and then incubated in the absence or presence of 10^{-7} M PACAP38 for 9 h (FIG. 1 A). At the end of the treatment, cells were washed twice with PBS and the proteins were extracted according to two different protocols to maximize the number of identified proteins. The first lysis buffer contained 1% Triton X-100, 50 mM Tris-HCl, and 10 mM EDTA. After sonication, the homogenate was centrifuged (9500 g, 4°C, 15 min) and the proteins contained in the supernatant were precipitated by the addition of 10% trichloroacetic acid. The samples were centrifuged (9500 g, 4°C, 15 min) and the pellet was washed three times with alcohol/ether. The extracted proteins were finally dried and kept frozen until use. The second extraction protocol was carried out according to the instructions provided with the Complete Mammalian Proteome Extraction Kit (Calbiochem, San Diego, CA).

For 2-DE, proteins were resuspended in 1 mL isoelectric focusing (IEF) buffer containing 7 M urea, 2 M thiourea, 0.5% ampholines (pH 3.5–10), 2 mM

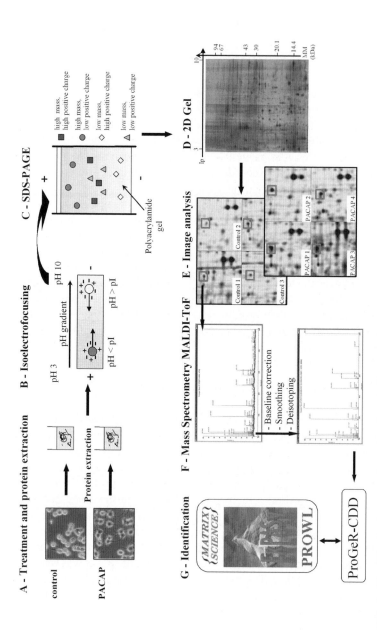

FIGURE 1. Identification process of proteins regulated by PACAP in PC12 cells. Cells were incubated in the absence or presence of PACAP (10^{-7} M) for 9 h before protein extraction (**A**). Proteins were separated according to their isoelectric point (**B**) and molecular weight (**C**) to generate 2-DE gels (**D**). The gels were digitized and analyzed with the PD-Quest software (**E**). After excision and tryptic digestion, a fingerprint of the proteins was obtained by MALDI-ToF mass spectrometry (**F**) and the identification was conducted on several proteomic servers (**G**).

tributyl phosphine, and 0.5% amidosulfobetaine 14 (ASB14). After quantification,[6] 200 μg of solubilized proteins were separated in the first dimension according to their isoelectric point (FIG. 1 B) along nonlinear immobilized pH-gradient strips (IPGS pH 3–10, 17 cm; Amersham Biosciences, Piscataway, NJ) in a Protean IEF-Cell (Bio-Rad, Hercules, CA). At the end of the first dimension, IPGS were reequilibrated (buffer containing 2% DTT and 2.5% iodoacetamide) and applied on top of a 12.5% SDS polyacrylamide gel for the second dimension (FIG.1 C, Protean II Xi; Bio-Rad). At the end of the migration, gels were fixed overnight in a solution of 10% acetic acid, 30% ethanol, washed three times in water, twice in 10% ethanol, and sensitized in a 0.02% sodium thiosulfate solution for 1 min. Finally, gels were stained in a 0.1% silver nitrate solution for 30 min and revealed in the presence of 0.12% sodium carbonate, 0.004% formaldehyde, and 0.008% sodium thiosulfate. Revelation was stopped with an 8% acetic acid solution (FIG. 1 D).

Gels were digitized using a computer-assisted densitometer (GS-800; Bio-Rad). Spot detection and statistical analysis were performed with the PD-Quest 7.01 software. For each experimental condition, spot quantities corresponded to the mean value of four independent gels (FIG. 1 E). For protein identification, 2-DE gels were performed by loading 500 μg proteins per gel and by using colloidal blue staining. Gels were fixed overnight in a solution containing 50% ethanol and 1.4% orthophosphoric acid. After rinsing in water, the gels were sensitized with 17% ammonium sulphate, 0.34% methanol, and 1.4% orthophosphoric acid. Staining was carried out for 3–4 days in the sensitization solution completed with 0.06% G 250 blue. Spots of interest were then excised (ProXcision; Perkin Elmer, Courtaboeuf, France), gel plugs were dried using a Speed-Vac evaporator and subjected to trypsin (MultiProbe II; Perkin Elmer). Peptide fragments were expulsed from the gels with acetonitrile, spotted on a MALDI plate in the presence of a cis-alpha cyano 4-hydroxy cinnamic acid matrix. Mass spectrometry was carried out on a Voyager DEPro MALDI-ToF spectrometer (Applied Biosystems, Courtaboeuf, France) in the reflectron mode (FIG. 1 F). Peptide fingerprints were matched against *in silico* digests of the SwissProt database using the Mascot (www.matrixscience.com), Aldente (www.expasy.org), and Profound (http://prowl.rockefeller.edu) servers (FIG. 1 G).

RESULTS AND DISCUSSION

For each extraction condition, protein regulations were investigated from the differential analysis of four control and PACAP-treated gels. Statistical analysis performed with the PD-Quest 7.01 software made it possible to identify 110 proteins significantly regulated two-fold or more after a 9-h treatment with PACAP. After staining with colloidal blue, 46 spots were excised and submitted to identification. MALDI-ToF mass spectra of the generated peptides were uploaded in the ProGeR-CDD database (http://proger-cdd.crihan.fr) and 22 of the proteins regulated by PACAP have already been identified. Ten of these regulated proteins are listed in TABLE 1.

One of the proteins repressed by PACAP is the p17 subunit of caspase-3. This protein, which plays a key role in apoptosis, has previously been shown to be inhibited by PACAP in cerebellar granule cells.[12] Inhibition of the active form of caspase-3 could be sufficient to protect PC12 cells from apoptosis induced by serum privation. Two heat shock proteins, i.e., hsp60 and hsp90β, were also repressed by PACAP. It is interesting to note that hsp60 is induced by neurotoxic agents, such as ethanol[13] or lipopolysaccharide[14] and repressed upon treatment with NGF.[15] The expression level of hsp90β is significantly higher in human breast cancer cells than in normal tissues but the role of this protein in the control of tumor cell proliferation or differentiation remains unclear.[16] Functional investigations should be conducted to clarify the involvement of hsp60 and/or hsp90β proteins in the antiapoptotic and neurotrophic effects of PACAP. Other proteins repressed by PACAP, such as the GTPase ran[17−19] and prohibitin[20,21] could also be involved in the control of PC12 cells proliferation. However, there is still considerable controversy concerning the role of these two proteins that can either induce or repress cell senescence or migration depending on the cell type.

PACAP stimulated the expression of hsp27, which is known to protect PC12 cells from apoptosis after heat shock and NGF withdrawal.[22] Recently, hsp27 has also been shown to inhibit c-Jun N-terminal kinase-induced neuronal death[23] and to prevent cytochrome c release and caspase-3 activation induced by 6-hydroxydopamine.[24] On the basis of these observations, the possible involvement of hsp27 in the antiapoptotic effect of PACAP should be considered. Expression of the tubulin β-5 subunit was found to be increased by PACAP in PC12 cells, which corroborates previous microarray results where the gene coding for this protein was induced during cell differentiation.[25] Another protein that appears to be regulated by several neurotrophic factors, including PACAP, during PC12 cell differentiation is the calcium-binding regulatory protein calmodulin.[26,27] This protein would at least participate to neuritogenesis as shown with cells treated with calmodulin antisense oligodeoxynucleotides[28,29] probably by interacting with the MEK/ERK transduction pathway.

For each of the identified proteins, various information including the putative functions and the availability of specific antibodies or siRNA have been repatriated in the ProGeR-CDD database. These data will now facilitate the investigation of the possible role of the proteins identified in the present study in the antiapoptotic and neurotrophic activities of PACAP.

ACKNOWLEDGMENTS

This work was supported by INSERM (U 413), CNRS (UMR 6522), the European Institute for Peptide Research (IFRMP23), the Association pour la

TABLE 1. Characterization of some of the proteins regulated by PACAP in PC12 cells

Name	Accession number	Molecular mass (kDa)	Ip	NMP	% of coverage	Regulation (% of control)
60S ribosomal protein L18a	P62718	21	10.7	5	26	Disappear
Caspase-3 p17 subunit	P55213	17	5.6	8	50	− 72
Calmodulin	P62161	17	4.1	4	40	+ 55
GTPase ran	Q8K586	24	6.6	4	13	− 55
Heat Shock protein (hsp27)	P69897	23	6.1	3	19	+ 50
Heat Shock protein (hsp60) precursor	P63039	61	5.9	5	18	− 55
Heat Shock protein 90β Nter (hsp84)	P11499	83	5.0	4	6	− 80
Prohibitin	P67779	30	5.6	15	50	− 72
Stress-induced-phosphoprotein 1	O35814	63	6.4	15	23	Appear
Tubulin β-5 chain.	P69897	50	4.8	10	21	+ 55

Ip = isoelectric point; NMP = number of matched peptides; Accession number from SwissProt DataBase.

Recherche sur le Cancer (to D.V.), an INSERM-FRSQ exchange program (to A.F. and H.V.), and the Conseil Régional de Haute-Normandie. H.V. is affiliated professor at the INRS-Institut Armand Frappier, University of Quebec.

REFERENCES

1. MIYATA, A., A. ARIMURA, R.R. DAHL, et al. 1989. Isolation of a novel 38 residue-hypothalamic polypeptide which stimulates adenylate cyclase in pituitary cels. Biochem. Biophys. Res. Commun. **164:** 567–574.
2. ARIMURA, A., A. SOMOGYVARI-VIGH, K. MIZUNO, et al. 1991. Tissue distribution of PACAP as determined by RIA: highly abundant in the rat brain and testes. Endocrinology **129:** 2787–2789.
3. VAUDRY, D., B.J. GONZALEZ, M. BASILLE, et al. 2000. Pituitary adenylate cyclase-activating polypeptide and its receptors: from structure to functions. Pharmacol. Rev. **52:** 269–324.
4. VAUDRY, D., B.J. GONZALEZ, M. BASILLE, et al. 1999. Neurotrophic activity of pituitary adenylate cyclase-activating polypeptide on rat cerebellar cortex during development. Proc. Natl. Acad. Sci. USA **96:** 9415–9420.
5. VAUDRY, D., A. FALLUEL-MOREL, S. LEUILLET, et al. 2003. Regulators of cerebellar granule cell development act through specific signaling pathways. Science **300:** 1532–1534.
6. VAUDRY, D., Y. CHEN, C.M. HSU & L.E. EIDEN. 2002. PC12 cells as a model to study the neurotrophic activities of PACAP. Ann. N. Y. Acad. Sci. **971:** 491–496.
7. VAUDRY, D., P.J. STORK, P. LAZAROVICI & L.E. EIDEN. 2002. Signaling pathways for PC12 cell differentiation: making the right connections. Science **296:** 1648–1649.
8. DEUTSH, P.J. & Y. SUN. 1992. The 38-amino acid form of pituitary adenylate cyclase-activating polypeptide stimulates dual signalling cascades in PC12 cells and promotes neurite outgrowth. J. Biol. Chem. **267:** 5108–5113.
9. LAZAROVICI, P., H. JIANG & D. FINK. 1998. The 38-amino-acid form of pituitary adenylate cyclase-activating polypeptide induces neurite outgrowth in PC12 cells that is dependent on protein kinase C and extracellular signal-regulated kinase but no on protein kinase A, nerve growth factor receptor tyrosine kinase, p21 (ras) G protein, and pp60 (c-src) cytoplasmic tyrosine kinase. Mol. Pharmacol. **54:** 547–558.
10. VAUDRY, D., Y. CHEN, A. RAVNI, et al. 2002. Analysis of the PC12 cell transcriptome after differentiation with pituitary adenylate cyclase activating polypeptide (PACAP). J. Neurochem. **83:** 1272–1284.
11. BARRIE, A.P., A.M. CLOHESSY, C.S. BUENSUCESO, et al. 1997. Pituitary adenylyl cyclase-activating peptide stimulates extracellular signal-regulated kinase 1 or 2 (ERK1/2) activity in a Ras-independent, mitogen-activated protein Kinase 1 or 2-dependent manner in PC12 cells. J. Biol. Chem. **272:** 19666–19671.
12. VAUDRY, D., T.F. PAMANTUNG, M. BASILLE, et al. 2002. PACAP protects cerebellar granule neurons against oxidative stress-induced apoptosis. Eur. J. Neurosci. **15:** 1451–1460.
13. DWYER, D.S., Y. LIU & R.J. BRADLEY. 1999. An ethanol-sensitive variant of the PC12 neuronal cell line: sensitivity to alcohol is associated with increased cell adhesion and decreased glucose accumulation. J. Cell. Physiol. **178:** 93–101.

14. HUANG, Y.H., A.Y. CHANG, C.M. HUANG, *et al.* 2002. Proteomic analysis of lipopolysaccharide-induced apoptosis in PC12 cells. Proteomics **2**: 1220–1228.
15. DWYER, D.S., Y. LIU, S. MIAO & R.J. BRADLEY. 1996. Neuronal differentiation in PC12 cells is accompanied by diminished inducibility of Hsp70 and Hsp60 in response to heat and ethanol. Neurochem. Res. **21**: 659–666.
16. YANO, M., Z. NAITO, S. TANAKA & G. ASANO. 1996. Expression and roles of heat shock proteins in human breast cancer. Jpn. J. Cancer. Res. **87**: 908–915.
17. FERRANDO-MAY, E., V. CORDES, I. BILLER-CKOVRIC, *et al.* 2001. Caspases mediate nucleoporin cleavage, but not early redistribution of nuclear transport factors and modulation of nuclear permeability in apoptosis. Cell Death Differ. **8**: 495–505.
18. KING, F.W. & E. SHTIVELMAN. 2004. Inhibition of nuclear import by the proapoptotic protein CC3. Mol. Cell. Biol. **24**: 7091–7101.
19. SCREATON, R.A., S. KIESSLING, O.J. SANSOM, *et al.* 2003. Fas-associated death domain protein interacts with methyl-CpG binding domain protein 4: a potential link between genome surveillance and apoptosis. Proc. Natl. Acad. Sci. USA **100**: 5211–5216.
20. FUSARO, G., P. DASGUPTA, S. RASTOGI, *et al.* 2003. Prohibitin induces the transcriptional activity of p53 and is exported from the nucleus upon apoptotic signaling. J. Biol. Chem. **278**: 47853–47861.
21. RAJALINGAM, K., C. WUNDER, V. BRINKMANN, *et al.* 2005. Prohibitin is required for Ras-induced Raf-MEK-ERK activation and epithelial cell migration. Nat. Cell Biol. **7**: 837–843.
22. MEAROW, K.M., M.E. DODGE, M. RAHIMTULA & C. YEGAPPAN. 2002. Stress-mediated signaling in PC12 cells—the role of the small heat shock protein, Hsp27, and Akt in protecting cells from heat stress and nerve growth factor withdrawal. J. Neurochem. **83**: 452–462.
23. NAKAGOMI, S., Y. SUZUKI, K. NAMIKAWA, *et al.* 2003. Expression of the activating transcription factor 3 prevents c-Jun N-terminal kinase-induced neuronal death by promoting heat shock protein 27 expression and Akt activation. J. Neurosci. **23**: 5187–5196.
24. GORMAN, A.M., E. SZEGEZDI, D.J. QUIGNEY & A. SAMALI. 2005. Hsp27 inhibits 6-hydroxydopamine-induced cytochrome c release and apoptosis in PC12 cells. Biochem. Biophys. Res. Commun. **327**: 801–810.
25. ISHIDO, M. & Y. MASUO. 2004. Transcriptome of pituitary adenylate cyclase-activating polypeptide-differentiated PC12 cells. Regul. Pept., **123**: 15–21.
26. BAI, G. & B. WEISS. 1991. The increase of calmodulin in PC12 cells induced by NGF is caused by differential expression of multiple mRNAs for calmodulin. J. Cell. Physiol. **149**: 414–421.
27. NATSUKARI, N., S.P. ZHANG, R.A. NICHOLS & B. WEISS. 1995. Immunocytochemical localization of calmodulin in PC12 cells and its possible interaction with histones. Neurochem. Int. **26**: 465–476.
28. HOU, W.F., S.P. ZHANG, G. DAVIDKOVA, *et al.* 1998. Effect of antisense oligodeoxynucleotides directed to individual calmodulin gene transcripts on the proliferation and differentiation of PC12 cells. Antisense Nucleic Acid Drug Dev. **8**: 295–308.
29. TASHIMA, K., H. YAMAMOTO, C. SETOYAMA, *et al.* 1996. Overexpression of Ca2+/calmodulin-dependent protein kinase II inhibits neurite outgrowth of PC12 cells. J. Neurochem. **66**: 57–64.

Identification of Repressor Element 1 in Secretin/PACAP/VIP Genes

LEO T.O. LEE, VIEN H.Y. LEE, PAULA Y. YUAN, AND BILLY K.C. CHOW

Department of Zoology, The University of Hong Kong, Hong Kong, China

ABSTRACT: Repressor element 1 (RE-1) is a negative, *cis*-acting regulatory element that interacts with the transcription factor RE-1-silencing transcription factor (REST). REST represses gene expression by two repressor domains that recruit other factors including mSin3 and CoREST. RE-1 has been identified in an increasing number of neuronal-specific genes, and recently, functional REST sites have also been discovered in VIP and PACAP genes. In the present article, we demonstrated for the first time that RE-1 sites are present in the 5′ flanking regions of several secretin/PACAP/VIP genes by *in silico* analysis. This observation suggests that RE-1/REST is a common negative regulatory pathway of this peptide family.

KEYWORDS: repressor element 1; secretin/PACAP/VIP peptide family; transcriptional regulation

FUNCTIONAL ROLE OF NRSF IN GENE REGULATION

Repressor element 1 (RE-1), also known as neuron-restrictive silencer element (NRSE), is a 21-bp long negative DNA regulatory element. As a *cis*-acting repressor element, RE-1 was initially defined as a motif that represses the expression of a number of neuronal genes in non-neural tissues with a proposed sequence "NNCAGCACCNNGCACAGNNNC".[1] Up-to-date, this RE-1 regulatory pathway has been identified in more than 50 genes, and most of them are neuronal-specific genes, such as type II sodium channel, μ-opioid receptor, and GABA$_A$ receptor $\gamma 2$,[2–4] while an increasing number of nonneuronal genes have also been reported to be controlled by RE-1. Knowing the importance of the RE-1 pathway, a RE-1 database (RE1db) was recently constructed by Bruce *et al.*,[5] in which search was performed by using an aligned sequence from 32 known RE-1 sequences (NTYAGMRCCNNRGMSAG). In human and mouse genomes, there are around 1800 putative RE-1 sites identified and 350 of them are located within 10 kb of the 5′ region of transcribed genes. Among these genes, more than 40% are known to be expressed in the nervous system.

Address for correspondence: Billy K.C. Chow, Department of Zoology, The University of Hong Kong, Pokfulam Road, Hong Kong, PRC. Voice: (852) 2299 0850; fax: (852) 2559 9114.
e-mail: bkcc@hkusua.hku.hk

Ann. N.Y. Acad. Sci. 1070: 388–392 (2006). © 2006 New York Academy of Sciences.
doi: 10.1196/annals.1317.050

The RE-1 sequences interact with a transcription factor, namely RE-1-silencing transcription factor (REST, which is also named as neuron-restrictive silencer factor, NRSF). REST is a Krippel-type zinc-finger transcriptional repressor for RE-1-dependent gene silencing. REST expression was detected in various nonneuronal tissues and in several vertebrate species, including human, mouse, rat, chicken, *Xenopus laevis,* and *Fugu rubripes.*[1,6] REST represses gene expression by two repressor domains that are located at the N- and C- termini.[7] The N-terminal repressor domain interacts with Sin3,[8] while the C-terminal repressor domain recruits the CoREST protein.[9] Both of these proteins are also associated with the histone deacetylase-containing complex[10] that deacetylates the N-terminal regions of the core histone proteins, resulting in an inactive and condensed chromatin structure. As a consequence, the target gene is silenced.

REST is also proposed as a modulator of gene expression. For example, although REST does not interact directly with the RE-1 sequence, the M4 muscarinic acetylcholine receptor gene could still be silenced in REST-expressing JTC-19 cells. In addition, treatment with Trichostatin A, an inhibitor of deacetylase, was unable to induce expression of this gene. It is therefore likely that other mechanism(s) should also exist in some REST-dependent gene-silencing event.[11,12] For instance, in contrast to gene silencing, transcriptional activator activity of the protein has been recently reported. When RE-1 was engineered within 50 nucleotides from the transcription start site of the neuronal nicotinic acetylcholine receptor b2 gene, an activation of gene expression was detected.[13] In the corticotropin-releasing hormone gene, REST could act as an activator via a mechanism that is independent of RE-1.[14] These data, therefore, suggest that the cellular context of the gene or the position of RE-1 site could determine the activity of REST.

IN SILICO PREDICTION OF RE-1 IN SECRETIN/PACAP/VIP PEPTIDE GENE FAMILY

The secretin/PACAP/VIP superfamily of brain-gut peptides contains at least nine members, including secretin, pituitary adenylate cyclase-activating polypeptide (PACAP), vasoactive intestinal polypeptide (VIP), glucagon, glucagon-like peptide-1 (GLP-1), glucagon-like peptide-2 (GLP-2), gastric inhibitory polypeptide (GIP), peptide histidine isoleucine (PHI), and growth hormone-releasing hormone (GHRH). They are encoded in six different genes, and among them, glucagon/GLP-1/GLP-2 and VIP/PHI are encoded in the same genes. These peptides are widely distributed throughout the body as well as the peripheral and central nervous systems. In the brain, majority of these peptides have already been identified as neurotransmitters, neuromodulators, neurotrophic factors, and/or neurohormones.[15,16] Recently, functional RE-1 sites have been located in VIP and PACAP

TABLE 1. Summary of the *in silico* analysis of the secretin/PACAP/VIP genes

	CpG island[c]	Predicted RE-1 in RE1-db[a]		Predicated RE-1 by MAPPER[b]		Reported RE-1
		Human	Mouse	Human	Mouse	
PACAP	+	−	−	+	+	+[18]
VIP	−	+	+	−	+	+[17]
Secretin	+	−	−	+	−	−
GIP	+	−	−	+	−	−
Glucagon/GLP-1/GLP-2	−	−	+	+	+	−
GHRH	+	−	−	−	−	−

+, RE-1 site identified or reported.

−, No significant finding or no related report.

[a]Based on the human and mouse RE-1 database (http://bioinformatics.leeds.ac.uk/group/online/RE1db/re1db_home.htm).

[b]Prediction by using the MAPPER (http://mapper.chip.org), with the following conditions: minimum score more than 2 and E-value not <0.1.

[c]Based on prediction from the human genome project, NCBI.

genes.[17,18] These findings and the neuronal specificity of this peptide family prompted us to investigate the presence of potential functional RE-1 sites in this peptide gene family. By searching the RE-1 database, RE1db (http://bioinformatics.leeds.ac.uk/group/online/RE1db/re1db_home.htm),[5] we could only locate the RE-1 sites in VIP and glucagon genes (TABLE 1). However, as many have reported functional RE-1 are not highly conserved when compared with the consensus, for example, in the PACAP gene, the RE-1-like element contains 10 mismatches out of 21-bp. For this reason, we employed a newly developed search engine "MAPPER" (from http://mapper.chip.org)[19] to identify putative RE-1 sites in these genes. This software has the advantage in allowing users to select a specific transcription factor in each search and to supply multiple sequences for alignment. In the study, we have used the defined NRSF-isoform 2 as the protein factor to search 1 kb upstream from the ATG codon as well as all the exon–intron sequences. Significant hits are defined if the score is more that 2.0 and E-value is <0.1. TABLE 1 listed the search results. RE-1 sites are found in both PACAP and glucagon genes in mouse and human. For VIP, secretin, and GIP genes, RE-1 sites are located in either species. As putative RE-1 sites are found in most of the genes in this family, hence, we propose here that the RE-1/REST regulation is a common regulatory pathway for this peptide family. In addition, RE-1 sites are also found in several receptor genes, including GLP-1, GLP-2, and VPAC2 receptors by the RE1db database, suggesting the possibility that both ligand and receptor genes share common mechanisms for gene expression. Interestingly, there is no significant hit for the GHRH gene, suggesting that the regulation of GHRH expression may not involve RE-1 and REST.

In summary, we have shown that putative RE-1 sites are present in the genes coding for the secretin/PACAP/VIP peptides. This cell-specific modulator should therefore play a regulatory role in the repression of these peptides in nonneural tissues and hence their expressions in the nervous system. This hypothesis opens up an entirely new research area to investigate the transcriptional regulation of this family of peptides and receptors in the future.

ACKNOWLEDGMENTS

This work was supported by research grants from RGC, HKU-7219/02M and CRCG 10203410 to Billy K.C. Chow.

REFERENCES

1. SCHOENHERR, C.J., A.J. PAQUETTE & D.J. ANDERSON. 1996. Identification of potential target genes for the neuron-restrictive silencer factor. Proc. Natl. Acad. Sci. USA **93:** 9881–9886.

2. MAUE, R.A., S.D. KRANER, R.H. GOODMAN & G. MANDEL. 1990. Neuron-specific expression of the rat brain type II sodium channel gene is directed by upstream regulatory elements. Neuron **4:** 223–231.

3. KIM, C.S., C.K. HWANG, H.S. CHOI, *et al.* 2004. Neuron-restrictive silencer factor (NRSF) functions as a repressor in neuronal cells to regulate the mu opioid receptor gene. J. Biol. Chem. **279:** 46464–46473.

4. MU, W. & D.R. BURT. 1999. Transcriptional regulation of GABAA receptor gamma2 subunit gene. Brain Res. Mol. Brain Res. **67:** 137–147.

5. BRUCE, A.W., I.J. DONALDSON, I.C. WOOD, *et al.* 2004. Genome-wide analysis of repressor element 1 silencing transcription factor/neuron-restrictive silencing factor (REST/NRSF) target genes. Proc. Natl. Acad. Sci. USA **101:** 10458–10463.

6. CHONG, J.A., J. TAPIA-RAMIREZ, S. KIM, *et al.* 1995. REST: a mammalian silencer protein that restricts sodium channel gene expression to neurons. Cell **80:** 949–957.

7. GRIFFITH, E.C., C.W. COWAN & M.E. GREENBERG. 2001. REST acts through multiple deacetylase complexes. Neuron **31:** 339–340.

8. ROOPRA, A., L. SHARLING, I.C. WOOD, *et al.* 2000. Transcriptional repression by neuron-restrictive silencer factor is mediated via the Sin3-histone deacetylase complex. Mol. Cell Biol. **20:** 2147–2157.

9. ANDRES, M.E., C. BURGER, M.J. PERAL-RUBIO, *et al.* 1999. CoREST: a functional corepressor required for regulation of neural-specific gene expression. Proc. Natl. Acad. Sci. USA **96:** 9873–9878.

10. BALLAS, N., E. BATTAGLIOLI, F. ATOUF, *et al.* 2001. Regulation of neuronal traits by a novel transcriptional complex. Neuron **31:** 353–365.

11. MIEDA, M., T. HAGA & D.W. SAFFEN. 1997. Expression of the rat m4 muscarinic acetylcholine receptor gene is regulated by the neuron-restrictive silencer element/repressor element 1. J. Biol. Chem. **272:** 5854–5860.

12. WOOD, I.C., A. ROOPRA & N.J. BUCKLEY. 1996. Neural specific expression of the m4 muscarinic acetylcholine receptor gene is mediated by a RE1/NRSE-type silencing element. J. Biol. Chem. **271:** 14221–14225.
13. BESSIS, A., N. CHAMPTIAUX, L. CHATELIN & J.P. CHANGEUX. 1997. The neuron-restrictive silencer element: a dual enhancer/silencer crucial for patterned expression of a nicotinic receptor gene in the brain. Proc. Natl. Acad. Sci. USA **94:** 5906–5911.
14. SETH, K.A. & J.A. MAJZOUB. 2001. Repressor element silencing transcription factor/neuron-restrictive silencing factor (REST/NRSF) can act as an enhancer as well as a repressor of corticotropin-releasing hormone gene transcription. J. Biol. Chem. **276:** 13917–13923.
15. SHERWOOD, N.M., S.L. KRUECKL & J.E. MCRORY. 2000. The origin and function of the pituitary adenylate cyclase-activating polypeptide (PACAP)/glucagon superfamily. Endocr. Rev. **21:** 619–670.
16. NYBERG, J., M.F. ANDERSON, B. MEISTER, *et al.* 2005. Glucose-dependent insulinotropic polypeptide is expressed in adult hippocampus and induces progenitor cell proliferation. J. Neurosci. **25:** 1816–1825.
17. HAMELINK, C., S.H. HAHM, H. HUANG & L.E. EIDEN. 2004. A restrictive element 1 (RE-1) in the VIP gene modulates transcription in neuronal and non-neuronal cells in collaboration with an upstream tissue specifier element. J. Neurochem. **88:** 1091–1101.
18. SUGAWARA, H., K. INOUE, S. IWATA, *et al.* 2004. Neural-restrictive silencers in the regulatory mechanism of pituitary adenylate cyclase-activating polypeptide gene expression. Regul. Pept. **123:** 9–14.
19. MARINESCU, V.D., I.S. KOHANE & A. RIVA. 2005. The MAPPER database: a multigenome catalog of putative transcription factor binding sites. Nucleic Acids Res. **33:** D91–D97.

Retinoic Acid-Induced Human Secretin Gene Expression in Neuronal Cells Is Mediated by Cyclin-Dependent Kinase 1

LEO T.O. LEE, K.C. TAN-UN, AND BILLY K.C. CHOW

Department of Zoology, The University of Hong Kong, Hong Kong, China

ABSTRACT: Previously, we found that secretin transcript levels were induced by *all-trans* retinoic acid (RA) in a neuroblastoma cell model, SH-SY5Y. In this article, this RA-dependent upregulation process was further investigated. In the cyclin-dependent kinase 1 (Cdk1) inhibitor-treated cells, the RA-dependent induction of secretin gene expression was inhibited. Together with our previous works, we propose here that the RA responsiveness of the secretin promotor is mediated by two different pathways. The first pathway is by changing the expression levels of NFI-C and Sp proteins while the second pathway is by modifying the phosphorylation status of both NFI-C and Sp proteins via Cdk1.

KEYWORDS: secretin; Cdk1; retinoic acid; promotor

INTRODUCTION

Recently, the regulation of the human secretin gene has been studied in several reports, and an E-box and two GC-boxes were found to be essential for activity in duodenum cells.[1] *In vivo*- (chromatin immunoprecipitation) and *in vitro* (gel mobility shift)-binding studies showed that these motifs were binding sites for transcription factors, NeuroD/E2A and Sp1/Sp3, respectively. As a neuropeptide, secretin gene expression in neuronal cells was studied in a well-established neuronal differentiation cell model, SH-SY5Y.[2] We found that the human secretin gene was upregulated upon *all-trans* retinoic acid (RA) treatment. This RA induction is regulated by the combined actions of reduced Sp3 and increased nuclear factor-I C (NFI-C) expressions. In the present article, we further investigated the mechanism by which RA induced human secretin gene expression. We found that this induction process was inhibited when cyclin-dependent kinase 1 (Cdk1) inhibitors were used to treat the cells. These data therefore suggested that the RA-dependent upregulation of the human secretin gene is mediated by protein modifications involving Cdk1.

Address for correspondence: Billy K.C. Chow, Department of Zoology, The University of Hong Kong, Pokfulam Road, Hong Kong, PRC. Voice: 852-2299-0850; fax: 852-2559-9114.
e-mail: bkcc@hkusua.hku.hk

Ann. N.Y. Acad. Sci. 1070: 393–398 (2006). © 2006 New York Academy of Sciences.
doi: 10.1196/annals.1317.051

MATERIALS AND METHODS

Cell Culture and Transient Transfection Assay

The neuroblastoma cell line, SH-SY5Y (kindly provided by Prof. J. Hugon, The University of Hong Kong), was cultured in MEM (Invitrogen, Carlsbad, CA, USA) with nonessential amino acids, 10% FBS, 100 U/mL penicillin, and 100 μg/mL streptomycin at 37°C and 5% CO_2. *All-trans* RA and 12-O-tetradecanoylphorbol-13-acetate (TPA) were purchased from Sigma (St. Louis, MO). Cdk1-inhibitor I [3-(2-chloro-3-indolymethylene)-1,3-dihydroindol-2-one] and III (Ethyl-(6-hydroxy-4-phenylbenzo[4,5]furo[2,3-b]) pyridine-3-carboxylate) were purchased from Calbiochem (Darmstadt, Germany).

In transient transfection assays, SH-SY5Y cells were plated at a density of 2.0×10^5 cells/35 mm well. After 2 days of incubation, 2 μg promotor-luciferase construct and 0.5 μg pCMV-β-gal were cotransfected into cells using 8 μL of LipofectAMINE 2000 reagent (Invitrogen) according to the manufacturer's protocol. For Cdk1-inhibitor treatment, cells were treated with the inhibitor (Cdk1 inhibitor I or III) 30 min before transfection. TPA (20 nM) or RA (10 μM) was incubated for the designated time after transfection. Afterward, the cells were harvested and cell extracts were assayed for luciferase and β-galactosidase activities. Data represent the mean ± SEM of three experiments performed in triplicates.

Real-Time Quantitative PCR Analysis

Total RNA from the cells was isolated using the TriPure reagent (Roche Molecular Biochemicals, Basel, Switzerland) and used (5 μg) for reverse-transcription with an oligo-dT primer and Superscript III reverse-transcriptase (Invitrogen). For quantitative real-time PCR, one-tenth of the first strand cDNA was used as the template. Secretin and GAPDH levels were detected by an assay on demand system (Applied Biosystems, Foster City, CA) with the Taqman Universal Master Mix (Applied Biosystems). Fluorescence signals were measured in real time during the extension step by the iCycler iQ Real Time Detection System (Bio-Rad, Hercules, CA). The ratio change in secretin gene relative to GAPDH was determined by the $2^{-\Delta\Delta Ct}$ method.[3]

RESULTS

Secretin is expressed endogenously in the neuroblastoma cell line SH-SY5Y, which can therefore serve as a cell model for studying secretin gene expression.[2] To unravel the mechanism that regulates human secretin gene during neuronal differentiation, RA and TPA were used to treat this SH-SY5Y cell model, and promotor activities were measured after drug treatment. Secretin promotor activity after RA treatment was increased in a time-dependent manner (108 % after 24 h and 270 % after 32 h). In contrast, no change was observed in TPA-treated samples (FIG. 1 A).

Cdk1-inhibitor I is a selective and ATP-competitive inhibitor of Cdk1/cyclin B by binding to the ATP pocket on the Cdk1 active site.[4] Cdk1-inhibitor III is also a highly selective inhibitor that displays little cross inhibitory activities against Cdk5/p25, Cdk2/E, and Cdk4/D1.[5] To investigate the *in vivo* roles of Cdk1 in regulating the core promotor construct, p341 construct, the transiently transfected SH-SY5Y cells were treated with 10 μM Cdk1-inhibitor I and 50 μM Cdk1-inhibitor III, and the effects were monitored by luciferase assays. As shown in FIGURE 1 B, Cdk1-inhibitor treatment caused a significant decrease in promotor function in RA-treated samples. In the presence of Cdk1 inhibitor, there was a significant reduction (45% for Cdk1-inhibitor I or 68% for Cdk1-inhibitor III) in promotor function but no change in luciferase activity was observed in samples with no RA treatment. In addition, endogenous secretin mRNA levels in SH-SY5Y cells were examined by real-time PCR analysis (FIG. 1 C). Similar to the results from luciferase assay, the endogenous secretin mRNA levels were significantly decreased from 2.3-fold (RA + control) to 0.4-fold after Cdk1-inhibitor I and to 0.6-fold after Cdk1-inhibitor III treatment. These data clearly indicated that the RA-induced secretin expression in neuroblastoma cells was mediated by Cdk1.

DISCUSSION

In previous studies, the SH-SY5Y neuroblastoma cell was used as a model to study the transcriptional regulation of the human secretin gene in neuronal cells.[2] Secretin transcript levels were induced by RA that also activated the process of neuronal differentiation. However, the use of another neuronal differentiating agent, TPA, resulted in no significant change of secretin level (FIG. 1 A). This indicated that the upregulation of secretin is direct and RA-dependent.

We have previously suggested the alteration of Sp proteins and NFI-C transcript levels by RA is an important component of this activation pathway. Besides the change in RNA and/or protein levels, it is also possible that RA activates secretin gene by modifying the structures of these transcription factors as NFI has previously been shown to be able to undergo protein phosphorylation.[6] Although no significant difference in DNA-binding affinity was observed in dephosphorylated and phosphorylated forms of NFI, NFI phosphorylation could still alter expressions of several genes.[7] Besides, it was proposed earlier that Cdk1 can phosphorylate NFI and Sp1 *in vitro*.[6,8] Other studies also suggested that the inhibition of Cdk1 activity by olomoucine could reduce

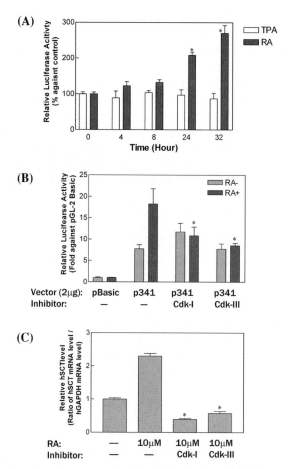

FIGURE 1. (**A**) Time dependence of TPA and RA on activating the human secretin promotor in SH-SY5Y cells. Two micrograms of promotor construct, p341, was transfected into SH-SY5Y cells. Transfected cells were treated with 10 μM RA or 20 nM TPA for different times. The relative promotor activity is shown as the percentage induction over the control (0-h treatment). $^{*}P < 0.001$ versus control. (**B**) Cdk1-inhibitor I or III can block the RA-induced human secretin gene promotor activity. In the transient transfection assays, Cdk1-inibitor I (10 μM) and Cdk1-inhibitor III (50 μM) were added to the p341-transfected SH-SY5Y cells. The relative promotor activity is shown as the fold change against the pGL2-basic control. $^{*}P < 0.001$ versus p-341-transfected cells without Cdk1-inhibitor treatment. (**C**) Effects of Cdk1-inhibitor treatment on the RA-induced human secretin gene expression in SH-SY5Y cells. $^{*}P < 0.001$ versus secretin level after RA treatment.

a Sp1-dependent reporter gene expression.[8] As activation of Cdk1 by RA is also observed in human hepatoma Hep3B cells,[9] it is therefore possible, as an alternative mechanism, that Cdk1 can modify the phosphorylation status of

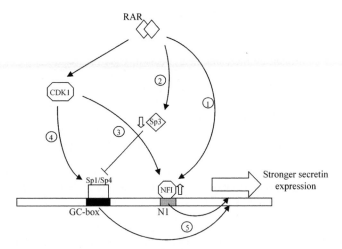

FIGURE 2. A proposed model for RA-induced human secretin gene expression. The numbers represent the steps for the RA-mediated effects: (1) RA upregulates NFI-C expression; (2) Sp3 expression levels is suppressed by RA to augment the cellular (Sp1 + Sp4)/Sp3 ratio; (3) RA, through Cdk1, changes the phosphorylation status of NFI proteins and (4) Sp proteins; (5) Sp proteins and NFI cooperatively bind onto the GC-box and N1 motif, respectively, leading to gene activation.

NFI and/or Sp1 in response to RA treatment and subsequently lead to secretin gene activation in neuroblastoma cells. In the present study, this hypothesis was studied by treating the cells with Cdk1 inhibitors during RA-induced neuronal differentiation. As shown in FIGURE 1 B and C, both inhibitors (I and III) inhibited the RA-induced promotor activity and *in vivo* transcript levels. These data indicated that the RA responsiveness of the secretin gene could also be mediated by Cdk1.

Together with our previous works,[5] we proposed that the RA responsiveness of the secretin promotor is regulated at transcription and protein levels (FIG. 2). At the transcription level, RA induces the expression of NFI-C and suppresses Sp3. At the protein level, RA modifies the phosphorylation status of both NFI-C and Sp proteins via Cdk1 resulting in the cooperative binding of Sp1/Sp4 and NFI-C with the GC-box 1 and N1 motif, respectively, to activate secretin gene expression.

ACKNOWLEDGMENTS

This work was supported by research grants from RGC, HKU-7219/02M, and CRCG 10203410 to Billy K.C. Chow.

REFERENCES

1. LEE, L.T., K.C. TAN-UN, R.T. PANG, et al. 2004. Regulation of the human secretin gene is controlled by the combined effects of CpG methylation, Sp1/Sp3 ratio, and the E-box element. Mol. Endocrinol. **18:** 1740–1755.
2. LEE, L.T., K.C. TAN-UN, M.C. LIN & B.K. CHOW. 2005. Retinoic acid activates human secretin gene expression by Sp proteins and nuclear factor I in neuronal SH-SY5Y cells. J. Neurochem. **93:** 339–350.
3. LIVAK, K.J. & T.D. SCHMITTGEN. 2001. Analysis of relative gene expression data using real-time quantitative PCR and the 2(-Delta Delta C(T)) method. Methods **25:** 402–408.
4. ANDREANI, A., A. CAVALLI, M. GRANAIOLA, et al. 2000. Imidazo[2,1 -b] thiazolylmethylene- and indolylmethylene-2-indolinones: a new class of cyclin-dependent kinase inhibitors. Design, synthesis, and CDK1/cyclin B inhibition. Anticancer Drug Des. **15:** 447–452.8.
5. BRACHWITZ, K., B. VOIGT, L. MEIJER, et al. 2003. Evaluation of the first cyto-statically active 1-aza-9-oxafluorenes as novel selective CDK1 inhibitors with P-glycoprotein modulating properties. J. Med. Chem. **46:** 876–879.
6. JACKSON, S.P., J.J. MACDONALD, S. LEES-MILLER & R. TJIAN. 1990. GC box binding induces phosphorylation of Sp1 by a DNA-dependent protein kinase. Cell **63:** 155–165.
7. BISGROVE, D.A., E.A. MONCKTON, M. PACKER & R. GODBOUT. 2000. Regulation of brain fatty acid-binding protein expression by differential phosphorylation of nuclear factor I in malignant glioma cell lines. J. Biol. Chem. **275:** 30668–30676.
8. HAIDWEGER, E., M. NOVY & H. ROTHENEDER. 2001. Modulation of Sp1 activity by a cyclin A/CDK complex. J. Mol. Biol. **306:** 201–212.
9. HSU, S.L., M.C. CHEN, Y.H. CHOU, et al. 1999. Induction of p21(CIP1/Waf1) and activation of p34(cdc2) involved in retinoic acid-induced apoptosis in human hepatoma Hep3B cells. Exp. Cell Res. **248:** 87–96.

Neuroendocrine Tumors Express PAC1 Receptors

SONG N. LIEU, DAVID S. OH, JOSEPH R. PISEGNA,
AND PATRICIA M. GERMANO

CURE: Digestive Diseases Research Center, VA Greater Los Angeles Healthcare System and Department of Medicine, University of California, Los Angeles, California 90073, USA

ABSTRACT: Neuroendocrine tumors (NETs) of the gastrointestinal tract can be grossly divided into two general types: carcinoid and pancreatic endocrine tumors. The former develop in the luminal intestine whereas the latter occur within the pancreas. To ascertain whether pituitary adenylate cyclase-activating polypeptide (PACAP) has a biological effect on the regulation of secretion or growth, we studied the well-established NET cell line, BON. BON cells have been shown previously to contain chromogranin A, neurotensin, and serotonin. In response to mechanical stimulation, BON cells have been demonstrated to release serotonin. The current article demonstrates that the high-affinity PAC1 receptor is expressed on the NET cell line BON. These results indicate that PACAP may regulate the biological release of peptides and serotonin from BON cells and that, like in solid tumors, PACAP could potentially stimulate the growth of BON cells.

KEYWORDS: BON; carcinoid tumor; PACAP

INTRODUCTION

Neuroendocrine tumors (NETs) include both carcinoid and pancreatic endocrine tumors. Originally these tumors were referred to as argentaffin tumors with intracytoplasmic granules owing to their staining characteristics against chromogranin A, neuron-specific enolase, or synaptophysin.[1] It is presumed that the majority of these tumors originate in the neuroendocrine cells of the gastrointestinal tract. Patients experience clinical syndromes that are related to the release of biologically active peptides and/or amines. For example, in patients presenting with carcinoid tumors, flushing and diarrhea are the characteristic symptoms.[1]

Address for correspondence: Patrizia M. Germano, M.D., David Geffen School of Medicine, CURE/UCLA/VA Bldg 115 Rm 313, 11301 Wilshire Blvd., Los Angeles, CA 90073. Voice: 310-268-4069; fax: 310-268-4096.
e-mail: pgermano@ucla.edu

Ann. N.Y. Acad. Sci. 1070: 399–404 (2006). © 2006 New York Academy of Sciences.
doi: 10.1196/annals.1317.052

A useful way to investigate the biology of NETs is to study the BON cell line. This cell line that was originally developed by C.M. Townsend Jr., (University of Texas, Galveston, TX) is derived from a metastatic lymph node in a patient with a pancreatic NET.[2] BON cells have been demonstrated to release a variety of peptides including transforming growth factor (TGF), fibroblast growth factor (FGF), and epidermal growth factor receptor (EGFR) as well as peptides, such as pancreastatin, neurotensin, and the amine 5-HT.[2] The release of secretory products, results in the clinical symptoms associated with these tumors.[1]

We have previously shown that receptors for pituitary adenylate cyclase-activating polypeptide (PACAP) can be identified on tumors originating from the lung, breast, and colon where stimulation by PACAP is linked to proliferative signaling pathways and tumor growth.[3–5] We therefore proposed that, similar to solid tumors, receptors for PACAP are expressed on NETs and may mediate biological effects induced by PACAP, such as secretion and growth.

MATERIALS AND METHODS

Materials

BON tumor cells were obtained from Dr. Courtney Townsend. Cells were cultured using DMEM F:12 (Hyclone, Logan, UT) supplemented with 10% (v/v) fetal bovine serum (FBS), kanamycin (Gibco, Carlsbad, CA), and gentamicin (Sigma, St. Louis, MO) antibiotics. Cells were grown to confluency in a humidified atmosphere of 5% CO_2 at 37°C.

Molecular Identification of PAC1

RT-PCR was used to demonstrate molecular expression of PAC1 R. These studies were performed on total RNA obtained from cells that were grown to confluency, using BD Biosciences' Nucleospin RNA II Kit (Palo Alto, CA) according to the manufacturer's instructions. PCR was performed in low salt Taq + DNA polymerase buffer and five units Taq + DNA polymerase (Stratagene, La Jolla, CA) in the presence of oligonucleotide primers under the following conditions: initial step (one cycle): 94°C for 2 min, 57°C for 1 min, and 72°C for 2 min; followed by 94°C for 1 min, 57°C for 1 min, and 72°C for 2 min (30 cycles); and a final extension step (one cycle) at 94°C for 1 min, 57°C for 1 min, and 72°C for 15 min. The sense primers used were: (SENSE 1) 5′CGAGTGGACAGTGGCAGGCGGTGA3′ (52–77); and (SENSE 2) 5′GCTCTCCCTGACTGCTCTCCTGCTG 3′ (145–170). The antisense primers used were: (ANTISENSE 1) 5′CAGTAGTGAGGGT-GGCGAGGGAAGT3′ (611–636) and (ANTISENSE 2) 5′CAGTAGGTGTC-CCCCAGCCGATGAT3′ (935–960).

Immunocytochemical Analysis of PAC1 R in BON Cells

In order to demonstrate the expression of PAC1 R on BON cells, we used immunofluorescent staining with antibodies specific to PAC1 R (CURE Antibodies Core), as described previously.[5] Confocal microscopy was used to characterize the expression of PAC1 R. To perform these studies, BON cells were plated on polylysine-treated coverslips overnight and fixed with 4% paraformaldehyde, permeabilized with 0.1% Triton-X, and then incubated at 4°C with polyclonal rabbit anti-PAC1 R (1:1000). To exclude nonspecific binding of the antibodies, BON cells were incubated with rabbit IgG (1 μg/mL) as a negative control overnight at 4°C. The following day, the cells were washed with PBS, and then incubated with Alexa-488 (Molecular Probes, Eugene, Oregon)-conjugated secondary antibodies—goat anti-rabbit IgG antibodies. Cells were visualized using a Zeiss LSM 510 Laser Scanning Microscope (Carl Zeiss, Thornwood, NY).

RESULTS

Molecular Determination of PAC1 Expression

Specific primer pairs to detect expression of PAC1 R and VPAC1 R were chosen for these studies as described previously.[6-8] RNA extracted from confluent BON cells only showed expression of PAC1 R and VPAC1-R. VPAC2 R was not detected in the test samples. These results demonstrate that high levels of PAC1 R and VPAC1 R mRNAs but not VPAC2 R mRNA are present in BON cells.

Immunocytochemistry

Immunocytochemistry and confocal microscopy were used to visualize PAC1 R in BON cells (FIG. 1). The cells were incubated with polyclonal rabbit anti-PAC1 antibodies or rabbit IgG as a negative control, followed by goat anti-rabbit FITC-conjugated antibodies. Expression of PAC1 R was detected at the surface of the BON cells. As a negative control, we used rabbit IgG followed by the secondary antibodies, to ascertain the specificity of the PAC1 R staining.

DISCUSSION

PACAP is a member of the vasoactive intestinal polypeptide (VIP) family of peptides.[6] The cloning of the high-affinity receptor for PACAP (PAC1 R), specific for PACAP-27 and PACAP-38, has identified this receptor as a member

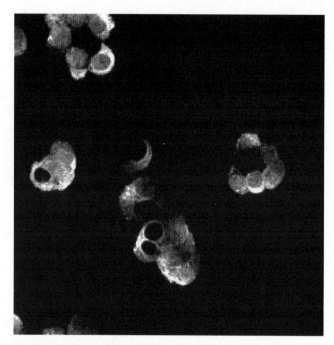

FIGURE 1. Immunocytochemistry demonstrating PAC1 R expression in BON cells. The cells were incubated with anti-PAC1 R rabbit polyclonal antibodies followed by Alexa-488-conjugated goat anti-rabbit IgG as a second antibody.

of the seven transmembrane G protein–coupled receptors family.[6,9,10] PACAP binds with a high affinity to PAC1 R and with a lower affinity to VPAC1 R and VPAC2 R.[11] Knowledge of the receptor structure has permitted the development of specific antibodies to detect receptors in native cell systems.[12] PAC1 has been shown to be expressed on a number of tumor cell lines derived from lung, breast, and colonic solid tumors.[3–5] In addition, PAC1 R has been shown to be expressed on the enterochromaffin-like cells of the gastric mucosa. In all these organs, PACAP stimulation leads to the proliferation of the cancer cells.[4,13,14] Expression of PAC1 R on neuroendocrine tumoral cells, such as BON, has not been shown previously.

In this study, we report the presence of PAC1 receptors in BON cells by means of RT-PCR and immunohistochemical staining. Immunohistochemical analysis using confocal laser scanning microscopy, performed in BON cells, using specific rabbit polyclonal anti-PAC1 R antibodies, has provided evidence of PAC1 R expression on the cell membranes. To determine whether activated PACAP receptors expressed on the surface of BON cells lead to the activation of intracellular signaling pathways, involved in cellular proliferation and secretion it will be necessary to confirm these results with functional pharmacological and biochemical assays. The precise signaling pathway involved in activating mitogen-activated protein kinases (MAPKs) would presumably originate from the activation of Gαq. It has been previously demonstrated in solid tumors that the Gαq pathway is involved in this process, because inhibition of phospholipase (PLC) or protein kinase C (PKC) results in blocking PACAP-induced activation of MAPKs.[13] These observations suggest that PAC1 R, expressed on BON cells, could result in the activation of PKC, and thereby activate MAPK leading to cell proliferation. This hypothesis needs to be tested in BON cells. The expression of PAC1 R on BON cells suggests that PACAP plays a role in stimulating the growth and survival of human NETs *in vivo*. These observations support the hypothesis that NETs cells can be stimulated by the neuropeptide PACAP.

In conclusion, the present data indicate that PAC1 R is expressed on neuroendocrine cells and suggests that PACAP may play a key role in the regulation of their growth. These data therefore may have important clinical implications by providing novel therapeutic targets to regulate the growth and development of NETs.

ACKNOWLEDGMENTS

This work was supported by the Department of Veterans Affairs Merit Review Grant (JRP) and National Institutes of Health DK37240 (HJC).

REFERENCES

1. PISEGNA, J.R. & M.P. SAWICKI. 2001. Neuroendocrine pancreas. *In* Cancer Treatment. C.M. HASKELL & J.S. BEREK Eds: 1065–1081. Saunders. Philadelphia, PA.
2. PAREKH, D., J. ISHIZUKA, C.M. TOWNSEND JR., *et al.* 1994. Characterization of a human pancreatic carcinoid in vitro: morphology, amine and peptide storage, and secretion. Pancreas **9:** 83–90.
3. ZIA, F., M. FAGARASAN, K. BITAR, *et al.* 1996. PACAP receptors regulate the growth of non-small cell lung cancer cells. Cancer Res. **55:** 4886–4891.
4. LEYTON, J., Y. GOZES, J.R. PISEGNA, *et al.* 1999. PACAP (6–38) is a PACAP receptor antagonist for breast cancer cells. Breast Cancer Res. Treat. **56:** 177–186.
5. LE, S.V., D.J. YAMAGUCHI, C.A. MC ARDLE, *et al.* 2002. PAC1 and PACAP expression, signaling, and effect in the growth of HCT8, human colonic tumor cells. Regul. Pept. **109:** 115–125.
6. PISEGNA, J.R. & S.A. WANK. 1996. Cloning and characterization of the signal transduction of four splice variants of the human pituitary adenylate cylcase activating polypeptide receptor. Evidence for dual coupling to adenylate cyclase and phospholipase C. J. Biol. Chem. **271:** 17267–17274.
7. ISHIHARA, T., R. SHIGEMOTO, K. MORI, *et al.* 1992. Functional expression and tissue distribution of a novel receptor for vasoactive intestinal polypeptide. Neuron **8:** 811–819.

8. LUTZ, E.M., W.J. SHEWARD, K.M. WEST, *et al.* 1993. The VIP2 receptor: molecular characterization of a cDNA encoding a novel receptor for vasoactive intestinal peptide. FEBS **334:** 3–8.

9. PISEGNA, J.R. & S.A. WANK. 1993. Molecular cloning and functional expression of the pituitary adenylate cylcase-activating polypeptide type I receptor. Proc. Natl. Acad. Sci. USA **90:** 6345–6349.

10. SPENGLER, D., C. WAEBER, C. PANTALONI, *et al.* 1993. Differential signal transduction by five splice variants of the PACAP receptor. Nature **365:** 170–175.

11. HARMER, A.J., A. ARIMURA, I. GOZES, *et al.* 1998. International Union of Pharmacology. XVIII. Nomenclature of receptors for vasoactive intestinal peptide and pituitary adenylate cyclase-activating polypeptide. Pharmacol. Rev. **50:** 265–270.

12. MIAMPAMBA, M., P.M. GERMANO, S. ARLI, *et al.* 2002. Expression of pituitary adenylate cyclase-activating polypeptide and PACAP type I receptor in the rat gastric and coloine myenteric neurons. Regul. Pept. **105:** 145–154.

13. PISEGNA, J.R., J. LEYTON, T. COELHO, *et al.* 1997. Differential activation of immediate-early gene expression by four splice variants of the human pituitary activating polypeptide receptor: evident for activation by PACAP hybrid and the phospholipase C inhibitor U73122. Life Sci. **61:** 631–639.

14. ZENG, N., C. ATHMANN, T. KANG, *et al.* 1999. PACAP type I receptor activation regulates ECL cells and gastric acid secretion. J. Clin. Invest. **104:** 1383–1391.

PAC1 Receptor

Emerging Target for Septic Shock Therapy

CARMEN MARTÍNEZ,[a] ALICIA ARRANZ,[b] YASMINA JUARRANZ,[b]
CATALINA ABAD,[b] MARÍA GARCÍA-GÓMEZ,[b]
FLORENCIA ROSIGNOLI,[b] JAVIER LECETA,[b] AND ROSA P. GOMARIZ[b]

[a]Department of Cell Biology, Faculty of Medicine, Complutense University,
28040 Madrid, Spain

[b]Department of Cell Biology, Faculty of Biology, Complutense University,
28040 Madrid, Spain

ABSTRACT: Septic shock is a systemic response to severe bacterial infections, generally caused by Gram-negative bacterial endotoxins, with
multiple manifestations such as hypotension, tissue injury, disseminated
intravascular coagulation, and multi-organ failure. All these effects, are
induced by the generation of pro-inflammatory and vasodilator mediators, cell adhesion molecules, coagulation factors, and acute-phase proteins. Vasoactive intestinal polypeptide (VIP) and pituitary adenylate
cyclase-activating polypeptide (PACAP) are two immunopeptides with
anti-inflammatory properties exerted through type 1 and 2 VIP receptors (VPAC1 and VPAC2, respectively), and PACAP receptor (PAC1). The
present results recapitulate the protective role of PAC1 in an experimental
model of lethal endotoxemia using a knockout for the PAC1 receptor. Our
results demonstrate that VIP and PACAP decrease lipopolysaccharide
(LPS)-induced interleukin-6 (IL-6) production, neutrophil infiltration
and intercellular adhesion molecule-1 (ICAM-1), vascular cell adhesion
molecule-1 (VCAM-1), and fibrinogen expression through PAC1 receptor, providing an advantage to design more specific drugs complementing
standard intensive care therapy in septic shock.

KEYWORDS: PAC1; endotoxemia; neutrophil infiltration; fibrinogen;
IL-6

INTRODUCTION

Septic shock is a systemic response to severe bacterial infections, generally
caused by Gram-negative bacterial endotoxins that trigger a cascade of biological events. During septic shock, the activation of coagulation and the release

Address for correspondence: Carmen Martínez, Ph.D., Department of Cell Biology, Faculty of
Medicine, Complutense University, Ciudad Universitaria, 28040 Madrid, Spain. Voice: 34-913944971;
fax: 34-913944981.
e-mail: cmmora@bio.ucm.es

Ann. N.Y. Acad. Sci. 1070: 405–410 (2006). © 2006 New York Academy of Sciences.
doi: 10.1196/annals.1317.053

of a large number of mediators of inflammation contribute to the pathogenesis of disseminated intravascular coagulation and multi-organ failure.[1] The recruitment of leukocytes into sites of inflammation by interactions with the vascular endothelium leads to the release of a wide range of pro-inflammatory cytokines such as interleukin-6 (IL-6) and tumor necrosis factor-α (TNF-α), which are pivotal mediators of septic shock. These cytokines initiate a wide range of systemic responses including a dramatic increase of acute-phase (AP) proteins such as serum amyloid A (SAA), which is synthesized in the liver. Coagulation factors as fibrinogen are also very rapidly released in response to endotoxins and inflammatory cytokines triggering blood clot formation. All these molecules associated with nitric oxide (NO), which leads to the hypotension that defines septic shock, are important markers of inflammation and belong to the growing list of potential therapeutic targets in the pathogenesis of sepsis. Though significant advances have been made in our understanding of the septic cascade, most of the early trials have failed to improve survival. Interference with adhesion molecules, inflammatory agents, and coagulation abnormalities as a whole might offer hopeful, therapeutic possibilities. The present article summarizes the protective role of the member of VIP/PACAP receptor family PAC1, mediating the blockade of several important markers into the complexities of sepsis.

VIP AND PACAP IN SEPTIC SHOCK

Vasoactive intestinal polypeptide (VIP) and pituitary adenylate cyclase-activating polypeptide (PACAP) are two immunopeptides that exert their immunological effects through binding to the family 2 of G protein–coupled receptors, VIP receptors type 1 and 2 (VPAC1 and VPAC2, respectively), and PACAP receptor (PAC1).[2] In recent years, VIP and PACAP effects inhibiting endotoxin-induced production of proinflammatory factors such as TNF-α, IL-6, IL-12, NO, and interferon-γ (IFN-γ), and stimulating anti-inflammatory cytokines, such as IL-10[3,4] have been described. Application of the anti-inflammatory properties of VIP and PACAP in a murine model,[5] has demonstrated that both immunopeptides protect from endotoxemia through inhibition of TNF-α and IL-6. Although our studies using specific VIP agonists suggested that VPAC1 was the main mediator,[6] PAC1 has recently emerged as an anti-inflammatory receptor improving survival in a murine model of lethal endotoxemia, as shown by using mice defective in PAC1 expression (PAC1$^{-/-}$) compared with their wild-type counterparts.[7]

HOW DOES PAC1 RECEPTOR PROTECT FROM ENDOTOXIC SHOCK?

The first findings on the role of PAC1 in endotoxic shock pointed out its involvement mediating the inhibition of VIP/PACAP on IL-6 production after a lethal dose (1 mg) of lipopolysaccharide[7] (LPS). However, it did not participate in the inhibition of TNF-α in accordance with previous studies using VPAC1 and VPAC2 agonists.[6] The best correlation of plasma cytokine levels with mortality from septic shock has been made with IL-6,[8] which appears to orchestrate a variety of inflammatory responses amplifying leukocyte recruitment through an increase of the local production of chemokines, and the expression of intercellular adhesion molecule-1 (ICAM-1) and vascular cell adhesion molecule-1 (VCAM-1), as well as inducing the hepatic AP response.[9,10] Thus, IL-6/IFN-γ activate nuclear factors for the regulation of ICAM-1 gene expression,[11] and an IL-6-response element has been described in the promotor of γ-chain of human fibrinogen.[12] Therefore, we speculated that PAC1 could modulate neutrophil recruitment, adhesion molecule expression, as well as AP proteins. Firstly, we evaluated the effect of VIP/PACAP (5 nmol) on myeloperoxidase activity (MPO), as a quantitative measure of neutrophil infiltration in liver and intestinal extracts. Both peptides inhibited MPO activity in wild-type but not in PAC1$^{-/-}$. This fact was confirmed by histopathological analysis. Transmigration of leukocytes involves endothelial cell adhesion, including ICAM-1 and vascular VCAM-1 which increase after LPS administration. VIP and PACAP reduced soluble ICAM-1 (ICAM) and the adhesion molecule transcripts in wild-type, whereas no differences were observed in PAC1$^{-/-}$ mice. Next, we showed the effects of VIP and PACAP on the

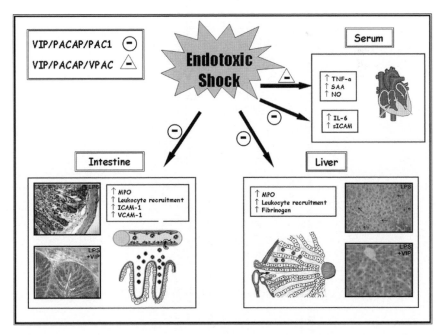

FIGURE 1. Endotoxic shock induces different pathological outcomes that VIP and PACAP abrogate through activation of VIP/PACAP receptors, notably of PAC1 receptor.

TABLE 1. Biological parameters modulated by PAC1 receptor in endotoxic shock

	Survival	Proinflammatory cytokines		Neutrophil infiltration		Adhesion			Coagulation factor fibrinogen	AP protein SAA	No production	DGA levels	cAMP levels
		IL-6	TNF-α	MPO	Histological scoring	sICAM	ICAM-1 VCAM-1						
WT	60%	↓	↓	↓	↓	↓	↓		↓	↓	↓	↑	↑
PAC1−/−	25–30%	=	↓	=	=	=	=		=	↓	↓	=	↑
PACAP (6–38)	ND	ND	ND	=	ND	ND	=		=	ND	ND	ND	ND

Note: Wild-type (WT) and PAC1−/− mice were injected with 1 mg of LPS plus VIP or PACAP (5 nmol). When indicated, wild-type mice were injected with a 40-nmol dose of PACAP (6–38) simultaneously with VIP/PACAP and LPS. Survival was monitored over the next 5 days. Survival and several mediators were measured. A downward arrow indicates inhibitory effects of VIP/PACAP, an upward arrow indicates an increase, and = symbol indicates no differences versus mice receiving LPS alone. DGA, 1,2 diacylglycerol; ND = not determined.

hepatic synthesis of the coagulation-related molecule β-fibrinogen during LPS-induced endotoxemia. Although treatment with VIP/PACAP resulted in a downregulation of β-fibrinogen in wild-type mice, it failed to reduce LPS-induced β-fibrinogen mRNA expression in PAC1$^{-/-}$ mice. Therefore, the effect of VIP and PACAP on neutrophil recruitment and ICAM-1, VCAM-1, and β-fibrinogen expression could be at least partly explained by their inhibitory effect on IL-6 production mediated through PAC1.[13] These results were corroborated using the PAC1 antagonist PACAP (6–38). In contrast, VIP and PACAP inhibited the production of the AP protein SAA and the reactive oxygen NO in a PAC1-independent way[4,14] (FIG. 1). The analysis of intracellular signal pathways involved in the inhibitory effects of VIP and PACAP suggests an important involvement of the inositol/phosphate/PLC cascade.[13] TABLE 1 summarizes information of advances in the identification of VIP/PACAP effects mediated through PAC1 in the experimental endotoxic shock model.

In conclusion, this study provides some important insights regarding the involvement of PAC1 into the complexities of sepsis and may suggest new approaches for the therapy of septic shock.

ACKNOWLEDGMENTS

This work was supported by grants BFI 2002-03489 from Ministerio de Ciencia y Tecnología (Spain) and a predoctoral fellowship from Ministerio de Ciencia y Tecnología (to A.A).

REFERENCES

1. RIEDEMANN, N.C., R.F. GUO & P.A. WARD. 2003. The enigma of sepsis. J. Clin. Invest. **112**: 460–467.
2. VAUDRY, D., B.J. GONZALEZ, M. BASILLE, *et al*. 2000. Pituitary adenylate cyclase-activating polypeptide and its receptors: from structure to function. Pharmacol. Rev. **52**: 269–324.
3. GOMARIZ, R.P., C. MARTÍNEZ, C. ABAD, *et al*. 2001. Immunology of VIP: a review and therapeutical perspectives. Curr. Pharm. Des. **7**: 89–111.
4. DELGADO, M., D. POZO & D. GANEA. 2004. The significance of vasoactive intestinal peptide in immunomodulation. Pharmacol. Rev. **56**: 249–290.
5. DELGADO, M., C. MARTINEZ, D. POZO, *et al*. 1999. Vasoactive intestinal peptide (VIP) and pituitary adenylate cyclase-activation polypeptide (PACAP) protect mice from lethal endotoxemia through the inhibition of TNF-alpha and IL-6. J. Immunol. **162**: 1200–1205.
6. DELGADO, M., R.P. GOMARIZ, C. MARTÍNEZ, *et al*. 2000. Anti-inflammatory properties of the type 1 and type 2 vasoactive intestinal peptide receptors: role in lethal endotoxic shock. Eur. J. Immunol. **30**: 3236–3246.
7. MARTÍNEZ, C., C. ABAD, M. DELGADO, *et al*. 2002. Anti-inflammatory role in septic shock of pituitary adenylate cyclase-activating polypeptide receptor. Proc. Natl. Acad. Sci. USA **99**: 1053–1058.

8. OBERHOLZER, A., S.M. SOUZA, S.K. TSCHOEKE, *et al.* 2005. Plasma cytokine measurements augment prognostic scores as indicators of outcome in patients with severe sepsis. Shock **23:** 488–493.

9. ROMANO, M., M. SIRONI, C. TONIATTI, *et al.* 1997. Role of IL-6 and its soluble receptor in induction of chemokines and leukocyte recruitment. Immunity **6:** 315–325.

10. HEINRICH, P.C., J.V. CASTELL & T. ANDUS. 1990. Interleukine-6 and acute phase response. Biochem. J. **265:** 621–636.

11. CALDENHOVEN, E., P. COFFER, J. YUAN, *et al.* 1994. Stimulation of the human intercellular adhesion molecule-1 promoter by interleukin-6 and interferon-gamma involves binding of distinct factors to a palindromic response element. J. Biol. Chem. **269:** 21146–21154.

12. RAY, A. 2000. A SAF binding site in the promoter region of human γ-fibrinogen gene functions as an IL-6 response element. J. Immunol. **165:** 3411–3417.

13. MARTÍNEZ, C., Y. JUARRANZ, C. ABAD, *et al.* 2005. Analysis of the role of PAC1 receptor in neutrophil recruitment, acute-phase response and nitric oxide production in septic shock. J. Leukoc. Biol. **77:** 729–738.

14. ABAD, C., C. MARTÍNEZ, Y. JUARRANZ, *et al.* 2003. Therapeutic effects of vasoactive intestinal peptide in the trinitrobenzene sulfonic acid mice model of Crohn's disease. Gastroenterology **124:** 961–971.

PACAP Stimulates Biosynthesis and Release of Endozepines from Rat Astrocytes

OLFA MASMOUDI-KOUKI,[a,b] PIERRICK GANDOLFO,[a]
JEROME LEPRINCE,[a] DAVID VAUDRY,[a] GEORGES PELLETIER,[c]
ALAIN FOURNIER,[d] HUBERT VAUDRY,[a]
AND MARIE-CHRISTINE TONON[a]

[a] Inserm U413, IFRMP23, University of Rouen, 76821 Mont-Saint-Aignan,
France

[b] 00/UR/08/01, University of Tunis, 2092 El Manar, Tunisia

[c] CHUL Research Center, Laval University Medical Center, Quebec,
G1V 4G2 Canada

[d] INRS-Institut Armand-Frappier, Pointe-Claire, H9R1G6 Canada

ABSTRACT: Astrocytes synthesize and release endozepines, a family of
neuropeptides related to diazepam-binding inhibitor (DBI). Astroglial
cells also express the receptors of pituitary adenylate cyclase-activating
polypeptide (PACAP) and vasoactive intestinal polypeptide (VIP). In the
present article, we show that PACAP dose dependently increases DBI
gene expression and stimulates endozepine release through activation
of PAC1-R. PACAP increases cAMP formation, enhances polyphosphoinositide turnover, and evokes calcium mobilization from intracellular
Ca^{2+} pools. The effect of PACAP on endozepine release is mediated
through the adenylyl cyclase/PKA pathway while the downregulation of
astrocyte response to PACAP can be ascribed to activation of the PLC/
PKC pathway.

KEYWORDS: astrocytes; endozepines; DBI mRNA

INTRODUCTION

The term endozepines designates a family of regulatory peptides that have
been initially isolated from the rat brain on the basis of their ability to dis-
place the binding of benzodiazepines from their receptors.[1] All endozepines
characterized so far derive from an 86-amino acid precursor called diazepam-
binding inhibitor (DBI) which generates, through proteolytic cleavage, several

Address for correspondence: Dr. Hubert Vaudry, Inserm U413, IFRMP 23, University of Rouen,
76821 Mont-Saint-Aignan, France. Voice: 33-235-14-6624; fax: 33-235-14-6946.
e-mail: hubert.vaudry@univ-rouen.fr

Ann. N.Y. Acad. Sci. 1070: 411–416 (2006). © 2006 New York Academy of Sciences.
doi: 10.1196/annals.1317.094

biologically active peptides including the triakontatetraneuropeptide DBI[17–50] (TTN) and the octadecaneuropeptide DBI[33–50] (ODN).[2,3] There is now clear evidence that, in the central nervous system (CNS), the DBI gene is primarily expressed in glial cells[4] and the presence of DBI-related peptides has been visualized by immunohistochemistry in astroglial cells in various brain regions.[5,6] *In vitro* studies have shown that cultured rat astrocytes contain and release substantial amounts of endozepines and that the secretion of endozepines is cAMP-dependent.[7]

Pituitary adenylate cyclase-activating polypeptide (PACAP) and its receptors are widely expressed in the CNS.[8] In particular, the presence of specific PACAP-binding sites has been demonstrated on astroglial cells[9] and it has been shown that PACAP stimulates adenylyl cyclase activity in cultured astrocytes.[10] The aim of the present article was to investigate the effects of PACAP on the synthesis and release of endozepines in cultured rat astrocytes and to determine the transduction pathways mediating the action of PACAP.

EFFECT OF PACAP ON DBI mRNA LEVEL

Although the DBI gene exhibits several typical features of housekeeping genes, it has been shown that the DBI promotor encompasses consensus sequence regulatory elements.[11] The expression of the DBI gene is clearly regulated by insulin and androgens,[12,13] and we have recently shown that somatostatin reduces DBI mRNA levels in cultured rat astrocytes.[14] Here, we demonstrate by quantitative RT-PCR that PACAP induces a concentration-dependent increase of DBI mRNA levels in cultured astroglial cells with an EC_{50} value of 0.5 nM. The stimulatory effect of PACAP is mimicked by forskolin and totally abrogated by the protein kinase A (PKA) inhibitor H89 (FIG. 1).

EFFECT OF PACAP ON ENDOZEPINE RELEASE

Exposure of cultured rat astrocytes to subnanomolar concentrations of PACAP-38 or PACAP-27 (10^{-13} to 10^{-6} M) induces a dose-dependent stimulation of endozepine release, while vasoactive intestinal polypeptide (VIP) is at least 1000-fold less potent.[15] The PAC1-R antagonist PACAP6-38 totally suppresses the stimulatory effects of both PACAP and VIP whereas the selective VPAC1/2 antagonist [4-Cl-D-Phe[6], Leu[17]]VIP is ineffective. Although high concentrations of VIP can stimulate endozepine release, the effects of PACAP and VIP are not additive.[15] Altogether, these data indicate that PACAP stimulates endozepine secretion from rat astrocytes through activation of PAC1-R. Reversed-phase high-performance liquid chromatography (HPLC) analysis of endozepines contained in conditioned media revealed that PACAP does not affect the processing of DBI.

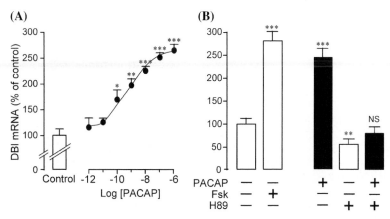

FIGURE 1. Effect of graded concentrations of PACAP and modulators of adenylyl cyclase/PKA pathway on DBI mRNA levels in cultured rat astrocytes. (A) Cells were incubated for 6 h with increasing concentrations of PACAP (10^{-12} to 10^{-6} M). (B) Cells were incubated for 6 h with forskolin (Fsk; 10^{-5} M), or with PACAP (10^{-7} M) in the absence or presence of the PKA inhibitor H89 (2×10^{-5} M). Each value, expressed as a percentage of the control, represents the mean ($\pm$SEM) of three independent experiments performed in quadruplicate. ANOVA followed by the Bonferroni's test: * $P < 0.05$; ** $P < 0.01$; *** $P < 0.001$; NS = not statistically different from the control.

SIGNAL TRANSDUCTION PATHWAYS INVOLVED IN PACAP-INDUCED ENDOZEPINE RELEASE

In agreement with previous reports, we found that incubation of astrocytes with PACAP (10^{-12}–10^{-6} M) provokes a dose-dependent increase in cAMP production in cultured rat astrocytes. PACAP also stimulated polyphosphoinositide turnover and induced calcium mobilization from IP$_3$-sensitive Ca^{2+} pools in astroglial cells. Forskolin and 8BrcAMP mimic the effect of PACAP on endozepine release, and the PKA inhibitor H89 totally suppresses PACAP-induced stimulation of endozepine secretion. In contrast, the PLC (phospholipase C) inhibitor U73122 and the PKC (protein kinase C) inhibitor chelerythrine do not affect basal and PACAP-evoked endozepine release.[15]

Time-course studies revealed that 10^{-10} M PACAP provokes a rapid and transient stimulation of cAMP formation and endozepine release, suggesting that PACAP, even at low concentrations, induces rapid desensitization of the glial cell response. Homologous desensitization following exposure to low doses of PACAP has already been reported in other cell types including cortical neurons and in PC12 cells.[16,17] Two observations indicate that, in glial cells, the down-regulation of the response observed during prolonged exposure to PACAP can be ascribed to activation of the PLC/PKC pathway: (*a*) the PLC inhibitor U73122 partially restores the stimulatory effect of PACAP on

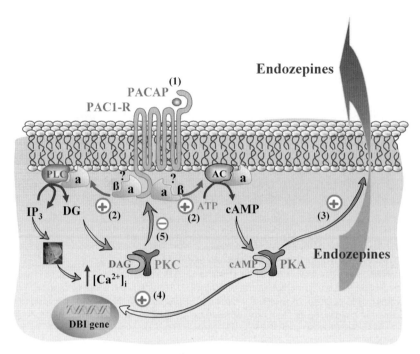

FIGURE 2. Schematic representation of the intracellular mechanisms involved in the effects of PACAP on endozepine biosynthesis and release in cultured rat astrocytes. PACAP, acting through PAC1-R (1), increases adenylyl cyclase (AC) and PLC activities (2). Activation of the PKA pathway stimulates endozepine release (3) and increases DBI mRNA levels (4). Concurrently, activation of the PLC/PKC pathway downregulates the effect of PACAP on endozepine secretion (5).

endozepine release and cAMP formation, and (b) the effects of U73122 are totally abolished by the phorbol ester PMA.[15]

CONCLUDING REMARKS

Taken together, these data indicate that (a) PACAP stimulates both DBI gene expression and endozepine secretion in cultured rat astrocytes, (b) PACAP acts through activation of PAC1-R receptors positively coupled to adenylyl cyclase and PLC, and (c) PACAP-evoked stimulation of endozepine release is mediated through the adenylyl cyclase/PKA pathway while desensitization of the effect of PACAP involves the PLC/PKC pathway (FIG. 2).

ACKNOWLEDGMENTS

This research was supported by Inserm (U413), the European Institute for Peptide Research and the Conseil Régional de Haute-Normandie. O.M. was the recipient of a fellowship from the French and Tunisian Ministries of Research. H.V. is affiliated professor at INRS-Institut Armand Frappier.

REFERENCES

1. GUIDOTTI, A., C.M. FORCHETTI, M.G. CORDA, *et al.* 1983. Isolation, characterization, and purification to homogeneity of an endogenous polypeptide with agonistic action on benzodiazepine receptors. Proc. Natl. Acad. Sci. USA. **80:** 3531–3535.
2. FERRERO, P., M.R. SANTI, B. CONTI-TRONCONI, *et al.* 1986. Study of an octadecaneuropeptide derived from diazepam binding inhibitor (DBI): biological activity and presence in rat brain. Proc. Natl. Acad. Sci. USA. **83:** 827–831.
3. SLOBODYANSKY, E., A. GUIDOTTI, C. WAMBEBE, *et al.* 1989. Isolation and characterization of a rat brain triakontatetraneuropeptide, a posttranslational product of diazepam binding inhibitor: specific action at the Ro 5-4864 recognition site. J. Neurochem. **53:** 1276–1284.
4. TONG, Y., D. TORANZO & G. PELLETIER. 1991. Localization of diazepam-binding inhibitor (DBI) mRNA in the rat brain by high resolution *in situ* hybridization. Neuropeptides **20:** 33–40.
5. TONON, M.C., L. DÉSY, P. NICOLAS, *et al.* 1990. Immunocytochemical localization of the endogenous benzodiazepine ligand octaecaneuropeptide (ODN) in the rat brain. Neuropeptides **15:** 17–24.
6. YANASE, H., H. SHIMIZU, K. YAMADA & T. IWANAGA. 2002. Cellular localization of the diazepam binding inhibitor in glial cells with special reference to its coexistence with brain-type fatty acid binding protein. Arch. Histol. Cytol. **65:** 27–36.
7. PATTE, C., P. GANDOLFO, J. LEPRINCE, *et al.* 1999. GABA inhibits endozepine release from cultured rat astrocytes. Glia **25:** 404–411.
8. VAUDRY, D., B.J. GONZALEZ, M. BASILLE, *et al.* 2000. Pituitary adenylate cyclase-activating polypeptide and its receptors: from structure to functions. Pharmacol. Rev. **52:** 269–324.
9. TATSUNO, I., P.E. GOTTSCHALL & A. ARIMURA. 1991. Specific binding sites for pituitary adenylate cyclase activating polypeptide (PACAP) in rat cultured astrocytes: molecular identification and interaction with vasoactive intestinal peptide (VIP). Peptides **12:** 617–621.
10. TATSUNO, I., H. MORIO, T. TANAKA, *et al.* 1996. Astrocytes are one of the main target cells for pituitary adenylate cyclase-activating polypeptide in the central nervous system. Astrocytes are very heterogeneous regarding both basal movement of intracellular free calcium ($[Ca^{2+}]_i$) and the $[Ca^{2+}]_i$ response to PACAP at a single cell level. Ann. N. Y. Acad. Sci. **805:** 613–619.
11. SWINNEN, J.V., P. ALEN, W. HEYNS & G. VERHOEVEN. 1998. Identification of diazepam-binding inhibitor/acyl-CoA-binding protein as a sterol regulatory element-binding protein-responsive gene. J. Biol. Chem. **273:** 19938–19944.
12. HANSEN, H.O., P.H. ANDREASEN, S. MANDRUP, *et al.* 1991. Induction of acyl-CoA-binding protein and its mRNA in 3T3-L1 cells by insulin during preadipocyte-to-adipocyte differentiation. Biochem. J. **277:** 341–344.
13. SWINNEN, J.V., I. VERCAEREN, M. ESQUENET, *et al.* 1996. Androgen regulation of the messenger RNA encoding diazepam-binding inhibitor/acyl-CoA-binding protein in the rat. Mol. Cell. Endocrinol. **118:** 65–70.

14. MASMOUDI, O., P. GANDOLFO, T. TOKAY, *et al.* 2005. Somatostatin down-regulates the expression and release of endozepines from cultured rat astrocytes via distinct receptor subtypes. J. Neurochem. **94:** 561–571.
15. MASMOUDI, O., P. GANDOLFO, J. LEPRINCE, *et al.* 2003. Pituitary adenylate cyclase-activating polypeptide (PACAP) stimulates endozepine release from cultured rat astrocytes via a PKA-dependent mechanism. FASEB J. **17:** 17–27.
16. TAUPENOT, L., M. MAHATA, S.K. MAHATA & D.T. O'CONNOR. 1999. Time-dependent effects of the neuropeptide PACAP on catecholamine secretion : stimulation and desensitization. Hypertension **34:** 1152–1162.
17. NIEWIADOMSKI, P., J.Z. NOWAK, P. SEDKOWSKA & J.B. ZAWILSKA. 2002. Rapid desensitization of receptors for pituitary adenylate cyclase-activating polypeptide (PACAP) in chick cerebral cortex. Pol. J. Pharmacol. **54:** 717–721.

Effects of Pituitary Adenylate Cyclase-Activating Polypeptide and Vasoactive Intestinal Polypeptide on Food Intake and Locomotor Activity in the Goldfish, *Carassius auratus*

KOUHEI MATSUDA,[a] KEISUKE MARUYAMA,[a] TOMOYA NAKAMACHI,[b] TOHRU MIURA,[a] AND SEIJI SHIODA[b]

[a]*Laboratory of Regulatory Biology, Graduate School of Science and Engineering, University of Toyama, Toyama 930-8555, Japan*

[b]*Department of Anatomy, Showa University School of Medicine, Shinagawa-ku, Tokyo 142-8555, Japan*

ABSTRACT: We investigated the effects of intracerebroventricular (ICV) and intraperitoneal (IP) administration of pituitary adenylate cyclase-activating polypeptide (PACAP) and vasoactive intestinal polypeptide (VIP) on food intake in the goldfish, *Carassius auratus*. Cumulative food intake was significantly decreased by ICV injection of PACAP or VIP. Similarly, IP administration of PACAP or VIP induced a significant decrease in food intake. ICV injection of PACAP or VIP also induced a significant decrease in locomotor activity. These results suggest that PACAP and VIP may be involved in the regulation of feeding in goldfish.

KEYWORDS: PACAP; VIP; food intake; goldfish; ICV and IP injections; hypomotility

INTRODUCTION

Pituitary adenylate cyclase-activating polypeptide (PACAP) and vasoactive intestinal polypeptide (VIP), share structural similarities and both polypeptides belong to the same molecular group, the secretin–glucagon superfamily.[1] PACAP and VIP act as hypophysiotropic factors mediating the release of pituitary hormones in the fish pituitary.[2,3] However, the roles of PACAP and VIP in the central nervous systems of fish have not yet been elucidated.

Address for correspondence: Kouhei Matsuda, Laboratory of Regulatory Biology, Graduate School of Science and Engineering, University of Toyama, 3190-Gofuku, Toyama 930-8555, Japan. Voice: +81-76-445-6638; fax: +81-76-445-6549.
e-mail: kmatsuda@sci.u-toyama.ac.jp

Ann. N.Y. Acad. Sci. 1070: 417–421 (2006). © 2006 New York Academy of Sciences.
doi: 10.1196/annals.1317.054

Intracerebroventricular (ICV) injection of PACAP and/or VIP has been shown to suppress food intake or feeding behavior in mammals[4] and chicks.[5] Therefore, we investigated the effects of ICV and intraperitoneal (IP) administration of synthetic PACAP and VIP on food intake and locomotor activity in the goldfish, *Carassius auratus*.

MATERIALS AND METHODS

Goldfish were purchased commercially and kept under controlled light–dark conditions (12 L/12 D) in a temperature-regulated room (20–24°C) for 2 weeks before use in the experiments. Fish were fed uniformly granular diets once a day. For 2 days before the experiment each fish was kept in a small experimental tank. Animal experiments were conducted in accordance with the Toyama University's guidelines for the care and use of animals. Synthetic PACAP and VIP were purchased from the Peptide Institute (Osaka, Japan), dissolved in saline at 1.0 mM, and then stored at −80°C until used.

Two hours before the experiments began, each fish was fed at a rate of at least 1% of its body weight (BW). For ICV administration of PACAP and VIP, each fish was placed in a stereotaxic apparatus under anesthesia with MS-222, and was treated with 1 μL of test solution into the third ventricle of the brain. Fish were injected with synthetic PACAP at 0.22, 2.2, 11, or 22 pmol/g BW or with synthetic VIP at 2.2, 11, or 22 pmol/g BW. Fish in the control group were given injections of the same volume of saline. Each fish received 3% mass of food/g BW and food intake was measured every 15 min during the 60-min period after treatment. For IP administration of PACAP or VIP, fish were injected with 100 μL of 22, 44, or 88 pmol of each peptide/g BW. Fish in the control group were injected IP with the same volume of saline. Each fish received 3% of its BW in food, and food intake was measured every 15 min during the 60 min after treatment.

Each fish that received ICV injection of PACAP or VIP at 20 pmol/g BW, enough to suppress food intake, or saline was also placed in a small experimental tank, and an area sheet marked into 24 separate regions by radial lines and concentric circles was laid under the tank. Each fish was automatically monitored using a fixed handycam, and locomotor activity was measured by counting the frequency of passes into the different regions on the area sheet every 5 min during the 60-min posttreatment observation period.

RESULTS

ICV or IP injection of synthetic PACAP or VIP inhibited the feeding behavior of goldfish (FIG. 1). A significant decrease in food intake was observed after

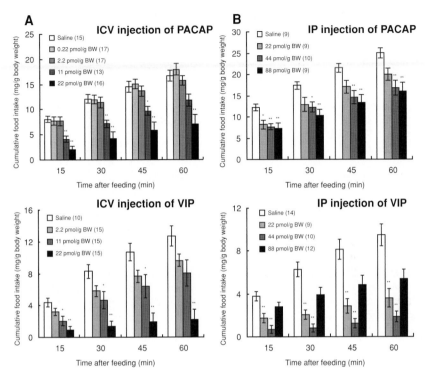

FIGURE 1. Effects of ICV (**A**) and IP (**B**) administrations of PACAP or VIP on food intake in the goldfish. Each column and bar is the mean ± SEM, respectively. The number of fish in each group is indicated between parentheses. Significant differences at each time point were evaluated by one-way ANOVA with Bonferroni's method as compared with a saline-injected group (* $P < 0.05$; ** $P < 0.01$).

ICV administration of PACAP or VIP at doses of 11 and 22 pmol/g BW, and cumulative food intake after these doses continued to decrease significantly at 45 and 60 min (Fig. 1 A). Suppression of food consumption was also observed after IP injection of PACAP at 22, 44, or 88 pmol/g BW or VIP at 22 or 44 pmol/g BW. Injection of PACAP at 44 or 88 pmol/g BW or VIP at 22 or 44 pmol/g BW continued to suppress cumulative food intake for 15 min to 60 min (Fig. 1 B). On the other hand, IP injection of VIP at 88 pmol/g BW did not affect cumulative food intake in the first 15 min to 60 min after injection (Fig. 1 B).

As shown in Figure 2, ICV injection of synthetic PACAP or VIP at 20 pmol/g BW inhibited the locomotor activity of the goldfish over the subsequent 60 min. The frequency of passes between the different regions on the area sheet remained significantly decreased between 30 min and 60 min.

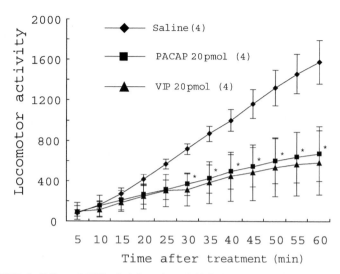

FIGURE 2. Effect of ICV administration of PACAP or VIP on locomotor activity in the goldfish. Each column and bar represents the mean ± SEM, respectively. The number of fish in each group is indicated between parentheses. One-way ANOVA with Bonferroni's test revealed significant differences at each time point between the PACAP or VIP group and the saline-injected control group (*$P < 0.05$).

DISCUSSION

In goldfish, the hypothalamic area is associated with control of food intake, and orexigenic and anorexigenic peptides regulate feeding behavior.[6] There are, however, no reports regarding the possible effects of PACAP or VIP on food intake in the goldfish. Our study is thus the first to show the inhibitory effects of ICV- or IP-administrated PACAP or VIP on food intake in the goldfish. PACAP and VIP have the potential to stimulate growth hormone, gonadotropin, and prolactin release from cultured pituitary cells of teleosts.[2,3] Our results suggest that PACAP and VIP may act centrally as feeding inhibitors in the goldfish, similarly to what has been described previously in the chick.[5] PACAP and VIP are also expressed in peripheral organs, especially in the gastrointestinal tract, of teleosts, but the peripheral effects of PACAP and VIP have not yet been investigated. Our results suggest that IP injection of PACAP and VIP may act to the brain to inhibit food consumption since fish were injected with relatively high doses of PACAP and VIP. IP injection of VIP at 88 pmol/g BW did not inhibit food intake, suggesting that IP injection of VIP at relatively high doses may induce downregulation or desensitization of the receptors.

Central injection of PACAP has also been found to influence other animal behaviors in rats and mice.[4,7] Furthermore, Hashimoto et al.[8] reported that adult mice that lack the PACAP gene display behavioral changes. These studies suggested that PACAP plays a role in the regulation of psychomotor behaviors in mammals. Our study is the first to show inhibitory effects of ICV-administered PACAP or VIP on locomotor activity in a nonmammalian animal, suggesting that hypomotility caused by PACAP or VIP may suppress food intake in the goldfish.

REFERENCES

1. VAUDRY, D., B.J. GONZALEZ, M. BASILLE, *et al*. 2000. Pituitary adenylate cyclase-activating polypeptide and its receptors from structure to functions. Pharmacol. Rev. **52:** 269–324.
2. KELLEY, K.M., R.S. NISHIOKA & H.A. BERN. 1988. Novel effect of vasoactive intestinal polypeptide and peptide histidine isoleucine: inhibition of in vitro secretion of prolactin in the tilapia, *Oreochromis mossambicus*. Gen. Comp. Endocrinol. **72:** 97–106.
3. MONTERO, M., L. YON, S. KIKUYAMA, *et al*. 2000. Molecular evolution of the growth hormone-releasing hormone/pituitary adenylate cyclase-activating polypeptide gene family. Functional implication in the regulation of growth hormone secretion. J. Mol. Endocrinol. **25:** 157–168.
4. MORLEY, J.E., M. HOROWITZ, P.M.K. MORLEY & J.F. FLOOD. 1992. Pituitary adenylate cyclase activating polypeptide (PACAP) reduces food intake in mice. Peptides **13:** 1133–1135.
5. TACHIBANA, T., S. SAITO, S. TOMONAGA, *et al*. 2003. Intracerebroventricular injection of vasoactive intestinal peptide and pituitary adenylate cyclase-activating polypeptide inhibits feeding in the chick. Neurosci. Lett. **339:** 203–206.
6. VOLKOFF, H., L.F. CANOSA, S. UNNIAPPAN, *et al*. 2005. Neuropeptides and the control of food intake in fish. Gen. Comp. Endocrinol. **142:** 3–19.
7. NORRHOLM, S.D., M. DAS & G. LEGRADI. 2005. Behavioral effects of local microinfusion of pituitary adenylate cyclase activating polypeptide (PACAP) into the paraventricular nucleus of the hypothalamus (PVN). Regul. Pept. **128:** 33–41.
8. HASHIMOTO, H., N. SHINTANI, K. TANAKA, *et al*. 2001. Altered psychomotor behaviors in mice lacking pituitary adenylate cyclase-activating polypeptide (PACAP). Proc. Natl. Acad. Sci. USA **98:** 13355–13360.

Functional Splice Variants of the Type II G Protein–Coupled Receptor (VPAC2) for Vasoactive Intestinal Peptide in Mouse and Human Lymphocytes

ALLISON L. MILLER, DEEPTI VERMA, CAROLA GRINNINGER, MEI-CHUAN HUANG, AND EDWARD J. GOETZL

Departments of Medicine and Microbiology-Immunology, University of California, San Francisco, California 94143-0711, USA

ABSTRACT: A PCR-based search for splice variants of the VPAC2 G protein–coupled receptor for vasoactive intestinal peptide (VIP) revealed: (*a*) a short-deletion variant in mouse lymphocytes termed *VPAC2de367–380*, that lacks 14 amino acids in the seventh transmembrane domain, and (*b*) a long-deletion variant in human lymphocytes termed *VPAC2de325–438(i325–334)*, that lacks 114 amino acids beginning with the carboxyl-terminal end of the third cytoplasmic loop and has 10 new carboxy-terminal amino acids. VPAC2de367–380 binds VIP normally, but shows reduced VIP-evoked signaling and effects on immune functions, whereas VPAC2de325–438(i325–334) shows reduced binding affinity for VIP and a complex pattern of functional differences. These splice variants may modify the immunoregulatory contributions of the VIP–VPAC2 axis.

KEYWORDS: neuropeptide; immunity; T cell; cytokine

Most of the vasoactive intestinal peptide (VIP) at sites of immune cell traffic and functional responses is generated and secreted by neurons and type 2 helper T lymphocytes, termed *Th2 cells*.[1] Many aspects of the specific effects of VIP on immune cells, including T cells and antigen-presenting cells, such as macrophages and dendritic cells, have been elucidated in several laboratories over the past two decades.[2] Expression of the types I (VPAC1) and II (VPAC2) G protein–coupled receptors for VIP by different immune cells has been documented and some mechanisms controlling their respective expression are clear determinants of the nature of immune responses. For example, stimulation of

Address for correspondence: Edward J. Goetzl, M.D., University of California Medical Center, Room UB8B, UC Box 0711, 533 Parnassus at 4th, San Francisco, CA 94143-0711. Voice: 415-476-5339; fax: 415-476-6915.

e-mail: edward.goetzl@ucsf.edu

Ann. N.Y. Acad. Sci. 1070: 422–426 (2006). © 2006 New York Academy of Sciences.
doi: 10.1196/annals.1317.055

T cells through their antigen receptor (TCR) downregulates VPAC1 but upregulates VPAC2. This shift in predominance of VPAC1 to VPAC2 on activated T cells permits VIP to control the balance between immediate-type or allergic inflammatory immune responses mediated by Th2 cells and delayed-type cellular protective or hypersensitivity responses mediated by Th1 cells. Further support for this role of the VIP–VPAC2 axis came from the results of immunological studies of VPAC2-knockout (KO) mice and VPAC2-transgenic (TG) mice with a constitutively high level of expression of VPAC2 selectively on T cells. VPAC2-TG mice have blood eosinophilia, elevated serum concentrations of IgE, and greater than normal IgE antibody responses and cutaneous immediate-type reactions to allergy-producing antigens, that all are typical of the allergic state.[3] VPAC2-TG mice also exhibit reduced cutaneous delayed-type reactions to antigens that is indicative of deviation from Th1- to Th2-mediated immunity. In contrast, VPAC2-KO mice manifest greater than normal cutaneous delayed-type reactions to antigens and reduced IgE antibody responses and cutaneous immediate-type reactions to allergy-producing antigens.[4] This immune deviation from Th1-mediated delayed-type cellular responses to Th2-mediated immediate-type allergic inflammatory responses is a function of the differential effects of the VIP–VPAC2 axis on T cell generation of cytokines that determine the relative numbers and activities of Th1 and Th2 cells.

The VIP–VPAC2 axis in Th cells signals expansion of the Th2 population and contraction of the Th1 set. As a result, stimulation of mixed Th cells from VPAC2-TG mice results in increased generation of interleukin-4 (IL-4) and, to a lesser extent, IL-5 and decreased generation of gamma-interferon. In contrast, similarly stimulated Th cells from VPAC2-KO mice generate less IL-4 and more gamma-interferon than Th cells from wild-type mice. VPAC2 signals these alterations in cytokine-secreting Th cell subsets through c-Maf and JunB, without the more usual involvement of GATA-3 and Tbet.[5] Other effects of VIP on T cells are mediated by both VPAC1 and VPAC2.[2] VIP is a chemotactic stimulus for T cells, in part through its capacity to enhance cell surface expression of the matrix metalloproteinase, MMP-9, that facilitates T cell movement across basement membranes and other tissues. The suppression of T cell proliferative responses by VIP is attributable largely to inhibition of generation of IL-2 and its actions as a T cell growth factor. These effects of VIP also reduce Th cell help to B cells that presumably underlies VIP suppression of T cell-dependent production of IgG antibodies. The coreceptor interactions necessary for effective antigen presentation to T cells and the binding of cell surface Fas/Fas ligand proteins that initiate apoptosis also are influenced by VIP-induced changes in expression of one or more such proteins.

The least well-understood aspect of VIP in immunity is the basis for immune regulation of expression of VPAC1 and VPAC2 and their signaling pathways in immune cells. It is generally assumed that some combination of cytokines transcriptionally regulates the constitutive levels of VPAC1 and VPAC2 and

TABLE 1. Natural carboxyl- terminal splice variants of the VPAC2

A. Mouse VPAC2

WT mVPAC2	^{350}ISSTYQILFELCVGSFQGLVVAVLYCFLNSEVQCELKRRWR^{390}
mVPAC2de367–380	^{350}ISSTYQILFELCVGSFQ————————VQCELKRRWR^{390}

B. Human VPAC2

WT hVPAC2	^{321}SQYKRLAK<u>STLLLIPLFGVHYMVFAVF</u>PISISSKYQIL<u>FELCLGS</u>^{365}
hVPAC2de325–438	^{321}SQYK*AWWWPSSTVS*———————————————————
WT hVPAC2	^{366}<u>FQGLVVAVLYCFLNSEVQCELK</u>RKWRSRCPTPSASRDYRVCG^{407}
hVPAC2de325–438	———————————————————————————————
WT hVPAC2	^{408}SSFSRNGSEGALQFHRGSRAQSFLQTETSVI^{438}
hVPAC2de325–438	———————————————————————————————

The sequence underlined in mVPAC2de367–380 is the seventh transmembrance domain and those underlined in hVPAC$_2$325–438 (i325–334) are the sixth and seventh transmembrance domains. Ten new amino acids (i325–334) at the carboxyl-terminus of hVPAC$_2$de325–438 (i325–334) are shown in italics. WT = wild type.

that changes in VIP concentration and the composition of the controlling cytokines is responsible for altered expression of VPAC1 and VPAC2 in immune responses. No investigations have addressed independent immune-specific alterations in signaling pathways. No one has considered the possible involvement of structural variants of VPAC receptors, hetero-oligomers of such variants with native wild-type VPAC receptors, or decoy or other signaling-impaired variants in immune-specific regulation of responsiveness of T cells to VIP. We therefore began a search for variants of VPAC2 in immune cells and tissues, using multiple sets of primers in a polymerase chain reaction (PCR)-based approach. Two deletion splice variants have been identified, one in mouse thymocytes and splenic T cells and the other in cultured lines of human malignant T cells and activated T cells of normal human blood. The mouse immune splice variant of VPAC2 is a simple deletion of the 42 base pairs of exon 12 and the resultant loss of 14 amino acids from 367–380 at the carboxyl-terminal end of the seventh transmembrane (TM) domain (TABLE 1).[6] In the human immune splice variant of VPAC2, the deletion of exon 11 results in loss of 114 amino acids from 325–438 by disruption of the codon sequence that creates a frame shift allowing transcription of mRNA encoding 10 different amino acids (i325–334) and introduces a premature stop codon with loss of sequence beyond amino acid 334 (TABLE 1). It is predicted that mouse VPAC2de367–380 will be a six TM domain protein and human VPAC2de325–438(i325–334) will be a five TM domain protein.

Similar levels of expression of mouse VPAC2de367–380 and mouse wild-type VPAC2 were established in human VPAC2 low Jurkat human T cell transfectants, which permitted comparative studies of their properties.[6] VPAC2de367–380 and wild-type VPAC2 in Jurkat T cells bound VIP with the same affinity, but only wild-type VPAC2-transduced VIP mediation of increases in [cAMP]i, chemotaxis, and decreases in IL-2 generation in the Jurkat T cells (FIG. 1). Similarly, only mouse wild-type VPAC2 and not VPAC2de 367–380 in mouse Th2 cell transfectants signaled. The results of preliminary studies of human VPAC2de325–438(i325–334) in contrast with

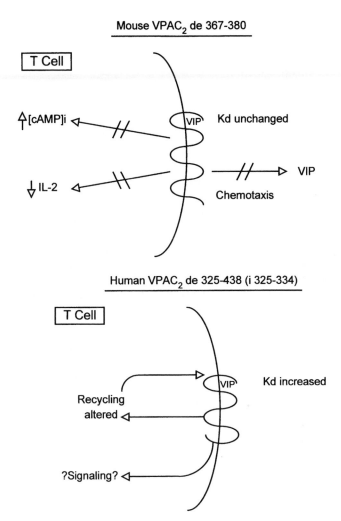

FIGURE 1. Distinctive characteristics of the carboxyl-terminal deletion splice variants of VPAC2. The mouse VPAC2de367–380 is shown as a six TM domain G protein–coupled receptor and human VPAC2de325–438(i325–334) is depicted as a five TM domain GPCR.

human wild-type VPAC2 suggest lower-affinity binding of VIP, decreased and less sustained VIP-induced downregulation, and diminished transduction of functional signals to T cells (FIG. 1).

Future goals include completion of analyses of the structures of mouse VPAC2de367–380 and human VPAC2de325–438(i325–334), elucidation of their distinctive mechanisms for regulation of expression and for cellular signaling, and delineation of functionally relevant interactions between these splice variants and their respective wild-type VPAC2 receptors.

ACKNOWLEDGMENT

The research described is supported by grant AI29912 from the National Institutes of Health.

REFERENCES

1. MARTINEZ, C. *et al.* 1999. Regulation of VIP production and secretion by murine lymphocytes. J. Neuroimmunol. **93:** 126–138.
2. DORSAM, G. *et al.* 2000. Vasoactive intestinal peptide mediation of development and functions of T lymphocytes. Ann. N. Y. Acad. Sci. **921:** 79–91.
3. VOICE, J.K. *et al.* 2001. Allergic diathesis in transgenic mice with constitutive T cell expression of inducible vasoactive intestinal peptide receptor. FASEB J. **15:** 2489–2496.
4. GOETZL, E.J. *et al.* 2001. Enhanced delayed-type hypersensitivity and diminished immediate-type hypersensitivity in mice lacking the inducible VPAC(2) receptor for vasoactive intestinal peptide. Proc. Natl. Acad. Sci. USA **98:** 13854–13859.
5. VOICE, J. *et al.* 2004. c-Maf and JunB mediation of Th2 differentiation induced by the type 2 G protein-coupled receptor (VPAC2) for vasoactive intestinal peptide. J. Immunol. **172:** 7289–7296.
6. GRINNINGER, C. *et al.* 2004. A natural variant type II G protein-coupled receptor for vasoactive intestinal peptide with altered function. J. Biol. Chem. **279:** 40259–40262.
7. HUANG, M.-C. *et al.* 2006. Differential signaling of T cell generation of IL-4 by wild-type and short-deletion variant of type 2 G protein-coupled receptor for vasoactive intestinal peptide (VPAC2). J. Immunol. **176:** 6640–6646.

Comparative Anatomy of PACAP-Immunoreactive Structures in the Ventral Nerve Cord Ganglia of Lumbricid Oligochaetes

LASZLO MOLNAR,[a] EDIT POLLAK,[a] AKOS BOROS,[a] DORA REGLÖDI,[b]
ANDREA TAMÁS,[b] ISRVAN LENGVARI,[b] AKIRA ARIMURA,[c]
AND ANDREA LUBICS[b]

[a]Department of General Zoology, University of Pécs, 7624 Pécs, Hungary

[b]Department of Anatomy, Neurohumoral Regulations Research Group of the
Hungarian Academy of Sciences, University of Pécs, 7624 Pécs, Hungary

[c]US-Japan Biomedical Research Laboratories, Tulane University, New Orleans,
Louisiana 70037, USA

ABSTRACT: By means of a whole mount immunocytochemical approach, the distribution patterns of pituitary adenylate cyclase-activating polypeptide (PACAP)-27 and PACAP-38 were identified in the ventral nerve cord (VNC) ganglia of the earthworms *Eisenia fetida* and *Lumbricus terrestris*. Each PACAP form appears to occur in a distinct neuron population. Positions of these populations, as well as numbers and sizes of the constituting neurons do not essentially differ between the two species. The data suggest that in Lumbricid Oligochaetes, PACAP-27 and PACAP-38 neuron populations may mediate distinct physiological processes.

KEYWORDS: earthworm; *Eisenia fetida*; *Lumbricus terrestris*

INTRODUCTION

Pituitary adenylate cyclase-activating polypeptide (PACAP) is a highly conserved neuropeptide, with only one to four amino acid differences between different vertebrate species.[1] Its presence has also been shown in invertebrates, from protozoa[2] to protochordata.[3] In the latter group PACAP-27 shows 90% similarity to human PACAP.[3] Except for Drosophila, in which PACAP may be involved in synaptic activity modulation[4,5] and memory processes,[6,7] there

Address for correspondence: Dora Reglodi, M.D., Ph.D., Department of Anatomy, University of Pécs, 7624 Pécs, Szigeti u 12. Hungary. Voice: +36-72-536001; ext.: 5398; fax: +36-72-536-393.
e-mail: dora.reglodi@aok.pte.hu

Ann. N.Y. Acad. Sci. 1070: 427–430 (2006). © 2006 New York Academy of Sciences.
doi: 10.1196/annals.1317.056

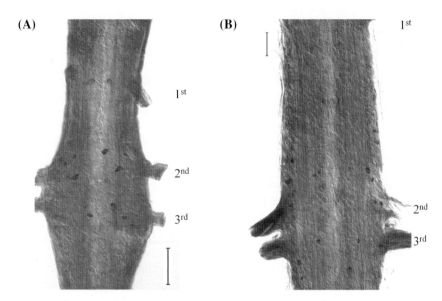

FIGURE 1. Whole mount immunocytochemistry of distribution pattern of PACAP-38 in VNC ganglia of *Eisenia fetida* (**A**) and *Lumbricus terrestris* (**B**). Anterior on the top. Scale bar 100 μm (**A**) and 250 μm (**B**). Note small numbers of neurons located at the level of the first, second, and third segmental nerves.

are relatively few data about PACAP functioning in invertebrates. PACAP-like immunoreactivity is present in annelids, and by means of radioimmunoassay, PACAP-27 has been proposed to represent the major form of the peptide in the oligochaete earthworm, *Lumbricus polyphemus*.[8,9] However, the exact anatomical positions of neurons expressing PACAP-38 or PACAP-27 have not been identified. The aim of the present article was to establish the three-dimensional distribution of PACAP forms in the ventral nerve cord (VNC) of oligochaete worms *Eisenia fetida* and *Lumbricus terrestris* by means of a whole mount immunocytochemical approach.

MATERIALS AND METHODS

All experiments were carried out on 5–5 sexually mature specimens of *Eisenia fetida* and *Lumbricus terrestris* (Annelida, Oligochaeta) kept in a standard breeding stock at our department. After anesthesia with chilling and carbon dioxide,[10] the VNC was dissected from the anterior 15–20 body segments and fixed in freshly prepared 4% ice-cold phosphate-buffered paraformaldehyde (pH 7.2) for 4 h. For whole mount immunocytochemistry, VNC ganglia were treated by immersion in 0.5% Triton-X 100 in 0.1 M phosphate-buffered saline (PBS), and then incubated with either PACAP-27 (No. 88,121–5)[11] or PACAP-38 (No. 92,112–6)[12] (1:500–1:1000) antisera, immunostained by avidin-biotin

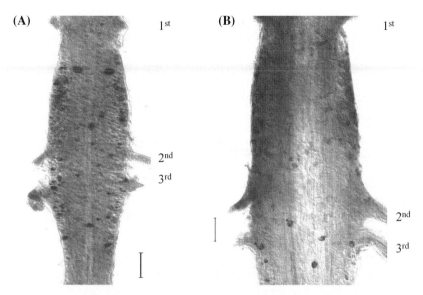

FIGURE 2. Whole mount immunocytochemistry of distribution pattern of PACAP27 in VNC ganglia of *Eisenia fetida* (**A**) and *Lumbricus terrestris* (**B**). Anterior on the top. Scale bar 100 μm (**A**) and 250 μm (**B**). Note high numbers of neurons at the ventrolateral and lateral parts of the ganglion and a pair of large neurons behind the first segmental nerves.

horse radish peroxidase staining kit (Sigma, Budapest, Hungary) and developed with 0.03% 3,3'-diaminobenzidine. After immunostaining and thorough washing in PBS, VNC ganglia were cleared in glycerol and observed with a Nikon Eclipse 80 microscope applying brightfield and/or Nomarski illumination.

RESULTS AND DISCUSSION

Results were essentially the same for each of the two earthworm species. Both PACAP-27 and PACAP-38 immunoreactivities were identified in the VNC, in fully separate neuron populations. PACAP-38-containing neurons were mainly located ventromedially and ventrolaterally at the level of segmental nerves. Neither somata of dorsal giant axons and giant motoneurons, nor neurosecretory cells[13] were found among stained cells (FIG.1). In contrast, PACAP-27-labeled neurons mainly occurred in the ventrolateral and lateral parts of ganglia, except for a pair of large somata behind the first segmental nerves and a few small neurons situated ventromedially (FIG. 2). The number of PACAP-27-labeled neurons was much higher (100–108 neuron per ganglion) than that of PACAP-38-stained ones (26–30 neuron per ganglion), which is consistent with previous data indicating that PACAP-27 is the predominant

PACAP form in the earthworm nervous system.[9] The size and the anatomical position of small-labeled neurons strongly suggest that they are interneurons forming fine fibers in polysegmental interneuronal tracts.[13]

The anatomical positions of PACAP-labeled somata do not resemble the pattern of any other neurons identified in *Eisenia fetida* and *Lumbricus terrestris*, suggesting that PACAP27 and PACAP38 are expressed in distinct neuron populations the members of which mediate specific physiological processes.

ACKNOWLEDGMENTS

This work was supported by the National Science Research Fund (OTKA T026652, T046589, F048908) and the Hungarian Academy of Sciences.

REFERENCES

1. VAUDRY, D. *et al*. 2000. Pituitary adenylate cyclase activating polypeptide and its receptors: from structure to functions. Pharmacol. Rev. **52:** 269–324.
2. MACE, S.R. *et al*. 2000. PACAP-38 is a chemorepellent and an agonist for the lysozyme receptor in *Tetrahymena thermophila*. J. Comp. Physiol. **A186:** 39–43.
3. MCRORY, J. & N.M. SHERWOOD. 1997. Two protochordate genes encode pituitary adenylate cyclase activating polypeptide and related family members. Endocrinology **138:** 2380–2390.
4. ZHONG, Y. & L.A. PENA. 1995. A novel synaptic transmission mediated by a PACAP-like neuropeptide in Drosophila. Neuron **14:** 527–536.
5. BHATTACHARYZ, A., S.S. LAKHMAN & S. SINGH. 2004. Modulation of L-type calcium channels in *Drosophila* via pituitary adenylyl cyclase activating polypeptide (PACAP)-mediated pathway. J. Biol. Chem. **279:** 37291–37297.
6. FEANY, M.B. & W.G. QUINN. 1995. A neuropeptide gene defined by the Drosophila memory mutant amnesiac. Science **268:** 869–873.
7. KEENE, A.C. *et al*. 2004. Diverse odor-conditioned memories require uniquely timed dorsal paired medial neuron output. Neuron **44:** 521–533.
8. REGLODI, D. *et al*. 2000. Distribution of PACAP-like immunoreactivity in the nervous system of oligochaeta. Peptides **21:** 183–188.
9. SOMOGYVARI-VIGH, A. *et al*. 2000. Tissue distribution of PACAP27 and -38 in oligochaeta: PACAP27 is the predominant form in the nervous system of Lumbricus polyphemus. Peptides **21:** 1185–1191.
10. ZSOMBOK, A. & L. MOLNAR. 2001. Patterns of NADPH-diaphorase containing structures in the subintestinal ganglion of *Lumbricus terrestris*. Neurobiology **9:** 67–69.
11. KÖVES, K. *et al*. 1990. Immunohistochemical demonstration of a novel hypothalamic peptide, pituitary adenylate cyclase activating polypeptide, in the ovine hypothalamus. Endocrinology **127:** 264–271.
12. YADA, T. *et al*. 1998. Autocrine action of PACAP in islets augments glucose-induced insulin secretion. Ann. N. Y. Acad. Sci. **865:** 451–457.
13. DORSETT, D.A. 1978. Organization of the nerve cord. *In* Physiology of Annelids. P.J. MILL Ed.: 115–159. Academic Press, London.

Involvement of the Adenylyl Cyclase/Protein Kinase A Signaling Pathway in the Stimulatory Effect of PACAP on Frog Adrenocortical Cells

MAITE MONTERO-HADJADJE,[a] CATHERINE DELARUE,[a]
ALAIN FOURNIER,[b] HUBERT VAUDRY,[a] AND LAURENT YON[a]

[a]INSERM U413, Laboratory of Cellular and Molecular Neuroendocrinology,
European Institute for Peptide Research (IFRMP23), University of Rouen,
76821 Mont-Saint-Aignan, France

[b]INRS, Institut Armand Frappier, Pointe-Claire, Quebec H9R 1G6, Canada

ABSTRACT: We have previously shown that PACAP stimulates in vitro the secretion of corticosteroids by frog adrenal explants and that PACAP increases cAMP formation and cytosolic calcium concentration ($[Ca^{2+}]_i$) in adrenocortical cells. The aim of the present study was to investigate the involvement of cAMP and $[Ca^{2+}]_i$ in the stimulatory effect of PACAP on steroid production. Incubation of adrenal explants with PACAP resulted in a significant increase in total inositol phosphate formation. Administration of the protein kinase A inhibitor, H89, markedly reduced the stimulatory effect of PACAP on corticosterone and aldosterone secretion by perifused adrenal slices. In contrast, chelation of intracellular or extracellular calcium, or incubation with calcium channel blockers, had no effect on PACAP-evoked steroid secretion. Incubation of the cells with BAPTA or thapsigargin totally suppressed the stimulatory effect of PACAP on $[Ca^{2+}]_i$. In contrast, suppression of extracellular calcium with EGTA or blockage of voltage-dependent Ca^{2+} channels did not impair PACAP-induced Ca^{2+} response. These data indicate that, in frog adrenocortical cells, the stimulatory effect of PACAP on steroid secretion is mediated through activation of the cAMP/PKA pathway. Concurrently, PACAP causes calcium mobilization from IP_3-dependent intracellular stores through activation of a phospholipase C, while the calcium response is not involved in the stimulatory effect of PACAP on corticosteroid secretion.

KEYWORDS: adrenal gland; amphibian; calcium; cAMP; steroids

Address for correspondence: Laurent Yon, INSERM U413, European Institute for Peptide Research (IFRMP 23), Laboratory of Cellular and Molecular Neuroendocrinology, University of Rouen, 76821 Mont-Saint-Aignan, France. Voice: 33-235-14-6945; fax: 33-235-14-6946.
e-mail: laurent.yon@univ-rouen.fr

Ann. N.Y. Acad. Sci. 1070: 431–435 (2006). © 2006 New York Academy of Sciences.
doi: 10.1196/annals.1317.057

INTRODUCTION

In the frog *Rana esculenta*, we have previously shown that frog PACAP-38 (fPACAP-38) exerts a direct stimulatory effect on corticosterone and aldosterone secretion[1] and we have found that the corticotropic action of fPACAP-38 is mediated through type I PACAP receptor.[2] The stimulatory effect of fPACAP-38 on frog adrenocortical cells is associated with activation of adenylyl cyclase and an increase in cytosolic calcium concentration ($[Ca^{2+}]_i$).[2] The aim of the present article was to investigate the respective contribution of Ca^{2+} and cAMP in fPACAP-38-induced steroid secretion. We have also investigated the source of calcium mobilized by the peptide in adrenocortical cells. Finally, cross-talks between Ca^{2+} and cAMP pathways were studied.

RESULTS

Effect of PACAP on IP formation

After an 18-h incubation of frog adrenal explants with *myo*-[^{3}H]inositol, incorporation of tritiated inositol into membrane phospholipids had reached equilibrium. Under these conditions, exposure of adrenal tissue to fPACAP-38 (1 μM) provoked a $40 \pm 13\%$ ($P < 0.01$) increase in total inositol phosphate formation.

Effect of a Protein Kinase A Inhibitor on PACAP-Induced Steroid Secretion

Prolonged infusion of the PKA inhibitor H89 (10 μM, 150 min) to perifused adrenal slices significantly reduced ($P < 0.05$) the effect of fPACAP-38 on corticosterone and aldosterone secretion ($-80 \pm 20\%$ and $-67 \pm 18\%$, respectively). At a concentration of 30 μM, H89 totally suppressed fPACAP-38-induced steroid secretion ($P < 0.01$).

Effect of Calcium Chelators and Calcium Channel Blockers on PACAP-Induced Steroid Secretion

Preincubation of adrenocortical explants with the intracellular calcium chelator BAPTA (50 μM; 100 min) did not significantly modify basal or fPACAP-38-induced secretion of corticosterone and aldosterone. Similarly, neither chelation of calcium in the perifusion medium with EGTA (1 mM; 240 min) nor administration of the L-type calcium channel blocker nifedipine (10 μM; 140 min) or the T-type calcium channel blocker pimozide (30 μM; 180 min) altered fPACAP-38-evoked corticosterone and aldosterone secretion.

Source of Calcium Involved in the fPACAP-38-Induced [Ca^{2+}]$_i$ Rise

Ejection of fPACAP-38 (1 μM; 2 s) in the vicinity of adrenocortical cells caused a rapid and transient mobilization of [Ca^{2+}]$_i$ (FIG. 1 A). Ejection of medium alone was performed to exclude nonspecific baroreceptor activation (FIG. 1 A). The addition of EGTA (10 mM; 30 min) to the medium reduced the extracellular-free calcium concentration from 1.3 mM to 8 nM. In these conditions, the kinetics of the fPACAP-38-induced calcium response remained unchanged (FIG. 1 B, F) compared to control conditions (FIG. 1 A). Similarly, in the presence of nifedipine (10 μM; 30 min) or pimozide (30 μM; 30 min), fPACAP-38 still caused [Ca^{2+}]$_i$ mobilization in adrenocortical cells (FIG. 1 F). Incubation of cultured adrenocortical cells with BAPTA (50 μM; 30 min) completely abolished the [Ca^{2+}]$_i$ rise in response to fPACAP-38 (FIG. 1 C, F; $P < 0.001$). Conversely, a brief ejection of the calcium ATPase inhibitor thapsigargin (10 μM; 2 s) resulted in a gradual and sustained rise in [Ca^{2+}]$_i$ (FIG. 1 D; $P < 0.001$). During the plateau phase, administration of fPACAP-38 had no effect on [Ca^{2+}]$_i$ (FIG. 1 D, F). Prolonged incubation of adrenocortical cells with ryanodine (10 μM; 30 min) did not modify the amplitude and the kinetics of the calcium response induced by fPACAP-38 (FIG. 1 E, F). Similarly, addition of the blocker of ryanodine-sensitive Ca^{2+} pools dantrolene (10 μM) to the incubation medium had no effect on fPACAP-38-provoked [Ca^{2+}]$_i$ rise (FIG. 1 F). Prolonged preincubation with pertussis toxin (250 ng/mL; 18 h) did not significantly inhibit nor potentiate the [Ca^{2+}]$_i$ response of adrenocortical cells to fPACAP-38 (FIG. 1 F). Preincubation of the cells with H89 (10 μM; 30 min) did not cause any changes in the calcium response induced by fPACAP-38 (FIG. 1 F). Similarly, incubation with the phosphodiesterase inhibitor IBMX (10μM; 30 min) did not potentiate the fPACAP-38-induced [Ca^{2+}]$_i$ mobilization (FIG. 1 F). Administration of the permeant analog 8-bromo-cAMP (100 μM; 2 s) to cultured adrenocortical cells induced a transient increase in [Ca^{2+}]$_i$. In contrast to fPACAP-38, the effect of 8-bromo-cAMP on [Ca^{2+}]$_i$ was abolished in calcium-free medium (10 mM EGTA).

DISCUSSION

In the presence of the PKA inhibitor, H89, the stimulatory effect of fPACAP-38 on corticosteroid secretion by perifused adrenal slices was markedly reduced suggesting that the steroidogenic effect of fPACAP-38 is mediated via the cAMP/PKA transduction pathway. In contrast, suppression of Ca^{2+} from the incubation medium, or addition of the calcium channel blockers, nifedipine or pimozide, or buffering intracellular calcium with BAPTA, did not impair the stimulatory effect of fPACAP-38 on corticosterone and aldosterone secretion, indicating that neither calcium influx through T- or L-type calcium channels nor intracellular calcium stores are involved in the corticotropic activity of

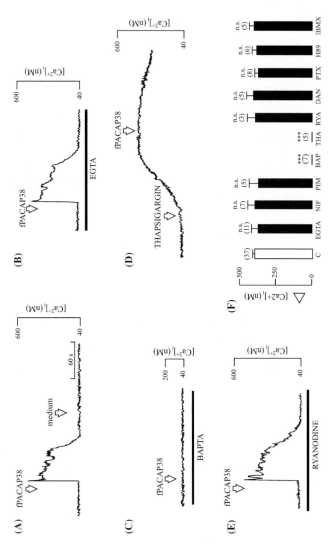

FIGURE 1. Effects of fPACAP-38 on $[Ca^{2+}]_i$ in cultured frog adrenocortical cells. (**A**) Typical profile illustrating the effect of a single application of fPACAP-38 (1 μM, 2 s). Ejection of medium alone (2 s) was used as a control. (**B**) Effect of fPACAP-38 in the presence of EGTA (10 mM). (**C**) Effect of fPACAP-38 in the presence of BAPTA (50 μM, 30 min). (**D**) Effect of a pulse of thapsigargin (10 μM, 2 s). Frog PACAP-38 was administered during the plateau phase. (**E**) Effect of fPACAP-38 in the presence of ryanodine (10 μM, 30 min). (**F**) Mean increase in $[Ca^{2+}]_i$ ($\Delta[Ca^{2+}]_i$) induced by fPACAP-38 in medium alone (control, C) and in the presence of EGTA, nifedipine (NIF), pimozide (PIM), BAPTA, thapsigargin (THA), ryanodine (RYA), dantrolene (DAN), pertussis toxin (PTX), H89, or IBMX. Profiles are representative of n experiments as indicated in parenthesis above histobars. *Arrows* indicate the onset of each pulse of fPACAP-38, medium or thapsigargin. *** $P < 0.001$ versus control; n.s. not statistically different versus control.

fPACAP-38. Microfluorimetric experiments performed on cultured adreno-cortical cells, revealed that reduction of the calcium concentration in the extracellular solution by adding 10 mM EGTA or incubation with nifedipine or pimozide had no effect on fPACAP-38-induced $[Ca^{2+}]_i$ rise. In contrast, chelation of intracellular calcium by BAPTA prevented the effect of the peptide on Ca^{2+} mobilization. Moreover, application of thapsigargin totally blocked the effect of fPACAP-38 on $[Ca^{2+}]_i$. Altogether, these results indicate that, in frog adrenocortical cells, the increase in $[Ca^{2+}]_i$ associated with type I PACAP receptor activation can be accounted for exclusive mobilization of calcium from intracellular pools. In support of this notion, we found that fPACAP-38 was able to stimulate inositol phosphate turnover in frog adrenal explants. Our data also suggest that the $[Ca^{2+}]_i$ rise induced by fPACAP-38 is attributable to IP_3-sensitive calcium stores since neither ryanodine nor dantrolene inhibited the $[Ca^{2+}]_i$ response to fPACAP-38. Finally, the effect of fPACAP-38 on $[Ca^{2+}]_i$ was not abolished by pertussis toxin pretreatment indicating that coupling of type I PACAP receptors to PLC may be mediated by G_q/G_{11} protein (PLC_β type) but not by G_i/G_o protein. The $[Ca^{2+}]_i$ mobilization induced by fPACAP-38 cannot be ascribed to an activation of the cAMP/PKA pathway since it was not blocked by H89 nor potentiated by IBMX. Furthermore, the $[Ca^{2+}]_i$ rise evoked by the cAMP analog 8-Br-cAMP, unlike that induced by fPACAP-38, was found to depend upon extracellular calcium. These results indicate that, in frog adrenocortical cells, activation of the cAMP/PKA pathway and $[Ca^{2+}]_i$ rise in response to fPACAP-38 are two independent mechanisms. It has previously been shown that, in the frog adrenal gland, fPACAP-38 is present in nerve endings in the vicinity of adrenocortical cells and stimulates type I PACAP adrenocortical receptors positively coupled to adenylyl cyclase and $[Ca^{2+}]_i$ mobilization.[1,2] The present results demonstrate that the stimulatory effect of fPACAP-38 on steroid secretion is mediated through activation of the cAMP/PKA pathway. Concurrently, fPACAP-38 causes calcium mobilization from IP_3-dependent intracellular stores through activation of a PCC via a pertussis toxin-insensitive G protein, while the calcium response is not involved in the stimulatory effect of PACAP on corticosteroid secretion.

REFERENCES

1. YON, L., M. FEUILLOLEY, N. CHARTREL, *et al.* 1993. Localization, characterization and activity of pituitary adenylate cyclase-activating polypeptide in the frog adrenal gland. J. Endocrinol. **139:** 183–194.
2. YON, L., N. CHARTREL, M. FEUILLOLEY, *et al.* 1994. Pituitary adenylate cyclase-activating polypeptide stimulates both adrenocortical and chromaffin cells in the frog adrenal gland. Endocrinology **136:** 1749–1758.

Breast Cancer VPAC1 Receptors

TERRY W. MOODY[a] AND ROBERT T. JENSEN[b]

[a]Department of Health and Human Services, Center for Cancer Research, National Cancer Institute, NIH, Bethesda, Maryland 20892, USA

[b]Digestive Diseases Branch, National Institute of Diabetes and Digestive and Kidney Diseases, NIH, Bethesda, Maryland 20892, USA

ABSTRACT: VIP receptors were investigated in breast cancer biopsy specimens. Twenty biopsy specimens bound ^{125}I-VIP with high affinity. Also, each of the 20 biopsy specimens had high amounts of VPAC1 receptor mRNA. MCF-7 cells have VPAC1 receptors that bound the VIP chemotherapeutic conjugate, (Ala^{2,8,9,19,24,25,27} Nle^{17}, Lys^{28})VIP-L2-camptothecin, with high affinity. VIP chemotherapeutic conjugates may be useful agents to inhibit the growth of breast cancer.

KEYWORDS: VIP; chemotherapeutical; conjugate; breast cancer

Over 200,000 new breast cancer cases are diagnosed annually in the United States and breast cancer causes the death of 40,000 U.S. women annually.[1] Traditionally, breast cancer has been treated with chemotherapeutic agents as well as surgery. In addition, breast cancer, which is estrogen receptor positive, is treated with selective estrogen receptor modifiers, such as tamoxifen. New therapeutic approaches are needed for estrogen receptor negative breast cancer patients.

Vasoactive intestinal polypeptide (VIP) receptors have been detected in human breast cancer cells. Addition of VIP to estrogen receptor positive MCF-7 cells, as well as estrogen receptor negative T47D cells, causes elevation of cAMP.[2] Subsequent studies showed that the VPAC1 receptor agonist, (Lys^{15}, Arg^{16}, Leu^{27})VIP$^{1-7}$GRF$^{8-27}$, but not the VPAC2 receptor agonist, R025-1553, caused cAMP elevation in MCF-7 and T47D cells.[3] The elevated cAMP may activate protein kinase A leading to phosphoryation of the CREB.[4] Phosphorylated CREB leads to increased expression of c-fos and c-jun. C-fos and c-jun form heterodimers and activate AP-1 sites on growth factor genes. VIP increases the clonal growth of breast cancer cells.[2] The actions of VIP on breast cancer cells are reversed by the VIP receptor antagonist, VIPhybrid (VIPhyb). VIPhyb and the structurally related (N-stearyl,

Address for correspondence: Dr. T. Moody, NCI Office of the Director, CCR, Bldg. 31, Rm. 4A48, 31 Center Dr. Bethesda, MD 20892. Voice: 301-451-9451; fax: 301-480-4323.
e-mail: moodyt@mail.nih.gov

Ann. N.Y. Acad. Sci. 1070: 436–439 (2006). © 2006 New York Academy of Sciences.
doi: 10.1196/annals.1317.058

TABLE 1. Breast cancer biopsy specimens

Specimen number	Cancer type	^{125}I-VIP binding (fmol/mg protein)	RT-PCR	
			VPAC1-R	VPAC2-R
1	DCIS	35	+	−
2	DCIS	25	+	−
3	DCIS	87	+	−
4	CARCINOMA	38	+	−
5	CARCINOMA	18	+	−
6	DCIS	24	+	−
7	DCIS	36	+	+
8	LOBULAR	18	+	+
9	DCIS	38	+	+
10	ADENO	25	+	+
11	DCIS	67	+	−
12	DCIS	14	+	+
13	DCIS	14	+	−
14	DCIS	9	+	+
15	DCIS	7	+	−
16	CARCINOMA	13	+	+
17	LOBULAR	20	+	+
18	CARCINOMA	10	+	−
19	DCIS	10	+	−
20	ADENO	16	+	+

Twenty breast cancer biopsy specimens were obtained, including 12 ductal carcinoma *in situ* (DCIS), 4 carcinoma (CARCINOMA), 2 adenocarcinoma (ADENO), and 2 lobular (LOBULAR). The specimens were assayed for receptor binding using 1 nM ^{125}I-VIP.

The mean value of three determinations repeated in triplicate is indicated (fmol bound/mg protein); the S.E. was <10% of the mean value. Messenger RNA was isolated from the breast cancer biopsy specimens and cDNA prepared. The RT-PCR was repeated four times for the analysis of VPAC1-R and VPAC2-R mRNA using receptor specific primers; mRNA present, +; mRNA absent, −.

Norleucine^{17})VIPhyb [(SN)VIPhyb] inhibit the proliferation of breast cancer cells *in vitro* and *in vivo*.[2] Here VIP receptors in breast cancer tumors were characterized.

Fresh frozen biopsy specimens were obtained from the Human Cooperative Tissue Network (Philadelphia, PA). The specimens were sectioned on a cryostat (10 μM), mounted on slides and analyzed for VIP receptors. TABLE 1 shows that using *in vitro* autoradiographic techniques, all of the biopsy specimens tested bound ^{125}I-VIP specifically. The density of binding sites was 26 ± 10 and ranged from 7 to 87 fmol/mg protein. Previously, we found that high densities of VIP receptors (100,000/cell) are present on MCF-7 cells. Due to the high densities of VIP receptors on cancer cells, it may be possible to image breast tumors in patients using VIP analogues. ^{99m}Tc-labeled VIP (TP-3645) has localized primary tumors in breast cancer patients.[5] A ^{18}F derivative of (Arg^{15}, Arg^{21})VIP was used to localize mammary cancer tumors in animal models of breast cancer.[6] Traditionally, breast cancer is detected by mammography and frequently it has undergone metastasis to lymph nodes.[1]

The density of specific ^{125}I-VIP binding sites was similar for all types of breast cancer biopsy specimens, including ductal carcinoma *in situ* (DCIS), breast carcinoma (CARCINOMA), lobular carcinoma (LOBULAR), and adenocarcinoma (ADENO). The VIP receptor subtypes were determined by RT-PCR. Breast carcinoma biopsy specimens had a major polymerase chain reaction product at 324 bp, indicative of VPAC1-R. In addition, 9 of 20 specimens had a minor band using a different set of primers indicative of VPAC2-R. These results indicate that VPAC1 receptor mRNA predominates in the breast cancer biopsy specimens.

The specificity of binding was investigated using MCF-7 cells. Specific ^{125}I-VIP binding to MCF-7 cells was inhibited by VIP, PACAP-27, and (Lys^{15}, Arg^{18}, Leu^{27})VIP$^{1-7}$GRF$^{8-27}$ with high affinity ($IC_{50} = 5, 2,$ and 3 nM, respectively). In contrast, (SN)VIPhyb and Ro25-1553 inhibited specific ^{125}I-VIP binding with IC_{50} values of 30 and >2000 nM, respectively. The VIP-ellipticine (E) conjugate VIP-LALA-E inhibited binding with IC_{50} value of 100 nM. Previously, we found that (Ala^{2,8,9,19,24,25,27,28})VIP was a long-lasting VPAC1 receptor agonist.[7] A VIP-camptothecin (CPT) conjugate was synthesized in which CPT was linked with N-methyl-amino-ethyl-glycine (L2) to the epsilon amino group of Lys^{28}. (Ala^{2,8,9,19,24,25,27}Nle^{17},Lys^{28})VIP-L2-CPT inhibited specific ^{125}I-VIP binding with an IC_{50} value of 1500 nM. Additional studies indicated that VIP-LALA-E bound with high affinity to cells enriched in VPAC1 or VPAC2 receptors. In contrast, (Ala^{2,8,9,19,24,25,27}Nle^{17},Lys^{28})VIP-L2-CPT bound with high affinity to cells enriched in VPAC1 but not VPAC2 receptors. Thus (Ala^{2,8,9,19,24,25,27}Nle^{17},Lys^{28})VIP-L2-CPT is a VIP-chemotherapeutic conjugate that is selective for VPAC1 receptors.

Previously, we demonstrated that ^{125}I-VIP-LALA-E was internalized by MCF-7 cells.[8] The receptor–agonist complex may undergo endocytosis in clathrin-coated pits accumulating in early endosomes. The VIP receptor may be recycled to the cell surface whereas the VIP agonist may be degraded in lysosomes. The VIP-LALA-E was metabolized by lysosomal enzymes and cytotoxic E released into the cytosol. E accumulated in the nucleus, inhibiting topoisomerase II, preventing the unwinding of DNA. Because the cancer cell cannot replicate DNA and proliferate, it ultimately undergoes apoptosis. Similarly, (Ala^{2,8,9,19,24,25,27}Nle^{17},Lys^{28})VIP-L2-CPT may be internalized by VPAC1 receptors. After (Ala^{2,8,9,19,24,25,27}Nle^{17},Lys^{28})VIP-L2-CPT degradation by lysosomal enzymes, CPT may be released. The CPT may accumulate in the cancer cell nucleus inhibiting topoisomerase I, leading to cancer cell apoptosis. Thus VPAC1 receptors are molecular targets that may be utilized to deliver chemotherapeutic drugs into cancer cells.

Almost all breast cancer tumors examined have VPAC1 receptors. Using autoradiographic techniques, VPAC1 receptors are abundant in bladder, breast, colon, liver, lung, prostate, stomach, thyroid, and uterine cancer tumors.[9] VPAC2 receptors are abundant in stomach leiomyoma cancer whereas PAC1 receptors predominate in glioblastoma, neuroblastoma, adrenal, and pituitary cancer. VPAC1, VPAC2 and PAC1 immunoreactivity has been detected in breast cancer biopsy specimens.[10] PACAP mRNA and immunoreactivity are present in breast cancer biopsy specimens. PreproPACAP (19.9 kDa) but not PACAP-38 was present in peritumoral and tumoral breast tissues as well as alveolar epithelial cells and leukocytes in connective tissue.[11] It is possible the PACAP-27, PACAP-38, and the precursor peptides have biological activity. Thus VIP and PACAP may function as autocrine growth factors in breast cancer tumors.

ACKNOWLEDGMENTS

The authors thank Drs. David Coy, Julius Leyton, Samuel Mantey, and Tapas Pradhan for assistance. This research is supported by the Intramural Research Program of the NIH and NIDDK.

REFERENCES

1. DeVita, V.T., S. Hellman & S.A. Rosenberg. 2001. Cancer: principles and Practice of Oncology, 6th ed. Lippincott, Williams and Wilkins. Philadelphia.
2. Zia, H., T. Hida, S. Jakowlew, et al. 1996. Breast cancer growth is inhibited by VIP hybrid, a synthetic VIP receptor antagonist. Cancer Res. **56:** 3486–3489.
3. Moody, T.W., J. Leyton, I. Gozes, et al. 1998. VIP and breast cancer. Ann. N.Y. Acad. Sci. **865:** 290–296.
4. Whitmarsh, A.J. & R.J. Davies. 1996. Transcription factor AP-1 regulation by mitogen activated protein kinase signal transduction pathways. J. Mol. Med. **74:** 589–607.
5. Pallela, V.R., M.L. Thakur, S. Chakder & S. Rattan. 1999. ^{99m}Tc-labeled vasoactive intestinal peptide receptor agonist. Functional studies. J. Nucl. Med. **40:** 357–360.
6. Moody, T.W., J. Leyton, E. Unsworth, C. John, et al. 1998. (Arg^{15}, Arg^{21})VIP: a potent VIP agonist for localizing breast cancer tumors. Peptides **19:** 585–592.
7. Igarishi, H., T. Ito, W. Hou, et al. 2002. Elucidation of vasoactive intestinal peptide pharmacophore for VPAC$_1$ receptors in human, rat and guinea pig. J. Pharm. Exp. Ther. **301:** 37–50.
8. Moody, T.W., G. Czerwinski, N.I. Tarasova & C.J. Michejda C. 2001. VIP ellipticine derivatives inhibit the growth of breast cancer cells. Life Sci. **19:** 1005–1014.
9. Reubi, J.C., U. Laderach, B. Waser, et al. 2000. Vasoactive intestinal peptide/pituitary adenylate cyclase-activating peptide receptor subtypes in human tumors and their tissues of origin. Cancer Res. **60:** 3105–3112.
10. Schulz, S., C. Rocken, C. Mawrin, et al. 2004. Immunocytochemical identification of VPAC$_1$, VPAC$_2$ and PAC$_1$ receptors in normal and neoplastic human tissues with subtype-specific antibodies. Clin. Cancer Res. **10:** 8235–8242.
11. Garcia-Fernandez, M.O., G. Bodega, A. Ruiz-Villaespesa, et al. 2004. PACAP expression and distribution in human breast cancer and healthy tissue. Cancer Let. **205:** 189–195.

PACAP and Type I PACAP Receptors in Human Prostate Cancer Tissue

COSTANZO MORETTI,[a] CATERINA MAMMI,[b] GIOVANNI VANNI
FRAJESE,[a] STEFANIA MARIANI,[c] LUCIO GNESSI,[c] MARIO ARIZZI,[c]
FRANCESCA WANNENES,[b] AND GAETANO FRAJESE[a]

[a]Department of Internal Medicine, Unit of Endocrinology, AfAR Fatebenefratelli
Hospital, University of Rome TorVergata, 00186 Rome, Italy

[b]IRCCS San Raffaele Pisana, 00163 Rome, Italy

[c]Department of Medical Pathophysiology, University of Rome "La Sapienza,"
00185 Rome, Italy

ABSTRACT: We characterized the expression and localization of pituitary adenylate cyclase-activating polypeptide (PACAP) and its specific type I receptor variants in prostatic, hyperplastic, and carcinomatous tissue collected from patients undergoing prostate biopsy and surgery for benign prostatic hyperplasia (BPH) and prostate cancer (PCa). The immunohistochemical studies using an indirect immunoperoxidase technique evidenced positive immunostaining for PACAP in the cytoplasm of epithelial cells of hyperplastic and carcinomatous prostate specimens and in some scattered cells of the stroma. Type I PACAP receptors (PAC1 R) in healthy and BPH tissues were localized in all epithelial cells lining the lumen of the acini and in some stromal cells, while in specimens from PCa the anti-PAC1 R antibody stained the apical portion of a large percentage of cells. Furthermore, our molecular studies provide evidence that several PAC1 R isoforms (null, SV1/SV2) are present in normal, hyperplastic, and neoplastic tissue, the null variant being the most intensely expressed in PCa. These observations provide additional evidence for a role of PACAP and PAC1 R in the events determining the outcome of PCa.

KEYWORDS: PACAP; PAC1 R; PACAP receptor variants; prostate; prostate cancer; BPH; SV1; SV2; SV3

INTRODUCTION

Pituitary adenylate cyclase-activating polypeptide (PACAP) is a hypothalamic neuropeptide that belongs to the secretin/glucagon/vasoactive intestinal

Address for correspondence: Costanzo Moretti, University of Rome TorVergata, Unit of Endocrinology, Fatebenefratelli Hospital Isola Tiberina, 00186 Rome, Italy. Voice: +390672596665; fax: +390672596663.

e-mail: moretti@med.uniroma2.it

Ann. N.Y. Acad. Sci. 1070: 440–449 (2006). © 2006 New York Academy of Sciences.
doi: 10.1196/annals.1317.059

polypeptide (VIP)/growth hormone-releasing hormone (GHRH) family of peptide hormones.[1] Two biologically active forms of PACAP, PACAP-38 and the C-terminally truncated PACAP-27, have been characterized, which derive from a 176 amino acid precursor protein by posttranslational cleavage.[2] PACAP exerts a potent stimulatory action on cyclic AMP production in anterior pituitary cells[1] and promotes the release of several pituitary hormones. It also stimulates phosphatidylinositol hydrolysis and increases cytosolic Ca^{2+} in several cell types.[3–6] PACAP is detectable and biologically active in many tissues, including pituitary, brain, adrenal, testis, and nerve fibers of both the gut and the lung,[7,8] where it is considered to act as a neurohormone, neurotransmitter, neuromodulator, and vasoregulator. Recently, the expression and distribution of PACAP in human prostate cancer (PCa) and healthy prostate tissue have been demonstrated by biochemical and morphological procedures.[9] Both PACAP-38 and PACAP-27 act through three classes of membrane G protein–coupled receptors: the specific PAC1 receptor (PAC1 R) and the common PACAP/VIP (VPAC1 and VPAC2) receptors.[10] An alternative splicing of two exons of the PAC1 R gene allows for four major splice variants: the PAC1 R *null*, the PAC1 SV1, the PAC1 SV2, and the PAC1 SV3.[11] PAC1 R and its isoforms are highly expressed in many areas of the central nervous system (CNS)[12,13] and moderately in the neurohypophysis[14] and the pituitary gland,[15] the male reproductive tract,[16] and the adrenal medulla.[17] The presence of functional PAC1, VPAC1, and VPAC2 receptors has been demonstrated in human benign prostatic hyperplasia (BPH),[18,19] and VPAC1 receptors have been detected by autoradiography in human normal prostate and prostate carcinoma.[20] Also, the expression of the mRNAs for the three receptor classes has been demonstrated in the androgen-dependent human LNCaP PCa cell line[21] where they display different abilities to activate adenylyl cyclase (AC) and phospholipase C.[22,23] We have recently demonstrated in LNCaP that the sustained versus transient intracellular cAMP increase induced by the binding of PACAP to its cognate PAC1 receptor is a crucial event determining the outcome of tumor progression.[24] In order to acquire more information about PACAP and PAC1 R variants in normal, hyperplastic, and neoplastic human prostate, we have performed expression studies and immunohystochemical experiments on tissues obtained from patients affected with different degrees of BPH and PCa.

MATERIALS AND METHODS

Patients and Tissue Procurement

We enrolled in our study 10 normal patients undergoing surgical treatment for pelvic and urogenital primary diseases, 10 patients affected with BPH, and 22 patients 50–70 years of age undergoing radical prostatectomy for previously untreated carcinoma of the prostate. After prostatectomy, a wedge-shaped specimen of the fresh prostate was cut. Samples were submitted for pathological examination to confirm the prostate origin, the diagnosis, and the absence of other diseases. Hematoxylin-eosin staining was used for the histopathological evaluation, diagnosis, and tumor grading. The TNM score referred to the pathological T stage was used to classify the tumors according to TNM score.[25] The carcinoma tissues were classified according to the Gleason grades and the epithelial/stroma proportion was evaluated. Tissue collected intrasurgically or from biopsies was immediately frozen on dry ice and subsequently kept at –80°C until processing. All patients had given their written informed consent before participating in this study.

Reagents and Antibodies

Maloney-murine leukemia virus (M-MLV) reverse transcriptase was purchased from Life Technologies (San Giuliano Milanese, Italy). Taq polymerase was purchased from Promega Corp. (Milano, Italy). Affinity-purified polyclonal rabbit anti-human PACAP serum was obtained from Peninsula Laboratories, Inc. (San Carlos, CA). Affinity-purified polyclonal rabbit anti-human PAC1 receptor primary antibody was procured by Prof. A. Arimura (Tulane University Hebert Center, New Orleans, LA).

Immunohistochemical Detection of PACAP and PAC1 R

The localization of hPACAP and PAC1 R was performed on 5-μm-thick sections of the fixed healthy prostate, BPH, and PCa tissues, and carried out by the streptavidin-biotin immunoperoxidase method, using a commercial kit (Zymed Laboratories Inc., San Francisco, CA). Sections were incubated overnight at 4°C with the following antisera: 1:50 dilution of the affinity-purified polyclonal rabbit anti-human PACAP, and 1:100 dilution for the affinity-purified polyclonal rabbit anti-human PAC1 R primary antibodies. For controls, the primary antiserum was omitted in some sections. Slides were developed using amino-ethylcarbazole as chromogenic substrate, which peroxidase converts into a red to brownish-red precipitate at the sites of antigen localization in the tissue. The preparations were counterstained with hematoxylin, dehydrated, mounted, and analyzed.

RNA Preparation and RT-PCR

The purity and integrity of the RNA, extracted by the single-step acid guanidinium thiocyanate-phenol-chloroform method,[26] were checked spectroscopically and by gel electrophoresis before carrying out the analytical procedures. First-strand cDNA synthesis was performed as follows: 1 μg total RNA was reverse transcribed by 200 U of M-MLV reverse transcriptase using 2.5 μM random hexamers in the presence of 250 μM deoxynucleotides triphosphate in a final volume of 20 μL. DNA contamination or PCR carry over controls were performed by omitting M-MLV during RT. The reaction mixture was heat denatured for 5 min at 75°C, and then incubated for 1 h at 42°C. Five microliters of the cDNA obtained was used to amplify hPAC1 or hPACAP. hPAC1 was PCR and nested-amplified as follows: the first round of PCR was carried out using primers designed to amplify a PAC1 cDNA sequence spanning the region 872–1626 [upstream Cat1: 5′-TGTATGCGGAGCAGGACAGC-3′; downstream Cat2: 5′-AGGCCAGACATGCGGATTTGGG-3′, amplified product 754 bp]. The PCR product was nested-amplified using a set of primers flanking the receptor-splicing site [upstream Don1: 5′-TTAACTTTGTGCTTTTTATTGG-3′; downstream Don2: 5′-GAGTCTTTCCCTTTTGCTGAC-3′]; multiple products were amplified relating to the splice variants expressed. cDNAs were amplified using Taq polymerase (2 U per tube) with 15 pmole of both upstream and downstream primers, 1.5 mM magnesium chloride in a final volume of 50 μL. Then, 35 cycles (94°C for 30 s, 60°C for 30 s, 72°C for 30 s, with 5 min final extension for the first round of PCR, and 94°C for 30 s, 48°C for 30 s, 72°C for 30 s, with 5 min final extension for the nested round of PCR) were applied. PACAP PCR was performed using primer designed for the amplification of the fragment 502–1078 spanning exon 5 of the human PACAP gene [upstream hP1: 5′-AAACAAAGGACGACGCCGATAG-3′; downstream hP2: 5′-AGACTCACTGGGAAAGAATGC-3′; amplified product 576 bp].[27] cDNAs were amplified using Taq polymerase (2 U per tube) with 15 pmole of both upstream and downstream primers, 1.5 mM magnesium chloride in a final volume of 50 μL. Then, 30 cycles (94°C for 30 s, 60°C for 30 s, 72°C for 30 s with 10-min final extension) were applied. Finally, a 15-μL aliquot of all the amplified products was analyzed on 2% (wt/vol) agarose gel (NuSieve 3:1, FMC, Rockland, ME) and stained with ethidium bromide. Quantitation of the signals was performed by densitometric analysis using densitometry computer software (Kodak Digital Science 1D Image Analysis software, Eastman Kodak Co., Rochester, NY).

RESULTS

Localization in Normal and Pathological Prostate Tissues

The localization of PACAP and PAC1 R proteins investigated in sections obtained from normal, BPH, and neoplastic prostate glands by an indirect immunoperoxidase technique with anti-PACAP and anti-PAC1 R antiserum demonstrated a specific positive immunostaining for PACAP in the cytoplasm

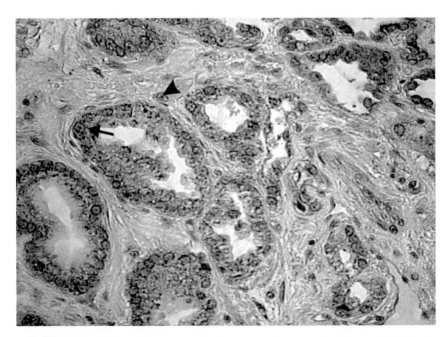

FIGURE 1. Section of human neoplastic prostate tissue showing immunohistochemical staining of PACAP in the PCa epithelium (arrow) and the stromal cells (arrow head).

of the epithelial cells lining the lumen of the acini and in some cells scattered in the stroma (FIG. 1), without significant differences in the distribution pattern between normal, hyperplastic, and tumoral tissues. PAC1 R was found in all epithelial cells and in some stromal cells in healthy and BPH tissues with the immunostaining mostly localized in the apical portion of the cells, while in the specimens from PCa, the anti-PAC1 R antibody stained the apical portion of a large percentage of cells (75 ± 10% positive cells).

Expression Studies

The expression of PACAP mRNA in normal, hyperplastic, and neoplastic human prostate examined in order to determine if these tissues synthesize PACAP and PAC1 R variants, demonstrated in all tissues the expression of PACAP and PAC1 R mRNAs. The nested RT-PCR of cDNA, in addition to the semiquantitative amplification from these tissues, identified the isoform without SV1/SV2 cassettes (PAC1 R *null*) as the predominant product (FIG. 2). The expected products corresponding to PAC1 R containing SV1 or SV2 cassettes were all present in the normal and hyperplastic tissues while in the PCa, the PAC1 R *null* isoform was the only one clearly expressed. The PAC1 SV3

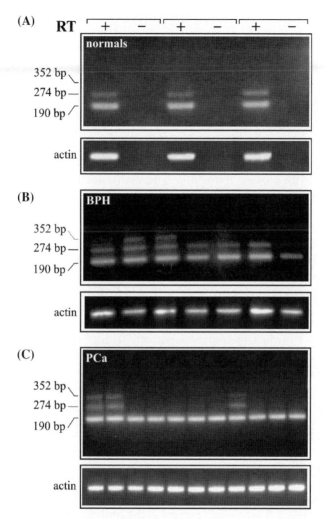

FIGURE 2. Analysis of type I PACAP R splice variants in normal (**A**), hyperplastic (BPH) (**B**), and neoplastic (PCa) (**C**). Reverse transcriptase (RT) was omitted in the control experiments (-RT). cDNAs were synthesized by RT-PCR using total RNA from tissues with cat1 and cat2 oligonucleotide as forward and reverse primers, respectively, followed by a second round of PCR products with don1 and don2 nested primers. PCR products were electrophoresed and stained with ethidium bromide. The product of expected size (190 bp) is related to the *null* variant. The product of expected size (274 bp) is related to the SV1/SV2 variant. The product of the expected size (352 bp) is related to the SV$_3$ variant. The β-actin transcript was analyzed as control.

isoform, a splice variant of the PAC1 receptor containing both the SV1 and SV2 boxes, was occasionally measurable with our method independently from the pathological state of the prostate tissue examined.

DISCUSSION

As in other human neoplasms, in the prostate gland, the process of forming a malignant tumor depends on a series of intermediate changes existing between the normal prostate epithelium and the newly formed prostate tumor. These changes involve a modification of the physiological reciprocal cellular interactions between stroma and epithelium that have been shown to accelerate local tumor growth[28,29] and distant metastases,[30] increase the genetic prostate cell instability,[31] and its subsequent androgen independent progression. The intricate intercellular communication between stromal and epithelial cells alters the physiology of the prostate epithelium in the malignant transformation of which two distinct circuits are involved: one operating in the nucleus and one in the cytoplasm, both of which are required for cell transformation.[32] Sex steroids are very important in the control of such mechanisms in that they alter within the prostate the main cell growth and survival as well as the action of several growth factors. The central characteristic of fatal PCa is androgen independence, a condition in which a perturbation of the androgen receptor signaling enables the androgen receptor to be activated by nonandrogenic steroid hormones and growth factors.[33] The androgen independence is markedly related to neuroendocrine (NE) cell activity and NE cells may represent an independent indicator of poor prognosis in patients with prostate carcinoma.[34] NE cells are dispersed throughout the prostate tumor and their number increases after long-term antiandrogen therapy.[35] They exhibit a fusiform morphology with neuritic processes and their ultrastructural characterization shows the presence of neurosecretory granules. Although they appear to be nonmitotic, proliferating carcinoma cells have been found in close proximity to them,[36] suggesting that NE cells provide paracrine stimuli for proliferation of the surrounding carcinoma cells. Among the factors that regulate NE differentiation from epithelially derived prostatic tumor cells, we investigated the localization and action of the PACAP and of the different variants of its specific PAC1 receptors. Our immunohistochemical studies show that, in the normal prostate, PACAP and PAC1 R are mainly present in the epithelial cells lining the lumen of the acini, while in the neoplastic tissue the anti-PAC1 R antibody stained the apical portion of a large percentage of cells, suggesting a different localization of the protein among tissues without variations in staining intensity. Furthermore, our expression studies show that the PAC1 R *null* isoform, which is related to the AC activation, is the most important receptor present in the neoplastic prostate. Considering the importance of the cAMP milieu in the mechanism of NE differentiation, the potential role of PACAP, as a natural cAMP inducer, may be reevaluated in the prostate neoplastic progression. We have recently demonstrated *in vitro* that PACAP induces the development of a NE morphology and NE differentiation in LNCaP cells[24]: LNCaP cells treated with different concentrations of PACAP-27 added at the beginning of the culture, showed a rapid but transient development of neuritic processes both in the presence and in the absence of 5% FBS; the acquisition of an NE morphology was evident after 15 min of treatment but had almost completely reverted after 24 h. This effect was partially prevented by pretreatment with PACAP6–27, indicating that PAC1 R is involved in this phenomenon. Chronic PACAP-27 treatment or co-treatment with PACAP-27 and IBMX, conditions that maintain a sustained cAMP accumulation, exerted the same effect as Fsk, and the NE morphology of LNCaP cells was still observable after 3 days. These data suggest that the inhibitory effect of the peptide on cell growth correlates with the maintenance of NE differentiation. Besides PACAP, several peptides produced by epithelial and prostate NE cells have been shown to possess the ability to elevate cAMP and to induce NE cell differentiation, capable of sensitizing the response of LNCaP PCa cells to growth factors.[24] A relevant consideration in the physiology of PCa is that cells with NE phenotype increase in number as cancer progresses to the androgen refractory condition. NE cells are particularly concentrated in the vicinity of proliferating cancer cells, on which they act in a paracrine fashion by secreting mitogenic factors.[30,36,37] It has been shown that the PACAP receptor antagonist PACAP6–27 suppresses the growth of a human PCa cell line, suggesting that this peptide could be an important prostate local regulator of cell growth and differentiation.[38] Thus, we propose that PACAP, through its specific receptors, may modulate the tissue availability of cAMP in the prostate. The evidence that the PAC1 R *null* variant, which is the predominant isoform expressed in human neoplastic tissue, is able to induce a cAMP rise may explain why in the neoplastic microenvironment it might be a determinant factor with the ability to change the fate of the prostate tumor. A large and sustained PAC1 R *null*–induced production of cAMP might be involved in the loss of androgen sensitivity. Further studies need to be performed in this direction.

REFERENCES

1. MIYATA, A. *et al.* 1989. Isolation of a novel 38 residue-hypothalamic polypeptide which stimulates adenylate cyclase in pituitary cells. Biochem. Biophys. Res. Commun. **164:** 567–574.
2. HOSOYA, M. *et al.* 1992. Structure of the human pituitary adenylate cyclase activating polypeptide (PACAP) gene. Biochem. Biophys. Acta **1129:** 199–206.
3. CANNY, B.J., S.R. RAWLINGS & D.A. LEONG. 1992. Pituitary adenylate cyclase activating polypeptide specifically increases cytosolic calcium ion concentration in rat gonadotropes and somatotropes. Endocrinology **130:** 211–215.
4. DEUTSCH, P.J. & Y. SUN. 1992. The 38-amino acid pituitary adenylate cyclase activating polypeptide stimulates dual signalling cascades in PC12 cells and promotes neurite outgrowth. J. Biol. Chem. **267:** 5108–5113.
5. TATSUNO, I., T. YADA, S. VIGH, *et al.* 1992. Pituitary adenylate cyclase activating polypeptide and vasoactive intestinal peptide increase cytosolic free calcium concentration in cultured rat hippocampal neurons. Endocrinology **31:** 73–81.

6. WATANABE, T. *et al.* 1992. Pituitary adenylate cyclase activating polypeptide provokes cultured rat chromaffin cells to secrete adrenaline. Biochem. Biophys. Res. Commun. **182:** 403–411.

7. ARIMURA, A. & S. SHIODA. 1995. Pituitary adenylate cyclase activating polypeptide (PACAP) and its receptors: neuroendocrine and endocrine interaction. Front. Neuroendocrinol. **16:** 53–88.

8. UDDMAN, R., A. LUTS, A. ARIMURA & F. SUNDLER. 1991. Pituitary adenylate cyclase activating polypeptide (PACAP), a new vasoactive intestinal peptide (VIP)-like peptide in the respiratory tract. Cell Tissue Res. **265:** 197–201.

9. GARCÌA-FERNÀNDEZ, M.O. *et al.* 2002. Expression and distribution of pituitary adenylate cyclase-activating peptide in human prostate and prostate cancer tissues. Regul. Pept. **110:** 9–15.

10. HARMAR, A.J. *et al.* 1998. International Union of Pharmacology. XVIII. Nomenclature of receptors for vasoactive intestinal peptide and pituitary adenylate cyclase-activating polypeptide. Pharmacol. Rev. **50:** 265–270.

11. PISEGNA, J.R. & S.A. WANK. 1996. Cloning and characterization of the signal transduction of four splice variants of the human pituitary adenylate cyclase activating polypeptide receptor. Evidence for dual coupling to adenylate cyclase and phospholipase C. J. Biol. Chem. **271:** 17267–17274.

12. HASHIMOTO, H. *et al.* 1996. Distribution of the mRNA for a pituitary adenylate cyclase activating polypeptide receptor in the rat brain: an in situ hybridization study. J. Comp. Neurol. **371:** 567–577.

13. SUDA, K., D.M. SMITH, M.A. GHATEI, *et al.* 1991. Investigation and characterization of receptors for pituitary adenylate cyclase-activating polypeptide in human brain by radioligand binding and chemical cross-linking. J. Clin. Endocrinol. Metab. **72:** 958–964.

14. LUTZ-BUCHER, B., D. MONNIER & B. KOCH. 1996. Evidence for the presence of receptors for pituitary adenylate cyclase-activating polypeptide in the neurohypophysis that are positively coupled to cyclic AMP formation and neurohypophyseal hormone secretion. Neuroendocrinology **64:** 153–161.

15. HART, G.R., H. GOWING & J.M. BURRIN. 1992. Effects of a novel hypothalamic peptide, pituitary adenylate cyclase-activating polypeptide, on pituitary release in rats. J. Endocrinol. **134:** 33–41.

16. ROMANELLI, F., S. FILLO, A. ISIDORI & D. CONTE. 1997. Pituitary adenylate cyclase-activating polypeptide regulates rat Leydig cell function in vitro. Neuropeptides **31:** 311–317.

17. WATANABE, T. *et al.* 1992. Pituitary adenylate cyclase-activating polypeptide provokes cultured rat chromaffin cells to secrete adrenaline. Biochem. Biophys. Res. Commun. **182:** 403–411.

18. SOLANO, R.M. *et al.* 1996. Characterization of vasoactive intestinal peptide/pituitary adenylate cyclase activating peptide receptors in human benign hyperplastic prostate. Endocrinology **137:** 2815–2822.

19. SOLANO, R.M. *et al.* 1999. Identification and functional properties of the pituitary adenylate cyclase activating peptide (PAC$_1$) receptor in human benign hyperplastic prostate. Cell. Signal. **11:** 813–819.

20. REUBI, J.C. 2000. In vitro evaluation of VIP/PACAP receptors in healthy and diseased human tissues. Clinical implications. Ann. N. Y. Acad. Sci. **921:** 1–25.

21. JUARRANZ, M.G., O. BOLAÑOS, I. GUTIERREZ-CAÑAS, *et al.* 2001. Neuroendocrine differentiation of the LNCaP prostate cancer cell line maintains the expression and function of VIP and PACAP receptors. Cell. Signal. **13:** 887–894.

22. PISEGNA, J.R. & S.A. WANK. 1996. Cloning and characterization of the signal transduction of four splice variants of the human pituitary adenylate cyclase activating polypeptide receptor. Evidence for dual coupling to adenylate cyclase and phospholipase C. J. Biol. Chem. **271:** 17267–17274.
23. VAUDRY, D., B.J. GONZALEZ, M. BASILLE, *et al.* 2000. Pituitary adenylate cyclase-activating polypeptide and its receptors: from structure to functions. Pharmacol. Rev. **52:** 269–324.
24. FARINI, D., A. PUGLIANIELLO, C. MAMMI, *et al.* 2003. Dual effect of pituitary adenylate cyclase activating polypeptide on prostate tumor LNCaP cells: short- and long-term exposure affect proliferation and neuroendocrine differentiation. Endocrinology **144:** 1631–1643.
25. MCNEAL, J.E., R.J. COHEN & J.D. BROOKS. 2001. Role of cytologic criteria in the histologic diagnosis of Gleason grade 1 prostatic adenocarcinoma. Hum. Pathol. **32:** 441–446.
26. CHOMCZYNSKI, P. & N. SACCHI. 1992. Single-step method of RNA isolation by acid guanidinium thiocyanate-phenol-chloroform extraction. Anal. Biochem. **162:** 156–159.
27. HOSOYA, M. *et al.* 1992. Structure of the human pituitary adenylate cyclase activating polypeptide (PACAP) gene. Biochem. Biophys. Acta **1129:** 199–206.
28. CAMPS, J.L. *et al.* 1990. Fibroblast-mediated acceleration of human epithelial tumor growth in vivo. Proc. Natl. Acad. Sci. USA **87:** 75–79.
29. CHUNG, L.W. & R. DAVIES. 1996. Prostate epithelial differentiation is dictated by its surrounding stroma. Mol. Biol. Rep. **23:** 13–19.
30. THALMANN, G.N. *et al.* 1994. Androgen-independent cancer progression and bone metastasis in the LNCaP model of human prostate cancer. Cancer Res. **54:** 2577–2581. Erratum in: Cancer Res. **54:** 3953.
31. TLSTY, T.D. 1998. Cell-adhesion-dependent influences on genomic instability and carcinogenesis. Curr. Opin. Cell. Biol. **10:** 647–653.
32. METTE, A.P. & E.A. OSTRANDER 2001. Prostate cancer: simplicity to complexity. Nature Genetics **27:** 134–135.
33. DEBES, J.D. & D.J. TINDALL. 2004. Mechanism of androgen-refractory prostate cancer. N. Engl. J. Med. **351:** 1488–1490.
34. YU, D.S., D.S. HSIEH, H.I. CHEN & S.Y. CHANG. 2001. The expression of neuropeptides in hyperplastic and malignant prostate tissue and its possible clinical implications. J. Urol. **166:** 871–875.
35. DI SANT'AGNESE, P.A. 1992. Neuroendocrine differentiation in human prostatic carcinoma. Hum. Pathos. **23:** 287–296.
36. BONKHOFF, H., U. STEIN, K. ROMBERG. 1995. Endocrine-paracrine cell types in the prostate and prostatic adenocarcinoma are postmitotic cells. Hum. Pathol. **26:** 167–170.
37. ABRAHAMSSON, P.T. 1999. Neuroendocrine cells in tumor growth of the prostate. Endocrine Relat. Cancer **6:** 503–519.
38. LEYTON, J., T. COELHO, D.H. COY, *et al.* 1998. PACAP(6–38) inhibits the growth of prostate cancer cells. Cancer Lett. **125:** 131–139.

Lack of Trimethyltin (TMT)-Induced Elevation of Plasma Corticosterone in PACAP-Deficient Mice

YOSHIKO MORITA,[a] DAISUKE YANAGIDA,[a] NORIHITO SHINTANI,[a] KIYOKAZU OGITA,[b] NORITO NISHIYAMA,[b] RIE TSUCHIDA,[a] HITOSHI HASHIMOTO,[a] AND AKEMICHI BABA[a]

[a]Laboratory of Molecular Neuropharmacology, Graduate School of Pharmaceutical Sciences, Osaka University, Suita, Osaka 565-0871, Japan

[b]Department of Pharmacology, Faculty of Pharmaceutical Sciences, Setsunan University, Hirakata, Osaka 573-0101, Japan

ABSTRACT: Accumulating evidence implicates pituitary adenylate cyclase-activating polypeptide (PACAP) in a number of stress responses. By using PACAP-deficient mice, PACAP has been shown to have an *in vivo* role in the regulation of the sympathoadrenal axis, but a role in regulating the hypothalamo-pituitary-adrenal (HPA) axis has not been fully addressed. To elucidate the role of endogenous PACAP in HPA axis regulation during pathological conditions, mice lacking the *Adcyap1* gene encoding the neuropeptide PACAP (*Adcyap1*$^{-/-}$) were injected with trimethyltin (TMT), a neurotoxin known to induce neuronal damage and several systemic responses including elevated plasma corticosterone levels. In wild-type controls, TMT induced transient decreases in water and food intake, with a concomitant decrease in body weight; however, no significant changes were observed in *Adcyap1*$^{-/-}$ mice. Basal corticosterone levels were not significantly different between the mutant and wild-type mice. TMT induced a marked elevation of plasma corticosterone above basal levels in wild-type mice but no significant increase was seen in *Adcyap1*$^{-/-}$ mice. The present article suggests that PACAP is involved in the corticosterone release in some pathological conditions but not in the basal state.

KEYWORDS: corticosterone; food intake; hypothalamo-pituitary-adrenal (HPA) axis; knockout mouse; PACAP; trimethyltin (TMT)

INTRODUCTION

Pituitary adenylate cyclase-activating polypeptide (PACAP) is a neuropeptide originally isolated from ovine hypothalamus based on its ability to

Address for correspondence: Norihito Shintani, Laboratory of Molecular Neuropharmacology, Graduate School of Pharmaceutical Sciences, Osaka University, 1-6 Yamadaoka, Suita, Osaka 565-0871, Japan. Voice: 81-6-6879-8182; fax: 81-6-6879-8184.
e-mail: shintani@phs.osaka-u.ac.jp

Ann. N.Y. Acad. Sci. 1070: 450–456 (2006). © 2006 New York Academy of Sciences.
doi: 10.1196/annals.1317.060

stimulate adenylate cyclase in cultured rat anterior pituitary cells, and it has a number of biologic activities.[1,2] A growing body of evidence suggests that PACAP plays a role in the control of the hypothalamo-pituitary-adrenal (HPA) axis. PACAP and its receptor subtypes are widely distributed in neuroendocrine tissues involved in HPA axis regulation, including different brain regions, the pituitary, as well as the adrenal gland. It is likely that, rather than directly stimulating secretion of HPA axis hormones, PACAP modulates different parts of the HPA axis function, acting mostly in a paracrine or autocrine manner.[1–3] PACAP has a direct stimulatory effect on catecholamine secretion coupled with its biosynthesis in the adrenal gland.[4] Recently, several groups, including our own, have independently produced mice lacking the *Adcyap1* gene encoding the neuropeptide PACAP (*Adcyap1*$^{-/-}$).[4–7] By using these mutant mice, PACAP has been clearly demonstrated to have an *in vivo* role in the sympathoadrenal axis and to be of pathophysiological importance.[4,5] However, the role of PACAP in the HPA axis has not been adequately addressed.

Trimethyltin (TMT) is a neurotoxin that induces *in vivo* neuronal damage,[8,9] accompanied by responses including temporal elevation of plasma corticosterone level,[10] body weight loss,[11] and behavioral changes, such as whole body tremor.[12] To elucidate the role of endogenous PACAP in HPA axis regulation under stress-related conditions, *Adcyap1*$^{-/-}$ mice were injected with TMT and examined for subsequent responses.

MATERIALS AND METHODS

All animal experiments were carried out in accordance with protocols approved by the Animal Care and Use Committee of Graduate School of Pharmaceutical Sciences, Osaka University, Japan. Generation of *Adcyap1*$^{-/-}$ mice by gene targeting has been reported previously.[6] The null mutation was backcrossed onto an Institute of Cancer Research (ICR) mouse background at least 10 times.

Adcyap1$^{-/-}$ and wild-type mice (all male, 7–12 weeks of age) obtained from heterozygote crosses were injected intraperitoneally with TMT (3 mg/kg; Strem Chemicals, Newburyport, MA) or vehicle (phosphate-buffered saline), and body weight changes at days 1, 2, 3, 4, and 7, and water and food intake for the first 24-h period were assessed. Plasma corticosterone levels were determined with an RIA kit (Rat Corticosterone ^{125}I Biotrack Assay System; GE Healthcare Bio-Sciences Corp., Piscataway, NJ).

Statistically significant differences were assessed by ANOVA using the Turkey–Kramer and paired *t*-tests, where applicable. All values were expressed as mean $\pm$ SE.

RESULTS AND DISCUSSION

Changes in Body Weight, and Water and Food Intake After TMT Injection

In accordance with a previous report,[11] TMT injection induced a transient decrease in body weight in wild-type mice, with the maximum reduction observed around 2–3 days after injection. In sharp contrast, TMT showed no significant effect on the body weight in $Adcyap1^{-/-}$ mice (FIG. 1A). Similarly, TMT significantly decreased water and food intake in wild-type mice, but no significant change was seen in $Adcyap1^{-/-}$ mice (FIG. 1B, C).

These results are somewhat unexpected since PACAP has been implicated in a protective role against various stresses, such as hypoglycemia[4] and cold exposure.[13] Intracerebroventricular injection studies and a knockout mouse study suggest that PACAP possesses both anorectic and orexigenic properties.[14,15] Although the mechanisms underlying TMT action remain poorly understood, PACAP deficiency might result in a suppression of the anorectic effect of TMT.

Lack of TMT-Induced Elevation of Plasma Corticosterone in $Adcyap1^{-/-}$ Mice

To address the question of HPA axis regulation by PACAP, we examined plasma corticosterone levels in $Adcyap1^{-/-}$ mice and wild-type controls under basal conditions and after TMT injection. In wild-type mice, TMT significantly increased plasma corticosterone level with a transient peak of approximately sixfold above basal level 24 h after injection (FIG. 2), followed by a lasting increase up to 96 h (data not shown). However, no significant increase in corticosterone level was seen in $Adcyap1^{-/-}$ mice. Basal corticosterone level was not significantly different between the two groups.

These results demonstrate that endogenous PACAP is crucial for plasma corticosterone release, at least under conditions of TMT-mediated stress. As mentioned above, it has been suggested that PACAP is involved with different parts of the HPA axis.[1–3] However, PACAP deficiency did not affect basal corticosterone levels. This result is in accord with previous observations that PACAP null pups are able to release corticosterone from the adrenal cortex,[5] and that the diurnal rhythm of plasma corticosterone at rest is equivalent in the mutant and wild-type mice.[4] In addition, it has been shown that corticosterone level is increased after insulin administration in PACAP null mice, indicating that the acute responsiveness of the HPA axis is unimpaired in these mice.[4] Thus, it is conceivable that PACAP is involved in corticosterone release as a consequence of the central stress response. TMT produced a similar degree of whole body tremor in $Adcyap1^{-/-}$ and wild-type mice (unpublished

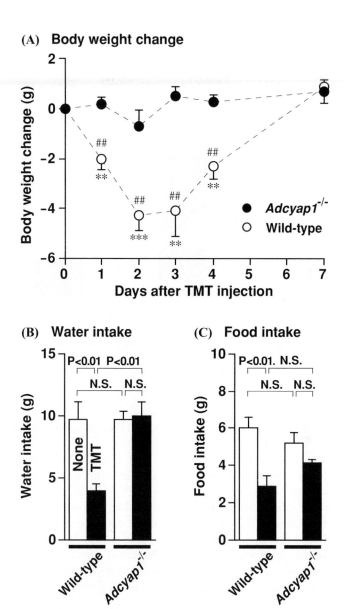

FIGURE 1. Body weight change, and water and food intake in *Adcyap1*$^{-/-}$ mice after TMT injection. (**A**) Body weight changes (relative to weight at day 0) were measured in *Adcyap1*$^{-/-}$ (closed circles) and wild-type (open circles) mice injected with TMT (3 mg/kg) at day 0. (**B**) and (**C**) Water intake (**B**) and food intake (**C**) were assessed in *Adcyap1*$^{-/-}$ (closed bars) and wild-type (open bars) mice during the 24-h period after TMT injection. $n = 6$–7 per group. **P < 0.01 and ***P < 0.001 compared with day 0; $^{##}P$ < 0.01 compared with *Adcyap1*$^{-/-}$ mice. N.S., not significant.

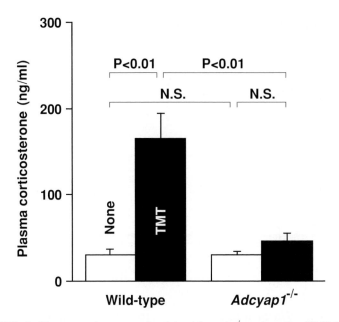

FIGURE 2. Plasma corticosterone level in *Adcyap1*$^{-/-}$ mice after TMT injection. Twenty-four hours after injection of TMT (3 mg/kg) or vehicle, plasma corticosterone level was assessed in *Adcyap1*$^{-/-}$ (closed bars; $n = 7$–10) and wild-type (open bars; $n = 13$–15) mice. N.S., not significant.

observation), indicating a similar distribution of this neurotoxin in the brain of both groups.

It is known that TMT causes neuronal cell death in hippocampus,[8,9] an important locus regulating the HPA axis, and that hippocampal lesions cause elevation of plasma corticosterone.[16] Because PACAP has been shown to exert neuroprotective effects, neural damage, as well as neurodegeneration in *Adcyap1*$^{-/-}$ mice after TMT injection, should be assessed in a future study. A plethora of knockout phenotypes has been shown for PACAP and its receptors, including psychomotor change and defects in circadian rhythm, as well as altered stress responses.[17] The mechanisms involved and their possible role in human diseases are of great interest and warrant further investigation.

ACKNOWLEDGMENTS

This research was supported, in part, by Grants-in-Aid for Scientific Research (A) and (B), and for Young Scientists (B) from Japan Society for the Promotion of Science, and by a grant from Taisho Pharmaceutical Co. Ltd.

REFERENCES

1. ARIMURA, A. 1998. Perspectives on pituitary adenylate cyclase activating polypeptide (PACAP) in the neuroendocrine, endocrine, and nervous systems. Jpn. J. Physiol. **48:** 301–331.

2. VAUDRY, D., B.J. GONZALEZ, M. BASILLE, *et al.* 2000. Pituitary adenylate cyclase-activating polypeptide and its receptors: from structure to functions. Pharmacol. Rev. **52:** 269–324.

3. NUSSDORFER, G.G. & L.K. MALENDOWICZ. 1998. Role of VIP, PACAP, and related peptides in the regulation of the hypothalamo-pituitary-adrenal axis. Peptides **19:** 1443–1467.

4. HAMELINK, C., O. TJURMINA, R. DAMADZIC, *et al.* 2002. Pituitary adenylate cyclase-activating polypeptide is a sympathoadrenal neurotransmitter involved in catecholamine regulation and glucohomeostasis. Proc. Natl. Acad. Sci. USA **99:** 461–466.

5. GRAY, S.L., K.J. CUMMINGS, F.R. JIRIK & N.M. SHERWOOD. 2001. Targeted disruption of the pituitary adenylate cyclase-activating polypeptide gene results in early postnatal death associated with dysfunction of lipid and carbohydrate metabolism. Mol. Endocrinol. **15:** 1739–1747.

6. HASHIMOTO, H., N. SHINTANI, K. TANAKA, *et al.* 2001. Altered psychomotor behaviors in mice lacking pituitary adenylate cyclase-activating polypeptide (PACAP). Proc. Natl. Acad. Sci. USA **98:** 13355–13360.

7. COLWELL, C.S., S. MICHEL, J. ITRI, *et al.* 2004. Selective deficits in the circadian light response in mice lacking PACAP. Am. J. Physiol. Regul. Integr. Comp. Physiol. **287:** R1194–R1201.

8. FIEDOROWICZ, A., I. FIGIEL, B. KAMINSKA, *et al.* 2001. Dentate granule neuron apoptosis and glia activation in murine hippocampus induced by trimethyltin exposure. Brain Res. **912:** 116–127.

9. OGITA, K., Y. NITTA, M. WATANABE, *et al.* 2004. In vivo activation of c-Jun N-terminal kinase signaling cascade prior to granule cell death induced by trimethyltin in the dentate gyrus of mice. Neuropharmacology **47:** 619–630.

10. CHANG, L.W., A.J. HOUGH, F.G. BIVINS & D. COCKERILL. 1989. Effects of adrenalectomy and corticosterone on hippocampal lesions induced by trimethyltin. Biomed. Environ. Sci. **2:** 54–64.

11. BUSHNELL, P.J. & H.L. EVANS. 1985. Effects of trimethyltin on homecage behavior of rats. Toxicol. Appl. Pharmacol. **79:** 134–142.

12. WENGER, G.R., D.E. MCMILLAN & L.W. CHANG. 1984. Behavioral effects of trimethyltin in two strains of mice. I. spontaneous motor activity. Toxicol. Appl. Pharmacol. **73:** 78–88.

13. GRAY, S.L., N. YAMAGUCHI, P. VENCOVA & N.M. SHERWOOD. 2002. Temperature-sensitive phenotype in mice lacking pituitary adenylate cyclase-activating polypeptide. Endocrinology **143:** 3946–3954.

14. MIZUNO, Y., K. KONDO, Y. TERASHIMA, *et al.* 1998. Anorectic effect of pituitary adenylate cyclase activating polypeptide (PACAP) in rats: lack of evidence for involvement of hypothalamic neuropeptide gene expression. J. Neuroendocrinol. **10:** 611–616.

15. NAKATA, M., D. KOHNO, N. SHINTANI, *et al.* 2004. PACAP deficient mice display reduced carbohydrate intake and PACAP activates NPY-containing neurons in the rat hypothalamic arcuate nucleus. Neurosci. Lett. **370:** 252–256.

16. JACOBSON, L. & R. SAPOLSKY. 1991. The role of the hippocampus in feedback regulation of the hypothalamic-pituitary-adrenocortical axis. Endocr. Rev. **12:** 118–134.
17. HASHIMOTO, H., N. SHINTANI & A. BABA. 2006. New insights into the central PACAPergic system from the phenotypes in PACAP- and PACAP receptor-knockout mice. Ann. N. Y. Acad. Sci. This volume.

Expression of PACAP Receptor mRNAs by Neuropeptide Y Neurons in the Rat Arcuate Nucleus

LOURDES MOUNIEN, PATRICE BIZET, ISABELLE BOUTELET,
GUILLAUME GOURCEROL, MAGALI BASILLE, BRUNO GONZALEZ,
HUBERT VAUDRY, AND SYLVIE JEGOU

*INSERM U413, Laboratory of Cellular and Molecular Neuroendocrinology,
European Institute for Peptide Research (IFRMP 23), University of Rouen,
76821 Mont-Saint-Aignan, France*

ABSTRACT: Neuropeptide Y (NPY) and pituitary adenylate cyclase-activating polypeptide (PACAP) exert opposite actions in energy homeostasis: NPY is a potent orexigenic peptide whereas PACAP reduces food intake. PAC1-R and VPAC2-R mRNAs are actively expressed in the arcuate nucleus of the hypothalamus which contains a prominent population of NPY neurons. By using a double-labeling *in situ* hybridization technique, we now show that a significant proportion of NPY neurons express PAC1-R or VPAC2-R mRNA. This observation indicates that PACAP may regulate the activity of NPY neurons, suggesting that the inhibitory effect of PACAP on food intake may be mediated, at least in part, through modulation of NPY neurotransmission.

KEYWORDS: arcuate nucleus; feeding behavior; PACAP; PAC1-R; VPAC2-R; neuropeptide Y

INTRODUCTION

The arcuate nucleus of the hypothalamus plays a pivotal role in the control of energy homeostasis and there is now clear evidence that several neuropeptides expressed by arcuate nucleus neurons are involved in this process.[1] In particular, neuropeptide Y (NPY) and pituitary adenylate cyclase-activating polypeptide (PACAP) appear to exert opposite effects on food intake and energy expenditure: when injected centrally, NPY induces a robust orexigenic response[1] while PACAP causes a long-lasting reduction of food consumption.[2] The mRNAs encoding the PACAP precursor, the PACAP-specific receptor

Address for correspondence: Dr. Hubert Vaudry, INSERM U413, Laboratory of Cellular and Molecular Neuroendocrinology, European Institute for Peptide Research (IFRMP 23), University of Rouen, 76821 Mont-Saint-Aignan, France. Voice: +33-235-14-6624; fax: +33-235-14-6946.
e-mail: hubert-vaudry@univ-rouen.fr

Ann. N.Y. Acad. Sci. 1070: 457–461 (2006). © 2006 New York Academy of Sciences.
doi: 10.1196/annals.1317.061

(PAC1-R), and the PACAP-vasoactive intestinal polypeptide (VIP) mutual receptor (VPAC2-R) are all expressed in the arcuate nucleus,[3] suggesting that
PACAP may act locally on other peptidergic neurons involved in the control
of appetite and energy expenditure. In the present article, we have investigated
the possible occurrence of PAC1-R and VPAC2-R mRNAs in NPY-expressing
neurons in the rat arcuate nucleus by means of double-labeling *in situ* hybridization histochemistry.

MATERIALS AND METHODS

Double-labeling *in situ* hybridization was performed as previously described.[4] Briefly, frontal brain sections were hybridized with the riboprobe-
hybridization buffer mix, containing the ^{35}S-labeled PAC1-R or VPAC2-R
riboprobe (1.5×10^7 cpm/mL) and the digoxigenin (DIG)-labeled NPY riboprobe (1:100, vol/vol). Tissue sections were incubated overnight at 55°C, and
then treated with RNase and submitted to a series of stringent washes, including
a high-stringency wash at 60°C. The sections were incubated with antidigox-
igenin Fab fragments conjugated to alkaline phosphatase, and stained with
4-nitroblue tetrazolium/5-bromo-4-chloro-3-indolyl-phosphate (NBT/BCIP).
Brain slices were then quickly dehydrated in graded alcohols containing 0.3 M
ammonium acetate, dipped into autoradiographic K5 emulsion and exposed
for 2 months. To calculate the percentage of NPY-positive neurons that express
either PAC1-R or VPAC2-R mRNA, the arcuate nucleus was arbitrarily sub-
divided into four areas (A–D) of equal length along the rostro-caudal axis.

RESULTS AND DISCUSSION

Double staining of rat brain sections with the DIG-labeled NPY probe and
the ^{35}S-labeled PAC1-R or VPAC2-R probe showed that a substantial proportion of NPY mRNA-containing neurons in the arcuate nucleus expressed
PAC1-R or VPAC2-R mRNAs. Several NPY neurons did not contain PAC1-R
or VPAC2-R mRNAs and, reciprocally, several PAC1-R or VPAC2-R mRNA-
expressing neurons did not contain NPY mRNA (Fig. 1). Quantitative analysis
of double-labeled neurons revealed that, in the whole arcuate nucleus, 20 ± 1%
of the NPY mRNA-positive perikarya expressed PAC1-R mRNA, and 30 ±
2% expressed VPAC2-R mRNA (Fig. 2). These observations strongly suggest
that PACAP may directly modulate the activity of NPY neurons through activation of PAC1-R and/or VPAC2-R. Consistent with this notion, it has been
shown that PACAP receptor agonists increase cytosolic calcium concentrations
in NPY neurons of the rat arcuate nucleus.[5] The fact that NPY mRNA expression was decreased in PACAP-deficient mice compared to wild-type animals[5]
provides additional support for a role of PACAP in the regulation of NPY neuron activity. It has been previously demonstrated that intracerebroventricular

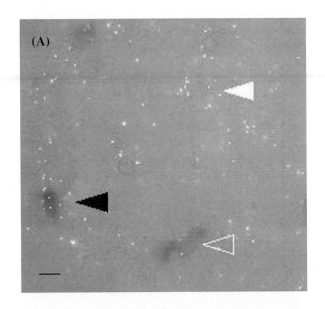

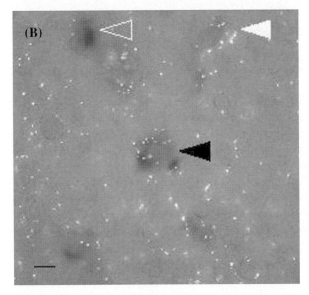

FIGURE 1. Bright-field microphotographs under epi-illumination of frontal sections at the level of the arcuate nucleus illustrating double *in situ* hybridization labeling with ^{35}S-labeled antisense probe to PAC1-R (**A**) or VPAC2-R (**B**) mRNA (silver grains) and digoxigenin-labeled antisense probe to NPY mRNA (dark precipitate). Black arrowheads point to NPY-positive neurons expressing PAC1-R or VPAC2-R mRNA. White arrowheads point to NPY-negative neurons expressing PAC1-R or VPAC2-R mRNA. Open white arrowheads point to NPY-positive neurons that do not express PAC1-R or VPAC2-R mRNA. Scale bars = 10 μM.

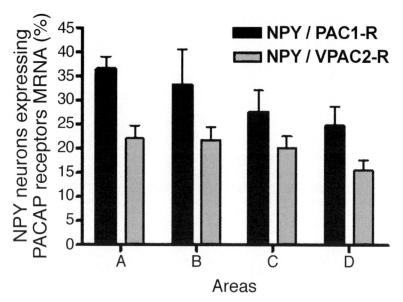

FIGURE 2. Quantitative analysis of the percentage of NPY-positive neurons expressing PAC1-R mRNA or VPAC2-R mRNA in each of the four antero-posterior subdivisions of the arcuate nucleus. For each subdivision, the data were obtained from five animals (six tissue sections each) and expressed as the mean value ± SEM. No significant differences were observed between the different areas.

administration of PACAP antagonizes NPY-induced feeding, suggesting that PACAP can modulate the action of NPY at a postsynaptic level.[6] This antagonistic effect probably occurs in the paraventricular nucleus which is innervated by arcuate NPY neurons and PACAP-nerve terminals and which express PAC1-R and VPAC2-R mRNAs.[3,7] The present observation that NPY neurons expressed PACAP mRNA receptors, indicates that the anorexigenic effect of PACAP may be regulated through modulation of NPY neuron activity. These results raise the question of the site of action of PACAP on NPY neurons which may occur either at the somato-dendritic or at the nerve terminal level.

ACKNOWLEDGMENTS

This work was supported by INSERM (U413), the European Institute for Peptide Research (IFRMP 23), the Regional Platform for Cell Imaging, and the Conseil Régional de Haute-Normandie.

REFERENCES

1. SCHWARTZ, M.W., S.C. WOODS, D. PORTE Jr. *et al.* 2000. Central nervous system control of food intake. Nature **404:** 661–671.

2. MIZUNO, Y., K. KONDO, Y. TERASHIMA, *et al.* 1998. Anorectic effect of pituitary adenylate cyclase activating polypeptide (PACAP) in rats: lack of evidence for involvement of hypothalamic neuropeptide gene expression. J. Neuroendocrinol. **10:** 611–616.
3. VAUDRY, D., B.J. GONZALEZ, M. BASILLE, *et al.* 2000. Pituitary adenylate cyclase-activating polypeptide and its receptors: from structure to functions. Pharmacol. Rev. **52:** 269–324.
4. JÉGOU, S., I. BOUTELET & H. VAUDRY. 2000. Melanocortin-3 receptor mRNA expression in pro-opiomelanocortin neurones of the rat arcuate nucleus. J. Neuroendocrinol. **12:** 501–505.
5. NAKATA, M., D. KOHNO, N. SHINTANI, *et al.* 2004. PACAP deficient mice display reduced carbohydrate intake and PACAP activates NPY-containing neurons in the rat hypothalamic arcuate nucleus. Neurosci. Lett. **370:** 252–256.
6. MORLEY, J.E., M. HOROWITZ, P.M. MORLEY, *et al.* 1992. Pituitary adenylate cyclase activating polypeptide (PACAP) reduces food intake in mice. Peptides **13:** 1133–1135.
7. BAI, F.L., M. YAMANO, Y. SHIOTANI, *et al.* 1985. An arcuato-paraventricular and -dorsomedial hypothalamic neuropeptide Y-containing system which lacks noradrenaline in the rat. Brain Res. **331:** 172–175.

Developmental Pattern of VIP Binding Sites in the Human Hypothalamus

MOHAMED NAJIMI, FATIMA RACHIDI, ABDELKRIM AFIF,
AND FATIHA CHIGR

*Laboratory of Functional and Pathologic Biology, Faculty of Sciences
and Techniques, 23000 Béni-Mellal, Morocco*

ABSTRACT: We have studied the developmental patterns of vasoactive intestinal polypeptide (VIP) binding sites in the human hypothalamus. VIP
recognition sites were widely distributed throughout the rostrocaudal extent of the hypothalamus. VIP binding was generally low in the fetal and
neonatal periods and a tendency in increasing densities was observed
during postnatal development. The age comparison of binding density
indicates variations in several structures. Thus, the densities were higher
in older infants in the preoptic area, lamina terminalis, and infundibular
(IN) nucleus. These differences suggest the implication of VIP receptors
in the development of this brain structure and the maintenance of its
various functions.

KEYWORDS: VIP binding sites; hypothalamus; development; human;
fetal and neonatal periods; postnatal period

INTRODUCTION

Vasoactive intestinal polypeptide (VIP) is a 28-amino acid polypeptide originally isolated from the porcine duodenum,[1] that is also present in the central
nervous system (CNS) of many vertebrates including mammals. Numerous
studies have shown that VIP exerts pleiotropic physiological functions: in addition to its role as a neurotransmitter/neuromodulator, VIP has also been found
to act as a hormone/neurohormone and, a neurotrophic or neuroprotective factor.[2] These various biological effects of VIP are mediated through interaction
with two receptor types, termed *VPAC1* and *VAPC2 receptors*, which recognize
with a similar high affinity both VIP and pituitary adenylate cyclase-activating
polypeptide (PACAP).[3] With regard to its neuroendocrine function and despite
the important role of VIP and its receptors in the hypothalamus during development, there is no information concerning the ontogeny of VIP binding sites

Address for correspondence: Mohamed Najimi, Laboratory of Functional and Pathologic Biology,
Faculty of Sciences and Techniques, 23000 Béni-Mellal, Morocco. Voice: +212-23485112; fax: +212-
23485201.
    e-mail: mnajimi1@fstbm.ac.ma

Ann. N.Y. Acad. Sci. 1070: 462–467 (2006). © 2006 New York Academy of Sciences.
doi: 10.1196/annals.1317.062

in the human hypothalamus. Thus, the aim of the present article was to provide baseline information about the quantitative distribution of VIP binding sites in human and their developmental pattern.

MATERIAL AND METHODS

Experiments were performed on brains from 2 fetuses (23- and 28-week-old postconceptional age), 4 neonates (2 h–3 days, postmortem delay [PMD]:10–34 h), 3 infants aged 1–2 months (PMD: 5–24 h), and 3 infants aged 4–12 months (PMD: 7–22 h). They presented no neurological or neuroendocrinological clinical signs and originate from the same brain collection used for our previous autoradiographic studies, for which the consent of parents has been given. For autoradiography, the coronal hypothalamic sections were incubated at 24°C for 3 h in 0.1 M HEPES/KOH buffer (pH 7.4) containing ^{125}I-VIP (60 pM) (2000 Ci/mmol) (Amersham, les Ulis France), NaCl (135 nM), KCl (7.4 mM), $MgCl_2$ (5 mM), EGTA (1 mM) (Sigma Chemical Co., St. Louis, MO), and bacitracin (1 mg/mL) to determine total binding. Nonspecific binding was determined on separate sections by the addition of 1 μM unlabeled VIP to the same incubation medium. The optical density of the autoradiographic images, obtained after development of the exposed films to labeled sections, was quantitated in terms of fmol/mg proteins using a computer-assisted image analysis system (RAG 200 Biocom, les Ulis, France). The difference of ^{125}I-VIP binding site densities between the different age periods analyzed was evaluated using the analysis of variance (ANOVA) with 99% significant level. When necessary *post hoc* tests were performed and the data were analyzed using the Scheffe' *F*-test of the Stat View 512^{+TM} computer program (Calabasas, CA).

RESULTS

The data obtained from quantitative autoradiographic measurements for ^{125}I-VIP binding in the different hypothalamic structures are reported in FIGURE 1. An illustration of total and nonspecific binding is shown in FIGURE 2 (anterior hypothalamic level). Autoradiographic labeling was seen throughout the rostrocaudal extent of the hypothalamic region. The highest densities were observed in the supraoptic (SON) and paraventricular nuclei (PVN), as well as in the organum vasculosum of the lamina terminalis (OVLT) (anterior hypothalamus) and in the infundibular (IN) and tuberal nuclei (TN) (mediobasal hypothalamus). Similar distribution patterns were found in all the cases studied. However, binding was generally low in the fetal and neonatal periods compared to infants and particularly to older infants (4–12 months). During the postnatal period, we observed quite similar results from newborn and infant hypothalamus, with no significant age relationship in the subregional distribution. In the first postnatal month (2 h–3 days and 1 month), as far as the

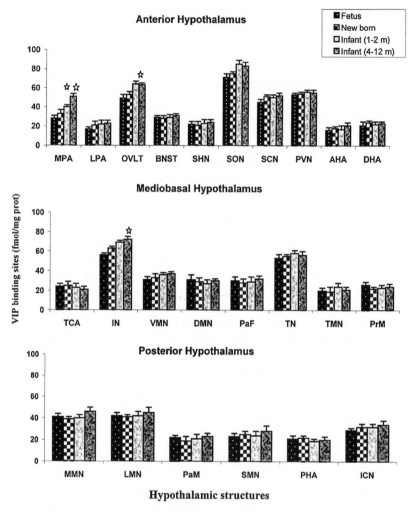

FIGURE 1. Age comparison of VIP receptor densities in the human hypothalamus. Abbreviations (not appearing in the text): AHA = anterior area; BNST = bed nucleus of stria terminalis; DHA = dorsal area; DMN = dorsomedial nucleus; ICN = intercalatus nucleus; LMN = lateral mammillary nucleus; LPA = lateral preoptic area; MMN = Medial mammillary nucleus; PaF = parafornical nucleus; PaM = paramammillary nucleus; PHA = posterior area; PrM = premammillary nucleus; SCN = suprachiasmatic nucleus; SHN = septohypothalamic nucleus; SMN = supramammillary nucleus; TCA = tuberocinereum area; TMN = tuberomammillary nucleus; VMN = ventromedial nucleus.

densities are concerned, there were no obvious quantitative changes in receptor binding between the cases examined. However, a tendency in increasing densities was observed during the following postnatal months (2–12 months). ^{125}I-VIP binding density was higher in several nuclei sampled in the older

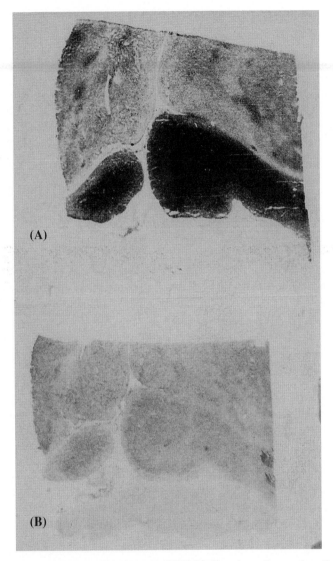

FIGURE 2. Autoradiographic images of VIP binding sites of coronal sections of the human infant hypothalamus as labeled by ^{125}I-VIP. (**A**) and (**B**) represent total and nonspecific binding, respectively, at the anterior hypothalamic level.

infants, although to different degrees. The most striking change we observed was the marked increase in ^{125}I-VIP binding in the medial preoptic area (MPA), the OVLT, and the IN nucleus. In the other hypothalamic structures, the density of binding sites was found to be generally constant between the different ages analyzed.

DISCUSSION

The results of the present study indicate that the level of VIP binding sites in some hypothalamic structures varies from fetal to the end of the first postnatal year. The low levels observed in these hypothalamic structures could be explained by the fact that their VIPergic system is not well developed at this postnatal period. We cannot exclude the possibility that these differences may reflect changes in presynaptic inputs into specific regions of these nuclei, although these events in the human hypothalamus during development are poorly documented. Taken together, the comparative results between fetuses, newborns, and infants, and with regard to the elevated densities reported in adult (data not shown), suggest that during the postnatal period, the full maturation of VIP binding sites is not achieved. Although the densities of ^{125}I-VIP binding sites show a tendency in increasing density from the fetal to the postnatal period, our sample size is too small to draw definitive conclusions about ontogenetically changes. Indeed, one can hypothesize that this developmental dynamics continue during the subsequent postnatal ages. The present study has shown a marked increase in VIP binding sites density during development, in several structures that contain neurons involved in neuroendocrine regulation, that is, the MPA, the OVLT, and the IN. Interestingly, these structures are known to contain high densities of luteinizing hormone-releasing hormone (LHRH) neurons in human newborn/infant.[4] This would indicate that the modulatory action of VIP on LHRH secretion[5] develops progressively in human suggesting that, during development, VIP may influence the secretion of pituitary hormones. The data reported in this study would be relevant with regard to a possible implication of VIP in some pathologies affecting neonate/infants such as sudden infant death syndrome.[6]

ACKNOWLEDGMENTS

This work was supported partially by the Grants N° 02–047 & N° 04–46 RG/BIO/AF/AC from the Third World Academy of Sciences (TWAS).

REFERENCES

1. SAID, S.I. & V. MUTT. 1970. Polypeptide with board biological activity: isolation from small intestine. Nature **225:** 863–864.
2. SHERWOOD, N.M., S.L. KRUECKL & J.E. McRORY. 2000. The origin and function of the pituitary adenylate cyclase-activating polypeptide (PACAP)/glucagons superfamily. Endocrine Rev. **21:** 619–670.
3. NAJIMI, M., F. CHIGR, D. JORDAN, *et al.* 1990. Anatomical distribution of LHRH-immunoreactive neurons in the human infant hypothalamus and extra hypothalamic regions. Brain Res. **516:** 280–291.

4. VAUDRY, D., B.J. GONZALEZ, M. BASILLE, *et al.* 2000. Pituitary adenylate cyclase-activating polypeptide and its receptors: from structure to function. Pharmacol. Rev. **52:** 269–324.
5. SMITH, M.J., L. JENNES & P.M. WISE. 2000. Localization of VIP$_2$ receptor protein on GnRH neurons in the female rat. Endocrinology **141:** 4317–4320.
6. STAINES, D.R. 2004. Is sudden infant death syndrome (SIDS) an autoimmune disorder of endogenous vasoactive neuropeptides? Med. Hypotheses **62:** 653–657.

Changes in PACAP Levels in the Central Nervous System after Ovariectomy and Castration

JOZSEF NEMETH,[a] ANDREA TAMAS,[b] RITA JOZSA,[b] JUDIT E. HORVATH,[b] BALAZS JAKAB,[a] ISTVAN LENGVARI,[b] AKIRA ARIMURA,[c] ANDREA LUBICS,[b] AND DORA REGLÓDI[b]

[a]Department of Pharmacology and Pharmacotherapy, Neuropharmacology Research Group of the Hungarian Academy of Sciences, Pécs University, 7624 Pécs, Hungary

[b]Department of Anatomy, Neurohumoral Regulations Research Group of the Hungarian Academy of Sciences, University of Pécs, 7624 Pécs, Hungary

[c]US–Japan Biomedical Research Laboratories, Tulane University, New Orleans, Louisana 70118, USA

ABSTRACT: The aim of the present article was to investigate the influence of gonadectomy on pituitary adenylate cyclase-activating polypeptide (PACAP) levels in different brain areas. In males, there seems to be an inverse relationship between gonadotropins and PACAP in the brain in the acute phase of castration: PACAP levels decreased in almost all brain areas examined within the first week after castration. In females, such pattern was observed in the hypothalamus, brain stem, and temporal cortex. In the pituitary, levels decreased only on the first day after ovariectomy, and later, as in the thalamus, increases were observed. Although the pattern of change showed gender differences, our results provide further evidence that levels of gonadotropins and possibly gonadotropin-releasing hormone influence PACAP levels and that PACAP is involved in the regulation of gonadal functions.

KEYWORDS: ovariectomy; castration; PACAP

INTRODUCTION

Pituitary adenylate cyclase-activating polypeptide (PACAP) was first isolated from ovine hypothalami based on its effect on pituitary adenylate cyclase. Several anatomical and functional evidences suggest that PACAP is involved

Address for correspondence: Dora Reglódi, M.D., Ph.D., Department of Anatomy, University of Pécs, 7624 Pécs, Szigeti u 12. Hungary. Voice: +36-72-536001 ext. 5398; fax: +36-72-536393.
e-mail: dora.reglodi@aok.pte.hu

Ann. N.Y. Acad. Sci. 1070: 468–473 (2006). © 2006 New York Academy of Sciences.
doi: 10.1196/annals.1317.063

in the regulation of hypothalamo-pituitary-gonadal axis.[1–5] PACAP has been shown to increase luteinizing hormone (LH) and follicle-stimulating hormone (FSH) secretion directly and by stimulating gonadotropin-releasing hormone secretion in the hypothalamus.[2–6] However, LH- and FSH-decreasing effects and no effect on gonadotropin secretion have also been demonstrated.[7–9] Similarly, both stimulating and inhibiting effects of PACAP on gonadal hormone secretion have been reported.[10–12] The reason for the existing contradictory results may be that several transduction pathways are involved in the effects of PACAP on pituitary gonadotropin release,[3] and these effects have also been shown to vary according to species, gender, endocrine status, circadian rhythm, and several other factors.[7,8,13–16] In spite of the vast amount of data on the effects of PACAP in gonadal functions, relatively little is known about endogenous PACAP levels after gonadectomy in mammals. The aim of the present article was to investigate the influence of gonadectomy on PACAP levels in different brain areas.

MATERIALS AND METHODS

Male and female Wistar rats of 2 months of age underwent gonadectomy. Rats were sacrificed at the following intervals after gonadectomy: 1 and 3 days, 1, 2, and 3 weeks, and 4 months. Different brain areas were removed and further processed for radioimmunoassay (RIA) analysis of the PACAP content. RIA procedure was performed as previously described.[17] Briefly, *Antiserum:* "88111-3" (working dilution 1:10,000).[18] *Tracer:* mono-^{125}I-labeled ovine PACAP 24–38 was prepared in our laboratory (5000 cpm/tube).[19] *Standard:* ovine PACAP-38 was used as an RIA standard ranging from 0 to 1000 fmol/mL. *Buffer:* the assay was prepared in 1 mL 0.05 mol/L (pH 7.4) phosphate buffer containing 0.1 mol/L sodium chloride, 0.25% (w/v) BSA and 0.05% (w/v) sodium azide. *Incubation time:* 48–72 h incubation at 4°C. *Separation solution:* charcoal/dextran/milk powder (10:1:0.5 g in 100-mL distilled water). Detection limit for PACAP-38 is 2 fmol/mL.

RESULTS AND DISCUSSION

Results are summarized in TABLE 1 (males) and TABLE 2 (females). In castrated males, PACAP levels significantly decreased in almost all brain areas examined within the first week after castration. In the hypothalamus, levels returned to normal by 3 weeks, while in the thalamus, temporal cortex, and pituitary by 2 weeks. Interestingly, PACAP concentrations showed a significantly higher level at 4 months after gonadectomy in the pituitary. In the brain stem, levels also decreased, reaching significance 1 day, 1 and 3 weeks after castration.

TABLE 1. PACAP concentration (fmol/mg tissue) after castration in male rats

Time	Hth	Thal	Pit	FrCtx	TemCtx	Bs	Cer
0 day	44.6 ± 3.9	36.8 ± 2.2	43.9 ± 1.4	26.9 ± 1.1	31.2 ± 1.7	25.6 ± 2.8	25.1 ± 1.4
1 day	20.9 ± 1.7***	18.5 ± 1.7***	13.4 ± 1.6***	29.5 ± 3.0	23.3 ± 2.6*	15.5 ± 2.1*	22.4 ± 2.9
3 days	20.9 ± 1.6***	25.1 ± 1.5***	19.9 ± 2.5***	27.0 ± 3.2	21.9 ± 1.3**	20.8 ± 1.4	28.3 ± 2.8
1 week	10.7 ± 4.0***	20.8 ± 1.0***	9.5 ± 2.5***	21.9 ± 2.6	23.5 ± 2.1*	14.8 ± 1.4**	22.0 ± 3.1
2 weeks	32.4 ± 2.4*	31.5 ± 3.1	41.5 ± 1.1	27.6 ± 3.2	29.6 ± 2.9	26.9 ± 2.6	33.2 ± 2.8
3 weeks	45.2 ± 4.4	32.5 ± 2.4	42.4 ± 3.1	23.5 ± 1.9	25.8 ± 1.7	18.0 ± 1.4*	23.0 ± 1.6
4 months	49.3 ± 1.8	32.0 ± 1.7	65.8 ± 2.6***	27.8 ± 2.2	32.3 ± 2.6	24.6 ± 2.6	30.3 ± 2.6

Hth = Hypothalamus; Thal = Thalamus; Pit = Pituitary; FrCtx = Frontal cortex; TemCtx = Temporal cortex; Bs = Brain stem; Cer = Cerebellum.
Values are given as mean ± SEM.
*$P < 0.05$, **$P < 0.01$, ***$P < 0.001$ vs. control.

TABLE 2. PACAP concentration (fmol/mg tissue) after ovariectomy in female rats

Time	Hth	Thal	Pit	FrCtx	TemCtx	Bs	Cer
0 day	38.7 ± 3.7	22.9 ± 1.8	27.8 ± 2.0	23.1 ± 1.9	25.8 ± 2.6	20.2 ± 1.7	18.2 ± 2.0
1 day	25.7 ± 3.0*	22.6 ± 1.4	12.1 ± 4.1**	21.8 ± 1.4	26.2 ± 3.5	19.9 ± 2.5	22.4 ± 2.2
3 days	39.4 ± 2.1	31.8 ± 2.6*	38.7 ± 2.6*	23.3 ± 3.9	21.9 ± 2.2	15.4 ± 0.8*	24.6 ± 3.2
1 week	40.8 ± 2.9	23.7 ± 1.4	24.9 ± 4.9	16.9 ± 2.4	15.8 ± 2.1*	12.8 ± 0.9**	18.8 ± 0.9
2 weeks	42.5 ± 1.9	33.6 ± 1.7***	43.7 ± 1.2***	22.1 ± 2.2	26.4 ± 3.1	19.2 ± 2.5	23.7 ± 2.7
3 weeks	46.0 ± 2.5	30.9 ± 1.2**	40.4 ± 2.8***	23.8 ± 2.2	28.2 ± 1.9	22.2 ± 1.8	26.4 ± 1.5
4 months	46.9 ± 2.9	25.5 ± 1.2	61.5 ± 6.5***	23.0 ± 2.6	27.5 ± 2.2	17.9 ± 0.6	18.3 ± 1.1

Hth = Hypothalamus; Thal = Thalamus; Pit = Pituitary; FrCtx,
Frontal cortex; TemCtx = Temporal cortex; Bs = Brain stem; Cer = Cerebellum.
Values are given as mean ± SEM.
$*P < 0.05, **P < 0.01, ***P < 0.001$ vs. control.

In females, a more complex pattern was observed in the changes of PACAP levels after gonadectomy (TABLE 2). A significant decrease in the hypothalamus was only observed 1 day after ovariectomy, then PACAP concentration returned to normal. In the thalamus, levels increased significantly 3 days, 2 and 3 weeks after gonadectomy. In the pituitary, there was an acute decrease at 1 day, but it was followed by a long-lasting significant increase. In the temporal cortex and brain stem, there was a significant decrease 1 week, and 3 days and 1 week after ovariectomy, respectively. In both males and females, there was no change in the frontal cortex and cerebellum at any examined time point.

In summary, our results show that the endogenous levels of PACAP sensitively react to gonadectomy in both sexes, not only in the hypothalamo-hypophyseal system, but also in other parts of the brain. In males, there seems to be an inverse relationship between gonadotropins and PACAP in the brain in the acute phase of castration. In females, such pattern was observed in the hypothalamus, brain stem, and temporal cortex. In the pituitary, the levels decreased only on the first day after ovariectomy, and later, as in the thalamus, increases were observed. This suggests that during the acute phase of gonadectomy, when LH/FSH levels are increased, PACAP levels decrease in most brain areas due to a yet unknown regulatory mechanism. Most studies show that PACAP increases gonadotropin secretion.[2–6] Based on these data we hypothesize a negative feedback mechanism that decreases endogenous PACAP levels when gonadotropins rise. Although the pattern of change showed gender differences, our results provide further evidence that the levels of gonadotropins and possibly GnRH influence PACAP levels and that PACAP is involved in the regulation of gonadal functions.

ACKNOWLEDGMENTS

This work was supported by the National Science Research Fund (OTKA T046589, F048908, T043467, ETT 82/2003, 03597/2003), the Hungarian Academy of Sciences, and Szechenyi Scholarship. The authors thank Brian K. Lucas and Dora Omboli for their help.

REFERENCES

1. KOVES, K. *et al*. 1998. PACAP colocalizes with luteinizing and follicle-stimulating hormone immunoreactivities in the anterior lobe of the pituitary gland. Peptides **19:** 1069–1072.
2. ARIMURA, A. 1998. Perspectives on pituitary adenylate cyclase activating polypeptide (PACAP) in the neuroendocrine, endocrine, and nervous systems. Jpn. J. Physiol. **48:** 301–331.
3. EVANS, J.J. 1999. Modulation of gonadotropin levels by peptides acting at the anterior pituitary gland. Endocr. Rev. **20:** 46–67.

4. VAUDRY, D. *et al.* 2000. Pituitary adenylate cyclase activating polypeptide and its receptors: from structure to functions. Pharmacol. Rev. **52:** 269–324.
5. ZHOU, C.J. *et al.* 2002. PACAP and its receptors exert pleiotropic effects in the nervous system by activating multiple signaling pathways. Curr. Prot. Pept. Sci. **3:** 423–439.
6. SZABO, E. *et al.* 2002. Cell immunoblot assay study demonstrating the release of PACAP from individual anterior pituitary cells of rats and the effect of PACAP on LH release. Regul. Pept. **109:** 75–81.
7. KOVES, K. *et al.* 2003. The role of PACAP in gonadotropic hormone secretion at hypothalamic and pituitary levels. J. Mol. Neurosci. **20:** 141–152.
8. SAWANGJAROEN, K. & J.D. CURLEWIS. 1994. Effects of pituitary adenylate cyclase activating polypeptide (PACAP) and vasoactive intestinal polypeptide (VIP) on prolactin, luteinizing hormone and growth hormone secretion in the ewe. J. Neuroendocrinol. **6:** 549–555.
9. TSUJII, T., K. ISHIKAZA & S.J. WINTERS. 1994. Effects of pituitary adenylate cyclase activating polypeptide on gonadotropin secretion and subunit messenger ribonucleic acids in perifused rat pituitary cells. Endocrinology **135:** 826–833.
10. CSABA, Z., V. CSERNUS & I. GERENDAI. 1997. Local effect of PACAP and VIP on testicular function in immature and adult rats. Peptides **18:** 1561–1567.
11. EL-GEHANI, F., M. TENA-SEMPERE & I. HUHTANIEMI. 2000. Evidence that pituitary adenylate cyclase activating polypeptide is a potent regulator of fetal rat testicular steroidogenesis. Biol. Reprod. **63:** 1482–1489.
12. ZHONG, Y. & B.G. KASSON. 1994. Pituitary adenylate cyclase activating polypeptide stimulates steroidogenesis and adenosine 3`,5`-monophosphate accumulation in cultured rat granulosa cells. Endocrinology **135:** 207–213.
13. CHIODERA, P. *et al.* 1996. Effects of intravenously infused pituitary adenylate cyclase activating polypeptide on adenohypophyseal hormone secretion in normal men. Neuroendocrinology **64:** 242–246.
14. KANTOR, O. *et al.* 2001. Study on the hypothalamic factors mediating the inhibitory effect of PACAP38 on ovulation. Peptides **22:** 2163–2168.
15. SAWISKY, G.R. & J.P. CHANG. 2005. Intracellular calcium involvement in pituitary adenylate cyclase activating polypeptide stimulation of growth hormone and gonadotrophin secretion in goldfish pituitary cells. J. Neuroendocrinol. **17:** 353–371.
16. SZABO, E. *et al.* 2004. Effect of PACAP on LH release studied by cell immunoblot assay depends on the gender, on the time of day and in female rats on the day of the estrus cycle. Regul. Pept. **123:** 139–145.
17. JAKAB, B. *et al.* 2004. Distribution of PACAP-38 in the central nervous system of various species determined by a novel radioimmunoassay. J. Biochem. Biophys. Methods **61:** 189–198.
18. ARIMURA, A. *et al.* 1991. Tissue distribution of PACAP as determined by RIA: highly abundant in the rat brain and testes. Endocrinology **129:** 2787–2789.
19. NÉMETH, J. *et al.* 2002. ^{125}I-labelling and purification of peptide hormones and bovine serum albumin. J. Radioanal. Nucl. Chem. **251:** 129–133.

Effects of Pituitary Adenylate Cyclase-Activating Polypeptide, Vasoactive Intestinal Polypeptide, and Somatostatin on the Release of Thyrotropin from the Bullfrog Pituitary

REIKO OKADA,[a] KAZUTOSHI YAMAMOTO,[a] YOICHI ITO,[a]
NICOLAS CHARTREL,[b] JEROME LEPRINCE,[b] ALAIN FOURNIER,[c]
HUBERT VAUDRY,[b] AND SAKAE KIKUYAMA[a]

[a]*Department of Biology, School of Education, Waseda University,
Tokyo 169-8050, Japan*

[b]*INSERM U413, Laboratory of Cellular and Molecular Neuroendocrinology,
European Institute for Peptide Research (IFRMP 23), University of Rouen,
76821 Mont-Saint-Aignan, France*

[c]*INRS–Institut Armand Frappier, Montreal, Canada*

ABSTRACT: The recent development of a specific radioimmunoassay for amphibian (bullfrog, *Rana catesbeiana*) thyrotropin (TSH) has made it possible to study the effects of various neuropeptides on the release of TSH from the pituitary *in vitro*. Up to now, corticotropin-releasing factor of bullfrog origin has been shown to have a potent TSH-releasing activity, whereas gonadotropin-releasing hormone and TSH-releasing hormone exhibit a moderate TSH-releasing effect on the adult, but not larval, pituitary. In the present study, the effects of pituitary adenylate cyclase-activating polypeptide (PACAP), vasoactive intestinal polypeptide (VIP), and somatostatin (SS) on the *in vitro* release of TSH from the bull-frog pituitary were investigated. Both frog (*R. ridibunda*) PACAP-38 and PACAP-27 caused a concentration-dependent stimulation of the release of TSH from dispersed pituitary cells during a 24-h culture. The PACAP-38- and PACAP-27-induced TSH release was suppressed by a simultaneous application of PACAP6–38. Application of high concentrations of PACAP6–38 alone caused a slight but significant stimulatory effect on the release of TSH. Frog VIP also stimulated TSH release from pituitary

Address for correspondence: Reiko Okada, Department of Biology, School of Education, Waseda University, Nishiwaseda 1-6-1, Shinjuku-ku, Tokyo 169-8050, Japan. Voice: +81-3-5286-1517; fax: +81-3-3207-9694.
e-mail: okada@aoni.waseda.jp

Ann. N.Y. Acad. Sci. 1070: 474–480 (2006). © 2006 New York Academy of Sciences.
doi: 10.1196/annals.1317.064

cells concentration-dependently. Frog SS1 (homologous to mammalian somatostatin-14) and SS2 (homologous to mammalian cortistatin) did not affect the basal release of TSH but caused a concentration-dependent suppression of the PACAP-38-induced release of TSH. These results suggest the involvement of multiple neuropeptides in the regulation of the release of TSH from the amphibian pituitary.

KEYWORDS: TSH; PACAP; VIP; somatostatin; amphibia

INTRODUCTION

Until recently, a radioimmunoassay (RIA) for amphibian thyrotropin (TSH) was not available because highly purified TSH preparations, necessary for antibody production, had never been obtained. We have recently developed an RIA for frog TSH using an antiserum against the C-terminal peptide of the bullfrog (*Rana catesbeiana*) TSH-β subunit deduced from the nucleotide sequence of the TSH-β complementary DNA.[1,2] Using this RIA, we have shown that corticotropin-releasing factor (CRF) stimulates the release of TSH from bullfrog pituitary cells more potently than thyrotropin-releasing hormone (TRH) and gonadotropin-releasing hormone (GnRH), and we have found that endogenous CRF accounts for approximately 50% of the total TSH-releasing activity in the frog hypothalamus.[3]

Pituitary adenylate cyclase-activating polypeptide (PACAP) and vasoactive intestinal polypeptide (VIP) belong to the secretin/glucagon/growth hormone (GH)-releasing hormone superfamily.[4] In amphibians as in mammals, PACAP exists in two molecular forms with 38 (PACAP-38) and 27 (PACAP-27) amino acids.[5–8] Three types of receptors for PACAP and VIP have been characterized so far: the PACAP-specific receptor, PAC1-R, that binds PACAP-38 and PACAP-27 with high affinity, and the PACAP/VIP mutual receptors, VPAC1-R and VPAC2-R, that bind PACAP-38, PACAP-27, and VIP with similar affinity.[9] PACAP and VIP have been shown to stimulate the release of various pituitary hormones in vertebrates.[10]

Somatostatin was first identified from the ovine hypothalamus on the basis of its GH release-inhibiting activity.[11] Two molecular forms of somatostatin (SS) have been characterized in the frog brain, that is, a variant called SS1 that possesses the same sequence as mammalian somatostatin-14, and a variant called SS2 ([Pro^2,Met^{13}]somatostatin-14) that is orthologous to cortistatin.[12–14] Evidence has recently shown that somatostatin inhibits the release of TSH from rat[15] and chicken[16] pituitaries.

In order to elucidate the possible involvement of hypothalamic peptides other than CRF, TRH, and GnRH in the release of TSH in amphibians, the effects of PACAP, VIP, and somatostatins on the release of TSH from bullfrog pituitary cells were investigated.

MATERIALS AND METHODS

Culture of dispersed distal lobe pituitary cells of adult bullfrogs was performed as described elsewhere.[1] Pituitary cells were suspended in 70% medium 199 (M199) containing 0.1% bovine serum albumin. An aliquot of the cell suspension was adjusted so that 1 mL contained 350,000 cells. The pituitary cells were incubated at 23°C in a humidified atmosphere of 95% air–5% CO_2 for 24 h. After preincubation, the culture medium was replaced with fresh medium containing frog (*R. ridibunda*) PACAP-38 (*f* PACAP-38),[7] PACAP-27, PACAP$_{6-38}$, frog VIP (*f* VIP),[17] and/or frog SS (SS1 or SS2).[11,12] Incubation was continued for 24 h. After incubation, the medium was collected from each well and centrifuged, and the supernatant was subjected to RIA for bullfrog TSH.[1] All experiments were approved by the Steering Committee for Animal Experimentation at Waseda University.

RESULTS

The release of TSH from dispersed pituitary cells was stimulated by both *f* PACAP-38 and PACAP-27 in a concentration (1–1000 nM)-dependent manner. During a 24-h incubation, a significant enhancement of the release of TSH (226–358% of the control value) was observed following application of 10, 100, and 1000 nM *f* PACAP-38. Likewise, 100 and 1000 nM PACAP-27 stimulated the release of TSH (420–568% of the control value) significantly. *f* VIP also exhibited a potent TSH-releasing activity: at concentrations of 1–100 nM, *f* VIP enhanced the release of TSH by 340–712% of the control value; at the concentration of 100 nM, the response of pituitary cells to *f* VIP was maximum (TABLE 1). On the other hand, the release of TSH enhanced

TABLE 1. Effects of frog PACAP-38 (*f* PACAP-38), PACAP-27, and frog VIP (*f* VIP) on the release of TSH from dispersed distal lobe pituitary cells of the bullfrog

	TSH released (% of control)				
	Concentration (nM)				
Treatment	0	1	10	100	1000
f PACAP-38	100 ± 8.6^a	$138 \pm 14.9^{a,b}$	$226 \pm 23.8^{b,c}$	$236 \pm 19.2^{c,d}$	358 ± 34.3^d
PACAP-27	100 ± 7.7^a	136 ± 12.1^a	178 ± 28.1^a	420 ± 35.7^b	568 ± 33.5^b
f VIP	100 ± 7.0^a	340 ± 46.5^b	503 ± 26.5^c	712 ± 34.5^d	705 ± 39.5^d

The amounts of TSH secreted into the medium by dispersed pituitary cells during a 24-h incubation are expressed as percentages of the mean control value. The values are expressed as means ± SEM of seven determinations. The mean values for TSH released in the absence of *f* PACAP-38, PACAP-27, and *f* VIP were 413 ± 36, 341 ± 26, and 356 ± 25 pg/10,000 cells, respectively. Within each treatment, values without a common letter in the superscripts are significantly different at the 5% level (ANOVA and Scheffe's test).

TABLE 2. Effects of PACAP$_{6-38}$, SS1, and SS2 on frog PACAP-38 (ƒPACAP-38)- or PACAP-27-induced TSH release from dispersed distal lobe pituitary cells of the bullfrog

	TSH released (% of control)			
	Concentration (nM) of PACAP$_{6-38}$, SS1, or SS2			
Treatment	0	10	100	1000
PACAP$_{6-38}$	100 ± 7.9^a	—	$131 \pm 14.4^{a,b}$	$151 \pm 9.0^{b,c}$
PACAP$_{6-38}$ + 100 nM ƒ PACAP-38	100 ± 8.7^a	—	$85 \pm 10.4^{a,b}$	46 ± 2.5^c
PACAP$_{6-38}$ + 100 nM PACAP-27	100 ± 7.5^a	—	103 ± 8.8^a	72 ± 5.2^b
SS1	100 ± 10.1^a	90 ± 13.3^a	69 ± 6.6^a	76 ± 2.4^a
SS1 + 1000 nM ƒ PACAP-38	100 ± 5.2^a	82 ± 3.3^b	58 ± 2.1^c	48 ± 3.9^c
SS2	100 ± 10.1^a	97 ± 4.6^a	85 ± 8.3^a	85 ± 8.3^a
SS2 + 1000 nM ƒ PACAP-38	100 ± 5.2^a	$79 \pm 3.5^{a,b}$	$69 \pm 6.7^{b,c}$	49 ± 3.2^c

The amounts of TSH secreted into the medium by dispersed pituitary cells during a 24-h incubation are expressed as percentages of the mean control value. The values are expressed as means ± SEM of seven determinations. The mean values for TSH released in the absence of PACAP$_{6-38}$, SS1, and SS2 (at concentration 0) from the top to the bottom were 424 ± 34, 1348 ± 101, 960 ± 83, 369 ± 37, 1175 ± 36, 369 ± 37, and 1175 ± 36 pg/10,000 cells, respectively. Within each treatment, values without a common letter in the superscripts are significantly different at the 5% level (ANOVA and Scheffe's test).

by 100 nM ƒPACAP-38 was reduced to 85–46% of the control value by simultaneous incubation with 100–1000 nM PACAP$_{6-38}$. Likewise, 100 nM PACAP-27-induced release of TSH was reduced to 72% of the control value by 1000 nM PACAP$_{6-38}$. On the other hand, PACAP$_{6-38}$ alone at concentrations of 100–1000 nM stimulated TSH release by 131–151% of the control value. SS1 and SS2 at concentrations of 10–1000 nM did not affect the basal release of TSH significantly, but SS1 at concentrations of 10–1000 nM and SS2 at concentrations of 100–1000 nM markedly attenuated the ƒPACAP-38-induced release of TSH to 82–48% and 69–49%, respectively (TABLE 2).

DISCUSSION

The occurrence of the mRNAs encoding PAC1-R,[18] VPAC1-R,[19] and VPAC2-R[20] has been reported in the frog pituitary gland, and we have previously shown that PACAP-38 causes an increase in intracellular calcium concentrations in frog TSH cells.[21] The present study has shown that ƒPACAP-38, PACAP-27, and ƒVIP stimulate the release of TSH from cultured bullfrog pituitary cells. In contrast, in mammals, PACAP stimulates the secretion of GH, adrenocorticotropic hormone (ACTH), and luteinizing hormone (LH), but has no effect on TSH release.[9] Likewise, VIP has no effect on the release of TSH from mammalian pituitaries although it stimulates the release of prolactin (PRL), LH, and GH.[22] The fact that fPACAP and fVIP could both enhance TSH secretion in frog strongly suggests the involvement of the VPAC1-R and/or the VPAC2-R in the response of thyrotrope cells to the peptide.

We have previously shown that the two frog somatostatin isoforms, SS1 and SS2, inhibit growth hormone-releasing hormone (GHRH)-induced GH secretion from bullfrog pituitary cells.[23] Here, we show that both SS1 and SS2 inhibit fPACAP-38-evoked TSH release. It thus appears that the two somatostatin isoforms may control the activity of both somatotrope and thyrotrope cells in amphibians. Consistent with this notion, it has been recently reported that the gene encoding the SS2 precursor is actively expressed in the pars intermedia of the frog pituitary,[24] suggesting that SS2, released by melanotrope cells, can diffuse toward the distal lobe to regulate the secretion of various adenohypophysial hormones.

In conclusion, the release of TSH in the amphibian pituitary is controlled by multiple hypothalamic neurohormones. The marked differences, between amphibians and mammals, in the hypothalamic neuropeptides regulating hormone secretion from each adenohypophysial cell type demonstrate the existence of phylogenetic variations in the expression pattern of the receptors for the hypothalamic factors in the different pituitary cells.

ACKNOWLEDGMENTS

This work was supported by Inamori Foundation (H17) JSPS (16370031, 15207007), Waseda University (2004B-839, 2003C-009), INSERM (U413), the European Institute for Peptide Research (IFRMP 23), and the Conseil Régional de Haute-Normandie.

REFERENCES

1. OKADA, R., K. YAMAMOTO, A. KODA, et al. 2004. Development of radioimmunoassay for bullfrog thyroid-stimulating hormone (TSH): effects of hypothalamic releasing hormones on the release of TSH from the pituitary in vitro. Gen. Comp. Endocrinol. 135: 42–50.
2. OKADA, R., Y. ITO, M. KANEKO, et al. 2005. Frog corticotropin-releasing hormone (CRH): isolation, molecular cloning, and biological activity. Ann. N. Y. Acad. Sci. 1040: 150–155.
3. ITO, Y., R. OKADA, H. MOCHIDA, et al. 2004. Molecular cloning of bullfrog corticotropin-releasing factor (CRF): effect of homologous CRF on the release of TSH from pituitary cells in vitro. Gen. Comp. Endocrinol. 138: 218–227.
4. ARIMURA, A. 1998. Perspectives on pituitary adenylate cyclase-activating polypeptide (PACAP) in the neuroendocrine, endocrine, and nervous systems. Jpn. J. Physiol. 48: 301–331.
5. MIYATA, A., A. ARIMURA, R.R. DAHL, et al. 1989. Isolation of a novel 38 residue-hypothalamic polypeptide which stimulates adenylate cyclase in pituitary cells. Biochem. Biophys. Res. Commun. 164: 567–574.

6. MIYATA, A., L. JIANG, R.D. DAHL, *et al.* 1990. Isolation of a neuropeptide corresponding to the N-terminal 27 residues of the pituitary adenylate cyclase-activating polypeptide with 38 residues (PACAP38). Biochem. Biophys. Res. Commun. **170:** 643–648.

7. CHARTREL, N., M.C. TONON, H. VAUDRY & J.M. CONLON. 1991. Primary structure of frog pituitary adenylate cyclase-activating polypeptide (PACAP) and effects of ovine PACAP on frog pituitary. Endocrinology **129:** 3367–3371.

8. YON, L., M. FEUILLOLEY, N. CHARTREL, *et al.* 1992. Immunohistochemical distribution and biological activity of pituitary adenylate cyclase-activating polypeptide (PACAP) in the central nervous system of the frog *Rana ridibunda*. J. Comp. Neurol. **324:** 485–499.

9. VAUDRY, D., B.J. GONZALEZ, M. BASILLE, *et al.* 2000. Pituitary adenylate cyclase-activating polypeptide and its receptors: from structure to function. Pharmacol. Rev. **52:** 269–324.

10. RAWLINGS, S.R. & M. HEZAREH. 1996. Pituitary adenylate cyclase-activating polypeptide (PACAP) and PACAP/VIP receptors: actions on the anterior pituitary gland. Endocr. Rev. **17:** 4–29.

11. BRAZEAU, P., W. VALE, R. BURGUS, *et al.* 1973. Hypothalamic polypeptide that inhibits the secretion of immunoreactive pituitary growth hormone. Science **179:** 77–79.

12. VAUDRY, H., N. CHARTREL & J.M. CONLON. 1992. Isolation of [Pro^2,Met^{13}]somatostatin-14 and somatostatin-14 from the frog brain reveals the existence of a somatostatin gene family in a tetrapod. Biochem. Biophys. Res. Commun. **188:** 477–482.

13. TOSTIVINT, H., I. LIHRMANN, C. BUCHARLES, *et al.* 1996. Occurrence of two somatostatin variants in the frog brain: characterization of the cDNAs, distribution of the mRNAs and receptor-binding affinities of the peptides. Proc. Natl. Acad. Sci. USA **93:** 12605–12610.

14. DE LECEA, L., J.R. CRIADO, O. PROSPERO-GARCIA, *et al.* 1996. A cortical neuropeptide with neuronal depressant and sleep-modulating properties. Nature **381:** 242–245.

15. REICHLIN, S. 2001. Neuroendocrine control of thyrotropin secretion. *In* The Thyroid L. Braverman, & R.D. Utiger, Eds.: 241–266. Lippincott Williams & Wilkins. Philadelphia.

16. GERIS, K.L., B. DE GROEF, E.R. KÜHN & V.M. DARRAS. 2003. *In vitro* study of corticotropin-releasing hormone-induced thyrotropin release: ontogeny and inhibition by somatostatin. Gen. Comp. Endocrinol. **132:** 272–277.

17. CHARTREL, N., Y. WANG, A. FOURNIER, *et al.* 1995. Frog vasoactive intestinal polypeptide and galanin: primary structures and effects on pituitary adenylate cyclase. Endocrinology **136:** 3079–3086.

18. ALEXANDRE, D., H. VAUDRY, V. TURQUIER, *et al.* 2002. Novel splice variants of type I pituitary adenylate cyclase-activating polypeptide receptor in frog exhibit altered adenylate cyclase stimulation and differential relative abundance. Endocrinology **143:** 2680–2692.

19. ALEXANDRE, D., Y. ANOUAR, S. JEGOU, *et al.* 1999. A cloned frog vasoactive intestinal polypeptide/pituitary adenylate cyclase-activating polypeptide receptor exhibits pharmacological and tissue distribution characteristics of both VPAC1 and VPAC2 receptors in mammals. Endocrinology **140:** 1285–1293.

20. Hoo, R.L.C., D. ALEXANDRE, S.M. CHAN, *et al*. 2001. Structural and functional identification of the pituitary adenylate cyclase-activating polypeptide receptor VPAC$_2$ receptor from the frog *Rana tigrina rugulosa*. J. Mol. Endocrinol. **27:** 229–238.

21. GRACIA-NAVARRO, F., M. LAMACZ, M.C. TONON, *et al*. 1992. Pituitary adenylate cyclase-activating polypeptide stimulates calcium mobilization in amphibian pituitary cells. Endocrinology **131:** 1069–1074.

22. SHERWOOD, N.M., L. SANDRA, L. KRUECKL, *et al*. 2000. The origin and function of the pituitary adenylate cyclase-activating polypeptide (PACAP)/glucagon superfamily. Endocr. Rev. **21:** 619–670.

23. JEANDEL, L., A. OKUNO, T. KOBAYASHI, *et al*. 1998. Effects of the two somatostatin variants somatostatin-14 and [Pro^2,Met^{13}]somatostatin-14 on receptor binding, adenylyl cyclase activity and growth hormone release from the frog pituitary. J. Neuroendocrinol. **10:** 187–192.

24. TOSTIVINT, H., D. VIEAU, N. CHARTREL, *et al*. 2002. Expression and processing of the [Pro^2,Met^{13}]somatostatin-14 precursor in the intermediate lobe of the frog pituitary. Endocrinology **143:** 3472–3481.

The Vasoactive Intestinal Peptide Receptor Turnover in Pulmonary Arteries Indicates an Important Role for VIP in the Rat Lung Circulation

VENTZISLAV PETKOV,[a] TEMENUSCHKA GENTSCHEVA,[b]
CHANTAL SCHAMBERGER,[b] INES HABERL,[b] ANDREAS ARTL,[c]
FRITZ ANDREAE,[c] AND WILHELM MOSGOELLER[b]

[a]Medical University Vienna, Pulmology, A-1090 Vienna, Austria

[b]Institute of Cancer Research, Borschkegasse 8a, A-1090 Vienna, Austria

[c]piCHEM Research & Development, A-8020 Graz, Austria

ABSTRACT: Vasoactive intestinal peptide (VIP) is a potent vasorelaxing peptide that plays a role in lung physiology and possibly in pulmonary hypertension. We investigated the turnover of the VIP receptors on rat pulmonary arteries *ex vivo*. There was evidence for a fast receptor turnover in pulmonary arteries, which underlines the important role of VIP for the regulation of pulmonary circulation and pulmonary pathology.

KEYWORDS: lung circulation; rat; VIP; therapy; receptor turnover

INTRODUCTION

The receptors for vasoactive intestinal peptide (VIP) have been remarkably well conserved during evolution, suggesting a high relevance of its biologic function. Because of a controversy over the VIP receptor distribution in human pulmonary artery smooth muscle cells (PASMCs),[1,2] we investigated the distribution of both known VIP receptor subtypes (VPAC1 and VPAC2) in human pulmonary arteries and cultured PASMCs. We found that cultured PASMCs expressed VIP receptors at the mRNA and protein levels (Petkov *et al.*, in preparation). Because the VIP receptors are internalized upon ligand binding and a rapid receptor turnover was described for human tumor cells[3] and rat cells,[4] we now used VIP-specific vasorelaxation to investigate the internalized VIP receptor turnover in cells of the rat pulmonary arterial wall.

Address for correspondence: Wilhelm Mosgoeller, Medical University Vienna, KIM-1, Institute of Cancer Research, Borschkegasse 8a, A-1090 Wien, Austria. Voice: +43 1 4277 65260; fax: +43 1 4277 65264.
e-mail: wilhelm.mosgoeller@meduniwien.ac.at

Ann. N.Y. Acad. Sci. 1070: 481–483 (2006). © 2006 New York Academy of Sciences.
doi: 10.1196/annals.1317.066

MATERIAL AND METHODS

Animals and Pulmonary Arteries

Six adult male Sprague-Dawley rats were used in agreement with the current legal and ethical national regulations. They were anesthetized, the heart and lungs were excised *en bloc*, and pulmonary arteries were excised in Krebs–Henseleit (KH) solution, in mmol: 118.40 NaCl; 5.01 KCl; 1.20 KH_2PPO_4; 2.5 $CaCl_2$; 1.2 $MgCl_2$; 10.1 Glucose; 25 $NaHCO_3$; 0°C).

The pulmonary arteries were divided into two pulmonary artery preparations (left or right pulmonary artery). The longitudinally opened arteries had a rectangular shape; they were cut as zigzag segments to obtain strips in which the circular musculature is linearized at large. In this way we produced 1.3- to 1.5- cm-long strips and mounted them in oxygenated organ baths at 37°C.

Changes in arterial tension were recorded on ink writers. The mounted 12 arterial strips were held extended with a weight load of 1 g, and were allowed to relax completely before the experiment started. Each artery ran through four to six experiment cycles and subsequent washes. One cycle consisted of a submaximal precontraction by phenylephrine (Phe, 0.1 μM), followed by submaximal relaxation by 100 nM VIP (Shahbazian *et al.*, in preparation) followed by acetylcholine (ACh, 1.0 μM), which was added to achieve further relaxation toward the base line. Each cycle with the subsequent washes took about 1 h or less. At the end of every experiment (four to six cycles) the viability of each artery was confirmed by contraction with Phe.

Data Analysis

On the records for each artery, the base line was drawn by a completely relaxed arterial strip. This and the deviation under precontraction provided the borders for a 100% contraction–relaxation range for each experiment. The actual distance of these borders was approximately 100 mm. Completed precontraction was set to 0% relaxation, return to the base line therefore was full (100%) relaxation. The relaxation by VIP or additional relaxation by ACh was expressed as part of the 100% range for each cycle.

RESULTS AND DISCUSSION

Following incubation with VIP, the addition of ACh during each cycle increased the average relaxation from 30% to about 60% (see TABLE 1). Because in our experiments the VIP receptors were still saturated with VIP when ACh was added, this result may be due to the fact that ACh and VIP have different targets. The values for ACh in the table indicate an additional vascular relaxation not triggered by VIP receptors. Most likely, the tendency of a relaxation

TABLE 1. *Ex vivo* pulmonary artery relaxation by VIP and ACh (reference substance)

	Cycle 1	Cycle 2	Cycle 3	Cycle 4	Mean of all cycles
VIP (100 nM)	36.6 ± 9.28	30.1 ± 12.78	26.9 ± 9.96	28.5 ± 6.00	30.5 ± 9.97
ACh(1 μM)	67.8 ± 10.71	60.8 ± 13.00	55.5 ± 10.74	51.8 ± 10.84	64.3 ± 18.19

The values are percent relaxation (mean ± standard deviations). The experimental cyles 1 to 4 reveal similar results with repeated submaximal VIP doses. Since one cycle and washes took less than 1 h, the values indicate a rapid restoration of the internalized VIP receptor to enable full VIP function after the prior cycle. (N = 12 arteries, each going through cycle 1 to 4, total number of cycles = 48.)

decrease from the cycle 1 to cycle 4 indicates some general function loss during the experimental time.

In previous experiments, we observed that the dose of VIP used resulted in submaximal relaxation, indicating almost complete VIP receptor saturation. The first cycle with VIP produced a vasorelaxation similar to the subsequent cycles (see TABLE 1). Since each cycle and triple wash lasted for about 1 h, we conclude that within this time the VIP-signaling function was fully restored to allow for full VIP function in subsequent cycles. Because the receptor VIP-complex is internalized by the cell,[4] we conclude that the saturated VIP receptors in the arterial wall must be fully replaced by the living cell in less than 1 h. In rat arteries, therefore, a rapid turnover of internalized receptors indicates an important role for VIP in the regulation of the pulmonary circulation.

REFERENCES

1. BUSTO, R. *et al.* 2000. Immunohistochemical and immunochemical evidence for expression of human lung PACAP/VIP receptors. Ann. N. Y. Acad. Sci. **921:** 308–311.
2. GRONEBERG, D.A. *et al.* 2001. Expression and distribution of vasoactive intestinal polypeptide receptor VPAC(2) mRNA in human airways. Lab. Invest. **81:** 749–755.
3. BOISSARD, C. *et al.* 1986. Vasoactive intestinal peptide receptor regulation and reversible desensitization in human colonic carcinoma cells in culture. Cancer Res. **46:** 4406–4413.
4. IZZO, R.S. *et al.* 1991. Binding and internalization of VIP in rat intestinal epithelial cells. Regul. Pept. **33:** 21–30.

A Splice Variant to PACAP Receptor That Is Involved in Spermatogenesis Is Expressed in Astrocytes

INBAR PILZER AND ILLANA GOZES

Department of Molecular Genetics and Biochemistry, Sackler Faculty of Medicine, Tel Aviv University, Tel Aviv 69978, Israel

ABSTRACT: The pituitary adenylate cyclase-activating polypeptide (PACAP) receptor, PAC1, recognizes PACAP with a higher affinity than it recognizes vasoactive intestinal peptide (VIP) and belongs to the subfamily G protein–coupled receptors. So far, more than 10 different splice variants of PAC1 have been cloned from rat tissue. Interestingly, the various PAC1 splice variants exhibit different signaling pathways. These splice variants are suggested to play a functional role mostly in the brain as well as in the testes. The present article introduces PAC1(3a) that was originally discovered in testes as another potential regulator in rat astrocytes.

KEYWORDS: PAC1; splice variants; astrocytes; VIP; PACAP

INTRODUCTION

The PAC1 receptor cDNA sequence was first cloned from a rat pancreatic acinar carcinoma cell line and later from human tissues.[1] Two types of pituitary adenylate cyclase-activating polypeptide (PACAP)-binding sites have been cloned: the PACAP type I receptor, PAC1, that binds PACAP with a higher affinity than vasoactive intestinal peptide (VIP), and the PACAP type II receptors, VPAC1 and VPAC2, which have an equal affinity for VIP and PACAP.[2] These receptors belong to a subfamily of the seven transmembrane-spanning G protein–coupled receptors.[3] Six splice variants in the third intracellular loop have been identified in rats.[4] The splice variants diverge by the absence or presence of either one or two cassettes of 84 bp (base pairs) (hip or hop1 variant) or 81 bp (hop2 variant).[5] The shortest variant PAC1 produces the same effect on cyclic adenosine 3′,5′-phosphate (cAMP) while binding to

Address for correspondence: Illana Gozes, Ph.D., Department of Molecular Genetics and Biochemistry, The Lily and Avraham Gildor Chair for the Investigation of Growth Factors, The Dr. Diana and Zelman Elton (Elbaum) Laboratory for Molecular Neuroendocrinology, Sackler Faculty of Medicine, Tel Aviv University, Tel Aviv 69978, Israel. Voice: 972-3-640-7240; fax: 972-3-640-8541.
e-mail: igozes@post.tau.ac.il

Ann. N.Y. Acad. Sci. 1070: 484–490 (2006). © 2006 New York Academy of Sciences.
doi: 10.1196/annals.1317.067

PACAP-38 or -27, whereas PACAP-38 shows higher potency than PACAP-27 in stimulating phospholipase C (PLC). The hop cassette shows similar effects. In contrast, the presence of the hip cassette abolishes PLC stimulation and impairs adenylyl cyclase stimulation. Using antisense oligodeoxynucleotides and receptor-binding assays on mixed glial-neuronal cells, the hop2 splice variant of PAC1 was suggested to be involved in neuronal survival.[6]

In human tissues, four splice variants coding the same domain as in rat PAC1 were identified. These variants were named null, SV-1, SV-2, and SV-3.[1] SV-2 is identical to the rat hop splice variant, while SV-1 displays similarities to the rat hip variant except for a difference in two amino acids. SV-3 contains both exons SV-1 and SV-2, thus displaying a great homology to the rat hip hop variant. In contrast, hPAC1 null does not contain any of the cassettes. All of these different splice variants affect both adenylate cyclase and PLC with similar potencies for PACAP-38 and -27. However, they present differences in the efficacy for PACAP-stimulated PLC with the SV-2 variant being more efficient than other splice variants.

Another PAC1 variant contains sequence substitutions that are located in the second and fourth transmembrane domains, hence is designated PACAP receptor (PACAPR) transmembrane domain (TM4). This variant has been cloned in rat cerebellum.[7] Interestingly, PACAPR TM4 does not induce adenylyl cyclase or PLC stimulation at all, but induces calcium influx through L-type calcium channels.

Alternative splicing of PAC1 was also found in the frog *Rana ridibunda*. Here three splice variants were characterized[8]: two of them, PAC1-R25 and PAC1-R41, contain additional amino acid cassettes in the third cytoplasmatic loop. The first one has an insertion of 25 amino acids that matches the hop cassette of the mammalian receptor, whereas the second one has a cassette with no homology to any other variant. The third variant displays a different cytoplasmic C-terminal domain, named PAC1-Rmc. These variants are abundant mostly in frog brain and spinal cord. Accumulation of cAMP activated by PACAP was similar in PAC1 and two of the frog splice variants, PAC1-R25 and PAC1-R41. However, the third receptor, PAC1-Rmc prevented cAMP formation.

Three other PAC1 splice variants that differ from the other variants in the region encoding the N-terminal extracellular domain were also cloned. The first one has a deletion of 21 amino acids (residues 89–109).[9] This variant was cloned from mouse brain and noncerebellum tissues derived from the human brain. The 21-amino acid domain determines receptor selectivity for PACAP-27 and -38 and affects the relative potencies of the two isoforms in PLC stimulation. The second variant, identified in the human cerebellum, was designated PAC1 very short (PAC1vs), because it has a deletion of 57 amino acid (residues 53–109).[10] Residues 53–88 were found to be important for high-affinity ligand binding and for VIP and PACAP-38 potency to stimulate cAMP accumulation (the deletion was associated with a decrease in cAMP accumulation). The third variant is characterized by an addition of 72-bp exon encoding 24 amino acids between coding exons 3 and 4, and is designated PAC1(3a).[11] PAC1 is alternatively spliced during the spermatogenesis cycle to this isoform. The 24-amino acid insertion results in a higher affinity for PACAP-38, however PAC1(3a) coupling to both cAMP and inositol phosphates were decreased.

Here, we demonstrate that the PAC1(3a) splice variant is also expressed in newborn rat cortex and in total brain, however, under the conditions used, this variant was not detected in RNA derived from the cerebral cortex of 4-month-old mouse. These new findings suggest age dependence and potential species specificity for the expression of PAC1(3a) and may contribute to the understanding of the PACAP signaling pathways in the brain.

MATERIALS AND METHODS

Cell Culture

Newborn rat cortical astrocytes were prepared as previously described.[12] Tissue was incubated for 20 min at 37°C in Hanks' balanced salt solution containing 15 mM HEPES, pH 7.3 (Biological Industries, Beit Haemek, Israel), and trypsin solution B without phenol red (Biological Industries). Dissociated cerebral cortical cells were added to the culture dish with 5% horse serum in Dulbecco's modified Eagle's medium (Biological Industries). Cells were plated in a ratio of one cortex per two 75-cm^2 cell culture flasks (polystyrene; Corning Glass, Corning, NY). The medium was changed 1 day after plating. Cells were split after 10 days of incubation and plated in new flasks for 6 additional days.

Oligonucleotides

Oligonucleotides specific for the 5'- and 3'-untranslated regions of PAC1 were:

354 Sense – 5 AGA GAA TTC CCG GAG ACC AGC AGC GAG TGG ACA GT 3'. 171 Antisense – 5' AGA CTC TAG AAA TAT CAG CCT ATC CCT ATC TCT CTC TTC CTT 3'. Oligonucleotides specific for the start and end sites of translation: 700 Sense – 5' ACC ATG GCC AGA GTC CT 3'. 699 Antisense – 5' TCA GGT GGC CAA GTT GT 3'. Oligonucleotides specific to PAC1(3a): Sense – 5' ATC TTC AAC CCG GAC CAA 3'. Antisense – 5' TGG TTT CTG TCA TCC AGA 3'.

Reverse Transcriptase Polymerase Chain Reaction

Total RNA was extracted from rat astrocytes using Tripure reagent (Roche Diagnostics, Indianapolis, IN) in accordance with the manufacturer's instructions. A total of 3 μg of RNA in H_2O were used for the reverse transcription (RT) reaction and were heated to 85°C for 2 min, then cooled down on ice. RT buffer, 0.5 μg oligo(dT), 2 units RNasin (Promega, Madison, WI), 10 units superscript II reverse transcriptase, 10 mM dithiothreitol, and 0.5 mM dNTPs mixture (Life Technologies, California, CA) were added. The reaction was performed in a final volume of 20 μL at 42°C for 90 min, and then was incubated at 85°C for 2 min followed by cooling. A total of 3 μL from the cDNA template were then taken for the polymerase chain reaction (PCR) with 1X PCR buffer (200 mM Tris-HCl, 500 mM KCl), 2.5 mM $MgCl_2$, 0.2 mM dNTPs mixture, 2.5 unit Taq DNA Polymerase (Life Technologies), and 100 pmol of primers (Sigma, Rehovot, Israel), in a total volume of 50 μL. For nested PCR, a second reaction with nested primers (start and end sites of translation) was carried out using 3 μL of the first reaction as template. The PCR was carried out in two rounds with primers 354 and 171 followed with primers 700 and 699. The first round of PCR used 30-s denaturation at 94°C, 30-s annealing at 55°C, and 2-min extension at 72°C for 35 cycles. The second round was identical except for an annealing at 65°C. For the 109-bp fragment that is specific to the PAC1(3a) sequence, the PCR method used a 30-s annealing time at 56°C and a 30-s extension time at 72°C for a total of 35 cycles.

Cloning of the RT-PCR Product

Products of the RT-PCR reactions were cloned into PGEM-T-Easy vector (Promega) and transformed into Xl-1-blue bacteria.

RESULTS AND DISCUSSION

PAC1 receptor cDNA was obtained from rat brain cortical astrocytes following a nested PCR protocol. The PCR product was then cloned. Surprisingly, in addition to the short PAC1 and the hop1 splice variant, one colony contained PAC1 which had an insertion of 72 bp at the N-terminal domain. Sequence analysis indicated that this PAC1 variant was identical to the PAC1(3a) splice variant (FIG. 1). Thus the formerly testes-related PAC1 splice variant was found here to be expressed in astrocytes derived from rat cerebral cortex. In order to verify this finding, either cortex or brain tissue that contained neither cortex nor cerebellum were dissected from newborn rats. RT-PCR using specific primers for the 72-bp spliced fragment resulted in a 109-bp product in both tissues (FIG. 2 A).

It was of interest to determine whether this splice variant is also expressed in other species such as mouse and whether the expression is age related. RT-PCR was performed using total RNA taken from cortex of 4-month-old mouse. The RT-PCR did not yield any product meaning that in contrast to

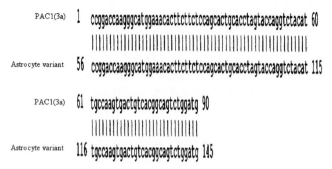

FIGURE 1. The sequence of exon 3a. Nucleotide sequence of exon 3a as obtained from sequence analysis of the astrocyte cDNA and comparing to the known sequence from testes.[11]

newborn rat cerebral cortex, 4-month-old mouse cortex does not express the PAC1(3a) splice variant (FIG. 2B).

Other species which were already found to express PAC1 splice variants, such as human and frog have not been checked yet for the expression of PAC1(3a).

It was previously reported that PAC1(3a) is expressed in testes and seminiferous tubules.[11] As indicated above, the 24-amino acid insertion was associated with spermatogenesis and resulted in a higher affinity for PACAP-38; however, PAC1(3a) coupling to both cAMP and inositol phosphates was decreased.[11] The current findings demonstrate that this variant is also expressed in rat

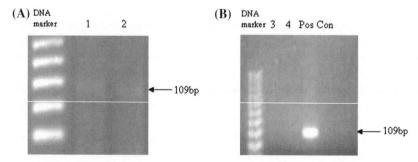

FIGURE 2. PAC1(3a) distribution in rat and mouse brain. **(A)** RT-PCR of PAC1(3a) from newborn rat shows the expression of this splice variant in brain tissue without cortex and cerebellum (1) and in the cerebral cortex (2). The PCR product is indicated by an arrow at 109 bp according to known DNA size markers (left lane). **(B)** PAC1(3a) is not expressed in the adult mouse cortex (1), (2) is a control to the RT step, and (3) is a positive control with PAC1(3a) from the rat cortex. The PCR product is indicated by an arrow at 109 bp.

cerebral cortex, and suggest a potential age-dependent and species-specific differences in the expression of this receptor.

Ligand-binding affinity was altered due to insertion or deletion of fragments in the PAC1 N-terminal domain.[9–11] Therefore, this region determines, in part, the affinity of PACAP to its receptors. The existence of the exon 3a-encoded amino acids in the receptor increases the affinity of the receptor for PACAP-38,[11] indicating that this region may be involved in important pathways in brain. It is thus interesting to investigate PAC1(3a) functions in the developing brain.

PACAP is present in the central and the peripheral nervous systems and has different functions. The peptide affects neurite outgrowth,[12] regulates neuropeptide and neurotransmitter production and secretion,[13,14] induces the transition from proliferation to differentiation of neurons,[15] and regulates neuronal survival.[16] The existence of many PACAP receptor splice variants with different functions may explain the ability of one peptide to induce multiple activities.[17,18]

ACKNOWLEDGMENTS

We would like to thank Dr. Eliezer Giladi and David Dangoor for their excellent advice. This study was supported, in part, by the US–Israel Binational Science Foundation, the Israel Science Foundation, and Allon Therapeutics, Inc.

REFERENCES

1. PISEGNA, J.R. & S.A. WANK. 1996. Cloning and characterization of the signal transduction of four splice variants of the human pituitary adenylate cyclase activating polypeptide receptor. Evidence for dual coupling to adenylate cyclase and phospholipase C. J. Biol. Chem. **271:**
2. SHIVERS, B.D. *et al.* 1991. Two high affinity binding sites for pituitary adenylate cyclase-activating polypeptide have different tissue distributions. Endocrinology **128:** 3055–3065.
3. HARMAR, A.J. *et al.* 1998. International Union of Pharmacology. XVIII. Nomenclature of receptors for vasoactive intestinal peptide and pituitary adenylate cyclase-activating polypeptide. Pharmacol. Rev. **50:** 265–270.
4. JOURNOT, L. *et al.* 1995. Differential signal transduction by six splice variants of the pituitary adenylate cyclase-activating peptide (PACAP) receptor. Biochem. Soc. Trans. **23:** 133–137.
5. SPENGLER, D. *et al.* 1993. Differential signal transduction by five splice variants of the PACAP receptor. Nature **365:** 170–175.
6. ASHUR-FABIAN, O. *et al.* 1997. Identification of VIP/PACAP receptors on rat astrocytes using antisense oligodeoxynucleotides. J. Mol. Neurosci. **9:** 211–222.
7. CHATTERJEE, T.K., R.V. SHARMA & R.A. FISHER. 1996. Molecular cloning of a novel variant of the pituitary adenylate cyclase-activating polypeptide (PACAP)

receptor that stimulates calcium influx by activation of L-type calcium channels. J. Biol. Chem. **271**: 32226–32232.

8. ALEXANDRE, D. *et al.* 2002. Novel splice variants of type I pituitary adenylate cyclase-activating polypeptide receptor in frog exhibit altered adenylate cyclase stimulation and differential relative abundance. Endocrinology **143**: 2680–2692.

9. PANTALONI, C. *et al.* 1996. Alternative splicing in the N-terminal extracellular domain of the pituitary adenylate cyclase-activating polypeptide (PACAP) receptor modulates receptor selectivity and relative potencies of PACAP-27 and PACAP-38 in phospholipase C activation. J. Biol. Chem. **271**: 22146–22151.

10. DAUTZENBERG, F.M. *et al.* 1999. N-terminal splice variants of the type I PACAP receptor: isolation, characterization and ligand binding/selectivity determinants. J. Neuroendocrinol. **11**: 941–949.

11. DANIEL, P.B. *et al.* 2001. Novel alternatively spliced exon in the extracellular ligand-binding domain of the pituitary adenylate cyclase-activating polypeptide (PACAP) type 1 receptor (PAC1R) selectively increases ligand affinity and alters signal transduction coupling during spermatogenesis. J. Biol. Chem. **276**: 12938–12944.

12. DEUTSCH, P.J., V.C. SCHADLOW & N. BARZILAI. 1993. 38-Amino acid form of pituitary adenylate cyclase activating peptide induces process outgrowth in human neuroblastoma cells. J. Neurosci. Res. **35**: 312–320.

13. MAY, V. & K.M. BRAAS. 1995. Pituitary adenylate cyclase-activating polypeptide (PACAP) regulation of sympathetic neuron neuropeptide Y and catecholamine expression. J. Neurochem. **65**: 978–987.

14. ZUSEV, M. & I. GOZES. 2004. Differential regulation of activity-dependent neuroprotective protein in rat astrocytes by VIP and PACAP. Regul. Pept. **123**: 33–41.

15. LU, N. & E. DICICCO-BLOOM. 1997. Pituitary adenylate cyclase-activating polypeptide is an autocrine inhibitor of mitosis in cultured cortical precursor cells. Proc. Natl. Acad. Sci. USA **94**: 3357–3362.

16. ARIMURA, A. *et al.* 1994. PACAP functions as a neurotrophic factor. Ann. N. Y. Acad. Sci. **739**: 228–243.

17. REGLODI, D. *et al.* 2002. Effects of pretreatment with PACAP on the infarct size and functional outcome in rat permanent focal cerebral ischemia. Peptides **23**: 2227–2234.

18. CHEN, W.H. & S.F. TZENG. 2005. Pituitary adenylate cyclase-activating polypeptide prevents cell death in the spinal cord with traumatic injury. Neurosci. Lett. **384**: 117–121.

Glucose Activation of the Glucagon Receptor Gene

Functional Dissimilarity with Several Other Glucose Response Elements

LAURENCE PORTOIS, MYRNA VIRREIRA, MICHÈLE TASTENOY, AND MICHAL SVOBODA

Department of Biochemistry and Nutrition, Medical School, Université Libre de Bruxelles, 1070 Brussels, Belgium

ABSTRACT: The glucagon receptor (GLR) expression is positively regulated by glucose. This regulation is allowed by the presence, in the promotor of the rat GLR gene, of a sequence feature similar to the two E-boxes motifs constituting the carbohydrate response elements (ChoRE) described for several glycolytic and lipogenic enzyme genes. Using reporter gene assays, we demonstrated here that, despite structural homologies with these ChoREs, the GLR gene glucose response element presents various functional dissimilarities. Testing glucose analogs, we demonstrated that, as for other genes, the glucose must be first phosphorylated. However, at variance with others homologue genes, our data showed the implication of the nonoxidative branch of the pentose phosphate pathway in the transmission of the glucose signal and lack of inhibition by adenosine monophosphate (AMP)-kinase. Furthermore, the activity of our reporter gene was strongly stimulated by butyrate, propionate, and acetate. This observation contrasts with fatty-acid-induced inhibition of the glucose activation, observed for all other genes containing homolog ChoREs. We also showed that glucose and butyrate influence the reporter gene expression via different features.

KEYWORDS: gene promotor; glucagon receptor gene; glucose activation; fatty acids activation; pentose phosphate

INTRODUCTION

Glucagon, together with insulin, are key hormones allowing glucose homeostasis. The glucagon receptor (GLR) is a member of the second (secretin-like)

Address for correspondence: M. Svoboda, Department of Biochemistry and Nutrition, Medical School, Université Libre de Bruxelles, CP 611, Route de Lennik 808, B-1070 Brussels, Belgium. Voice: 32-2-555-62-25; fax: 32-2-555-62-30.
e-mail: msvobod@ulb.ac.be

Ann. N.Y. Acad. Sci. 1070: 491–499 (2006). © 2006 New York Academy of Sciences.
doi: 10.1196/annals.1317.068

group of the seven transmembrane domains receptor family. The GLR gene is composed of 13 exons in the coding domain, and of 2 alternatively spliced exons in the 5′ noncoding domain.[1] The GLR mRNA level is positively regulated by extracellular glucose.[2] This regulation is allowed by the presence, in the promotor the rat GLR gene, of a 19 nucleotide (nt) quasipalindromic sequence containing two E-boxes.[1,3] This region, which is the core of glucose activation, is similar to the two E-boxes motifs constituting the carbohydrate response elements (ChoRE) reported for several glycolytic and lipogenic genes, such as liver pyruvate kinase (L-PK),[4] fatty acid synthase (FAS),[5] Spot 14 (S14),[6] and acetyl CoA carboxylase (ACC)[7] genes. However, some notable differences exist: for example, a mandatory separation of 5 nt between the two E-boxes was observed in the ChoRE of L-PK, S14, and FAS genes,[4] whereas only 3 nt separate the two E-boxes in the promotor of the GLR gene.[1]

The mechanism of activation of the promotors of the glycolytic and lipogenic enzyme genes has been largely studied. First, the glucose effect is independent of insulin.[8,9] Second, upstream stimulatory factor (USF) transcription factors, which strongly bind to E-boxes, are not implicated in this glucose regulation.[10,11] Third, carbohydrate response element binding protein (ChoREBP), another nuclear factor belonging to the basic helix-loop-helix/leucin zipper (bHLH/LZ) family, has been proposed as the key nuclear regulator. This factor is phophorylated and cytosolic in the resting state, and can be dephosphorylated by a protein phosphatase 2B. Dephosphorylated ChoREBP enters into the nucleus and then binds to the ChoRE.[12] The protein phosphatase 2B is activated by xylulose 5′-phosphate, a glucose metabolite generated through the pentose phosphate pathway.[13] More recently, the need for the dimerization of ChoREBP with another nuclear factor coined Max-like protein (Mlx) was described.[14,15]

It was reported that, in the liver, the inhibition of glucose metabolism by fatty acids ("glucose sparing effect") is mediated by a decrease in xylulose-5 phosphate resulting in a lower protein phosphates 2A activity, an increased amount of the phosphorylated form of the bidirectional enzyme Fru-6-P,2-kinase:Fru-2,6-Pase, and finally a lowered concentration of the fructose-2,6 bisphosphate (Fru-2,6P_2).[16] In another hand, fatty acids activate adenosine monophosphate (AMP)-dependent kinase which phosphorylates ChoREBP and thus inhibit this transcription factor activity.[17,18]

In this article, we compare the glucose activation reporter gene drove by GLR promotor with the glucose activation of the homologous ChoRE located within the glycolytic and lipogenic enzyme genes.

MATERIALS AND METHODS

Insulin-secreting INS-1 cell line was kindly provided by Professor Wollheim (University Medical Center, Geneva, Switzerland). Transfections were performed as described previously.[1] Transfections were usually performed on cells 4–5 days after seeding in six-well plates, in order to achieve roughly 60%

cell confluence. The day before transfection, the usual cell culture medium was replaced by RPMI 1640 with a low glucose concentration (3.5 mM instead of 11 mM). The cationic liposomes (Dosper (1,3-di-oleyloxy-2-(6-carboxy-spermyl)-propylamid) [Roche, Mannheim, Germany] or Lipofectamin[TM] 2000 Transfection Reagent [Invitrogen, Carlsbad, CA]) was used according to the manufacturer's instructions. After 5 h, the transfection medium was replaced by RPMI 1640, with variable glucose concentrations, and the various tested molecules (butyric acid, drugs) were added. The cells were then incubated for 48 h. Finally, the chloramphenicol acetyltransferase (CAT) activity of trans-fected cells was assayed by acylation of ^{14}C-chloramphenicol using n-butyryl-CoA as described previously.[1] Products of acylation were separated by thin layer chromatography and autoradiographied. Total proteins were assayed in cell extracts in order to check that cell amount was similar in various wells.

RESULTS

Effect of Glucose Metabolism Modulators

We tested the effect of different molecules, described as interfering with the glucose metabolism, on the activation of the reporter gene in INS-1 cells incubated in either low or high glucose concentration. Three glucose analogs in-hibited the glucose-induced activation: [α]-methylglucose (which inhibits the glucose entry), mannoheptulose (a hexokinase inhibitor), and 2-deoxyglucose (a nonphosphorylable glucose analog). The results demonstrated that, for the induction of the promotor activity, the extracellular glucose must enter into the cell and must be phosphorylated. In contrast, glycerin, used at different concentrations ranging from 2 to 20 mM, and xylitol, from 0.5 mM to 20 mM, were without noticeable effect on the glucose activation. Dihydroxyacetone 5 mM and peroxyvanadate 30 μM were toxic to the INS-1 cells and therefore not usable in our system. Were also tested the effect of two potential inhibitors of the xylulose 5'-phosphate production in the cell: 6-aminonicotinamide, an inhibitor of the oxidative branch of the pentose phosphate pathway, and oxythi-amine, an inhibitor of the transketolase implicated in the nonoxidative branch of this pathway. At 200 μM, 6-aminonicotinamide was without effect on the glu-cose stimulation, whereas a strong inhibition of the glucose activation was ob-served with oxythiamine 250 μM (FIG. 1 A). 5-Amino4-imidazolcarboxamide ribotide (AICAR), an AMP-kinase activator, did not inhibit the glucose acti-vation in INS-1 cells (FIG. 1 B).

Effect of Aliphatic Acids

We observed that 2 mM butyrate induced a strong activation of the reporter gene. The effect of butyrate could be more pronounced than the glucose stim-ulation, especially on older INS-1 cells (cells with high number of passages).

(A)

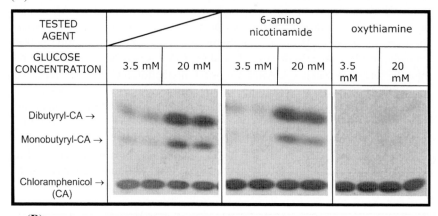

(B)

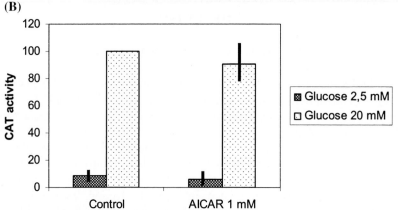

FIGURE 1. Role of the pentose phosphate pathway in the glucose regulation of the GLR gene. (**A**) INS-1 cells were transfected by the plasmid (pTkCd) containing the promotor domain (−2 kb to −287 pb) bearing the ChoRE of GLR gene. Transfected cells were grown in either 3.5 mM glucose or 20 mM glucose, and in the presence of 200 μM 6-aminonicotinamide (center part) or 250 μM oxythiamine (right part). Products of acylation were separated by thin layer chromatography and autoradiographied. Positions of mono- and di-butyrylchloramphenicol are indicated. The experiment was performed in duplicate. One of three experiments with similar results is represented. (**B**). INS-1 cells were transfected by the plasmid pTkCd and cells were grown in either 3.5 mM glucose or 20 mM glucose, in control condition or in the presence of 1 mM AICAR. (mean ± SEM of three experiments.)

This activation was partially additional with the glucose stimulation. Acetate and propionate had a similar but less pronounced effect (not shown).

In order to test if butyrate and glucose act via the same regulatory element, plasmids bearing a mutated glucose regulatory element were tested. These plasmids have been described previously.[3] The first one contains a mutation in the first part of the first E-box (mutation mE1-AC), the other one bears a mutation in the second part of the first E-box (mE1-TG). In the third mutated plasmid, the two E-boxes of the ChoRE are inverted. The reporter gene activity of cells transfected by plasmids in the first E-box were no more activated by glucose, whereas butyrate activation was still present (FIG. 2A for mE1-TG, mE1-AC not shown).

In contrast, the plasmids with inverted E-boxes are any more stimulable by neither glucose nor butyrate (FIG. 2 B). Furthermore, CAT activity of INS-1 cells transfected by control Rous Sarcoma Virus promotor (pRSV)-CAT plasmid is also not activable by glucose and butyrate (not shown). This demonstrates that glucose and butyrate do not activate the reporter gene expression by the same regulatory mechanism.

DISCUSSION

Our data suggest that, despite the similar structure of their glucose response element, glucose-induced activation of the GLR gene, in one hand, and of glycolytic and lipogenic enzyme gene, in another hand, could not be mediated by identical mechanisms. We observed the inhibition of glucose activation by oxythiamine, this suggesting an implication of the nonoxidative branch of the pentose phosphate pathway. In contrast, 6-aminonicotinamide, inhibitor of oxidative branch, was without effect.

Our results contrasts with those observed for the glucose-6-phosphatase gene: inhibition of glucose activation by 6-aminonicotinamide, and absence of effect for oxythiamine (up to 500 μM).[19] In our knowledge, no data are available for other glucose-activated genes. However, a part of the glucose activation mechanism (glycolytic and lipogenic enzyme genes versus GLR gene) must be similar: the key molecule of regulation of the precited genes is xylulose 5-P, and our data suggest implication of the pentose phosphate pathway.

Others experiments demonstrate differences in the glucose activation of glycolytic and lipogenic enzyme genes versus GLR gene: in our hands, the AMP-kinase activator, AICAR, which inhibits glucose activation of various glycolytic and lipogenic enzyme genes,[20,21] did not inhibit the glucose induction in INS-1 cells.

Moreover, we did not observe any effect of xylitol, a precursor of xylitol 5-P, whereas it has been reported that xylitol activates many ChoRE.[19,22,23] Nevertherless, it has been reported that the GLR mRNA is increased by xylitol.[2] However, all these experiments were realized in hepatic cells: it is conceivable that xylitol is not efficiently converted into xylulose 5-P in INS-1 cells.

Glycerin, which activates the glucose-6-phosphatase gene,[19] was without effect in our system.

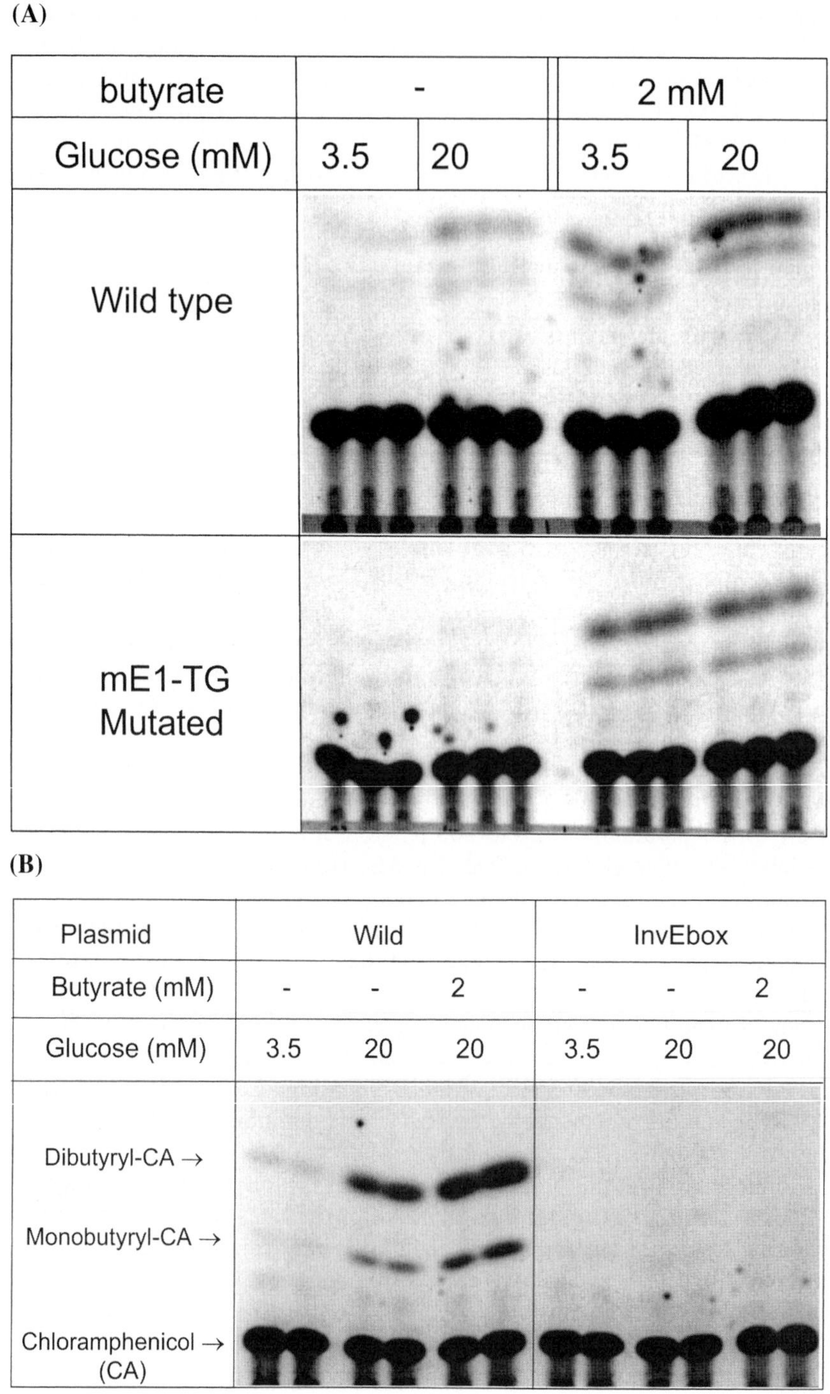

(A)

butyrate	-		2 mM	
Glucose (mM)	3.5	20	3.5	20
Wild type				
mE1-TG Mutated				

(B)

Plasmid	Wild			InvEbox		
Butyrate (mM)	-	-	2	-	-	2
Glucose (mM)	3.5	20	20	3.5	20	20
Dibutyryl-CA →						
Monobutyryl-CA →						
Chloramphenicol → (CA)						

Finally, short aliphatic acids (acetate, propionate, and butyrate) activate the reporter gene expression. The effect of fatty acids, and glucose, are partially additional. This contrasts with the generally observed inhibition produced by aliphatic acids on the glucose-activated lipogenic enzyme gene expression.[18,24]

The effect of butyrate appeared as an effect on a specific regulatory element, as the plasmid bearing reporter gene droved by inverted ChoRE or by RSV were not stimulated by butyrate. However, our experiments demonstrated that glucose and butyrate did not act via the same regulatory element, as mutation of the first of the two E-boxes constituting the ChoRE suppressed the glucose stimulation, but left the butyrate activation unmodified.

Several type of consensus sequences for butyrate regulatory elements were described (for review see Ref. 25). Such types of sequence are encountered in several folds in sequence of used plasmid. However actual identification of the active element requires additional experiments.

ACKNOWLEDGMENTS

This work was supported by a grant from the Fonds de la Recherche Scientifique Médicale 3.4511.00. L. Portois was a recipient for a doctoral fellowship from the Interuniversity Poles of Attraction Program, Belgian State Prime Minister's Office, Federal Office for Scientific, Technical and Cultural Affairs. M. Virreira is a recipient for a doctoral fellowship of Found Xenophilia of Université Libre de Bruxelles.

REFERENCES

1. PORTOIS, L., B. MAGET, M. TASTENOY, *et al.* 1999. Identification of a glucose response element in the promotor of the rat glucagon receptor gene. J. Biol. Chem. **274:** 8181–8190.

←————————————————————————————————

FIGURE 2. Effect of mutation in the ChoRE of the GLR gene on its stimulation by glucose or butyrate. (**A**) INS-1 cells were transfected by the wild-type plasmid pTkCd, or by a mutated plasmid in which the TG have been replaced by the AC in the first E-box (mE1-TG). Transfected cells were grown in either 3.5 mM glucose or 20 mM glucose, in the absence (left part) or presence (right part) of 2 mM butyrate. Experiments were performed in triplicate. One of four experiments with similar results is represented. (**B**) INS-1 cells were transfected by the wild-type plasmid (pTkCd), or by a mutated plasmid in which the first and second E-box in ChoRE were inverted (InvE-box). Transfected cells were grown in either 3.5 mM glucose or 20 mM glucose alone, in presence of 20 mM glucose and 2 mM butyrate. The experiment was performed in duplicate. Positions of mono- and di-butyrylchloramphenicol are indicated. One of two experiments with similar results is represented.

2. BURCELIN, R., C. MREJEN, J.F. DECAUX, *et al*. 1998. In vivo and in vitro regulation of hepatic glucagon receptor mRNA concentration by glucose metabolism. J. Biol. Chem. **273:** 8088–8093.
3. PORTOIS, L., M. TASTENOY, B. VIOLLET & M. SVOBODA. 2002. Functional analysis of the glucose response element of the rat glucagon receptor gene in insulin-producing INS-1 cells biochim. Biophys. Acta **1574:** 175–186.
4. SHIH, H.M., Z. LIU & H.C. TOWLE. 1995. Two CACGTG motifs with proper spacing dictate the carbohydrate regulation of hepatic gene transcription. J. Biol. Chem. **270:** 21991–21997.
5. RUFO, C., M. TERAN-GARCIA, M.T. NAKAMURA, *et al*. 2001. Involvement of a unique carbohydrate responsive factor in the glucose regulation of rat liver fatty acid synthase gene transcription. J. Biol. Chem. **276:** 21969–21975.
6. KOO, S.H. & H.C. TOWLE. 2000. Glucose regulation of mouse S(14) gene expression in hepatocytes. Involvement of a novel transcription factor complex. J. Biol. Chem. **275:** 5200–5207.
7. O'CALLAGHAN, B.L., S.H. KOO, Y. WU, *et al*. 2001. Glucose regulation of the acetyl-CoA carboxylase promotor PI in rat hepatocytes. J. Biol. Chem. **276:** 16033–16039.
8. LEFRANCOIS-MARTINEZ, A.M., M.J. DIAZ-GUERRA, V. VALLET, *et al*. 1994. Glucose-dependent regulation of the L-pyruvate kinase gene in a hepatoma cell line is independent of insulin and cyclic AMP. FASEB J. **8:** 89–96.
9. KOO, S.H., A.K. DUTCHER & H.C. TOWLE. 2001. Glucose and insulin function through two distinct transcription factors to stimulate expression of lipogenic enzyme genes in liver. J. Biol. Chem. **276:** 9437–9445.
10. KAYTOR, E.N., H. SHIH & H.C. TOWLE. 1997. Carbohydrate regulation of hepatic gene expression. Evidence against a role for the upstream stimulatory factor. J. Biol. Chem. **272:** 7525–7531.
11. WANG, H. & C.B. WOLLHEIM. 2002. ChREBP rather than USF2 regulates glucose stimulation of endogenous L-pyruvate kinase expression in insulin-secreting cells. J. Biol. Chem. **277:** 32746–32752.
12. KAWAGUCHI, T., M. TAKENOSHITA, T. KABASHIMA & K. UYEDA. 2001. Glucose and cAMP regulate the L-type pyruvate kinase gene by phosphorylation/dephosphorylation of the carbohydrate response element binding protein. Proc. Natl. Acad. Sci. USA **98:** 13710–13715.
13. KABASHIMA, T., T. KAWAGUCHI, B.E. WADZINSKI & K. UYEDA. 2003. Xylulose 5-phosphate mediates glucose-induced lipogenesis by xylulose 5-phosphate-activated protein phosphatase in rat liver. Proc. Natl. Acad. Sci. USA **100:** 5107–5112.
14. MA, L., N.G. TSATSOS & H.C. TOWLE. 2005. Direct role of ChREBP/Mlx in regulating hepatic glucose-responsive genes. J. Biol. Chem. **280:** 12019–12027.
15. STOECKMAN, A.K., L. MA & H.C. TOWLE. 2004. Mlx is the functional heteromeric partner of the carbohydrate response element-binding protein in glucose regulation of lipogenic enzyme genes. J. Biol. Chem. **279:** 15662–15669.
16. LIU, Y.Q. & K. UYEDA. 1996. A mechanism for fatty acid inhibition of glucose utilization in liver. Role of xylulose 5-P. J. Biol. Chem. **271:** 8824–8830.
17. UYEDA, K., H. YAMASHITA & T. KAWAGUCHI. 2002. Carbohydrate responsive element-binding protein (ChREBP): a key regulator of glucose metabolism and fat storage. Biochem. Pharmacol. **63:** 2075–2080.

18. KAWAGUCHI, T., K. OSATOMI, H. YAMASHITA, *et al.* 2002. Mechanism for fatty acid "sparing" effect on glucose-induced transcription: regulation of carbohydrate-responsive element-binding protein by AMP-activated protein kinase. J. Biol. Chem. **277:** 3829–3835.

19. MASSILLON, D. 2001. Regulation of the glucose-6-phosphatase gene by glucose occurs by transcriptional and post-transcriptional mechanisms. Differential effect of glucose and xylitol. J. Biol. Chem. **276:** 4055–4062.

20. LECLERC, I., A. KAHN & B. DOIRON. 1998. The 5′-AMP-activated protein kinase inhibits the transcriptional stimulation by glucose in liver cells, acting through the glucose response complex. FEBS Lett. **431:** 180–184.

21. FORETZ, M., D. CARLING, C. GUICHARD, *et al.* 1998. AMP-activated protein kinase inhibits the glucose-activated expression of fatty acid synthase gene in rat hepatocytes. J. Biol. Chem. **273:** 14767–14771.

22. DOIRON, B., M.H. CUIF, R. CHEN & A. KAHN. 1996. Transcriptional glucose signaling through the glucose response element is mediated by the pentose phosphate pathway. J. Biol. Chem. **271:** 5321–5324.

23. MOURRIERAS, F., F. FOUFELLE, M. FORETZ, *et al.* 1997. Induction of fatty acid synthase and S14 gene expression by glucose, xylitol and dihydroxyacetone in cultured rat hepatocytes is closely correlated with glucose 6-phosphate concentrations. Biochem. J. **326:** 345–349.

24. DENTIN, R., J. GIRARD & C. POSTIC. 2005. Carbohydrate responsive element binding protein (ChREBP) and sterol regulatory element binding protein-1c (SREBP-1c): two key regulators of glucose metabolism and lipid synthesis in liver. Biochimie **87:** 81–86.

25. PEGORIER, J.P., C. LE MAY & J. GIRARD. 2004. Control of gene expression by fatty acids. J. Nutr. **134:** 2444S–2449S.

NAP, a Peptide Derived from the Activity-Dependent Neuroprotective Protein, Modulates Macrophage Function

FRANCISCO J. QUINTANA,[a] ROY ZALTZMAN,[b]
RAFAEL FERNANDEZ-MONTESINOS,[c] JUAN LUIS HERRERA,[c]
ILLANA GOZES,[b] IRUN R. COHEN,[a] AND DAVID POZO[a,c]

[a]Department of Immunology, The Weizmann Institute of Science, Rehovot
76100, Israel

[b]Department of Human Molecular Genetics and Biochemistry, Sackler School of
Medicine, Tel Aviv University, Tel Aviv 69978, Israel

[c]Department of Medical Biochemistry and Molecular Biology, The University of
Seville Medical School, Avda. Sánchez Pizjuan, 4, 41009 Sevilla, Spain

ABSTRACT: NAP is an eight-amino acid neuroprotective peptide
NAPVSIPQ; it is the smallest active element derived from the recently
cloned activity-dependent neuroprotective protein (ADNP). NAP readily
enters the brain from the blood. It will be important to learn whether
NAP, in addition to its neuroprotective activity, also might influence
immune-mediated inflammation. Here, we report that: (a) macrophages
express ADNP; (b) expression of ADNP in macrophages responds to VIP;
and (c) NAP downregulates the key inflammatory cytokines tumor necrosis factor (TNF-α), interleukin-16 (IL-16), and IL-12 in macrophages.
These findings indicate that ADNP/NAP can play an important role in
immune regulation as well as in neuroprotection, which may be mutually
related processes.

KEYWORDS: vasoactive intestinal polypeptide (VIP); activity-dependent
neuroprotective protein (ADNP); macrophages; neuroimmunology; gene
expression

INTRODUCTION

Neurotrophic proteins and neuropeptides have important regulatory functions and are a focus of intensive research in rational drug design.[1,2] Previous

Address for correspondence: Dr. David Pozo, Department of Medical Biochemistry and Molecular
Biology, The University of Seville School of Medicine, Avda. Sanchez Pizjuan, 4, 41009 Sevilla, Spain.
Voice: +34-95-4559852; fax: +34-95-4907048.
e-mail: dpozo@us.es

Ann. N.Y. Acad. Sci. 1070: 500–506 (2006). © 2006 New York Academy of Sciences.
doi: 10.1196/annals.1317.069

studies have identified several components by sequential chromatographic methods within the neurotrophic milieu produced by astroglia, activity-dependent neurotrophic factor (ADNF) being one of the most potent.[3] The active peptide fragment of ADNF is ADNF-14 (VLGGGSALLRSIPA)[3]; ADNF-9 (SALLRSIPA), a shorter C-terminal peptide, retains full biological activity.[4] Antibodies to ADNF-14 or to ADNF-9 were used to identify the activity-dependent neuroprotective protein (ADNP), and the cDNA has been cloned from mouse neuroglial cells and human fetal brain.[5,6] Structure/activity screening of several peptides derived from ADNP identified a potent octapeptide, NAP (NAPVSIPQ).[5] NAP has a greater *in vivo* neuroprotective efficacy than ADNF-9.[5,7,8]

The neuroprotective activities of NAP have been studied in a wide variety of systems. NAP induces neuroprotection against the β-amyloid peptide's toxicity involved in the onset of Alzheimer's disease,[5,9,10] oxidative stress,[11] NMDA excitotoxicity,[5] tumor necrosis factor-α (TNF-α) toxicity,[12] transient glucose deprivation,[9] dopamine toxicity, and decreased glutathione.[13] NAP's biological properties related to neuroprotection have been demonstrated in *in vivo* models of closed head injury, fetal alcohol syndrome, and stroke[7,14] and it also been involved in neurodevelopment.[15] NAP is under phase I clinical trials in the United States.

Neuropeptides and neurotrophic proteins perform a broad array of seemingly unrelated functions. Vasoactive intestinal polypeptide (VIP), for example, promotes neuronal survival,[16,17] but is also a potent immunomodulator[18,19] and is under clinical trials. Remarkably, VIP inhibits the acute inflammatory response that follows spinal cord injury[20] and prevents activated microglia-induced neurodegeneration under inflammatory conditions[21] while increasing the synthesis of the NAP-containing protein ADNP in astroglia.[5] Given the breath of NAP's neuroprotective activities and the fact that ADNP is a VIP-responsive gene, we were interested to examine the direct consequences of NAP on the immune system. The present article shows for the first time direct effects of NAP on the macrophage, a cell with a critical role in the initiation and coordination of the immune response. Mindful that VIP acts on activated macrophages as a potent, endogenous anti-inflammatory neuropeptide and that ADNP is a VIP-responsive gene, the current study was performed to investigate whether ADNP mRNA expression can be detected in a mouse macrophage cell line and whether VIP is able to increase the steady-state levels ADNP mRNA.

MATERIAL AND METHODS

Synthetic VIP was purchased from Calbiochem-Novabiochem (Laufelfingen, Switzerland). NAP was used as before.[22] The mouse macrophage cell line RAW 264.7 was obtained from the American Type Tissue Collection (Rockville, MD). These cells were maintained in RPMI 1640 supplemented with 25 mM HEPES, 10% (v/v) heat-inactive fetal calf serum (FCS) (Biowhittaker, Wokingham, UK), 10 mM glutamine, 100 U/mL penicillin, and 100 μg/mL streptomycin (components from Sigma Chemical Co., St. Louis, MO). For mRNA analysis cells, total RNA was extracted and DNAse-treated after TriPure isolation reagent (Roche Diagnostics GmbH, Mannheim, Germany) following manufacturer's instructions. Murine ADNP primers were derived from the published sequence of murine ADNP and reverse transcription polymerase chain reaction (RT-PCR) experimental conditions were previously reported.[23] cDNA was previously titrated to amplify in the linear range. Cytokine levels were determined by enzyme-linked immunosorbent assay (ELISA) according to manufacture's instructions (BD-Pharmingen, San Diego, CA).

RESULTS AND DISCUSSION

RT-PCR of mRNA using the ADNP primers from RAW 264.7 macrophages resulted in single DNA band when analyzed by agarose gel electrophoresis (FIG. 1 A). RT-PCR reactions were also processed with control β-actin house-keeping gene primers. These RT-PCR reactions correspond to the predicted size for PCR amplification using the ADNP primers, and nucleotide sequences of the amplified fragments showed an identical sequence to the mouse ADNP gene.[5] Thus, we report for the first time that ADNP mRNA is expressed in immune system cells, namely, macrophages in a resting state. To further investigate the physiological role of ADNP in this context, we tested whether VIP treatment might influence ADNP gene expression in macrophages. FIGURE 1 A shows that ADNP mRNA levels were increased after 24 h of VIP treatment. The highest increase in ADNP mRNA was produced at nanomolar concentrations of VIP, with a slightly increased level at 10^{-12} M VIP. The effect of VIP was not dose dependent, most probably due to VIP receptor desensitization in macrophages.[24] Thus, ADNP is a VIP-responsive gene in macrophages at concentrations that can be sensed by VIP receptors (VPAC) on immunocompetent cells. VIP production and secretion are elevated after immunological stimuli[25] and, therefore, some of VIP's immunomodulatory properties might be mediated in part by ADNP. To know whether ADNP mRNA levels are modified by immunological stimuli, we incubated RAW 264.7 cells in the presence of increasing concentrations of lipopolysaccharide (LPS) (0.1–10 mg/mL) for 24 h. ADNP gene expression was not altered after LPS treatment (FIG. 1 B). Toll-like receptors (TLRs) are key regulators of innate immunity, sensing and responding to invading microorganisms. LPS is the main ligand of the TLR-4 and TLRs include up to 10 different gene products.

So, at this point, we cannot rule out whether ADNP is regulated by other TLR-ligands. Nevertheless, these data should be taken as qualitative, taking into consideration the limitations of the RT-PCR approach to quantify

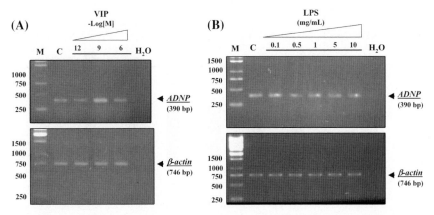

FIGURE 1. ADNP is expressed in macrophages. **(A)** RT-PCR analysis of ADNP in the macrophage-like RAW 264.7 cell line. Arrows indicate the PCR product amplified with specific ADNP primers (390 bp [base pairs]) and β-actin primers (746 bp). **(B)** Effect of LPS treatment on ADNP mRNA levels in RAW 264.7 cells. Results are representative of four independent experiments.

accurately mRNA. In this sense, differential expression of ADNP mRNA is currently being studied in our lab to assess VIP and known TLR-ligands effects by real-time PCR.

Given the breath of NAP's neuroprotective activities and the fact that ADNP is a VIP-responsive gene expressed in macrophages, we were interested to examine direct consequences of NAP treatment on key cytokines involved in the inflammatory response such as TNF-α, interleukin-6 (IL-6), and IL-12. RAW 264.7 cells treated with 0.1 mg/mL LPS in the presence of increasing concentrations of NAP for 24 h showed an inhibition of TNF-α, IL-6, and IL-12 secretion (FIG. 2). Although the mechanism of action involved is not yet known, our results support the role of NAP as a potent immunomodulator. Several questions related to the mechanism of action are under current investigation; we wish to learn whether ADNP could act in a paracrine and/or autocrine fashion under different circumstances. A study on NAP potential new functions can define novel mechanisms that modulate immune responses, and might lead to the development of new therapies for immune-mediated disorders, particularly, for neurodegenerative diseases in which neuronal defense mechanisms and immunomodulation represent innovative approaches.

ACKNOWLEDGMENTS

This research was supported in part by The Weizmann Institute Exchange Fellowship Fund from Cambridge University (to David Pozo) and grants from Fondo de Investigación Sanitaria, Spanish Ministry of Health (PI 030359 to

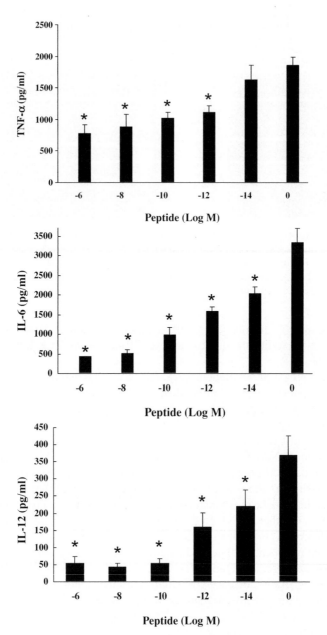

FIGURE 2. Effect of NAP on cytokine release (TNF-α, IL-6, and IL-12) by RAW 264.7 cells. Cytokine levels were determined in supernatants by ELISA after activation of RAW cells with 0.1 mg/mL LPS for 24 h in the presence of different concentrations of NAP. Cytokine basal levels for TNF-α, IL-6, and IL-12 were 485 $\pm$ 83 pg/mL, 337 $\pm$ 101 pg/mL, and 60 $\pm$ 27 pg/mL, respectively. Statistical significance was determined by ANOVA followed by a Student-Newman-Keuls test.

David Pozo), VI European Framework Program UEMERG (CT2004–00638 to David Pozo), Minerva Foundation (to Irun R. Cohen), and the Center for the Study of Emerging Diseases (to Irun R. Cohen). Francisco J. Quintana was funded by a fellowship from the Feinberg Graduate School, The Weizmann Institute of Science. Rafael Fernandez-Montesinos and Juan Luis Herrera were funded by fellowships from Junta de Andalucia. Professor Illana Gozes is the incumbent of the Lily and Avraham Gildor Chair for the Investigation of Growth Factors and heads the Dr. Diana and Zelman Elton (Elbaum) Laboratory for Molecular Neuroendocrinology. This study was supported by Allon Therapeutics, ISF, and BSF. Irun R. Cohen is the Meuerberger Professor of Immunology at the Weizmann Institute of Science, the Director of the Center for the Study of Emerging Diseases, Jerusalem, and the Director of the National Center for Biotechnology in the Negev, at the Ben-Gurian University of the Negev.

REFERENCES

1. NGUYEN, M.D., J.P. JULIEN & S. RIVEST. 2002. Innate immunity: the missing link in neuroprotection and neurodegeneration? Nat. Rev. Neurosci. **3:** 216–227.
2. GOZES, I. 2001. Neuroprotective peptide drug delivery and development: potential new therapeutics. Trends Neurosci. **24:** 700–705.
3. BRENNEMAN, D.E. & I. GOZES. 1996. A femtomolar-acting neuroprotective peptide. J. Clin. Invest. **97:** 2299–2307.
4. BRENNEMAN, D.E. *et al.* 1998. Activity-dependent neurotrophic factor: structure-activity relationships of femtomolar-acting peptides. J. Pharmacol. Exp. Ther. **285:** 619–627.
5. BASSAN, M. *et al.* 1999. Complete sequence of a novel protein containing a femtomolar-activity-dependent neuroprotective peptide. J. Neurochem. **72:** 1283–1293.
6. ZAMOSTIANO, R. *et al.* 2001. Cloning and characterization of the human activity-dependent neuroprotective protein. J. Biol. Chem. **276:** 708–714.
7. GOZES, I. *et al.* 2003. From vasoactive intestinal peptide (VIP) through activity-dependent neuroprotective protein (ADNP) to NAP: a view of neuroprotection and cell division. J. Mol. Neurosci. **20:** 315–322.
8. GOZES, I., R.A. STEINGART & A.D. SPIER. 2004. NAP mechanisms of neuroprotection. J. Mol. Neurosci. **24:** 67–72.
9. ZEMLYAK, I. *et al.* 2000. A novel peptide prevents death in enriched neuronal cultures. Regul. Pept. **96:** 39–43.
10. ASHUR-FABIAN, O. *et al.* 2003. The neuroprotective peptide NAP inhibits the aggregation of the beta-amyloid peptide. Peptides **24:** 1413–1423.
11. STEINGART, R.A. *et al.* 2000. VIP and peptides related to activity-dependent neurotrophic factor protect PC12 cells against oxidative stress. J. Mol. Neurosci. **15:** 137–145.
12. BENI-ADANI, L. *et al.* 2001. A peptide derived from activity-dependent neuroprotective protein (ADNP) ameliorates injury response in closed head injury in mice. J. Pharmacol. Exp. Ther. **296:** 57–63.

13. OFFEN, D. *et al*. 2000. Vasoactive intestinal peptide (VIP) prevents neurotoxicity in neuronal cultures: relevance to neuroprotection in Parkinson's disease. Brain Res. **854:** 257–262.

14. GOZES, I. *et al*. 2000. Activity-dependent neurotrophic factor: intranasal administration of femtomolar-acting peptides improve performance in a water maze. J. Pharmacol. Exp. Ther. **293:** 1091–1098.

15. PINHASOV, A. *et al*. 2003. Activity-dependent neuroprotective protein: a novel gene essential for brain formation. Dev. Brain Res. **144:** 83–90.

16. SAID, S.I. 1996. Molecules that protect: the defense of neurons and other cells. J. Clin. Invest. **97:** 2163–2164.

17. GRESSENS, P. *et al*. 1997. Vasoactive intestinal peptide prevents excitotoxic cell death in the murine developing brain. J. Clin. Invest. **100:** 390–397.

18. DELGADO, M., D. POZO & D. GANEA. 2004. The significance of vasoactive intestinal peptide in immunomodulation. Pharmacol. Rev. **56:** 249–290.

19. POZO, D. & M. DELGADO. 2004. The many faces of VIP in neuroimmunology: a cytokine rather a neuropeptide? FASEB J. **18:** 1325–1334.

20. KIM, W.-K. *et al*. 2000. Vasoactive intestinal peptide and pituitary adenylate cyclase-activating polypeptide inhibit tumor necrosis factor-a production in injured spinal cord and in activated microglia via a cAMP-dependent pathway. J. Neurosci. **20:** 3622–3630.

21. DELGADO, M. & D. GANEA. 2003. Vasoactive intestinal peptide prevents activated microglia-induced neurodegeneration under inflammatory conditions: potential therapeutic role in brain trauma. FASEB J. **17:** 1922–1924.

22. ALCALAY, R.N. *et al*. 2004. Intranasal administration of NAP, a neuroprotective peptide, decreases anxiety-like behavior in aging mice in the elevated plus maze. Neurosci. Lett. **361:** 128–131.

23. POGGI, S.H. *et al*. 2002. Differential expression of embryonic and maternal activity-dependent neuroprotective protein during mouse development. Am. J. Obstet. Gynecol. **187:** 973–976.

24. POZO, D., J.M. GUERRERO & J.R. CALVO. 1995. Homologous regulation of vasoactive intestinal peptide (VIP) receptors on rat peritoneal macrophages. Peptides **16:** 313–318.

25. MARTINEZ, C. *et al*. 1999. Regulation of VIP production and secretion by murine lymphocytes. J. Neuroimmunol. **93:** 126–138.

Involvement of ERK and CREB Signaling Pathways in the Protective Effect of PACAP in Monosodium Glutamate-Induced Retinal Lesion

BOGLARKA RÁCZ,[a] ANDREA TAMÁS,[b] PETER KISS,[b] GABOR TÓTH,[c] BALAZS GASZ,[a] BALAZS BORSICZKY,[a] ANDREA FERENCZ,[a] FERENC GALLYAS Jr.,[d] ERZSEBET RŐTH,[a] AND DORA REGLŐDI[b]

[a]Department of Surgical Research and Techniques, University of Pecs, 7624 Pécs, Hungary

[b]Department of Anatomy, University of Pecs, (MTA-TKI Neurohum. Regul. Res. Group), 7624 Pécs, Hungary

[c]Department of Medical Chemistry, University of Szeged, 6701 Szeged, Hungary

[d]Department of Biochemistry and Medical Chemistry, University of Pecs, 7624 Pécs, Hungary

ABSTRACT: Pituitary adenylate cyclase-activiting polypeptide (PACAP) has well-documented neuroprotective actions, which have also been shown in retinal degeneration induced by monosodium glutamate (MSG) in neonatal rats. The aim of this article was to investigate the activation of extracellular signal-regulated kinase (ERK1/2) and cyclic adenosine 3',5'-phosphate (cAMP)-responsive element binding protein (CREB) signaling pathways by Western blot analysis in retinal degeneration induced by MSG. We found that intravitreal administration of PACAP preceding the MSG treatments induced significant increases in the phosphorylation, that is, the activation of ERK1/2 and its downstream target, CREB, 12 h after the treatment compared to the contralateral untreated eye during the first two treatments, with no further elevations 24 h after treatments. These results demonstrate that the degenerative effect of MSG and the protective effect of PACAP involve complex kinase signaling pathways and are related to cAMP/ERK/CREB activation.

KEYWORDS: neuroprotection; intravitreal; neonatal rat

Address for correspondence: Dora Reglődi, M.D., Ph.D., Department of Anatomy, University of Pécs, 7624 Pécs, Szigeti u 12. Hungary. Voice: +36-72-536001 ext. 5398; fax: +36-72-536393.
e-mail: dora.reglodi@aok.pte.hu

Ann. N.Y. Acad. Sci. 1070: 507–511 (2006). © 2006 New York Academy of Sciences.
doi: 10.1196/annals.1317.070

INTRODUCTION

Pituitary adenylate cyclase-activiting polypeptide (PACAP) and its receptors can be found in the retina[1,2] and it is an important transmitter in the retino–hypothalamic tract.[3] The well-known neuroprotective effects of PACAP[4] have also been shown *in vitro* in the retina in several pathological conditions, such as hypoxia[5] and optic nerve transection.[6] Elevated glutamate levels lead to retinal damage and PACAP has been shown to be protective against glutamate-induced cell death in the retina *in vitro*.[7] Recently, we have shown that this protective effect is also present *in vivo*, in monosodium glutamate (MSG)-induced retinal degeneration.[8] The underlying molecular mechanism of this protective effect is not yet known. The aim of the present article was to study the activation of extracellular signal-regulated kinase (ERK1/2) and cyclic adenosine 3',5'-phosphate (cAMP)-responsive element binding protein (CREB), both of which have been shown to be involved in the protective effects exerted by PACAP in neurons.[4,9,10]

MATERIALS AND METHODS

Wistar rat pups were used from the first postnatal day. All procedures were performed in accordance with the ethical guidelines approved by the University of Pécs. Pups were injected s.c. with 2 mg/g bodyweight MSG on postnatal days 1, 5, and 9.[8] Preceding each MSG treatment, 100 pmol PACAP in 5 µL saline was injected unilaterally in the vitreous body. Retinas were removed 12 and 24 h following each MSG treatment and were processed for Western blot analysis. Samples were homogenized in ice-cold Tris buffer (50 mM, pH = 8.0), containing Protease Inhibitor Coctail (1:1000, Sigma-Aldrich Co., Budapest, Hungary) and harvested in $2\times$ concentrated sodium dodecyl sulfate (SDS)-polyacrilamide gel electrophoretic sample buffer. Proteins were separated on 12% SDS-polyacrilamide gel and transferred to nitrocellulose membranes. After blocking (2 h with 3% nonfat milk in Tris-buffered saline), membranes were probed overnight at 4°C with antibodies recognizing the following antigens: Phospho-specific ERK1/2, phospo-specific CREB (Cell Signaling Technology, Beverly, MA). Membranes were washed 6 times for 5 min in Tris-buffered saline (pH 7.5) containing 0.2% Tween (TBST) before addition of goat anti-rabbit horseradish peroxidase-conjugated secondary antibody (1:3000 dilution, Bio-Rad, Budapest, Hungary). Peroxidase labeling was visualized using a Super Signal West Pico Chemiluminescent Substrate (Kvalitex Hungary). The developed films were scanned and the pixel volumes of the bands were determined by using HIN Image J software. Each experiment was repeated a minimum of three times.

RESULTS AND DISCUSSION

In a previous study, we have shown that the presently applied MSG treatment protocol led to a severe degeneration of the inner retinal layers which could be significantly ameliorated by local pre-MSG PACAP administration.[8] In the present study, the phosphorylation, that is, the activation of ERK1/2 and CREB (P-ERK1/2 and P-CREB) was monitored 12 and 24 h after treatments. There was a gradual increase in both proteins following MSG treatments. PACAP treatments led to further, significant increases in the level of P-ERK1/2 (FIG. 1) and P-CREB (FIG. 2) 12 h after the first and the second PACAP treatments. There was no further elevation after the third treatment in either control or PACAP-treated retinas (data not shown). Interestingly, PACAP increased the ERK1/2 and CREB phosphorylation 12 h, but not 24 h after the first and second treatments. It suggests that during the first two MSG challenges PACAP induced a transient increase in the activation of ERK1/2 and of its downstream target CREB, which disappeared by 24 h after the treatment. Then the phosphorylation reached a plateau value, and did not increase any further. These results are in accordance with our earlier histological observations using the same experimental setup.[11] In that study, we showed that in order to achieve significant amelioration in MSG-induced degeneration at least two PACAP treatments are necessary, but there was no further amelioration after the third treatment. Also, our present and earlier observations indicate that repeated application of PACAP may lead to a primed state promoting a long-lasting protection, which has also been observed by others in cerebellar neurons.[12]

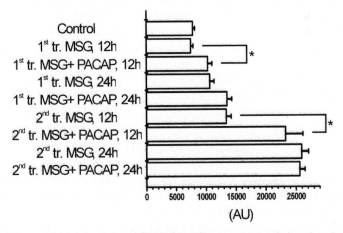

FIGURE 1. Effects of MSG and PACAP on the expression of phosphorylated ERK in rat retina. Western blot analysis of ERK phosphorylation in rat retina after unilateral injection of 100 pmol PACAP to the vitreous body on postnatal days 1 and 5. Bands were determined by using NIH Image J software, demonstrated in arbitrary unit (AU). $^*P < 0.05$ compared to the MSG-treated control. (tr = treatment.)

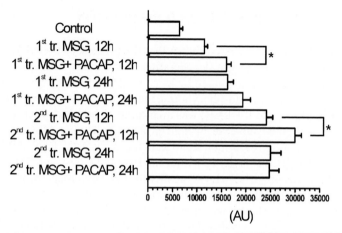

FIGURE 2. Effects of MSG and PACAP on the expression of phosphorylated CREB in rat retina. Western blot analysis of CREB phosphorylation in rat retina after unilateral injection of 100 pmol PACAP to the vitreous body on postnatal days 1 and 5. Bands were determined by using NIH Image J software, demonstrated in arbitrary unit (AU). *$P < 0.05$ compared to the MSG-treated control. (tr = treatment.)

The protective effect of PACAP against glutamate toxicity may involve several mechanisms, including antiapoptotic effects and regulating effects on glutamate transporters and metabolism.[4,13] The PACAP-induced activation of the ERK-type of MAP kinases through a cAMP/PKA-dependent pathway has been shown in numerous *in vitro* studies.[4,7] Activation of ERK1/2 results in phosphorylation, that is, activation of its downstream target, CREB, which forms complex with cAMP-responsive element (CRE), which in turn induces expression of CRE-regulated genes. Under our experimental conditions, PACAP induced transient activation of the aforementioned signaling sequence which is important in neuronal survival. Although further studies are necessary to elucidate other signaling mechanisms, our results demonstrate that the degenerative effect of MSG and the protective effect of PACAP involve complex kinase signaling pathways and are related to cAMP/ERK/CREB activation.

ACKNOWLEDGMENTS

The present work was supported by Grants OTKA T046589, T048851, T048848, F046504, ETT 596/2003 and the Hungarian Academy of Sciences.

REFERENCES

1. SEKI, T. *et al.* 1997. Distribution and ultrastructural localization of a receptor for pituitary adenylate cyclase activating polypeptide and its mRNA in the rat retina [abstract]. Neurosci. Lett. **238:** 127–130.

2. Izumi, S. *et al.* 2000. Ultrastructural localization of PACAP immunoreactivity in the rat retina. Ann. N. Y. Acad. Sci. **921:** 317–320.
3. Hannibal, J. 2002. Neurotransmitters of the retino-hypothalamic tract. Cell Tissue Res. **309:** 73–88.
4. Somogyvari-Vigh, A. & D. Reglodi. 2004. Pituitary adenylate cyclase activating polypeptide: a potential neuroprotective peptide. Review. Curr. Pharm. Des. **10:** 2861–2889.
5. Rabl, K. *et al.* 2002. PACAP inhibits anoxia-induced changes in physiological responses in horizontal cells in the turtle retina. Regul. Pept. **109:** 71–74.
6. Seki, T. *et al.* 2003. Pituitary adenylate cyclase activating polypeptide (PACAP) protects against ganglion cell death after cutting of optic nerve in the rat retina [abstract]. Regul. Pept. **115:** 55.
7. Shoge, K. *et al.* 1999. Attenuation by PACAP of glutamate-induced neurotoxicity in cultured retinal neurons. Brain Res. **839:** 66–73.
8. Tamas, A. *et al.* 2004. Effects of pituitary adenylate cyclase activating polypeptide in retinal degeneration induced by monosodium-glutamate. Neurosci. Lett. **372:** 110–113.
9. Vaudry, D. *et al.* 2002. PACAP protects cerebellar granule neurons against oxidative stress-induced apoptosis. Eur. J. Neurosci. **15:** 1451–1460.
10. Shioda, S. *et al.* 1998. PACAP protects hippocampal neurons against apoptosis: involvement of JNK/SAPK signaling pathway. Ann. N. Y. Acad. Sci. **865:** 111–117.
11. Babai, N. *et al.* 2005. Degree of damage compensation by various PACAP treatments in monosodium glutamate-induced retina degeneration. Neurotox. Res. **8:** 227–233.
12. Vaudry, D. *et al.* 1998. Pituitary adenylate cyclase activating polypeptide stimulates both c-fos gene expression and cell survival in rat cerebellar granule neurons through activation of the protein kinase A pathway. Neuroscience **84:** 801–812.
13. Figiel, M. & J. Engele. 2000. Pituitary adenylate cyclase activating polypeptide (PACAP), a neuron-derived peptide regulating glial glutamate transport and metabolism. J. Neurosci. **20:** 3596–3605.

Mechanisms of VIP-Induced Neuroprotection against Neonatal Excitotoxicity

CLAIRE-MARIE RANGON,[a] ELENI DICOU,[b] STÉPHANIE GOURSAUD,[c]
LOURDES MOUNIEN,[d] SYLVIE JÉGOU,[d] THIERRY JANET,[c]
JEAN-MARC MULLER,[c] VINCENT LELIÈVRE,[a]
AND PIERRE GRESSENS[a]

[a]INSERM U 676—Université Paris 6 & Service de Neurologie Pédiatrique,
Hôpital Robert Debré, 75019 Paris, France

[b]IPMC du CNRS, 06560 Valbonne, France

[c]IPBC CNRS-UMR 6187 Pôle Biologie Santé, 86000 Poitiers, France

[d]INSERM U 413, Laboratory of Cellular and Molecular Neuroendocrinology,
European Institute for Peptide Research (IFRMP23), University of Rouen,
76821 Mont St Aignan, France

ABSTRACT: Two VIP receptors, shared with a similar affinity by pituitary
adenylate cyclase-activating polypeptide (PACAP), have been cloned:
VPAC1 and VPAC2. PHI binds to these receptors with a lower affinity.
We previously showed that VIP protects against excitotoxic white matter
damage in newborn mice. This article aimed to determine the receptor involved in VIP-induced neuroprotection. VIP effects were mimicked with
a similar potency by VPAC2 agonists and PHI but not by VPAC1 agonists,
PACAP 27 or PACAP 38. VIP neuroprotective effects were lost in mice
lacking VPAC2 receptor. In situ hybridization confirmed the presence
of VPAC2 mRNA. These data suggest that, in this model, VIP-induced
neuroprotection is mediated by VPAC2 receptors. The pharmacology of
this VPAC2 receptor seems unconventional as PACAP does not mimic
VIP effects and PHI acts with a comparable potency.

KEYWORDS: VPAC2; PACAP; PKC; BDNF; PHI

INTRODUCTION

Prepro-vasoactive intestinal peptide (VIP) mRNA codes for two neuropeptides: VIP and peptide histidine isoleucine (PHI) in rodents or VIP and peptide

Address for correspondence: Pierre Gressens, INSERM U 676, Hôpital Robert Debré, 48 Blvd
Sérurier, 75019 Paris, France. Voice: 33-1-40-03-47-83; fax: 33-1-40-03-47-74.
e-mail: gressens@rdebre.inserm.fr

Ann. N.Y. Acad. Sci. 1070: 512–517 (2006). © 2006 New York Academy of Sciences.
doi: 10.1196/annals.1317.071

histidine methionine (PHM) in humans. Two VIP receptors, shared with a similar affinity by pituitary adenylate cyclase-activating polypeptide (PACAP), have been cloned: VPAC1 and VPAC2 (for a review, see Ref. 1). PHI binds to these receptors with a lower affinity. Furthermore, PACAP-27 and PACAP-38, but not VIP, bind with high affinity to a specific PACAP receptor called the PAC1 receptor. VPAC receptors are preferentially coupled to Gαs protein that stimulates adenylate cyclase activity and induces cAMP increase (for a review, see Ref. 1). VPAC receptors can also be coupled to Gαq and Gαi proteins that stimulate the inositol phosphate/calcium/protein kinase C (PKC) pathways.

VIP BUT NOT PACAP PROTECTS THE NEWBORN BRAIN AGAINST EXCITOTOXIC LESIONS

In line with its previously reported neurotrophic properties,[2] VIP potently protects the developing brain against an excitotoxic insult in newborn mice.[3]

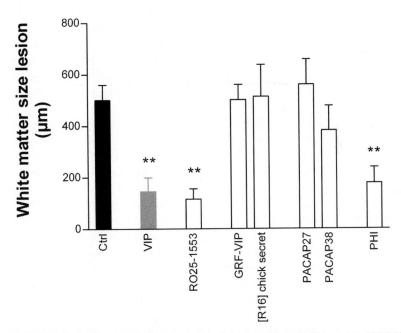

FIGURE 1. Effects of VIP, VPAC2 agonist (RO25-1553), VPAC1 agonists (GRF-VIP and [R16] chicken secretin), PACAP-27, PACAP-38, and PHI on white matter lesions induced by ibotenate injected to P5 mice. One microgram of each peptide was administered. Bar represents mean length of the lesions ± SEM. Asterisks indicate difference from phosphate-buffered saline solution (PBS) control group (black bar); $**P < 0.01$ in ANOVA with Dunnet's multiple comparison test.

In this *in vivo* model, VIP, coinjected with the glutamatergic agonist, ibotenate, in the brain of 5-day-old (P5) pups, reduces ibotenate-induced white matter lesions by up to 85 % when compared to controls. These ibotenate-induced white matter lesions mimic several aspects of periventricular leukomalacia, a white matter lesion often observed in human preterm infants. VIP effects are mimicked with a similar potency by VPAC2 agonists and PHI but not by VPAC1 agonists (FIG. 1).[4,5] Surprisingly, VIP-induced neuroprotection is not mimicked by a large range of doses of PACAP-27 or PACAP-38 (FIG. 1).[3,5]

VPAC2 RECEPTORS MEDIATE VIP-INDUCED NEUROPROTECTION

This atypical pharmacology of VIP-induced neuroprotection in newborn mice raised several hypotheses: (*a*) Activation of PAC1 receptors could have a toxic effect on the excitotoxic lesions while activation of VPAC receptors could be neuroprotective, leading to a lack of detectable effect for PACAP. (*b*) During some stages of brain development, the binding of VIP or PACAP to VPAC receptors leads to activation of separate transduction pathways. (*c*) VIP acts through a yet-to-be-identified specific VIP receptor which is not recognized by PACAP. Indeed, Ekblad *et al.*[6] characterized a PACAP-27 preferring receptor and a VIP-specific receptor, distinct from those that have been cloned (VPAC1, VPAC2, and PAC1 receptors), in intestine of rat and PAC1$^{-/-}$ mice.

The first stated hypothesis that activation of PAC1 receptors could have a toxic effect on the excitotoxic lesions while activation of VPAC receptors could be neuroprotective, leading to a lack of detectable effect for PACAP-38, can be ruled out by the lack of protective effects of PACAP-38 in PAC1$^{-/-}$ mice.[5] In contrast, VIP neuroprotective effects are completely abolished in mice lacking VPAC2 receptor.[5] *In situ* hybridization confirms the presence of VPAC2 mRNA in the postnatal day 5 white matter.[5] When analyzed between embryonic life and adulthood, VIP-specific binding site density peaks at postnatal day 5.[5] These data suggest that, in this model, VIP-induced neuroprotection is mediated by VPAC2 receptors. The pharmacology of this VPAC2 receptor seems unconventional as (*a*) PACAP does not mimic VIP effects, (*b*) PHI acts with a comparable potency, and (*c*) PACAP-27 modestly inhibits the VIP-specific binding while for PHI or VIP, inhibition is complete.

Furthermore, supporting an atypical pharmacological profile of this VPAC2 receptor, stearyl norleucine VIP, a specific VIP agonist that does not activate adenylate cyclase, mimics VIP effects and treatment with forskolin, an adenylate cyclase activator, fails to provide a VIP-like protection.[3] In contrast, VIP protective effects are abolished by a PKC inhibitor and a mitogen-associated protein kinase (MAPK) inhibitor in a dose-dependent manner.[3,7]

POTENTIAL MECHANISMS UNDERLYING THE ATYPICAL PHARMACOLOGY OF VIP EFFECTS

In order to explain the observed characteristics of VPAC2 receptors in this model of neuroprotection, some hypotheses can be formulated: (*a*) During some stages of brain development, the binding of VIP or PACAP to VPAC2 receptors leads to activation of separate transduction pathways. This differential coupling could be secondary to VPAC2 receptors dimerization (homo- or heterodimers) or to their interaction with larger oligomeric complexes, as demonstrated for other types of G protein–coupled receptors (GPCRs) (for a review, see Ref. 8). A variant of this hypothesis would be a developmental change in the G proteins available for the receptor to couple to in the relevant cells. (*b*) An alternative hypothesis has been suggested by recent studies. A first study identified a deletion variant of the mouse VPAC2 receptor in immune cells.[9] This natural deletion abrogates VIP-induced cAMP production without apparent alterations of expression or ligand binding. Second, Langer and Robberecht[10] showed that mutations in the proximal domain of the third intracellular loop of the VPAC1 receptor reduced the capability of VIP to increase adenylate cyclase activity without any change in the calcium response, whereas mutations in the distal part of the loop markedly reduced the calcium

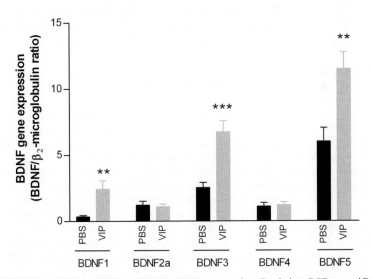

FIGURE 2. Effects of VIP on BDNF mRNA expression. Real-time PCR quantification of BDNF-1 to -5 mRNA variants in brains treated with ibotenate + 1 μg VIP or with ibotenate + PBS. Intracerebral injections were performed on P5 and tissues were collected at the site of injection 4 h later. Data are presented as mean BDNF variant/ß$_2$-microglobulin ratios ± SEM. Asterisks indicate statistically significant differences between PBS-treated controls and pups treated with VIP (**$P < 0.01$ in a *t*-test; ***$P < 0.001$ in a *t*-test).

increase and Gαi coupling but only weakly reduced the adenylate cyclase activity. Based on these studies, we can hypothesize that a yet-to-be-identified substitution or deletion in the newborn mouse VPAC2 receptor transcript, through RNA editing for instance, might be able to induce VIP specificity and modulate the coupling with different G proteins.

BRAIN-DERIVED NEUROTROPHIC FACTOR (BDNF) MEDIATES VIP-INDUCED NEUROPROTECTION

Interestingly, VIP neuroprotective effects against neonatal excitotoxic damage have recently been shown to be mediated by BDNF production (as demonstrated by BDNF-5 primers which recognize all BDNF mRNA variants) which induces a secondary axonal sprouting and white matter repair through MAPK pathway activation.[11] In particular, VIP induces through activation of this atypical VPAC2 receptor an increased expression of BDNF-1 and -3 mRNA variants (FIG. 2). These BDNF-1 and -3 mRNA transcripts have been recently shown to be decreased in brains of Alzheimer patients.[12]

CONCLUSION

Altogether, these data strongly support the hypothesis that, in newborn mice, VIP neuroprotective effects against an excitotoxic insult are mediated by VPAC2 receptors showing atypical pharmacological properties.

REFERENCES

1. VAUDRY, D. *et al*. 2000. Pituitary adenylate cyclase-activating polypeptide and its receptors: from structure to functions. Pharmacol. Rev. **2**: 269–324.
2. GRESSENS, P. *et al*. 1993. Growth factor function of vasoactive intestinal peptide in whole cultured mouse embryos. Nature **362**: 155–158.
3. GRESSENS, P. *et al*. 1997. Vasoactive intestinal peptide prevents excitotoxic cell death in the murine developing brain. J. Clin. Invest. **100**: 390–397.
4. GRESSENS, P. *et al*. 1999. Systemic administration of VIP derivatives and neuroprotection of the developing brain against excitotoxic lesions. J. Pharmacol. Exp. Ther. **288**: 1207–1213.
5. RANGON, C.M. *et al*. 2005. VPAC2 receptors mediate VIP-induced neuroprotection against neonatal excitotoxic brain lesions in mice. J. Pharmacol. Exp. Ther. **314**: 745–752.
6. EKBLAD, E. *et al*. 2000. Characterization of intestinal receptors for VIP and PACAP in rat and in PAC1 receptor knockout mouse. Ann. N. Y. Acad. Sci. **921**: 137–147.
7. GRESSENS, P. *et al*. 1998. Regulation of neuroprotective action of vasoactive intestinal peptide in the murine developing brain by protein kinase C and mitogen-activated protein kinase cascades: in vivo and in vitro studies. J. Neurochem. **70**: 2574–2584.

8. MILLIGAN, G. 2004. G protein-coupled receptor dimerization: function and ligand pharmacology. Mol. Pharmacol. **66:** 1–7.

9. GRINNINGER, C. *et al.* 2004. A natural variant type II G protein-coupled receptor for vasoactive intestinal peptide with altered function. J. Biol. Chem. **279:** 40259–40262.

10. LANGER, I. & P. ROBBERECHT. 2005. Mutations in the carboxy-terminus of the third intracellular loop of the human recombinant VPAC1 receptor impair VIP-stimulated [Ca2+]i increase but not adenylate cyclase stimulation. Cell. Signal. **17:** 17–24.

11. HUSSON, I. *et al.* 2005. BDNF-induced white matter neuroprotection and stage-dependant neuronal survival following a neonatal excitotoxic challenge. Cereb. Cortex **15:** 250–261.

12. GARZON, D., G. YU & M. FAHNESTOCK. 2002. A new brain-derived neurotrophic factor transcript and decrease in brain-derived neurotrophic factor transcripts 1, 2 and 3 in Alzheimer's disease parietal cortex. J. Neurochem. **82:** 1058–1064.

Comparative Study of the Effects of PACAP in Young, Aging, and Castrated Males in a Rat Model of Parkinson's Disease

D. REGLÓDI,[a] A. TAMÁS,[a] I. LENGVÁRI,[a] G. TOTH,[b] L. SZALONTAY,[a] AND A. LUBICS[a]

[a]Department of Anatomy, Pécs University (Neurohumoral Regulations Research Group of the Hungarian Academy of Sciences), 7624 Pécs, Hungary

[b]Department of Medical Chemistry, University of Szeged, Szeged Hungary

ABSTRACT: We have previously shown that PACAP ameliorates the neurological symptoms and reduces the dopaminergic cell loss in young male rats, in a 6-hydroxydopamine (6-OHDA)-induced lesion of the substantia nigra, a model of Parkinson's disease. In the present study, we compared the effects of PACAP in young, aging, and castrated males. Our results show that PACAP significantly reduced the dopaminergic cell loss in young and aging males. In castrated males, 6-OHDA did not induce such a severe cell loss, and it was not altered by PACAP. However, PACAP effectively ameliorated behavioral symptoms in all groups, with a degree of recovery depending on age and endocrine status.

KEYWORDS: PACAP; Parkinson; age; gender; neuroprotection

INTRODUCTION

Pituitary adenylate cyclase-activating polypeptide (PACAP) has been shown to be neuroprotective in animal models of different brain pathologies and nervous injuries, including focal and global cerebral ischemia, optic and facial nerve injuries, and traumatic brain and spinal cord injuries.[1–3] In mesencephalic cultures, PACAP is protective against 6-OHDA-induced toxicity.[4] Recently, we have shown that PACAP ameliorates the neurological symptoms and reduces the dopaminergic cell loss in young male rats, in a 6-OHDA-induced lesion of the substantia nigra (SN), a model of Parkinson's disease.[5–7] It is known that Parkinson's disease mainly affects the elderly and also shows gender differences, which may affect the possible therapeutic interventions.[8,9] Therefore,

Address for correspondence: Dora Reglódi, M.D., Ph.D., Department of Anatomy, University of Pécs, 7624 Pécs, Szigeti u 12. Hungary. Voice: +36-72-536001; fax: +36-72-536393.
e-mail: dora.reglodi@aok.pte.hu

Ann. N.Y. Acad. Sci. 1070: 518–524 (2006). © 2006 New York Academy of Sciences.
doi: 10.1196/annals.1317.072

the aim of the present article was to compare the effects of PACAP in young, aging, and castrated males in a 6-OHDA-induced unilateral SN lesion.

MATERIALS AND METHODS

Male Wistar rats were housed under standard laboratory conditions. Procedures were performed according to ethical guidelines (No: BA02/2000-31/2001). The following groups were used: 2-month-old young males ($n = 24$), 18 to 20-month-old aging males ($n = 20$), and castrated males ($n = 20$) that underwent castration at 2 month of age, 3 wk before the lesion. Drug administrations were performed as previously described.[5-8] Briefly, all animals received 2-μL 6-OHDA dissolved in saline (4 μg/μL containing 0.2% ascorbic acid) into the left SN. Half of the animals in each group received 0.1-μg PACAP38 in 0.5-μL saline preceding the 6-OHDA lesion, the other half received the same volume of saline. Behavioral testing in an open-field of 45 × 45 × 50 cm was performed 1 day before the operation (0 day) and 1 and 10 days following treatments according to previous descriptions.[5-8] Briefly, two groups of signs were evaluated from the 15-minute video recordings: activity signs (ambulation time, distance traveled, and the number of rearings), and asymmetrical signs due to the unilateral lesion (asymmetrical turning, rearing, and thigmotaxis).

Following the behavioral testing, analysis of dopaminergic cell survival was performed with tyrosine-hydroxylase (TH) immunohistochemistry on serial sections from the mesencephalon as previously described.[5-8] TH-positive cells in each section on both contralateral and ipsilateral sides of the SN pars compacta were counted using the Scion Image computerized image analysis system (Scion Corp., Frederick, MD) on digital photomicrographs. Statistical analysis was done by analysis of variance followed by Neuman Keuls *post hoc* analysis.

RESULTS AND DISCUSSION

Results of the activity signs are summarized in TABLE 1. Briefly, all three hypokinetic measures were severely reduced in young control animals 1 day after the operation when compared to 0 day values, and these signs did not show significant recovery 10 days later. In accordance with our previous observations,[5,6] PACAP-treated young males displayed no hypokinesia. Aging animals were less active in horizontal movements even before the operation, and it was further reduced 1 day after the lesion. Also, in the aging group, PACAP-treated animals showed hypokinesia 1 day after the lesion, but they recovered by 10 days. Interestingly, control aging rats showed significant recovery in activity by 10 days. Castrated males also moved less than normal young animals before the operation, but it was not further reduced after the lesion. Only the number

TABLE 1. Activity measures in the open-field before the injury (0 day) and 1 and 10 days after the injury

	Young control	Young PACAP	Aging control	Aging PACAP	Castrated control	Castrated PACAP
Ambulation						
0 day	266 ± 14		196 ± 13		158 ± 16	
1 day	192 ± 37*	285 ± 38	61 ± 10*	87 ± 15*	139 ± 40	113 ± 40
10 days	166 ± 27*	293 ± 43	106 ± 29	147 ± 30	109 ± 34	183 ± 06
Distance						
0 day	2101 ± 167		1084 ± 67		1768 ± 139	
1 day	742 ± 267**	1782 ± 289	179 ± 125***	652 ± 110	1204 ± 756	618 ± 125*
10 days	1001 ± 263*	2409 ± 369	934 ± 339	1339 ± 125	974 ± 300	2278 ± 140
Rearing						
0 day	36.6 ± 3.5		25.2 ± 3.9		43.4 ± 8.8	
1 day	9.1 ± 3.8***	23.6 ± 5.1	0 ± 0**	4 ± 2*	4.2 ± 1.9**	4.5 ± 3.5*
10 days	10.2 ± 4.2*	39.7 ± 7.7	9.6 ± 6.0*	15.0 ± 3.7	13.7 ± 5.2*	32.2 ± 6.0

*$P < 0.05$, **$P < 0.01$, ***$P < 0.001$ vs. 0 day values.

NOTE: All values are given as mean ± SEM. Ambulation is given in seconds, distance in cm traveled, and rearing in number of rearings during the 15-minute open-field test.

of rearings was significantly less 1 and 10 days after the lesion in the control group, while PACAP-treated animals showed complete recovery in rearings by 10 days. Surprisingly, castrated PACAP-treated animals traveled significantly less distance 1 day after the lesion, but it returned to normal 10 days later.

Asymmetrical signs are summarized in TABLE 2. All animals had balanced turning, rearing, and thigmotaxis before the operation, and all signs showed left bias 1 day after the lesion (except for the thigmotaxis in the castrated control group). PACAP-treated rats had significantly better recovery in almost all signs. Ten days after the lesion, PACAP-treated rats ceased to show biased movements in all groups (except for the slight asymmetry in thigmotaxis in young males). In contrast, control lesioned animals still displayed asymmetrical signs 10 days after the lesion.

In summary, behavioral signs show that PACAP treatment ameliorated the deficits caused by 6-OHDA. Best effect was observed in young males, where PACAP-treated rats showed no hypokinesia. A less striking effect was observed in aging males, where PACAP treatment improved recovery. In castrated rats, only the recovery of rearing activity was better in the PACAP-treated group. In the asymmetrical signs, PACAP-treated animals showed a better recovery in all groups. It has been described that the observed behavioral symptoms are most severe during the first few days after the lesion, and animals may show partial or complete recovery depending on the degree of dopaminergic neuronal loss.[10] Our results regarding the TH-immunopositive neurons in the SN are in accordance with the behavioral symptoms (TABLE 2). Most severe (more than 90%) reduction of the dopaminergic neurons was observed in young and aging males, and a less severe (about 70%) reduction was observed in castrated males. PACAP treatment significantly decreased the cell loss in young and aging males, while there was no significant difference in the cell numbers in the castrated groups.

The possible protective mechanisms have been previously discussed.[5–7] Briefly, 6-OHDA treatment leads to oxidative stress-induced apoptosis of the dopaminergic neurons and the degeneration is also associated with inflammatory processes, similar to human Parkinson's disease. PACAP is a potent anti-apoptotic agent, as shown in various neuronal cultures against different pro-apoptotic agents, including oxidative stress.[3,11,12] PACAP also acts as a protective mediator in inflammatory processes, mainly through attenuating the aggravating effects of microglia, which is known to play a role in Parkinson's disease by the release of proinflammatory and cytotoxic mediators.[13] Vasoactive intestinal peptide, the closest homologue peptide of PACAP, has been found to be protective in a mouse model of Parkinson's disease by blocking microglial activation.[14] Other effects of PACAP may also play a role in the observed neuroprotection and less severe acute behavioral symptoms: PACAP upregulates other neurotrophic factors, increases spontaneous motor activity, and increases dopamine levels.[3,15,16]

TABLE 2. Asymmetrical behavioral signs before the operation (0 day) and 1 and 10 days after the injury

	Young control	Young PACAP	Aging control	Aging PACAP	Castrated control	Castrated PACAP
Turning						
0 day	46.4 ± 4.1		50.8 ± 4.0		50.7 ± 5.2	
1 day	96.6 ± 1.7***	88.7 ± 3.3***	82.4 ± 10.7**	71.4 ± 3.2*	80.8 ± 4.1**	94.6 ± 3.4***
10 days	83.9 ± 5.3***	42.8 ± 8.5	70.7 ± 8.0*	63.4 ± 5.2	76.9 ± 7.1*	53.9 ± 4.5
Rearing						
0 day	41.5 ± 4.5		49.7 ± 9.3		42.6 ± 12.0	
1 day	79.6 ± 9.5***	80.0 ± 7.7***	—	75.0 ± 12.5*	74.9 ± 7.8*	75.0 ± 13.0*
10 days	76.5 ± 11.7**	42.4 ± 5.7	100 ± 0.0**	58.3 ± 8.3	76.6 ± 13.0*	54.8 ± 15.2
Thigmotaxis						
0 day	47.1 ± 7.5		62.3 ± 6.3		45.7 ± 7.9	
1 day	95.2 ± 3.8***	89.7 ± 3.0***	74.6 ± 6.7*	90.4 ± 2.5**	61.9 ± 15.2	88.4 ± 6.1**
10 days	95.0 ± 2.5***	76.8 ± 4.5**	77.5 ± 7.6*	61.7 ± 13.5	72.1 ± 9.8*	52.6 ± 7.5
Cell loss	6.4 ± 2.7***	53.8 ± 5.5**,#	2.9 ± 0.7***	34.2 ± 1.3***,#	33.4 ± 8.9**	30.1 ± 10.0**

NOTE: All values are given as mean percentage of left turning, rearing, or thigmotaxis compared to total numbers $\pm$ SEM. $* P < 0.05$, $** P < 0.01$, $*** P < 0.001$ vs. 0 day values. Note that asymmetrical rearing in the aging male group 1 day after the operation was not assessible due to the lack of rearings (see TABLE 1). Cell loss is given as percentage of TH-positive cells on the injured side compared to the normal, right side. $* P < 0.05$, $** P < 0.01$, $*** P < 0.001$ vs. normal animals (102.3 $\pm$ 21.9%), $\# P < 0.01$ vs. control animals.

In summary, our present study shows that PACAP is effective in a rat model of Parkinson's disease not only in young males but in aging and castrated males, but the degree of this effect differs depending on age and endocrine status.

ACKNOWLEDGMENTS

This work was supported by the grants OTKA T046589, T048848, and F048908 and the Hungarian Academy of Sciences.

REFERENCES

1. ARIMURA, A. 1998. Perspectives on pituitary adenylate cyclase activating polypeptide (PACAP) in the neuroendocrine, endocrine, and nervous systems. Jpn. J. Physiol. **48:** 301–331

2. WASCHEK, J.A. 2002. Multiple actions of pituitary adenylyl cyclase activating peptide in nervous system development and regeneration Dev. Neurosci. **24:** 14–23.

3. SOMOGYVARI-VIGH, A. & D. REGLODI. 2004. Pituitary adenylate cyclase activating polypeptide: a potential neuroprotective peptide. Rev. Curr. Pharm. Des. **10:** 2861–2889.

4. TAKEI, N., Y. SKOGLOSA & D. LINDHOLM. 1998. Neurotrophic and neuroprotective effects of pituitary adenylate cyclase activating polypeptide (PACAP) on mesencephalic dopaminergic neurons. J. Neurosci. Res. **54:** 698–706.

5. REGLODI, D., A. LUBICS, A. TAMAS, *et al.* 2004. Pituitary adenylate cyclase activating polypeptide protects dopaminergic neurons and improves behavioral deficits in a rat model of Parkinson's disease. Behav. Brain Res. **151:** 303–312.

6. REGLODI, D., A. TAMAS, A. LUBICS, *et al.* 2004. Morphological and functional effects of PACAP in 6-hydroxydopamine-induced lesion of the substantia nigra in rats. Regul. Pept. **123:** 85–94.

7. REGLODI, D., A. TAMÁS & A. SOMOGYVARI-VIGH. 2005. Pituitary adenylate cyclase activating polypeptide in animal models of neurodegenerative disorders— implications for Huntington and Parkinson's diseases. Lett. Drug Des. Discov. **2:** 239–244

8. TAMAS, A., A. LUBICS, L. SZALONTAY, *et al.* 2005. Age and gender differences in behavioral and morphological outcome after 6-hydroxydopamine-induced lesion of the substantia nigra in rats. Behav. Brain Res. **158:** 221–229.

9. COLLIER, T.J., C.E. SORTWELL & B.F. DALEY. 1999. Diminished viability, growth, and behavioral efficacy of fetal dopamine neuron grafts in aging rats with long-term dopamine depletion: an argument for neurotrophic supplementation. J. Neurosci. **19:** 5563–5573.

10. SCHWARTING, R.K. & J.P. HUSTON. 1996. The unilateral 6-hydroxydopamine lesion model in behavioral brain research: analysis of functional deficits, recovery and treatments. Prog. Neurobiol. **50:** 275–331.

11. VAUDRY, D., C. ROUSSELLE, M. BASILLE, *et al.* 2002. PACAP protects cerebellar granule neurons against oxidative stress-induced apoptosis. Eur. J. Neurosci. **15:** 1451–1460.

12. Ito, Y., M. Arakawa, K. Ishige, *et al.* 1999. Comparative study of survival signal withdrawal- and 4-hydroxynonenal-induced cell death in cerebellar granule cells. Neurosci. Res. **35:** 321–327.
13. Delgado, M., J. Leceta & D. Ganea. 2003. Vasoactive intestinal peptide and pituitary adenylate cyclase activating polypeptide inhibit the production of inflammatory mediators by activated microglia. J. Leukoc. Biol. **73:** 155–164.
14. M. Delgado & D. Ganea. 2003. Neuroprotective effect of vasoactive intestinal peptide (VIP) in a mouse model of Parkinson's disease by blocking microglial activation. FASEB J. **17:** 944–946.
15. Masuo, Y., J. Noguchi, S. Morita, *et al.* 1995. Effects of intracerebroventricular administration of pituitary adenylate cyclase activating polypeptide (PACAP) on the motor activity and reserpine-induced hypothermia in murines. Brain Res. **700:** 219–236.
16. Lee, F.S., R. Rajagopal, A.H. Kim, *et al.* 2002. Activation of trk neurotrophin receptor signaling by pituitary adenylate cyclase activating polypeptides. J. Biol. Chem. **277:** 9096–9102.

VIP and Tolerance Induction in Autoimmunity

F. ROSIGNOLI,[a] M. TORROBA,[a] Y. JUARRANZ,[a] M. GARCÍA-GÓMEZ,[a] C. MARTINEZ,[b] R.P. GOMARIZ,[a] C. PÉREZ-LEIRÓS,[c] AND J. LECETA[a]

[a]Dept. Biología Celular, Facultad de Biologia, UCM, Madrid, Spain

[b]Dept. Biología Celular, Facultad de Medicina, UCM, Madrid, Spain

[c]Dept. Quimica Biologica, Facultad de Ciencias Naturales y Exactas, UBA-CONICET, Buenos Aires, Argentina

ABSTRACT: Vasoactive intestinal peptide (VIP) is a potent anti-inflammatory agent with immunoregulatory properties, skewing the immune response to a Th2 pattern of cytokine production. Here, we studied the effect of treatment with VIP in the development of diabetes in nonobese diabetic (NOD) mice, an animal model of type 1 diabetes. Mice treated with VIP from 4 weeks of age did not develop diabetes and showed milder insulitis than nontreated mice. The protective mechanism of VIP was associated with a reduction in the circulating levels of Th1 cytokines. In the pancreas of VIP-treated animals, regulatory T cell markers predominate, as indicated by the upregulation of FoxP3 and transforming growth factor-β (TGF-β), and the downregulation of the transcription factor, T-bet. These findings indicate that VIP restores tolerance to pancreatic islets by promoting the local differentiation and function of regulatory T cells.

KEYWORDS: VIP; autoimmune; FoxP3; TGF-β

INTRODUCTION

The NOD strain of mouse is a useful model of autoimmune type 1 diabetes. The pathological mechanisms that lead to disease development are not known but the disease is associated with loss of immunological tolerance, in which autoreactive T cells play a major role in the pathogenesis. Th1-dependent cellular immune response and its associated cytokine secretion profile are thought to be responsible for the destruction of β cells.[1] *In vivo* systemic or local delivery of Th2 and regulatory cytokines[2], as well as diverting the immune response, have been reported to prevent disease development and have been proposed as effective immunotherapeutic strategies.[3]

Address for correspondence: Javier Leceta, Dept. Biología Celular, Facultad de Biologia, UCM 28040 Madrid, Spain. Voice: 34-913-944-971; fax: 34-913-944-981.

e-mail: jleceta@bio.ucm.es

Ann. N.Y. Acad. Sci. 1070: 525–530 (2006). © 2006 New York Academy of Sciences.

doi: 10.1196/annals.1317.073

Vasoactive intestinal peptide (VIP) is widely distributed in the central and peripheral nervous system as well as in the immune system, where it is produced by different cell types and exhibit important immunoregulatory functions,[4] inhibiting Th1 and promoting Th2 immune responses.[5] In this sense, VIP has been proposed as a therapeutic candidate for autoimmune diseases such as rheumatoid arthritis, multiple sclerosis, or autoimmune diabetes.[6] In the present study, we show that treating NOD mice with VIP from 4 weeks of age protects them against diabetes development. The *in vivo* effect is correlated with a decrease in Th1/Th2 balance and the induction of markers of T-regulatory function in the pancreas.

RESULTS

VIP Prevents Spontaneous Diabetes and Insulitis

Overt diabetes, as determined by the circulating levels of glucose, was apparent from 18 weeks onward in female NOD mice and augmented gradually to reach 70% of diabetic mice at 30 weeks of age. None of the NOD mice treated from 4 weeks of age with 2.5 nmoles/animal of VIP (i.p.) every other day developed diabetes during the duration of the study (FIG. 1A). Since insulitis is a key feature in the development of type 1 diabetes, we studied the infiltration of the islets by histological methods from 10 weeks of age. Infiltration was evident in 80% of the islets in NOD mice from 10 weeks of age and its severity augmented progressively to the highest grade of infiltration apparent in 60–80% of the islets from 20 to 30 weeks of age (FIG. 1B). Insulitis was not observed in NOD mice treated with VIP at 10 weeks of age. Although it was apparent from 15 to 30 weeks of age, insulitis in VIP-treated NOD mice appeared in <50% of the islets and was mostly restricted to the periphery (FIG. 1B).

VIP Decreases the Th1/Th2 Balance and Enhances the Expression of FoxP3 and TGF-β in the Pancreas

In NOD mice, a generalized intrinsic T cell defect is manifested in a dysregulated cytokine effector function,[1] and a marker of diabetes progression is the increase in the balance of circulating levels of Th1/Th2 cytokines.[7] To find out if the protective effect of VIP against the development of insulitis and diabetes is associated with changes in circulating levels of Th1/Th2 cytokines we have determined the levels of interleukin (IL)-12, IL-4, and IL-10 in serum by enzyme-linked immunosorbent assay (ELISA). As shown in FIGURE 2A, circulating levels of both, IL-4 and IL-12 were readily detected in NOD mice but do not change significantly along the duration of the study. However, the

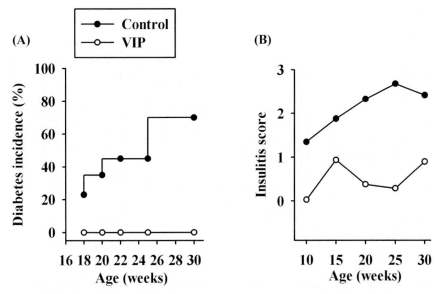

FIGURE 1. Protection against diabetes development and decreased insulitis in VIP-treated NOD mice. Each point represents the mean value ($n = 6$). (**A**) Cumulative incidence of diabetes in vehicle and VIP-treated NOD mice. (**B**) Histological scoring of insulitis was performed on pancreatic sections stained with hematoxylin/eosin. Islets were scored for absence of insulitis (grade 0), peri-insulitis (grade 1), moderate insulitis (grade 2), and severe insulitis (grade 3).

circulating levels of IL-10 were almost undetectable until 30 weeks of age. VIP treatment had no significant effects on the circulating levels of IL-4, but reduced the IL-12 serum levels and greatly increased the IL-10 levels. Since a breakdown of immunoregulatory balance in the pancreas has been implicated in the development of diabetes in NOD mice, we paid attention to the Th1/Th2 deviation in the pancreas at 15 weeks of age, a time point close to the onset of diabetes, as well as to the presence of regulatory T cells that have been involved in the control of autoimmunity. Markers for the lineage commitment to Th1 and Th2 are the transcription factors T-bet and GATA3, respectively,[8] while the expression of FoxP3 and transforming growth factor (TGF)-β is associated with the commitment to regulatory T cells. The quantitative determination of these markers by reverse-transcription polymerase chain reaction (RT-PCR) in mRNA extracted from the pancreas showed that the transition from the prediabetic stage to diabetes onset was marked by an increase in the Th1 commitment marker, T-bet, while markers for Th2 and T-regulatory subsets remained low (FIG. 2B). VIP treatment significantly reduced the expression of T-bet and increased the expression of the Th2 commitment marker, GATA-3. Furthermore, the markers for T-regulatory function, FoxP3, and TGF-β, were significantly upregulated in the pancreas of VIP-treated mice.

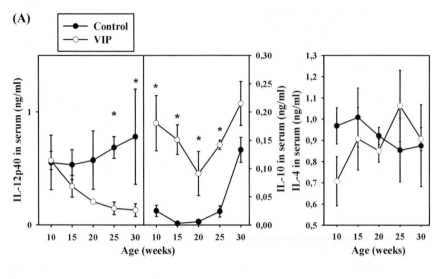

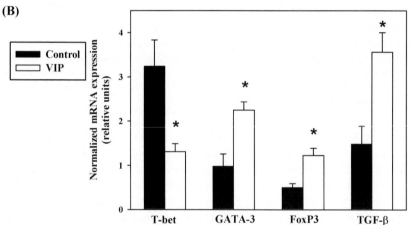

FIGURE 2. (**A**) Circulating cytokine levels. Circulating levels of IL-12, IL-10, and IL-4 were obtained by sandwich ELISA on serum samples. *$P < 0.05$. (**B**) mRNA expression of T-bet, GATA-3, FoxP3, and TGF-β in the pancreas of vehicle and VIP-treated mice at 15 weeks of age. Expression of the corresponding mRNA was measured by quantitative RT-PCR and corrected by mRNA expression for β-actin in each sample. *$P < 0.05$.

DISCUSSION

This study shows that treatment of NOD mice with VIP inhibits diabetes development. The data indicate that the protective mechanism may be mediated by the induction of Th2-like immune response and the differentiation of regulatory T cells counteracting the Th1-dependent cellular immune response in the pancreas.

Central and peripheral tolerance defects are responsible for diabetes in NOD mice. Th1 effectors for autoantigens expressed in the pancreatic islets are associated with disease progression. During its development, the dominance of Th2 cytokines appears at the early stage of nondestructive insulitis that is followed by a predominance of Th1 cytokines during destructive insulitis.[7] Conversely, it appears that deviation to Th2 responses helps protection against diabetes. Elevation of circulating levels of IL-4 and IL-10 by injection or gene therapy has been shown to prevent or ameliorate autoimmune diabetes in NOD mice in a dose-dependent manner, inducing changes in the proportion of T cell subsets in the periphery and the islet lesions, without reduction of islet infiltration.[9] The protective effect of VIP in diabetes development may be due partially to its effect in the Th1/Th2 balance. A shift in this balance was evident in the pancreas after VIP treatment that may prevent islet destruction in spite of a patent infiltration, which nevertheless does not reach the core of the islet. VIP shift toward a Th2-like response locally in the pancreas is indicated by the upregulation of GATA3 and the downregulation of T-bet. In the periphery, however, although VIP decreases IL-12 and increases IL-10 circulating levels, there is no effect in IL-4 levels in serum. However, some authors have questioned the role of the Th1/Th2 balance as a determinant of susceptibility to develop diabetes based on the fact that both IL-12 and interferon-γ knockout NOD mice develop diabetes.[10] Some experimental approaches indicate that cytokine shift is an outcome rather than a cause of diabetes protection and may be attributed to the higher resistance of activated Th2 cells to apoptosis in NOD mice.[11] Instead of that, they point to multiple immunoregulatory cell defects.[12] Prominent among the cellular component of this regulatory network are $CD4^+CD25^+$ cells in which expression of the transcription factor FoxP3 is necessary for the development and function of the different subsets of regulatory T cells.[13] Although the mode of action of these cells is not well understood, they express a number of cell surface markers as well as a particular cytokine secretion pattern.[14] In line with these observations, maximal contribution of VIP to prevent diabetes development seems to be mediated by the generation of regulatory T cells. The upregulation of FoxP3 and TGF-β indicate a potentiation of regulatory functions in the pancreas that may lead to tolerance restoration to pancreatic autoantigens. Given the action of VIP in the development of regulatory T cells reported here and the efficacy of these cells to control autoimmune process, this peptide arises as a promising candidate for an effective treatment of type 1 diabetes. Further studies are needed to identify these cells in the islet infiltrates or the pancreatic lymph node and their efficacy to suppress the action of effector diabetogenic cells in transfer experiments.

ACKNOWLEDGMENTS

This work was supported by grants BFI 2002-03489 from Ministerio de Ciencia y Tecnología (Spain) and a postdoctoral fellowship from Fundacion Antorcha (Argentina).

REFERENCES

1. KOARADA, S., Y. WU, G. OLSHANSKY, et al. 2002. Increased nonobese diabetic Th1:Th2 (IFN-:IL-4) ratio is CD4 T cell intrinsic and independent of APC genetic background. J. Immunol. **169:** 6580–6587.
2. MATHISEN, P.M. & V.K. TUOHY. 1998. Gene therapy in the treatment of autoimmune disease. Immunol. Today **19:** 103–105.
3. RYU, S., S. KODAMA, K. RYU, et al. 2001. Reversal of established autoimmune diabetes by restoration of endogenous b cell function. J. Clin. Invest. **108:** 63–72.
4. GOMARIZ, R.P., C. MARTINEZ, C. ABAD, et al. 2001. Immunology of VIP: a review and therapeutical perspectives. Curr. Pharm. Des. **7:** 89–111.
5. DELGADO, M., J. LECETA & D. GANEA. 2002. Vasoactive intestinal peptide and pituitary adenylate cyclase-activating polypeptide promote in vivo generation of memory Th2 cells. FASEB J. **16:** 1844–1846.
6. DELGADO, M., C. ABAD, C. MARTINEZ, et al. 2002. Vasoactive intestinal peptide in the immune system: potential therapeutic role in inflammatory and autoimmune diseases. J. Mol. Med. **80:** 16–24.
7. SCHLOOT, N.C., P. HANIFI-MOGHADDAM, C. GOEBEL, et al. 2002. Serum IFN-g and IL-10 levels are associated with disease progression in non-obese diabetic mice. Diabetes Metab. Res. Rev. **18:** 64–70.
8. LIEW, F.Y. 2002. TH1 and TH2 cells: a historical perspective. Nat. Rev. Immunol. **2:** 55–60.
9. TOMINAGA, Y., M. NAGATA, H. YASUDA, et al. 1998. Administration of IL-4 prevents autoimmune diabetes but enhances pancreatic insulitis in NOD mice. Clin. Immunol. Immunopathol. **86:** 209–218.
10. TREMBLEAU, S., G. PENNA, S. GREGORI, et al. 2003. IL-12 administration accelerates autoimmune diabetes in both wild-type and IFN-γ-deficient nonobese diabetic mice, revealing pathogenic and protective effects of IL-12-induced IFN-γ. J. Immunol. **170:** 5491–5501.
11. SERREZE, D.V., H.D. CHAPMAN, C.M. POST, et al. 2001. Th1 to Th2 cytokine shifts in nonobese diabetic mice: sometimes an outcome, rather than the cause, of diabetes resistance elicited by immunostimulation. J. Immunol. **166:** 1352–1359.
12. KUKREJA, A., G. COST, J. MARKER, et al. 2002. Multiple immuno-regulatory defects in type-1 diabetes. J. Clin. Invest. **109:** 131–140.
13. HORI, S., T. NOMURA & S. SAKAGUCHI. 2003. Control of regulatory T cell development by the transcription factor FoxP3. Science **299:** 1057–1061.
14. ZHENG, S.G., J.H. WANG, J.D. GRAY, et al. 2004. Natural and induced CD4$^+$CD25$^+$ cells educate CD4$^+$CD25$^-$ cells to develop suppressive activity: the role of IL-2, TGF-β, and IL-10. J. Immunol. **172:** 5213–5221.

Neuroprotective Effect of PACAP against Kainic Acid-Induced Neurotoxicity in Rat Retina

TAMOTSU SEKI,[a,b,c] MASAYOSHI NAKATANI,[d] CHISATO TAKI,[d]
YUKO SHINOHARA,[d] MOTOKI OZAWA,[d] SHIGERU NISHIMURA,[d]
HIROYUKI ITO,[a] AND SEIJI SHIODA[b]

[a]Department of Ophthalmology, Kozawa Eye Hospital Eye Research Center,
Mito 3100063, Japan

[b]Department of Anatomy, Showa University School of Medicine,
Tokyo 1428555, Japan

[c]Department of Ophthalmology, Showa University School of Medicine,
Tokyo 1478666, Japan

[d]Bioengineering Institute, Assessment Research, Nidek Co., Ltd.,
Gamagohri Aichi 4430022, Japan

ABSTRACT: Pituitary adenylate cyclase-activating polypeptide (PACAP) is well known to protect delayed neuronal cell death in the brain of rodents. In order to investigate the neuroprotective action of PACAP in the retina, we examined the effects of PACAP on kainic acid (KA)-induced neurotoxicity in the rat retina. Many ganglion cells in the retina died after KA injection in the control group and PACAP treatment significantly promoted cell survival. These findings strongly suggest that PACAP plays very important roles in preventing cell death in the retina.

KEYWORDS: PACAP (pituitary adenylate cyclase-activating polypeptide); kainic acid; rat retina; ganglion cell

INTRODUCTION

Pituitary adenylate cyclase-activating polypeptide (PACAP) and its receptors are widely expressed in the central nervous system and eye.[1–4] In the retina, PACAP has been reported to attenuate glutamate-induced delayed neurotoxicity in cultured retinal neurons and in the neonatal rat *in vivo*.[5,6] We have already reported that intravitreal injection of PACAP increases the number of

Address for correspondence: Dr. Tamotsu Seki, Department of Ophthalmology, Kozawa Eye Hospital Eye Research Center, 2-2-11 Gokencho, Mito, Ibaraki 3100063, Japan. Voice: +81-29-224-5722; fax: +81-29-225-5721.
e-mail: t.seki@kozawa-ganka.or.jp

Ann. N.Y. Acad. Sci. 1070: 531–534 (2006). © 2006 New York Academy of Sciences.
doi: 10.1196/annals.1317.074

surviving retinal ganglion cells (RGCs) after transient ischemia or transection of the optic nerve.[7] However, it has not been shown clearly whether PACAP has a neuroprotective effect against retinal damages induced by kainic acid (KA), a non-N-methyl-D-aspartate (NMDA) glutamate receptor agonist. Therefore, in the present article we have investigated the effect of PACAP on delayed neuronal cell death.

MATERIALS AND METHODS

Adult Wistar rats were anesthetized by intraperitoneal injection of 45 mg/kg sodium pentobarbital. Ten picomol of PACAP, 5 nmol of KA (Sigma, S. Louis, MO), or 100 nmol of 6-cyano-7-nitroquinoxaline-2,3,-dione (CNQX, Sigma), a non-NMDA receptor antagonist diluted in 3 μL of 0.9% saline, were coinjected into the vitreous body using a 30-gauge needle via a corneal limbus. Saline alone administered in the same way was used as a negative control. Apoptotic-like cell death was detected with a terminal deoxynucleotidyl transferase-mediated dUTP nick end-labeling (TUNEL) assay. KA-induced neuronal damage was evaluated 7 days after KA injection by determining phosphorylated neurofilament heavy chain subunit (pNF-H) content in the retina with an enzyme-linked immunosorbent assay (ELISA) system. The protein concentration of each sample was measured with the Bio-Rad DC protein assay (Bio-Rad Laboratories, Hercules, CA). Animals were sacrificed 1, 2, 3, or 7 days after treatment and the eyes were dissected. Frozen retinal sections (10-μ thick) were cut. Double immunostaining of PAC-1 receptor (PAC1-R) and glial fibrillary acidic protein (GFAP) was carried out. The sections were also immunostained with a monoclonal antibody against interleukin-6 (IL-6) (1:50; Biosource Inc., Worchester, MA) to detect the localization of IL-6 in the rat retina. The excitotoxicity of KA was examined in the ganglion cell layer (GCL). Differences were analyzed using Mann-Whitney U test and Student's t-test.

RESULTS AND DISCUSSION

Many TUNEL-positive cells were observed in the inner nuclear layer (INL) especially abundant in its inner part. In the retinas, 7 days after preinjection of saline 2 days prior to the injection of 5 nmol of KA, which were used as the control group, the mean pNF-H contents were decreased to 31% of the normal retina. Preinjection of 10 pmol of PACAP at the second day before KA injection significantly inhibited the reduction of pNF-H content in the retina, compared with the saline-injected control group. Coinjection of 10 pmol of PACAP with KA did not cause a significant inhibition, while coinjection of 100 nmol of CNQX significantly inhibited the reduction of pNF-H. In the eyes

7 days after KA injection, cell loss in the GCL and INL, and thinning of the inner plexiform layer (IPL) were observed. The numbers of RGCs were 55.4 ± 4.9 cells/mm in the normal eyes, 24.0 ± 6.7 cells/mm in saline-injected eyes, and 34.8 ± 9.4 cells/mm in PACAP-injected eyes (mean $\pm$ SD, $n = 8$). These data demonstrate that PACAP significantly inhibits KA-induced cell loss in the GCL (FIG. 1).

PACAP significantly inhibited the loss of pNF-H content in the retina. However, PACAP coinjected with KA did not cause any significant changes in the retina, while CNQX treatment almost completely inhibited KA-induced cell death. We have already shown that PAC_1-R-like immunoreactivity is localized in neuronal cell bodies in the GCL and INL.[2] These results suggest that PACAP protects RGC death through PAC1-R as shown in the hippocampus.[1]

We also investigated the changes in pNF-H in the retina. It has been reported that the loss of nerve fiber protein in RGC axons corresponds to the extent of excitotoxicity induced by NMDA and KA, while the dephosphorylation status of nerve fibers in RGC axons corresponds to glaucomatous damage.[8,9] It has been reported that endogenous IL-6 is upregulated in the IPL and INL after ischemic injury.[10] Furthermore, the addition of IL-6 enhances the survival of RGCs *in vitro*.[11] In our immunohistochemical studies, longitudinal IL-6 expression was observed throughout the length from the NFL to the outer plexiform layer (OPL) at the second and third days after intravitreal injection of PACAP in the untreated retina. There is little information on the neuroprotective mechanism involving IL-6 in the retina. We have already reported that intracerebroventricular injection of PACAP stimulates astrocytes to secrete

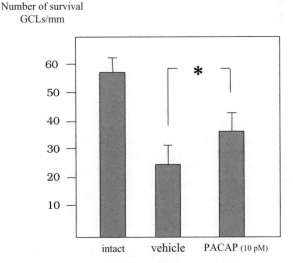

FIGURE 1. Numbers of survival ganglion cells after KA injection. *$P < 0.05$ versus saline group (Mann-Whitney U test) Data expressed as the mean $\pm$ SD ($n = 8$/group)

IL-6 into the cerebrospinal fluid.[1] We have also reported that PACAP inhibits the activation of Jun N-terminal kinase/stress-activated protein kinase in neurons in the CA1 region after ischemic stress.[12] We have shown in another paper published in this volume that PACAP stimulates the release of IL-6 in cultured rat Müller cells *in vitro*.[13] These findings, together with previous reports, suggest that PACAP prevents retinal cell death induced by KA and that alterations in the expression of PAC_1-R in GFAP-positive cells, presumably Müller cells, may play important roles to prevent neurotoxicity induced by KA.

REFERENCES

1. SHIODA, S. 2000. Pituitary adenylate cyclase-activating polypeptide (PACAP) and its receptors in the brain. Kaibogaku Zasshi **75:** 487–507.
2. VAUDRY, D., B.J. GONZALEZ, M. BASILLE, ET AL. 2000. Pituitary adenylate cyclase-activating polypeptide and its receptors: from structure to function. Pharmacol. Rev. **52:** 269–324.
3. SEKI, T., S. SHIODA, *et al.* 1997. Distribution and ultrastructural localization of a receptor for pituitary adenylate cyclase activating polypeptide and its mRNA in the rat retina. Neurosci. Lett. **238:** 127–130.
4. SEKI, T., S. IZUMI, *et al.* 2000. Gene expression for PACAP receptor mRNA in the rat retina by *in situ* hybridization and *in situ* RT-PCR. Ann. N. Y. Acad. Sci. **921:** 366–369.
5. SHOGE, K. *et al.* 1999. Attenuation by PACAP of glutamate-induced neurotoxicity in cultured retinal neurons. Brain Res. **839:** 66–73.
6. TAMAS, A., R. GABRIEL, *et al.* 2004. Effects of pituitary adenylate cyclase activating polypeptide in retinal degeneration induced by monosodium-glutamate. Neurosci. Lett. **372:** 110–113.
7. SEKI, T., S. IZUMI, *et al.* 2003. Pituitary adenylate cyclase-activating polypeptide (PACAP) protects ganglion cell death against cutting of optic nerve in the rat retina. Regul. Pept. **115:** 55.
8. KASHIWAGI, K., B. OU, S. NAKAMURA, *et al.* 2003. Increase in dephosphorylation of the heavy neurofilament subunit in the monkey chronic glaucoma model. Invest. Ophthalmol. Vis. Sci. **44:** 154–159.
9. ZHANG, X., M. CHENG, *et al.* 2004. Kainic acid-mediated upregulation of matrix metalloproteinase-9 promotes retinal degeneration. Invest. Ophthalmol. Vis. Sci. **45:** 2374–2383.
10. MENDONCA TORRES, P.M. & E.G. DE ARAUJO. 2001. Interleukin-6 increases the survival of retinal ganglion cells *in vitro*. J. Neuroimmunol. **117:** 43–50.
11. CHILDOW, G. & N.N. OSBORNE. 2003. Rat retinal ganglion cell loss caused by kainate, NMDA and ischemia correlates with a reduction in mRNA and protein of Thy-1 and neurofilament light. Brain Res. **963:** 298–306.
12. SHIODA, S., H. OZAWA, K. DOHI, *et al.* 1998. PACAP protects hippocampal neurons against apoptosis: involvement of JNK/SAPK signaling pathway. Ann. N. Y. Acad. Sci. **865:** 111–117.
13. SEKI, T., Y. SHINOHARA, C. TAKI, *et al.* 2006. PACAP stimulates the release of Interleukin-6 in cultured rat Müller cells. Ann. N. Y. Acad. Sci. In press.

PACAP Stimulates the Release of Interleukin-6 in Cultured Rat Müller Cells

T. SEKI,[a,b,c] Y. HINOHARA,[d] C. TAKI,[d] M. NAKATANI,[d] M. OZAWA,[d]
S. NISHIMURA,[d] A. TAKAKI,[e] H. ITHO,[a,c] F. TAKENOYA,[c]
AND S. SHIODA[c]

[a]Ophthalmology, Kozawa Eye Hospital Eye Research Center,
Mito 310-0063, Japan

[b]Ophthalmology and [c]Anatomy, Showa University School of Medicine,
Tokyo 142-8666, Japan

[d]Bioengineering Institute, Assessment Research, Nidek Co. Ltd.,
Aichi 443-0038, Japan

[e]Integrative Physiology, Graduate School of Medical Sciences,
Kyushu University, Fukuoka 812-8581, Japan

ABSTRACT: We have investigated the *in vivo* effect of PACAP on rat
Müller cells that are the predominant glial element in the retina. Müller
cells were treated with PACAP38, either alone or in the presence of the
PACAP-selective antagonist, PACAP6–38. Cellular proliferation was determined by measuring the incorporation of bromodeoxyuridine, while
interleukin-6 (IL-6) levels in the culture medium were examined using a
B9 cell bioassay. In cultured rat Müller cells, the expression of PACAP receptor (PAC1-R) was assessed with immunohistochemistry using a PAC1-
R-specific antiserum. PACAP stimulated IL-6 production in Müller cells
at a concentration as low as 10^{-12} M, which was not sufficient to induce
cell proliferation. This elevation of IL-6 production was significantly inhibited by PACAP6–38. These data suggest that Müller cells are one of
the target cells for PACAP, stimulating the release of IL-6, and providing
a mechanism whereby PACAP exerts a significant neuroprotective effect
in the retina.

KEYWORDS: PACAP;IL-6;rat retina;Müller cell;PAC1-R; neuroprotective effect

Address for correspondence: Dr. Tamotsu Seki, Ophthalmology, Kozawa Eye Hospital Eye Research
Center, 2-2-11 Gokencho, Mito, Ibaraki 310-0063, Japan. Voice: +81-29-224-5722; fax: +81-29-225-
5721.
e-mail: t.seki@kozawa-ganka.or.jp

Ann. N.Y. Acad. Sci. 1070: 535–539 (2006). © 2006 New York Academy of Sciences.
doi: 10.1196/annals.1317.043

INTRODUCTION

Numerous eye diseases, including glaucoma,[1] involve the death of retinal ganglion cells. However, despite recent advances, this cell death remains relatively intractable to treatment. Pituitary adenylate cyclase-activating polypeptide (PACAP) is known to prevent delayed neuronal death following a hippocampal ischemic insult and has also been reported to stimulate the protective release of interleukin-6 (IL-6) from astrocytes.[2] On the basis of these findings, we investigated the mitogenic effects and the changes in IL-6 production induced by PACAP in cultured rat retinal Müller cells that comprise the major glial population in the retina.

MATERIALS AND METHODS

Müller cells were isolated from 14- or 15-day-old Wistar rats,[3] and cultured in 25 mM HEPES and DMEM (Dullbecco's modified minimum essential medium), containing 10% fetal bovine serum at 37°C in 5% CO_2. The cells were treated with different concentrations (10^{-14} to 10^{-6} M) of PACAP38 in serum-free medium for 48 h. Müller cells were identified immunocytochemically by the expression of glutamine synthetase (GS, 1:800 dilution; Chemicon, Temecula, CA), glial fibrillary acidic protein (GFAP, 1:1,000 dilution; Sigma Chemical Company, St. Louis, MO), and cellular retinaldehyde-binding protein (CRALBP, Santa Cruz Biotechnology, Santa Cruz, CA). CRALBP is expressed in Müller cells but not in astrocytes.[4,5] Cell proliferation was assayed by the incorporation of the thymidine analog bromodeoxyuridine (BrdU) using an enzyme-linked immunosorbent assay (ELISA) kit. IL-6 levels in the supernatants, obtained from 24-h cultures, were determined using a B9 cell assay. We also determined whether the increase in IL-6 was PACAP specific in 48-h cultured Müller cells exposed to the specific antagonist, PACAP6–38. At the end of the incubation period, proliferation was determined colorimetrically using 3-(4,5-dimethylthiazol-2-yl)-2,5-diphenyl tetrazolium bromide (MTT).[6] The cells were solubilized with acidic sodium dodecyl sulfate and the plates read at 570 nm with a microplate reader.

RESULTS AND DISCUSSION

The incorporation of BrdU into Müller cells was significantly stimulated by PACAP at 10^{-9}, 10^{-8}, and 10^{-7} M, with maximum cellular proliferation (2.23-fold above the control) in response to 10^{-9} M PACAP. In the absence of PACAP, cultured Müller cells produced very little IL-6. However, even at 10^{-12} M of PACAP treatment, IL-6 levels in the treated medium were 40-fold higher than in the control, with increased levels recorded from 10^{-12} M to

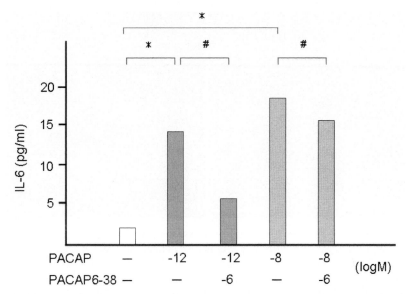

FIGURE 1. Effect of PACAP6–38 on the production of IL-6 in cultured rat Müller cells. Müller cells were treated with PACAP6–38 in the absence and presence of PACAP38 for 24 h. Data represent the mean ± SD ($n = 4$) *$P < 0.01$ versus no treatment; #$P < 0.01$ versus treatment with 10^{-12} M or 10^{-8} M PACAP38.

10^{-8} M (FIG. 1) and with 10^{-6} M.[7] Treatment with PACAP6–38 (a potent antagonist of the PAC1-R) alone at a concentration of 10^{-6} M did not affect the production of IL-6 (data not shown). However, PACAP6–38 significantly inhibited the action of PACAP in a dose-dependent fashion. When treated with both PACAP6–38 (10^{-6} M) and PACAP (10^{-12} M and 10^{-8} M), the mean IL-6 level was significantly decreased (FIG. 1).

The expression and localization of PAC1-R in cultured rat Müller cells were assessed by immunohistochemistry by the use of a specific antiserum. No obvious immunoreactivity was detected in the control that was incubated with normal serum. However, PAC1-R-like immunoreactivity was intensely observed in Müller cells (FIG. 2). On the basis of these and other morphological findings, we believe that IL-6 is produced in Müller cells, which may express PAC1-R. These results strongly suggest that Müller cells represent the target of PACAP in the retina, and that PACAP may play a neuroprotective role, mediating the release of IL-6 from Müller cells, and thus providing protection for neurons, including retinal ganglion cells.

In this study, we found that PACAP-stimulated IL-6 production in cultured Müller cells at the lowest concentration tested, 10^{-12} M, which was not sufficient to induce cellular proliferation. Paradoxically, PACAP has previously been reported to induce cellular proliferation at a concentration as low as 10^{-12} M, a concentration that did not stimulate the production of IL-6

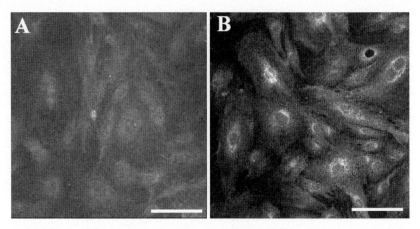

FIGURE 2. Light photomicrographs showing immunohistochemical labeling for PAC1-R in cultured rat Müller cells. (**A**) normal serum, (**B**) PAC1-R antiserum bar = 100 μm.

in cultured astrocytes.[8] The difference between astrocytes and Müller cells can be explained on the basis of the intrinsic mitogenic ability of each cell type. Gottschall *et al.* have reported that PACAP stimulates IL-6 secretion in astrocytes across a wide range of concentrations (from 10^{-11} M to 10^{-7} M).[9] In contrast to the mitogenic responses, our results regarding IL-6 production are very similar to those reported by these authors. There are two types of PACAP receptors: PAC1-R, which binds to PACAP with much higher affinity than to VIP, and VPAC1 and VPAC2 receptors, which bind to VIP and PACAP with similar affinities.[10] PAC1-R is known to be coupled to several signal transduction pathways that activate adenylate cyclase, phospholipase C, extracellular signal-regulated kinases, mitogen-activated protein (MAP) kinase, and p38 MAP kinase, whereas the VPAC receptors are mainly coupled to the adenylate cyclase pathway.[10,11] Surprisingly, the elevation in IL-6 production observed with PACAP at the low concentration of 10^{-12} M was inhibited by PACAP6–38 to a greater extent than when PACAP was given at a higher concentration (10^{-8} M). We have previously reported on the localization and gene expression of PAC1-R in the rat retina.[12] Strong PAC1-R mRNA expression and immunoreactivity were detected in retinal ganglion cells, amacrine cells, and in the inner plexiform layer. We have also reported a significant decrease in retinal ganglion cell death compared with vehicle 7 days after PACAP treatment, both in an ischemia reperfusion model based on intraocular hypertension and in an optic nerve transection model.[13] This effect was bimodal, with peaks at PACAP concentrations of 10^{-14} M and 10^{-11}–10^{-10} M. These results suggest that functional PAC1-R and/or VPAC2 R are expressed on Müller cells and that they could be involved in IL-6-mediated neuroprotection in response to low concentrations of PACAP. The increased cellular proliferation and elevated IL-6 production observed at higher concentrations of PACAP may be involved in the activation of VPAC1-R in addition to PAC1-R. However, in the present study, we were unable to identify the exact type of PACAP receptors found in Müller cells. Similarly, the underlying signaling mechanisms remain unknown, and form the basis of ongoing studies.

REFERENCES

1. GARCIA-VALENZUELA, E., S. SHAREEF, *et al.* 1995. Programmed cell death of retinal ganglion cells during experimental glaucoma. Exp. Eye Res. **61:** 33–44.
2. SHIODA, S., H. OZAWA, K. DOHI, *et al.* 1998. PACAP protects hippocampal neurons against apoptosis: involvement of JNK/SAPK signaling pathway. Ann. N. Y. Acad. Sci. **865:** 111–117.
3. HICKS, D. & Y. COURTOIS. 1990. The growth and behavior of rat retinal Müller cells in vitro. 1. An improved method for isolation and culture. Exp. Eye Res. **51:** 119–129.
4. TOIMELA, T., H. MAENPAA, *et al.* 2000. Retinal Müller cell culture. ATLA **28:** 477–482.
5. SARTHY, V.P., S.J. BRODJIAN, K. DUTT, *et al.* 1998. Establishment and characterization of a retinal Muller cell line. Invest. Ophthalmol. Vis. Sci. **39:** 212–216.
6. TAKAKI, A., Q.H. HUANG, *et al.* 1994. Immobilization stress may increase plasma interleukin-6 via central and peripheral catecholamines. Neuroimmunomodulation **1:** 335–342.
7. SHINOHARA, Y., T. SEKI, *et al.* 2005. Pituitary adenylate cyclase-activating peptide (PACAP) stimulates proliferation and production of interleukin-6 in cultured rat Muller cells. Invest. Ophthalmol. Vis. Sci. **46:** E-Abstract 181.
8. TATSUNO, I., H. MORIO, *et al.* 1999. Pituitary adenylate cyclase-activating polypeptide (PACAP) is a regulator of astrocytes: PACAP stimulates proliferation and production of interleukin 6 (IL-6), but not nerve growth factor (NGF), in cultured rat astrocyte. Ann. N. Y. Acad. Sci. **805:** 482–488.
9. GOTTSCHALL, P.E., I. TATSUNO, *et al.* 1994. Regulation of interleukin-6 (IL-6) secretion in primary cultured rat astrocytes: synergism of interleukin-1 (IL-1) and pituitary adenylate cyclase activating polypeptide (PACAP). Brain Res. **637:** 197–203.
10. VAUDRY, D., B.J. GONZALEZ, *et al.* 2000. Pituitary adenylate cyclase-activating polypeptide and its receptors: from structure to functions. Pharmacol. Rev. **52:** 269–324.
11. HASHIMOTO, H., A. KUNUGI, *et al.* 2003. Possible involvement of a cyclic AMP-dependent mechanism in PACAP-induced proliferation and ERK activation in astrocytes. Biochem. Biophys. Res. Commun. **311:** 337–343.
12. SEKI, T., S. SHIODA, D. OGINO, *et al.* 1997. Distribution and ultrastructural localization of a receptor for pituitary adenylate cyclase activating polypeptide and its mRNA in the rat retina. Neurosci. Lett. **238:** 127–130.
13. SEKI, T., S. IZUMI, *et al.* 2003. Pituitary adenylate cyclase-activating polypeptide (PACAP) protects ganglion cell death against cutting of optic nerve in the rat retina. Regul. Pept. **115:** 55.

VIP Protects Th2 Cells by Downregulating Granzyme B Expression

VIKAS SHARMA,[a] MARIO DELGADO,[b] AND DOINA GANEA[a]

[a]Department of Biological Sciences, Rutgers University, Newark, New Jersey 07102, USA

[b]Instituto de Parasitologia y Biomedicina, CSIC, Granada 18001, Spain

ABSTRACT: Selective differentiation of Th1/Th2 effectors contributes to cell- or antibody-mediated immunity. Vasoactive intestinal peptide (VIP) induces Th2 responses by promoting Th2 differentiation and survival. Here we investigate the mechanisms of VIP-induced Th2 survival. Microarray and protein data indicate that VIP prevents the upregulation of granzyme B (GrB) in Th2, but not Th1 effectors. This is the first report of GrB expression and of its involvement in activation-induced apoptosis of T helper cells. The enhanced responsiveness of Th2 cells to VIP is probably due to the higher expression of VIP receptors and alternative signaling pathways. This study identifies GrB as a new significant player in Th1/Th2 activation-induced cell death, and characterizes the mechanisms for the protective effect of VIP on Th2 survival.

KEYWORDS: Th1/Th2 differentiation; vasoactive intestinal peptide; granzyme B; T cell apoptosis

INTRODUCTION

Following antigenic stimulation, CD4$^+$ T cells differentiate into Th1 and Th2 effectors, with different cytokine profiles and different physiological functions.[1] The differentiation into Th1/Th2 effectors is controlled by various factors, such as the nature of the antigen-presenting cells (APCs), the nature and amount of antigen, the genetic background of the host, and particularly the cytokine microenvironment.[2]

Clonally expanded CD4$^+$ T cells are eliminated primarily through activation-induced cell death (AICD), and the major mechanism involves signaling through the death receptor CD95 (Fas).[3] Although both Th1 and Th2 effectors are ultimately eliminated, Th1 cells are more susceptible to AICD.[4,5] Recent studies indicate that Th1 cells are more susceptible to death induced by either

Address for correspondence: Doina Ganea, Temple University School of Medicine, Dept. Physiology, 3420 N. Broad St., Philadelphia, PA 19140. Voice: 215-707-9921; fax: 215-707-4003.
e-mail: dganea@temple.edu

Ann. N.Y. Acad. Sci. 1070: 540–544 (2006). © 2006 New York Academy of Sciences.
doi: 10.1196/annals.1317.077

FasL or TNF-related apoptosis-inducing ligand (TRAIL), and that both Th1 and Th2 cells can kill Th1, but not Th2 targets.[6]

RESULTS AND DISCUSSION

Endogenous factors such as progesterone, glucocorticoids, and neuropeptides, such as vasoactive intestinal peptide (VIP) and pituitary adenylate cyclase-activating polypeptide (PACAP), have been reported to favor Th2 differentiation.[7,8] VIP/PACAP was shown to affect Th1/Th2 differentiation through effects on APCs and direct effects on the master transcriptional factors c-Maf and JunB.[9–11] In addition, we reported previously that VIP/PACAP promotes the specific survival of Th2 effectors *in vivo*.[12] In this study, we investigated the mechanisms involved in the VIP-induced survival of Th2 effectors. Our experiments reveal a surprising new mechanism involved in Th1/Th2 AICD, i.e., the induction of enzymatically active granzyme B (GrB) upon CD3 restimulation.

To investigate the molecular mechanisms by which VIP promotes the survival of Th2 cells, we focused on *in vitro* experiments using Th1 and Th2 effectors. Th1 and Th2 effectors, generated following anti-CD3/anti-CD28 stimulation, were restimulated with immobilized anti-CD3 and treated with or without VIP for 24 h. As reported earlier, VIP protected Th2 cells and not Th1 from undergoing AICD (Fig. 1A).

To investigate the molecular mechanism of Th2 protection, we analyzed the differential gene expression by macroarray analysis (results not shown). GrB was one of the genes downregulated by VIP. GrB has been associated previously with cytotoxic T cells and NK cells and its function in $CD4^+$ T cells has not yet been defined. Therefore, we investigated if GrB can induce apoptosis in $CD4^+$ Th2 cells. Effector Th2 cells were restimulated with immobilized anti-CD3 in the presence of a GrB inhibitor or blocking anti-Fas antibodies. The GrB inhibitor protected >50% Th2 cells from undergoing AICD, which was higher than the protection from Fas. The combination of the GrB inhibitor and anti-Fas Abs completely inhibited apoptosis in Th2 cells. This is the first report of GrB inducing apoptosis in $CD4^+$ T effector cells (Fig. 1B).

Since we observed downregulation of GrB mRNA expression by VIP in Th2 cells, we investigated if the downregulation was consistent at protein and enzymatic activity level. Th1 and Th2 cells were restimulated with immobilized anti-CD3 in the presence or absence of VIP for 6 h. Cells were fixed and stained for GrB-protein expression (Fig. 1C). Cell lysates were also extracted from restimulated Th2 cells and analyzed for GrB-enzyme activity (Fig. 1D). Consistent with the mRNA data, GrB protein and enzyme activity were downregulated by VIP in Th2 cells and not in Th1 cells.

This study defines the function of GrB, i.e., to induce apoptosis in $CD4^+$ T effectors. VIP protects Th2 cells from undergoing apoptosis by downregulating

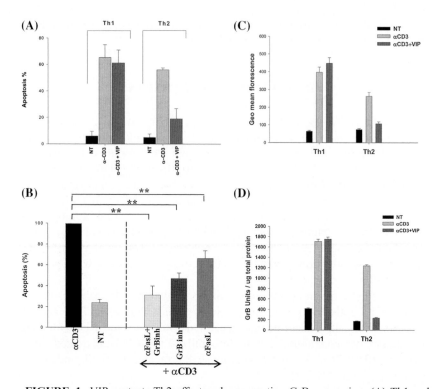

FIGURE 1. VIP protects Th2 effectors by preventing GrB expression. **(A)** Th1 and Th2 effectors were preincubated with VIP followed by restimulation with immobilized anti-CD3 and apoptosis was determined by annexin V/PI staining. **(B)** Th2 effectors were preincubated with the GrB inhibitor I (20 μM), the neutralizing anti-FasL Ab, or both, followed by restimulation. Apoptosis was determined 24 h later by TUNEL assay. The results are the mean ± SD of three independent experiments. **statistically significant ($P < 0.005$). **(C)** Th1 and Th2 effectors were preincubated with VIP, followed by restimulation. Cells were fixed and permeabilized, followed by staining with rabbit anti-mouse GrB Ab. The secondary reagent was FITC-conjugated F(ab)'2 goat anti-rabbit IgG. Normal goat IgG was used as control. The cells were analyzed by FACS. **(D)** Th1 and Th2 effectors were preincubated with VIP, followed by restimulation. GrB activity was determined in cell lysates by using the synthetic colorimetric GrB substrate Ac-IEPD-pNA. The GrB activity was calculated based on a standard curve generated with recombinant murine GrB and normalized to the amounts of total protein in lysates. The results are the mean ±SD of three independent experiments.

GrB expression. In lpr (Fas mutant) Th1 and Th2 effectors, CD3 restimulation results in apoptosis mediated solely by GrB (results not shown), suggesting that GrB induction is independent of Fas signaling. VIP also downregulates FasL expression in Th2 cells, but not in Th1 cells.

We conclude that VIP induces the specific survival of wild-type Th2 effectors by preventing the induction of GrB and the upregulation of FasL expression

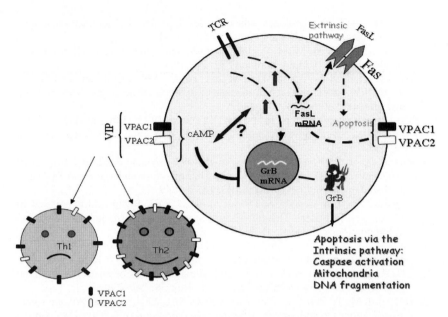

FIGURE 2. TCR restimulation induces FasL and GrB expression in both Th1 and Th2 effectors, leading to apoptosis. Th2 effectors express higher levels of VPAC1 and VPAC2, being more responsive to VIP. VIP receptor signaling activates the adenylate cyclase, ultimately preventing FasL and GrB expression in Th2 effectors. Whether the cAMP signaling pathway acts directly on the expression of the GrB gene, or interferes with the TCR-induced GrB expression remains to be determined.

in Th2, but not Th1 cells (FIG. 2). The VIP-induced survival of Th2 effectors is in agreement with its general anti-inflammatory function. By tilting the balance in favor of Th2 cells through effects on both Th1/Th2 differentiation and survival, VIP contributes to the reduction of the pro-inflammatory potential of the immune response. This is particularly relevant for sites with high abundance of VIP sources, such as the GI tract and the immune-privileged organs, where acute inflammatory processes are particularly harmful.

ACKNOWLEDGMENTS

This work was supported by grants AI 47325 and AI 52306 (DG), and the Johnson & Johnson Neuroimmunology Fellowships (VS).

REFERENCES

1. MOSSMAN, T.R., H. CHERWINSKI, M.W. BOND, *et al.* 1986. Two types of murine helper T cell clone. I. Definition according to profiles of lymphokine activities and secreted proteins. J. Immunol. **136:** 2348–2357.

2. O'GARRA, A. 1998. Cytokines induce the development of functionally heterogeneous T helper cell subsets. Immunity **8**: 275–283.
3. ASNAGLI, H. & K.M. MURPHY. 2001. Stability and commitment in T helper cell development. Curr. Opin. Immunol. **13**: 242–247.
4. BRUNNER, T., R.J. MOGIL, D. LAFACE, *et al.* 1995. Cell-autonomous Fas (CD95)/Fas-ligand interaction mediates activation-induced apoptosis in T-cell hybridomas. Nature **373**: 441–444.
5. VARADHACHARY, A.S., S.N. PERDOW, C. HU, *et al.* 1997. Differential ability of T cell subsets to undergo activation-induced cell death. Proc. Natl. Acad. Sci. USA **94**: 5778–5783.
6. ZHANG, X., T. BRUNNER, L. CARTER, *et al.* 1997. Unequal death in T helper cell (Th)1 and Th2 effectors: Th1, but not Th2, effectors undergo rapid Fas/FasL-mediated apoptosis. J. Exp. Med. **185**: 1837–1849.
7. ZHANG, X.R., L.Y. ZHANG, S. DEVADAS, *et al.* 2003. Reciprocal expression of TRAIL and CD95L in Th1 and Th2 cells: role of apoptosis in T helper subset differentiation. Cell Death. Differ. **10**: 203–210.
8. MIYAURA, H & M. IWATA. 2002. Direct and indirect inhibition of Th1 development by progesterone and glucocorticoids. J. Immunol. **168**: 1087–1094.
9. DELGADO, M., D. POZO & D. GANEA. 2004. The significance of vasoactive intestinal peptide in immunomodulation. Pharmacol. Rev. **56**: 249–290.
10. DELGADO, M., A. REDUTA, V. SHARMA & D. GANEA. 2004. VIP/PACAP oppositely affects immature and mature dendritic cell expression of CD80/CD86 and the stimulatory activity for CD4 (+) T cells. J. Leukoc. Biol. **75**: 1122–1130.
11. GOETZL E.J., J.K. VOICE, S. SHEN, *et al.* 2001. Enhanced delayed-type hypersensitivity and diminished immediate-type hypersensitivity in mice lacking the inducible VPAC(2) receptor for vasoactive intestinal peptide. Proc. Natl. Acad. Sci. USA **98**: 13854–13859.
12. DELGADO, M., J. LECETA & D. GANEA. 2002. Vasoactive intestinal peptide and pituitary adenylate cyclase-activating polypeptide promote in vivo generation of memory Th2 cells. FASEB J. **16**: 1844–1846.

Serotonergic Inhibition of Intense Jumping Behavior in Mice Lacking PACAP (*Adcyap1*$^{-/-}$)

NORIHITO SHINTANI,[a] HITOSHI HASHIMOTO,[a] KAZUHIRO TANAKA,[a] NAOFUMI KAWAGISHI,[a] CHIHIRO KAWAGUCHI,[a] MICHIYOSHI HATANAKA,[a] YUKIO AGO,[a,b] TOSHIO MATSUDA,[b] AND AKEMICHI BABA[a]

[a]*Laboratory of Molecular Neuropharmacology, Graduate School of Pharmaceutical Sciences, Osaka University, Suita, Osaka 565-0871, Japan*

[b]*Laboratory of Medicinal Pharmacology, Graduate School of Pharmaceutical Sciences, Osaka University, Suita, Osaka 565-0871, Japan*

ABSTRACT: Genetic manipulation of pituitary adenylate cyclase-activating polypeptide (PACAP) in mice has uncovered its involvement in psychomotor function. We previously observed that mice lacking the *Adcyap1* gene encoding the neuropeptide PACAP (*Adcyap1*$^{-/-}$) displayed intense jumping behavior when placed in a novel environment such as an open field. Here, we show that *Adcyap1*$^{-/-}$ mice manifest jumping behavior as early as at least 6 weeks of age when compared with wild-type mice and that the selective serotonin (5-HT) reuptake inhibitor, fluoxetine, as well as the serotonin precursor, 5-hydroxytryptophan, suppress jumping behavior. Our previous study showed a slight decrease in 5-HT metabolite, 5-hydroxyindoleacetic acid (5-HIAA) in *Adcyap1*$^{-/-}$ mouse brain. Taken together, these results suggest that there is a developmental aspect to the jumping behavior seen in *Adcyap1*$^{-/-}$ mice, and that jumping behaviour may involve the serotonergic system.

KEYWORDS: *Adcyap1*; behavior; PACAP; psychiatric disorder; selective serotonin reuptake inhibitor; serotonin

INTRODUCTION

Controlled psychomotor function is provided not only by the regulatory influence of small-molecule neurotransmitters, but also by some regulatory neuropeptides having important physiological significance. One of them is the

Address for correspondence: Hitoshi Hashimoto, Laboratory of Molecular Neuropharmacology, Graduate School of Pharmaceutical Sciences, Osaka University, 1-6 Yamadaoka, Suita, Osaka 565-0871, Japan. Voice: 81-6-6879-8181; fax: 81-6-6879-8184.
e-mail: hasimoto@phs.osaka-u.ac.jp

Ann. N.Y. Acad. Sci. 1070: 545–549 (2006). © 2006 New York Academy of Sciences.
doi: 10.1196/annals.1317.079

neuropeptide, pituitary adenylate cyclase-activating polypeptide (PACAP), a member of the vasoactive intestinal peptide (VIP)/secretin/glucagon family.[1–3]

Because of the lack of selective low-molecular-weight PACAP antagonists suitable for *in vivo* study, we have previously generated mice lacking the *Adcyap1* gene encoding the neuropeptide PACAP (*Adcyap1*$^{-/-}$).[4] *Adcyap1*$^{-/-}$ mice show marked central nervous system (CNS) phenotypes: they have a high early mortality rate before weaning, and the surviving *Adcyap1*$^{-/-}$ females exhibit reduced fertility, which is partly due to reduced mating frequency.[5] Furthermore, *Adcyap1*$^{-/-}$ mice display remarkable behavioral changes, including hyperactive and explosive jumping behavior in an open field, and increased novelty seeking behavior.[4] The hyperactive behavior can be ameliorated by the antipsychotic drug haloperidol. In *Adcyap1*$^{-/-}$ mouse brain, the serotonin (5-HT) metabolite 5-hydroxyindoleacetic acid (5-HIAA) level is slightly decreased in the cerebral cortex and striatum. These findings suggest possible involvement of altered monoaminergic neurotransmission in the CNS phenotypes of *Adcyap1*$^{-/-}$ mice, however, their mechanistic basis and pathophysiological significance remain largely unclear.

Three different knockout lines[6–8] developed separately from our colony have been reported to show dysfunction of lipid and carbohydrate metabolism, cold hypersensitivity, impaired catecholamine regulation in the sympathoadrenal axis, and deficits in the circadian light response (for reviews, see Hashimoto *et al.*, this volume).[9]

In the present article, intense jumping behavior in *Adcyap1*$^{-/-}$ mice was investigated in relation to age-related changes and the effect of elevation of extracellular 5-HT by the selective serotonin reuptake inhibitor, fluoxetine, and the 5-HT precursor, 5-hydroxytryptophan (5-HTP).

MATERIALS AND METHODS

Animals

All animal experiments were carried out in accordance with protocols approved by the Animal Care and Use Committee of Graduate School of Pharmaceutical Sciences, Osaka University, Japan. Generation of *Adcyap1*$^{-/-}$ mice by a gene-targeting technique has been reported previously.[4] The null mutation was backcrossed onto an Institute of Cancer Research (ICR) mouse background for at least five times. Three genotypes—wild-type, heterozygous (*Adcyap1*$^{+/-}$)—, and homozygous (*Adcyap1*$^{-/-}$)—mice were obtained from heterozygote crosses.

Behavioral Study

The number of jumps by mice in a $30 \times 30 \times 30$-cm Plexiglas open field was scored for 90 min using video recordings by experienced observers blinded

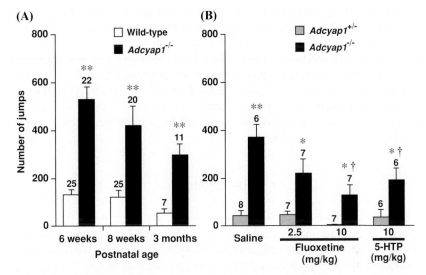

FIGURE 1. Jumping behavior in an open field. (**A**) Number of jumps for 90 min was determined in wild-type (open bars) and *Adcyap1*$^{-/-}$ (solid bars) at postnatal weeks 6 and 8 and 3 mo. (**B**). Effects of fluoxetine and 5-HTP at the indicated doses on the number of jumps were examined in 5-week-old *Adcyap1*$^{+/-}$ (shaded bars) and *Adcyap1*$^{-/-}$ (solid bars). The numbers of mice are indicated above the bars. Data are given as mean + S.E. *$P < 0.05$ and **$P < 0.01$ compared with wild-type (**A**) or *Adcyap1*$^{+/-}$ (**B**); $^{†}P < 0.05$ compared with saline; two-way ANOVA with the Turkey–Kramer multiple comparisons test.

to the genotypes of mice, as previously described.[4] Drugs were injected intraperitoneally just prior to placing mice in the open field.

RESULTS

Developmental Changes of Jumping Behavior

To examine the developmental changes of jumping behavior in *Adcyap1*$^{-/-}$ mice, jumping was assessed at postnatal weeks 6 and 8 and 3 mo (FIG. 1A). *Adcyap1*$^{-/-}$ mice exhibited a significantly increased number of jumps compared with wild-type mice at all postnatal ages tested (6–8 week and 3 month) [two-way ANOVA, $F (1, 109) = 47.1, P < 0.0001$]. There was no significant difference in the number of jumps between *Adcyap1*$^{+/-}$ mice and wild-type mice (data not shown).

Effects of Fluoxetine and 5-HTP on Jumping Behavior in Adcyap1$^{-/-}$ *Mice*

We previously showed that 5-HIAA levels are slightly decreased in the cerebral cortex and striatum in *Adcyap1*$^{-/-}$ mice,[4] implying that altered 5-HT neurotransmission is involved in some behavioral phenotypes in these mutant mice. Therefore, we examined the effects of elevating extracellular 5-HT by systemic administration of the selective serotonin reuptake inhibitor, fluoxetine, or the 5-HT precursor, 5-HTP (FIG. 1B). Fluoxetine [2.5 and 10 mg/kg, intraperitoneal (i.p.)] attenuated the number of jumps in $Adcyap1^{-/-}$ mice in a dose-dependent manner [$F(2, 36) = 3.84, P = 0.031$]. 5-HTP (10 mg/kg, i.p.) similarly attenuated the number of jumps in $Adcyap1^{-/-}$ mice [$F(1, 22) = 4.79, P = 0.040$]. Neither fluoxetine nor 5-HTP significantly affected jumping in $Adcyap1^{+/-}$ control mice.

DISCUSSION

The present findings support our previous study showing that PACAP plays a pivotal role in the regulation of psychomotor behavior. In this study, $Adcyap1^{-/-}$ mice manifested increased jumping behavior as early as at least 6 weeks of age when compared with wild-type mice. $Adcyap1^{-/-}$ mice also showed a high degree of novelty-induced hyperactivity from a similar early age (unpublished data). Several lines of evidence suggest that PACAP acts as a neurotrophic factor and plays diverse roles in mammalian neurogenesis and regeneration.[1,2,10,11] In contrast, it has been shown that PACAP depletion does not affect early (embryonic days 10.5 and 12.5) development of the monoaminergic nervous system.[12] Therefore, it is conceivable that late developmental deficits and/or defects in the acute role of PACAP contribute to the intense jumping as well as hyperactive behaviors in $Adcyap1^{-/-}$ mice.

The present results, taken together with our previous findings, suggest that an alteration in 5-HT neurotransmission or the relative balance between 5-HT and other neurotransmitters, such as dopamine, may contribute to abnormal behavior seen in $Adcyap1^{-/-}$ mice. Although the pathophysiological significance of increased jumping behavior in $Adcyap1^{-/-}$ mice is unclear, it can be speculated that the behavior and its responsiveness to fluoxetine and 5-HTP are relevant to certain aspects of antidepressant therapy.

A plethora of knockout phenotypes have been shown for PACAP and its receptors, including altered stress responses and defects in circadian rhythm, as well as psychomotor change.[9] The mechanisms involved and their possible role in human diseases are of great interest and warrant further investigation.

ACKNOWLEDGMENTS

This research was supported, in part, by Grants-in-Aid for Scientific Research (A) and (B), and for Young Scientists (B) from Japan Society for the Promotion of Science, and by a grant from Taisho Pharmaceutical Co. Ltd.

REFERENCES

1. ARIMURA, A. 1998. Perspectives on pituitary adenylate cyclase activating polypeptide (PACAP) in the neuroendocrine, endocrine, and nervous systems. Jpn. J. Physiol. **48:** 301–331.

2. VAUDRY, D., B.J. GONZALEZ, M. BASILLE, *et al.* 2000. Pituitary adenylate cyclase-activating polypeptide and its receptors: from structure to functions. Pharmacol. Rev. **52:** 269–324.

3. HASHIMOTO, H., N. SHINTANI & A. BABA. 2002. Higher brain functions of PACAP and a homologous Drosophila memory gene amnesiac: insights from knockouts and mutants. Biochem. Biophys. Res. Commun. **297:** 427–431.

4. HASHIMOTO, H., N. SHINTANI, K. TANAKA, *et al.* 2001. Altered psychomotor behaviors in mice lacking pituitary adenylate cyclase-activating polypeptide (PACAP). Proc. Natl. Acad. Sci. USA **98:** 13355–13360.

5. SHINTANI, N., W. MORI, H. HASHIMOTO, *et al.* 2002. Defects in reproductive functions in PACAP-deficient female mice. Regul. Pept. **109:** 45–48.

6. GRAY, S.L., K.J. CUMMINGS, F.R. JIRIK, *et al.* 2001. Targeted disruption of the pituitary adenylate cyclase-activating polypeptide gene results in early postnatal death associated with dysfunction of lipid and carbohydrate metabolism. Mol. Endocrinol. **15:** 1739–1747.

7. HAMELINK, C., O. TJURMINA, R. DAMADZIC, *et al.* 2002. Pituitary adenylate cyclase-activating polypeptide is a sympathoadrenal neurotransmitter involved in catecholamine regulation and glucohomeostasis. Proc. Natl. Acad. Sci. USA **99:** 461–466.

8. COLWELL, C.S., S. MICHEL, J. ITRI, *et al.* 2004. Selective deficits in the circadian light response in mice lacking PACAP. Am. J. Physiol. Regul. Integr. Comp. Physiol. **287:** R1194–R1201.

9. HASHIMOTO, H., N. SHINTANI & A. BABA. 2006. New insights into the central PACAPergic system from the phenotypes in PACAP- and PACAP receptor-knockout mice. Ann. N. Y. Acad. Sci. This volume.

10. SUH, J., N. LU, A. NICOT, *et al.* 2001. PACAP is an anti-mitogenic signal in developing cerebral cortex. Nat. Neurosci. **4:** 123–124.

11. WASCHEK, J.A. 2002. Multiple actions of pituitary adenylyl cyclase activating peptide in nervous system development and regeneration. Dev. Neurosci. **24:** 14–23.

12. OGAWA, T., T. NAKAMACHI, H. OHTAKI, *et al.* 2005. Monoaminergic neuronal development is not affected in PACAP-gene-deficient mice. Regul. Pept. **126:** 103–108.

Pleiotropic Functions of PACAP in the CNS

Neuroprotection and Neurodevelopment

SEIJI SHIODA, HIROKAZU OHTAKI, TOMOYA NAKAMACHI,
KENJI DOHI, JUN WATANABE, SHIGEO NAKAJO, SATORU ARATA,
SHINJI KITAMURA, HIROMI OKUDA, FUMIKO TAKENOYA,
AND YOSHITAKA KITAMURA

Department of Anatomy, Showa University School of Medicine, Shinagawa-ku, Tokyo 142-8555, Japan

ABSTRACT: Pituitary adenylate cyclase-activating polypeptide (PACAP) is a pleiotropic neuropeptide that belongs to the secretin/glucagon/vasoactive intestinal peptide (VIP) family. PACAP prevents ischemic delayed neuronal cell death (apoptosis) in the hippocampus. PACAP inhibits the activity of the mitogen-activated protein kinase (MAPK) family, especially JNK/SAPK and p38, thereby protecting against apoptotic cell death. After the ischemia-reperfusion, both pyramidal cells and astrocytes increased their expression of the PACAP receptor (PAC1-R). Reactive astrocytes increased their expression of PAC1-R, released interleukin-6 (IL-6) that is a proinflammatory cytokine with both differentiation and growth-promoting effects for a variety of target cell types, and thereby protected neurons from apoptosis. These results suggest that PACAP itself and PACAP-stimulated secretion of IL-6 synergistically inhibit apoptotic cell death in the hippocampus. The PAC1-R is expressed in the neuroepithelial cells from early developmental stages and in various brain regions during development. We have recently found that PACAP, at physiological concentrations, induces differentiation of mouse neural stem cells into astrocytes. Neural stem cells were prepared from the telencephalon of mouse embryos and cultured with basic fibroblast growth factor. The PAC1-R immunoreactivity was demonstrated in the neural stem cells. When neural stem cells were exposed to PACAP, about half of these cells showed glial fibrillary acidic protein (GFAP) immunoreactivity. This phenomenon was significantly antagonized by a PAC1-R antagonist (PACAP6-38), indicating that PACAP induces differentiation of neural stem cell into astrocytes. Other our physiological studies have demonstrated that PACAP acts on PAC1-R in mouse neural stem cells and its

Address for correspondence: Dr. Seiji Shioda, Department of Anatomy, Showa University School of Medicine, 1-5-8 Hatanodai, Shinagawa-ku, Tokyo 142-8555, Japan. Voice: +81-3-3784-8103; fax: +81-3-3784-6815.

e-mail: shioda@med.showa-u.ac.jp

Ann. N.Y. Acad. Sci. 1070: 550–560 (2006). © 2006 New York Academy of Sciences.
doi: 10.1196/annals.1317.080

signal is transmitted to the PAC1-R-coupled G protein Gq but not to Gs. These findings strongly suggest that PACAP plays very important roles in neuroprotection in adult brain as well as astrocyte differentiation during development.

KEYWORDS: PACAP; brain ischemia; delayed cell death; neural stem cells

INTRODUCTION

Pituitary adenylate cyclase-activating polypeptide (PACAP) belongs to the secretin/glucagon/vasoactive intestinal peptide (VIP) family and exists in two α-amidated forms, PACAP-38 and PACAP-27, that share the same N-terminal 27 amino acids and are derived from a precursor of 176 or 175 amino acids.[1] PACAP was first isolated from the hypothalamus and thought to be a hypophysiotropic hormone, but its specific receptor, PAC1-R, has been demonstrated in the brain, testis, pancreas, adrenal gland, and other organs or tissues as well as the anterior pituitary gland. In addition, PACAP has pleiotropic functions, such as the regulation of neurodevelopment and protection against delayed neuronal cell death (apoptosis) in the brain and spinal cord. Thus, this article focuses on the functional significance of PACAP and the PAC1 receptor in the brain, especially during neurodevelopment and neurodegeneration. We will present here our recent findings and discuss the possibility that PACAP could be used for the clinical treatment of brain ischemia and injury, and further to make an important role of PACAP in regenerative medicine in near future.

PACAP IN NEURODEVELOPMENT

PACAP has been found to stimulate the outgrowth of neurites from PC12 cells and to enhance the survival of sympathetic ganglion cells.[2,3] The effect of PACAP on naturally occurring neuronal cell death was examined in the chick embryo. The daily administration of PACAP-38 from embryonic day (E)3.5 to E8.5 significantly reduced the number of pyknotic cells in the dorsal root ganglion on E9.[4] PACAP-38 also functions as a neurotrophic factor during the development of motoneurons in the chick spinal cord.[4] Cultured cerebellar granule cells undergo apoptosis in the absence of fetal calf serum within 48 h. The addition of PACAP-38 or PACAP-27 dose dependently rescued the cells from apoptosis. PACAP-38 was effective at 0.1 nM, and PACAP-27 suppressed apoptosis at 1 nM or higher concentrations.[5] These findings suggest that the neuroprotective effect of PACAP is mediated through PACAP receptors.

We have shown that PACAP affects the neuroepithelial cells isolated from the neural fold (headfold) at the primitive streak stage as early as E9.[6] In sequential studies, we have shown that PAC1-R, the specific receptor for PACAP, is expressed in the neuroepithelium at this early developmental stage, and we have also determined that two splice variants (PAC1-R-s and PAC1-R-hop) are expressed in the same region.[7] These data support the hypothesis that the protein kinase A (PKA) or PKC signaling cascades (which are coupled to PAC1-R) are activated by their agonists and inhibited by antagonists in the PACAP-responsive neuroepithelial cells.

PACAP has been shown to induce cell cycle withdrawal and promote the transition from proliferation to differentiation in cultured cortical precursor cells from E13 rats.[8] PACAP potently increases cyclic AMP (cAMP) levels in cultured hindbrain neuroepithelial cells from E10.5 mice and downregulates the expression of the *sonic hedgehog*-gene and the PKA-dependent target gene *gli*-1.[9] The rat at E13 or mouse at E10.5 have completed the process of turning in which the extra-embryonic membranes surround the headfolds and then change their conformation dynamically. We have revealed that before this process, PACAP affects the neuroepithelial cells at the primitive streak stage at E9. Furthermore, PKA has been demonstrated as a common negative regulator of Hedgehog signaling in vertebrate embryos,[10] and we[11] have shown that PKA is the main signaling cascade (and PKC is the minor signaling cascade) in PACAP-responsive neuroepithelial cells. This should be considered in relation to our previous results that PAC1-R-s (without any cassette) was the main receptor isoform expressed and PAC1-R-hop (which contains a "hop" cassette which encodes a consensus motif for phosphorylation by PKC) was the minor receptor isoform.[7] However, the differences between these two splice variants in signaling pathways and related physiological functions still remain to be demonstrated.

It is well established that neural stem cells have pluripotent and self-renewing properties.[12–14] Neural progenitor cells prepared from embryonic brain undergo mitosis and subsequent differentiation into neurons and glial cells. However, the molecular mechanism of the differentiation of neural stem cells is less well characterized. It has been demonstrated that neural stem cells can be differentiated into astrocytes via the action of ciliary neurotrophic factor (CNTF), into oligodendrocytes by the thyroid hormone T3, and into neurons by brain-derived neurotrophic factor (BDNF) and platelet-derived growth factor (PDGF).[15] Furthermore, the participation of bone morphogenetic protein 2 and leukemia inhibitory factor in the differentiation of neural stem cells to astrocytes,[16] of Mash1, Math3,[17–19] neurogenin,[20] neurogenesin-1,[21] and REN[22] in differentiation to neurons, and of hes5 in differentiation to oligodendrocytes[23] has been reported. In contrast, the basic region helix-loop-helix (bHLH) gene *Hes1* prevents neural differentiation in the central nervous system (CNS).[24,25] The differentiation of neural stem cells *in vivo* as well as *in vitro* is believed to be regulated by such transcription factors.

When isolated cells prepared from telencephalons of E14.5 mouse embryos were cultured in the presence of bFGF, they proliferated in a time-dependent manner. Immunostaining was performed to determine the cell types present, with the results showing that about 85% of the cells in the primary culture were immunopositive for nestin at day 4. The typical cell shape was morphologically very similar to that of neural stem cells as reported previously.[15] In the absence of bFGF, these cells differentiated into neurons, astrocytes, and oligodendrocytes following the addition of PDGF isoform BB, CNTF, and T3, respectively.[15] The cells prepared during these experiments were characterized by self-renewal and multiple lineage properties, indicating that they were neural stem cells.

During the exposure of neural stem cells to various substances instead of bFGF in an attempt to seek a novel differentiation factor, we found that physiological concentrations (0.1–2 nM) of PACAP evoked morphological changes in these cells.[26] The results indicated that PACAP induced an increase in the number of glial fibrillary acidic protein (GFAP)-positive cells, which was significant at PACAP concentrations as low as 0.2 nM, and maximal at 2 nM PACAP. Approximately 55%, 30%, and 16% of the cells treated with 2 nM PACAP for 7 days were immunopositive for GFAP, nestin, and microtubule-associated protein-2, respectively, while Gal-C-immunopositive cells accounted for just 4% of all cells.[26] These results suggest that PACAP mainly induces the differentiation of neural stem cells into astrocytes. This induction of differentiation was significantly antagonized by the PACAP antagonist PACAP$_{6-38}$, while PACAP$_{6-38}$ used alone had virtually no effect on the differentiation process.[26] The number of GFAP-positive cells increased about sixfold in the presence of PACAP. PACAP was found to influence not only cell differentiation but also cell proliferation. Coupled with the recent report that PACAP promotes neural stem cell proliferation in adult mouse brain,[27] these results are noteworthy for their demonstration of a new function for PACAP.

The signal transduction pathway for PACAP in neural stem cells was also evaluated. First, PACAP receptor gene expression was analyzed by RT-PCR. This is the first report which describes the expression of PACAP receptor protein in mouse neural stem cells.[26] We subsequently examined the role that cAMP plays in the differentiation of neural stem cells by PACAP. Neural stem cells were exposed to cAMP analogs, such as 8-bromo-cAMP and dibutyryl-cAMP, with neither of these compounds inducing any change in the cells.[26] In addition, the PACAP-induced differentiation of neural stem cells could not be blocked by the PKA inhibitor, Rp-cAMP.[26] Moreover, Rp-cAMP did not influence the PACAP-induced cell proliferation.[26] Neural stem cells prepared from mouse embryonic telencephalons have been shown to differentiate into astrocytes in the presence of PACAP.[26] It has been demonstrated that nanomolar concentrations of PACAP stimulate the growth of neural stem cells in adult mouse brain.[27] The involvement of PKC in the signaling pathway was presumed because the effect could be blocked by the PKC inhibitor Gö6976. In our study, the participation of PKC in the proliferation of embryonic mouse neural stem cells was suggested because the growth of mouse neural stem cells was also observed in the presence of phorbol 12-myristate 13-acetate (FIG. 1).[26]

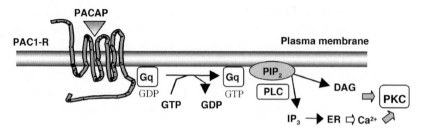

FIGURE 1. Signaling pathway of PACAP in the differentiation of neural stem cells into astrocytes. PACAP acts with the PAC1-R of neural stem cells. The signal, transmitted via PAC1-R, activates Gq protein but not Gs and it promotes the activation of phospholipase C and then PKC.

These findings indicate that PACAP might be a novel mediator of neural stem cell proliferation.

Three types of receptors, PAC1, VPAC1, and VPAC2, are well known to mediate the actions of PACAP.[28] In our experiment, the PAC1 receptor, which possesses the highest affinity for PACAP, was shown to be expressed on the plasma membrane of neural stem cells as well as in the adult mouse cerebral cortex as reported previously.[29] From our result, we postulate that PACAP binds to PAC1, and then the signal is transmitted intracellularly via a receptor-coupled G protein. Recently, it has been reported that PACAP induces differentiation of rat cortical precursor cells into astrocytes.[30] These investigators have also demonstrated that elevation of intracellular cAMP is responsible for the differentiation of cortical precursor cells into astrocytes since the induction by PACAP was influenced by a cAMP analog and a PKA inhibitor.[30] Recently, it has been reported that VIP and PACAP induce selective neuronal differentiation of mouse embryonic stem cells through VPAC2 and PAC1 receptors.[31] We showed, however, that neither cAMP analogs nor a PKA inhibitor had any effect on the differentiation of neural stem cells into astrocytes, suggesting that a signaling pathway that involves another receptor-coupled G protein is responsible rather than Gs.[26] We have also shown that the differentiation was suppressed by BAPTA-AM, an intracellular calcium chelator and chelerythrine, a selective inhibitor of PKC. These findings strongly suggest that Ca^{2+}, phospholipid-dependent PKC (cPKC) is responsible for the differentiation. PACAP has been shown to promote neural stem cell proliferation in adult mouse brain through the PKC pathway as the principal signaling pathway.[27] It is considered that PACAP acts on its receptors on plasma membranes of mouse neural stem cells and the signal is transmitted to the PAC1-R-coupled G protein Gq, which then leads to activation of cPKC which may play a crucial role in the differentiation of neural stem cells into astrocytes. It remains to be determined whether PACAP itself is expressed in embryos or transported from the maternal plasma across the placental barrier in the same manner as the transport of insulin-like growth factor and epidermal growth factor.[32] Further experiments are needed to clarify the *in vivo* role of PACAP in the differentiation of neural stem cells into astrocytes.

PACAP IN NEUROPROTECTION

VIP and PACAP were induced in several types of neurons after diverse types of experimental injury, suggesting that these peptides are involved in the reaction to nerve injury.[33,34] Moreover, PAC1-R was found to be upregulated in the cortex and caudate putamen after transient focal cerebral ischemia in mice.[35] VIP and PACAP have been reported to be neuroprotective in several *in vivo* models of brain injury. For example, neurons in the CA1 region of the hippocampus are vulnerable to global forebrain ischemia, and this model has been widely used for evaluating neuroprotective agents.[36,37] Intracerebroventricular infusion of PACAP-38 into ischemic animals prevented the otherwise total loss of pyramidal cells and their dendritic processes throughout CA1 and cerebral cortex.[38–40] PACAP-38 is neuroprotective at concentrations as low as 0.1 pM *in vitro*, and it can cross the blood–brain barrier by a saturable mechanism.[38] PACAP-38 has been reported to be effective even when administered intravenously.[39] Thus, PACAP may act on the hippocampal neurons by crossing the blood–brain barrier. Although PAC1-R gene expression is detected in rat CA1 pyramidal and nonpyramidal neurons,[41] only a small amount of PAC1-R immunoreactivity is detected in the CA1 region in the normal rat brain.[41]

Evidence from both *in vitro* and *in vivo* experiments suggests that VIP and PACAP may be useful for enhancing neuronal survival and neuronal regeneration in the CNS. PACAP has been reported to prevent programmed cell death in cultured cerebellar granule cells[42,43] and basal forebrain cholinergic neurons.[44] VIP and PACAP have also been shown to exhibit a variety of trophic or growth factor-like actions on other populations of cultured neurons and glia subtypes.[45] For example, in cultured rat astrocytes, PACAP mobilizes intracellular free calcium,[46] increases interleukin-6 (IL-6) secretion,[47] and regulates glial glutamate transport and metabolism through activation of PKA and PKC.[48] Early studies have also shown that PACAP stimulates the outgrowth of neurites from PC12 cells and enhances the survival of sympathetic ganglion cells.[3] More recently, it has been shown that PACAP induces cell cycle withdrawal and promotes the transition from proliferation to differentiation in cultured cortical precursor cells from E13 rats.[8] This antimitotic action of PACAP appears to be mediated by an increase in the activity of the cyclin-dependent kinase inhibitor p57.[49] PACAP potently increased cAMP levels in cultured hindbrain neuroepithelial cells from E10.5 mice and downregulated the expression of the *sonic hedgehog*-dependent and PKA-dependent target gene *gli-1*.[9] More recently, it was reported that PACAP increased mitosis in cultured embryonic superior cervical ganglion (SCG) precursors and potently enhanced precursor survival.[50] In addition, PACAP promoted neuronal differentiation, increased neurite outgrowth, and enhanced expression of the neurotrophin receptors trkC and trkA. Most or all of the physiological effects of PACAP on SCG precursors were shown to be mediated by the PAC1-R, via increased intracellular second messengers, including cAMP and phosphatidylinositol as well as Ca^{2+}.[50] Early studies demonstrated that high concentrations of VIP stimulate mitosis, promote neurite outgrowth, and enhance survival of sympathetic neuroblasts in culture.[51] However, these effects of VIP have now been attributed to action on the PAC1-R.[50] In other studies, lower concentrations of VIP increased neuronal survival during a critical period of development in dissociated spinal cord cultures,[52] and VIP administration reduced the size of excitotoxin-induced lesions in brains of neonatal mice.[53] VIP and PACAP were also reported to increase BDNF mRNA in cultured cortical neurons through an action of glutamate on NMDA receptors.[54]

In the ischemic hippocampus, including the CA1 region, both an in-gel kinase assay and an immunocomplex kinase assay indicated that both c-Jun N-terminal kinase (JNK) and extracellular signal-regulated kinase (ERK) activities increased during the first 6 h after ischemia-reperfusion.[55] During the first 6 h, p38 was also stimulated after ischemia-reperfusion.[55] Taken together, these observations suggest that the activation of JNK or p38 MAPK, or both, contribute to the induction of apoptosis in CA1 neurons. In contrast, in PACAP-treated animals, no significant increase in JNK activity was detected within the first 6 h.[56] This indicates that PACAP inhibited the activation of JNK after ischemia-reperfusion stress.

In intact animals, there were no cells that showed immunoreactivity for IL-6, while a small number of IL-6-positive cells were visible 12 h after ischemia. A peak of IL-6-LI in the CA1 region was found at 1–2 days after the ischemia and the strong immunoreactivity was found in the reactive astrocytes. The number of IL-6-positive cells was decreased at 4 and 7 days after the ischemia. IL-6-LI was not detected in microglial cells during ischemia-reperfusion. Considerably high concentrations of IL-6 (pg range) in the cerebrospinal fluid (CSF) were detected at 6 and 24 h after the ischemia.[56] We measured the IL-6 concentration in the CSF of the ischemic animals with or without the infusion of PACAP (1 pmol/h). The secretion of IL-6 into the CSF was stimulated slightly by the vehicle. Unexpectedly, the secretion of IL-6 into the CSF was stimulated dramatically (ng range) after the infusion of PACAP (1 pmol/h).[56] We have also determined that PACAP participated in the expression and generation of IL-6, and the neuroprotection by PACAP injection after ischemia is decreased in IL-6 gene-deficient mice, but not in the wild-type mice (unpublished observations). However, the neuroprotective mechanism of PACAP has not been elucidated in detail. Moreover, the infarct volume in PACAP Knockout (KO) mice is significantly larger than in wild-type mice. The cytochrome C release from

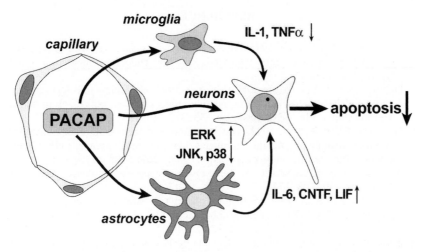

FIGURE 2. Putative mechanism of PACAP on suppression of apoptosis in neurons; PACAP, crossing blood–brain barrier, binds to PAC1-R in neurons, astrocytes and microglia, and suppresses neuronal apoptosis through direct and indirect pathways. In neurons, PACAP regulates MAPK pathway with activation of ERK and inhibition of JNK and p38. PACAP decreases deleterious cytokine, such as IL-1 and tumor necrosis factor-α (TNF-α) in microglia and increases beneficial cytokines, such as IL-6, CNTF, and LIF in astrocytes.

the mitochondria to the cytoplasm that is a trigger for apoptosis is increased in the ipsilateral hemisphere of hetero- and homozygous PACAP KO mice. Therefore, it is possible that PACAP itself and IL-6, whose secretion has been stimulated by PACAP, act synergistically to inhibit the JNK signaling pathway, and thereby protecting neurons against cell death. Very low concentrations of PACAP, which cross the blood–brain barrier, may stimulate neurons directly and may also stimulate astrocytes, and possibly microglia, to secrete neuroprotective factors, which affect neurons (FIG. 2). During the process of the cytoprotective action of PACAP, PACAP may regulate the dynamic balance between the growth factor-activated ERK and stress-activated JNK-p38 pathways. The significant neuroprotection with intravenous infusions of PACAP begun 24 h after the ischemic injury suggests a potential clinical importance for PACAP.

ACKNOWLEDGMENTS

This study was supported in part by grants to S.S. from the Ministry of Education, Science, Sports, and Culture of Japan, and the High-Technology Research Center Project from the Ministry of Education, Science, Sports, and Culture of Japan.

REFERENCES

1. ARIMURA, A. & S. SHIODA. 1995. Pituitary adenylate cyclase activating polypeptide (PACAP) and its receptors: neuroendocrine and endocrine interaction. Front. Neuroendocrinol. **16:** 53–88.
2. DEUTSCH, P.J., V. SCHADLOW & N. BARZILAI. 1992. 38-Amino acid form of pituitary adenylate cyclase activating peptide induces process outgrowth in human neuroblastoma cells. J. Neurosci. Res. **35:** 312–320.
3. DEUTSCH, P.J. & Y. SUN. 1992. The 38 amino acid form of pituitary adenylate cyclase activating polypeptide stimulates dual signalling cascades in PC12 cells and promotes neurite outgrowth. J. Biol. Chem. **267:** 5108–5113.
4. ARIMURA, A. et al. 1994. PACAP functions as a neurotrophic factor. Ann. N. Y. Acad. Sci. **739:** 228–243.
5. CANONICO, P.L. et al. 1996. Activation of pituitary adenylate cyclase-activating polypeptide (PACAP) is a neurotrophic factor for cultured rat cortical neurons. Ann. N. Y. Acad. Sci. **805:** 476–481.
6. ZHOU, C-J. et al. 1999. Pituitary adenylate cyclase-activating polypeptide receptors during development: expression in the rat embryo at primitive streak stage. Neuroscience **93:** 376–391.
7. ZHOU, C-J. et al. 2000. Cellular distribution of the splice variants of the receptor for pituitary adenylate cyclase-activating polypeptide (PAC1-R) in the rat brain by in situ RT-PCR. Mol. Brain Res. **75:** 150–158.
8. LU, N. & E. DICICCO-BLOOM. 1997. Pituitary adenylate cyclase-activating polypeptide is an autocrine inhibitor of mitosis in cultured cortical precursor cells. Proc. Natl. Acad. Sci. USA **94:** 3357–3362.
9. WASCHEK, J. A. et al. 1998. Neural tube expression of pituitary adenylate cyclase-activating peptide (PACAP) and receptor: potential role in patterning and neurogenesis. Proc. Natl. Acad. Sci. USA **95:** 9602–9607.
10. HAMMERSCHMIDT, M. et al. 1996. Protein kinase A is a common negative regulator of Hedgehog signaling in the vertebrate embryo. Genes Dev. **10:** 647–658.
11. ZHOU, C-J. et al. 2001. PACAP activates PKA, PKC, and Ca signaling cascades in rat neuroepithelial cells. Peptides **22:** 1111–1117.
12. GAGE, F.H. et al. 1995. Isolation, characterization, and use of stem cells from the CNS. Annu. Rev. Neurosci. **18:** 159–192.
13. KUHN, H.G. & C.N. SVENDSEN. 1999. Origins, functions, and potential of adult neural stem cells. Bioessays **21:** 625–630.
14. MOMMA, S. et al. 2000. Get to know your stem cells. Curr. Opin. Neurobiol. **10:** 45–49.
15. JOHE, K.K. et al. 1996. Single factors direct the differentiation of stem cells from the fetal and adult central nervous system. Genes Dev. **10:** 3129–3140.
16. NAKASHIMA, K. et al. 1999. Synergistic signaling in fetal brain by Stat3-Smad1 complex bridged by p300. Science **284:** 479–482.
17. LEE, J.E. 1997. Basic helix-loop-helix genes in neural development. Curr. Opin. Neurobiol. Dev. **7:** 13–20.
18. KEGEYAMA, R. & S. Nakanishi. 1997. Helix-loop-helix factors in growth and differentiation of the vertebrate nervous systems. Curr. Opin. Genet. Dev. **7:** 659–665.
19. TORII, M. et al. 1999. Transcription factors Mash-1 and Prox-1 delineate early steps in differentiation of neural stem cells in the developing central nervous system. Development **126:** 443–456.

20. SUN, Y. *et al.* 2001. Neurogenin promotes neurogenesis and inhibits glial differentiation by independent mechanisms. Cell **104:** 365–376.
21. UEKI, T. *et al.* 2003. A novel secretory factor, *neurogenesin-1*, provides neurogenic environmental cues for neural stem cells in the adult hippocampus. J. Neurosci. **23:** 11732–11740.
22. GALLO, R. *et al.* 2002. REN: a novel, developmentally regulated gene that promotes neural cell differentiation. J. Cell Biol. **158:** 731–740.
23. WANG, S. *et al.* 1998. Notch recepter activation inhibits oligodendrocyte differentiation. Neuron **21:** 63–75.
24. ISHIBASHI, M. *et al.* 1994. Persistent expression of helix-loop-helix factors *HES-1* prevents mammalian neural differentiation in the central nervous system. EMBO J. **13:** 1799–1805.
25. NAKAMURA, Y. *et al.* 2000. The bHLH *Hes1* as a repressor of the neuronal commitment of CNS stem cells. J. Neurosci. **20:** 283–293.
26. OHNO, F. *et al.* 2005. Pituitary adenylate cyclase-activating polypeptide promotes differentiation of mouse neural stem cells into astrocytes. Regul. Pept. **126:** 115–122.
27. MERCER, A. *et al.* 2004. PACAP promotes neural stem cell proliferation in adult mouse brain. J. Neurosci. Res. **76:** 205–215.
28. SPENGLER, D. *et al.* 1993. Differential signal transduction by five splice variants of the PACAP receptor. Nature **365:** 170–175.
29. SUZUKI, R *et al.* 2003. Expression of the receptor for pituitary adenylate cyclase-activating polypeptide (PAC1-R) in reactive astrocytes. Brain Res. Mol. Brain Res. **115:** 10–20.
30. VALLEJO, I. & M. VALLEJO. 2002. Pituitary adenylate cyclase-activating polypeptide induces astrocyte differentiation of precursor cells from developing cerebral cortex. Mol. Cell Neurosci. **21:** 671–683.
31. CAZILLIS, M. *et al.* 2004. VIP and PACAP induce selective neuronal differentiation of mouse embryonic stem cells. Eur. J. Neurosci. **19:** 798–808.
32. GARNICA, A.D. & W.Y. CHAN. 1996. The role of the placenta in fetal nutrition and growth. J. Am. Coll. Nutr. **15:** 206–222.
33. ZIGMOND, R.E. 1997. LIF, NGF, and the cell body response to axotomy. New Sci. **3:** 176–185.
34. WASCHEK, J.A. 2002. Multiple actions of pituitary adenylyl cyclase activating peptide (PACAP) in nervous system development and regeneration. Dev. Neurosci. **24:** 14–23.
35. GILLARDON, F. *et al.* 1998. Delayed up-regulation of Zac1 and PACAP type I receptor after transient focal cerebral ischemia in mice. Brain Res. Mol. Brain Res. **61:** 207–210.
36. PULSINELLI, W.A. & J.B. BRIERLY. 1979. A new model of bilateral hemispheric ischemia in the unanesthetized rat. Stroke **10:** 267–272.
37. DOHI, K. *et al.* 1998. Delayed neuronal cell death and microglial cell reactivity in the CA1 region of the rat hippocampus in the cardiac arrest model. Med. Electron Microsc. **31:** 85–89.
38. BANKS, W.A. *et al.* 1996. Transport of adenylate cyclase-activating polypeptide across the blood-brain barrier and the prevention of ischemia-induced death of hippocampal neurons. Ann. N. Y. Acad. Sci. **805:** 270–279.
39. UCHIDA, A. *et al.* 1996. Prevention of ischemia-induced death of hippocampal neurons by pituitary adenylate cyclase activating polypeptide. Brain Res. **736:** 280–286.

40. REGLODI, D. *et al.* 2000. Delayed systemic administration of PACAP-38 is neuroprotective in transient middle cerebral artery occlusion in the rat. Stroke **31:** 1411–1417.
41. SHIODA, S. *et al.* 1997. Localization and gene expression of the receptor for pituitary adenylate cyclase-activating polypeptide in the rat brain. Neurosci. Res. **28:** 345–354.
42. VILLALBA, M. *et al.* 1997. Pituitary adenylate cyclase-activating polypeptide (PACAP) protects cerebellar granule neuron from apoptosis by activating the mitogen-activated protein kinase (MAP kinase) pathway. J. Neurosci. **17:** 83–90.
43. VAUDRY, D. *et al.* 2000. The neuroprotective effect of pituitary adenylate cyclase-activating polypeptide on cerebellar granule cells is mediated through inhibition of the CED3-related cysteine protease caspase-3/CPP32. Proc. Natl. Acad. Sci. USA **97:** 13390–13395.
44. TAKEI, N. *et al.* 2000. Pituitary adenylate cyclase-activating polypeptide promotes the survival of basal forebrain cholinergic neurons in vitro and in vivo: comparison with effects of nerve growth factor. Eur. J. Neurosci. **12:** 2273–2280.
45. WASCHEK, J.A. 1995. Vasoactive intestinal peptide: an important trophic factor and developmental regulator? Dev. Neurosci. **17:** 1–7.
46. TATSUNO, I. & A. ARIMURA. 1994. Pituitary adenylate cyclase-activating polypeptide (PACAP) mobilizes intracellular free calcium in cultured type-2, but not type-1, astrocytes. Brain Res. **662:** 1–10.
47. GOTTSCHALL, P.E. *et al.* 1994. Regulation of interleukin-6 (IL-6) secretion in primary cultured rat astrocytes: synergism of interleukin- (IL-1) and pituitary adenylate cyclase-activating polypeptide (PACAP). Brain Res. **637:** 197–203.
48. FIGIEL, M. & J. ENGELE. 2000. Pituitary adenylate cyclase-activating polypeptide (PACAP): a neuron-derived peptide regulating glial glutamate transport and metabolism. J. Neurosci. **20:** 3596–3605.
49. CAREY, R.G. *et al.* 2002. Pituitary adenylate cyclase activating polypeptide antimitogenic signaling in cerebral cortical progenitors is regulated by p57Kip2-dependent CDK2 activity. J. Neurosci. **22:** 1583–1591.
50. DiCICCO-BLOOM, E. *et al.* 2000. Autocrine expression and ontogenetic functions of the PACAP ligand/receptor system during sympathetic development. Dev. Biol. **219:** 197–213.
51. PINCUS, D.W. *et al.* 1990. Vasoactive intestinal peptide regulates, differentiation and survival of cultures sympathetic neuroblasts. Nature **343:** 564–567.
52. BRENNEMAN, D. & L.D. EIDEN. 1986. Vasoactive intestinal peptide and electrical activity influence neuronal survival. Proc. Natl. Acad. Sci. USA **83:** 1159–1162.
53. GRESSENS, P. *et al.* 1997. Vasoactive intestinal peptide prevents excitotoxic cell death in the murine developing brain. J. Clin. Invest. **100:** 390–397.
54. PELLEGRI, G. *et al.* 1998. VIP and PACAP potentiate the action of glutamate on BDNF expression in mouse cortical neurons. Eur. J. Neurosci. **10:** 272–280.
55. OZAWA, H. *et al.* 1999. Delayed neuronal cell death in the rat hippocampus is mediated by the mitogen-activated protein kinase signal transduction pathway. Neurosci. Lett. **262:** 57–60.
56. SHIODA, S. *et al.* 1998. PACAP protects hippocampal neurons against apoptosis. Involvement of JNK/SAPK signaling pathway. Ann. N. Y. Acad. Sci. **865:** 111–117

The Prenatal Expression of Secretin Receptor

FRANCIS K.Y. SIU,[a] M.H. SHAM,[b] AND BILLY K.C. CHOW[a]

[a] Department of Zoology, The University of Hong Kong, Hong Kong SAR, People's Republic of China

[b] Department of Biochemistry, The University of Hong Kong, Hong Kong SAR, People's Republic of China

ABSTRACT: Secretin is a classical gastrointestinal peptide while its neuroactive functions in the central nervous system have recently been consolidated. In the past, there was little information regarding the expression of secretin receptor in prenatal development. In this article, using mouse embryos and by *in situ* hybridization, secretin receptor transcripts were detected in several developing brain regions including the cerebellar primordium and choroid plexus. In the developing intestine, secretin receptor is present in the epithelial lining of the villi and the inner circular muscle. Interestingly, the transcripts for secretin receptor were also detected in the epicardium and myocardium of the developing heart as well as the glomerulus and collecting duct in the developing kidney. Taken together, our data suggest a potential pleiotrophic role of secretin during embryonic development.

KEYWORDS: secretin receptor; expression; prenatal; embryo; development

INTRODUCTION

Secretin is a 27-amino acid peptide hormone discovered in 1902 by Bayliss and Starling.[1] The major function of secretin in the periphery is to neutralize the acidic chyme entering into the duodenum by stimulating the release of bicarbonate, water, and electrolytes from the pancreas.[2,3] Secretin exerts its biological functions via its interaction with a specific cell surface receptor, the secretin receptor. In this article, we investigated the expression of secretin receptor in two embryonic stages (E10.5 and E14.5) of mouse by *in situ* hybridization.

Address for correspondence: Billy K.C. Chow, Department of Zoology, The University of Hong Kong, Pokfulam Road, Hong Kong SAR, PRC. Voice: +852-22990850; fax: +852-28574672.
e-mail: bkcc@hkusua.hku.hk

Ann. N.Y. Acad. Sci. 1070: 561–565 (2006). © 2006 New York Academy of Sciences.
doi: 10.1196/annals.1317.081

MATERIALS AND METHODS

Probe Preparation

The mouse secretin receptor cDNA (GeneBank Accession No: NM_001012322, nucleotide 247-748) was initially subcloned into the vector pBlueScript II KS+ (Stratagene, La Jolla, CA). Anti-sense DIG-labeled riboprobe was synthesized by T7 RNA polymerase using a DIG RNA labeling kit (Roche Diagnostics, Mannheim, Germany). To show the specificity of hybridization, unlabeled anti-sense riboprobe was also generated and used as a competitor (20X) of the DIG-labeled probe in the controls.

In Situ *Hybridization*

In situ hybridizations using whole-mount embryos and paraffin sections were performed. In summary, the fixed whole-mount embryos or paraffin sections were rehydrated in a graded series of ethanol followed by proteinase K digestion. After rinsing with PBS, the samples were incubated with the hybridization buffer (50% formamide, 2X SSC, 1% SDS) at 70°C (whole-mount) or 50°C (sections) for 1 h, and then hybridized with the probe at 65°C (whole-mount) or 50°C (sections) overnight with or without 20X excess of the unlabeled competitor. The next day, samples were washed at 65°C (whole-mount) or 37°C (sections) with 2X, 1X, and 0.5X SSC, and finally PBS. Before incubating the samples with 1:2000 (whole-mount) or 1:500 (sections) alkaline phosphatase-conjugated anti-DIG antibody (Roche Diagnostics) at 4°C overnight, the samples were blocked by 10% serum for 1 h at room temperature. After washing with PBS, positive signals (purple) were produced by using the NBT/BCIP substrate solution (Sigma, St. Louis, MO).

RESULTS AND DISCUSSION

Secretin Receptor Expression in CNS Regions

During mouse brain development, secretin receptor was expressed as early as at E10.5 in the mid- and hide-brain regions (FIG. 1A). At E14.5, the mid-brain region also showed secretin receptor expression. It was expressed in the tegmentum and mesenchyme flexure (FIG. 1B). In the hide-brain region of E14.5 embryo, secretin receptor was expressed in the cerebellar primordium and choroid plexus. The expression pattern of secretin receptor overlaps with that of secretin at E10.5 and E14.5.[4,5] The presence of secretin receptor in the cerebellum is consistent with previous studies, both secretin and its receptor are expressed in distinct neurons in the cerebellum of adult rat and secretin

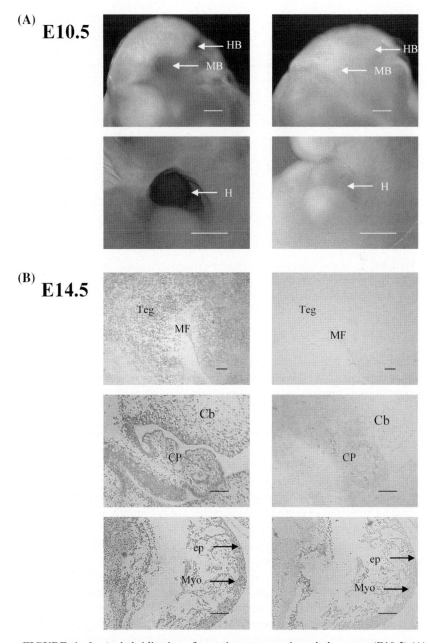

FIGURE 1. *In situ* hybridization of secretin receptor using whole-mount (E10.5) **(A)** and sectioned (E14.5) **(B)** mouse embryos. All the positives are in left panel while all the controls are in the right panel. MB: mid-brain region; HB: hide-brain region; H: heart; Cb: cerebellar primordium; CP: choroid plexus; Teg: tegmentum; MF: mesencephalic flexure; ep: epicardium; Myo: myocardium. Scale bars: 40 μm.

E14.5

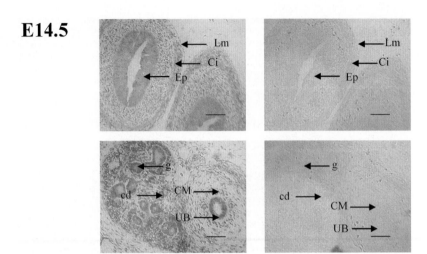

FIGURE 2. *In situ* hybridization of secretin receptor on mouse embryos sections. All the positive sections are in left panel while all the control sections are in the right panel. Lm: longitudinal muscle layer; Ci: inner circular muscle; Ep: epithelial lining of the villi; cd: collecting duct; g: glomerulus; UB: ureteric bud; CM: condensing mesenchyme. Scale bars: 40 μM.

is neuroactive.[6] Although the function of secretin in the brain remains largely unknown, the presence of both secretin and its receptor in similar areas suggests the potential of secretin in modulating brain development.

Secretin Receptor Expression in Peripheral Tissues

Secretin receptor transcripts are expressed in the heart at E10.5 and E14.5 (FIG. 1). More specifically, the expression was found in the epicardium and myocardium of the developing heart at E14.5. This expression pattern is again consistent with that of secretin.[5] As secretin can potentially stimulate the contraction of cardiac ventricular cardiomyocytes and increase cAMP accumulation,[7–9] the presence of secretin and its receptor in developing heart suggests that secretin has the same function in prenatal stages.

At E14.5, secretin receptor was found in the epithelial lining of the intestinal villi and the inner circular muscle layer (FIG. 2). The expression in the epithelial lining overlapped with that of secretin.[5] However, in the inner circular muscle layer, no secretin transcript expression could be detected,[5] suggesting that in mouse embryo, secretin receptor in the inner circular muscle layer may receive signals from the epithelial layer to regulate intestinal contraction.[10,11]

In the developing kidney at E14.5, secretin receptor transcripts co-localized with secretin[5] and were detected in the ureteric bud, condensing mesenchyme, collecting duct, and glomerulus (FIG. 2). Secretin was previous found to increase the nephron filtration rate, glomerular filtration rate, and reabsorption rate of water and bicarbonate.[12–14] These actions of secretin are therefore consistent with the presence of secretin receptor in the kidney as previously predicted.[15]

ACKNOWLEDGMENTS

The work was supported by the HK government RGC HKU 7384/04M and CRCG 10203915 to Billy K.C. Chow.

REFERENCES

1. BAYLISS, W.M. & E.H. STARLING. 1902. The mechanism of pancreatic secretion. J. Physiol. (Lond.) **28**: 325–353.
2. MEYER, J.H., L.W. WAY & M.I. GROSSMAN. 1970. Pancreatic response to acidification of various lengths of proximal intestine in the dog. Am. J. Physiol. **219**: 971–977.
3. WATANABE, S., W.Y. CHEY, K.Y. LEE, *et al.* 1986. Secretin is released by digestive products of fat in dogs. Gastroenterology **90**: 1008–1017.
4. LOSSI, L., L. BOTTARELLI, M.E. CANDUSSO, *et al.* 2004. Transient expression of secretin in serotoninergic neurons of mouse brain during development. Eur. J. Neurosci. **20**: 3259–3269.
5. SIU, F.K., M.H. SHAM & B.K. CHOW. 2005. Secretin, a known gastrointestinal peptide, is widely expressed during mouse embryonic development. Gene Expr. Patterns **5**: 445–451.
6. YUNG, W.H., P.S. LEUNG, S.S.M. NG, *et al.* 2001. Secretin facilitates GABA transmission in the cerebellum. J. Neurosci. **21**: 7063–7068.
7. CHIBA, S. 1976. Effect of secretin on pacemaker activity and contractility in the isolated blood-perfused atrium of the dog. Clin. Exp. Pharmacol. Physiol. **3**: 167–172.
8. GUNNES, P., H.L. WALDUM, K. RASMUSSEN, *et al.* 1983. Cardiovascular effects of secretin infusion in man. Scand. J. Clin. Lab. Invest. **43**: 637–642.
9. BELL, D. & B.J. McDERMOTT. 1994. Secretin and vasoactive intestinal peptide are potent stimulants of cellular contraction and accumulation of cyclic AMP in rat ventricular cardiomyocytes. J. Cardiovasc. Pharmacol. **23**: 959–969.
10. HERMON-TAYLOR, J. & C.F. CODE. 1970. Effect of secretin on small bowel myoelectric activity of conscious healthy dogs. Am. J. Dig. Dis. **15**: 545–550.
11. RAMIREZ, M. & J.T. FARRAR. 1970. The effect of secretin and cholecystokinin-pancreozymin on the intraluminal pressure of the jejunum in the unanesthetized dog. Am. J. Dig. Dis. **15**: 539–544.
12. VITERI, A.L., J.W. POPPELL, J.M. LASATER, *et al.* 1975. Renal response to secretin. J. Appl. Physiol. **38**: 661–664.
13. MARCHAND, G.R. 1986. Effect of secretin on glomerular dynamics in dogs. Am. J. Physiol. **250**: F256–F260.
14. ROMANO, G., P. GIAGU, G. FAVRET, *et al.* 2000. Dual effect of secretin on nephron filtration and proximal reabsorption depending on the route of administration. Peptides **21**: 723–728.
15. CHARLTON, C.G., R. QUIRION, G.E. HANDELMANN, *et al.* 1986. Secretin receptors in the rat kidney: adenylate cyclase activation and renal effects. Peptides **7**: 865–871.

Cyclic AMP Formation in C6 Glioma Cells

Effect of PACAP and VIP in Early and Late Passages

PAULINA SOKOLOWSKA[a] AND JERZY Z. NOWAK[a,b]

[a]Centre for Medical Biology, Polish Academy of Sciences, Lodz 93-232, Poland

[b]Department of Pharmacology, Medical University of Lodz, Lodz, Poland

ABSTRACT: Pituitary adenylate cyclase-activating polypeptide (PACAP) and vasoactive intestinal peptide (VIP) exert their actions via common receptors, VPAC1 and VPAC2, which are equally sensitive to both peptides, while PACAP stimulates its specific PAC1-type receptors. Both peptides potently stimulate cAMP production in different biological systems. In the present article, we examined the effects of PACAP and VIP on cAMP formation in C6 rat glioma cells used between passages 12-28 (early) and 120-136 (late). In the presence of the PDE inhibitor IBMX (0.1 mM), PACAP (0.1 μM) and VIP (1 μM) strongly stimulated cAMP synthesis in C6 cells in early passages, but not in C6 cells in late passages. In contrast, forskolin (10 μM), a direct activator of adenylyl cyclase, and isoprenaline (10 μM), a β-adrenergic receptor agonist, strongly stimulated cAMP production in both early and late C6 cell passages. Concentration-dependent studies carried out in early passages with PACAP-38, PACAP-27, mammalian and chicken VIPs, and PHI/PHM peptides (1–5 μM) revealed that both forms of PACAP produced strong cAMP accumulation, VIP peptides were less effective than PACAP, while the cAMP effects of PHI/PHM peptides were noticeable only at the highest doses tested. These results suggest that C6 glioma cells in early passages possess functional PAC1 and possibly VPAC-type receptors, but either the density of PACAP/VIP receptors progressively declines or the PACAP/VIP receptor-Gs protein coupling becomes less effective through culture passages.

KEYWORDS: PACAP; VIP; cAMP; C6 cells; glioma; rat

Address for correspondence: Paulina Sokolowska, Centre for Medical Biology, Polish Academy of Sciences, 106 Lodowa Road, Lodz 93-232, Poland. Voice: +48-42-681-51-01; fax: +48-42-272-36-30. e-mail: psokolowska@cbm.pan.pl

Ann. N.Y. Acad. Sci. 1070: 566–569 (2006). © 2006 New York Academy of Sciences. doi: 10.1196/annals.1317.082

INTRODUCTION

On the basis of their structural similarity, pituitary adenylate cyclase-activating polypeptide (PACAP) and vasoactive intestinal peptide (VIP) have been classified to the same superfamily of polypeptides, embracing also peptide histidine-isoleucine (PHI) and its human analog peptide histidine-methionine (PHM), secretin, glucagon, and helodermin.[1,2] PACAP exists in an organism in two forms: a longer one possessing 38 amino acids (PACAP-38), and a shorter one consisting of 27 amino acids (PACAP-27).[1,2] PACAP shows 70% sequence identity with VIP which is present in the body in only one form composed of 28 amino acids.[1] PACAP and VIP share affinity for common receptors, named VPAC1 and VPAC2, which recognize similarly and with high affinity VIP, PACAP-38 and PACAP-27, and with lower affinity PHI/PHM.[2] PACAP may also exert its biological effects via PAC1 receptors showing selective affinity for both forms of the peptide, recognizing poorly VIP and PHI/PHM.[2] The main signal transduction pathway of VIP/PACAP receptors is the adenylyl cyclase (AC)-cAMP cascade.[1]

In this article, we examined the effects of PACAP (both forms), VIP (mammalian and chicken), and PHI/PHM peptides on cAMP formation in the C6 rat glioma cell line in early (12-28) and late (120-136) passages. We compared the cAMP effects of the studied peptides in the C6 cell line with those of a direct AC activator, forskolin, and a potent cAMP stimulator, isoprenaline (β-adrenergic agonist).

MATERIALS AND METHODS

Experiments were carried out on the C6 rat glioma cell line (ECACC, Salisbury, UK). Cells were incubated with [^{3}H]adenine. The formation of [^{3}H]cAMP was measured with the method of Shimizu et al.[3] The formed [^{3}H]cAMP was isolated by sequential Dowex-alumina column chromatography.[4] Results were evaluated by analysis of variance (ANOVA) followed by the Newman–Keuls test.

RESULTS AND DISCUSSION

Studies carried out in C6 cells in early passages with PACAP-38, PACAP-27, mammalian VIP (mVIP), and chicken VIP (cVIP) revealed the ability of all tested peptides to stimulate cAMP formation. Both forms of PACAP stimulated cAMP synthesis in a dose-dependent manner effectively acting at wide range of concentrations (1 nM–1 μM). Unexpectedly, PACAP-38 was clearly less effective than PACAP-27 (323 versus 703% of control, at 1 μM). Other tested peptides, i.e., mVIP and cVIP were effective only at higher doses (0.1–5 μM) (FIG. 1). Comparative studies showed that at a concentration of

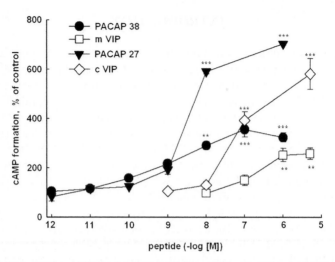

FIGURE 1. Effects of PACAP-38, PACAP-27, mVIP, and cVIP on cAMP formation in C6 rat glioma cells in early passages. Data are means ± SEM of 4–6 values per group. The results are expressed as percent of respective control values (each taken as 100%). **$P < 0.01$, ***$P < 0.001$.

10 μM, isoprenaline and forskolin produced much stronger cAMP responses than the peptides (TABLE 1). The effects of isoprenaline and forskolin remained unchanged in late passages of C6 glioma cells, whereas PACAP (0.1 μM) and VIP (1 μM) were ineffective in stimulating cAMP formation in these cells (TABLE 1). A time-course study (2–60 min) conducted in early C6 cell passages with 0.1 μM PACAP-38 revealed that the peptide evoked its action rapidly, reaching a maximum after 15 min of incubation (data not shown). Both forms of PHI (porcine and rat) and PHM appeared to be the least effective among the peptides studied in early C6 cell passages, showing their effects on cAMP production (202, 254, 197% of control, respectively) only at the highest dose tested, i.e., 5 μM (data not shown). The rank order of potency seen in that study, i.e., PACAP-27 ≈ PACAP-38 >> cVIP > mVIP, suggests that the PACAP actions on the cAMP generating system in early C6 cell passages were mediated through the PAC1 type receptor. On the other hand, the significant differences

TABLE 1. Effects of PACAP-38, mVIP, forskolin, and isoprenaline on cAMP formation in C6 rat glioma cells in early and late passages. The results are expressed as percent of respective control values (each taken as 100%)

		% of control	
Drug	μM	Early passages	Late passages
PACAP-38	0.1	355.5 ± 29.8	99.2 ± 9.4
mVIP	1	252.2 ± 26.6	96.6 ± 6.1
Forskolin	10	1227.0 ± 49.3	1356.3 ± 53.2
Isoprenaline	10	1437.2 ± 147.8	1260.6 ± 65.7

in maximal cAMP responses evoked by PACAP-27 and PACAP-38 may be ascribed to the existence of atypical receptors in C6 cells, characterized by diverse interaction with shorter and longer forms of PACAP. In support of such a possibility, recent findings indicate the presence of PACAP-27-preferring receptors in the rat intestine.[5] In respect of a low affinity of PAC1-type receptors for VIP and PHI/PHM peptides,[2] the occurrence of functional VPAC-type receptors (VPAC1 and/or VPAC2) in C6 cells seems likely to occur. Recent data, based on [^{125}I]PACAP-27 and [^{125}I]VIP binding studies, suggest the expression in C6 cells of high-affinity VIP/PACAP binding sites as well as the presence of an unusual subclass of very high-affinity VIP binding sites.[6]

The disappearance of PACAP- and VIP-induced cAMP responses in C6 cells in late passages, despite an active AC-cAMP system in these cells (which was proved in studies with forskolin and isoprenaline), deserves attention. It has been demonstrated that even small alterations in the primary structure of the PAC1 and VPAC receptor proteins lead to changes in ligand binding and/or G protein coupling.[7,8] Based on these findings, we suggest that rapidly proliferating C6 cells may undergo spontaneous mutations in PACAP/VIP receptors transcripts resulting in a lack of functional receptors on C6 cells in late passages; however this issue needs further investigations.

In conclusion, the present results show that PACAP and VIP, acting via PAC1 and possibly VPAC-type receptors, stimulate cAMP formation only in early passages of C6 rat glioma cells.

REFERENCES

1. SHERWOOD, N.M., S.L. KRUECKL & J.E. MCRORY. 2000. The origin and function of the pituitary adenylate cyclase-activating polypeptide (PACAP)/glucagon super-family. Endocr. Rev. **21:** 619–670.
2. VAUDRY, D., B.J. GONZALEZ, M. BASILLE, *et al.* 2000. Pituitary adenylate cyclase-activating polypeptide and its receptors: from structure to function. Pharmacol. Rev. **52:** 269–324.
3. SHIMIZU, H., J.W. DALY & C.R. CREVELING. 1969. A radioisotopic method for mea-suring the formation of adenosine 3′,5′-cyclic monophosphate in incubated slices of brain. J. Neurochem. **16:** 1609–1619.
4. SALOMON, Y., C. LONDOS & M. RODBELL. 1974. A highly sensitive adenylate cyclase assay. Analyt. Biochem. **58:** 541–548.
5. EKBLAD, E. *et al.* 2000. Characterization of intestinal receptors for VIP and PACAP in rat and PAC1 receptor knockout mouse. Ann. N. Y. Acad. Sci. **921:** 137–147.
6. DUFES, C. *et al.* 2003. Effects of the vasoactive intestinal peptide (VIP) and related peptides on glioblastoma cell growth in vitro. J. Mol. Neurosci. **21:** 97–108.
7. PISEGNA, J.R., R.M. LYU & P.M. GERMANO. 2000. Essential structural motif in the C terminus of the PACAP type I receptor for signal transduction and internalization. Ann. N. Y. Acad. Sci. **921:** 195–201.
8. NICOLE, P. *et al.* 1998. Site-directed mutagenesis of human vasoactive intestinal peptide receptor subtypes VIP1 and VIP2: evidence for difference in the structure-function relationship. J. Pharmacol. Exp. Ther. **284:** 744–750.

Protective Effects of PACAP in Excitotoxic Striatal Lesion

A. TAMÁS, A. LUBICS, I. LENGVÁRI, AND D. REGLÓDI

Department of Anatomy, Neurohumoral Regulations Research Group of the Hungarian Academy of Sciences, Pécs University, 7624 Pécs, Hungary

ABSTRACT: The present article investigated the effects of pituitary adeny-late cyclase-activating polypeptide (PACAP) treatment in a quinolinic acid (QA)-induced unilateral lesion of the striatum, a model of Huntington's disease (HD). PACAP was given locally, preceding the lesion. Behavioral analysis was performed after 1, 10, and 30 days, when motor activity and asymmetrical signs were evaluated. Three weeks after the treatment, a catalepsy test was performed by haloperidol administration, and finally histological assessment of the striatum was done. Our results show that PACAP treatment attenuated the behavioral deficits and reduced the number of lesioned neurons in the striatum.

KEYWORDS: PACAP; Huntington's disease; quinolinic acid; striatum; neuroprotection

INTRODUCTION

Huntington's disease (HD) is a progressive neurodegenerative disorder, characterized by severe degeneration of basal ganglia neurons.[1] Oxidative stress, inflammation, and toxic agents are involved in the pathways leading to mainly apoptotic neuronal cell death.[1,2] Numerous studies show that pituitary adenylate cyclase-activating polypeptide (PACAP) has protective effects against all these factors both *in vitro* and *in vivo* experimental conditions.[3-5] Recently, we have shown that PACAP reduced dopaminergic cell loss and improved behavioral deficits after 6-hydroxydopamine lesion of the substantia nigra, a model of Parkinson's disease.[6-8] These results imply the possibility of PACAP being protective in other neurodegenerative diseases. Therefore, the aim of the present article was to investigate the effects of PACAP in a quinolinic acid (QA)-induced lesion of the striatum, a model of Huntington's disease.

Address for correspondence: Andrea Tamás, M.D., Department of Anatomy, University of Pécs, 7624 Pécs, Szigeti u 12. Hungary. Voice: +36-72-536001 ext. 1805; fax: +36-72-536393.
e-mail: andrea.tamas@aok.pte.hu

Ann. N.Y. Acad. Sci. 1070: 570–574 (2006). © 2006 New York Academy of Sciences.
doi: 10.1196/annals.1317.083

MATERIALS AND METHODS

Male Wistar rats (200–250 g) were housed under standard laboratory conditions. Procedures were performed in accordance with the ethical guidelines. Drug administration was performed as previously described.[8] Briefly, rats were treated with 30 μg QA/2 μL saline (sigma) injected into the left striatum. Rats received 0.2-μg PACAP ($n = 7$) or 2-μg PACAP ($n = 8$) in 0.5-μL saline *in loco*, preceding the quinolinic acid (QA) treatment, while control animals received the same volume of saline ($n = 7$). Rats were videorecorded during 15 minutes in an open field 1, 10, and 30 days after the lesion as previously described.[6,7] Briefly, motor activity (total ambulation time and total number of turning) and asymmetrical signs (number of left and right turns, forelimb use in rearing and runs along the walls) were measured.

Three weeks after the treatment we evaluated changes in haloperidol-induced catalepsy[9] (0.5 mg/kg i.p.). The typical catalepsy test consists of placing the animal into an unusual posture and recording the time taken to correct this posture within 3 min. Three different bar tests were evaluated. We placed the forepaws of the rats on two different round bars (7 and 10 cm high), or on the side of the glass (7 cm high) and the hindpaws remained on the floor.

Following the behavioral testing, the number of neurons was assessed by nicotinamide adenine dinucleotide phosphate (NADPH)-diaphorase staining 30 days after the injury as previously described.[8] NADPH-d-positive cells in each section on both contralateral and ipsilateral sides of the striatum were counted using the Scion Image computerized image analysis system, and data are expressed as percentage of NADPH-d-positive cells in the lesioned side compared to the contralateral, undamaged side. Statistical analysis was done by ANOVA followed by Neuman Keuls *post hoc* analysis.

RESULTS AND DISCUSSION

Results of the motor activity test are summarized in TABLE 1. QA-treated control animals showed significant hyperkinesia in motor activity tests 1 day

TABLE 1. Activity Measures in the Open-Field Before the Injury (Normal Animals) and 1, 10, and 30 Days After the Injury

	Normal	Control	PACAP 0.2 μg	PACAP 2 μg
Ambulation				
1 day	162 ± 86	$231.4 \pm 30.1^*$	149.2 ± 21.5	$248.5 \pm 34.3^*$
10 days		176 ± 7.8	159 ± 14.5	$282.2 \pm 27.6^*$
30 days		192.5 ± 27.6	182 ± 18.8	192.4 ± 17
Turning				
1 day	91 ± 7.2	$171.4 \pm 41.1^*$	129.1 ± 26.5	$184.5 \pm 26^*$
10 days		106.2 ± 8.6	119.7 ± 10.8	110.4 ± 17.9
30 days		92.1 ± 15.4	107.1 ± 11.7	101.4 ± 9.5

All values are given as mean $\pm$ SEM. Ambulation is given in seconds and turning in the number of turns during the 15-min open-field test. $^*P < 0.05$ vs. normal animals.

TABLE 2. Asymmetrical Behavioral Signs Before the Operation (Normal Animals) and 1, 10, and 30 days After the Injury

	Normal	Control	PACAP 0.2 µg	PACAP 2 µg
Turning				
1 day	40.48 ± 6.4	95.5 ± 2.2***	81.7 ± 4.6***	95.4 ± 2.4***
10 days		74 ± 5**	61.4 ± 6.6	66.4 ± 4.6**
30 days		72.3 ± 5.5**	57.1 ± 4.4	62.6 ± 3.4**
Rearing				
1 day	32.8 ± 8.9	97.2 ± 2.7***	75.4 ± 12.2*	75.9 ± 12.12*
10 days		88.1 ± 3.8***	77.6 ± 8.4**	71.7 ± 7.6**
30 days		72 ± 10.4*	69 ± 7.2*	65 ± 4.7*
Thigmotaxis				
1 day	45.6 ± 8.7	80.3 ± 6.5**	65.4 ± 6*	81.9 ± 6.4**
10 days		81.7 ± 3.7**	72.8 ± 5.6*	75.9 ± 2.8**
30 days		77.5 ± 5.4**	76 ± 6.2*	78 ± 3.7**
Catalepsy				
high bar	136 ± 18.6	62.2 ± 16.3*	116.6 ± 17.8	47.7 ± 25.6*
small bar	180	93.8 ± 21.8*	116.3 ± 19.8*	103.6 ± 19.9*
glass test	173.8 ± 3.8	93.5 ± 20.3*	154.5 ± 23.3	113.4 ± 17.2

All values are given as mean percentage of left turning, rearing, or thigmotaxis compared to total number $\pm$ SEM. *$P < 0.05$, **$P < 0.01$; ***$P < 0.001$ vs. normal animals. All values of the catalepsy test show the time during which the animals maintained this posture in seconds $\pm$ SEM. *$P < 0.05$ vs. normal animals.

after the injury compared to normal animals, but the time spent with ambulation and the total number of turns recovered to normal levels by 10 days. Animals treated with 2-µg PACAP showed similar tendency in the turning activity, but these animals spent significantly more time with ambulation both 1 and 10 days after the injury. However, rats treated with 0.2-µg PACAP displayed no hyperkinesia at any time point.

Asymmetrical signs are summarized in TABLE 2. Normal animals showed no significant asymmetry, but severe left-biased turning, rearing, and wall runs were observed in all injured animals after the lesion. Asymmetry in turning was markedly present in QA-treated animals and in rats treated with 2-µg PACAP 1 day following the injury, with partial but not complete recovery by 10 days and no further recovery by 30 days. Animals treated with 0.2-µg PACAP also showed significantly increased left-biased turning activity 1 day after the injury. However, these animals ceased to display turning asymmetry by 10 days. Rearing asymmetry was markedly expressed 1 and 10 days after the injury in QA-lesioned animals, but significantly improved by 30 days. There was no significant difference between the PACAP-treated groups that showed less left-biased rearing activity at 1 day after the injury. Similarly to other signs of asymmetry, left bias in wall runs could also be observed in all injured animals. Asymmetrical active thigmotaxis was markedly present in QA- and 2-µg PACAP-treated animals throughout the whole observation period. Although asymmetry was present in animals treated with 0.2-µg PACAP, they showed significantly less left-biased wall runs than the other groups 1 day after the injury.

In summary, the unilateral QA lesion of the striatum produced hyperactivity and behavioral asymmetries most pronounced during the first few days after the operation, and animals showed partial recovery after 10 days with no further recovery by 30 days. PACAP treatment attenuated these behavioral deficits.

Results of the catalepsy test are summarized in TABLE 2. Three weeks after the lesion the QA-treated animals and 2-μg PACAP showed significantly less degree of catalepsy: they spent less time in the bars than the normal animals. However, there was no significant difference between the normal animals and rats treated with 0.2-μg PACAP in the high bar test and glass test.

NADPH-diaphorase staining revealed severe loss of neurons in the striatum ipsilateral to the QA injection in the control animals. The loss of neurons on the lesioned side was $18.14 \pm 2.74\%$. There was a significant cell loss also in the 0.2-μg PACAP and 2-μg PACAP-treated animals, but it was only $8.16 \pm 3.74\%$ and $8.22 \pm 2.05\%$, respectively. The average cell loss was significantly different between the control and PACAP-treated animals.

Our results show that pretreatment with PACAP significantly attenuated the deficits caused by QA injection. Best effect was observed in 0.2-μg PACAP treated animals, while the animals treated with 2-μg PACAP spent significantly more time with ambulation and the asymmetrical signs and rigidity were markedly present after the injury. Several studies have demonstrated the dose-dependent effect of PACAP and in these experiments, better results were achieved by the lower doses of PACAP.[10] PACAP has also been shown to enhance locomotor activity at higher doses.[11] It is believed that different doses of PACAP act through different signal transduction pathways.[10] The exact mechanism of the neuroprotective effect of PACAP in the model used in this study is not known. The possible protective mechanisms have been previously discussed.[5,8] It has been reported that QA-induced lesions closely mimic the striatal neuropathology seen in the human disease.[2,12] Although there are no data on the effect of PACAP in QA-induced lesions, the peptide has been shown to protect neurons against other types of toxicity. The neuroprotective effect of PACAP against glutamate-, ethanol-, oxidative stress-, ceramide-, β-amyloid-, and 6-OHDA-induced toxicity has been shown in different cell cultures.[5]

Our present results show that pretreatment with PACAP significantly attenuated the behavioral deficits and the number of lesioned neurons in the striatum in QA-induced injury. These results provide further evidence for the neuroprotective effects of PACAP in animal models of neurodegenerative diseases.

ACKNOWLEDGMENTS

This work was supported by the Grants OTKA T046589 and F048908 and the Hungarian Academy of Sciences.

REFERENCES

1. MARTIN, J.B. & J.F. GUSELLA. 1986. Huntington's disease. Pathogenesis and management. N. Engl. J. Med. **315:** 1267–1276.
2. ROBERTS, R.C., A. AHN, K.J. SWARTZ, *et al.* 1993. Intrastriatal injections of quinolinic acid or kainic acid: differential patterns of cell survival and the effects of analysis on outcome. Exp. Neurol. **124:** 274–282.
3. VAUDRY, D., B.J. GONZALEZ, M. BASILLE, *et al.* 2000. Pituitary adenylate cyclase activating polypeptide and its receptors: from structure to functions. Pharmacol. Rev. **52:** 269–324.
4. WASCHEK, J.A. 2002. Multiple actions of pituitary adenylyl cyclase activating peptide in nervous system development and regeneration. Dev. Neurosci. **24:** 14–23.
5. SOMOGYVÁRI-VÍGH, A. & D. REGLŐDI. 2004. Pituitary adenylate cyclase activating polypeptide: a potential neuroprotevtive peptide. Rev. Curr. Pharm. Des. **10:** 2861–2889.
6. REGLÓDI, D., A. LUBICS, A. TAMÁS, *et al.* 2004. Pituitary adenylate cyclase activating polypeptide protects dopaminergic neurons and improves behavioral deficits in a rat model of Parkinson's disease. Behav. Brain Res. **151:** 303–312.
7. REGLÓDI, D., A. TAMÁS, A. LUBICS, *et al.* 2004. Morphological and functional effects of PACAP in a 6-hydroxydopamine-induced lesion of the substantia nigra in rats. Regul. Pept. **123:** 85–95.
8. REGLÓDI, D., A. TAMÁS & A. SOMOGYVÁRI-VIGH. 2005. Pituitary adenylate cyclase activating polypeptide in animal models of neurodegenerative disorders—implications for Huntington and Parkinson's diseases. Lett. Drug. Des. Disc. **2:** 311–315.
9. SANBERG, P.R., S.F. CALDERON, M. GIORDANO, *et al.* 1989. The quinolinic acid model of Huntington's disease: locomotor abnormalities. Exp. Neurol. **105:** 45–53.
10. ARIMURA, A. 2003. Perspectives on the development of a neuroprotective drug based on PACAP [abstract]. Regul. Peptides **115:** 40.
11. MASUO, Y., J. NOGUCHI, S. MORITA, *et al.* 1995. Effects of intracerebroventricular administration of pituitary adenylate cyclase activating polypeptide (PACAP) on the motor activity and reserpine-induced hypothermia in murines. Brain Res. **700:** 219–236.
12. BEAL, M.F., N.W. KOWALL, D.E. ELLISON, *et al.* 1986. Replication of the neurochemical characteristics of Huntington's disease by quinolinic acid. Nature **321:** 168–171.

Characterization of the New Photoaffinity Probe (Bz$_2$-K^{24})-VIP

YOSSAN-VAR TAN, ALAIN COUVINEAU, JEAN JACQUES LACAPERE, AND MARC LABURTHE

INSERM U773-Faculté de Médecine X. Bichat, 75018 Paris, France

ABSTRACT: Site-directed mutagenesis and molecular modeling demonstrated that the N-terminal ectodomain of the VPAC1 receptor is a major site of vasoactive intestinal peptide (VIP) binding. Previous studies with the [Bpa^6]-VIP and [Bpa^{22}]-VIP probes (substitution with the photoactivable Bpa for the residues 6 and 22 in VIP) showed spatial approximation between the amino acids 6 and 22 of VIP and the 104-108 and 109-119 sequences within the N-terminal ectodomain of the receptor, respectively. Here, we characterize the new probe (Bz$_2$-K^{24})-VIP (substitution with the photoreactive Bz$_2$-K for the residue 24 in VIP). After photolabeling and sequential digestions of the receptor, the 121-133 sequence of the N-terminal ectodomain was identified as the site of interaction. The N-terminal ectodomain of the VPAC1 receptor is therefore an affinity trap for the central part of VIP, at least between residues 6 and 24.

KEYWORDS: GPCR; VIP; photoaffinity

INTRODUCTION

The neuropeptide vasoactive intestinal peptide (VIP)[1] is present in both central and peripheral nervous systems as well as in immune cells.[1-3] In consonance with its ubiquitous distribution, it controls a large array of biological functions including exocrine secretions, release of hormones, growth control of fetuses, and embryonic brain development.[1-5] VIP was shown recently to exert potent anti-inflammatory actions.[2] VIP actions are mediated by two serpentine G protein–coupled receptors, VPAC1 and VPAC2.[6] They belong to the class II subfamily within the superfamily of G protein–coupled receptors including receptors for peptide hormones, such as secretin, parathyroid hormone, or glucagon.[6-8] The VPAC1 receptor is a prototypical class II G protein–coupled receptor that has been extensively studied by site-directed mutagenesis and molecular chimerism.[6] These studies showed that the N-terminal extracellular

Address for correspondence: Marc Laburthe, INSERM U773, Faculté de Médecine Bichat, 16 rue Henri Huchard, 75018 Paris, France. Voice: 33-1-44856135; fax: 33-1-42288765.
e-mail: laburthe@bichat.inserm.fr

Ann. N.Y. Acad. Sci. 1070: 575–580 (2006). © 2006 New York Academy of Sciences.
doi: 10.1196/annals.1317.084

domain of hVPAC1 receptor plays a crucial role for VIP binding, although it is not sufficient to ensure high affinity.[6] Despite these studies of the structure–function relationship of hVPAC1 receptor, the physical sites of interaction between VIP and its receptors had remained elusive until the direct approach of intrinsic photoaffinity labeling. Photoaffinity experiments evidenced physical interactions between VIP and the N-terminal extracellular domain of hVPAC1 receptor. Indeed, the side chains of positions 6^9 and 22^{10} of VIP are in direct contact with receptor segments localized in the N-terminal ectodomain of the receptor. However, it is well known that VIP has diffuse pharmacophoric domains, with the amino acid residues important for biological activity being distributed along the whole 28-amino acid peptide chain.[11] In this context, it is clear that the synthesis of new probes with photolabile residues in the N-terminal and the C-terminal ends of VIP will be essential to further understand the molecular nature of the interaction between VIP and its receptor. In this article, we developed a new photoaffinity probe of VIP by substituting benzoylbenzoyl-L-Lys (Bz$_2$-K) for Asn^{24} residue. We report here the development and characterization of the (Bz$_2$-K^{24})-VIP probe for covalent labeling of the human VPAC1 receptor.

MATERIALS AND METHODS

Ligand Binding and Adenylyl Cyclase Activity Assays

Membranes from Chinese Hamster Ovary (CHO) cells stably expressing hVPAC1 receptor fusioned to the green fluorescent protein[12] were incubated with ^{125}I-VIP in the presence of increasing concentrations of VIP or (Bz$_2$-K^{24})-VIP probe. The radioactivity was then assayed in a γ-counter. Specific binding was calculated as the difference between the amount of ^{125}I-VIP bound in the absence (total binding) and the presence (nonspecific binding) of 1 μM unlabeled VIP. The Ki was determined by computer analysis. Adenylyl cyclase activity in cell membranes was assayed in the presence of increasing concentrations of VIP or (Bz$_2$-K^{24})-VIP. Dose-response curves were fitted and EC$_{50}$ were calculated using the Prism software suite (Graph Pad, San Diego, CA).

Receptor Photoaffinity Labeling

A total of 50 μg of enriched receptor-bearing transfected cells were incubated in darkness with ^{125}I-[(Bz$_2$-K^{24})-VIP] and photolyzed ($\lambda = 365$ nm). The photolabeled receptors were then analyzed directly by SDS-NuPAGE electrophoresis under reducing conditions or after chemical (CNBr) and/or enzymatic (PNGase F, endoproteinase Glu-C) cleavages as described.[9,10] The apparent molecular masses of radiolabeled receptor fragments were determined by interpolation on a plot of the mobility of protein molecular weight markers from Amersham (Buckinghamshire, UK) or Invitrogen (Carlsbad, CA) versus the log values of their masses.

RESULTS AND DISCUSSION

In this study, the synthetic (Bz$_2$-K^{24})-VIP probe with a substitution of Asn^{24} residue by the photolabile residue benzoylbenzoyl-L-Lys was used to define the interaction between the position 24 of VIP and the human VPAC1 receptor. In order to determine the biological activity of the probe, it was tested for its ability to bind the human VPAC1 receptor stably expressed in CHO-F7 cells and to stimulate adenylyl cyclase activity (FIG. 1). (Bz$_2$-K^{24})-VIP is a full VPAC1 receptor agonist with a potency similar to that of native VIP in stimulating cAMP accumulation with EC$_{50}$ of 0.1 nM and 0.03 nM, respectively (FIG. 1). Its binding affinity for the VPAC1 receptor is also similar to that of VIP (Ki of 0.6 nM) (FIG. 1). Incubation of the ^{125}I-[(Bz$_2$-K^{24})-VIP] probe with CHO-F7 cells expressing the VPAC1 receptor was followed by UV exposure. NuPAGE analysis revealed the existence of a single band that completely disappeared in the presence of 1μM of native VIP. This band migrated at 95 kDa and corresponded to ^{125}I-[(Bz$_2$-K^{24})-VIP]/hVPAC1-R complex. This band represents the glycosylated VPAC1 receptor (64 kDa) fusioned to the GFP protein (25 kDa) and covalently associated to the radioiodinated probe (3.3 kDa). Altogether, these data indicated that the photolabile ^{125}I-[(Bz$_2$-K^{24})-VIP] probe was able to efficiently and specifically photolabel the human VPAC1 receptor. To identify the VPAC1 receptor domain to which the probe is covalently

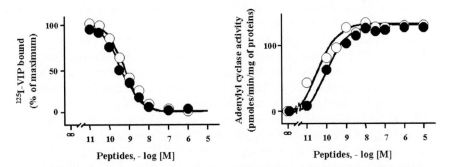

FIGURE 1. Binding and adenylyl cyclase activity assays of the (Bz$_2$-K^{24})-VIP probe. *Left panel*, competitive inhibition of ^{125}I-VIP binding to CHO-F7 cell membranes expressing wild-type hVPAC1 receptor by VIP and (Bz$_2$-K^{24})-VIP. The data are expressed as percentage of initial specific binding in the absence of competitor. *Right panel*, effect of increasing concentrations of VIP and (Bz$_2$-K^{24})-VIP on adenylyl cyclase activity in the same membranes. The data are expressed as percentage of maximal stimulation above basal obtained with 1 μM native VIP. The symbols are: (o) VIP and (•) (Bz$_2$-K^{24})-VIP.

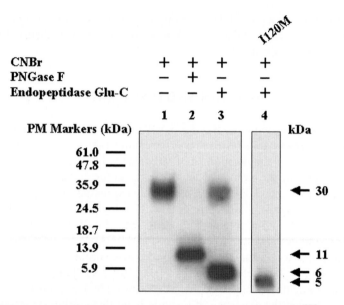

FIGURE 2. Photoaffinity labeling of the hVPAC1 receptor with the ^{125}I-[Bz$_2$-K^{24}]-VIP probe. Autoradiography of SDS-NuPAGE gel (+DTT) reveals digestion products of covalent complex ^{125}I-[(Bz$_2$K^{24})-VIP]/VPAC1 receptor, after treatment with CNBr, PNGase F or endopeptidase Glu-C for the wt (left) and the I120M mutant (right) receptors. See "Materials and Methods" for details. MW = molecular mass.

linked, the 95-kDa complex ^{125}I-[(Bz$_2$-K^{24})-VIP]/hVPAC1-R was subjected to sequential chemical and enzymatic digestions. The cyanogen bromide (CNBr) cleavage generated a single 30-kDa labeled band that shifted after deglycosylation to a 11-kDa band indicating that it is heavily glycosylated (FIG. 2). Previous work demonstrated the presence of 9-kDa carbohydrate moiety on each of the three consensus N-glycosylation sites in the N-terminal domain of the hVPAC1 receptor.[13] In this context, our data are consistent with the covalent attachment of the probe to the Trp^{67}-Met^{137} glycosylated fragment present in the N-terminal ectodomain of the receptor. To further reduce the size of the labeled fragment after CNBr action, the endoproteinase Glu-C cleavage was realized after electroelution of the 30-kDa band, generating a major 6-kDa band that represents the receptor segment 109-133 (FIG. 2). In order to further narrow the fragment covalently linked to the ^{125}I-[(Bz$_2$-K^{24})-VIP] probe, a mutant was produced by substitution of Ile^{120} by a methionine residue, which creates a new CNBr cleavage site in this position. The mutant I120M was constructed and stably expressed in CHO cells. It bound VIP with an affinity similar to that of the wild-type receptor. After incubation of the I120M mutant with the iodinated probe and CNBr treatment of proteins, a 5-kDa labeled band was observed instead of a 30-kDa band for the wild-type receptor (FIG. 2).

TABLE 1. VPAC1 receptor fragments covalently attached to photoaffinity probes

Photoaffinity probe	Labeled receptor sequence	Ref.
^{125}I-[Bpa^6-VIP]	104-108	9
^{125}I-[Bpa^{22}-VIP]	109-120	10
^{125}I-[Bz$_2$-K^{24}-VIP]	121-133	This paper

This labeled protein represents the segment 121-133 covalently attached to the ^{125}I-[(Bz$_2$-K^{24})-VIP] probe. Thus, our data show that there is a spatial approximation between Asn^{24} of VIP and the nonglycosylated 121-133 sequence within the N-terminal ectodomain of the human VPAC1 receptor.

In conclusion, the results with (Bz$_2$-K^{24})-VIP show that the 121-133 region of the N-terminal ectodomain is in contact with the side chain at position 24 of VIP. In previous reports, the use of Bpa^6-VIP and Bpa^{22}-VIP photoaffinity probes identified covalent interactions with 104-108[9] and 109-120[10] segments in the amino-terminal ectodomain of the human VPAC1 receptor, respectively. Therefore, our data are consistent with a binding model in which the central part of VIP (6-24) interacts with the N-terminal ectodomain of the receptor (TABLE 1). The development of new probes, in which photolabile residues are inserted in the N-terminal and the C-terminal ends of VIP, will be essential to strengthen this binding model.

ACKNOWLEDGMENTS

Y.-V. Tan is a Ph.D. student supported by the Ministère de l'Enseignement Supérieur et de la Recherche and by grant FDT 20041202869 from the Fondation pour la Recherche Médicale. This research was supported by the Institut National de la Santé et de la Recherche Médicale, the Centre National de la Recherche Scientifique, the Université Paris 7, grant ACIM-2-18, 2003-5, Programmes de Microbiologie: Microbiologie Fondamentale et Appliquée, Maladies Infectieuses, Environnement et Bioterrorisme, and a grant in 2005 from Association de Recherche sur la Polyarthrite.

REFERENCES

1. SAID, S.I. 1991. VIP biologic role in health and disease. Trends Endocrinol. Metab. **2:** 107–122.
2. KLIMASCHEWSKI, L. 1997. VIP, a 'very important peptide' in the sympathetic nervous system? Anat. Embryol. **196:** 269–277.
3. GANEA, D. & M. DELGADO. 2002. Vasoactive intestinal peptide (VIP) and pituitary adenylate cyclase-activating polypeptide (PACAP) as modulators of both innate and adaptive immunity. Crit. Rev. Oral Biol. Med. **13:** 229–237.

4. LABURTHE, M., M. ROUSSET, C. BOISSARD, *et al*. 1978. Vasoacive intestinal peptide: a potent stimulator of adenosine 3':5'-cyclic monophosphate accumulation in gut carcinoma cell lines in culture. Proc. Natl. Acad. Sci. USA **75:** 2772–2775.

5. GRESSENS, P., J.M. HILL, I. GOZES, *et al*. 1993. Growth factor function of vasoactive intestinal peptide in whole cultured mouse embryos. Nature **362:** 155–158.

6. LABURTHE, M., A. COUVINEAU & J.C. MARIE. 2002. VPAC receptors for VIP and PACAP. Receptors Channels **8:** 137–153.

7. LABURTHE, M., A. COUVINEAU, P. GAUDIN, *et al*. 1996. Receptors for VIP, PACAP, secretin, GRF, glucagon, GLP-1, and other members of their new family of G protein-linked receptors: structure-function relationship with special reference to the human VIP-1 receptor. Ann. N. Y. Acad. Sci. **805:** 94–111.

8. LABURTHE, M. & A. COUVINEAU. 2002. Molecular pharmacology and structure of VPAC Receptors for VIP and PACAP. Regul. Peptides **108:** 165–173.

9. TAN, Y.V., A. COUVINEAU & M. LABURTHE. 2004. Diffuse pharmacophoric domains of vasoactive intestinal peptide (VIP) and further insights into the interaction of VIP with the N-terminal ectodomain of human VPAC1 receptor by photoaffinity labeling with [Bpa6]-VIP. J. Biol. Chem. **279:** 38889–38894.

10. TAN, Y.V., A. COUVINEAU, J. VAN RAMPELBERGH & M. LABURTHE. 2003. Photoaffinity labeling demonstrates physical contact between vasoactive intestinal peptide and the N-terminal ectodomain of the human VPAC1 receptor. J. Biol. Chem. **278:** 36531–36536.

11. NICOLE, P., L. LINS, C. ROUYER-FESSARD, *et al*. 2000. Identification of key residues for in eraction of vasoactive intestinal peptide with human VPAC1 and VPAC2 receptors and development of a highly selective VPAC1 receptor agonist. Alanine scanning and molecular modeling of the peptide. J. Biol. Chem. **275:** 24003–24012.

12. GAUDIN, P., J.J. MAORET, A. COUVINEAU, *et al*. 1998. Constitutive activation of the human vasoactive intestinal peptide 1 receptor, a member of the new class II family of G protein-coupled receptors. J. Biol. Chem. **273:** 4990–4996.

13. COUVINEAU, A., C. FABRE, P. GAUDIN, *et al*. 1996. Mutagenesis of N-glycosylation sites in the human vasoactive intestinal peptide 1 receptor. Evidence that asparagine 58 or 69 is crucial for correct delivery of the receptor to plasma membrane. Biochemistry **35:** 1745–1752.

PACAP Receptor (PAC1-R) Expression in Rat and Rhesus Monkey Thymus

N. TOKUDA,[a] Y. ARUDCHELVAN,[a] T. SAWADA,[a] Y. ADACHI,[a]
T. FUKUMOTO,[a] M. YASUDA,[b] H. SUMIDA,[b] S. SHIODA,[c,d] T. FUKUDA,[e]
A. ARIMA,[e] AND S. KUBOTA[f]

[a]Department of Human Science, Yamaguchi University School of Medicine,
Ube 755-8505, Japan

[b]Department of Clinical Radiology, Faculty of Health Sciences, Hiroshima
International University, Higashi-Hiroshima, Japan

[c]Department of Anatomy, Showa University School of Medicine, Tokyo, Japan

[d]CREST of JST, Kagoshima, Japan

[e]Shin Nippon Biomedical Laboratories, Ltd., Drug Safety Research Laboratories,
Kagoshima, Japan

[f]Department of Life Sciences, Graduate School of Arts and Sciences, University
of Tokyo, Tokyo, Japan

ABSTRACT: The expression of PACAP receptor (PAC1-R) was investigated in the thymus of rats and rhesus monkeys. In the rat thymus, PAC1-R positive cells were found in the intermediate type of thymic epithelial cells of the medulla. PAC1-R-positive cells were also seen in the thymic medulla of the rhesus monkey. The thymus showed unusual structures in some rhesus monkey dams (F0) and offspring (F1) exposed to 2, 3, 7, 8-tetrachlorodibenzo-p-dioxin (TCDD). Additionally, in these rhesus monkeys, PAC1-R expression was different from that in the control thymus.

KEYWORDS: rat thymus; rhesus monkey thymus; TCDD

INTRODUCTION

Pituitary adenylate cyclase-activating polypeptide (PACAP) is a multifunctional and pleiotropic signal molecule[1] with important activities related to immunity. We previously reported that the PACAP receptor (PAC1-R) is strongly expressed in stromal cells in the medulla of the rat thymus and that its expression is affected by irradiation.[2] In the present article we present a detailed

Address for correspondence: Nobuko Tokuda, Department of Anatomy, Department of Human Science, Yamaguchi University School of Medicine, Ube 755-8505, Japan. Voice: +81-836-22-2202; fax: +81-836-22-2203.

e-mail: toku@yamaguchi-u.ac.jp

Ann. N.Y. Acad. Sci. 1070: 581–585 (2006). © 2006 New York Academy of Sciences.
doi: 10.1196/annals.1317.085

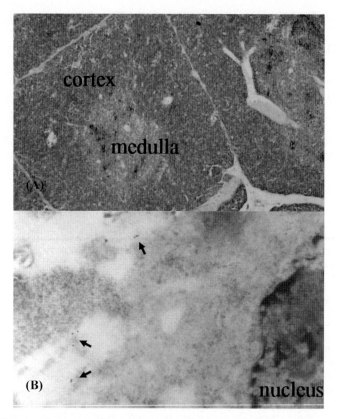

FIGURE 1. (**A**) Immunohistochemical analysis of PAC1-R in the rat thymus (x40). (**B**) Immunoelectron microscopy analysis of PAC1-R. Arrows indicate PAC1-R-immunoreactivity in an intermediate type of thymic epithelial cell magnification (16,000x).

investigation of PAC1-R-positive cells in the rat thymus and examine thymus cells of rhesus monkeys exposed to tetrachlorodibenzo-*p*-dioxin (TCDD), a chemical known to impair thymus and T cells.[3]

MATERIALS AND METHODS

The cell type expressing PAC1-R in the thymus of young-adult Dark Agouti (DA) rats was determined immunohistochemically. TCDD (300 ng/kg) was administered subcutaneously to pregnant rhesus monkeys. Five percent of the

---

FIGURE 2. Immunohistochemical analysis of PAC1-R in the rhesus monkey thymus (x100). (**A**) normal. (**B, C**) TCDD-exposed (**B**: dam, **C**: offspring). In TCDD-exposed rhesus thymuses (**B, C**), discrete cortex and medulla were not seen.

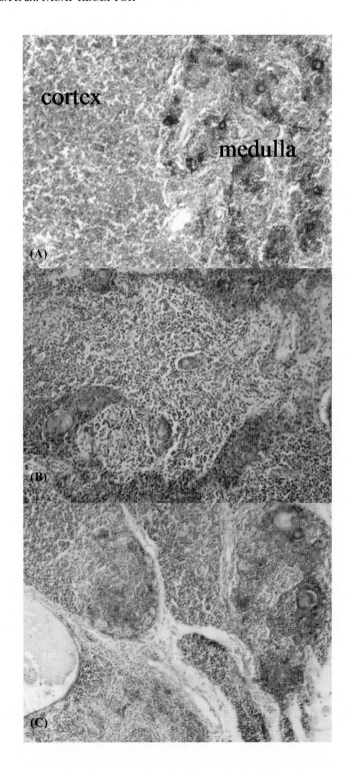

initial dose was then administered every 30 days until day 90 after delivery.[4] Thymuses from dams (F0) and offspring (F1) exposed to TCDD were analyzed immunohistochemically using the antibody G51 to detect the intercellular region of PAC1-R.[2,5]

RESULTS

PAC1-R expression in the thymus was similar in rats and rhesus monkeys (FIGS. 1A and 2A). In the rat, immunoelectron microscopic analysis detected PAC1-R in the intermediate type of thymic epithelial cells of the thymic medulla (FIG. 1B). Thymocytes and other types of stromal cells were not positive for PAC1-R. Similarly, in the rhesus monkeys, PAC1-R was strongly expressed in the epithelial cells of the medulla (FIG. 2A). However, in some thymuses from TCDD-exposed F0 and F1 rhesus monkeys, a discrete cortex and medulla was not seen. In these instances fewer PAC1-R-positive cells were seen than in the control thymuses (FIGS. 2B and 2C).

DISCUSSION

PACAP is thought to regulate stromal cells in the central nervous system.[5] In the rat thymus, PAC1-R was detected in the intermediate type of thymic epithelial cells (FIGS. 1A and 1B). Our results suggest that PACAP may have an important role in the regulation of thymocyte maturation and/or proliferation.

Similarly, PAC1-R-positive cells were also seen in the rhesus monkey thymuses (FIG. 2A), and PACAP may also work as a regulator of thymocyte functions in rhesus monkeys. PAC1-R expression in TCDD-exposed thymuses appeared to be different from those of controls (FIG. 2B), suggesting that TCDD affects the thymus. Furthermore, TCDD may affect offspring thymuses via the placenta and/or through milk (FIG. 2C).

ACKNOWLEDGMENTS

This study was supported in part by a grant-in-aid for Scientific Research from the Ministry of Education, Science of Japan (No. 17590164) and Health Science Research Grants for Research on Environmental Health from the Ministry of Health, Welfare, and Labor of Japan.

REFERENCES

1. VAUDRY, D., B.J. GONZALEZ, M. BASILLE, et al. 2000. Pituitary adenylate cyclase-activating polypeptide and its receptors: from structure to function. Pharmacol. Rev. **52:** 269–324.

2. TOKUDA, N. *et al.* 2004. Expression of PAC1 receptor in rat thymus after irradiation. Regul. Pept. **123:** 167–172.
3. CAMACHO, I.A *et al.* 2004. Evidence for induction of apoptosis in T cells from murine fetal thymus following perinatal exposure to 2, 3, 7, 8-tetrachlorodibenzo-p-dioxin (TCDD). Toxicol. Sci. **78:** 96–106.
4. YASUDA, I. *et al.* 2005. In utero and lactational exposure to 2,3,7,8-tetrachlorodibenzo-p-dioxin (TCDD) affects tooth development in rhesus monkeys. Reprod. Toxicol. **20:** 21–30.
5. SHIODA, S. *et al.* 2003. PACAP Receptor signaling. *In* Pituitary Adenylate Cyclase-Activating Polypeptide. H. Vaudry & A. Arimura, Eds.: 95–124. Kluwer Academic Publishers. Norwell, MA.

Characterization of the PAC1 Variants Expressed in the Mouse Heart

MINA USHIYAMA,[a,b] HIDEKI SUGAWARA,[a,b] KAZUHIKO INOUE,[a] KENJI KANGAWA,[c] KATSUSHI YAMADA,[b] AND ATSURO MIYATA[a]

[a]Department of Pharmacology, Graduate School of Medical and Dental Sciences, Kagoshima University, Kagoshima 890-8544, Japan

[b]Department of Clinical Pharmacy, Graduate School of Medical and Dental Sciences, Kagoshima University, Kagoshima 890-8544, Japan

[c]Department of Biochemistry, National Cardiovascular Center Research Institute, Suita, Osaka 565-8565, Japan

ABSTRACT: Pituitary adenylate cyclase-activating polypeptide (PACAP), a pleiotropic neuropeptide, exerts a variety of physiological functions through three types of G protein–coupled receptors, PAC1, VPAC1, and VAPC2. Characterization of the molecular forms of PAC1 in mouse heart revealed the presence of four types of variant receptors harboring the N or S variant in the first extracellular domain (EC1 domain) with or without the HOP1 insert in the third intracellular cytoplasmic loop (IC3 loop). Then, we assessed the binding affinity and ability to stimulate adenylyl cyclase of the PCA1 variant-expressing cells for PACAP. Adenylyl cyclase activation by PACAP was markedly influenced with the variant in the EC1 domain as well as that in the IC3 loop, in spite of a little difference in their binding properties. These data suggest that the combination of EC1 domain variants and IC3 loop variants might account for the diversity of intracellular signaling, which might contribute to multiple functions of PACAP including a role in the cardiovascular system.

KEYWORDS: PACAP; PAC1; splice variant; RT-PCR; cyclic AMP

INTRODUCTION

Pituitary adenylate cyclase-activating polypeptide (PACAP) exerts pleiotropic functions through three G protein–coupled receptors, PAC1, VPAC1, and VPAC2, which are preferentially linked to stimulation of adenylate cyclase activity.[1,2] In addition, some of PAC1 receptors have also been reported to couple with intracellular calcium mobilization through phospholipase C activation.

Address for correspondence: A. Miyata, Department of Pharmacology, Graduate School of Medical and Dental Sciences, Kagoshima University, 8-35-1 Sakuragaoka, Kagoshima 890-8544, Japan. Voice: +81-99-275-5256; fax: +81-99-265-8567.

e-mail: amiyata@m3.kufm.kagoshima-u.ac.jp

Ann. N.Y. Acad. Sci. 1070: 586–590 (2006). © 2006 New York Academy of Sciences.

doi: 10.1196/annals.1317.086

As to the molecular forms of PAC1, the presence of six splice variants, R (no insertion) and HIP, HOP1, HOP2, HIP-HOP1, HIP-HOP2 which differ from R by insertion of specific splice cassettes hip and hop, was initially reported in the third intracellular cytoplasmic loop (IC3 loop) of PAC1.[1,2] Then, in the first extracellular domain (EC1 domain) of PAC1, the presence of three splice variants was reported as the subtype S and VS forms, which lack 21 and 57 amino acids, respectively, whereas the N form has no deletion.[3,4] Although the effect of individual variant on the intracellular signaling has been demonstrated, the effect produced by the combination of variants in the EC1 domain and those in the IC3 loop have not been investigated so far.

Recently, Otto *et al.* reported that genetic disruption of PAC1 in mice leads to postnatal pulmonary hypertension followed by right heart failure, while VPAC2-deficient mice do not suffer from pulmonary hypertension.[5] Together with the report of positive inotropic and cardioprotective effects of PACAP,[6] these findings indicate that PACAP could play a key role in the cardiovascular system through a distinct PACAP preferring receptor PAC1. Little is known, however, about the molecular forms of PAC1 in heart. Therefore, we attempted to characterize PAC1 splice variants in the mouse heart and assess their binding affinity for PACAP and the ability of PACAP to activate adenylyl cyclase.

MATERIALS AND METHODS

Reverse Transcription Polymerase Chain Reaction (RT-PCR) Analysis of PAC1 Variants

Total RNAs were isolated from various mouse tissues with TRIZOL LS (Invitrogen, Carlsbad, CA) according to the manufacturer's protocol. The first strand cDNA was synthesized from total RNA, using random hexamers and Superscript H⁻ reverse transcriptase (Invitrogen). The resulting cDNAs were subjected to PCR with two sets of forward/reverse primers corresponding to amino acids [KKEQAMC]/[ACGFDDY] and [GPVVGSI]/[ILSKSS] of mouse PAC1, which were designed for amplification of EC1 domain variants and IC3 loop variants, respectively.

Preparation of the Cells Expressing PAC1 Variants

cDNAs of N/R, N/HOP1, S/R, and S/HOP1 were inserted into pcDNA3.1, and then transfected into Chinese Hamster Ovary (CHO) cells using FuGENE6 (Roche, Basel, Switzerland) according to the manufacturer's protocol. Clonal cell lines expressing PAC1 variants were isolated by selection in medium supplemented with G418.

Binding of ^{125} I-PACAP-27

The crude membrane fractions were prepared from PAC1 variant-expressing cells. Radioreceptor binding assay was performed using ^{125}I-PACAP27 as described previously.[6]

Assay of Intracellular cAMP in PAC1 Variant-Expressing Cells

After stimulation of PACAP for 1 h, the accumulated cAMP in PAC1 variant-expressing cells was measured by using cAMP-specific RIA kit (YAMASA, Tokyo, Japan) as described previously.[6]

RESULTS AND DISCUSSION

As shown in FIGURE 1, RT-PCR analysis demonstrated the tissue distribution of PAC1 splice variants in mice. As regards the EC1 variant of PAC1, two splice variants, N form (337 bp) and S form (274 bp), were amplified, and the N form was found to be predominant in mouse heart. Also in other tissues including thymus, lung, liver, spleen, colon, kidney, and prostate, the N form and S form were amplified, and the N form was dominantly expressed as well. Incidentally, no evidence for the expression of the VS form, corresponding to a 169-bp product, was found in any of the examined tissues.

With regard to the IC3 loop variants, the expression of two bands, 430 bp and 514 bp, was observed (FIG. 2). One of these bands, corresponding to the 430-bp product was identified as the R form. The other band, corresponding to the 514-bp product, was confirmed by DNA sequencing as carrying the HOP1 cassette insert. In the mouse heart, the R form and HOP1 form were amplified in the same extent. However, the R form was found to be predominant or exclusive in most other tissues. According to the combination of the EC1 domain variants and IC3 loop variants, four variant receptors harboring the

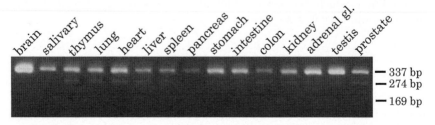

FIGURE 1. Tissue distribution of splice variants in the first extracellular domain of PAC1 as determined by RT-PCR. The arrows indicate the bands of 337 bp, 274 bp, and 169 bp corresponding to the N form (no deletion), S form, and VS form, respectively.[4]

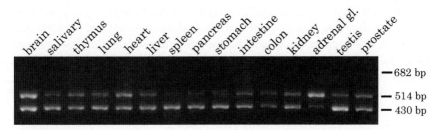

FIGURE 2. Tissue distribution of splice variants in the third intracellular loop of PAC1 as determined by RT-PCR. The arrows indicate the bands of 682 bp, 514 bp, and 430 bp corresponding to the HIP/HOP1 insertion, HOP1 or HIP insertion, and R form (no insertion), respectively.[8]

N or S variants of the EC1 domain with or without the HOP1 insert in the IC3 loop were identified as expected theoretically. We finally isolated full-length cDNAs of these PAC1 variants and transfected all the constructs in CHO cells. Subsequently, stable cell lines for the respective receptors were generated and designated as N/R, N/HOP1, S/R, and S/HOP1.

Utilizing these cells, we performed radioreceptor-binding assay with ^{125}I-PACAP-27 and assessed the effect of PACAP on cyclic AMP accumulation. The following rank order of binding and agonist potency profiles were obtained with the four different receptors; N/HOP1 >N/R>S/R=S/HOP1 in regard to the binding affinity, N/R= S/R=S/HOP1>>N/HOP1 in regard to adenylyl cyclase activation (TABLE 1). Stimulation of cAMP accumulation by PACAP was greatly influenced by the variant subtype, although the binding affinities for PACAP were not significantly different among the cell lines. In the case of the HOP1 insertion in the IC3 loop, the potency of PACAP on the S form of EC1 domain was more greatly facilitated than that on the N form. These data suggest that the EC1 domain as well as the IC3 loop can influence not only the ligand binding but also the intracellular signaling. Regarding the effects of IC3 variants, McCulloch *et al.* revealed that the R and HOP1 forms potently activate adenylate cyclase and phospholipase C by PACAP.[7] On the other hand, Pantaloni *et al.*[3] and Dautzenberg *et al.*[4] demonstrated that the

TABLE 1. IC_{50} values for PACAP-38 displacement of $[^{125}I]$-PACAP-27 and EC_{50} values for PACAP-38 stimulation of cAMP accumulation in the PAC1 variant-expressing cells. The values are the mean $\pm$ SEM

Expressed receptor	IC_{50} (nM) $\pm$ SEM (displacement of ^{125} I-PACAP-27)	EC_{50} (nM) $\pm$ SEM (cAMP formation)
N/R	0.25 ± 0.003	0.07 ± 0.003
N/HOP1	0.12 ± 0.02	0.67 ± 0.34
S/R	0.39 ± 0.009	0.10 ± 0.002
S/HOP1	0.41 ± 0.04	0.05 ± 0.25

EC1 domain plays a critical role in conferring ligand selectivity or regulation of cellular signaling. Taken together these findings indicate that the combination of EC1 variants and IC3 variants might give arise to multiple responses of intracellular signaling, contributing to multiple functions of PACAP, including its role in the cardiovascular system.

REFERENCES

1. VAUDRY, D., B.J. GONZALEZ, et al. 2000. Pituitary adenylate cyclase-activating polypeptide and its receptors: from structure to functions. Pharmacol. Rev. **52:** 269–324.
2. SHIODA, S. & J.A. WASCHEK. 2002. VIP and PACAP receptors. In Understanding G Protein-Coupled Receptors and Their Role in the CNS. M.N. Pangalos & C.H. Davies, Eds.: 527–545. Oxford University Press. Oxford, UK.
3. PANTALONI, C., P. BRABET, et al. 1996. Alternative splicing in the N-terminal extracellular domain of the pituitary adenylate cyclase-activaing polypeptide (PACAP) receptor modulates receptor selectivity and relative potencies of PACAP-27 and PACAP-38 in phospholipase C activation. J. Biol. Chem. **271:** 22146–22151.
4. DAUTZENBERG, F.M., G. MEVENKAMP, et al. 1999. N-terminal splice variants of the type 1 PACAP receptor: isolation, characterization and ligand binding/selectivity determinants. J. Neuroendocrinol. **11:** 941–949.
5. OTTO, C., L. HEIN, et al. 2004. Pulmonary hypertension and right heart failure in pituitary adenylate cyclase-activating polypeptide Type I receptor-deficient mice. Circulation **110:** 3245–3251.
6. SANO, H., A. MIYATA, et al. 2002. The effect of pituitary adenylate cyclase activating polypeptide on cultured rat cardiocytes as a cardioprotective factor. Regul. Pept. **109:** 107–113.
7. MCCULLOCH, D.A., E.M. LUTZ, et al. 2005. ADP-ribosilation factor-dependent phospholipase D activation by VPAC receptors and PAC1 receptor splice variant. Mol. Pharmacol. **59:** 1523–1532.
8. SPENGLER, D., C. WAEBER, et al. 1993. Differential signal transduction by five splice variants of the PACAP receptor. Nature **365:** 170–175.

Distribution of PACAP in the Brain of the Cartilaginous Fish *Torpedo Marmorata*

SALUATORE VALIANTE, MARINA PRISCO, LOREDANA RICCHIARI, VINCENZA LAFORGIA, LORENZO VARANO, AND PIERO ANDREUCCETTI

Department of Biological Sciences, Section of Evolutionary and Comparative Biology, University of Naples, 80134 Naples, Italy

ABSTRACT: In this article, we investigated the distribution of pituitary adenylate cyclase-activating polypeptide (PACAP) and its mRNA by immunohistochemistry, *in situ* hybridization, and RT-PCR techniques, in the central nervous system of the elasmobranch *Torpedo marmorata*. RT-PCR analysis showed that the CNS of *T. marmorata* expresses a messenger encoding PACAP. The immunohistochemistry and *in situ* hybridization patterns were partly overlapping, with a major expression in the hypothalamo–pituitary region and, surprisingly, in the saccus vasculosus. Our results show that, in *T. marmorata*, PACAP is synthesized and widely distributed in the CNS, suggesting an as yet unidentified role for this peptide in elasmobranch brain physiology.

KEYWORDS: central nervous system; elasmobranch; RT-PCR; immunohistochemistry; *in situ* hybridization

INTRODUCTION

Pituitary adenylate cyclase-activating polypeptide (PACAP) has been shown to have a broader range of actions than the merely hypophysiotropic role.[1] The primary structure of PACAP is thoroughly preserved across vertebrates, indicating a strong selective pressure in the molecular evolution of this neuropeptide.[2] In mammals as in nonmammalian vertebrates, PACAP is widely distributed in the central nervous system (CNS).[3–7] Although vertebrates have been widely studied in this respect, elasmobranch information is very scant, *Dasyatis akajei* being the only species studied so far.[8] Here, we investigated the presence and distribution of PACAP mRNA by RT-PCR and *in situ* hybridization, and PACAP localization by immunohistochemistry in the CNS of the elasmobranch *Torpedo marmorata*.

Address for correspondence: S. Valiante, Department of Biological Sciences, Section of Evolutionary and Comparative Biology, University of Naples, Via Mezzocannone, 8, 80134 Naples, Italy. Voice: +390812535038; fax: 390812535035.
e-mail: valiante@unina.it

Ann. N.Y. Acad. Sci. 1070: 591–596 (2006). © 2006 New York Academy of Sciences.
doi: 10.1196/annals.1317.087

MATERIALS AND METHODS

Six animals were captured in the Bay of Naples (southern Italy). The experiments were admitted by institutional committees (Italian Ministry of Health). Fish brains were fixed for 24 h in Bouin's fluid or paraformaldehyde 4% with PBS and embedded in paraffin wax. All results were obtained through three independent experiments. Two RNA probes were synthesized using turkey PACAP cDNA, kindly provided by Prof. Foster. The plasmid was recovered with Ultraclean miniplasmid kit (MoBio Laboratories Carlsbad, CA) after transfection and used for *in vitro* transcription reaction to synthesize both digoxigenin-labeled sense and antisense RNA probes of 300 bp each. The procedure for *in situ* hybridization has been described in detail elsewhere.[9] Slides were hybridized overnight at 60°C with 2 ng/µL of PACAP sense or antisense riboprobes. Anti-DIG alkaline phosphatase-conjugated antibody (Roche, Mannheim Germany) and BM Purple substrate (Roche) were used to detect the hybridization signal. Antigen unmasking was carried out for immunohistochemistry in 10 mM citrate buffer pH 6.0 at 95°C and 1:100 normal goat serum step was performed. Overnight incubation at 4°C with a 1:300-diluted antibody raised against PACAP of *D. akajei*, kindly provided by Prof. Matsuda, was carried out. Avidin-biotin peroxidase complex (ABC) (Pierce, Rockford, IL) and diaminobenzidine (Sigma, St. Louis, MO) in 0.05 M Tris-HCl pH 7.4, 3% H_2O_2 were used to reveal the immunocomplex. Both *in situ* and immunohistochemical signals were analyzed with Axioskop System (Zeiss, Oberkochen, Germany) and images were acquired by KS 300 software (Zeiss). For RT-PCR, brains were removed, and total RNA was extracted using SV Total RNA Isolation System (Promega, Madison, WI) following the manufacturer's instructions. cDNA synthesis was carried out from 4 µg of total RNA with ImProm-II™ Reverse Transcription System (Promega) using oligo (dt)$_{12-18}$ as primers at 37°C for 1.5 h. Polymerase chain reaction (PCR) amplification was executed using 3 µL of the cDNA, 1.25 U of Taq DNA polymerase (Promega), 0.4 mM dNTP, 2.5 mM $MgCl_2$, and 0.4 µM of each primer. Primers were designed by comparing the PACAP coding sequence of vertebrates. The amplification product was analyzed on 2% agarose gel.

RESULTS

PCR amplification revealed the presence of a 250-bp band corresponding to the sequence flanked by the forward and reverse primers (FIG. 1 a). In the forebrain, PACAP immunoreactivity (IR) was mainly found in the hypothalamus region, where numerous PACAP-positive nerve fibers were seen (FIG. 1 b). The pituitary showed PACAP IR only in the neurointermediate lobe (FIG. 1 c), and strong staining was observed in the saccus vasculosus (FIG. 1 d). In the hindbrain, the cerebellum was labeled for PACAP (FIG. 1 e); negative controls showed no PACAP IR (FIG. 1 f). A number of brain areas synthesized PACAP:

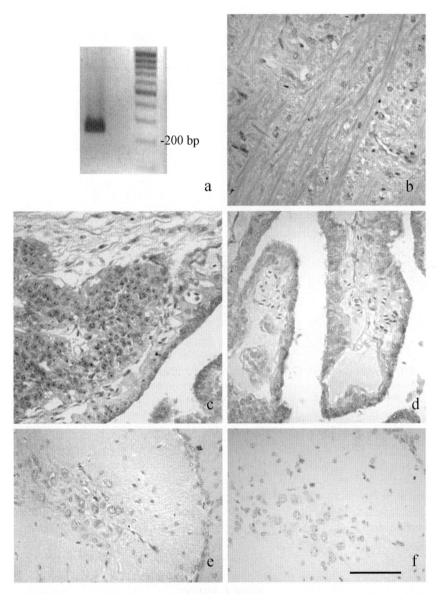

FIGURE 1. (a) RT-PCR analysis showed that PACAP mRNA is expressed in the brain at appreciable levels; (b) the hypothalamus of *Torpedo* showed immunoreactivity toward PACAP in many nerve fibers; (c) neurointermediate lobe of the pituitary immunoreactive to PACAP; (d) saccus vasculosus showed strong labeling for PACAP antibody, in epithelial cells exclusively; (e) labeling in the cerebellum was observed in the cell cytoplasm; and (f) negative control showing no labeling. Scale bars correspond to: (b) 40 μm; c,d 20 μm; e,f 33 μm.

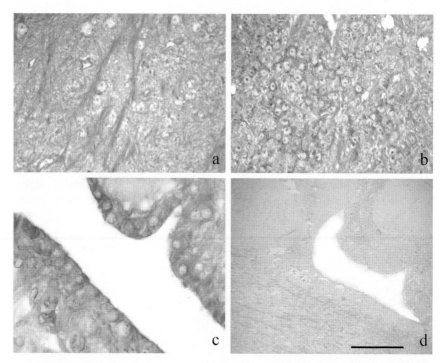

FIGURE 2. (a) Nerve fibers were labeled for PACAP antisense riboprobe in the hypothalamus area; (b) hypophysis of *Torpedo*: cells were labeled for PACAP messenger; (c) cells of the saccus vasculosus synthesized PACAP as demonstrated by *in situ* hybridization; and (d) no labeling was seen in sections treated with sense riboprobe, counterstained with nuclear Fast Red (Vector, Burlingame, CA). Scale bars correspond to: a,b 20 μm; (c) 14 μm (d) 100 μm.

the main regions where PACAP mRNA was detected were the hypothalamus and the neurointermediate lobe of the pituitary (FIG. 2 a,b); moreover the saccus vasculosus expressed PACAP mRNA in the cellular portion while the vascular portion was devoid of PACAP mRNA (FIG. 2 c). Sense riboprobe controls showed no labeling (FIG. 2 d).

DISCUSSION

Here, we report the localization of PACAP in the brain of *T. marmorata*. In general, it can be stated that there is a wide expression of PACAP in the brain of this elasmobranch, suggesting a major role for this peptide in the regulation of cartilaginous fish physiology. Interestingly, the pituitary of *T. marmorata* was immunoreactive to PACAP only in the neurointermediate lobe, so PACAP could be involved in the endocrine regulation of the pituitary activity of this cartilaginous fish as occur in other vertebrate species,[1] especially the elasmo-branch *D. akajei* where PACAP IR was localized mainly in the hypothalamo–pituitary region.[8] The evidence of mRNA in hypothalamic nerve fibers is consistent with previous reports showing that mRNAs encoding for several neuropeptides are located in the distal portion of hypothalamic axons.[10,11] Since neuronal mRNAs have important role in many cell functions, such as neuronal differentiation and in maintaining cell polarity,[12] PACAP could be implicated in these basic functions of nerve cell biology. The cells of the saccus vasculosus were labeled for both PACAP and its mRNA, suggesting that this characteristic organ of fishes is probably implicated in endocrine functions rather than homeostasis of cerebrospinal fluid and osmoregulation. Whether PACAP can be considered as one of these factors remains to be determined, but our data support this hypothesis. However, it has been previously demon-strated that the epithelial cells of the saccus vasculosus of both cartilaginous and bony fishes are immunoreactive to neuropeptide-Y,[13,14] suggesting that neuroendocrine regulation of saccus vasculosus can occur. In the cerebellum, considered to be mainly involved in motor and sensory processing in verte-brates,[15] PACAP IR could suggest a role for this neuropeptide in the different functions of this organ, such as the modulation of motor output of the CNS and/or visual and tactic input to the cerebellum. In conclusion, PACAP is a highly conserved molecule that has wide expression in the brain of the carti-laginous fish *T. marmorata*, suggesting that this peptide may have a key role in elasmobranch brain functions.

ACKNOWLEDGMENTS

We are indebted to both Prof. Foster (University of Minnesota) and Prof. Matsuda (University of Toyama) for the kind gift of PACAP plasmid and PACAP antibody, respectively.

REFERENCES

1. VAUDRY, D., B.J. GONZALEZ, M. BASILLE, *et al.* 2000. Pituitary adenylate cyclase-activating polypeptide and its receptors: from structure to functions. Pharmacol. Rev. **52:** 269–324.
2. SHERWOOD, N.M., S.L. KRUECKL & J.E. McRORY. 2000. The origin and function of the pituitary adenylate cyclase-activating polypeptide (PACAP)/glucagon su-perfamily. Endocr. Rev. **21:** 619–670.
3. HANNIBAL, J. 2002. Pituitary adenylate cyclase-activating peptide in the rat central nervous system: an immunohistochemical and in situ hybridization study. J. Comp. Neurol. **453:** 389–417.
4. YON, L., M. FEUILLOLEY, N. CHARTREL, *et al.* 1992. Immunohistochemical distribu-tion and biological activity of pituitary adenylate cyclase-activating polypeptide

(PACAP) in the central nervous system of the frog Rana ridibunda. J. Comp. Neurol. **324:** 485–499.

5. MONTERO, M., L. YON, K. ROUSSEAU, *et al.* 1998. Distribution, characterization, and growth hormone-releasing activity of pituitary adenylate cyclase-activating polypeptide in the European eel, Anguilla anguilla. Endocrinology **139:** 4300–4310.

6. YON, L., D. ALEXANDRE, M. MONTERO, *et al.* 2001. Pituitary adenylate cyclase-activating polypeptide and its receptors in amphibians. Microsc. Res. Tech. **54:** 137–157.

7. MATSUDA, K., Y. NAGANO, M. UCHIYAMA, *et al.* 2005. Pituitary adenylate cyclase-activating polypeptide (PACAP)-like immunoreactivity in the brain of a teleost, Uranoscopus japonicus: immunohistochemical relationship between PACAP and adenohypophysial hormones. Regul. Pept. **126:** 129–136.

8. MATSUDA, K., Y. ITOH, T. YOSHIDA, *et al.* 1998. The localization of pituitary adenylate cyclase-activating polypeptide (PACAP)-like immunoreactivity in the hypothalamo-pituitary region of an elasmobranch, stingray, *Dasyatis akajei*. Peptides **19:** 1263–1267.

9. VALIANTE, S., A. CAPALDO, F. VIRGILIO, *et al.* 2004. Distribution of alpha7 and alpha4 nicotinic acetylcholine receptor subunits in several tissues of *Triturus carnifex* (Amphibia, Urodela). Tissue Cell **36:** 391–398.

10. MOHR, E. & D. RICHTER. 2000. Axonal mRNAs: functional significance in vertebrates and invertebrates. J. Neurocytol. **29:** 783–791.

11. MOHR, E. & D. RICHTER. 2003. Molecular determinants and physiological relevance of extrasomatic RNA localization in neurons. Front. Neuroendocrinol. **24:** 128–139.

12. KINDLER, S., E. MOHR & D. RICHTER. 1997. At the cutting edge. Quo vadis: extrasomatic targeting of neuronal RNA in mammals. Mol. Cell. Endocrinol. **128:** 7–10.

13. YANEZ, J., M. RODRIGUEZ, S. PEREZ, *et al.* 1997. The neuronal system of the saccus vasculosus of trout (*Salmo trutta fario* and *Oncorhynchus mykiss*): an immunocytochemical and nerve tracing study. Cell Tissue Res. **288:** 497–507.

14. CHIBA, A., S. OKA & E. SAITOH. 2002. Ontogenetic changes in neuropeptide Y-immunoreactive cerebrospinal fluid-contacting neurons in the hypothalamus of the cloudy dogfish, *Scyliorhinus torazame* (Elasmobranchii). Neurosci. Lett. **329:** 301–304.

15. NEW, J.G. 2001. Comparative neurobiology of the elasmobranch cerebellum: theme and variations on sensorimotor interface. Environ. Biol. Fishes **60:** 93–108.

Involvement of Protein Kinase C in the PACAP-Induced Differentiation of Neural Stem Cells into Astrocytes

JUN WATANABE,[a,b] FUSAKO OHNO,[c] SEIJI SHIODA,[a]
SAKAE KIKUYAMA,[b] KAZUYASU NAKAYA,[d] AND SHIGEO NAKAJO[c]

[a]Department of First Anatomy, School of Medicine, Showa University,
Tokyo 142-8555, Japan

[b]Department of Biology, School of Education, Waseda University,
Tokyo 169-8050, Japan

[c]Laboratory of Biological Chemistry, School of Pharmaceutical Sciences,
Showa University, Tokyo 142-8555, Japan

[d]Faculty of Applied Life Sciences, Niigata University of Pharmacy and Applied
Life Sciences, Niigata 956-8603, Japan

ABSTRACT: Expression of members of the conventional protein kinase C
(cPKC) family in the differentiation of mouse neural stem cells (NSCs)
induced by pituitary adenylate cyclase-activating polypeptide (PACAP)
was investigated. In particular, expression of the α and β subtypes of
cPKC in NSCs was observed. In response to activation by PACAP, cPKCβ
transiently increased twofold by day 2 and returned to basal levels by day
4, suggesting that cPKCβ might be responsible for the differentiation
process.

KEYWORDS: neural stem cell; progenitor cell; differentiation; astrocyte;
pituitary adenylate cyclase-activating polypeptide

INTRODUCTION

It has been shown that the neuropeptide pituitary adenylate cyclase-
activating polypeptide (PACAP), which belongs to the secretin/glucagon family, has pleiotropic functions not only in the central nervous system (CNS) but
also in peripheral tissues.[1,2] Recent reports demonstrated that PACAP induced
neuronal differentiation of embryonic stem cells[3] and differentiation of neural stem cells (NSCs) into astrocytes.[4,5] Our data (manuscript in preparation)

Address for correspondence: Dr. Shigeo Nakajo, Laboratory of Biochemistry, Yokohama College of
Pharmacy, 601 Matanocho, Totsuka-Ku, Yokohama, Kanagawa 245-0066, Japan. Voice: 81-045-859-
1300; fax: 81-045-859-1301.
e-mail: s.nakajou@hamayaku.ac.jp

Ann. N.Y. Acad. Sci. 1070: 597–601 (2006). © 2006 New York Academy of Sciences.
doi: 10.1196/annals.1317.090

showed that phorbol 12-myristate 13-acetate (PMA) is sufficient to mimic the effect produced by PACAP in NSCs; in addition, the differentiation was inhibited by the PKC inhibitor, chelerythrine, and by 1,2-bis (o-aminophenoxy) ethane-N,N,N′,N′-tetraacetic acid tetra (acetoxymethyl) ester (BAPTA/AM). These results indicated that members of the conventional protein kinase C (cPKC) family, which are Ca^{2+}- and phospholipid-dependent forms of PKC, might be responsible for the PACAP-induced differentiation of NSCs into astrocytes. Furthermore, it has been shown that PACAP promotes NSC proliferation in adult mouse brain and that PKC is implicated in the signaling pathway of the proliferation process.[6] Although the involvement of PKC in the signaling pathway of PACAP-induced differentiation was shown, the nature of the PKC subtype[7,8] involved was not determined.

In this article, gene and protein expression levels of the different cPKC subtypes, cPKCα, cPKCβ, and cPKCγ, in NSCs were investigated to determine which cPKC subtype is involved in the PACAP-induced differentiation of these cells.

MATERIALS AND METHODS

Materials

Antibodies (sc-208, sc-210G, sc-211) against cPKCα, cPKCβ, and cPKCγ were purchased from Santa Cruz Biotechnology, Inc. (Santa Cruz, CA), and GAPDH antibody was obtained from Chemicon International Inc. (Temewla, CA).

Isolation and Culture of NSCs

The cell culture was essentially carried out as described previously.[5,9] Telencephalons were prepared from embryos (E14.5) of ICR mice and gently dissociated by pipetting. The dissociated cells ($2 - 2.5 \times 10^7$ cells/10 cm dish) were cultured for 4 days in DMEM/F12 medium containing 10 ng/mL of recombinant human basic fibroblast growth factor (bFGF), and were then exposed to 2 nM PACAP for 8 days in the absence of bFGF.

Reverse Transcription-Polymerase Chain Reaction (RT-PCR) Analysis

Total RNA was isolated from NSCs, E14.5 whole brain (containing telencephalon, diencephalon, and mesencephalon), and adult cerebral cortex, using the RNeasy Mini Kit (Qiagen Co., Tokyo, Japan). Total RNA (1 μg) was subjected to reverse transcription reaction, and then 1 μL of the reaction mixture was used for PCR. Primers used were as follows: for PKCα, 5′-primer,

5′-GAACCATGGCTGACGTTTAC-3′, and 3′-primer 5′-GCAAGATTGGG
TGCACAAAC-3′; for PKCβ (I and II), 5′-primer 5′-TTCAAGCAGCCCA-
CCTTCTG-3′ and 3′-primer 5′-AAGGTGGCTGAATCTCCTTG-3′; and for
PKCγ, 5′-primer 5′-GACCCCTGTTTTGCAGAAAG-3′ and 3′-primer 5′-
GTAAAGCCCTGGAAATCAGC-3′. For amplification, an initial denatura-
tion step of 2 min at 94°C was followed by 30 cycles of 1 min at 94°C, 1 min at
65°C, and 2 min at 72°C, and the last cycle was extended for 10 min at 72°C.

Immunoblot Analysis

NSCs, E14.5 whole brain (containing telencephalon, diencephalon, and mesencephalon), and adult cerebral cortices washed with PBS were lysed with 10
volumes of lysis buffer containing 1% Triton X-100 and the lysate was centrifuged as described previously.[5] The clear supernatant (10 μg of protein) was
subjected to SDS-10% polyacrylamide gel electrophoresis and transferred onto
a nitrocellulose membrane. The membrane was incubated overnight at 4°C with
antibodies against cPKCs (1:2000 PKCα, 1:400 PKCβ, 1:800 PKCγ), washed
6 times with 0.2% Tween 20/Tris-buffered saline, and the protein bands visualized as described previously.[5] Antibody against GAPDH was used at 1:1000.

RESULTS AND DISCUSSION

To our knowledge, there have been no reports in which expression levels
of cPKC genes and proteins have been examined in embryonic NSCs and in
which differentiation was induced by PACAP. Here, expression of the genes
coding for the different cPKCs was investigated by RT-PCR (Fig. 1A). Although expression of the gene coding for cPKCα in the E14.5 whole brain
was low, expression of the gene in NSCs and in the adult cerebral cortex was
present at higher levels. Expression of the gene coding for cPKCβ was found
in NSCs, E14.5 whole brain, and adult brain tissues examined. In contrast,
expression of the gene coding for cPKCγ was only observed in adult brain
tissues. Levels of expression of protein corresponding to the different cPKC
subtypes were investigated by immunoblot analysis. As shown in Figure 1B,
expression levels of cPKC proteins followed a similar pattern to that of gene
expression. It is thus demonstrated that NSCs and E14.5 whole brain express
the α and β cPKC subtypes (mainly PKCβ and to a lower extent α, but not the
γ form).

Protein expression levels of cPKCα and cPKCβ were then investigated in
NSCs exposed to 2 nM PACAP (Fig. 2). The expression of cPKCα showed
a linear increase up to day 6, whereas cPKCβ underwent a transient twofold
increase up to day 2, then returned to basal levels by day 4 of exposure of NSCs
to PACAP. Morphological alterations observed in NSCs 4 days after addition

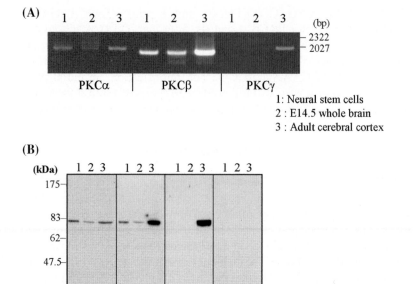

FIGURE 1. Gene and protein expressions of cPKC subtypes in NSCs. **(A)** Purified total RNA (1 μg) were subjected to RT-PCR as described in "MATERIALS AND METHODS," and the products were analyzed by agarose gel electrophoresis. **(B)** Lysates (10 μg) prepared from NSCs and brain tissues were analyzed by immunoblot. GAPDH: glyceraldehyde-3-phosphate dehydrogenase.

of PACAP were as described previously.[5] It is considered that these alterations might reveal differentiation into astrocytes. The maximum level of expression of cPKCβ was found to occur prior to that of the morphological changes. Therefore, while cPKCα may be involved, it is considered that the increased cPKCβ could have a crucial role in the PACAP-induced differentiation of NSCs into astrocytes. Further study will be necessary to elucidate the roles of the cPKC subtypes and substrate proteins participating in the PACAP-induced differentiation process.

ACKNOWLEDGMENTS

This study was supported in part by a Showa University Grant-in-Aid for Innovative Collaborative Research Projects.

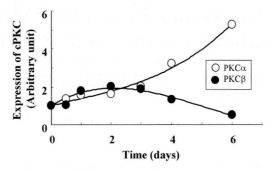

FIGURE 2. Expression of cPKCα and cPKCβ during PACAP-induced differentiation of NSCs. NSCs (2.5 × 10^7 cells/10 cm dish) were exposed to 2 nM PACAP for the indicated times, and cell lysates were prepared. Lysates (10 μg) were subjected to immunoblot analysis. The extent of expression was quantified using NIH image software after scanning the protein bands on X ray films. Very similar results were obtained in another independent experiment. Each value normalized to GAPDH is represented.

REFERENCES

1. ARIMURA, A. 1998. Perspectives on pituitary adenylate cyclase activating polypeptide (PACAP) in the neuroendocrine, endocrine, and nervous systems. 1998. Jpn. J. Physiol. **48:** 301–331.
2. VAUDRY, D., B.-J. GONZALEZ, M. BASILLE, *et al.* 2000. Pituitary adenylate cyclase-activating polypeptide and its receptors: from structure to functions. Pharmacol. Rev. **52:** 269–324.
3. CAZILLIS, M., B.J. GONZALEZ , C. BILLARDON, *et al.* 2004. VIP and PACAP induce selective neuronal differentiation of mouse embryonic stem cells. Eur. J. Neurosci. **19:** 798–808.
4. VALLEJO, I. & M. VALLEJO. 2002. Pituitary adenylate cyclase-activating polypeptide induces astrocyte differentiation of precursor cells from developing cerebral cortex. Mol. Cell. Neurosci. **21:** 671–683.
5. OHNO, F., J. WATANABE, H. SEKIHARA, *et al.* 2005. Pituitary adenylate cyclase-activating polypeptide promotes differentiation of mouse neural stem cells into astrocytes. Regul. Peptides **126:** 115–122.
6. MERCER, A., H. RÖNNHOLM, J. HOLMBERG, *et al.* 2004. PACAP promotes neural stem cell proliferation in adult mouse brain. J. Neurosci. Res. **76:** 205–215.
7. NISHIZUKA, Y. 1992. Intracellular signalling by hydrolysis of phospholipids and activation of protein kinase C. Science **258:** 607–614.
8. NISHIZUKA, Y. 1995. Protein kinase C and lipid signaling for sustained cellular responses. FASEB J. **9:** 484–496.
9. JOHE, K.-K., T.-G. HAZEL, T. MULLER, *et al.* 1996. Single factors direct the differentiation of stem cells from the fetal and adult central nervous system. Genes Dev. **10:** 3129–3140.

Role of Two Genes Encoding PACAP in Early Brain Development in Zebrafish

SHENG WU, BRUCE A. ADAMS, ERICA A. FRADINGER, AND NANCY M. SHERWOOD

Department of Biology, University of Victoria, Victoria, B.C. V8W 3N5, Canada

ABSTRACT: To study the role of pituitary adenylate cyclase-activating polypeptide (PACAP) in early brain development, we examined PACAP and its receptors for first expression and then separately knocked down the two forms of PACAP in zebrafish where development is rapid and observable. We injected morpholinos (antisense oligonucleotides) into fertilized eggs to block PACAP. Morphological changes in the brain were observed in embryos at 27 h post fertilization (hpf). Using *in situ* hybridization of early brain marker genes, we found that the most striking effects were an increase in *pax2.1* expression in eye stalks associated with absence of either form of PACAP or an increase in *eng2* and *fgf8* in the midbrain–hindbrain boundary after loss of PACAP2. These marker genes are among the earliest factors in the formation of the midbrain–hindbrain boundary, an early organizing center. We suggest that PACAP is a target gene with feedback inhibition on *pax2.1*, *eng2*, or *fgf8* in specific brain areas. In the hindbrain, the absence of either form of PACAP had little effect, as shown by expression of *ephA4* and *meis1.1*. During midbrain development, our evidence suggests that PACAP1 can activate *mbx*. In both the diencephalon and/or forebrain, lack of PACAP1 or PACAP2 led to an increase in *fgf8*, again suggesting a suppressive effect of PACAP during development on these important genes that help to define cells in the forebrain. The early expression of transcripts for PACAP and its receptors by 0.5–6 hpf make both PACAP1 and PACAP2 candidates for factors that influence brain development.

KEYWORDS: PACAP gene knockdown; Morpholinos; brain development; brain marker genes (*fgf8*; *eng2*; *pax2.1*; *ephA4*; *mab21l2*; *mbx*; *meis1.1*); zebrafish

INTRODUCTION

Pituitary adenylate cyclase-activating polypeptide (PACAP) and its PAC1 receptor are expressed in the neural tube of mouse at embryonic day 10 suggesting

Address for correspondence: Nancy M. Sherwood, Ph.D., Department of Biology, University of Victoria, 3800 Finnerty Road, Victoria, B.C. V8W 3N5, Canada. Voice: 250-721-7143; fax: 250-721-7120.

e-mail: nsherwoo@uvic.ca

Ann. N.Y. Acad. Sci. 1070: 602–621 (2006). © 2006 New York Academy of Sciences.
doi: 10.1196/annals.1317.091

that PACAP might be important in development.[1] Although our group[2] and others[3-5] have created lines of mice with the PACAP gene disrupted, the zebrafish (*Danio rerio*) offers advantages in the study of development. In zebrafish, it is possible to avoid the influence of maternal PACAP and to use a nearly transparent embryo for constant observation. The method of gene knockdown offers only a temporary blocking of the PACAP translation, but is sufficient for developmental studies in zebrafish where hatching occurs at 72 h post fertilization (hpf). Although PACAP is highly conserved between mammals and fish, there are some interesting differences between them. First, PACAP is encoded on two separate genes in zebrafish,[6,7] unlike the single copy gene in mice.[8] In zebrafish, the two forms of PACAP (PACAP1 and PACAP2) are each 38 amino acids in length and the amino acids are 82% identical, but we do not know if the peptides have distinct functions. Therefore, each of the two PACAPs would have to be eliminated to examine their separate effects. Second, there is alternative splicing in the zebrafish PACAP2 transcript in which exon 4 is cut out of the mRNA. Third, the PACAP genes of all fish studied to date encode a growth hormone-releasing hormone (GHRH)-like peptide on exon 4, which would be blocked also if the mRNA is knocked down. However, GHRH in fish appears to have minor functional effects even for release of growth hormone compared to PACAP.[9]

To date, the temporal expression of PACAP1 and PACAP2 in zebrafish are known through expression studies using reverse transcription-polymerase chain reaction (RT-PCR).[6,7] More specific localization in the brain and eye was provided by an immunocytochemistry study in zebrafish from the 24-hpf embryo through to the adult, but the antibody could not distinguish the two PACAP forms.[10] The functions of the two zebrafish PACAPs are not known, but in mice and other mammals PACAP has a role in the regulation of other endocrine systems,[11] in carbohydrate and protein metabolism,[2] in the stress response,[2,3] in the immune system,[11] as a neuromodulator in the sympathetic and parasympathetic nervous system[12,13]; and in the brain as a neurotrophic, neuroprotective, or proliferative factor.[11]

The structures of zebrafish receptors that bind PACAP exclusively, PAC1, or bind both PACAP and vasoactive intestinal peptide (VIP), VPAC1 and VPAC$_2$, have been fully or partially (VPAC$_2$ receptor) identified.[7,14] When the zebrafish PAC1 receptor was expressed in monkey COS cells, it responded equally to zebrafish PACAP1 or PACAP2 as measured by an increase in cAMP or inositol phosphate (IP).[14] In contrast, only cAMP signaling was activated when the zebrafish VPAC1 receptor was tested with PACAP1 or PACAP2.[14] In conclusion, zebrafish have two genes encoding PACAP1 and PACAP2, which can be distinguished as to structure, but not yet as to exact brain location and function in development or in adults. Evidence to date suggests that the two zebrafish PACAPs activate their three shared receptors in the same way.

Here, we examine early brain development and the role of PACAP using zebrafish, where rapid development occurs outside of the mother in contrast to mammals. The first appearance of mRNA for PACAP1, PACAP2, and their receptors, PAC1, VPAC1, VPAC$_2$, and GHRH, is determined with RT-PCR and for PACAP2 using *in situ* hybridization. The effects of morpholino (MO)-induced knockdown of PACAP1 or PACAP2 is determined initially by screening morphological defects in the embryos at 27 hpf and subsequently by detailed examination of the expression pattern of brain genes that can be used as molecular markers for early stages of development.

MATERIALS AND METHODS

Zebrafish

All zebrafish were kept in dechlorinated water at 28°C with 14 h of light and 10 h of dark. Embryos were collected in the morning after natural spawning of adult fish. All procedures were approved by the Animal Care Committee at the University of Victoria.

RT-PCR of PACAP and its Receptors

To examine the onset of *PACAP* expression, mRNA from 30 embryos at developmental stages ranging from 30 min to 25 hpf was extracted with the RNeasy mini-kit (Qiagen, Mississauga, ON). The samples were treated with Turbo RNase-free DNase (Ambion, Austin, TX). Single stranded cDNA was synthesized using oligo (dT) and Superscript II RT (Invitrogen, Burlington, ON) according to the manufacturer's instructions. The primers for PCR of *PACAP1* and *PACAP2* mRNA are shown in TABLE 1. A control reaction was prepared using 1 μL of cDNA amplified with primers for RNA polymerase II (TABLE 1). PCR reactions were carried out for 35 cycles at 55°C annealing temperature with a 10-min extension at 72°C on the last cycle. The PCR products were normalized to the expression of RNA polymerase II. Reaction products were separated by 1.5% agarose gel electrophoresis and visualized using ethidium bromide staining.

To examine the onset of receptor expression, mRNA from five embryos at developmental stages ranging from 1 to 18 hpf was extracted with the Poly (A) Pure kit (Ambion). Single-stranded cDNA was synthesized using oligo (dT) and Superscript II RT (Invitrogen) according to the manufacturer's instructions. The transcripts encoding the PAC1, VPAC1, and GHRH receptors, and tubulin control were amplified from 2 μL of cDNA by PCR and detected by Southern analysis with ^{32}P-dCTP-labeled receptor specific probes as previously described.[14]

The sequence of the VPAC$_2$ receptor was not identified until after the other three receptors had been studied. To examine expression of VPAC2 receptor, the same procedure was followed as for RT-PCR of *PACAP1* and *PACAP2* mRNA, described above. The primers are listed (TABLE 1).

TABLE 1. Primer sequences

Primer name	Direction-use	Sequence 5′ to 3′
RT-PCR		
PACAP1	F	CGCCTCTGAGTTACCCGAAAA
PACAP1	R	TAGCGAGCCGCCGTCCTTTG
PACAP2	F	TCAGGGAAGAGGTGCTGTGAGGA
PACAP2	R	CATCTGTTTTCGGTAGCGACTGT
VPAC$_2$ receptor	F	GTCCTGCTGCCCTCAAAA
VPAC$_2$ receptor	R	AGATGGATGTAGTTCCTGGTG
RNA polymerase II	F	CCCTTTAACATTGACA
RNA polymerase II	R	CACAGCAATAACCTGAGAAA
tubulin	F	CAGGTGTCCACGGCTGTGGTG
tubulin	R	AGGGCTCCATCGAAACGCAG
Probes		
PACAP2	F	AGACAAACGAGTGCTGCTGA
PACAP2	R	CAAAAGCCAGGTCGCTTAAC
mab	F	ATTCGCTCCCGCTTTCAG
mab	R	TCGTCCCAGTCAGTCTCCC
fgf8	F	TCATGGACAGCTCGGGATT
fgf8	R	GTCCGTCACCTTACTTTGCTCACT
meis1.1	F	ACCACCCTGCTCAGTTGC
meis1.1	R	TATGTCCTCATCCCTCCC
eng2	F	GACGCAGCAATCGTTCGG
eng2	R	GTGGCGCTTACCCTGGCTT
ephA4	F	GTGTTCACAGCGACTTTTGCG
ephA4	R	AGCCGCAATGGTGTCG
pax2.1	F	CCGCACGAATCTGACTGT
pax2.1	R	ATGGGATAGCAAAGGAAGGA
mbx	F	CAAGCGATAAGGATGAAGC
mbx	R	CAGCGGAGACGAGGAGTA

Microinjection of MO Antisense Oligonucleotides

MOs are antisense oligonucleotides that are designed to block a specific mRNA from being translated into a protein.[15] The morpholine ring modifies the nucleotides so that degradation is prevented,[16] although the original quantity injected is diluted at each cell division. In developmental studies of zebrafish, MOs are effective blockers of translation, but are mainly restricted to the period before hatching. The methods for knockdown in zebrafish have been described by others.[17]

MOs, including those for control injections, were designed and synthesized by Gene Tools (Eugene, OR). The design of the MOs against the RNA region around the AUG translation start site (ATG) and the 5′ untranslated region (UTR) ensured that the two MOs did not overlap (TABLE 2). Also, this design resulted in blockage of translation of both the full length and splice variants lacking exon 4 of PACAP. In addition, a PACAP splicing MO[17] was used to confirm the specificity of the PACAP1-UTR-MO. MOs were dissolved in

TABLE 2. MO sequences and rescue primers

MO name	Location	Sequence 5′ to 3′
PACAP1 UTR	5′ UTR	CGGACGGATGCTGTCCAATGGAGGC
PACAP1 ATG	Start codon	GAGTCGTTTTGCTGCTCGTAATCAT
PACAP2 UTR	5′ UTR	GAAATGCTGTTGGAATGCGACTCGGG
PACAP2 ATG	Start codon	GCCATGCTATTGCAGAGTAGGTAGA
Rescue primers		
PACAP1	F	AGAATGATTGCGAGCAGC
PACAP1	R	TTGGAGCAAGCATGATAAAC
PACAP2	F	TATCTACCTACTCTGCAATAGCATG
PACAP2	R	ATATCAGGCGCATAACGT

filtered ddH$_2$O to a working concentration of 2–12 μg/μL. Zebrafish embryos were injected with 1–1.5 nL of the MO solution into the yolk at the one-cell stage.[15,18] Pictures of morphants were captured by an Olympus (SZX9) dissection microscope with a Sony color video camera (DXC-950P). Images were imported into Northern Eclipse software (Empix Imaging Inc, Mississauga, ON).

Rescue of MO-Induced Effects by Injection of PACAP mRNA

To rescue MO-treated embryos, mRNA was prepared by amplification using PCR of the corresponding *PACAP2* cDNA containing part of the 5′ UTR plus the coding sequence and 3′ UTR. The rescue mRNA did not overlap with the MO *PACAP2*-UTR sequence. The primers for PCR amplification of rescue mRNA are shown (TABLE 2). The PCR products were ligated into pGEM-T vector and sequenced. The vector with insert was linearized with NdeI, and synthetic mRNA was transcribed from the T7 promotor using the T7 mMESSAGE mMACHINE kit (Ambion) following the manufacturer's directions. After synthesis, all capped RNAs were purified by MEGAclear kit (Ambion) and precipitated by ammonium acetate and ethanol. Capped mRNA was mixed with the MO solution just before microinjection and co-injected into the one-cell stage embryos. The co-injection concentrations for rescue were 2 ng *PACAP2*-UTR-MO + 50 pg *PACAP2* mRNA of either long (5 exons) mRNA or short (lacking exon 4) mRNA. Embryos were scored for defects 27 hpf.

Whole Mount In Situ Hybridization

RNA probes were prepared from a linearized template using the Digoxigenin-RNA labeling kit (Roche, Germany) according to the manufacturer's instructions. *In situ* hybridization and signal detection with anti-digoxigenin antibody coupled to alkaline phosphatase (Roche) were performed as described[19] with a few modifications. Anti-dig-alkaline phosphatase antibody was diluted 1:3000 in blocking solution. After a blue product was visible, the reaction was stopped by washing twice in PBS. Embryos were dehydrated in a series of solutions: 70% PBS-30% methanol, 50% PBS-50% methanol, 30% PBS-70% methanol, and 100% methanol. Expression patterns were photographed after clearing in benzyl benzoate:benzyl alcohol (1:2). The primers used to prepare probes for the brain gene markers *eng2, fgf8, pax2.1, meis1.1, ephA4, mab21l2,* and *mbx* are listed in TABLE 1. To analyze expression patterns for *PACAP2*, the probe was prepared using primers listed in TABLE 1 and the procedure was modified from the one listed above by storing embryos in a 10% glycerol in PBS solution, instead of dehydrating in methanol, before photographing them.

RESULTS

Expression of PACAP1 and PACAP2 Transcripts in Early Embryos

To determine the earliest expression of *PACAP* mRNA in the embryo, we used RT-PCR. Our analysis showed the presence of both *PACAP1* and *PACAP2* cDNA beginning at 0.5 hpf and continuing until the last measurement at 25 hpf (FIG. 1 A). However, the pattern of expression was quite different for the two mRNAs. The *PACAP1* cDNA was always expressed as a single band of the same length with the strongest expression at 0.5–1.5 hpf and 18–25 hpf; expression was least intense at 4 and 14 hpf compared to the control expresssion of RNA polymerase II. In contrast, *PACAP2* cDNA always appears as a double band with the strongest expression at 14–25 hpf compared to the control cDNA.

First Expression of Transcripts Encoding PAC1, VPAC$_1$, VPAC$_2$, and GHRH Receptors

The cDNAs encoding the PAC1 or GHRH receptors were expressed beginning at 1–2 hpf, whereas the cDNA encoding VPAC1 was detected at 5–6 hpf (FIG. 1 B). Once the expression began, it was continuous throughout the early period measured until 18 hpf. The expression pattern showed double bands at 1–8 hpf for the PAC1 receptor and single bands of either the long or short form thereafter. Sequence analysis demonstrated that the shorter PAC1 receptor was a full-length receptor without inserts, whereas the longer PAC1 receptor had a hop1 cassette inserted into its third intracellular loop. The other receptors were expressed as single-size transcripts only. The VPAC$_2$ receptor was analyzed at a latter time using regular RT-PCR rather than the Southern method used for the other three receptors. The VPAC$_2$ receptor is detected as a single band beginning at 0.5 hpf after fertilization and is detected

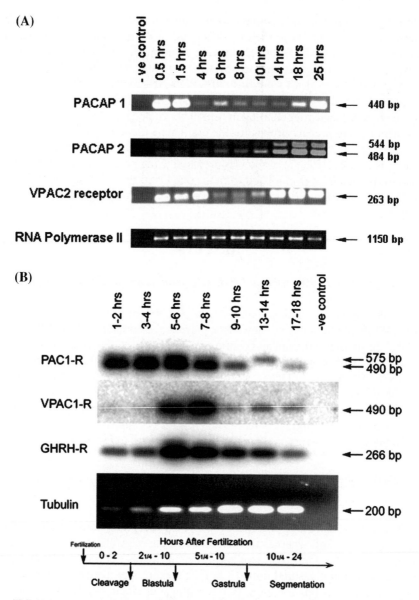

FIGURE 1. Expression of zebrafish PACAP and receptors at different stages of development. **(A)** PACAP1, PACAP2, and VPAC$_2$ receptor cDNA are shown in comparison to control cDNA amplified from RNA polymerase II in zebrafish embryos at times from 0.5 to 25 hpf. RT-PCR was used to amplify the mRNA. **(B)** PAC1 receptor (R), VPAC1-R, and GHRH-R cDNA converted to Southern blots are shown in comparison to control cDNA amplified from tubulin in zebrafish embryos at times from 1 to 18 hpf. The length of each band is shown at the right in base pairs (bp). The developmental stage represented by hours after fertilization is shown at the bottom.

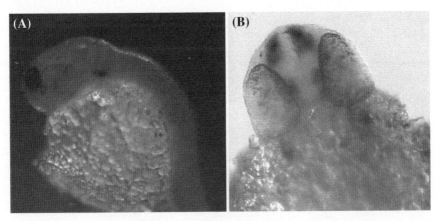

FIGURE 2. *In situ* hybridization of PACAP2 in a zebrafish embryo at 24 hpf. **(A)** Lateral view. **(B)** Dorsal view.

thereafter with varying intensity using RNA polymerase II as the control for comparison (FIG. 1 A).

In Situ *Hybridization of PACAP mRNA in 24-hpf Embryos*

Previously we showed that *PACAP1* mRNA is expressed in 24-hpf embryos in the eye, midbrain, midbrain–hindbrain boundary (MHB), and hindbrain.[19] In the present study, *PACAP2* mRNA at 24 hpf is expressed extensively in the forebrain, in the diencephalon near the eyes, and in the hindbrain as shown from both the lateral and dorsal views (FIG. 2).

Brain and Eye Morphology After Knockdown of PACAP1 or PACAP2

A series of MO injections at 2, 4, and 8 ng were tested to determine the threshold dose that hybridizes to *PACAP* mRNA resulting in blockage of translation of PACAP1 or PACAP2 as determined by altered brain morphology, especially in the MHB where the fold can be easily viewed in whole embryos. For *PACAP1* mRNA, the threshold was 4 ng MOs whereas for *PACAP2* mRNA hybridization, the threshold was 2 ng MOs to alter brain morphology. These are low doses and far below any reported to cause toxic effects. In this study, we only used 2 or 4 ng of MOs. The effects caused by using either UTR- or ATG-directed MOs were similar and suggested that the MOs were not misdirected. Injections of unrelated MOs into control embryos did not result in any developmental delays with 2–6% showing some altered morphology at 24 hpf.

Wild type   KD- PACAP1   KD-PACAP2

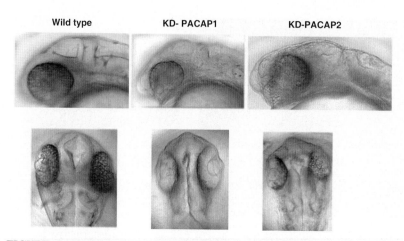

FIGURE 3. Morphological changes in 27-hpf zebrafish embryos, which are wild type or injected with 2–4 ng of MOs to knockdown (KD) PACAP1 or PACAP2 in the one-celled stage. Lateral view is above; dorsal view is below.

The screening of embryos ($n = 80$–90/ injection group) at 27 hpf after injection of MOs is shown in representative embryos in FIGURE 3. The knockdown or blocking of PACAP1 peptide resulted in morphological changes in which the most obvious were the distortion of the forebrain and mid-hindbrain boundary. The knockdown of PACAP2 at only 2 ng MO always produced more severe changes than knockdown of PACAP1 with 4 ng MO. The lateral view shows an aberrant MHB and the dorsal view shows distortion of the forebrain in the embryos in which PACAP2 was knocked down (FIG. 3). Eye size was measured at its greatest diameter in 10 embryos in each of three groups to verify our initial observations: wild type (156 μM), PACAP1-UTR MO of 4 ng (124 μM), and PACAP2-UTR MO of 2 ng (110 μM). The two knockdown groups had eyes that were significantly smaller than wild-type embryos. Statistical analysis ($P < 0.05$) was performed by Prism using one-way ANOVA and Tukey's Multiple Comparison test.

To further demonstrate that the MOs were hybridizing to the correct mRNA, rescue experiments with exogenous mRNA were done. In 82 embryos, 2 ng MO directed against *PACAP2* mRNA resulted in 8.7% normal embryos; in 100 embryos treated in the same amount and type of MO along with 50 pg of the long form of *PACAP2* mRNA including all five exons, 22.1% of the embryos were normal; and in 83 embryos treated with the same MO along with 50 pg of the short form (lacking exon 4) of *PACAP2*, 34.8% of the embryos were normal. The results in both rescued groups were significantly changed compared to the group without mRNA rescue. Statistical analysis ($P < 0.05$) was Tukey's Multiple Comparison test as above.

The screening process showing morphological defects after knockdown of either PACAP1 or PACAP2 with low doses of MOs and the partial rescue of the morphants provided evidence that further testing after PACAP knockdown was warranted. Therefore, marker gene expression for specific brain areas was evaluated after PACAP knockdown as described below.

Change in Expression of Markers for Brain and Eye After Knockdown of PACAP1 or PACAP2

Seven different brain markers were selected for their expression in different brain regions including the MHB, the hindbrain, midbrain, diencephalon–midbrain boundary (DMB), diencephalon, forebrain, and eyes. All markers were examined in the 24-hpf embryos, which were wild type or treated with 2–4 ng of MOs against *PACAP1* or *PACAP2* mRNA.

The MHB is an important organizing center for this region of the brain and hence three markers were selected to study the region. Marker *eng2* is expressed in the MHB in a similar pattern for both the wild-type embryo and PACAP1-blocked embryo (FIG. 4 A). In contrast, *eng2* is not only overexpressed in the MHB as judged by its intensity of labeling, but the expression shows a striking increase in the posterior midbrain and rostral hindbrain (presumptive cerebellum) in embryos in which PACAP2 is knocked down compared to the wild type. The *fgf8* gene marker also has more intense labeling of the MHB, especially after PACAP2 knockdown, than the wild-type embryo (FIG. 4 A). The third MHB marker, *pax2.1,* had stronger labeling intensity in the MHB after knockdown compared to that for the wild-type embryo (FIG. 4 A).

To analyze alterations in the hindbrain after PACAP knockdown, two markers, *ephA4* and *meis1.1*, were used. The expression of *ephA4* in specific rhombomeres appears to be almost identical in the wild type and in PACAP-blocked embryos (FIG. 4 B). Changes of expression of *ephA4* in other parts of the brain are discussed below. Marker *meis1.1* expression in hindbrain is decreased after PACAP1 or PACAP2 are blocked (FIG. 4 A).

Three midbrain markers include *mbx, mab21l2,* and *meis1.1*. There is a severe reduction in the midbrain for expression of marker *mbx* but only a modest reduction in *mab21l2 and meis1.1* in a PACAP1-blocked embryo (FIG. 4 A, B). In PACAP2-blocked embryos, expression of the same gene markers is not reduced but the shape of the midbrain is altered in the embryo (FIG. 4 A, B).

In the DMB, *ephA4* was used as a marker. There is a marginal increase in *ephA4* expression in embryos after PACAP knockdown compared to wild-type embryos, but other markers are needed to confirm this result (FIG. 4). In the diencephalon, *fgf8* shows an increase in expression in the DMB in PACAP1- and PACAP2-blocked embryos.

Eye stalk development is visualized with markers *pax2.1* and *fgf8*. Marker *fgf8* showed an increased expression in the absence of either PACAP1

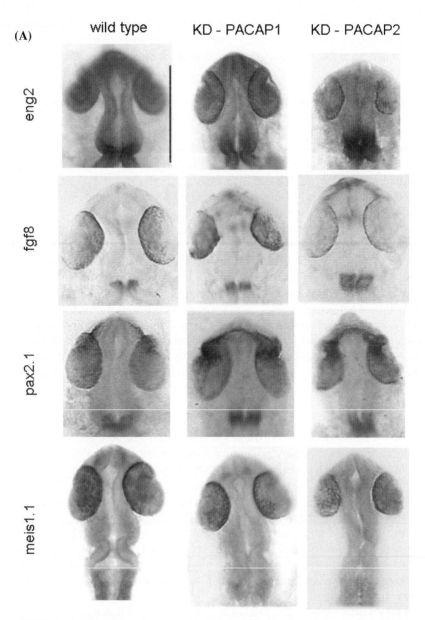

FIGURE 4. Zebrafish embryos at 24 hpf after fertilization in which *in situ* hybridization has been used to show gene expression of specific factors listed on the left. The vertical column on the left-hand side has wild-type embryos; the column in the middle has embryos injected with 4 ng of MOs directed against PACAP1 to knockdown translation of the mRNA; and the column on the right-hand side is the same as the middle column except that PACAP2 was knocked down. The scale bar in each picture represents 250 μM. All embryos are dorsal view.

(B)

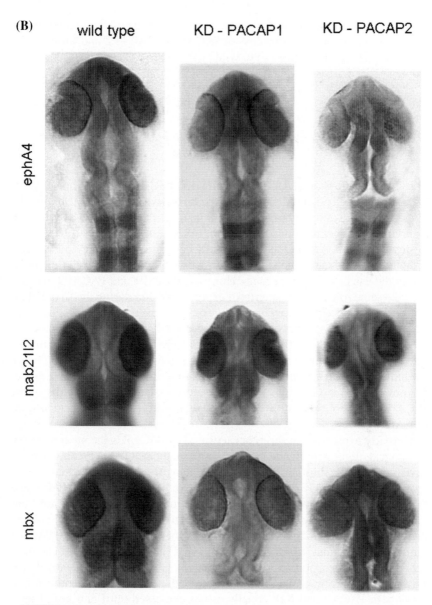

FIGURE 4. *continued.*

or PACAP2 compared to wild-type zebrafish. A striking change in the pattern of expression is seen with marker *pax2.1* in which overexpression in cells of the eye cup and optic stalk in the forebrain occurs after knockdown of either form of PACAP.

Further, a decrease in eye size and a decrease in gene marker expression in the eye occurred for *mbx* after loss of PACAP1 or PACAP2 and for *mab21l2* after loss of PACAP2 compared to wild-type embryos (FIG. 4 B).

DISCUSSION

Early Expression of Transcripts Encoding PACAP Peptides and Their Receptors

The early expression of *PACAP1 and PACAP2* mRNA at only 0.5 hpf suggests that PACAP could potentially influence the early development of the brain. The transcripts at 0.5 and 1.5 hpf are likely to be maternal transcripts as zebrafish embryos do not begin to make their own mRNA until 3–4 hpf.[20] The earlier expression of PACAP1 in the present study compared to our previous study[19] is thought to be due to the use of 30 rather than 5 embryos at each time point and to the use of different primers. However, PACAP1 and PACAP2 may not have the same role as their patterns of expression differ during the first 25 hpf. Transcripts for *PACAP1* are constantly expressed except for a decrease at 4–14 hpf, whereas transcripts of two sizes for *PACAP2* are always present except for lower levels at 0.5–10 hpf. The shorter form is the result of alternative splicing of the full-length form, which removes exon 4 encoding GHRH (1-32), the biologically active portion.[21] The functional implication is that a higher ratio of PACAP to GHRH is produced in any tissue with both forms of the *PACAP2* transcript. Also, growth hormone release in fish depends primarily on PACAP.[9]

The early expression of receptors activated by PACAP is the other essential component along with ligand expression that indicates PACAP is involved in development. In zebrafish, only one gene copy of each receptor that is activated by PACAP or GHRH is present. The expression of PAC1, VPAC$_2$, and GHRH receptors occurs at 1–2 hpf, which implies they are maternal transcripts, whereas the expression of VPAC1 receptor begins at 5–6 hpf implying that it is an embryonic transcript. PAC1 receptor binds exclusively to PACAP. Earlier we found that zebrafish PACAP1 (1-27) and PACAP2 (1-27) activate zebrafish PAC1 receptor expressed in monkey COS cells with identical profiles as measured by cAMP and IP accumulation.[14] VPAC1 receptor, which binds both PACAP and VIP equally, is not expressed until 5–6 hpf. Thereafter, VPAC1 receptor continues to be expressed as a single form for at least 25 hpf. The zebrafish VPAC1 receptor was expressed also in COS cells and found to be activated by PACAP1 and PACAP2 in an almost identical manner with an increase in cAMP but not IP accumulation. In regard to early brain development, all of the components, two ligands and four receptors, were expressed no later than 5–6 hpf indicating that PACAP may be involved in early development.

Location of PACAP1 and PACAP2 mRNA Expression in 24-hpf Embryo

Previously we showed that *PACAP1* mRNA is expressed in the eye, midbrain, MHB, and hindbrain.[19] Clearly, the location of the two *PACAP* mRNAs is different in that *PACAP2* is shown here to be widely expressed in the forebrain, diencephalon, and hindbrain, but not in the eye. With two PACAPs, large areas of the brain and eye could be affected. *In situ* hybridization shows only the location of cell bodies, so the influence via axonal extensions is even greater. Although we only tested *in situ* hybridization at 24 hpf, the mRNA is present much earlier as shown by the RT-PCR.

The translation of the mRNA to protein is important as developmental effects depend on the protein expression of the PACAP peptides. A new study has examined the presence of PACAP peptides by immunocytochemistry from 24 hpf to adulthood.[10] The antisera used in the study did not distinguish between PACAP1 and PACAP2. Nonetheless, PACAP is present in the telencephalon, diencephalon, eye, and rhombencephalon in the 24-hpf embryo. The protein may be present even earlier, but was not checked and, indeed, may be present even though it cannot be detected by immunocytochemistry. The receptors have not been studied for protein expression because antibodies are not available. In comparing zebrafish to a mammal such as mouse, the single copy gene in mouse is expressed early in development at embryonic day 10.5 and the protein is detected at embryonic day 14; the PAC1 receptor mRNA is expressed on embryonic day 10.[1] Thus, the zebrafish embryo, which has no maternal protection, appears to express PACAP at an earlier stage of development than mammals suggesting PACAP is essential whether obtained from embryonic translation in zebrafish or the placenta in mice.

Morphological Changes in Brain and Eye After Knockdown of PACAP1 or PACAP2

The MO doses of 2–4 ng are very low for threshold doses that are effective in terms of morphological changes in brain and eye morphology. A number of controls, including the injection of an unrelated MO, the injection of MOs directed against the UTR or ATG regions for comparison of effects, the injection of a PACAP1 splicing MO, and the injection of exogenous mRNA along with MOs to partially rescue embryos provide evidence that the MOs were targeting the correct mRNA and that the morphological defects were not due to the injection process or toxicity of the MO. That rescue is partial and not complete is known from the literature[22–24] and is thought to reflect the difficulty of injecting mRNA at exactly the needed dose or location.[25]

The observation of morphological changes in the 27-hpf brain and eye provides a screening method to determine if knockdown of a factor expressed early in development has a critical role that needs to be examined in detail.

Both PACAP1 and PACAP2 produced observable changes in the size and shape of the brain and eye. PACAP 2 knockdown at a lower threshold dose (2 ng) is more effective than PACAP1 with a threshold dose of 4 ng. However, both produced similar morphological effects in reduction of brain size and distortion of the MHB, which usually results from changes in the midbrain, hindbrain or both. The size of the eye was significantly reduced by loss of both forms of PACAP. Although we tested higher doses of MOs (8 ng), which produced the same type of defects with greater severity, these doses were not used further to ensure that toxic effects were not a factor in the observed neural changes.

Selective Changes in Expression of Gene Markers in Brain and Eye After PACAP Knockdown

Development involves cascades of factors, both transcription factors and secreted factors, in which sequential events must occur in a strict spatial and temporal pattern to produce a normal brain. Furthermore, some steps involve feedback loops so that it is more difficult to determine the exact sequence of events. Brain marker genes can be used to visualize the developmental events as they are expressed at a specific time and in a specific pattern within the three dimensional space of the brain during development. We selected seven marker genes that have distinct expression patterns in brain regions of interest. Any change in marker expression suggests PACAP is important at some point in the cascade, although not necessarily immediately before marker gene expression. Lack of PACAP can reduce or enhance marker gene expression. The markers we chose, *eng2, fgf8, pax2.1, meis1.1, ephA4, mab21l2*, and *mbx* are all zebrafish homologs of similar proteins in mice and humans.

The *pax2.1* marker is one of the earliest acting genes needed for the development of the midbrain, MHB (isthmus), and cerebellum from gastrulation onward. The loss of *pax2.1* protein leads to loss of the midbrain tectum, MHB, and cerebellum in zebrafish.[26] In our embryos, *pax2.1* is overexpressed in the MHB after PACAP knockdown suggesting that PACAP has an influence in the midbrain–hindbrain region. The expression of *pax2.1* in eyes and forebrain is discussed below.

Downstream genes of *pax2.1* include *eng2* and *eng3* with a feedback loop among the three.[27] Expression of *eng2* is strongly dependent on *pax2.1*.[28] Others have shown that a MO knockdown of *eng2* and *eng3* results in an absence of *pax 2.1* and that the MHB does not develop.[27] In our experiments, the absence of PACAP1 had little effect on the expression of *eng2*, but the knockdown of PACAP2 had a striking effect in that *eng2* expression was substantially increased in the MHB. This suggests that PACAP2 acts in the cascade involving *eng2* and normally suppresses the expression. PACAP2 is a secreted protein and hence acts in a different way from *pax2.1*, which is thought to act as an activating transcription factor that directly binds to the promotor region of *eng2*.

In contrast, PACAP2, which activates seven-transmembrane receptors in the outer membrane, has been shown to increase both cAMP and IP accumulation in zebrafish.[14] The PAC1, VPAC1, and VPAC2 receptors may be available for PACAP activations, as shown here the mRNA is expressed by 1–6 hpf, although their expression as protein has not yet been studied.

Further support for an early role of PACAP as a suppressor in the cascade involving formation of the MHB, is the overexpression of *fgf8* in embryos that have MO-induced knockdown of PACAP2. The *fgf8* gene is expressed as early as *pax2.1* and its expression is independent of *pax2.1*.[28] *Fgf8* is important in the MHB organizer[28] and if *fgf8* is mutated, the zebrafish lack a cerebellum.[27,29] After PACAP2 knockdown, the cerebellum is present and *fgf8* expression is increased. One hypothesis is that PACAP is a target gene for *fgf8*, *eng2*, and *pax2.1* and expression of PACAP feeds back to inhibit *fgf8*, *eng2* and *pax2.1*. Rhinn and Brand[28] have clearly stated the problem: "Given the potency as a signaling molecule, the activity of *Fgf8* must be carefully controlled in the embryo. An emerging theme for several signaling pathways is that extracellular or intracellular inhibitors control their activity" (p. 38). PACAP is a candidate as an extracellular inhibitor of this pathway.

In hindbrain formation, there is some indication that PACAP1 and PACAP2 may normally influence the pathway regulating the expression of *meis1.1* in the hindbrain. The expression of the *meis1.1* gene in wild-type zebrafish has been described.[30] Another gene marker for the hindbrain is *ephA4*, which is an ephrin receptor; *ephA4* is clearly expressed in specific rhombomeres in the wild-type embryo.[27] However, the expression of *ephA4* and the morphology of the hindbrain are not altered by knockdown of either PACAP1 or PACAP2 suggesting that PACAP does not act in this cascade of factors, although PACAP1 and PACAP2 may act in a minor way through *meis1.1*.

PACAP1 and PACAP2 both influence the formation of the midbrain. The expression of the *mab21l2* gene[25,31] was affected, although to a lesser extent than *mbx* expression. The reduction of *mbx* expression in the midbrain was even more evident than *mab21l2* after PACAP1 was blocked. Both the intensity of expression and morphology of tissue were altered after knockdown. Thus PACAP affects the midbrain by activating *mbx*, *mab21l2,* and *meis1.1* expression; the three genes appear later in development than *pax2.1*, *fgf8,* or *eng2*.

Our data on the diencephalon and the DMB are both derived from our *in situ* hybridization study of *ephA4* and *fgf8* genes. *Fgf8* shows an increase in expression if PACAP1 or PACAP2 are absent, but the greatest increase in *ephA4* is after PACAP1 knockdown. Again, the suggestion is that PACAP normally acts to suppress these genes. *Fgf8* is thought to be one of several factors that are involved in setting the boundary between the diencephalon and midbrain. *Fgf8, pax2.1,* and *eng2* genes are all thought to be involved in repressing forebrain fate,[27] so it is of interest that PACAP normally acts to inhibit the expression of *fgf8* and *eng2*. There are other studies in mice that show an interaction of PACAP with other genes expressed in the diencephalon, such as sonic hedgehog (*shh*), and these genes will be of interest in future studies.

Our evidence suggests that PACAP peptides affect the formation of the forebrain. Patterning of the forebrain in zebrafish begins in gastrulation, and as in the midbrain–hindbrain region, *fgf8* and *pax2.1* are known to play a role.[32,33] Gene *fgf8* showed a substantial increase in expression, after blockage of either form of PACAP.

In forebrain and eye, PACAP1 and PACAP2 appear to influence the cascade that includes the *pax2.1* gene. After knockdown of either form of PACAP, expression of *pax2.1* in the eye stalk and eye cup is greatly increased. Therefore, PACAP appears to be an inhibitor of *pax2.1*.

The *mab21l2* gene is expressed in eye primordia and midbrain at 11 hpf; others showed that knockdown of this gene resulted in a 20% decrease in eye size at 24 hpf in zebrafish, and in mice was embryonically lethal.[25,31,34,35] Although our *in situ* hybridization studies showed *PACAP1* location in the eye and midbrain, *PACAP2* fibers may be distributed in the midbrain so that the knockdown of either peptide could explain the results. Other evidence that PACAP affects development of the eyes is the decreased expression of *mab21l2* in the absence of PACAP and our measurement of eye diameter that was significantly reduced after knockdown of either form of PACAP.

The strategy of examining the role of a hormone by eliminating it in a living organism is a long-established method to elucidate function and identify novel actions. Knockdown of either PACAP1 or PACAP2 has shown that both secreted proteins have a role in early development. The most dramatic results are the increase of *pax2.1* in the eye and eye stalk in the absence of both PACAPs, the increase of *eng2*, *fgf8*, and *pax2.1* in the absence of PACAP2, and the decrease in *mbx* in the absence of PACAP1. The result of the combined gene markers is that PACAP is shown to have an effect in early development in the midbrain, hindbrain, and their boundary and on the diencephalon and forebrain. The changes in marker gene expression and in brain morphology imply that PACAP is normally translated into a protein early in development to act on the brain and eye. A study by others[10] substantiates this claim by immunocytochemistry. The authors do not test for protein detection until the 24-hpf embryo, but already PACAP is detected in the telencephalon, diencephalon, retina, and rhombencephalon.

The zebrafish offers a model for the study of PACAP on the expression of specific genes during early development. Currently, microarray projects to identify target genes of PACAP are likely to produce long lists of genes. The zebrafish offers a method to study the effects of PACAP on these genes in a living organism in which the spatial and temporal aspects of gene expression can be observed after the absence of PACAP. We examined seven marker genes in the absence or PACAP1 or PACAP2, but other genes known to be interrelated in the formation of a brain region or pituitary could be examined.

ACKNOWLEDGMENTS

We thank the Canadian Natural Sciences and Engineering Research Council (NSERC) and the Canadian Institutes of Health Research (CIHR) for financial support.

REFERENCES

1. SHERWOOD, N.M., S.L. KRUECKL & J.E. MCRORY. 2000. The origin and function of the pituitary adenylate cyclase-activating polypeptide (PACAP)/glucagon superfamily. Endocr. Rev. **21:** 619–670.
2. GRAY, S.L. *et al.* 2001. Targeted disruption of the pituitary adenylate cyclase-activating polypeptide gene results in early postnatal death associated with dysfunction of lipid and carbohydrate metabolism. Mol. Endocrinol. **15:** 1739–1747.
3. HAMELINK, C. *et al.* 2002. Pituitary adenylate cyclase-activating polypeptide is a sympathoadrenal neurotransmitter involved in catecholamine regulation and glucohomeostasis. Proc. Natl. Acad. Sci. USA **99:** 461–466.
4. HASHIMOTO, H. *et al.* 2001. Altered psychomotor behaviors in mice lacking pituitary adenylate cyclase-activating polypeptide (PACAP). Proc. Natl. Acad. Sci. USA **98:** 13355–13360.
5. COLWELL, C.S. *et al.* 2004. Selective deficits in the circadian light response in mice lacking PACAP. Am. J. Physiol. Regul. Integr. Comp. Physiol. **287:** R1194–R1201.
6. FRADINGER, E.A. & N.M. SHERWOOD. 2000. Characterization of the gene encoding both growth hormone-releasing hormone (GRF) and pituitary adenylate cyclase-activating polypeptide (PACAP) in zebrafish. Mol. Cell. Endocrinol. **165:** 211–219.
7. WANG, Y., A.O.L. WONG & W. GE. 2003. Cloning, regulation of messenger ribonucleic acid expression, and function of a new isoform of pituitary adenylate cyclase-activating polypeptide in the zebrafish ovary. Endocrinology **144:** 4799–4810.
8. CUMMINGS, K.J. *et al.* 2002. Mouse pituitary adenylate cyclase-activating polypeptide (PACAP): gene, expression and novel splicing. Mol. Cell. Endocrinol. **192:** 133–145.
9. PARKER, D.B. *et al.* 1997. Exon skipping in the gene encoding pituitary adenylate cyclase-activating polypeptide (PACAP) in salmon alters the expression of two hormones that stimulate growth hormone release. Endocrinology **138:** 414–423.
10. MATHIEU, M. *et al.* 2004. Pituitary adenylate cyclase-activating polypeptide in the brain, spinal cord and sensory organs of the zebrafish, *Danio rerio*, during development. Dev. Brain Res. **151:** 169–185.
11. VAUDRY, H. & A. ARIMURA, Eds. 2003. Pituitary Adenylate Cyclase-Activating Polypeptide. Kluwer Academic Publishers. Dordrecht, The Netherlands.
12. MAY, V. *et al.* 2000. PACAP modulates rat sympathetic neuron depolarization through IP_3. Ann. N. Y. Acad. Sci. **921:** 186–194.
13. PARSONS, R.L. *et al.* 2000. PACAP peptides modulate guinea pig cardiac neuron membrane excitability and neuropeptide expression. Ann. N. Y. Acad. Sci. **921:** 202–210.

14. FRADINGER, E.A. *et al.* 2005. Characterization of four receptor cDNAs: PAC1, VPAC1, a novel PAC1 and a partial GHRH in zebrafish. Mol. Cell. Endocrinol. **231:** 49–63.
15. NASEVICIUS, A. & S.C. EKKER. 2000. Effective targeted gene "knockdown' in zebrafish. Nat. Genet. **26:** 216–220.
16. SUMMERTON, J. & D. WELLER. 1997. Morpholino antisense oligomers: design, preparation, and properties. Antisense Nucleic Acid Drug Dev. **7:** 187–195.
17. EKKER, S.C. & J.D. LARSON. 2001. Morphant technology in model developmental systems. Genesis **30:** 89–93.
18. XU, Q. 1999. Microinjection into zebrafish embryos. Methods Mol. Biol. **127:** 125–132.
19. KRUECKL, S.L., E.A. FRADINGER & N.M. SHERWOOD. 2003. Developmental changes in the expression of growth hormone-releasing hormone and pituitary adenylate cyclase-activating polypeptide in zebrafish. J. Comp. Neurol. **455:** 396–405.
20. KANE, D.A. & C.B. KIMMEL. 1993. The zebrafish midblastula transition. Development **119:** 447–456.
21. GUILLEMIN, R. 1986. Hypothalamic control of pituitary functions. *In* The Growth Hormone Releasing Factor, pp. 173. The Sherrington Lectures XVIII, Liverpool University Press. Liverpool, UK.
22. CUI, Z. *et al.* 2001. Inhibition of *ski*A and *ski*B gene expression ventralizes zebrafish embryos. Genesis **30:** 149–153.
23. HASHIGUCHI, A., K. OKABAYASHI & M. ASASHIMA. 2004. Role of TSC-22 during early embryogenesis in *Xenopus laevis*. Dev. Growth Differ. **46:** 535–544.
24. KANZLER, B. *et al.* 2003. Morpholino oligonucleotide-triggered knockdown reveals a role for maternal E-cadherin during early mouse development. Mech. Dev. **120:** 1423–1432.
25. KENNEDY, B.N. *et al.* 2004. Zebrafish *rx3* and *mab21l2* are required during eye morphogenesis. Dev. Biol. **270:** 336–349.
26. BRAND, M. *et al.* 1996. Mutations in zebrafish genes affecting the formation of the boundary between midbrain and hindbrain. Development **123:** 179–190.
27. SCHOLPP, S. & M. BRAND. 2003. Integrity of the midbrain region is required to maintain the diencephalic-mesencephalic boundary in zebrafish *no isthmus/pax2.1* mutants. Dev. Dyn. **228:** 313-322.
28. RHINN, M. & M. BRAND. 2001. The midbrain-hindbrain boundary organizer. Curr. Opin. Neurobiol. **11:** 34–42.
29. REIFERS, F. *et al.* 1998. Fgf8 is mutated in zebrafish *acerebellar* (*ace*) mutants and is required for maintenance of midbrain-hindbrain boundary development and somitogenesis. Development **125:** 2381–2395.
30. WASKIEWICZ, A.J. *et al.* 2001. Zebrafish Meis functions to stabilize Pbx proteins and regulate hindbrain patterning. Development **128:** 4139–4151.
31. KUDOH, T. & I.B. DAWID. 2001. Zebrafish *mab21l2* is specifically expressed in the presumptive eye and tectum from early somitogenesis onwards. Mech. Dev. **109:** 95–98.
32. WALSHE, J. & I. MASON. 2003. Unique and combinatorial functions of Fgf3 and Fgf8 during zebrafish forebrain development. Development **130:** 4337–4349.

33. MACDONALD, R. *et al.* 1995. Midline signalling is required for Pax gene regulation and patterning of the eyes. Development **121:** 3267–3278.
34. WONG, Y-M. & K.L. CHOW. 2002. Expression of zebrafish *mab21* genes marks the differentiating eye, midbrain and neural tube. Mech. Dev. **113:** 149–152.
35. YAMADA, R. *et al.* 2003. Cell-autonomous involvement of *Mab21l1* is essential for lens placode development. Development **130:** 1759–1770.

A Role for Pituitary Adenylate Cyclase Activating Polypeptide (PACAP) in Detrusor Hyperreflexia after Spinal Cord Injury (SCI)

PETER ZVARA,[c] KAREN M. BRAAS,[a] VICTOR MAY,[a,d]
AND MARGARET A. VIZZARD[a,b]

[a]University of Vermont College of Medicine, Department of Anatomy and Neurobiology, Burlington, Vermont 05405, USA

[b]University of Vermont College of Medicine, Department of Neurology, Burlington, Vermont 05405, USA

[c]University of Vermont College of Medicine, Department of Surgery, Burlington, Vermont 05405, USA

[d]University of Vermont College of Medicine, Department of Pharmacology, Burlington, Vermont 05405, USA

ABSTRACT: Intrathecal administration of the PAC1 receptor antagonist, PACAP6-38 (10 nM), significantly ($P \leq 0.05$) reduced intermicturition, threshold and micturition pressures in chronic (3–6 weeks) spinal cord injured rats but intravesical administration (100–300 nM) was without effect. Intrathecal PACAP6-38 reduced the number and amplitude of nonvoiding bladder contractions observed after spinal cord injury (SCI). PACAP may contribute to detrusor hyperreflexia induced by SCI and PACAP antagonists may be a novel approach to reduce detrusor hyper-reflexia after SCI.

KEYWORDS: urinary bladder; cystometry; intrathecal; intravesical; lumbosacral spinal cord

INTRODUCTION

Micturition is sensitive to a wide variety of injuries, diseases, and chemicals that affect the nervous system.[1,2] Complete transection of the spinal cord, rostral to the lumbosacral level, eliminates voluntary supraspinal control of voiding.[1,2] This is followed by the emergence of automatic, involuntary

Address for correspondence: Margaret A. Vizzard, Ph.D., University of Vermont College of Medicine, Department of Neurology, D411 Given Building, Burlington, VT 05405. Voice: 802-656-3209 fax: 802-656-8704.
e-mail: margaret.vizzard@uvm.edu

Ann. N.Y. Acad. Sci. 1070: 622–628 (2006). © 2006 New York Academy of Sciences.
doi: 10.1196/annals.1317.092

reflex micturition with detrusor hyperreflexia and detrusor-sphincter dyssynergia (DSD).[1,2] The neuropeptide, pituitary adenylate cyclase-activating peptide (PACAP), may play a role in peripheral and central neurons after peripheral nerve injury[3-5] or peripheral[6,7] or central inflammatory states.[8] A subpopulation of PACAP-immunoreactive (IR) sensory neurons in the dorsal root ganglia (DRG) is co-localized with calcitonin gene-related peptide and substance P and exhibits capsaicin sensitivity.[9] Intrathecal administration of PACAP induces bladder hyperreflexia in rats.[10] PACAP expression is increased in micturition reflex pathways after cystitis[7] or spinal cord injury (SCI)[11] and PACAP antagonists reduce bladder hyperreflexia after cystitis.[12] This article determined if intrathecal or intravesical administration of the PAC1 selective antagonist, PACAP6-38, reduces detrusor hyperreflexia in rats after SCI.

Experimental Animals

Adult female Wistar rats (150–200 g; spinal cord intact (control; $n = 9$); SCI ($n = 9$) were used. Complete spinal cord transection was performed at T8-T10.[13] Animals were studied 3–6 weeks after SCI. Animal use was approved by The University of Vermont Institutional Animal Care and Use Committee.

Intrathecal Catheter Placement

An intrathecal catheter (PE-10, Clay Adams, Parsippany, NJ) was passed caudally to the lumbosacral spinal cord (L6-S1) and placement confirmed after euthanasia. The volume of fluid in the intrathecal catheter was 9 μL. At the time of study, 9 μL of the PAC1 selective antagonist, PACAP6-38, dissolved in artificial cerebrospinal fluid (ACF),[14] was injected through the intrathecal catheter followed by a 9-μL ACF flush. Control animals received ACF alone. The concentrations of PACAP6-38 selected for intrathecal or intravesical administration were based upon previous studies.[15]

Intravesical Catheter Placement

Polyethylene tubing (PE-50, Clay Adams) was inserted into the dome of the bladder and secured.[16] PACAP6-38 (0.5 mL; 100–300 nM) was injected into the bladder of an anesthetized animal to prevent elimination of the antagonist. PACAP6-38 remained in the bladder for 30 min.

Cystometry

A rat was placed in a Small Animal Cystometry Lab Station (MED Associates, Inc., St. Albans, VT) for urodynamic measurements.[17] Isotonic saline

(0.9% NaCl) was infused into the bladder (10 mL/h). These cystometric parameters were determined: filling, intermicturition, threshold and micturition pressure, micturition interval, void volume, and presence of nonvoiding bladder contractions (NVC) defined as increases in bladder pressure of at least 7 cm H_2O without urination.[16] Intermicturition pressure was a measure of average NVC pressure.

RESULTS

SCI rostral to the lumbosacral spinal cord significantly increased threshold, intermicturition and micturition pressure, resulted in the appearance of NVC and increased bladder capacity (two- to threefold) (FIG. 1A, B). Intrathecal administration of PACAP6-38 (10 nM) significantly reduced intermicturition, threshold, and micturition pressure in SCI rats compared to SCI rats treated with vehicle (FIG. 1A, B). Subsequent intrathecal administration of agonist (PACAP, 50 nM) reversed these changes (FIG. 1B). Intrathecal administration of PACAP6-38 also significantly reduced the number and amplitude of NVC and reduced the intercontraction interval (i.e., decreased bladder capacity) in rats with SCI compared to SCI rats treated with vehicle (FIG. 2A, B). The duration of effect of PACAP6-38 was 30–60 min. In contrast, intravesical administration of PACAP6-38 (100–300 nM) was without effect on any cystometric parameter examined.

DISCUSSION

SCI rostral to the lumbosacral spinal cord alters the coordination between the urinary bladder and external urethral sphincter (EUS) in many species and results in DSD that interferes with efficient voiding and results in urinary retention, bladder hypertrophy, increased voiding pressures, increased bladder capacity, and numerous NVC during bladder filling.[1,2] PACAP is upregulated in micturition reflex pathways after SCI.[11] These studies demonstrate that intrathecal (L6-S1) administration of the PAC1 receptor antagonist, PACAP6-38 (10 nM), significantly reduced intermicturition, threshold, and micturition pressure, and number and amplitude of NVC after SCI. In addition, PACAP6-38 increased voiding frequency (i.e., decreased bladder capacity) after SCI. In contrast, intravesical administration of PACAP6-38 was without any effect, possibly due to lack of penetration of the PAC1 antagonist through the urothelium. The presence of PAC1 receptor transcript in the lumbosacral spinal cord and DRG has been demonstrated.[12] Thus, intrathecal administration of PACAP6-38 may act at the lumbosacral spinal cord or DRG. The effects of PACAP6-38 after SCI are consistent with PACAP-27 facilitation of micturition in rats.[10] The facilitatory effects of PACAP are consistent with the

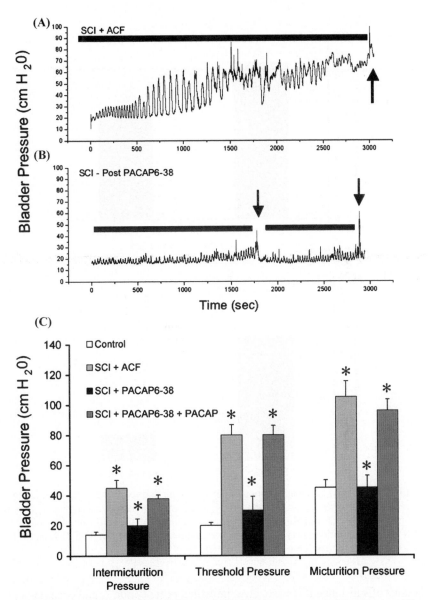

FIGURE 1. Intrathecal PACAP6-38 reduces bladder pressure and increases voiding frequency in SCI rats. Bladder pressure recordings from a SCI rat with intrathecal administration of ACF (**A**) and after intrathecal administration of PACAP6-38 (10 nm) (**B**). NVC occur during bladder filling (black bar) prior to a micturition event (arrow). NVC were still present (black bar) after PACAP6-38 but were reduced in number and amplitude. Reductions in bladder pressure are also noted. PACAP6-38 also increased voiding frequency (B, arrows) in the SCI rat. (**C**) Histogram of reductions in intermicturition, threshold and micturition pressure after intrathecal administration of PACAP6-38 in SCI rats. $^*P \leq 0.05$.

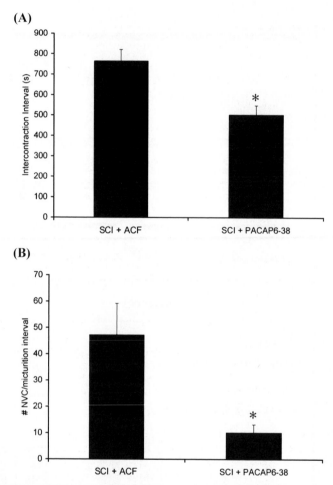

FIGURE 2. (A) Summary histogram demonstrating the reduction in intercontraction interval after intrathecal PACAP6-38 (10 nM) administration in SCI rats compared to SCI rats treated with ACF. **(B)** Summary histogram demonstrating the reduction in the number of NVC per micturition interval after intrathecal administration of PACAP6-38 (10 nM) in SCI rats compared to SCI rats treated with ACF. *$P \leq 0.05$.

actions of substance P[18], vasoactive intestinal polypeptide[19], and cocaine- and amphetamine-regulated transcript peptide (CARTp)[20] on lower urinary tract tissues. The present study is consistent with the effects of tachykinin antagonists on detrusor hyperreflexia after SCI.[21] PACAP6-38 reduced the number and amplitude of NVC after SCI. This may be attributed to a reduction in DSD or through an effect on C-fiber bladder afferents.[22] An effect of PACAP6-38 on urinary bladder C-fiber afferents is consistent with previous studies that demonstrate that capsaicin depletes PACAP-IR in the lower urinary tract[23] and that PACAP-IR nerve fibers in the bladder express the vanilloid receptor.[11] PACAP expression is upregulated in micturition pathways after SCI[12] and the present studies demonstrated improved bladder function after intrathecal PACAP antagonist administration. PACAP may contribute to detrusor hyper-reflexia after SCI.

ACKNOWLEDGMENTS

The support of NIH DK051369, DK060481, DK065989, NS040796 is acknowledged.

REFERENCES

1. DE GROAT, W.C. & M.N. KRUSE. 1993. Central processing and morphological plasticity in lumbosacral afferent pathways from the lower urinary tract. *In* Basic and Clinical Aspects of Chronic Abdominal Pain. Pain Research and Clinical Management, Vol. 9. E.A. Mayer & H.E. Raybould, Eds.: 219–235. Elsevier Science Publishers. Amsterdam.
2. YOSHIMURA, N. 1999. Bladder afferent pathways and spinal cord injury: possible mechanisms inducing hyperreflexia of the urinary bladder. Prog. Neurobiol. **57:** 583–606.
3. ZHANG, Q., T.J. SHI, R.R. JI, *et al.* 1995. Expression of pituitary adenylate cyclase-activating polypeptide in dorsal root ganglia following axotomy: time course and coexistence. Brain Res. **705:** 149–158.
4. ZHANG, Y.-Z., J. HANNIBAL, Q. ZHAO, *et al.* 1996. Pituitary adenylate cyclase activating peptide expression in the rat dorsal root ganglia: up-regulation after peripheral nerve injury. Neuroscience **74:** 1099–1110.
5. SUNDLER, F., E. EKBLAD, J. HANNIBAL, *et al.* 1996. Pituitary adenylate cyclase-activating peptide in sensory and autonomic ganglia: localization and regulation. Ann. N. Y. Acad. Sci. **805:** 410–426.
6. ZHANG, Y.Z., N. DANIELSON, F. SUNDLER, *et al.* 1998. Pituitary adenylate cyclase-activating peptide is upregulated in sensory neurons by inflammation. Neuroreport. **9:** 2833–2836.
7. VIZZARD, M.A. 2000. Up-regulation of pituitary adenylate cyclase-activating polypeptide in urinary bladder pathways after chronic cystitis. J. Comp. Neurol. **420:** 335–348.
8. HANNIBAL, J., D.S. JESSOP, J. FAHRENKRUG, *et al.* 1999. PACAP gene expression in neurons of the rat hypothalamo-pituitary-adrenocortical axis is induced by endotoxin and interleukin-1B. Neuroendocrinology **70:** 73–82.
9. MOLLER, K., M. REIMER, J. HANNIBAL, *et al.* 1993. Pituitary adenylate cyclase activating peptide is a sensory neuropeptide: immunocytochemical and immuno-chemical evidence. Neuroscience **57:** 725–732.
10. ISHIZUKA, O., P. ALM, B. LARSSON, *et al.* 1995. Facilitatory effect of pituitary adenylate cyclase-activating polypeptide on micturition in normal, conscious rats. Neuroscience **66:** 1009–1014.
11. ZVAROVA, K., J.D. DUNLEAVY & M.A. VIZZARD. 2005. Changes in pituitary adenylate cyclase activating polypeptide expression in urinary bladder pathways after spinal cord injury. Exp. Neurol. **192:** 46–59.

12. BRAAS, K.M., V. MAY, P. ZVARA, *et al.* 2005. Role for pituitary adenylate cyclase activating polypeptide (PACAP) in cystitis-induced plasticity of micturition reflexes. *In* VIIth International Symposium on VIP, PACAP and Related Peptides. Regul. Pept. **130:** 157–158.

13. QIAO, L.Y. & M.A. VIZZARD. 2002. Up-regulation of tyrosine kinase (Trka, Trkb) receptor expression and phosphorylation in lumbosacral dorsal root ganglia after chronic spinal cord (T8-T10) injury. J. Comp. Neurol. **449:** 217–230.

14. IZQUIERDO, I. & J.H. MEDINA. 1995. Correlation between the pharmacology of long-term potentiation and the pharmacology of memory. Neurobiol. Learn. Mem. **63:** 19–32.

15. BEAUDET, M.M., R.L. PARSONS, K.M. BRAAS, *et al.* 2000. Mechanisms mediating pituitary adenylate cyclase-activating polypeptide depolarization of rat sympathetic neurons. J. Neurosci. **20:** 7353–7361.

16. HU, V.Y., S. MALLEY, A. DATTILIO, *et al.* 2003. COX-2 and prostanoid expression in micturition pathways after cyclophosphamide-induced cystitis in the rat. Am. J. Physiol. **284:** R574–R585.

17. HU, V.Y., P. ZVARA, A. DATTILIO, *et al.* 2005. Decrease in bladder overactivity with REN1820 in rats with cyclophosphamide induced cystitis. J. Urol. **173:** 1016–1021.

18. CHIEN, C.T., H.J. YU, T.B. LIN, *et al.* 2003. Substance P via NK1 receptor facilitates hyperactive bladder afferent signaling via action of ROS. Am. J. Physiol. **284:** F840–F851.

19. IGAWA, Y., K. PERSSON, K.E. ANDERSSON, *et al.* 1993. Facilitatory effect of vasoactive intestinal polypeptide on spinal and peripheral micturition reflex pathways in conscious rats with and without detrusor instability. J. Urol. **149:** 884–889.

20. ZVAROVA, K. & M.A. VIZZARD. 2005. Distribution and fate of cocaine- and amphetamine-regulated transcript peptide ($CART_p$)-expressing cells in rat urinary bladder: a developmental study. J. Comp. Neurol. **489(4):** 501–517.

21. ABDELGAWAD, M., S.B. DION & M.M. ELHILALI. 2001. Evidence of a peripheral role of neurokinins in detrusor hyperreflexia: a further study of selective tachykinin antagonists in chronic spinal injured rats. J. Urol. **165:** 1739–1744.

22. CHENG, C.L., C.P. MA & W.C. DE GROAT. 1995. Effect of capsaicin on micturition and associated reflexes in chronic spinal rats. Brain Res. **678:** 40–48.

23. FAHRENKRUG, J. & J. HANNIBAL. 1998. Pituitary adenylate cyclase activating polypeptide immunoreactivity in capsaicin-sensitive nerve fibres supplying the rat urinary tract. Neuroscience **83:** 1261–1272.

Index of Contributors